Edited by
Javier García-Martínez and
Elena Serrano-Torregrosa

Chemistry Education

Related Titles

García-Martínez, J., Li, K. (eds.)

Mesoporous Zeolites

Preparation, Characterization and Applications

2015
ISBN: 978-3-527-33574-9

García-Martínez, J. (ed.)

Nanotechnology for the Energy Challenge

2nd Edition

2013
ISBN: 978-3-527-33380-6

García-Martínez, J., Serrano-Torregrosa, E. (eds.)

The Chemical Element

Chemistry's Contribution to Our Global Future

2011
ISBN: 978-3-527-32880-2

Armaroli, N., Balzani, V., Serpone, N.

Powering Planet Earth

Energy Solutions for the Future

2013
ISBN: 978-3-527-33409-4

Quadbeck-Seeger, H.-J.

World of the Elements

Elements of the World

2007
ISBN: 978-3-527-32065-3

Ebel, H.F., Bliefert, C., Russey, W.E.

The Art of Scientific Writing

From Student Reports to Professional Publications in Chemistry and Related Fields
2nd Edition

2004
ISBN: 978-3-527-29829-7

*Edited by Javier García-Martínez and
Elena Serrano-Torregrosa*

Chemistry Education

Best Practices, Opportunities and Trends

With a Foreword by Peter Atkins

Verlag GmbH & Co. KGaA

The Editors

Prof. Dr. Javier García-Martínez
University of Alicante
Department of Inorganic Chemistry
Campus de San Vicente del Raspeig
03690 San Vincente del Raspeig
Alicante
Spain

Dr. Elena Serrano-Torregrosa
University of Alicante
Department of Inorganic Chemistry
Campus de San Vicente del Raspeig
03690 San Vincente del Raspeig
Alicante
Spain

Cover: (c) fotolia

All books published by **Wiley-VCH** are carefully produced. Nevertheless, authors, editors, and publisher do not warrant the information contained in these books, including this book, to be free of errors. Readers are advised to keep in mind that statements, data, illustrations, procedural details or other items may inadvertently be inaccurate.

Library of Congress Card No.: applied for

British Library Cataloguing-in-Publication Data
A catalogue record for this book is available from the British Library.

Bibliographic information published by the Deutsche Nationalbibliothek
The Deutsche Nationalbibliothek lists this publication in the Deutsche Nationalbibliografie; detailed bibliographic data are available on the Internet at <http://dnb.d-nb.de>.

© 2015 Wiley-VCH Verlag GmbH & Co. KGaA, Boschstr. 12, 69469 Weinheim, Germany

All rights reserved (including those of translation into other languages). No part of this book may be reproduced in any form – by photoprinting, microfilm, or any other means – nor transmitted or translated into a machine language without written permission from the publishers. Registered names, trademarks, etc. used in this book, even when not specifically marked as such, are not to be considered unprotected by law.

Print ISBN: 978-3-527-33605-0
ePDF ISBN: 978-3-527-67933-1
ePub ISBN: 978-3-527-67932-4
Mobi ISBN: 978-3-527-67931-7
oBook ISBN: 978-3-527-67930-0

Cover Design Grafik-Design Schulz
Typesetting Laserwords Private Limited, Chennai, India

Printed on acid-free paper

Contents

Foreword *XXI*
Preface *XXV*
List of Contributors *XXXIII*

Part I: Chemistry Education: A Global Endeavour *1*

1	**Chemistry Education and Human Activity** *3*	
	Peter Mahaffy	
1.1	Overview *3*	
1.2	Chemistry Education and Human Activity *3*	
1.3	A Visual Metaphor: Tetrahedral Chemistry Education *4*	
1.4	Three Emphases on Human Activity in Chemistry Education *5*	
1.4.1	The Human Activity of Learning and Teaching Chemistry *6*	
1.4.1.1	Atoms or Learners First? *6*	
1.4.1.2	Identifying Learners and Designing Curriculum to Meet Their Needs *7*	
1.4.1.3	Effective Practices in the Human Activity of Learning and Teaching Chemistry *8*	
1.4.1.4	Identifying and Eliminating Worst Practices as a Strategy? *8*	
1.4.1.5	Exemplar: Emphasizing the Human Activity of Learning and Teaching Chemistry *9*	
1.4.2	The Human Activity of Carrying Out Chemistry *10*	
1.4.2.1	Explicit and Implicit Messages about the Nature of Chemistry *10*	
1.4.2.2	Breathing the Life of Imagination into Chemistry's Facts *11*	
1.4.2.3	Exemplars: Emphasizing the Human Activity of Carrying Out Chemistry *13*	
1.4.3	Chemistry Education in the Anthropocene Epoch *14*	
1.4.3.1	Planetary Boundaries: A Chemistry Course Outline? *15*	
1.4.3.2	Steps toward Anthropocene-Aware Chemistry Education *16*	
1.4.3.3	Exemplars: Anthropocene-Aware Chemistry Education *17*	
1.5	Teaching and Learning from Rich Contexts *18*	
1.5.1	Diving into an Ocean of Concepts Related to Acid–Base Chemistry *18*	

1.5.2	What Is Teaching and Learning from Rich Contexts? *20*
1.5.3	Teaching and Learning from Rich Contexts – Evidence for Effectiveness *21*
1.5.4	From "Chemical" to "Chemistry" Education – Barriers to Change *22*
	Acknowledgments *23*
	References *24*
2	**Chemistry Education That Makes Connections: Our Responsibilities** *27*
	Cathy Middlecamp
2.1	What This Chapter Is About *27*
2.2	Story #1: Does This Plane Have Wings? *28*
2.3	Story #2: Coaching Students to "See" the Invisible *30*
2.4	Story #3: Designing Super-Learning Environments for Our Students *34*
2.5	Story #4: Connections to Public Health (Matthew Fisher) *37*
2.6	Story #5: Green Chemistry Connections (Richard Sheardy) *39*
2.7	Story #6: Connections to Cardboard (Garon Smith) *41*
2.8	Story #7: Wisdom from the Bike Trail *44*
2.9	Conclusion: The Responsibility to "Connect the Dots" *46*
	References *48*
3	**The Connection between the Local Chemistry Curriculum and Chemistry Terms in the Global News: The Glocalization Perspective** *51*
	Mei-Hung Chiu and Chin-Cheng Chou
3.1	Introduction *51*
3.2	Understanding Scientific Literacy *52*
3.3	Introduction of Teaching Keywords-Based Recommendation System *55*
3.4	Method *56*
3.5	Results *57*
3.5.1	Example 1: Global Warming *57*
3.5.2	Example 2: Sustainability *57*
3.5.3	Example 3: Energy *58*
3.5.4	Example 4: Acid *59*
3.5.5	Example 5: Atomic Structure *60*
3.5.6	Example 6: Chemical Equilibrium *61*
3.5.7	Example 7: Ethylene *62*
3.5.8	Example 8: Melamine *63*
3.5.9	Example 9: Nano *64*
3.6	Concluding Remarks and Discussion *65*
3.7	Implications for Chemistry Education *68*
	Acknowledgment *70*
	References *70*

4	**Changing Perspectives on the Undergraduate Chemistry Curriculum** *73*	
	Martin J. Goedhart	
4.1	The Traditional Undergraduate Curriculum *73*	
4.2	A Call for Innovation *74*	
4.2.1	Constructivism and Research on Student Learning *74*	
4.2.2	New Technologies *76*	
4.2.3	The Evolving Nature of Chemistry *77*	
4.2.4	Developments in Society and Universities *77*	
4.3	Implementation of New Teaching Methods *78*	
4.3.1	The Interactive Lecture *79*	
4.3.2	Problem- and Inquiry-Based Teaching *80*	
4.3.3	Research-Based Teaching *80*	
4.3.4	Competency-Based Teaching *81*	
4.4	A Competency-Based Undergraduate Curriculum *83*	
4.4.1	The Structure of the Curriculum *84*	
4.4.2	Competency Area of Analysis *86*	
4.4.3	Competency Area of Synthesis *88*	
4.4.4	Competency Area of Modeling *89*	
4.4.5	The Road to a Competency-Based Curriculum *90*	
4.5	Conclusions and Outlook *92*	
	References *93*	
5	**Empowering Chemistry Teachers' Learning: Practices and New Challenges** *99*	
	Jan H. van Driel and Onno de Jong	
5.1	Introduction *99*	
5.2	Chemistry Teachers' Professional Knowledge Base *102*	
5.2.1	The Knowledge Base for Teaching *102*	
5.2.2	Chemistry Teachers' Professional Knowledge *103*	
5.2.3	Development of Chemistry Teachers' Professional Knowledge *105*	
5.3	Empowering Chemistry Teachers to Teach Challenging Issues *107*	
5.3.1	Empowering Chemistry Teachers for Context-Based Teaching *107*	
5.3.2	Empowering Chemistry Teachers to Teach about Models and Modeling *109*	
5.3.3	Empowering Chemistry Teachers to Use Computer-Based Technologies for Teaching *111*	
5.4	New Challenges and Opportunities to Empower Chemistry Teachers' Learning *113*	
5.4.1	Becoming a Lifelong Research-Oriented Chemistry Teacher *113*	
5.4.2	Learning Communities as a Tool to Empower Chemistry Teachers' Learning *114*	
5.5	Final Conclusions and Future Trends *116*	
	References *118*	

6	Lifelong Learning: Approaches to Increasing the Understanding of Chemistry by Everybody *123*
	John K. Gilbert and Ana Sofia Afonso
6.1	The Permanent Significance of Chemistry *123*
6.2	Providing Opportunities for the Lifelong Learning of Chemistry *123*
6.2.1	Improving School-Level Formal Chemistry Education *123*
6.2.2	Formal Lifelong Chemical Education *125*
6.2.3	Informal Chemical Education *126*
6.2.4	Emphases in the Provision of Lifelong Chemical Education *127*
6.3	The Content and Presentation of Ideas for Lifelong Chemical Education *129*
6.3.1	The Content of Lifelong Chemical Education *129*
6.3.2	The Presentation of Chemistry to Diverse Populations *130*
6.4	Pedagogy to Support Lifelong Learning *131*
6.5	Criteria for the Selection of Media for Lifelong Chemical Education *133*
6.6	Science Museums and Science Centers *133*
6.6.1	Museums *133*
6.6.2	Science Centers *134*
6.7	Print Media: Newspapers and Magazines *134*
6.8	Print Media: Popular Books *135*
6.9	Printed Media: Cartoons, Comics, and Graphic Novels *136*
6.9.1	Three Allied Genre *136*
6.9.2	The Graphic Novel *137*
6.9.3	The Educational Use of Graphic Novels in Science Education *138*
6.9.4	Case Study: A Graphic Novel Concerned with Cancer Chemotherapy *140*
6.10	Radio and Television *140*
6.11	Digital Environments *141*
6.12	Citizen Science *143*
6.13	An Overview: Bringing About Better Opportunities for Lifelong Chemical Education *144*
	References *146*

Part II: Best Practices and Innovative Strategies *149*

7	Using Chemistry Education Research to Inform Teaching Strategies and Design of Instructional Materials *151*
	Renée Cole
7.1	Introduction *151*
7.2	Research into Student Learning *153*
7.3	Connecting Research to Practice *154*
7.3.1	Misconceptions *154*
7.3.2	Student Response Systems *157*

7.3.3	Concept Inventories *158*	
7.3.4	Student Discourse and Argumentation *159*	
7.3.5	Problem Solving *161*	
7.3.6	Representations *161*	
7.3.7	Instruments *163*	
7.4	Research-Based Teaching Practice *165*	
7.4.1	Interactive Lecture Demonstrations *166*	
7.4.2	ANAPOGIL: Process-Oriented Guided Inquiry Learning in Analytical Chemistry *167*	
7.4.3	CLUE: Chemistry, Life, the Universe, and Everything *169*	
7.5	Implementation *171*	
7.6	Continuing the Cycle *172*	
	References *174*	
8	**Research on Problem Solving in Chemistry** *181*	
	George M. Bodner	
8.1	Why Do Research on Problem Solving? *181*	
8.2	Results of Early Research on Problem Solving in General Chemistry *184*	
8.3	What About Organic Chemistry *186*	
8.4	The "Problem-Solving Mindset" *192*	
8.5	An Anarchistic Model of Problem Solving *193*	
8.6	Conclusion *199*	
	References *200*	
9	**Do Real Work, Not Homework** *203*	
	Brian P Coppola	
9.1	Thinking About Real Work *203*	
9.1.1	Defining Real Work: Authentic Learning Experiences *203*	
9.1.2	Doing Real Work: Situated Learning *206*	
9.2	Attributes of Real Work *209*	
9.2.1	Balance Convergent and Divergent Tasks *209*	
9.2.1.1	Convergent Assignments *212*	
9.2.1.2	Divergent Assignments *213*	
9.2.1.3	Balancing Convergent and Divergent Assignments *214*	
9.2.1.4	Convergent Assignments in Team Learning *215*	
9.2.1.5	Divergent Assignments in Team Learning *216*	
9.2.2	Peer Presentations, Review, and Critique *218*	
9.2.2.1	Calibrated Peer Review *221*	
9.2.2.2	Guided Peer Review and Revision *221*	
9.2.2.3	Argumentation and Evidence *222*	
9.2.3	Balance Teamwork and Individual Work *222*	
9.2.3.1	Team-Based Learning: Face-to-Face Teams *222*	
9.2.3.2	Team-Based Learning: Virtual Teams *223*	
9.2.3.3	Team-Based Learning: Laboratory Projects *223*	

9.2.3.4	Team-Based Learning: Collaborative Identification	223
9.2.3.5	Team-Based Learning: Experimental Optimization	224
9.2.4	Students Use the Instructional Technologies	224
9.2.4.1	Learning by Design	224
9.2.4.2	Electronic Homework System: In the Classroom	225
9.2.4.3	Student-Generated Videos	225
9.2.4.4	Student-Generated Animations	225
9.2.4.5	Student-Generated Video Blogs	226
9.2.4.6	Wikipedia Editing	227
9.2.4.7	Wiki Environment	227
9.2.4.8	Student-Generated Metaphors	227
9.2.5	Use Authentic Texts and Evidence	228
9.2.5.1	Literature Summaries	228
9.2.5.2	Literature Seminars	229
9.2.5.3	Public Science Sources	230
9.2.5.4	Generating Questions	230
9.2.5.5	Course-Based Undergraduate Research Experiences (CURE)	230
9.2.5.6	Interdisciplinary Research-Based Projects	231
9.2.6	As Important to the Class as the Teacher's Work	232
9.2.6.1	Student-Generated Instructional Materials	232
9.2.6.2	Wiki Textbooks	232
9.2.6.3	Print and Web-Based Textbooks	233
9.2.6.4	Electronic Homework Systems	235
9.2.6.5	Podcasts	236
9.2.6.6	Classroom: Active-Learning Assignments	238
9.2.6.7	Laboratory: Safety Teams	239
9.3	Learning from Real Work	239
9.3.1	Evidence of Creativity through the Production of Divergent Explanations	240
9.3.2	Peer Review and Critique Reveal Conceptual Weaknesses	240
9.3.3	Team Learning Produces Consistent Gains in Student Achievement	241
9.3.4	Students Use Instructional Technologies	242
9.3.5	Using Authentic Materials Result in Disciplinary Identification and Socialization	243
9.3.6	Student-Generated Instructional Materials Promotes Metacognition and Self-Regulation	244
9.4	Conclusions	245
	Acknowledgments	247
	References	247

10	**Context-Based Teaching and Learning on School and University Level** *259*	
	Ilka Parchmann, Karolina Broman, Maike Busker, and Julian Rudnik	
10.1	Introduction *259*	
10.2	Theoretical and Empirical Background for Context-Based Learning *260*	
10.3	Context-Based Learning in School: A Long Tradition with Still Long Ways to Go *261*	
10.4	Further Insights Needed: An On-Going Empirical Study on the Design and Effects of Learning from Context-Based Tasks *263*	
10.4.1	Strategies to Approach Context-Based Tasks *265*	
10.4.2	Application of Chemical Knowledge *267*	
10.4.3	Outlook on the Design of Tasks and Research Studies *269*	
10.5	Context-Based Learning on University Level: Goals and Approaches *269*	
10.5.1	Design of Differentiated CBL-Tasks *271*	
10.5.2	Example 1 Physical and Chemical Equilibria of Carbon Dioxide – Important in Many Different Contexts *272*	
10.5.3	Example 2 Chemical Switches – Understanding Properties like Color and Magnetism *273*	
10.5.4	Feedback and Implications *275*	
10.6	Conclusions and Outlook *275*	
	References *276*	
11	**Active Learning Pedagogies for the Future of Global Chemistry Education** *279*	
	Judith C. Poë	
11.1	Problem-Based Learning *280*	
11.1.1	History *281*	
11.1.2	The Process *281*	
11.1.3	Virtual Problem-Based Learning *283*	
11.1.4	The Problems *285*	
11.1.4.1	Selected Problems for Introductory Chemistry at the UTM *285*	
11.1.4.2	Project for a UTM Upper Level Bioinorganic Chemistry Course *288*	
11.1.5	PBL – Must Content Be Sacrificed? *289*	
11.2	Service-Learning *290*	
11.2.1	The Projects *291*	
11.2.1.1	Selected Analytical/Environmental Chemistry Projects *291*	
11.2.1.2	Selected Projects in Chemistry Education *292*	
11.2.1.3	Project for an Upper Level Bioinorganic Chemistry Course at UTM *293*	
11.2.2	Benefits of Service-Learning *294*	

11.3	Active Learning Pedagogies 296
11.4	Conclusions and Outlook 297
	References 297

12	**Inquiry-Based Student-Centered Instruction** 301
	Ram S. Lamba
12.1	Introduction 301
12.2	Inquiry-Based Instruction 303
12.3	The Learning Cycle and the Inquiry-Based Model for Teaching and Learning 304
12.4	Information Processing Model 308
12.5	Possible Solution 308
12.6	Guided Inquiry Experiments for General Chemistry: Practical Problems and Applications Manual 310
12.7	Assessment of the Guided-Inquiry-Based Laboratories 314
12.8	Conclusions 316
	References 317

13	**Flipping the Chemistry Classroom with Peer Instruction** 319
	Julie Schell and Eric Mazur
13.1	Introduction 319
13.2	What Is the Flipped Classroom? 320
13.2.1	Three Big Ideas about Flipped Classrooms 321
13.2.2	Blended Learning and Flipped Classrooms 322
13.2.3	A Brief History of the Flipped Classroom 323
13.2.4	Traditional versus a Flipped Chemistry Classroom 323
13.2.5	Flipped Classrooms and Dependency on Technology 324
13.3	How to Flip the Chemistry Classroom 325
13.3.1	Common Myths about Flipped Classrooms 326
13.3.1.1	Myth 1: Flipped Classrooms are Just Video Lectures 326
13.3.1.2	Myth 2: Flipped Classrooms Have No Lectures 326
13.3.1.3	Myth 3: Students Won't Be Prepared for Class 327
13.3.1.4	Myth 4: Flipping Your Classroom Means Changing Everything You Do 327
13.3.1.5	Myth 5: Flipped Classrooms Solve All Students' Problems Immediately 328
13.3.2	FLIP 329
13.3.3	Student Attitudes toward Flipping General Chemistry 329
13.4	Flipping Your Classroom with Peer Instruction 329
13.4.1	What Is Peer Instruction? 330
13.4.2	What Is a ConcepTest? 331
13.4.3	Workflow in a Peer Instruction Course 332
13.4.4	ConcepTest Workflow 333
13.4.5	Peer Instruction and Classroom Response Systems 333
13.4.6	The Instructional Design of a Peer Instruction Course 334

13.4.7	Research on Peer Instruction	*336*
13.4.8	Strategies for Avoiding Common Pitfalls of Flipping the Classroom with Peer Instruction	*336*
13.4.8.1	Effective Grouping	*337*
13.4.8.2	Response Opportunities	*337*
13.4.8.3	Peer Discussion Opportunities	*337*
13.4.8.4	Response Sharing	*338*
13.4.9	Flipping the Chemistry Classroom with Peer Instruction	*338*
13.5	Responding to Criticisms of the Flipped Classroom	*339*
13.6	Conclusion: The Future of Education	*341*
	Acknowledgments	*341*
	References	*341*

14 **Innovative Community-Engaged Learning Projects: From Chemical Reactions to Community Interactions** *345*
Claire McDonnell

14.1	The Vocabulary of Community-Engaged Learning Projects	*345*
14.1.1	Community-Based Learning	*346*
14.1.2	Community-Based Research	*346*
14.1.3	Developing a Shared Understanding of CBL and CBR	*347*
14.2	CBL and CBR in Chemistry	*349*
14.2.1	Chemistry CBL at Secondary School (High School) Level	*352*
14.2.2	Chemistry Projects Not Categorized as CBL or CBR	*352*
14.2.3	Guidelines and Resources for Getting Started	*352*
14.3	Benefits Associated with the Adoption of Community-Engaged Learning	*353*
14.3.1	How Do Learners Gain from CBL and CBR?	*354*
14.3.1.1	Personal Development and Graduate Attributes	*354*
14.3.1.2	High-Impact Educational Practices	*354*
14.3.2	How Do HEIs and Schools Gain from CBL and CBR?	*356*
14.3.3	How Do Communities Gain from CBL and CBR?	*359*
14.3.3.1	Reciprocity	*359*
14.3.3.2	Maximizing Impact for Community Partners	*359*
14.4	Barriers and Potential Issues When Implementing Community-Engaged Learning	*360*
14.4.1	Clarity of Purpose	*360*
14.4.2	Regulatory and Ethical Issues	*360*
14.4.3	Developing Authentic Community Partnerships	*361*
14.4.3.1	Useful Frameworks	*361*
14.4.3.2	Case Studies on Developing Authentic Community Partnerships	*361*
14.4.4	Sustainability	*362*
14.4.5	Institutional Commitment and Support	*363*
14.4.6	An Authentic Learning Environment	*363*
14.4.7	Reflection	*363*

14.5	Current and Future Trends	364
14.5.1	Geographic Spread	364
14.5.2	Economic Uncertainty	364
14.5.3	The Scholarship of Community-Engaged Learning	365
14.5.4	Online Learning	365
14.5.5	Developments in Chemistry Community-Engaged Learning	366
14.6	Conclusion	366
	References	367

15 The Role of Conceptual Integration in Understanding and Learning Chemistry *375*
Keith S. Taber

15.1	Concepts, Coherence, and Conceptual Integration	375
15.1.1	The Nature of Concepts	375
15.1.2	Concepts and Systems of Public Knowledge	377
15.1.3	Conceptual Integration	378
15.2	Conceptual Integration and Coherence in Science	381
15.2.1	Multiple Models in Chemistry	383
15.3	Conceptual Integration in Learning	385
15.3.1	The Drive for Coherence	386
15.3.2	Compartmentalization of Learning	387
15.3.3	When Conceptual Integration Impedes Learning	388
15.3.4	Conceptual Integration and Expertise	389
15.4	Conclusions and Implications	390
15.4.1	Implications for Teaching	390
15.4.2	Directions for the Research Programme	391
	References	392

16 Learners Ideas, Misconceptions, and Challenge *395*
Hans-Dieter Barke

16.1	Preconcepts and School-Made Misconceptions	395
16.2	Preconcepts of Children and Challenge	396
16.3	School-Made Misconceptions and Challenge	396
16.3.1	Ions as Smallest Particles in Salt Crystals and Solutions	397
16.3.1.1	Challenge of Misconceptions	398
16.3.2	Chemical Equilibrium	401
16.3.2.1	Most Common Misconceptions	402
16.3.2.2	Challenge of Misconceptions	402
16.3.3	Acid–Base Reactions and Proton Transfer	405
16.3.4	Redox Reactions and Electron Transfer	411
16.4	Best Practice to Challenge Misconceptions	415
16.4.1	Misconceptions	416
16.4.2	Integrating Misconceptions into Instruction	417
16.5	Conclusion	419
	References	419

17	**The Role of Language in the Teaching and Learning of Chemistry** *421*	
	Peter E. Childs, Silvija Markic, and Marie C. Ryan	
17.1	Introduction *421*	
17.2	The History and Development of Chemical Language *423*	
17.2.1	Chemical Symbols: From Alchemy to Chemistry, from Dalton to Berzelius *423*	
17.2.2	A Systematic Nomenclature *425*	
17.3	The Role of Language in Science Education *428*	
17.4	Problems with Language in the Teaching and Learning of Chemistry *430*	
17.4.1	Technical Words and Terms *432*	
17.4.2	Nontechnical Words *433*	
17.4.3	Logical Connectives *434*	
17.4.4	Command Words *435*	
17.4.5	Argumentation and Discourse *436*	
17.4.6	Readability of Texts *436*	
17.5	Language Issues in Dealing with Diversity *437*	
17.5.1	Second Language Learners *437*	
17.5.2	Some Strategies for Improving Language Skills of SLLs *440*	
17.5.3	Special-Needs Students *440*	
17.6	Summary and Conclusions *441*	
	References *442*	
	Further Reading *445*	
18	**Using the Cognitive Conflict Strategy with Classroom Chemistry Demonstrations** *447*	
	Robert (Bob) Bucat	
18.1	Introduction *447*	
18.2	What Is the Cognitive Conflict Teaching Strategy? *448*	
18.3	Some Examples of Situations with Potential to Induce Cognitive Conflict *449*	
18.4	Origins of the Cognitive Conflict Teaching Strategy *451*	
18.5	Some Issues Arising from *A Priori* Consideration *453*	
18.6	A Particular Research Study *455*	
18.7	The Logic Processes of Cognitive Conflict Recognition and Resolution *459*	
18.8	Selected Messages from the Research Literature *461*	
18.9	A Personal Anecdote *465*	
18.10	Conclusion *466*	
	References *467*	
19	**Chemistry Education for Gifted Learners** *469*	
	Manabu Sumida and Atsushi Ohashi	
19.1	The Gap between Students' Images of Chemistry and Research Trends in Chemistry *469*	

19.2	The Nobel Prize in Chemistry from 1901 to 2012: The Distribution and Movement of Intelligence *470*
19.3	Identification of Gifted Students in Chemistry *472*
19.3.1	Domain-Specificity of Giftedness *472*
19.3.2	Natural Selection Model of Gifted Students in Science *474*
19.4	Curriculum Development and Implementation of Chemistry Education for the Gifted *477*
19.4.1	Acceleration and Enrichment *477*
19.4.2	Higher Order Thinking and the Worldview of Chemistry *478*
19.4.3	Promoting Creativity and Innovation *479*
19.4.4	Studying Beyond the Classrooms *480*
19.4.5	Can the Special Science Program Meet the Needs of Gifted Students? *482*
19.5	Conclusions *484*
	References *486*

20	**Experimental Experience Through Project-Based Learning** *489*
	Jens Josephsen and Søren Hvidt
20.1	Teaching Experimental Experience *489*
20.1.1	Practical Work in Chemistry Education *489*
20.1.2	Why Practical Work in Chemistry Education? *490*
20.1.3	Practical Work in the Laboratory *491*
20.2	Instruction Styles *492*
20.2.1	Different Goals and Instruction Styles for Practical Work *492*
20.2.2	Emphasis on Inquiry *493*
20.3	Developments in Teaching *494*
20.3.1	Developments at the Upper Secondary Level *494*
20.3.2	Trials and Changes at the Tertiary Level *495*
20.3.3	Lessons Learned *497*
20.4	New Insight and Implementation *498*
20.4.1	Curriculum Reform and Experimental Experience *498*
20.4.1.1	Problem-Based Group-Organized Project Work *498*
20.4.1.2	Second Semester Project Work *499*
20.4.2	Analysis of Second Semester Project Reports *502*
20.4.2.1	Analysis of Reports from a Chemistry Point of View *503*
20.4.2.2	Elements of Experimental Work *503*
20.5	The Chemistry Point of View Revisited *511*
20.6	Project-Based Learning *512*
	References *514*

21	**The Development of High-Order Learning Skills in High School Chemistry Laboratory: "Skills for Life"** *517*
	Avi Hofstein
21.1	Introduction: The Chemistry Laboratory in High School Setting *517*

21.2	The Development of High-Order Learning Skills in the Chemistry Laboratory *519*
21.2.1	Introduction *519*
21.2.2	What Are High-Order Learning Skills? *520*
21.3	From Theory to Practice: How Are Chemistry Laboratories Used? *522*
21.4	Emerging High-Order Learning Skills in the Chemistry Laboratory *523*
21.4.1	First Theme: Developing Metacognitive Skills *523*
21.4.2	Second Theme: Scientific (Chemical) Argumentation *527*
21.4.2.1	The Nature of Argumentation in Science Education *527*
21.4.2.2	Argumentation in the Chemistry Laboratory *528*
21.4.3	Asking Questions in the Chemistry Laboratory *531*
21.5	Summary, Conclusions, and Recommendations *532*
	References *535*
22	**Chemistry Education Through Microscale Experiments** *539*
	Beverly Bell, John D. Bradley, and Erica Steenberg
22.1	Experimentation at the Heart of Chemistry and Chemistry Education *539*
22.2	Aims of Practical Work *540*
22.3	Achieving the Aims *540*
22.4	Microscale Chemistry Practical Work – "The Trend from Macro Is Now Established" *541*
22.5	Case Study I: Does Scale Matter? Study of a First-Year University Laboratory Class *542*
22.6	Case Study II: Can Microscale Experimentation Be Used Successfully by All? *543*
22.7	Case Study III: Can Quantitative Practical Skills Be Learned with Microscale Equipment? *544*
22.7.1	Volumetric Analysis – Microtitration *544*
22.7.2	Gravimetric Measurements *546*
22.7.3	The Role of Sensors, Probes, and the Digital Multimeter in Quantitative Microscale Chemistry *548*
22.7.3.1	Cell Potential Measurements *549*
22.7.3.2	Electrical Conductivity, Light Absorption, and Temperature Measurements *551*
22.8	Case Study IV: Can Microscale Experimentation Help Learning the Scientific Approach? *554*
22.9	Case Study V: Can Microscale Experimentation Help to Achieve the Aims of Practical Work for All? *555*
22.9.1	The UNESCO-IUPAC/CCE Global Microscience Program and Access to Science Education for All *555*

22.9.2	The Global Water Experiment of the 2011 International Year of Chemistry – Learning from the Experience *556*
22.10	Conclusions *559*
	References *559*

Part III: The Role of New Technologies *563*

23 Twenty-First Century Skills: Using the Web in Chemistry Education *565*
Jan Apotheker and Ingeborg Veldman

23.1	Introduction *565*
23.2	How Can These New Developments Be Used in Education? *567*
23.3	MOOCs (Massive Open Online Courses) *572*
23.4	Learning Platforms *574*
23.5	Online Texts versus Hard Copy Texts *575*
23.6	Learning Platforms/Virtual Learning Environment *577*
23.7	The Use of Augmented Reality in (In)Formal Learning *579*
23.8	The Development of Might^y/Macht^{ig} *580*
23.9	The Evolution of MIGHT-^{y} *580*
23.10	Game Play *581*
23.11	Added Reality and Level of Immersion *582*
23.12	Other Developments *586*
23.13	Molecular City in the Classroom *587*
23.14	Conclusion *593*
	References *593*

24 Design of Dynamic Visualizations to Enhance Conceptual Understanding in Chemistry Courses *595*
Jerry P. Suits

24.1	Introduction *595*
24.1.1	Design of Quality Visualizations *595*
24.1.2	Mental Models and Conceptual Understanding *596*
24.2	Advances in Visualization Technology *598*
24.3	Dynamic Visualizations and Student's Mental Model *603*
24.4	Simple or Realistic Molecular Animations? *607*
24.5	Continuous or Segmented Animations? *608*
24.6	Individual Differences and Visualizations *609*
24.6.1	Self-Explanations and Spatial Ability *609*
24.6.2	Individual Differences and Visualization Studies *610*
24.7	Simulations: Interactive, Dynamic Visualizations *611*
24.7.1	Pedagogic Simulations *611*
24.7.2	An Organic Pre-Lab Simulation *613*
24.8	Conclusions and Implications *615*
	Acknowledgments *616*
	References *616*

25	**Chemistry Apps on Smartphones and Tablets** *621*	
	Ling Huang	
25.1	Introduction *621*	
25.2	Operating Systems and Hardware *625*	
25.3	Chemistry Apps in Teaching and Learning *626*	
25.3.1	Molecular Viewers and Modeling Apps *626*	
25.3.2	Molecular Drawing Apps *629*	
25.3.3	Periodic Table Apps *631*	
25.3.4	Literature Research Apps *633*	
25.3.5	Lab Utility Apps *634*	
25.3.5.1	Flashcard Apps *635*	
25.3.5.2	Dictionary/Reference Apps *636*	
25.3.5.3	Search Engine Apps *637*	
25.3.5.4	Calculator Apps *639*	
25.3.5.5	Instrumental Apps *640*	
25.3.6	Apps for Teaching and Demonstration *641*	
25.3.7	Gaming Apps *642*	
25.3.8	Chemistry Courses Apps *644*	
25.3.9	Test-Prep Apps *644*	
25.3.10	Apps are Constantly Changing *645*	
25.4	Challenges and Opportunities in Chemistry Apps for Chemistry Education *646*	
25.5	Conclusions and Future Perspective *647*	
	References *649*	
26	**E-Learning and Blended Learning in Chemistry Education** *651*	
	Michael K. Seery and Christine O'Connor	
26.1	Introduction *651*	
26.2	Building a Blended Learning Curriculum *652*	
26.3	Cognitive Load Theory in Instructional Design *654*	
26.4	Examples from Practice *655*	
26.4.1	Podcasts and Screencasts *656*	
26.4.2	Preparing for Lectures and Laboratory Classes *657*	
26.4.3	Online Quizzes *659*	
26.4.4	Worked Examples *661*	
26.4.5	Clickers *662*	
26.4.6	Online Communities *663*	
26.5	Conclusion: Integrating Technology Enhanced Learning into the Curriculum *665*	
	References *666*	
27	**Wiki Technologies and Communities: New Approaches to Assessing Individual and Collaborative Learning in the Chemistry Laboratory** *671*	
	Gwendolyn Lawrie and Lisbeth Grøndahl	
27.1	Introduction *671*	

27.2	Shifting Assessment Practices in Chemistry Laboratory Learning *672*	
27.3	Theoretical and Learning Design Perspectives Related to Technology-Enhanced Learning Environments *675*	
27.4	Wiki Learning Environments as an Assessment Platform for Students' Communication of Their Inquiry Laboratory Outcomes *678*	
27.4.1	Co-Construction of Shared Understanding of Experimental Observations *679*	
27.4.2	Enhancing the Role of Tutors in the Wiki Laboratory Community *679*	
27.5	Practical Examples of the Application of Wikis to Enhance Laboratory Learning Outcomes *681*	
27.5.1	Supporting Collaborative Discussion of Experimental Data by Large Groups of Students during a Second-Level Organic Chemistry Inquiry Experiment *681*	
27.5.2	Virtual Laboratory Notebook Wiki Enhancing Laboratory Learning Outcomes from a Collaborative Research-Style Experiment in a Third-Level Nanoscience Course *682*	
27.5.3	Scaffolding Collaborative Laboratory Report Writing through a Wiki *682*	
27.6	Emerging Uses of Wikis in Lab Learning Based on Web 2.0 Analytics and Their Potential to Enhance Lab Learning *684*	
27.6.1	Evaluating Student Participation and Contribution as Insight into Engagement *684*	
27.6.2	Categorizing the Level of Individual Student Understanding *686*	
27.7	Conclusion *688*	
	References *689*	

28 New Tools and Challenges for Chemical Education: Mobile Learning, Augmented Reality, and Distributed Cognition in the Dawn of the Social and Semantic Web *693*
Harry E. Pence, Antony J. Williams, and Robert E. Belford

28.1	Introduction *693*	
28.2	The Semantic Web and the Social Semantic Web *694*	
28.3	Mobile Devices in Chemical Education *702*	
28.4	Smartphone Applications for Chemistry *706*	
28.5	Teaching Chemistry in a Virtual and Augmented Space *708*	
28.6	The Role of the Social Web *717*	
28.7	Distributed Cognition, Cognitive Artifacts, and the Second Digital Divide *721*	
28.8	The Future of Chemical Education *726*	
	References *729*	

Index *735*

Foreword

What is it about chemistry? Why do so many students, having tasted it in high school, turn away from it with distaste and remember only the horror of their experience? Why, on the other hand, are other students immediately hooked on it and want it to lie at the core of their studies and subsequent careers? The issue is plainly important, for chemistry touches us all, like it or not, and everyone's role in and interaction with society depends on at least an appreciation of what chemists and the chemical industry achieve, especially in the light of dangers to the environment that it presents and the extraordinary positive contribution it makes to everyday and ever-longer life. Moreover, those who turn their back on chemistry are closing their minds to its cultural contribution to understanding the nature of the world around them. Motivation is plainly important, and there is plenty of it lying around, as the contributors emphasize: their message is that if you seek motivation, then look around, for chemistry deepens our understanding of the natural world, be it through our natural environment or the artifacts of the industry. Once motivated, there is an obligation, as the authors rightly argue, for that enthusiasm to be encouraged throughout life, not merely at the incubators of school and college.

Why does chemical education play such a pivotal role? I think the essence of the difficulty of learning chemistry is the combination of the perceived abstraction of its concepts and the fact that (unlike so often in physics) there is such a tension between possible explanations that judgment is needed to arrive at the true explanation. The abstraction, of course, is perceived rather than real. We educated chemists *all* know that atoms and molecules are real, and we are confident about our reasoning about energy and entropy; however, the neophyte has no such confidence and needs to come to terms with the reality of the infrastructure of our explanations. A part of this volume is the exploration of how to convey our concepts in an accessible way, in part planting but also dispelling misconception, perhaps by using that powerful entry into the brain, visualization. Furthermore, there is the question of judgment: chemistry is, in fact, a multidimensional tug-of-war, with rival influences in perplexing competition. Is it ionization energy that should be dominant in an explanation or is it some other aspect of structure or bulk matter? How can the starting student learn to judge what is dominant and retain self-confidence?

Pervading these problems is the perennial problem of problem-solving. How can this most inductive of activities be ingrained into the thinking of our students? I frankly do not know; however, the authors struggle here with the challenge. It probably comes down to ceaseless demonstration of how we practitioners of chemistry practice our profession: a ceaseless Confucian exposure to the actions of masters in the hope that skill will emerge through observation and emulation. We see a little of what is involved in this text; however, it is central to education, and perhaps there should have been more of it. Volume 2, should it ever emerge, might take up that theme and explore another omission, the role of mathematics in science in a universe where confident deployment is in decline in many countries and is a source of worry to us all. Mathematics adds spine to otherwise jelly-like qualitative musings, enabling them to stand up to quantitative exploration and is absolutely central to the maintenance of chemistry as a part of the physical sciences. How can students be led from the qualitative into the quantitative, and how can they distil the meaning of, not merely derive, an equation? There is little of that here; however, it is crucial to the future of our subject and is related to the formulation of solutions to problems.

In short, the concepts of chemistry at first sight are abstract, its arguments intricate, its formulation sometimes mathematical, and its applications spanning widely between the horizons of physics and biology. This perfect storm of aspects can be overwhelming and, unless handled with the utmost care and professional judgment, results in confusion and disaffection. The responsibility of educators is to calm this storm.

The improvement of chemical education, to ensure not only a progression of specialists but also an appreciation of its content, role, and attitude among that most elusive but vital entity, the general public, is of paramount importance in the modern world. Collected in this volume are contributions from many notable thinkers and writers who have devoted their intellectual life to seeking ways to advance society by improving chemical education at all levels. Thus, they need to confront the identification of the central concepts of chemistry and how they can be rendered familiar and concrete. How do the ways that chemists think become deconstructed, then repackaged for transmission? How should the central importance of mathematics be illustrated, and how does quantitative reasoning get conveyed convincingly and attractively? How should the intricacies of applications be presented such that they do not overwhelm the simplicities of the underlying ideas? In all these considerations, where does the balance lie between the education of a specialist and the well-informed member of general society?

It is also not as though there is a shortage of ideas about how to proceed. This timely volume displays the current vigor of research into chemical education and the range of approaches being explored to carry out this most valuable and important of tasks. Should social conscience be deployed to motivate, as in concern for the environment, or should motivation be sought it an appreciation of the material fruits of chemistry? As an academic and probably out-of-touch purist, I wonder whether elaborately contrived motivation is helpful, believing that an emerging sunrise of intellectual love of understanding should be motivation enough. Should classrooms be inverted, as some authors argue, to generate more involvement in the process of learning, or should downward projection authority-to-student succeed more effectively in the transmission of learning? These matters are discussed here by those who have explored their efficacy in practice. I suppose the issue is whether learning can be democratized, with instructor and student as equal partners, or whether a touch of the whip of authority is advantageous.

The authors of this collection of essays are sensitive to the problems of introducing the young to the special language of chemistry. Common sense is all very well; however, a great deal of science is concerned with looking under everyday perceptions of the world and identifying their infrastructure, which at first sight sometimes seems to run against common sense and opens the door to misconception. Science, in fact, deepens common sense. The central point, apart from the precision that comes from careful definition, is to show how a new language is needed when entering any new country, in this case a country of the intellect.

With the language in place, or at least emerging, it is necessary to turn to a consideration of what is in effect its syntax: the stringing together of concepts and techniques to solve problems. Problem-solving is perhaps the most troublesome aspect of chemical education, being largely inductive, and a huge amount of attention is rightly directed at its techniques, including the roles of instructors and peers.

Crucial to this endeavor is the demonstration that the concepts and calculations of chemistry relate to actual physical phenomena (or should) and that experiment and observation, not ungrounded algebra, lie at the heart of science. The contributions acknowledge this core feature of science, and although microscale experiments, which are discussed here, are not to everyone's taste, they are far better than unsupported printed assertion and unadorned abstraction.

Many of the problems of chemical education have been around for decades, perhaps a century or more, ever since chemistry became a rational subject and numbers were attached to matter. New problems and concomitantly extraordinary opportunities are now emerging as new technologies move to an educator's reach. The later sections of this book are like the emergence of mammals in the world of dinosaurs (I do not intend to be in the least disrespectful to my wonderful colleagues, but merely to draw an analogy!): new technologies are the future, possible savior, and, undoubtedly, enhancer of chemical education. They do not simply enhance our present procedures; they have the potential to be transformative in the same way that plastics have replaced wood.

Almost by definition, "new technologies" are in their infancy, with even the farsighted seeing only dimly the extraordinary opportunities that they will bring to chemical education. However, the crucial point is that those opportunities must not run wild: they must build on the extraordinary insights and expertise of the extant practitioners of chemical education, developing securely on a strong foundation. This collection of chapters contributes substantially to that strong foundation and will provide inspiration and insight for old-timers and newcomers alike.

For me, the most exciting chapter of this collection is the one that peers into the future to explore the consequences of the ubiquity of devices that tap into that store of universal knowledge we know as the Internet. We are all currently groping to find ways to employ this extraordinary resource, currently standing on the shore of the ocean of opportunity that it represents, still unaware of what lies over the horizon. It is already influencing publishing and the dissemination of knowledge, and it is facilitating the involvement of the entire academic community in corporate activity, transforming the attitude to personally stored information as distinct from publicly available data, affecting the deployment of information, and encouraging interpersonal accessibility and cooperation. The future of chemical education lies here, and this volume provides a glimpse of what it might bring.

I am not a chemical educator in the professional sense of the term; however, I am deeply involved in the deployment of its activities. As such, I welcome a volume that brings together in a single source so many different, multiple facets of this intricate and rewarding exercise. The authors and their editors should be congratulated on the timeliness of this publication, acting as a pivot between good practice in the present and opportunity in the future.

Peter Atkins
University of Oxford

Preface

The Science of Teaching and Learning Chemistry

The world we live is increasingly complex and interconnected; a world where an event in a corner of a remote country can rapidly grow and affect millions of people in places thousands of kilometers away. Both globalization and technology provide us with great opportunities and also with enormous challenges. Our planet is becoming increasingly crowded and interdependent. From climate change to access to water and from food security to new pandemics, the number of global challenges and their implications on our future is truly daunting.

But as US President John F. Kennedy said in 1963: "Our problems are man-made; therefore they may be solved by man." Many solutions, from new vaccines to cleaner ways to produce energy, will only be made possible by the right science and technology. As in the past, mankind has overcome its problems through science: terrible illnesses and poor living conditions have been overcome through the ingenuity and hard work of great men and women. From the artificial synthesis of ammonia, which allowed the green revolution, to the discovery of antibiotics, the breakthroughs of a few have improved the lives of many.

But with all that science has achieved to date, technological advances alone are insufficient to continue to address mankind's challenges. The *human* drive for improvement, the attitude, the willingness to contribute, and the desire to help solve problems is at least as important as having the right tools. Therefore, investment in education is also an essential component of any attempt to build a better and more sustainable future, as education interconnects the human desire to help with the science that creates solutions. Science and education are two of the most common elements discussed when talking about how to build a better future. Part of this "investment" is exactly allocating enough resources to make sure that long-term objectives are possible. Financial investment is not enough; we need to be able to teach science in the most effective way to create a new generation of scientists who are able to find the solutions to our global challenges and then take those solutions from the laboratory to the market place.

Both the teaching and learning of science in general – and chemistry in particular – are not easy tasks. Each requires hard work, dedication, and practice. There is definitely a component of "art" (one could even say craftsmanship) in effectively communicating complex chemistry concepts, many related to the molecular world. But there is much more science in chemistry education than many teachers and students appreciate. Years of research in chemistry education have provided clear and well-established results in terms of best practices, common mistakes, and which tools are most effective.

Despite the decades of research on chemistry education, the authors of this book were moved by how little the broad chemistry community knows about the results of this work. We felt it was about time to invite some of the world's leading experts to contribute an original piece to a compendium of the most effective ways to teach and learn chemistry. Obviously, no single person could write such a book. This book is therefore a diverse, sometimes controversial, but always interesting collection of chapters written by leading experts in chemistry education.

Learning and teaching chemistry is far from an exact science, but there are plenty of lessons to take from the research done so far. In fact, it is quite surprising how little has changed the way chemistry is taught in the last century despite all the recent advances in chemistry and the numerous possibilities that information technologies offer. A typical vision of a general chemistry course will still be an image of a large classroom packed with students who passively listen to a single person.

Some of the most interesting research in chemistry education deals with the way we learn: how we grasp new concepts and connect the macro with the micro world. The three traditional thinking levels of chemistry: macroscopic, molecular, and symbolic, all require a different way to communicate, visualize, and comprehend new concepts. Another critically important topic in chemistry education is the role of misconceptions. Every student enters the classroom with his or her own bag of ideas about "how the world works." Many of these come from the way previous teachers have taught them key concepts. Other preconceptions come from students' personal interpretations of their experience. Identifying these misconceptions and knowing how to challenge them is critically important, but rarely done in a chemistry course.

In addition to all the opportunities that the years of research conducted on how to efficiently teach and learn chemistry offer to the those interested in chemistry education, technology itself is also bringing a whole set of opportunities (and of course challenges) to both educators and learners. The easy and immediate access to chemistry courses through different Internet-based platforms is radically changing the way our students study, expand their own interests, and interact with their teachers and peers.

And of course, in addition to all of this, the more fundamental fact remains that every single student is a different person. Although there are many things we can do to improve the way chemistry is taught, there are no silver bullets. Our students are evolving individuals, with their own personalities, interests, and challenges.

This book consists of 28 chapters grouped into three parts: *Chemistry Education: A Global Endeavor*, *Best Practices and Innovative Strategies*, and *The Role of the New Technologies*.

The first part, covering Chapters 1–6, provides a broad introduction to the book and touches on critically important aspects of chemistry education. The opening chapter introduces the reader to the scope and the context of the book. In this chapter, Prof. Peter Mahaffy of the King's University College provides an excellent analysis of the connection between human activity and education in general, and then in chemistry in particular. Prof. Mahaffy asserts that the difference between "Chemical Education" and "Chemistry Education" is human activity. The tetrahedral chemistry education metaphor, an extension of the triangle of thinking levels that includes the focus on human activity in their three dimensions in learning and teaching chemistry, is nicely reviewed to give some keys to overcome the barriers to change from "Chemical" to "Chemistry" education.

Chapter 2, by Prof. Cathy Middlecamp of the University of Wisconsin-Madison, is focused on the connection between chemistry education and "the real world" as a high-level thinking skill. As pointed out by the author, "if we can better see the connections, we have set the stage for transforming the way we think. In turn, we can better recognize and meet our responsibilities." Further on, the connection between chemistry curriculum and the content of chemistry news is addressed by Prof. Mei-Hung Chiu and Prof. Chin-Cheng Chou in Chapter 3, where a deep analysis of the need to bridge formal school chemistry with chemistry in everyday life is carried out.

In Chapter 4, Prof. Goedhart of the University of Groningen sketches how curricula in universities transformed as a result of a changing environment and the effectiveness of the new pedagogical approaches, based on the combination of pedagogical ideas and the use of authentic learning environments on the teaching and learning of chemistry. Finally, a new division of chemistry from a competency-based perspective, which can be used as the basis for the structure of a new curriculum, is proposed.

Chapters 5 and 6 are written based on the idea that chemistry teachers need to develop their professional knowledge and practice throughout their entire career, a field closely related to the main focus of this book. Chapter 5, by Prof. Jan H. van Driel of the University of Leiden and Prof. Onno de Jong of Utrecht University, focuses on empowering chemistry teachers' professional learning, identifying successful approaches to promote chemistry teacher learning and the specific areas that present challenges to chemistry teachers. In particular, the authors address context-based teaching, teaching about models and modeling, and the use of computer-based technologies. Chapter 6, by Prof. John K. Gilbert of King's College London and Dr. Ana Sofia Afonso of the University of Minho, discusses the need for increased efforts to both revise the school chemistry curriculum, so that more students are encouraged to persist in the study of the subject, and make the ideas of chemistry more readily available and appealing to adults.

The second part of the book (Chapters 7–22) deals with the most innovative practices and strategies derived from years of research in chemistry education for efficacious learning and teaching of chemistry at different levels. Chapter 7, by Prof. Renée Cole of the University of Iowa, gives an excellent survey of the general field and a comprehensive introduction of teaching strategies and the design of instructional materials (research-based materials) developed so far to improve chemistry education. In Chapter 8, Prof. George M. Bodner of Purdue University focuses on problem solving in chemistry, describing the model developed by the author's research group and their more than 30 years of research in this content domain. Chapter 9, by Prof. Brian P. Coppola of the University of Michigan, deals with the design of real work for a successful learning of chemistry based on a six-part framework of tenets: (i) use of authentic texts; (ii) a balance of team and individual work; (iii) peer presentation, review, and critique; (iv) student-generated instructional material; (v) a balance of convergent and divergent tasks against the traditional homework; and (vi) as important to the class as the teacher's work.

Active learning pedagogies such as the so-called context-based learning (CBL), problem-based learning (PBL), and inquiry-based student-centered instruction are carefully reviewed in Chapters 10–12, respectively. Chapter 10, by Prof. Ilka Parchman of the University of Kiel *et al.*, focuses on CBL pedagogy. As pointed out by the authors, chemistry seems to be an interesting and encouraging area for some students, while others do not see relevance for it to their own life and interests. The CBL pedagogy aims to overcome this challenge by not only linking chemistry to applications that often refer to daily life or societal issues but also linking chemistry to modern research and development. In a similar way, in Chapter 11 Prof. Judith C. Pöe of the University of Toronto Mississauga carefully reviews the use of PBL, a process by which the content and methods of a discipline are learned in an environment in which they are to be used to address a real-world problem, on the learning and teaching of chemistry. In Chapter 12, Prof. Ram S. Lamba of Carlos Albizu University describes the most recent advances in student-centered inquiry-based instruction, giving guidance to instructors on how to interact with students during instruction, how to design activities for classroom use and what to emphasize, as the goal of instruction is to enable students to think like scientists do. These active learning pedagogies, all of them recommended to reach beyond the front rows of our classes, allow the students to develop an enhanced sense of responsibility for their learning and for the applications of their learning, a key point in global learning communities. The implementation of an efficacious flipped classroom as a model based on a student-centered learning environment, and the use of those and the related active learning pedagogies as a part of the flipped-class process, is then discussed in Chapter 13 by Dr. Julie Shell and Prof. Eric Mazur of Harvard University.

A critical review of developments in community-based learning and community-based research in chemistry education at second and third levels is provided in Chapter 14 by Prof. Claire McDonnell of Dublin Institute of Technology.

Chapter 15, by Prof. Keith S. Taber of the University of Cambridge, is aimed to highlight the importance of the notion of conceptual integration in teaching and learning chemistry from two perspectives: (i) the theory of learning (the linking of concepts within current understanding is considered to facilitate further learning and later accessing of that learning); (ii) the nature of science (NOS) – increasingly considered a central curricular aim – for helping students to relate ideas about the submicroscopic realm of molecules, ions, and electrons to the macroscopic description of the subject. Related to conceptual integration, Chapter 16, by Prof.

Hans-Dieter Barke of the University of Münster, is centered on the most representative student's preconcepts and student's misconceptions related to chemistry, giving some instructions on how to prevent it and to overcome them during the teaching of chemistry at different levels. In a broader way, Chapter 17, by Prof. Peter E. Childs *et al.* of the University of Limerick, looks at the role of language in the teaching and learning of chemistry, focusing not only on the typical problems related to terminology and symbols but also on other language-related problems such as the use of nontechnical terms in chemistry which have a different meaning to their use in everyday discourse, for example, to students that are non-native speakers, suggesting some teaching strategies to reduce the barrier and facilitate a novice's mastery of chemical language.

Chapter 18, by Prof. Robert Bucat of the University of Western Australia, deals with the use of the cognitive conflict strategy in classroom chemistry demonstrations. This chapter, oriented to secondary school teachers and university lecturers, concerns the use of discrepant events to induce cognitive conflict in students' understanding of chemistry, with references to particular experiences and some theoretical references, and consideration of the conditions under which they may (or may not) be effective.

Chapter 19, by Prof. Manabu Sumida and Dr. Atsushi Ohashi of Ehime University, outlines the characteristics of gifted learners in science, focusing on identification, curriculum development, and the implementation of gifted education in chemistry from diverse contexts. In this chapter, Prof. Manabu Sumida also illustrates how giftedness in chemistry is required in the new century by analyzing the world trends of Nobel Laureates in chemistry from 1901 to 2012. According to *The New York Times* (December 15, 2013), "even gifted students can't keep up, in math and science, the best fend for themselves. The nation (US) has to enlarge its pool of the best and brightest science and math students and encourage them to pursue careers that will keep the country competitive."

Chapter 20, by Prof. Jens Josephsen and Prof. Søren Hvidt of Roskilde University, discusses the outcomes of the use of different types and aims of experimental work in chemistry education, including the project-based learning pedagogy as an effective tool for students' experience with scientific inquiry processes and obtains practical laboratory skills, experimental experience, and other skills needed by an experimental chemist. In the same vein, research-based evidence showing that high-order learning skills can be developed by involving the students in inquiry-oriented high school laboratories in chemistry is discussed by Prof. Avi Hofstein of Weizmann Institute of Science in Chapter 21.

The second part of the book concludes with a chapter on microscale experiments, by Prof. John D. Bradley *et al.* of the University of Witwatersrand (Chapter 22), where different case studies are analyzed. This chapter is dedicated to Prof. Erica Steenberg (1953–2013), whose valuable contributions to chemistry education were many, especially through microscale experimentation.

In the third part of this book (Chapters 23–28), the central question is focused on the role of new technologies on learning and teaching of chemistry. This part begins with an introductory chapter by Prof. Jan Apotheker and Ingeborg Veenstra of the University of Groningen on several resources on the Internet that can be used in education, introducing the concept of technological pedagogical content knowledge as a condition for the design of instructional materials, and giving some recommendations derived from a particular case study on augmented reality developed at this university (Chapter 23). Chapter 24, by Prof. Jerry P. Suits of the University of Northern Colorado, is more focused on the design of dynamic visualizations to enhance conceptual understanding in chemistry courses. Recent advances in visualization technologies (good multimedia software) and research studies in this field applied to chemistry education are carefully reviewed to analyze how students use dynamic visualizations to internalize concepts and imagery and to explore chemical phenomena.

From students in universities, high school, college, and graduate school, to chemical professionals, and teachers, everyone has a mobile computing device such as smartphone or/and a tablet. So, many chemistry-related apps are seeing dramatic growth with increasing adoption rates to enhance the chemistry teaching and learning experience in the classrooms and laboratories. Chapter 25, by Prof. Ling Huang of Hofstra University, covers the use of these chemistry apps in different teaching contexts, analyzing the pros and cons of using them in chemistry education and extracting conclusions about future trends.

The shift from the picture of a general chemistry course composed by a large classroom packed with students, who passively listen to a single person, to learner-centered collaborative based on active learning environments has provoked a parallel increase in the use of Web 2.0/3.0 technologies in active learning pedagogies. Two closely related approaches on this topic are included in this book, Chapter 26, by Dr. Michael K. Seery and Dr. Christine O'Connor of Dublin Institute of Technology, and Chapter 27, by Prof. Gwendolyn Lawrie and Prof. Lisbeth Grøndahl of the University of Queensland. Chapter 26 is more focused on e-learning and blended learning in chemistry education, while Wiki technologies and communities as a part of e-learning and blended learning approaches are covered by Chapter 27.

Finally, this book is concluded with Chapter 28 by Prof. Robert E. Belford *et al.* by the University of Arkansas, which attempts to contextualize contemporary Information and Communication Technologies (ICT) challenges to education and the practice of science in a perspective of relevance to the twenty-first-century chemical educators.

This book was inspired by many interactions with members of the IUPAC Committee on Chemistry Education, a truly dedicated group of educators, and it is the result of several years of work conducted by a large number of experts in the field, many chemistry professors with decades of experience. The final product is a fascinating read, covering a wide range of topics. But let us be clear, there are no magic solutions. However, if you are interested in knowing what years of research on how to best teach and best learn chemistry has produced, and how to take these lessons into your own classroom, this book is a great source of information and, we hope, of inspiration too. All of the contributing authors have put a significant amount of time aside from their daily work to produce this collective work, in order to help others to teach and learn chemistry more effectively. We want to thank all and each of them for their work and invaluable contributions. Hopefully, this book will help you learn about the best practices, opportunities, and trends that years of research in chemistry education has to offer to anyone involved in the teaching or learning of chemistry.

Javier Garcia-Martinez and
Elena Serrano-Torregrosa
University of Alicante
December 2014

List of Contributors

Ana Sofia Afonso
University of Minho
Department of Education
Braga
Portugal

Jan Apotheker
Science LinX
Faculty of Mathematics and
Natural Sciences
University of Groningen
Nijenborgh 9
9717 CG Groningen
The Netherlands

Hans-Dieter Barke
University of Münster
Institut für Didaktik der Chemie
Fliednerstraße 21
48149 Münster
Germany

Robert E. Belford
University of Arkansas at Little
Rock, Department of Chemistry
2801 South University Avenue
Little Rock, AR 72204
USA

Beverly Bell
University of the Witwatersrand
RADMASTE Centre
27 St Andrews Road
Parktown
Johannesburg
South Africa

George M. Bodner
Purdue University
560 Oval Drive
West Lafayette, IN 47907
USA

John D. Bradley
University of the Witwatersrand
RADMASTE Centre
27 St Andrews Road
Parktown
Johannesburg
South Africa

Karolina Broman
Umeå University
Department of Science and
Mathematics Education
Umeå 90187
Sweden

Robert (Bob) Bucat
The University of Western Australia
School of Chemistry and Biochemistry
Crawley Campus
Bayliss Building, Room 127
35 Stirling Highway
Crawley, WA 6009
Australia

Maike Busker
University of Flensburg
Department of Chemistry Education
Flensburg
Germany

Peter E. Childs
University of Limerick
Department of Chemical and Environmental Sciences
National Centre for Excellence in Mathematics and Science Teaching and Learning
Plassey Park
Limerick
Ireland

Mei-Hung Chiu
National Taiwan Normal University
Graduate Institute of Science Education
88, Sec 4 Ting-Chou Road
Taipei, 11678
Taiwan

Chin-Cheng Chou
National Taipei University of Education
Department of Science Education
No. 134, He-ping E. Rd., Sec. 2.
Taipei, 10671
Taiwan, R.O.C.

Renée Cole
The University of Iowa
Department of Chemistry
W331 Chemistry Building
Iowa City, IA 52242-1294
USA

Brian P. Coppola
University of Michigan
College of Literature, Science, and the Arts
Department of Chemistry
930 N. University
Ann Arbor, MI 48109-1055
USA

Jan H. van Driel
ICLON – Leiden University
Graduate School of Teaching
PO Box 905
2300 AX Leiden
The Netherlands

John K. Gilbert
The University of Reading
King's College London
Australian National Unversity
International Journal of Science Education
Australia

and

The University of Reading
Reading RG6 1HY
UK

Martin J. Goedhart
University of Groningen
Faculty of Mathematics and
Natural Sciences
Nijenborgh 9
9747 AG Groningen
The Netherlands

Lisbeth Grøndahl
The University of Queensland
School of Chemistry and
Molecular Biosciences
Brisbane
St Lucia QLD 4072
Australia

Avi Hofstein
Department of Science Teaching
Weizmann Institute of Science
Rehovot 76100
Israel

Ling Huang
Hofstra University
Chemistry Department
151 Hofstra University
Hempstead, NY 11549
USA

Søren Hvidt
Roskilde University
Department of Science
Systems and Models, NSM
Universitetsvej 1, 28
DK 4000 Roskilde
Denmark

Onno de Jong
FISME Institute
Utrecht University
PO Box 80 000
3508 TA Utrecht
The Netherlands

Jens Josephsen
Roskilde University
Department of Science
Systems and Models, NSM
Universitetsvej 1, 28
DK 4000 Roskilde
Denmark

Ram S. Lamba
Carlos Albizu University
PO Box 9023711
San Juan, PR 00902-3711
USA

Gwendolyn Lawrie
The University of Queensland
School of Chemistry and
Molecular Biosciences
Brisbane
St Lucia QLD 4072
Australia

Peter Mahaffy
The King's University
King's Centre for Visualization in
Science
9125 50th Street
Edmonton, AB T6B 2H3
Canada

and

The King's University
Department of Chemistry
9125 50th Street
Edmonton, AB T6B 2H3
Canada

Silvija Markic
University of Bremen
Institute for Didactics of Science
(IDN) – Chemistry Department
28359 Bremen
Germany

Eric Mazur
Harvard University
Harvard School of Engineering
and Applied Sciences
29 Oxford Street
Cambridge, MA 02138
USA

Claire McDonnell
Dublin Institute of Technology
School of Chemical and
Pharmaceutical Sciences
Kevin Street
Dublin 8
Ireland

Cathy Middlecamp
University of
Wisconsin-Madison
Nelson Institute for
Environmental Studies
Science Hall
550 North Park Street
Madison, WI 53706
USA

Christine O'Connor
School of Chemical and
Pharmaceutical Sciences
Dublin Institute of Technology
DIT Kevin Street
Dublin 8
Ireland

Atsushi Ohashi
Ehime University
Faculty of Education
Department of Science
Education
3, Bunkyo-cho
Matsuyama City 790-8577
Japan

Ilka Parchmann
Kiel University and
Leibniz-Institute for Science and
Mathematics Education (IPN)
Olshausenstraße 62
24118 Kiel
Germany

and

IPN
Department of Chemistry
Education
Kiel
Germany

Harry E. Pence
Department of Chemistry and
Biochemistry
SUNY Oneonta
Oneonta, NY
USA

Judith C. Poë
University of Toronto
Mississauga
Department of Chemical and
Physical Sciences
Room DV 4048
3359 Mississauga Rd. N.
Mississauga, ON L5L 1C6
Canada

Julian Rudnik
University of Kiel
Leibniz Institute for Science and
Mathematics Education (IPN)
Olshausenstraße 62
24118 Kiel
Germany

and

IPN
Department of Chemistry
Education
Kiel
Germany

Marie C. Ryan
University of Limerick
Department of Chemical and
Environmental Sciences
National Centre for Excellence in
Mathematics and Science
Teaching and Learning
Plassey Park
Limerick
Ireland

Julie Schell
Harvard University
Harvard School of Engineering
and Applied Sciences
29 Oxford Street
Cambridge, MA 02138
USA

Michael K. Seery
School of Chemical and
Pharmaceutical Sciences
Dublin Institute of Technology
DIT Kevin Street
Dublin 8
Ireland

Erica Steenberg
University of the Witwatersrand
RADMASTE Centre
27 St Andrews Road
Parktown
Johannesburg
South Africa

Jerry P. Suits
University of Northern Colorado
Department of Chemistry and
Biochemistry
UNC Campus Box 98
Greeley, CO 80639
USA

Manabu Sumida
Ehime University
Faculty of Education
Department of Science
Education
3, Bunkyo-cho
Matsuyama City 790-8577
Japan

Keith S. Taber
University of Cambridge
Faculty of Education
184 Hills Road
Cambridge CB2 8PQ
UK

Ingeborg Veldman
Science LinX
Faculty of Mathematics and
Natural Sciences
University of Groningen
Nijenborgh 9
9717 CG Groningen
The Netherlands

Antony J. Williams
Cheminformatics Department
904 Tamaras Circle
Wake Forest, NC 27587
USA

Part I
Chemistry Education: A Global Endeavour

1
Chemistry Education and Human Activity
Peter Mahaffy

1.1
Overview

The context for the book *Chemistry Education: Best Practices, Opportunities, and Trends* is set by this opening chapter, which asserts that the difference between historical "chemical education" and contemporary "chemistry education" is human activity. Tetrahedral chemistry education is reviewed as a visual and conceptual metaphor that was created to emphasize the need to situate chemical concepts, symbolic representations, and chemical substances and reactions in important human contexts. Three dimensions of human activity that require strong emphasis for educational practice to meet the learning needs of students are developed: (i) the human activity of learning and teaching chemistry; (ii) the human activity of carrying out chemistry; and (iii) the human activity that has imprinted itself in such a substantial way on the chemistry of our planet that it has defined a new geological epoch. Introducing chemistry content through rich contexts is proposed as one evidence-based approach for weaving all three of these dimensions of human activity into the practice of teaching and learning chemistry at secondary and post-secondary levels.

1.2
Chemistry Education and Human Activity

The term "chemical" education, which I encounter every day, has a long and storied history. I belong to the "chemical" education divisions of both the Chemical Institute of Canada and the American Chemical Society (ACS). On my bookshelf is the Journal of "Chemical" Education, and I access resources from the "Chemical" Education Digital Library. I regularly attend "chemical" education conferences and visit "chemical" education centers. In my professional circles, research and practice is supported by "chemical" education foundations, and exemplary practitioners of the art, science, and craft of teaching chemistry receive awards for contributions to "chemical" education.

Chemistry Education: Best Practices, Opportunities and Trends, First Edition.
Edited by Javier García-Martínez and Elena Serrano-Torregrosa.
© 2015 Wiley-VCH Verlag GmbH & Co. KGaA. Published 2015 by Wiley-VCH Verlag GmbH & Co. KGaA.

Yet, by design, the title of both this chapter and this book uses the word "chemistry" instead of "chemical" education. Should the two terms be used interchangeably, as is so often done?

The difference between chemical education and chemistry education is human activity.

How should the modern profession of "chemistry" education differ from historic "chemical" education? The term "chemical" education accurately conveys that at the heart of this domain of education are substances: their structures and properties, and the reactions that change them into other substances. But, beyond chemicals, human activity is central to (i) teaching and learning chemistry, (ii) the practice of chemistry in laboratories and industry, and (iii) the use and reactions of chemical substances by ordinary people. This opening chapter in *Chemistry Education: Best Practices, Opportunities, and Trends* asserts that chemistry educators should embed an understanding of all three of these different types of human activity into their practices of teaching and learning about the structures, properties, and reactions of chemical substances. And consistently using the term "chemistry education" as a more authentic descriptor than "chemical education" is a good starting point in conveying to students and the public the centrality of human activity in our professional domain.

1.3
A Visual Metaphor: Tetrahedral Chemistry Education

As chemistry educators, are we stuck in some of the historic practices of "chemical" education that we may have experienced as students? Have we narrowed our field of vision to presenting the intricate details of chemical substances and their reactions? Do our course and program learning objectives sufficiently incorporate students' need to understand why they should care about the "chemical" content they receive? Understanding how to effectively present "chemistry" authentically to students, including the multifaceted human connections of the discipline, has motivated an important thread of my research and practice for over a decade. Knowing that metaphors can influence as well as reflect practice, I have encouraged stronger emphasis on human activity in chemistry education through a new visual and conceptual metaphor – tetrahedral chemistry education [1].

How does a tetrahedral shape relate to the move from "chemical" to "chemistry" education? Chemistry educators have shown that students need to encounter chemistry at different thinking levels to obtain a rich understanding of chemical substances and reactions. To address human learning patterns, Johnstone, Gabel, and others [2] have proposed three widely accepted thinking levels needed to learn chemistry: the symbolic or representational (symbols, equations, calculations), the macroscopic (tangible, visible, laboratory), and the molecular or submicroscopic. These are often represented as a triangle of thinking levels required for mastery of chemistry. As shown in Figure 1.1, the visual metaphor of tetrahedral chemistry education extends the triangle of levels of engaging chemistry into a third dimension, in which the fourth vertex represents the human

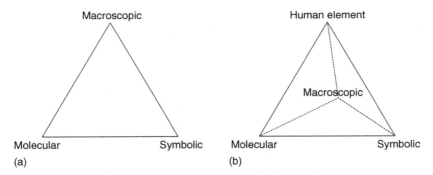

Figure 1.1 Tetrahedral chemistry education (b), as an extension of the triangle of thinking levels (a), making the focus on human activity in learning and teaching chemistry more visible and intentional.

contexts for chemistry. This new visual dimension emphasizes the need to situate chemical concepts, symbolic representations, and chemical substances and processes in the authentic contexts of the human beings who create substances, the culture that uses them, and the students who try to understand them. The tetrahedral chemistry education metaphor has been adapted and extended in various initiatives to articulate and support approaches to curriculum that foreground the human contexts for chemistry [3].

1.4
Three Emphases on Human Activity in Chemistry Education

What sorts of human activities are implied by changing the description of "chemical" to chemistry education, and emphasized by invoking the metaphor of tetrahedral chemistry education? What implications might more formal and systematic emphases on the human element have on learning through and about chemistry? How does emphasizing the human activity of chemistry flow from and inform research findings? Is our developing understanding of how the scale of human activity impacts the chemistry of our planet's life support systems adequately reflected in curriculum and pedagogy?

In this opening chapter, we take a 10-km high view of chemistry education to articulate three dimensions of human activity that should receive strong emphases in our professional efforts to ensure that our practice meets the learning needs of chemistry students: (i) the human activity of learning and teaching chemistry; (ii) the human activity of carrying out chemistry; and (iii) the human activity that has imprinted itself in such a substantial way on the chemistry of our planet that it has defined a new geological epoch. Our analysis will focus on chemistry education at the upper secondary and introductory post-secondary levels, with examples of effective practices that weave these emphases through both curriculum and pedagogy.

1.4.1
The Human Activity of Learning and Teaching Chemistry

Johnstone [4] highlights some of the results of paying too much attention to the "chemical" and not enough to the "education" part of chemical education. He suggests that current educational practice often clusters ideas into indigestible bundles, and that theoretical ideas are not linked to the reality of students' lives. The result: "chemical" education that is irrelevant, uninteresting, and indigestible, leading to student attitudes that range from not being able to understand to indifference about arriving at understanding.

Gilbert's [5] review of the interrelated problems facing chemical education over the past two decades reinforces Johnstone's critique, suggesting that students experience (i) an overload of content, (ii) numerous isolated facts that make it difficult for students to give meaning to what they learn, (iii) lack of ability to transfer conceptual learning to address problems presented in different ways, (iv) lack of relevance of knowledge to everyday life, and (v) too much emphasis on preparation for further study in chemistry rather than for development of scientific literacy.

Tetrahedral chemistry education implies identifying and meeting the needs of the diverse groups of students we serve with chemistry courses, and a transition from an emphasis on teaching to what research has to say about effective strategies and approaches to help students learn, and to learn chemistry.

What aspects of the human activity of teaching and learning chemistry need ongoing attention? Consider an example.

1.4.1.1 Atoms or Learners First?

Fifteen years after Johnstone's call to "begin where students are [2a]," vestiges of "chemical" rather than "chemistry" education remain. One example can be found when educators take quite literally the "atoms-first" approach to teaching chemistry. While it is difficult to find consistent definitions of this "new" approach, and the research evidence supporting it is very limited [6], the term is often used to describe a flow of ideas that begins with introducing the simplest building blocks of matter, and then assembles those first blocks of knowledge into more complex pieces, to eventually reach the point where the relevance of that understanding becomes evident to a student. The approach is summarized in the promotion for a 2013 chemistry textbook:

> The atoms-first approach provides a consistent and logical method for teaching general chemistry. This approach starts with the fundamental building block of matter, the atom, and uses it as the stepping stone to understanding more complex chemistry topics. Once mastery of the nature of atoms and electrons is achieved, the formation and properties of compounds are developed. Only after the study of matter and the atom will students have sufficient background to fully engage in topics such as stoichiometry, kinetics, equilibrium, and thermodynamics ... [7]

Atoms-first may have roots over a half-century old in the work of Linus Pauling, who, in the first edition (1950) of his much-emulated *College Chemistry* suggests a similar flow of ideas:

> In this book I begin the teaching of chemistry by discussing the properties of substances in terms of atoms and molecules ... [8]

The flow of ideas in putting atoms first is logical, consistent, and perhaps even elegant to the instructor who is an expert in chemistry and who already sees in his/her mind's eye important and motivating applications that will provide the reward for obtaining and stacking the first blocks of knowledge. But to a novice learner who is asked to wait to see the beauty and significance of the whole until the key pieces of knowledge are in place, the approach easily leads to fragmented understanding and difficulty in seeing the relevance of the knowledge learned. A parallel to "atoms first" in architecture education might be a deferred-gratification "sand-first" approach, where beginning architecture students study in sequence the details of sand, mortar, aggregate, rebar, and slabs of concrete, before finally seeing, perhaps half-way through a course, the exquisite building that motivates the vision and passion of an architect [9]. Perhaps atoms and other isolated "chemical" building blocks need to come second, after first motivating learners with the beauty and importance of the whole, based on an understanding of their diverse needs for learning chemistry.

Science is built up with facts, as a house is with stones. But a collection of facts is no more a science than a heap of stones is a house.

Henri Poincaré, La Science et l'hypothèse [10]

1.4.1.2 Identifying Learners and Designing Curriculum to Meet Their Needs

The learning needs of post-secondary chemistry students cannot possibly be met without first identifying who populates chemistry courses at the first-year university level. In first-year university chemistry courses in North America, an overemphasis is often placed on providing all of the foundational pieces for the few students who major in chemistry, rather than for the majority of students who will pursue careers in health professions, engineering, or other areas. Perhaps, practice here, too, has been shaped by Linus Pauling's influential approach in his 1950 textbook, who seems to have considered those who weren't majoring in chemistry as a bit of an after-thought:

> Although General Chemistry was written primarily for use by students planning to major in chemistry and related fields it has been found useful also by students with primary interest in other subjects ... [8]

Effective educational practice requires understanding who the students are who take chemistry, and ensuring that learning objectives are formulated to meet the needs of the many students who won't again darken the door of a chemistry course or lab, as well as those going on to study chemistry.

1.4.1.3 Effective Practices in the Human Activity of Learning and Teaching Chemistry

Re-hybridizing learning toward tetrahedral chemistry education that attends thoughtfully to the human activity of learning and teaching chemistry requires much more than tinkering with curriculum. Rather, systemic efforts to deliberately design learning environments, curriculum, pedagogy, and physical spaces are all needed to enrich the experiences of learners. In the past several decades, the community of educators has taken monumental strides to pay more attention to the "education" part of "chemical education." This includes efforts to identify and understand the learning needs of all students studying chemistry, to create learning communities, and to implement both curriculum and pedagogical strategies that lead to more active and engaged learning. It would be impossible to adequately summarize here the approaches and initiatives that have emerged, but there is now substantial literature supporting effective practices on the human activity of learning and teaching chemistry.

A review of that literature suggests helpful practices to enrich experiences of learning chemistry [11], including (i) understanding the student's prior conceptual understanding and developing validated inventories and strategies to identify and address misconceptions; (ii) using models for learning that account for different learning styles and limits to cognitive load; (iii) motivating students; (iv) engaging students with active and collaborative instruction and building and supporting intentional learning communities; (v) developing curriculum that connects to the lived experience of students and societal needs; (vi) implementing strategies for faculty professional development; and (vii) integrating into education the responsible and ethical practice of science. Many of these strategies and practices are the focus of later chapters of this book.

1.4.1.4 Identifying and Eliminating Worst Practices as a Strategy?

A U.S. National Academies National Research Council report on linking evidence and promising practices in reforming Science, Technology, Engineering, and Mathematics (STEM) education [12] reinforces effective practices in many of the areas listed above. The report identifies challenges in disseminating best practices beyond individual faculty members in undergraduate institutions. It suggests that, in addition to improving student learning and faculty teaching, it may be helpful to focus on improving student learning productivity. The greatest gain in aggregate student learning in STEM might be achieved, suggests the report, not by insisting on adopting optimal teaching practices in every classroom, but by identifying and eliminating the worst practices in each classroom. For example, substantial gains in student learning might result from encouraging the majority of STEM faculty members who only lecture to use *any form* of active learning, rather than unrealistically insisting that the optimal practices of these instructional approaches be adopted.

1.4.1.5 Exemplar: Emphasizing the Human Activity of Learning and Teaching Chemistry

Visualizing the Chemistry of Climate Change (VC3) [13] is one example of an evidence-based approach to implementing reform for introductory university chemistry courses, based on an analysis of the motivational and learning needs and conceptual understanding of students. Starting with the recognition that interdisciplinary understanding of complex systems is fundamental to understanding modern science, the end goal of VC3 is to provide tested interactive digital learning resources to support chemistry instructors in adopting active-learning pedagogies that situate cognition in authentic science practice and a particularly important context – global climate change. VC3 has developed an interactive set of resources, targeting first-year university chemistry students and teachers, with a triptych of goals, to (i) exemplify science education for sustainability, (ii) improve the understanding of climate change by both undergraduate students and faculty members, and (iii) provide resources to support pedagogical reform by modeling how chemistry topics can be contextualized to enhance student motivation and learning.

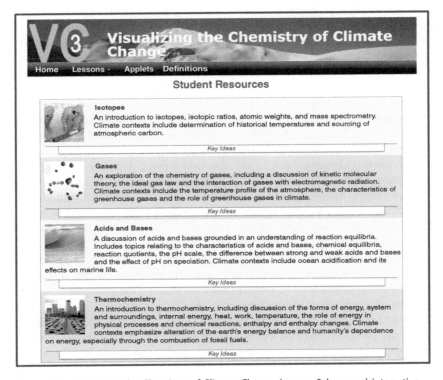

Figure 1.2 Visualizing the Chemistry of Climate Change (*www.vc3chem.com*) interactive electronic resources to introduce topics in general chemistry through climate contexts. (Figure courtesy of the King's Centre for Visualization in Science.)

The VC3 initiative (Figure 1.2) has been implemented in five phases: (i) mapping the correlation between climate literacy principles and core first year university chemistry content; (ii) documenting underlying science preconceptions and misconceptions, developing an inventory of chemistry concepts related to climate change, and validating instruments that make use of the inventory to assess understanding; (iii) developing and testing peer-reviewed interactive digital learning objects related to climate literacy principles with particular relevance to undergraduate chemistry; (iv) piloting the materials with first-year students and measuring the change in student understanding of both chemistry and climate science concepts, relative to control groups not using the materials; and (v) disseminating the digital learning objects for use by chemistry educators and students. An overview of the VC3 approach and a detailed example of one of the four VC3 topics developed to date at the King's Centre for Visualization in Science is given in Section 1.5.1 of this chapter.

1.4.2
The Human Activity of Carrying Out Chemistry

A second way for chemistry educators to emphasize the human element is by attending to the scholarship that asks whether the chemistry taught and learned in classrooms authentically reflects the practice of chemistry. Research on portrayals of science in formal curricula has documented student misconceptions about scientists, how science develops over time, and the nature of scientific knowledge [14]. The stakes are high in addressing these misconceptions, as chemistry students' understanding about the nature of science will influence their attitudes toward learning chemistry and their ability to react thoughtfully and critically to scientific claims. Talanquer [15] suggests that the unique features of chemistry as a discipline add complexity to the efforts to categorize the authenticity of portrayals of how chemistry is carried out. In addition to observing, explaining, and modeling, as has been the case for many other sciences, chemistry is also about creating new substances, designing new synthetic and analytical processes, and analyzing and transforming material systems. Deep understanding of science, including chemistry, requires understanding the evidence for theories and the discipline's underlying assumptions and methods [14].

Tetrahedral chemistry education emphasizes the coherence between the rich human activity of carrying out chemistry and the portrayals of that activity in classrooms and laboratories. Chemistry students should have an authentic understanding of where ideas and theories come from, how they develop over time, and how they connect to observations about the world. They should frequently engage the question: "How do we know what we know?" in addition to "What do we know?"

1.4.2.1 Explicit and Implicit Messages about the Nature of Chemistry
Without overt attention to the authenticity of how chemistry is portrayed, "chemical" education can introduce misconceptions about science as an intellectual and

social endeavor. But one challenge in analyzing the authenticity of portrayals of the human activity of carrying out chemistry is that implicit, as well as explicit, messages about the nature of science are communicated to students as they learn the "facts" of chemistry. By recognizing and countering unauthentic messages, chemistry educators can seize opportunities to paint a picture of chemistry as a creative science [16]. Non-authentic portrayals are introduced or reinforced in a variety of unexamined and implicit ways including static, contrived, and predetermined laboratory exercises; presentation of chemistry as isolated facts to be remembered, without a genuine understanding of how chemists develop explanations; lack of attention to where ideas come from and how they change over time; insufficient attention to the processes and tools chemists use to analyze, interpret, and apply data; neglect to highlight the imaginative process that is such a central part of "thinking like a chemist"; and failure to mention the ethical choices chemists and chemistry students make about how knowledge is used [17].

While practice has improved over the past decade, some textbooks still present the naïve and distorted caricature of a single hypothetico-deductive method used to carry out chemistry, often referred to as *the scientific method*. More authentic portrayals of how chemistry is carried out will leave students with an understanding that science grows through communities of practice that stand on the shoulders of prior understanding and that are influenced by a wide variety of human influences, including societal pressures and the availability of research funding. Understanding in chemistry develops in fits and starts, involves a mix of inductive, deductive, and abductive [18] methods, and at times is moved dramatically forward by chemists willing to challenge existing paradigms, and occasionally by serendipitous discoveries. Simplistic or distorted caricatures of science not only create misconceptions about the nature of science, but also make it difficult for human learners to see themselves as meaningful participants in carrying out science [19].

> The images that many people have of science and how it works are often distorted. The myths and stereotypes that young people have about science are not dispelled when science teaching focuses narrowly on the laws, concepts, and theories of science. Hence, the study of science as a way of knowing needs to be made explicit in the curriculum… not all of the historical emphasis should be placed on the lives of great scientists, those relatively few figures who, owing to genius and opportunity and good fortune, are best known. Students should learn that all sorts of people, indeed, people like themselves, have done and continue to do science.
>
> American Association for the Advancement of Science, Project 2061 Benchmarks [20]

1.4.2.2 Breathing the Life of Imagination into Chemistry's Facts

Implicit messages that convey less-than-authentic understandings of science are ubiquitous, and are found beyond the opening chapters of chemistry texts that outline the methods of science. But they are sometimes difficult to spot, due to entrenched patterns for sequencing instruction in "chemical" education. The flow of ideas in many learning resources at both the secondary and first-year postsecondary levels starts with facts and concepts to be learned – often presented in isolation from the evidence that underlies those facts, and then moves to applications of those concepts. A good example is found in treatment of structure and bonding of molecular substances, where the sequence of learning often *begins* with theories of bonding, such as hybridization and Valence Shell Electron Pair Repulsion (VSEPR) theory, *before* any evidence of experimental geometries, and without discussion of the nature and complementarity of different theories and models to explain that experimental evidence. As a result of such sequencing of ideas, and sometimes because of explicit language to that effect, students develop misconceptions. They may come to believe, for example, that carbon atoms in molecules of alkanes are tetrahedral *because* they are sp^3-hybridized. Assessment questions often ask students to list the hybridization of certain atoms in molecules or their "VSEPR geometries" without overtly referencing hybridization and VSEPR geometries as powerful, but limited models for making sense of experimental data.

A learning sequence for an introductory university chemistry course that presents a more authentic view of how chemists arrive at their understanding is to start with the activity of human beings who provide experimental evidence for structure and bonding, using techniques such as infrared spectroscopy (evidence for connectivity patterns in functional groups), mass spectrometry (evidence for molecular formulas), X-ray crystallography (bond lengths and bond angles), and nuclear magnetic resonance (NMR) spectroscopy (map of the C–H framework of organic compounds), and then to convey a sense of how chemists *imagine complementary scientific models* to explain that evidence [21]. Such a sequence can help students see both the power and limitations of models: the imaginative and creative processes that lead to robust explanations, and to avoid equating models with reality. In his 1951 Tilden Lecture, Oxford University Chemist Charles Coulson, whose work played an important role in developing our current theories of chemical bonding, describes the result of conflating models with experimental evidence, when considering a simple chemical bond, such as the C–H bond in methane:

> Sometimes it seems to me that a bond between two atoms has become so real, so tangible, (and) so friendly that I can almost see it. And then I awake with a little shock: for a chemical bond is not a real thing; it does not exist; no-one has ever seen it, no-one ever can ... Hydrogen I know, for it is a gas and we keep it in large cylinders; benzene I know, for it is a liquid and we keep it in bottles. The tangible, the real, the solid, is explained by the intangible, the unreal, (and) the purely mental. Yet that is what chemists are always doing ... [22]

Coulson goes on to articulate the importance of recognizing the human imagination as an integral part of chemistry sense-making.

> With us, as Mendeleev said, the facts are there and are being steadily accumulated day by day. Chemistry certainly includes all the chemical information and classification with which most school test-books are cluttered up. But it is more; for, because we are human, we are not satisfied with the facts alone; and so there is added to our science the sustained effort to correlate them and breathe into them the life of the imagination.
>
> Charles A. Coulson, 1951 Tilden Lecture [22]

All chemistry educators, knowingly and unknowingly, communicate messages about the nature of science. However, the messages students receive are often unrecognized or unexamined [14]. Substantial efforts are being taken in several countries to ensure that students develop an authentic understanding of science as a human endeavor [23]. The United States Next Generation Science Standards elaborate on this with recommendations that the following aspects of the nature of science should be communicated implicitly and explicitly in science classrooms [24]:

- Scientific investigations use a variety of methods.
- Scientific knowledge is based on empirical evidence.
- Scientific knowledge is open to revision in light of new evidence.
- Scientific models, laws, mechanisms, and theories explain natural phenomena.
- Science is a way of knowing.
- Scientific knowledge assumes an order and consistency in natural systems.
- Science is a human endeavor.
- Science addresses questions about the natural and material world.

1.4.2.3 Exemplars: Emphasizing the Human Activity of Carrying Out Chemistry

- McNeil [25] uses an innovative pedagogical strategy for moving university students from algorithmic application of "rules" for structure and bonding to deeper conceptual understanding that emphasizes the strength and limitations of complementary models to explain a large set of experimental observations. He divides students into two or more learning communities (e.g., a valence bond theory community and a molecular orbital theory community), each of which is required to persuasively explain pertinent data using *only* their assigned bonding theory. "Dueling bonding theories" result, as members of each learning community try to convince the others that their theory will better explain particular examples of data such as experimental geometries, bond strengths, magnetic properties, chemical reactivity, spectroscopic data, and chemical reactivity. Problem-solving, communication, and higher order skills are demonstrated as the groups attain deeper conceptual understanding, and the strength and limitations of models to explain evidence.
- The first-year university chemistry learning resource, *Chemistry: Human Activity, Chemical Reactivity* [21], uses two opening-chapter narratives to

introduce the power of modern chemistry to solve important problems and improve the quality of life while giving an authentic glimpse into the way modern chemistry is carried out in research groups and laboratories. Students are introduced to chemistry through the stories of David Dolphin, a Canadian chemist who has designed and made new substances that have improved the quality of life for over a million people suffering from cancer or eye disease, and Gavin Flematti, an Australian chemist who, while he was a postgraduate student, identified a compound in smoke that causes plant seeds to germinate after a forest fire. Flematti then found a way to make this compound in the laboratory. Modern techniques such as spectroscopy and chromatography are introduced through these human activity stories, and students obtain a feel for the time scale involved in chemical discoveries, as well as the role of both prior knowledge and serendipity. Then, each subsequent unit in the textbook begins with a relevant "rich context" that is designed to trigger student interest and the need to know more about the concepts covered in that chapter. Section 1.5.1 provides a detailed example.

1.4.3
Chemistry Education in the Anthropocene Epoch

> Two billion years ago, cyanobacteria oxygenated the atmosphere and powerfully disrupted life on earth … But they didn't know it. We're the first species that's become a planet-scale influence and is aware of that reality.
>
> Andrew Revkin, New York Times [26]

With awareness of the reality of the planet-scale influence of our species comes responsibility by educators to communicate, over educational levels and across disciplines, fundamental ideas about the fit between humans and our habitat. A compelling case for a new emphasis on human activity in chemistry education comes from considering the paradoxical ways in which chemistry has affected (usually for the better) virtually every aspect of human life, while at the same time comprehending that the scale and nature of modern human activity since the Industrial Revolution, aided by those very developments in chemistry, has fundamentally changed the chemistry of planet Earth.

What is the scope and magnitude of this human activity, linked to chemistry? The scientific community is moving toward accepting the term "Anthropocene Epoch" as an appropriate chronological term on the geological time scale to describe the transition from the Holocene Epoch to a new epoch that is defined by the imprint of global human activity. The term "anthropocene" [Greek "anthropo-" (human), and "-cene" (new)] was coined by ecologist Eugene Stoermer and popularized by chemist Paul Crutzen to emphasize the scale of the impact of human activity on the chemistry, biology, and geology of Earth's life support systems. Still to be determined, based on evidence about the human imprint on Earth's geology, is the appropriate time period for the beginning of the Anthropocene. Candidates include the beginnings of agriculture (~900 AD), the Nuclear Age (1950s), and the Industrial Revolution (~1750).

Tetrahedral chemistry education implies that students and teachers see how chemistry affects virtually every aspect of modern life – usually, but not always, for the better; that human activity is fundamentally altering our planetary boundaries; and that knowledge of chemistry is crucial in developing strategies to tackle global sustainability challenges.

1.4.3.1 Planetary Boundaries: A Chemistry Course Outline?

A series of seminal research publications, beginning in 2009, have set out to define and quantify the boundaries to our planet that should not be crossed, if humans are to prevent unacceptable human-induced global environmental change [27]. When reading, for the first time, the list of nine planetary boundaries (Figure 1.3) within which humans can safely operate, I was struck by the coherence between these boundaries and numerous underlying chemistry concepts presented to secondary and first-year post-secondary students. The list

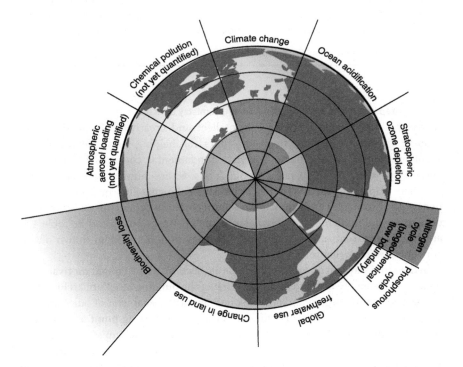

Figure 1.3 Illustration of the change to seven planetary boundaries from preindustrial levels to the present. The planetary boundaries attempt to quantify the limits within which humanity can safely operate without causing unacceptable global environmental change. (Photo credit: Azote Images/Stockholm Resilience Centre.)

of boundaries includes climate change, ocean acidification, interference with the nitrogen and phosphorus biogeochemical cycles, stratospheric ozone depletion, global freshwater use, changes in land use, atmospheric aerosol loading, rate of biodiversity loss, and (the poorly defined) chemical pollution.

Does secondary and post-secondary chemistry education, as currently practiced, reflect to any substantial effect our understanding of the fundamental role chemistry plays in altering Earth's boundaries? What are the implications of teaching students at more advanced levels who have learned, starting in elementary school, that we now live in the Anthropocene Epoch – defined in large part by changes to the chemistry of our lithosphere and atmosphere [28]?

> To master this huge shift, we must change the way we perceive ourselves and our role in the world. Students in school are still taught that we are living in the Holocene, an era that began roughly 12,000 years ago at the end of the last Ice Age. But teaching students that we are living in the Anthropocene ... could be of great help. Rather than representing yet another sign of human hubris, this name change would stress the enormity of humanity's responsibility as stewards of the Earth. It would highlight the immense power of our intellect and our creativity, and the opportunities they offer for shaping the future.
>
> Nobel Laureate Chemist Paul Crutzen [29]

1.4.3.2 Steps toward Anthropocene-Aware Chemistry Education

A starting point in highlighting the intellect and creativity of chemists, and probing the opportunities chemistry offers for helping to shape the future (Crutzen, above), would be to correlate fundamental concepts in chemistry curriculum with those planetary boundaries where human activity is substantially impacting earth systems processes. Following this mapping activity, pedagogical strategies and curriculum that make these connections overt can be developed and implemented. Examples of such connections, listed in Table 1.1, are readily apparent upon even cursory examination, but seldom drawn out in core chemistry curriculum.

In the same way that an architect-educator (Section 1.4.1.1) might motivate students to develop the detailed understanding of the building blocks of architecture by introducing an elegant and complex building, the visual icon of our planetary boundaries could helpfully become one meaningful starting point and point of reference integrated throughout an introductory university chemistry course. The chemistry educator might motivate by starting, not with atoms, but with a view of the intricate and elegant chemical structures and processes that are found in every part of everyday life, including those that define our place in geological time. Keeping relevant human contexts in sight, students can then be guided through the details of structures, properties, and reactions of substances.

Table 1.1 Examples of relevant connections between six of the planetary boundaries and chemistry concepts.

Planetary boundary	Examples of underlying chemistry concepts
Climate change	Interaction of electromagnetic radiation with matter, infrared spectroscopy, thermochemistry, aerosols, isotopes, states of matter, combustion reactions, stoichiometry, hydrocarbons, and carbohydrates
Ocean acidification	Acid–base chemistry, equilibria, solubility, chemistry in and of water, chemical speciation, stoichiometry, and models
Stratospheric ozone depletion	Photochemistry, interaction of electromagnetic radiation with matter, ultraviolet spectroscopy, free-radical reactions, reaction mechanisms, thermochemistry, and kinetics
Nitrogen and phosphorus biogeochemical cycles	Main group chemistry, chemical speciation, stoichiometry, atom economy and atom efficiency, thermochemistry, and kinetics
Global freshwater use	Chemistry in and of water, chemical speciation, solubility and precipitation, equilibria, and states of matter
Atmospheric aerosol loading	States of matter and phase changes, thermochemistry, and acid–base chemistry

1.4.3.3 Exemplars: Anthropocene-Aware Chemistry Education

- The United Nations resolution declaring 2011 the International Year of Chemistry (IYC-2011) placed strong emphasis on the role chemistry plays in building a sustainable future [30]. During the year, educational and outreach activities focused on climate science, water resources [31], and energy gave further momentum to the link between chemistry and sustainability. One IYC-2011 legacy resource is a comprehensive set of free, interactive, critically reviewed, and Web-based learning tools to help students, teachers, science professionals, and the general public make sense of the underlying science of climate change. *www.explainingclimatechange.com* builds on and integrates connections to concepts in chemistry and physics, and is being used in chemistry and other courses at both the secondary and post-secondary level [32].
- National chemical societies have created programs and committees to raise the profile of chemistry education initiatives that address the human imprint on our planet. The Royal Society of Chemistry (RSC, UK) has teamed up with the ACS to form a sustainability alliance to help people understand the basic chemistry behind our global challenges and potential solutions. The ACS Committee on Environmental Improvement has instituted an annual award for exemplary incorporation of sustainability into Chemistry Education.
- As early as 1993, the ACS *Chemistry in Context* textbook for teaching chemistry to university students majoring in disciplines other than science has taught chemistry through real-world examples that engage students on multiple levels: their individual health and well-being, the health of their local communities, and the health of wider ecosystems that sustain life on Earth. Despite the success of this initiative, large inertia barriers have been experienced in extending similar approaches into courses for science majors.

1.5
Teaching and Learning from Rich Contexts

> … Now, for the first time in history, we are educating students for life in a world about which we know very little, except that it will be characterized by substantial and rapid change, and is likely to be more complex and uncertain than today's world … 'What kind of science education is appropriate as preparation for this unknown world?'
>
> Derek Hodson [33]

Are there approaches, whose effectiveness is supported by evidence, in which all three of these human activity dimensions to chemistry education (human activity of learning and teaching chemistry, human activity of carrying out chemistry, and Anthropocene-aware chemistry education) can be woven seamlessly into the practice of teaching and learning chemistry?

1.5.1
Diving into an Ocean of Concepts Related to Acid–Base Chemistry

Consider an example of traditional "chemical" education curriculum that has seen little evolution over three or more decades. Almost every final-year secondary and first-year post-secondary chemistry course introduces a set of concepts that many students find challenging, and with a history of robust misconceptions, related to acid–base chemistry and solution equilibria and precipitation. Often taught as isolated concepts, these topics are introduced with a strong emphasis on the symbolic level of understanding. Coverage of these complex topics are highly mathematical, and algorithmic questions related to chemical equilibria, including acids and bases, feature prominently in classroom and standards exams. Yet, Yaron [34] reports that interviews with students only a few months after completing successfully a chemistry course with strong emphasis on acid–base equilibria, reveal that very little of the knowledge is retained. He suggests that a likely cause of this poor retention is that mathematical procedures related to equilibria are learned as procedures, with little connection to underlying concepts.

An alternative approach, which can incorporate all three dimensions of human activity described in Sections 1.4.1–1.4.3, is to introduce these core acid–base and equilibrium concepts through a compelling narrative, whose importance is or becomes evident to students. In VC3 (Section 1.4.1.5) [13] and in the learning resource *Chemistry: Human Activity, Chemical Reactivity* [21], we start with thought-questions that convey the urgency of the ocean acidification planetary boundary, rather than beginning with the detailed building blocks of knowledge related to mathematical and chemical reaction equations and chemical speciation. The conceptual building blocks are not neglected, but are carefully introduced *after* students understand both the importance of the global challenge and that they require knowledge about acids and bases and equilibria to make sense of the chemistry underlying the challenge to Earth's oceans.

Ocean acidification is one of the seven planetary boundaries that have been quantified [27b] with a global scale threshold. The 2009 analysis shows that Earth is approaching the proposed boundary of sustaining ≥80% of preindustrial aragonite mean surface seawater saturation state levels. About 70% of Earth is covered by oceans, and the oceans play a critical role in regulating earth's radiation balance, as well as providing habitat for more than a million species of plants and animals [35]. Mean ocean pH has already dropped from a preindustrial value of 8.2 to 8.1, and data on natural CO_2 seeps in tropical waters show that biodiversity of coral species is substantially diminished when ocean acidity increases [36]. While awareness is increasing of the damage to ocean coral reefs and other components of marine ecosystems from increasing hydronium ion concentration, little is generally known about another facet of the underlying chemistry – the dependence of speciation of dissolved CO_2 on pH. As pH drops, the speciation of carbon oxides from the initial dissolved CO_2 shifts significantly, with an increasing percentage of aquated hydrogen carbonate ions and lower availability of the aquated carbonate ions, which are particularly important in the formation of many marine exoskeletons.

In developing this approach to introducing acid–base chemistry and solution equilibria and precipitation, the VC3 research team started with a review of research literature of prior knowledge and documented student misconceptions related to the chemical concepts. Learning objectives were then developed to address core concepts related to the chemistry of acids and bases and equilibria, as well as the principles of climate literacy related to the role of the oceans. Learning objectives included not only what students should know about particular concepts but also what the evidence for that knowledge is, and the climate contexts that are related to those concepts. A set of interactive Web-based lessons were then created, based on effective practices for the creation of electronic simulations [37]. Integrated throughout the lessons are new digital learning objects, such as one that invites students to interrogate the ocean system with a model that correlates atmospheric CO_2 with surface ocean pH, and the relationship between pH and speciation of carbonic acid, bicarbonate, and carbonate ions (Figure 1.4). To provide evidence about the effectiveness of this approach, an assessment of learning gains of chemistry concepts related to acid–base chemistry and relevant climate science concepts is being carried out, comparing students using the VC3 resources with control groups using conventional approaches.

It can readily be seen how each of the three dimensions of human activity described in Sections 1.4.1–1.4.3 is emphasized in this approach, with careful implementation. (i) *The human activity of learning and teaching chemistry.* Materials are developed after consideration of student conceptions and documented student misconceptions to achieve both lower and higher order learning objectives. A story is used to introduce an important set of concepts through interactive learning tools that probe student understanding, provide feedback, and encourage deep and active learning. Evidence about the effectiveness of

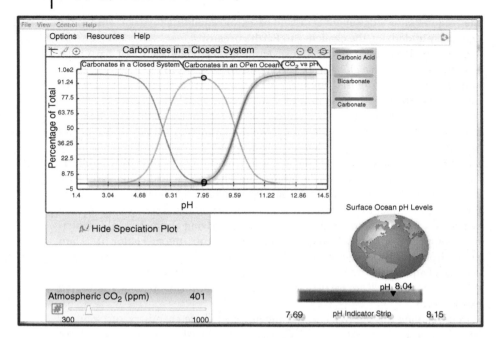

Figure 1.4 Screen capture from interactive digital learning object that models the correlations between atmospheric CO_2 and ocean pH, as well as the speciation of carbon in the ocean. (Figure courtesy of the King's Centre for Visualization in Science.) (*www.kcvs.ca*).

the approach in achieving learning gains is obtained. (ii) *The human activity of carrying out chemistry.* The nature of scientific evidence as well as the application of conceptual and mathematical models to analyze data and explain observations is demonstrated, and students are introduced to complexity in a contemporary scientific global challenge. Complexity is a feature of the nature of science that is often given insufficient attention in conventional approaches. (iii) *The human activity that has imprinted itself in such a substantial way on the chemistry of our planet that it has defined a new geological epoch.* Chemistry students learn about ocean acidification, one of the planetary boundaries where human-induced global environmental change is becoming increasingly evident on a relatively short time scale.

1.5.2
What Is Teaching and Learning from Rich Contexts?

Introducing acid–base chemistry and solution equilibria through ocean acidification is an example of an approach we describe as *teaching from a rich context*, and that has potential to effectively model science education as preparation for the complexity and uncertainty of our world (Hodson, Section 1.5). Context-based approaches, which are also considered by Parchmann in Chapter 10 of this title [38], have been described as "approach(es) adopted in science teaching where context and applications of science are used as the starting point for the development of scientific ideas. This contrasts with more traditional approaches that cover scientific ideas first, before looking at applications [39]." Through the context, students are expected to give meaning to the chemical concepts they learn [40]. The learning theory concept of situated cognition provides one theoretical framework for using contexts to scaffold the development of chemical concepts [41]. Situated cognition assumes that learning is embedded in social, cultural, and physical contexts [42], and when meaningful contexts are used to introduce concepts, content will be more firmly anchored in memory and more easily applied to new contexts. Meaningful learning occurs when incoming information can be linked to and interpreted by what the learner already knows or by what comes to have meaning for him/her [41].

Context-based learning shares features and aspects of historical evolution with other approaches, such as case studies, problem-based learning, and Science–Technology–Society (STS) [43] and Science–Technology–Society–Environment (STSE) teaching. DeJong [44] identifies four common domains for contexts: the personal, social and societal, professional practice, and scientific and technological domains. He traces the evolution of the use of contexts to teach chemistry from initiatives in the 1980s such as ChemCom and Chemistry in Context in the United States, Salters Chemistry in the United Kingdom, Chemie in Kontext in Germany, and Chemistry in Practice in the Netherlands. He suggests that in some of the most recent and effective implementation of contextual learning, contexts initially precede concepts, providing both an orienting function and increasing motivation for learning new concepts. In some cases, a follow-up inquiry context is used to help students see the need to apply their knowledge to new situations. We propose that the term "rich context" appropriately applies to implementation that provides deep and rich opportunities for learning concepts through contexts and applying that knowledge to new contexts.

Using insights from linguistics and learning theories, Gilbert [5] suggests ways in which context-based approaches have the potential to address the five challenges of chemical education identified in Section 1.4.1 (content overload, numerous isolated facts, difficulty in transferring learning to problems presented in different ways, lack of relevance of knowledge to everyday life, and too much emphasis on preparation for further study in chemistry). Pilot and Bulte [45] analyze significant new initiatives to develop context-based approaches to chemistry curriculum in five different countries, and describe ways in which each initiative has contributed to the five challenges listed above. They also identify the importance of assessment for context-based chemistry education that is coherent with the learning goals, approaches, and vision of the approaches.

1.5.3
Teaching and Learning from Rich Contexts – Evidence for Effectiveness

Several challenges become apparent when seeking to obtain research evidence about the effectiveness of context-based approaches relative to long-standing practices of "chemical" education. One challenge is a common understanding of key terms – all learning occurs within multiple contexts – the system contexts of a classroom and curriculum, the socio-cultural context for learning, and the internalized context – and practitioners use the term "context" in diverse ways. Context-based approaches also have a wide range of goals and approaches, and they are implemented across a range of educational levels to both science and non-science majors, and in very diverse school cultures. It can be difficult to measure gains from any learning intervention when learning outside of formal curriculum dominates [46]. However, evidence for effectiveness of context-based learning is emerging from assessment [45] of implementations such as the five international approaches described in Section 1.5.2, and several recent large-scale reviews converge on some important conclusions, while pointing the way toward areas where further research is needed. Bennett *et al.* [39] presents a synthesis of the research evidence from 17 experimental studies in eight countries on the effects of context-based and STS approaches. This work draws on the findings of two systematic reviews of the research literature, including a previous survey of 220 context-based and conventional approaches [47]. Overton [41] gives an overview of both context-based approaches and problem-based learning in both chemistry and physics, and Ültay and Çalik [46] has recently reviewed 34 context-based chemistry studies.

Perhaps the most important finding from this review of assessments of implementations is that context-based learning results in positive effects on attitudes. Students using context-based curricula view chemistry as more motivating, interesting, and relevant to their lives [46, 47]. Students develop a range of transferrable and intellectual skills [41], including higher order thinking skills. Conclusions about the important question of gains in student's understanding/cognition are more tenuous, as relatively few well-designed, comparative studies have been done with different contemporary learning models. Bennett's systematic review of experimental studies [39], using methods from the UK Evidence, Policy, and Practice Initiative, concludes (from 12 studies considered medium to high quality) that the understanding of scientific ideas through context-based approaches is comparable to that of conventional approaches. Bennett concludes that "reliable and valid evidence is available to support the use of contexts as a starting point in science teaching: there are no drawbacks in the development of understanding of science, and considerable benefits in terms of attitudes to 'school' science [39]."

1.5.4
From "Chemical" to "Chemistry" Education – Barriers to Change

Educational change from "chemical" toward more tetrahedrally shaped "chemistry" education pre-requires both a vision for identifying and meeting the learning needs of students in diverse cultural and educational settings, and a critical mass of professional educators willing to step back from historical practices to examine evidence for what works and what doesn't in current practice.

Necessary also are a keen awareness of inertia and other barriers to change, along with healthy doses of both time and patience. Communities of learning and professional practice which place high priority on effective practices for learning and teaching need to support and guide innovators and galvanize others into examining and trying out more effective approaches. Finally, a commitment to an on-going process of assessment to provide evidence for the effectiveness of new approaches is needed, and a willingness of proponents of educational reform to dynamically adapt new approaches in response to that evidence.

Professional development to support chemistry educators is crucial. Stolk has recently carried out iterative empirical studies in which a professional development framework to empower chemistry teachers for context-based education is designed, implemented, and evaluated [48]. Building on Stolk's three-phase framework (preparation to teach context-based units, instruction, and reflection and evaluation), Dolfing [49] reports on four empirical studies to better understand what strategies within a professional development program are needed to support teachers in developing domain-specific expertise in teaching context-based chemistry education.

Effective teaching of chemistry does not develop in the abstract. It needs to be grounded in the discipline of chemistry and in the many interfaces where chemistry is practiced. So, in addition to actions of individuals, communities of learning, and professional teacher development programs, disciplinary scientific societies also have an important role to play. Recognizing that most STEM faculty at the post-secondary level begin their careers with little or no professional training in teaching and little or no knowledge about effective teaching practices, a collaborative initiative of the U.S. Council of Scientific Society Presidents and other partners have produced a report [50] on the importance of disciplinary societies in stimulating and supporting faculty to implement successful teaching strategies. The report highlights effective practices which include new faculty workshops, annual disciplinary teaching workshops and education sessions, teaching fellowships, teaching institutes, and other strategies for professional development to facilitate more widespread change in STEM learning and teaching.

Acknowledgments

Coauthors on the *Chemistry: Human Activity, Chemical Reactivity Learning Resources* are B. Bucat (University of Western Australia) and R. Tasker (University of Western Sydney). Collaborators on the VC3 project are B. Martin (King's University College), M. Towns, A. Versprille, and P. Shepson (Purdue University), M. Kirchhoff and L. McKenzie (American Chemical Society), C. Middlecamp (University of Wisconsin), and T. Holme (Iowa State University, evaluator). Undergraduate King's University College students M. Price, D. Vandenbrink, T. Keeler, and D. Eymundson contributed significantly to the ocean acidification resources, and T. VanderSchee carried out a comprehensive literature review of context-based learning. Funding has been provided by the Natural Sciences and Engineering Research Council of Canada through the CRYSTAL Alberta and

Undergraduate Student Research Award programs and by the National Science Foundation (CCLI Award #1022992 for VC3).

References

1. Mahaffy, P. (2006) Moving chemistry education into 3D: a tetrahedral metaphor for understanding chemistry. *J. Chem. Educ.*, **83** (1), 49–55.
2. (a) Johnstone, A.H. (2000) Teaching of chemistry – logical or psychological? *Chem. Educ.: Res. Pract. Eur.*, **1**, 9–15; (b) Gabel, D. (1999) Improving teaching and learning through chemistry education research: a look to the future. *J. Chem. Educ.*, **76**, 548.
3. (a) Lewthwaite, B.E. and Wiebe, R. (2011) Fostering teacher development towards a tetrahedral orientation in the teaching of chemistry. *Res. Sci. Educ.*, **40** (11), 667; (b) Sjöström, J. (2013) Towards *bildung*-oriented chemistry education. *Sci. Educ.*, **22** (7), 1873–1890; (c) Lewthwaite, B., Doyle, T., and Owen, T. (2013) Tensions in intensions for chemistry education. *Chem. Educ. Res. Pract.* doi: 10.1039/C3RP00133D (d) Sileshi, Y. (2011) Chemical reaction: diagnosis and towards remedy of misconceptions. *Afr. J. Chem. Educ.*, **1**, 10–28; (e) Apotheker, J. (2004) Viervlakkig chemie-onderwijs in Groningen. *NVOX*, **9**, 488–492; (f) Savec, V.F., Sajovic, I., and Wissiak, K.S. (2009) in *Multiple Representations in Chemical Education, Models and Modeling in Science Education*, vol. 4 (eds J. Gilbert and D. Treagust), Springer, London, p. 309.
4. Johnstone, A. (2010) You can't get there from here. *J. Chem. Educ.*, **87** (1), 22–29.
5. Gilbert, J.K. (2006) On the nature of "context" in chemical education. *Int. J. Sci. Educ.*, **28** (9), 958.
6. Esterling, K.M. and Bartels, L. (2013) Atoms-first curriculum: a comparison of student success in general Chemistry. *J. Chem. Educ.*, **90**, 1433–1436.
7. Burdge, J. and Overby, J. (2012) *Chemistry: Atoms First*, McGraw-Hill, Columbus, OH.
8. Pauling, L. (1950) *College Chemistry: An Introductory Textbook of General Chemistry*, W.H. Freeman, New York.
9. Bucat, R. (2011) University of Western Australia, personal communication.
10. Poincaré, H. (1905), Dover Abridged edition 1952), *Science and Hypothesis*, Chapter IX (translated by, G.B. Halsted), Dover Books, New York.
11. (a) Mahaffy, P. (2011) The human element: chemistry education's contribution to our global future, in *The Chemical Element: Chemistry's Contribution to Our Global Future* (ed. J. Garcia), Wiley-VCH Verlag GmbH, Weinheim; (b) Byers, B. and Eilks, I. (2009) in *Innovative Methods of Teaching and Learning Chemistry in Higher Education* (eds I. Elks and B. Byers), Royal Society of Chemistry, Cambridge, p. 17.
12. Fairweather, J. (2008) Linking evidence and promising practices in Science, Technology, Engineering, and Mathematics (STEM) undergraduate education: a status report for the National Academies National Research Council Board of Science Education. Commissioned Paper for the National Academies Workshop: Evidence on Promising Practices in Undergraduate Science, Technology, Engineering, and Mathematics (STEM) Education, http://sites.nationalacademies.org/DBASSE/BOSE/DBASSE_071087 (accessed 29 April 2014).
13. Mahaffy, P. G.; Martin, B.E; Kirchhoff, M.; McKenzie, L.; Holme, T.; Versprille, A.; Towns, M. (2014) Infusing Sustainability Science Literacy Through Chemistry Education: Climate Science as a Rich Context for Learning Chemistry. *ACS Sustainable Chemistry & Engineering*, **2** (11), 2488–2622.
14. Clough, M.P. (2008) We all teach the nature of science – whether accurately or not. *Iowa Sci. Teachers J.*, **35** (2), 2–3.

15. Talanquer, V. (2013) School chemistry: the need for transgression. *Sci. Educ.*, **22**, 1757–1773.
16. Christensson, C. and Sjöström, J. (2014) Chemistry in context: analysis of thematic chemistry videos available online. *Chem. Educ. Res. Pract.* **15**, 59–69.
17. Mahaffy, P., Zondervan, J., Hay, A., Feakes, D., and Forman, J. (2014) Multiple Uses of Chemicals - IUPAC and OPCW Working Together Toward Responsible Science. *Chemistry International*, **36** (5), 9–13.
18. Chamizo, J.A. (2013) A new definition of models and modeling in chemistry's teaching. *Sci. Educ.*, **22**, 1613–1632.
19. Martin, B.E., Kass, H., and Brouwer, W. (1990) Authentic science: a diversity of meanings. *Sci. Educ.*, **74** (5), 541–554.
20. Benchmarks Project 2061, http://www.project2061.org/publications/bsl/online/index.php?chapter=1 (accessed 29 April 2014).
21. Mahaffy, P., Bucat, B., Tasker, R. et al. (2015) *Chemistry: Human Activity, Chemical Reactivity*, Chapters 9-10, 2nd edn, Nelson, Toronto.
22. Coulson, C.A. (1955) The Tilden Lecture: contributions of wave mechanics to chemistry. *J. Chem. Soc.*, 2069.
23. Rasmussen, S.C., Glunta, C., and Tomchuk, M.R. (2008) Content standards for the history and nature of science, in *Chemistry in the National Science Education Standards*, Chapter 9, 2nd edn (ed. S.L. Bretz), American Chemical Society, Washington, DC, http://www.acs.org/content/acs/en/education/policies/hsstandards.html (accessed 29 April 2014).
24. National Academy of Sciences (2013) Appendix H – Understanding the Scientific Enterprise: The Nature of Science in the Next Generation Science Standards. A Framework for K-12 Science Education. http://www.nextgenscience.org/sites/ngss/files/Appendix H -The Nature of Science in the Next Generation Science Standards4.15.13.pdf (accessed 29 April 2014).
25. Kohout, J.D. and McNeil, W.S., (2012) Duelling bonding theories: using student engagement to address misconceptions of chemical bonding. 2012 Biennial Conference on Chemical Education, University Park, PA, Abstract, p. 499.
26. Revkin, A., quoted by Stromberg, J. (2013) What is the Anthropocene and are We in It? Smithsonian Magazine, http://www.smithsonianmag.com/science-nature/What-is-the-Anthropocene-and-Are-We-in-It-183828201.html#ixzz2liqVbpaz (accessed 05 January 2014).
27. (a) Rockström, J. et al. (2009) A safe operating space for humanity. *Nature*, **461**, 472–475; (b) Rockström, J. et al. (2009) Planetary boundaries: exploring the safe operating space for humanity. *Ecol. Soc.*, **14** (2), 32, http://www.ecologyandsociety.org/vol14/iss2/art32/ (accessed 29 April 2014).
28. Mahaffy, P. (2014) Telling time: chemistry education in the anthropocene epoch. *J. Chem. Educ.*, **91** (4), 463–465.
29. Crutzen, P.J. and Schwägerl, C. (2011) Living in the Anthropocene: Toward a New Global Ethos. Yale Environment360, http://e360.yale.edu/feature/living_in_the_anthropocene_toward_a_new_global_ethos/2363/ (accessed 29 April 2014).
30. Vilches, A. and Gil-Perez, D. (2013) Creating a sustainable future: some philosophical and educational considerations for chemistry teaching. *Sci. Educ.*, **22**, 1857–1872.
31. Serrano, E., Sigamoney, R., and Garcia-Martinez, J. (2013) ConfChem conference on a virtual colloquium to sustain and celebrate IYC 2011 initiatives in global chemical education – The global experiment of IYC2011: creating online communities for education and science. *J. Chem. Educ.*, **90** (11), 1544–1546.
32. Mahaffy, P., Martin, B.E., Schwalfenberg, A., Vandenbrink, D., and Eymundson, D. (2013) ConfChem conference on a virtual colloquium to sustain and celebrate IYC 2011 initiatives in global chemical education: visualizing and understanding the science of climate change. *J. Chem. Educ.*, **90** (11), 1552–1553.
33. Hodson, D. (2003) Time for action: science education for an alternative future. *Int. J. Sci. Educ.*, **25** (6), 648.

34. Yaron, D., Karabinos, K., Evans, K., Davenport, J., Cuadros, J., and Greeno, J. (2010) in *Instructional Explanations in the Disciplines* (eds M. Stein and L. Kucan), Springer, New York, pp. 41–50.
35. Appeltans, W. et al. (2012) The magnitude of global marine species diversity. *Curr. Biol.*, **22**, 2189–2202.
36. Service, R. (2012) Rising acidity brings an ocean of trouble. *Science*, **337**, 146–148.
37. Martin, B.E. and Mahaffy, P.G. (2013) in *Pedagogic Roles of Animations and Simulations in Chemistry Courses*, ACS Symposium Series, vol. 1142 (ed. J. Suits), American Chemical Society, Washington, DC, pp. 411–440.
38. Parchmann, I. (2014) Context-based teaching and learning on school and university level, in *Chemistry Education: Best Practices, Innovative Strategies and New Technologies*, Chapter 5 (eds J. Garcia-Martinez and E. Serrano), Wiley-VCH Verlag GmbH, Weinheim.
39. Bennett, J., Lubben, F., and Hogarth, S. (2007) Bringing science to life: a synthesis of the research evidence on the effects of context-based and STS approaches to science teaching. *Sci. Educ.*, **91** (3), 347–370.
40. Bulte, A.M.W., Westbroek, H.B., de Jong, O., and Pilot, A. (2007) A research approach to designing chemistry education using authentic practices as contexts. *Int. J. Sci. Educ.*, **28** (9), 1063–1086.
41. Overton, T., Byers, B., and Seery, M.K. (2009) in *Innovative Methods in Teaching and Learning Chemistry in Higher Education* (eds I. Eilks and B. Byers), Royal Society of Chemistry, Cambridge, pp. 43–60.
42. Greeno, J.G. (1998) The situativity of knowing, learning, and research. *Am. Psychol.*, **53**, 5–26.
43. Fensham, P.J. (1985) Science for all. *J. Curriculum Stud.*, **17**, 415–435.
44. DeJong, O. (2006) Context-based chemical education: how to improve it? Plenary Lecture, 19th International Conference on Chemistry Education, Seoul, Korea, August 12–17, 2006, http://old.iupac.org/publications/cei/vol8/0801xDeJong.pdf (accessed 06 January 2014).
45. Pilot, A. and Bulte, A.M.W. (2006) The use of "contexts" as a challenge for the chemistry curriculum: its successes and the need for further development and understanding. *Int. J. Sci. Educ.*, **28** (9), 1087–1112.
46. Ültay, N. and Çalik, M. (2012) A thematic review of studies into the effectiveness of context-based chemistry curricula. *J. Sci. Educ. Technol.*, **21**, 686–701.
47. Bennett, J. (2005) Context-based and conventional approaches to teaching chemistry: comparing teachers' views. *Int. J. Sci. Educ.*, **27**, 1521–1547.
48. Stolk, M. (2013) Empowering chemistry teachers for context-based education. Utrecht University Dissertation, Freudenthal Institute for Science and Mathematics Education, Faculty of Science, No. 74.
49. Dolfing, R. (2013) Teacher's professional development in context-based chemistry education. Utrecht University Dissertation, Freudenthal Institute for Science and Mathematics Education, Faculty of Science, No. 78.
50. Hilborn, R.C. (2012) The Role of Scientific Societies in STEM Faculty Workshops, Meeting of the Council of Scientific Society Presidents, May 3, 2012, http://www.aapt.org/Conferences/newfaculty/upload/STEM_REPORT-2.pdf (accessed 29 April 2014).

2
Chemistry Education That Makes Connections: Our Responsibilities

Cathy Middlecamp

2.1
What This Chapter Is About

The universe is made of stories, not of atoms [1].

This chapter tells stories aimed at helping college chemistry instructors better "connect the dots" as they teach. These stories involve people, places, and events both on a university campus and in the wider communities in which we work and live.

This chapter also tells stories from a national curriculum reform project in the United States, SENCER (Science Education for New Civic Engagements and Responsibilities) [2]. Funded by the U.S. National Science Foundation (NSF), the SENCER project is now well into its second decade. Hundreds of college faculty members have contributed to this project by developing science courses that teach "through" real-world issues such as obesity, coal mining, stem cells, and water quality "to" the scientific principles that underlie these issues [3, 4]. In essence, SENCER courses use the power of a good story to engage students in learning. The scientific knowledge that students gain is seamlessly linked with topics of immediate interest, thus inextricably linking context and content.

In telling stories, this chapter invites conversations about the "R" in SENCER: *Responsibilities*. These include those to ourselves, to our colleagues, and to others in our discipline. They also include our responsibilities as teachers to frame content in ways consistent with how people learn. And they include those of students to dedicate themselves to learning; in fact, the SENCER project squarely locates the responsibilities (the burdens and the pleasures) of discovery with the students [5].

We also have responsibilities to those in future generations, which is why ideas relating to sustainability underlie this chapter. Although sustainability can be articulated in many ways [6], it almost always connects to our *responsibilities* to each other and to future generations:

Sustainability may be described as our responsibility to proceed in a way that will sustain life that will allow our children, grandchildren and great-grandchildren to live comfortably in a friendly, clean, and healthy world [7].

It is nearly impossible to feel a responsibility toward something to which one has no connection!

Given the competing demands on our time in teaching introductory chemistry courses, why take the time to tell a story? Chemistry, of course, is a physical science, sometimes called the *central science*, that studies matter. But in the best sense of the word, chemistry is a *story*, one that connects us to the fascinating world atoms and molecules as well as to the larger world in which we live. If a picture is worth 1000 words, then telling a story is worth 1000 minutes of class time. Years later, when the facts and figures are long forgotten, students may remember the stories, perhaps ones that captured their hearts, minds, and imagination. As several students have pointed out to the author of this chapter years later about their experiences in a college chemistry classroom, "I remembered the stories" [8].

The stories that follow span three decades of teaching and learning by the author. They largely are drawn from a keynote address, "How We Live & Learn on Campus Tells Us Much about Life (and Death) Issues in the Wider World," which was presented at the 2013 SENCER Summer Institute, funded in part by a grant from the NSF [9]. The next section opens with the first story. When told to students, it quickly makes the point that important connections in our lives – perhaps ones of life and death – may be out of our line of sight.

2.2
Story #1: Does This Plane Have Wings?

Connections are what transform the way we think [10].

Some connections are separated from us in time or space. Such was the case with the elderly nun in the window seat who sat next to me on a DC-9. These noisy old jets have carried millions of passengers over the years, although some airlines now are selling them in favor of quieter and more fuel-efficient planes. Eventhough the interaction with this woman took place over a decade ago, I still clearly remember it.

For passengers seated in the first few rows of the longer "stretch" version of the DC-9, the wings are well to the rear. The elderly nun next to me was seated at the window in one of these first few rows. The two of us, both on our way home to Madison, Wisconsin, were chatting. Suddenly, though, she stopped talking and lines of anxiety crossed her face.

"Dearie," she asked, "Does this plane have wings? I don't see any."

Really? A plane without wings? Her words sorely tested my belief that there's no such thing as a stupid question. Even so, I quickly reassured the nun that the wings of the plane were in the back, hidden from her view. She relaxed and continued chatting. She had made the connection between her safety and the ability of the plane to fly.

Only upon later reflection, did I realize the wisdom of her question. In truth, things *are* hidden from us. Not being able to see certain things in our daily surroundings leaves us unable to make connections. Consider switching on a light. Not only are the wires that carry the electric current hidden inside a wall, but also the plant that generated the electricity most likely is miles away. The people who supplied the fuel for the electric power plant are even more distant. Just as the nun could not see the swept-back wings that would support her flight on a DC-9, we cannot see the electrical infrastructure that supports our use of electricity.

Disconnects! Pump gas into your car. The 50 000-gal tank beneath the filling station is not visible, nor is the gasoline refinery from which the petroleum was distilled. Eat a hamburger. Most likely, the cow was raised and butchered far away. Flush a toilet. The waste moves swiftly from sight and smell. We are disconnected in time or space from our energy supply, from our food supply, and from our waste stream.

Disconnects! Through the process of photosynthesis, green plants release oxygen, a colorless, tasteless, and odorless gas that is invisible to our senses. Without our perceiving it, plants are providing the life-support system for our planet. Through the combustion of coal, we release gigatons of carbon dioxide, another colorless, tasteless, and odorless gas. Again without perceiving it, we are engaging in activities that are changing the composition of the atmosphere.

Also consider what students are able to perceive in our classrooms when they learn about the submicroscopic world of atoms and molecules. They cannot watch chemical bonds breaking and forming in a chemical reaction. With their eyes, they cannot read a molecule of DNA to see how much we have in common with each other. They neither can hear nor see the α-particles, β-particles, and γ-rays that zip through our world. The limitations of our senses have consequences in the wider world as well. For example, if humans were born equipped to detect nuclear radiation, then those who mined uranium [11] and who painted watch dials with radium [12] might not have died in their workplaces.

Disconnects! We are blind to some things because they are too small to see (e.g., atoms). We are blind to others because our senses have no way to detect them (e.g., colorless odorless gases, nuclear radiation). We cannot see still other things because they are separated from us in time and space (e.g., waste water, coal mining). It's relatively easy to ask questions about what is right before our eyes. In contrast, it is harder to question what is missing. The remarkable thing about the elderly nun was that *she inquired about what she could not see*. Clearly, it was of concern to her. Once she made the connection, this transformed her thinking.

Can we transform the way our students think about the world around them? The next section offers stories for those attempting to do just that.

2.3
Story #2: Coaching Students to "See" the Invisible

> You could say paradigms are harder to change than anything else about a system … But there's nothing physical or expensive or even slow about paradigm change. In a single individual it can happen in a millisecond. All it takes is a click in the mind, a new way of seeing [13].

A click in the mind. A new way of seeing. These words should sound familiar to chemistry instructors. The ability to "see" the invisible always has been at the heart of learning about atoms, molecules, and chemical reactions.

For example, in the United States, the College Board recently released a list of the Big Ideas in chemistry, biology, physics, and environmental science. Chemists named five Big Ideas, leading off with the following:

> Big Idea 1: The chemical elements are fundamental building materials of matter, and all matter can be understood in terms of arrangements of atoms. These atoms retain their identity in chemical reactions [14].

Of the five Big Ideas in chemistry, four speak to visualizing the particulate nature of matter: atoms, ions, and molecules together with the forces between them, their arrangements, and the transfer of electrons.

Clearly, our students cannot learn chemistry *without* powers of our imagination: in essence, the ability to see the invisible. As a result, models, representations, and visualizations are used by chemistry instructors at all levels, with some quite elegant ones now available for wider audiences [15].

Consider, for example, that the air we breathe is about four-fifths nitrogen (N_2) and one-fifth oxygen (O_2). Students need this information to build their understanding of combustion, a process that takes place in the lab with a Bunsen burner, on nearby streets in an automobile engine, or across the globe with a cooking fire. Both O_2 and N_2 play important roles, the former supporting combustion and the latter preventing a worldwide conflagration because it does not readily ignite. The oxygen in our atmosphere also underlies discussions of respiration and photosynthesis. Similarly, the "unreactive" nitrogen in our atmosphere underlies the nitrogen cycle, including the formation of N_2O, a greenhouse gas.

To help students imagine ("see") the atoms and molecules in the air they breathe, consider Figure 2.1. On a submicroscopic scale, it shows a sample of dry air containing N_2 and O_2 molecules, which can be colored blue and red consistent with atom color conventions. "1% Other" primarily is argon but notably includes about 400 ppm CO_2.

Figure 2.1 offers an opportunity for students to practice their critical thinking skills. For example, an instructor might invite students to turn to their neighbor and come up with answers to this question: *What's wrong with this representation?* Students quickly can point out that, in many ways, the portrayal falls short of the reality. The gas molecules should be in motion at different speeds, these molecules should be colliding, and the space between them should be adjusted relative to

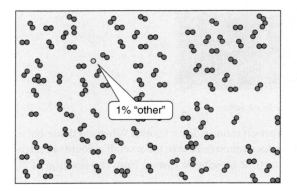

Figure 2.1 A representation of dry air that shows N_2 and O_2 molecules plus 1% other gases.

their size. Furthermore, this representation could include water vapor as well as dust particles and pollen. And, of course, oxygen and nitrogen molecules are not really red and blue.

The subsequent question for class discussion is *What's right with this representation?* Several responses are possible. Figure 2.1 depicts the components of dry air in their correct proportions. This quickly conveys the idea that our atmosphere contains much more nitrogen than oxygen. The spacing of the molecules and their random orientations indicate the gas phase.

Both the shortcomings and the strengths of a representation such as Figure 2.1 offer opportunities for learning. So do the pair of questions: what's right and what's wrong with the representation. In fact, these questions are useful to ask about *any* representations instructors employ to help students learn chemistry, including chemical equations.

Consider, for example, this chemical equation that represents the combustion of methane in oxygen:

$$CH_4(g) + 2O_2(g) \rightarrow CO_2(g) + 2H_2O(g)$$

What's right with this representation? Students readily can provide several answers. For example, they can describe how the chemical equation properly shows the two combustion products for the combustion of methane, carbon dioxide, and water. Similarly, they can see that it is correctly balanced and provides useful information about the physical states of the reactants and products. With coaching, students also can describe how chemical equations are useful symbolic representations.

However, the question *What's wrong with this representation?* is more difficult for students to answer. One reason is that it requires them to make connections to what is *not* shown by a chemical equation. For example, equations do not show what a chemical reaction looks like: in this case, the "visual" fireball produced when methane is burned in sufficient quantity. Another reason is that chemical equations are simplifications. In this case, several products are missing, those of

Figure 2.2 A three-part carbon cycle message.

incomplete combustion (e.g., carbon monoxide and soot). When methane burns in sufficient quantity, it also produces enough heat to form small amounts of nitrogen monoxide (NO, a precursor to tropospheric ozone) from the reaction of N_2 with O_2.

Chemical equations simultaneously reveal information and, by their simplifications, hide it from view. In the case of methane, the missing information is part of a story that connects to air quality and to global climate change. In turn, these stories connect to our responsibilities to each other and to those in the generations that will follow us.

In that carbon dioxide plays a character in so many stories on our planet, it is worth a mention in its own right. However, as pointed out earlier in this chapter, CO_2 is a colorless, odorless, and tasteless gas that we cannot perceive with our senses. We are blind to its presence when it is released into the air via combustion. Similarly, we are unable to see plants taking CO_2 up via photosynthesis. Earlier in this chapter, we remarked that human history might have taken a different course had people been able to detect nuclear radiation with their senses. Perhaps the same is true had people been able to detect the carbon dioxide released when fuels are burned.

Carbon is found many places ("reservoirs") on our planet. The fact that it is building up in the atmosphere and in the oceans should be of concern to citizens of all nations. Figure 2.2 summarizes a three-part message about the carbon cycle that can launch a set of stories about carbon on our planet, all of which involve making connections across time and space.

How might students better perceive the connections between carbon dioxide and combustion? Figure 2.3 shows one possibility. Using presentation software such as PowerPoint (Microsoft Office 2010), instructors can insert colored "clouds" to make the invisible gas visible. In Figure 2.3, purple clouds represent the CO_2 emitted by vehicles on a campus street. The person on the sidewalk also is exhaling CO_2.

The color of the "cloud" holds no meaning other than to signal students of the presence of an invisible gas. Although instructors may worry that students would learn that CO_2 was a purple gas, this fear appears ungrounded. In a quiz, the author provided a photograph similar to Figure 2.3 and asked the meaning of the purple clouds. Almost without exception, students responded that the clouds represented a gas that they were unable to see because it was colorless.

In recent years, elegant ways to visualize carbon dioxide emissions have been created by combining the expertise of scientists and with that of graphic artists.

Figure 2.3 Using "clouds" to indicate the release of CO_2, a colorless gas.

Consider, for example, the work of Carbon Visuals, a project dedicated to helping everyone better understand carbon emissions [16]. One of this project's more compelling videos makes visible the 6204 metric tons of CO_2 emitted each hour by vehicle traffic in New York City. Rather than using purple clouds, the graphic artists employed 10-m turquoise spheres, each containing a metric ton of the invisible gas. In the animation, the pile of spheres soon buries a section of Manhattan [17, 18].

Carbon Visuals aims to "help everyone on the planet make more sense of the invisible." CEO Ted Turner explains why this is important: " … Up to now we've only been able to understand the primary cause of climate change through numbers. The numbers are important but we also need to incorporate more direct and sensory ways of experiencing the world [19]."

Who better to join in this effort than chemists? At every level of instruction, those of us who teach chemistry have opportunities to help people make more sense of the invisible. In engaging students in learning about carbon, our responsibility is to enable them to look at carbon differently, making connections.

Connections can transform the way we think. The combustion of coal, natural gas, and petroleum moves carbon on our planet, invisibly changing the concentration of CO_2 in the atmosphere. In turn, this connects to ocean acidification and the melting of icecaps. As Donella Meadows pointed out in the quote that opened this section, all it takes is "a click in the mind [13]."

Ideally, chemistry instructors provide their students with learning goals at the start of a semester. Can *making connections* be framed as one of these learning goals? How about *using powers of the imagination*? Indeed these can, as a story from a course taught by the author points out.

2.4
Story #3: Designing Super-Learning Environments for Our Students

> With great power comes great responsibility.[1]

Chemistry in Context, a project of the American Chemical Society, is an undergraduate textbook that entered its eighth edition in 2014. It was designed with the goal of engaging non-majors in learning chemistry [20, 21]. The word "context," derived from the Latin word to weave, reflects the fact that every chapter of the book weaves chemistry into a societal issue such as the air we breathe, the water we drink, and the food we eat.

Each chapter of *Chemistry in Context* includes a set of learning goals. For example, Chapter 4, The Energy of Combustion, includes these as part of a longer list:

- Name the fossil fuels, describe the characteristics of each, and compare them in terms of how cleanly they burn and how much energy they produce.
- Explain how fossil fuels, photosynthesis, and the Sun are connected.
- Compare and contrast biodiesel with diesel fuel in terms of chemical composition, energy released on combustion, and energy required to produce.
- Explain the concept of a functional group and give three examples of compounds that are alcohols [22].

The second item in the list is an example of a learning goal that requires students to make connections across time and space. Through the process of photosynthesis, green plants capture the energy of sunlight. During the growing season, these plants slowly move carbon dioxide from the atmosphere to their leaves, stems, and bark. Over millions of years, the biomass of green plants became the fossil fuels we take out of the earth today. The carbon atoms in gasoline not only have a connection to the past but also to our future.

What about powers of imagination? Can these also get framed as a learning goal as well? This task is more difficult, because learning goals tend not to be written with verbs such as "to imagine" or "to use a power." But in 2012, a UW-Madison graduate student, Travis Blomberg, suggested a creative way to get around this when he suggested that we might frame the learning goals as the ability to use *superpowers*.

Superpowers? In essence, Travis was suggesting that students employ powers of the mind to "see" water molecules, branched hydrocarbons, and the DNA molecule in the submicroscopic world. Similarly, in the macroscopic world they could employ superpowers and to "see" the hidden connections among air, water, food, and energy.

---

1) Most recently, this line has been attributed to Uncle Ben in the Spider-Man series. However, it also can be traced back to Voltaire (Francois-Marie Arouet).

Making a connection to contemporary culture, Travis pointed out that the enhanced sensory perception that contemporary superheroes employ (e.g., Batman, Superman, Wonder Woman, and Spider-Man) has a role in the classroom. In film and in comic books, superheroes save their families, their neighborhoods, or even their planet by making use of their powers. For example, the power to leap tall buildings in a single bound is a power of enhanced strength. Flying is the power to overcome gravity. X-ray vision and the ability to listen in on conversations are powers of enhanced perception.

Travis clearly articulated the need for powers of the imagination in the large introductory environmental science course taught by the author. The result was a set of three new learning goals, each one framed as a superpower:

- The power of X-ray vision in order to perceive what is hidden, such as in the infrastructures that supply energy and transport food to campus.
- The power to travel through time in order to learn from people who lived in the past as well as from people in future generations.
- The power to hold conversations with animals and plants in order to more easily take different perspectives on issues [23].

Why not superpowers? If our students need the ability to think across time and space in order to gain multiple points of view, and if our students need to connect the dots both globally and locally, tracing the pathways of substances from cradle to grave, why not explicitly build these powers of imagination into the learning goals of a course? If we are asking our students to imagine atoms and molecules, surely we also can ask them to imagine hidden infrastructures in the world in which they live.

Were the story to stop here, framing learning goals as superpowers might be viewed simply as a curiosity, perhaps an amusing way to liven up an otherwise boring list of what students should know and be able to do. But Travis took the idea of power a step further, pointing out that there was more to the superhero narratives than feats of strength or feats of the mind. Power is accompanied by *a responsibility to use that power wisely*. He produced a short video for students to drive home his point [24].

Did the idea of superpowers work with students enrolled in a large introductory environmental science course taught by the author? Spring semester 2013 was the first time the course was taught with some of the learning goals framed as superpowers. At the end of the semester, students were asked to respond to this prompt: *In this course, we've tried to come up with learning goals that go beyond "learn this" or "understand that". To do this, we have framed these learning goals in terms of superpowers.* Table 2.1 shows the responses.

The Tables 2.2 and 2.3 show responses to two additional questions administered to students in spring semesters 2013 and 2014. *In this course, we framed some of your learning goals in terms of "super powers." To what extent did using a superpower (or two) in this course stimulate you to think more creatively? Here is a theme from several superhero tales: "With power comes responsibility." To what extent do you agree with this theme?*

Table 2.1 Class responses to framing learning goals in terms of superpowers ($n = 112$).

	2013
The idea works for me	49%
Bad plan! I don't like it	1%
Some parts I like; others I don't	26%
I know it's a new idea. Keep working on it!	26%

Table 2.2 Class responses to being stimulated to think more creatively ($n = 112$ in 2013 and $n = 93$ in 2014).

	2013	2014
Not at all (%)	12	10
A little bit (%)	22	28
Somewhat (%)	34	30
Quite a bit (%)	28	27
A great deal (%)	4	4

Table 2.3 Class responses to connecting power with responsibility ($n = 112$ in 2013 and $n = 93$ in 2014).

	2013	2014
Not at all (%)	0	0
A little bit (%)	4	2
Somewhat (%)	11	11
Quite a bit (%)	42	29
A great deal (%)	43	58

Finally, in 2013 students were asked to write short essays to convey their point of view about setting learning goals in the course that included superpowers. Here is one response, typical of many we received. It is reprinted with the student's permission:

> I think using a superpower constantly reminded me to think past what is given to me. In the past, learning has been more of a memorization game rather than understanding the how and why of a subject. When I use my superpower, I have the ability to seek out information that changes the way I see the world.

If our students can benefit when they change the way in which they see the world, so can we as their instructors. The next three sections each tell a story of *change*.

If chemistry instructors change their instructional focus to more broadly make connections in the world, they find that new ways of teaching and learning become possible.

2.5
Story #4: Connections to Public Health (Matthew Fisher)

Why do only our non-majors get the good stuff?[2]

Success in teaching a course for non-majors can open pathways for instructors to change courses for their majors as well. In 2003, this is exactly what happened when Matthew Fisher Footnote [3]first taught a course on science and global sustainability for non-science majors [25]. He structured this course using the approach taken by the SENCER national curriculum reform project, teaching through real-world topics to the underlying science. This approach was described earlier in Section 2.1.

Figure 2.4 represents the SENCER approach, one that positions the learner in a real-world issue such as global sustainability. Students then have the need of the underlying scientific facts, concepts, and problem-solving skills in order to grapple with the larger issue.

The real-world issues of climate change, feeding the world's population, and threats to world health were the starting points for students – all non-science majors – to learn basic principles in chemistry, biology, and earth science. At the same time, the course was designed to reflect that part of Saint Vincent's goal for

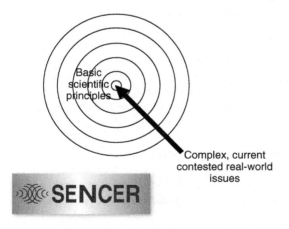

Figure 2.4 A representation of the SENCER approach to teaching and learning.

2) Many have posed this question, including several of my graduate student Teaching Assistants in a chemistry course for non-science majors focused on real-world issues that was taught using Chemistry in Context [20].
3) Text for this section was provided by Matthew Fisher, Department of Chemistry, Saint Vincent College.

the natural sciences focused on understanding "the impact science has had on daily life and the human condition." So the course used these topics as starting points for conversations about the responsibility that comes with knowledge about these real-world issues.

Using the SENCER approach, Fisher created a course that simultaneously was content-rich and context-rich. As he compared the enthusiasm of the students in the sustainability course for making a difference in the world with what he saw in his biochemistry students, Fisher realized that a conventional approach to teaching biochemistry did not develop in students a significant ability to "use scientific knowledge and ways of thinking for personal and social purposes" [26]. So why not approach the content differently? In fact, why not use this content to enable students to look at their world differently?

In biochemistry, a natural place to make broad connections to the world is with topics relating to public health. Many professional programs such as medical, dental, and veterinary school require biochemistry as a prerequisite. With his audience in mind, Fisher restructured a two-semester upper-level biochemistry course so that each section of the course was set in the context of a public health issue. These issues and their related biochemical content included Alzheimer's disease/influenza/vaccines (protein structure and function), HIV/AIDS (enzyme function), malnutrition and diabetes (metabolism), microbial drug resistance (membranes and transport), neuropsychiatric conditions (signal transduction), and the relationship between cancer and the environment (signal transduction, replication, DNA repair, transcription).

Public health issues were incorporated in three ways:

- *Examples used in class were drawn from topics in public health.*
  For example, Aβ or influenza proteins were used as examples of protein structure and folding; HIV protease was used as the central example for enzyme mechanisms; malaria and diabetes served as examples for glycolysis, and multidrug-resistant TB served as an example for secondary active transport.
- *Readings were added on the broader societal context of public health issues.*
  Students read and responded to articles published in the *New Yorker, Atlantic Monthly, Nature,* or *Science* as well as to excerpts from books, for example, *28: Stories of AIDS from Africa*.
- *Group work offered opportunities for integrative learning.*
  In small groups, students developed a presentation on the biochemistry of a public health topic of their choice. The presentation was created using the KEEP Toolkit developed by staff at the Carnegie Foundation for the Advancement of Teaching [27].

As shown in Figure 2.5, there was a natural progression from starting with the global AIDS epidemic and ending at enzyme kinetics and catalytic strategies, both widely regarded as fundamental biochemistry concepts.

Changing the course structure in this way to incorporate real-world connections and the stories and concerns of many individuals resulted in a significant shift in affective engagement on the part of students [28–30]. At the same time, students

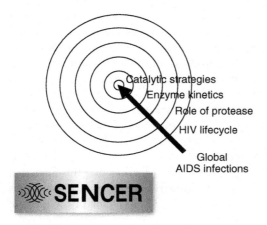

Figure 2.5 Teaching "through" global AIDS infections "to" the underlying science.

were able to demonstrate the same level of content mastery as what was routinely observed prior to these changes. Furthermore, with the changes he made, Fisher more closely aligned his course with the same phrase from the goals for the natural sciences at Saint Vincent College that had initially been viewed as "more relevant" to courses for non-science majors.

The impact on students was more than Fisher initially anticipated. Several students have chosen, after taking one or both biochemistry course, to pursue postgraduate plans to medical or graduate school that have a significant focus on public health. These include medical students pursuing a Master's in Public Health, a graduate student pursuing a certificate in public health, and several students seriously reflecting on what they wanted to pursue after graduating from Saint Vincent.

If our non-majors can benefit by having connections to "the good stuff," surely our majors can as well. We have the responsibility to engage all of our students in learning topics of interest both now and in their future professions.

2.6
Story #5: Green Chemistry Connections (Richard Sheardy)

> How can people pass the carbon and water molecules along, benefiting all the generations that will follow us? [31]

In referring to carbon and water molecules, the authors of *The Upcycle: Beyond Sustainability* more generally are pointing their readers to a key question: *How can the people of our generation upcycle to the benefit those in future generations?* Just as the connections we make need to span time and space, so do our responsibilities.

Michael Braungart, a chemist and one of the authors of *The Upcycle*, points out that in the process of going from a tree ... to a wooden table ... to particle board ... to a piece of paper, carbon atoms should not become contaminated with toxic glues, dyes, or inks. Rather, these carbon atoms need to return to the soil or atmosphere free of contaminants. We need designers who can create pathways free of contaminants during the life cycle of a product.

At Texas Woman's University (TWU), Richard Sheardy describes how instructors bring questions relating to upcycling to their undergraduate chemistry students.[4] In the spring of 2012, these instructors redesigned their organic chemistry laboratory experiments in order to minimize contaminants or toxins produced as waste or by-products. Chemists may recognize that this approach fits under the umbrella of green chemistry, summarized by a set of 12 principles that allow for more environmental accountability [32–34]. One of the goals of the Department of Chemistry and Biochemistry at TWU is to make the dream of sustainable green chemistry a reality in its teaching and training of undergraduates. To echo Matthew Fisher (Section 2.5, Story #4), our majors *can* get the good stuff.

And at TWU they are getting it. In 2012, the concepts of green chemistry were introduced into the second-semester organic chemistry laboratory. During lab recitation, students discuss "The Twelve Principles of Green Chemistry" [35] and the ramifications of continuing to do things the old-fashioned way. Instructors emphasize the importance of using sustainable practices to our students – some of whom will be future chemists. This approach is a form of civic engagement because the products and by-products of the chemical industry will find their ways into our local environments. "As chemists," writes Dick Sheardy, "we must be responsible for the stewardship of the planet and be willing and able to minimize the environmental impact of what we do on our local communities."

This conversation continues with the laboratory experiments. One experiment, the dehydration of cyclohexanol, was redesigned to use phosphoric acid instead of sulfuric acid. Another experiment, a Diels–Alder coupling, now employs water rather than an organic solvent. A third experiment, the photo-initiated synthesis of benzopinacol, exposes reaction flasks to sunlight to initiate the reaction. Sheardy comments, "Our organic chemistry teaching laboratories have been the forerunners that will lead the rest of TWU into a golden age of green chemistry through sustainability, better, more environmentally friendly experiments, and improved waste disposal."

It's not just the faculty. Undergraduates also share the excitement generated when making changes that "connect the dots." Meet David, one of the undergraduate teaching assistants for TWUs organic chemistry laboratory. He is a business major pursuing a minor in chemistry and performed so well in organic chemistry lab that TWU hired him to help teach the course. Students like David also become ambassadors and advocates for change. In David's case, he joined

---

4) Text for this section was provided by Richard Sheardy, Department of Chemistry and Biochemistry, Texas Women's University.

an undergraduate research project on expanding green chemistry in the organic lab and introducing the concepts into the general chemistry labs. In the summer of 2013, he was awarded an NSF fellowship to attend the Green Chemistry and Engineering Conference in Washington, DC. He came back from that meeting full of ideas and resources to help us out. David clearly connected the dots when it comes to sustainability, civic responsibility, and chemistry education. He explains:

> After attending the 17th annual Green Chemistry and Engineering Conference, I have gained a greater appreciation for how connected and important civic engagement, as well as environmental consciousness is in all facets of the world from academia to the business sector. Everything is connected, and we need to treat our Earth in a manner that will foster synergy and cooperation between everyone to make our planet a better place in which to live.

In going forward, the plan at TWU is to make all of the general and organic chemistry laboratory experiments greener. The goal is twofold. First, instructors at TWU hope to apply the methods and lessons learned in the organic chemistry laboratories to all the science labs. Second, they hope to educate future scientists about the challenge and promise of green chemistry. By using the application of these experiments and through teaching, they can begin to train more environmentally conscious scientists who could one day take these concepts and make great strides in the fields of chemistry and biochemistry.

2.7
Story #6: Connections to Cardboard (Garon Smith)

> ... an urgent call for chemistry educators to catalyze change in chemistry education, making it relevant, interesting, and digestible ... [36]

Many of our purchases come in boxes or packages. How much thought do we give to a box or package? Do we connect it to its point of manufacture? Do we question whether the packaging is needed or not?

Add packaging to the list of items right in front of us yet possibly hidden in plain sight. Also add packaging to the list of real-world topics that can catalyze change in chemistry education, making it relevant, interesting, and digestible.

In Missoula, home to the University of Montana, Garon Smith notes that boxes have been manufactured right under the noses of residents for more than 50 years.[5] He quips, "We also could perceive this using our noses!" Until recently, the mill in Missoula produced linerboard, the outside and inside surfaces of boxes via the kraft process, sometime called *kraft pulping* or the *sulfate process*.

---

5) Text for this section was provided by Garon Smith, Department of Chemistry and Biochemistry, University of Montana, Missoula.

Kraft is the German word for strong. Using a chemical procedure based on sulfur chemistry, the kraft process converts wood into cellulose pulp by dissolving the lignin that binds the fibers of cellulose.

Many sulfur compounds stink! Thus, it is hard for people to miss the presence of a chemical reaction involving sulfur. For example, Smith points out to his students that they can detect the difference between good meat and rotting meat by using their sense of smell. Why? Because protein contains two sulfur-bearing amino acids: cysteine and methionine. As the proteins in a piece of meat break down, the sulfur is incorporated into molecules that volatilize and can be detected by most people at a concentration of 5 ppb. He also points out that decaying onions contain organo-sulfur compounds. Then, of course, consider hydrogen sulfide, the odorant in rotten eggs.

Similarly, most Missoula residents are unlikely to miss a bad air day. If a layer of cold air sinks into the Missoula valley overnight, emissions become trapped in this layer, causing the city to smell of putrefying meat, decomposing onions, and rotten eggs. In large part, these emissions were courtesy of the kraft mill.

Missoula residents have stories to tell. For example, a woman who moved to town was applying for a position on the Air Quality Advisory Council of Missoula. At her new home, she called a plumber to identify the source of a bad smell. She was aghast to learn that there was nothing wrong with her septic system. Unfortunately for her, she had purchased a house downwind of the odor plume released by the mill.

A large plant is required to meet the demand for cardboard for consumer products (Figure 2.6). The former pulp mill included a 250-ft digestion tower, four 150-ft chemical recovery boilers, and 800 acres of wastewater treatment ponds.

Figure 2.6 Missoula's former pulp mill (1959–2009), showing the digestion tower (upper left).

For Garon Smith's Introduction to College Chemistry course at the University of Montana, the pulp mill became a frequently used example of the connections between chemistry and daily life. Annually, his course enrolls about 1000 students in fields of applied science, most notably forestry, wildlife biology, environmental studies, and allied health care. These students represent the wide range of issues generated by the presence of the mill.

The mill provides a rich array of topics to fuel discussions of real-world chemistry in the classroom. For example, the first stoichiometry problem is based on the lime kiln operation at the mill. As another example, the last step in the kraft process chemical recovery loop is a double displacement reaction. The list continues! The formation of the linerboard sheet is a large-scale illustration of the attractive power of numerous hydrogen bonds. The odors from the mill serve as a perfect example of teaching concentration, including how parts per billion by volume, when applied to gas molecules, is a molecular ratio.

The mill also demonstrates how partnering with a perceived environmental liability can lead to a much more rapid resolution of a problem. Between 1993 and 2009 (when the corporate office decided to close the mill for economic reasons), the mill supported four of Smith's doctoral students and invested almost a million dollars in research to improve its environmental performance. Reduction of the odor from the mill was one of the most dramatic ones.

As part of the odor reduction, the Montana Department of Environmental Quality imposed an esthetic-based emission standard on the mill of 50 ppb for hydrogen sulfide (H_2S). For each 1-hour average that exceeded this level, the mill was subject to a \$25 000 fine. In August 2000, the mill personnel called Smith's office in a panic. They had recorded 156 hours with levels of H_2S above 50 ppb. Because Montana was in the midst of a particularly severe wild fire season, they were hoping something in the smoke was affecting the H_2S measurements. Smith quickly determined that this was not the case. So the personnel at the mill asked Smith and his graduate students to research the project.

What they found rivals a Who Dunnit story. The culprit? Microbes! The odors from the mill were mostly due to sulfur-reducing bacteria living in an emergency overflow pond that produced hydrogen sulfide. Daily weather patterns brought these microbes into play.

For years, the pulp mill had been paying fines for exceeding allowable limits of hydrogen sulfide emissions. As the cause was unknown, these were just chalked up to "plant upsets." With the research done by Smith's graduate students, the connection between microbes and air-flow patterns became clear. They averaged 10 years of data, creating maps that were color-keyed according to the 1-hour H_2S readings. Immediately, it became obvious that something was happening in summer at the start of the day.

The map created for August showed that an entire band of peaks was shifted in time. The cause was the later sunrise in August. When sunlight hit the dark soil, it created convectional updrafts south of the ponds that dragged the overnight accumulation of H_2S in the surface layer right past the monitors.

Why were there so many violations in the summer of 2000? The mill had redirected the flow pattern in their treatment ponds and had inadvertently fed the microbes exactly the nutrients what they loved. Smith then was able to work with mill technical staff to eventually feed the microbes a nitrate alternative that brought the mill back into compliance. For the 1-hour standard for H_2S (50 ppb), the violations dropped from a high of 716 in 1 year to less than 20 in recent years. The happy ending to the story was that the Montana Department of Environmental Quality forgave the permit violations, respecting that the operator had investigated the issue and come back with a solution.

As a chemist, Smith heeds the call for chemistry educators cited in the words that open this section: "to catalyze change in chemistry education, making it relevant, interesting, and digestible." In his case, the wood actually is digestible, to a pulp. As evidenced by the responses of his students, the stories of the kraft mill are interesting and relevant as well. Garon Smith embodies the R in SENCER. The responsibility includes going out into the community as well as returning to bring the story back into the classroom.

The next section tells a personal story of the author who, as a bicycle commuter out in the community, found a story that she brings into her chemistry classroom.

2.8
Story #7: Wisdom from the Bike Trail

The stories provided by Matt Fisher, Dick Sheardy, and Garon Smith have a common theme: *responsibility*. So do the stories in a best seller released in 1986 with the title *Everything I Really Need to Know I Learned in Kindergarten* [37]. Its words of wisdom included to play fair, share, don't hit people, put things back where you found them, and clean up your own mess. Good advice!

For many of us in Madison, these same words of wisdom could have been learned on our city bike trails. Largely, these trails were built on former railroad right-of-ways and accommodate the needs of a variety of users. These include cyclists who commute daily, speed racers, joggers, those pushing infants in strollers, and folks simply out for a stroll. People must play fair and share the trail. *Don't hit people* seems an especially appropriate piece of wisdom.

As any cyclist knows, it requires one's full concentration to navigate the trail when it gets crowded. Although cyclists largely are kept separate from automobile traffic, the situation changes the instant a cyclist leaves the trail. For example, those who leave the trail near my campus find themselves on a busy city street with vehicles, cyclists, and pedestrians. As if these weren't enough for a cyclist to watch out for, a set of railroad tracks cut diagonally across the street. The bike lane curves to allow cyclists to hit the tracks at a right angle, helping them avoid getting a wheel caught in the tracks (Figure 2.7).

And now the story. One beautiful fall day, I had successfully cycled through this intersection, watching out for the cars and pedestrians. In addition, I managed to cross the bed of railroad tracks with my wheels perpendicular to it, so as not

Figure 2.7 The intersection where train tracks cross a bike path. This photo was taken at a rare moment when no vehicles, bicycles, or pedestrians were present.

to snag a wheel and take a spill. While intently focusing on my path *across* the tracks, I missed seeing what was *on* the tracks. A locomotive! It was slowly chugging along at an ungated intersection, pulling several cars of coal to our nearby campus heating plant. Once I cleared the tracks, the engineer sounded the train whistle. I looked up in panic, but there really was no danger to me. Fortunately, the engineer had his eye on me, the unwitting cyclist who somehow had missed all cues that a train was coming. Kindly, he did not frighten me with a blast of the horn until I was safely clear of the intersection and thus wouldn't tumble off my bicycle while crossing the tracks.

How could I miss the train? In hindsight, it is no surprise. Perhaps the simplest explanation is sensory overload. As I narrowed my focus so as not to hit pedestrians and nearby cyclists, I failed to look up. It simply was not possible for me to pay attention to everything at once.

Similarly, when we are busily engaged in our day-to-day teaching activities, it is not possible to pay attention to everything. For our students, the same holds true. But we cannot afford to keep our heads down very long. Our lives and well-being ultimately depend on our ability to see the bigger picture.

We need to look up! The metaphorical locomotives coming down the train tracks in our world surely are visible, but may not be equipped with whistles. Some changes occur silently. For example, discarded prescription drugs make their way into the drinking water. Plastic bottles degrade and the tiny pieces accumulate in marine wildlife. The fuels we burn silently add carbon dioxide to the atmosphere. For decades, the ozone thinned silently in the stratosphere above us as a result of the chlorofluorocarbons we released. Today, the tundra quietly thaws and releases methane. At increasing rates, species blink out of existence with neither a bang nor a whimper.

As busy instructors, we are well acquainted with sensory overload. We have materials to prepare and exams to grade. In the classroom, we have important facts and concepts to "cover" so that students will be prepared for future chemistry courses, if not future for professions in science, engineering, and medicine. As we narrow our focus to stay safe in one regard, we miss things that threaten us, our students, and possibly even our discipline in another. Not only do we need to look up, we have a responsibility to do so.

2.9
Conclusion: The Responsibility to "Connect the Dots"

Genchem is the first and the last chemistry course that many students take.

Although the quote that opens this section is attributed to Melanie Cooper, who spoke recently in the chemical education lecture series at the University of Wisconsin–Madison [38], she is not the first person to have made this remark. Others have noted that introductory college courses are the *last* formal instruction in chemistry that many of our students ever will receive.

As part of her talk, Cooper described common practices in introductory college chemistry courses. She commented: "If you think about a traditional course, you move to a new topic, and a new topic, and a new topic. The dots are not connected." This chapter underscores this point. We and our students can benefit by attending more carefully to connecting the dots. In fact, we have a responsibility to do so.

Faculty members at the college level who teach introductory ("general") chemistry classes usually serve a wide range of students. In the United States, if not in classrooms across the world, we've successfully trained the next generation of chemists and chemical engineers. In industry, government, and academia, chemists have a demonstrated track record in synthesis, innovation, and creativity. Examples of their success are featured at annual meetings of the American Chemical Society as well as at other national and international chemistry gatherings across the globe.

However, chemistry instructors have not served the general public nearly as well. In the author's experience, evidence of the shortcomings can be found in casual conversation with people who took general chemistry courses decades ago. The line "I hated chemistry" is all too common. Equally importantly, the evidence is documented in high-profile national reports on science education in the United States [39, 40]. Killing a love of chemistry rather than kindling it is a disservice to individual students, if not to the wider society to which they will contribute.

How can we best make our choices in preparing the syllabus for an introductory college chemistry course? What really is important for our students to learn, especially when our discipline contains more knowledge than anybody possibly could hope to master? Indeed we must make choices.

Depending on whom you ask, you will get a longer or a shorter list of these choices: that is, the topics to "cover." In the author's experience, instructors readily can produce a long list of topics likely to resemble the table of contents in a general chemistry book.

Given that students in a general chemistry course range from those who will never see college chemistry again to those for whom it is the gateway to a life-long career in the chemical sciences, it makes sense to find strategies that benefit both ends of the spectrum. This chapter makes the case that one of these strategies is to "connect the dots" between chemistry and the world in which our students will live and seek a profession. In essence, this means connecting chemistry with individual and societal approaches to sustainability.

Does "connecting the dots" favor taking *a macroscopic* or *a microscopic* approach to teaching and learning chemistry? Both! Metaphorically, these are two sides of the chemical coin. Setting a microscopic approach to chemistry in opposition to a macroscopic one (or the other way around) establishes a false dichotomy. For example, instructors who strictly follow the "atoms first" approach easily can miss one side of the coin. Of course atoms are important. The problem is the word *first*. In essence, teaching atoms first privileges microscopic over macroscopic, the two sides of the chemical coin. To connect the dots, simultaneously a view of the submicroscopic world and the macroscopic world around us is necessary. The "real world" sets up a need to know for the world of atoms and molecules; the microscopic world set us a deeper understanding of the issues in our world today. Both are necessary.

Does "connecting the dots" mean taking *a content-based* or a *context-based approach* to teaching? Again both. Metaphorically, these are two sides of the teaching and learning coin. Setting learning chemical content in opposition to learning a real-world context (or the other way around) establishes a false dichotomy. For example, instructors who first need to "cover content" miss one side of the coin. In essence, privileging content over context disconnects students from real-world issues for which a knowledge of chemistry is essential. It is not content first, later weaving in a bit of context as time permits. Similarly, it is not context first without the chemical science needed to understand this context. To connect the dots, both content and context are necessary.

What can empower us to connect the chemical dots, in essence to rethink our chemistry curriculum for twenty-first century students in a twenty-first century world? Perhaps it is useful to recall the message of contemporary superheroes that resonates so well with my students: with power comes responsibility.

Actually it is not one responsibility, but many. The R word in SENCER is *responsibilities*: Science Education for New Civic Engagements and Responsibilities. David Burns, the founder and chief architect of the SENCER national curriculum project explains his choice of this word: "My notion of "responsibilities" was a projection (and a prediction that there would be many things to which we might be called upon to act responsibly toward/about). Along with being a conscientious citizen in a democracy comes the responsibility to know something about the things about which you are "governing" ... so it was precisely the responsibility toward the commonweal (D. Burns, personal communication via email, 2013)."

His words underlie the message of this chapter – our responsibility to "connect the dots" for students in our chemistry classrooms. Doing so is not burdensome; rather, our responsibilities lend meaning to what we teach and to what end. In essence, we have responsibilities that we can embrace. These include the following:

- Responsibilities to "get our heads up," even while we are busily enmeshed in everyday tasks. In our world, metaphorical trains are coming down the tracks;
- Responsibilities to our current students, to send them forth from our classrooms with the curiosity and intellectual habits of mind to put to use the chemistry that they learned;
- Responsibilities to the future citizen that our students will become, to set them on a path of becoming life-long learners of chemistry that they will enable them to both ask and answer questions of civic engagement;
- Responsibilities to the future professionals that our students will become. We need the very best chemical scientists to responsibly lead us into the future. We need lawyers, doctors, and business executives who are curious and motivated to keep learning chemistry, even though it is not their primary field of expertise;
- Responsibilities to the future parents and grandparents that our students will become, to tell stories to children of why their study of chemistry was worthwhile and helped make the world a better place;
- Responsibilities to those we will never meet but surely will follow us. We hope to be able to say that our generation has used its knowledge wisely to live sustainably and to make the world a better place.

Not one responsibility, but many. Perhaps in the simplest of terms, we have a responsibility *to each other* as we engage in the struggle to do our very best as teachers (and learners) of chemistry. Our responsibilities lend meaning – if not higher purpose – to what we teach and why we teach it.

References

1. Rukeyser, M. (1992) The speed of darkness, in *Out of Silence: Collected Poems*, Triquarterly Books, Northwestern University, Evanston, IL.
2. SENCER (2014) Science Education for New Engagements and Responsibilities, http://www.sencer.net/ (accessed 25 May 2014).
3. Sheardy, R. and Burns, W.D. (eds) (2012) *Science Education and Civic Engagement: The Next Level*, Symposium Series, vol. **1121**, American Chemical Society, Washington, DC.
4. Middlecamp, C.H., Jordan, T., Schlacter, A., Lottridge, S., and Oates, K.K. (2006) Chemistry, society and civic engagement, part I: the SENCER project. *J. Chem. Educ.*, **83**, 1301–1307.
5. SENCER (2014) The SENCER Ideals, http://www.sencer.net/About/ sencerideals.cfm (accessed 25 May 2014).
6. Sustainable Measures (2014) Definitions of Sustainability and Sustainable Development, http://www.sustainablemeasures.com/node/36 (accessed 25 May 2014).
7. Thomas Jefferson Planning District Council, Virginia (2014) Although this definition is credited to the Thomas Jefferson Planning District Council, it no longer appears on its website.

http://www.tjpdc.org/environment/ index.asp (accessed 25 May 2014).
8. Middlecamp, C.H. (2009) The old woman and the rug: the wonder and pain of teaching (and learning) chemistry. *Fem. Teach.*, **19** (2), 134–149.
9. SENCER (2013) 2013 SENCER Summer Institute, Santa Clara University, Santa Clara, CA, August 1–5, 2013, *http://www.sencer.net/Symposia/ summerinstitute2013.cfm* (accessed 25 May 2014).
10. Meine, C. (2013) *Sustainability in Time and Space*, Climate, People, and the Environment Program Seminar Series, University of Wisconsin-Madison, Madison, WI, September 13, 2013.
11. Eichstaedt, P. (1994) *If You Poison Us: Uranium and Native Americans*, Red Crane Books, Santa Fe, NM.
12. Mullner, R. (1999) *The Deadly Glow: The Radium Dial Worker Tragedy*, American Public Health Association, Washington, DC.
13. Meadows, D. (1999) *Leverage Points: Places to Intervene in a System*, The Sustainability Institute, Hartland, VT, *http://www.sustainer.org/pubs/Leverage_ Points.pdf* (accessed 25 May 2014).
14. College Board (2009) Science: College Board Standards for College Success.
15. The King's Centre for Visualization *http://www.kcvs.ca/site/index.html* (accessed 25 May 2014).
16. Carbon Visuals *http://www. carbonvisuals.com/* (accessed 25 May 2014).
17. Visualizing carbon (2013) Conservation Magazine, University of Washington, *http://conservationmagazine.org/2013/ 06/visualizing-carbon/* (accessed 25 May 2014).
18. (2014) New York City's Greenhouse Gas Emissions as One-ton Spheres of Carbon Dioxide Gas, *http://www. youtube.com/watch?v=DtqSIplGXOA& feature=player_embedded* (accessed 25 May 2014).
19. Carbon Visuals (2012) (2012) New York City's CEO Anthony Turner TEDx Talk, *http://www.carbonvisuals.com/work/ceo- antony-turner-tedx-talk* (accessed 25 May 2014).
20. American Chemical Society (2014) Chemistry in Context, *http://www.acs. org/content/acs/en/education/resources/ undergraduate/chemistryincontext.html* (accessed 25 May 2014).
21. Middlecamp, C. et al (2015) *Chemistry in Context*, 8th edn, American Chemical Society and McGraw-Hill, Dubuque, IA.
22. Middlecamp, C. et al (2015) *Chemistry in Context*, 8th edn, American Chemical Society and McGraw Hill, Dubuque, IA, p. 196.
23. Middlecamp, C. (2014) Principles of Environmental Science *http://faculty. nelson.wisc.edu/middlecamp/2014/*. (accessed 30 November 2013).
24. Blomberg, T. (2013) With Great Power, *http://www.youtube.com/watch?v= ceLUtjfkbl8* (accessed 30 November 2013).
25. Davis, B.A. and Fisher, M.A. (2011) Science and global sustainability as a course context for non-science majors, in *Sustainability in the Chemistry Curriculum*, ACS Symposium Series, vol. 1087 (eds C. Middlecamp and A. Jorgensen), American Chemical Society, Washington, DC.
26. Rutherford, F.J. and Ahlgren, A. (1994) *Science for All Americans*, Oxford University Press, New York, p. 18.
27. Iiyoshi, T. and Richardson, C.R. (2008) in *Opening Up Education: The Collective Advancement of Education Through Open Technology, Open Content, and Open Knowledge* (eds T. Iiyoshi and M.S. Vijay Kumar), Cambridge, MIT Press, pp. 337–355.
28. Fisher, M.A. (2010) Educating for scientific knowledge, awakening to a citizen's responsibility, in *Citizenship Across the Curriculum* (eds M.B. Smith, R.S. Nowacek, and J.L. Bernstein), Indiana University Press, Bloomington, IN.
29. Fisher, M.A. (2011) Sustainability and the pedagogical perspective of connected science, in *Sustainability in the Chemistry Curriculum*, ACS Symposium Series, vol. 1087 (eds C. Middlecamp and A. Jorgensen), American Chemical Society, Washington, DC.
30. Fisher, M.A. (2013) Public health and biochemistry: connecting content, issues,

and values for majors, in *Connected Science: Strategies for Integrative Learning in College* (eds T.A. Ferrett, D. Geelan, W.M. Schlegel, and J.L. Stewart), Indiana University Press, Bloomington, IN.

31. McDonough, W. and Braungart, M. (2013) *The Upcycle, Beyond Sustainability*, North Point Press, New York.

32. Doxsee, K.M. and Hutchison, J.E. (2004) *Green Organic Chemistry: Strategies, Tools, and Laboratory Experiments*, Thomson-Brooks/Cole Publishing Co, London.

33. Anastas, P.T., Levy, I.J., and Parent, K.E. (eds) (2009) *Green Chemistry Education*, ACS Symposium Series, vol. 1011, American Chemical Society, Washington, DC.

34. Dicks, A.P. (ed.) (2012) *Green Organic Chemistry in Lecture and Laboratory*, CRC Press and Taylor & Francis, Boca Raton, FL.

35. ACS (2014) *The Twelve Principles of Green Chemistry*, Green Chemistry Institute, American Chemical Society, http://www.acs.org/content/acs/en/greenchemistry/about/principles/12-principles-of-green-chemistry.html (accessed 25 May 2014).

36. Mahaffy, P. (2011) in *The Chemical Element, Chemistry's Contribution to Our Global Future* (eds J. Caria-Martinez and E. Serrano-Torregrosa), Wiley-VCH Verlag GmbH, Weinheim, p. 137.

37. Fulghum, R. (2004) *Everything I Need to Know I Learned in Kindergarten*, Ballantine Books.

38. Cooper, M. (2013) *Discipline-Based Education Research: An NRC Report from an Emerging Field*, Chemical Education Seminar Series, Department of Chemistry, University of Wisconsin-Madison, Madison, WI, September 13, 2013.

39. Seymour, E. and Hewitt, N. (1997) *Talking About Leaving: Why Undergraduates Leave the Sciences*, Westview Press, Boulder, CO.

40. Advisory Committee to the National Science Foundation (1998) *Shaping the Future*, Volume II: Perspectives on Undergraduate Education in Science, Mathematics, Engineering, and Technology, National Science Foundation, http://www.nsf.gov/pubs/1998/nsf98128/contents.pdf (accessed 25 May 2014).

3
The Connection between the Local Chemistry Curriculum and Chemistry Terms in the Global News: The Glocalization Perspective

Mei-Hung Chiu and Chin-Cheng Chou

3.1
Introduction

"Science for all" has been advocated since the 1980s (e.g., [1, 2]). Enhancing the students' scientific literacy is important in school teaching. Appropriate instruction that promotes the students' competence in the science domain in terms of cognitive understanding of science contents, practical skills, reasoning of scientific phenomena, and attitude toward science has become essential in school science. To extend the impact of scientific literacy, Miller [3] stated that civic-minded citizens should have scientific literacy with the understanding of scientific terms and should be able to read daily newspapers and generate competing arguments for a given context or issue. Collins [4] also argued that: "A literate citizen should be able to evaluate the quality of scientific information on the basis of its source and the methods used to generate it. Scientific literacy also implies the capacity to pose and evaluate arguments based on evidence and to apply conclusions from such arguments appropriately" (p. 22). However, the link between school science and daily life remains incomplete and requires more research to uncover the reason for this disparity. Furthermore, to extend the impact of science education in society, the public should also be investigated carefully. Shen [5] pointed out that, despite the growing level of concern regarding civic literacy, there has been little debate or agreement about the best methods to measure scientific literacy (cited in [3], p. 204). Moreover, newspapers have the capacity to draw attention to important issues and concepts in science, but they do not purport to consider these topics in depth nor do they consider it their job to empower or educate citizens [6].

According to the report of the Programme for International Student Assessment (PISA; [7]), students in Taiwan lack competence in identifying scientific issues and scientifically explaining phenomena even though Taiwan is ranked at the top of the participating countries. Also, The Organization for Economic Cooperation and Development (OECD) [8] reported that students in Taiwan dislike chemistry the most among the science subjects (e.g., physics, biology, earth science) and the value of the likeness was even below OECD average scores in the PISA 2006. These results highlight areas of concern for science education in Taiwan.

Chemistry Education: Best Practices, Opportunities and Trends, First Edition.
Edited by Javier García-Martínez and Elena Serrano-Torregrosa.
© 2015 Wiley-VCH Verlag GmbH & Co. KGaA. Published 2015 by Wiley-VCH Verlag GmbH & Co. KGaA.

Lin *et al.* [9] investigated 209 citizens in Taiwan and found that 70% of the respondents did not understand most of the major concepts associated with nanotechnology. As a result, researchers, educators, and policymakers struggle with deciding what should be taught in school science classes. In order to better inform this decision-making process, we must first understand the relationship between school science and daily life experiences.

In this chapter, we will describe how selected keywords from chemistry were depicted in local daily newspapers and textbooks compared to how these same keywords were used in two well-known international newspapers to show the frequency trends of selected keywords locally and globally across 5 years. Therefore, we have adopted the term "glocalization" to express the importance of linking local and global issues in science learning.

3.2
Understanding Scientific Literacy

Recently, the United States established the *National Science Education Standards* (NSES) which illustrate the specific skills and knowledge students should possess at each grade level. These science education standards are benchmarks of what is to be expected from school teaching and learning [10]. Among the standards are explicit content standards that must be fulfilled by students in order for them to be considered scientifically literate. These content standards include (i) using scientific information to make every day choices, (ii) engaging in public discourse intelligently and debate about important issues that involve science and technology, (iii) sharing the excitement and personal fulfillment that can arise from understanding and learning about the natural world, (iv) being able to learn, reason, think creatively, make decisions, and solve problems, and (v) keeping pace with the global marketplace [10, pp. 1–2]. As stated in the NSES, modern-day citizens need to be literate in science. Therefore, schools need to prepare students to face the future and to competently engage with science in their everyday encounters so as to be able to lead productive and fulfilling lives. OECD [11] defined scientific literacy as "the ability to engage with science-related issues, and with the ideas of science, as a reflective citizen" (p. 7). The scientifically literate person, therefore, is willing to engage in reasoned discourse about science and technology, which requires the competencies to (i) explain the phenomena scientifically, (ii) evaluate and design scientific inquiry, and (iii) interpret data and evidence scientifically. Miller [12, p. 4] explicitly pointed out that, given the economy of the twenty-first century, countries the world over will need a higher proportion of scientifically literate consumers.

Scientific literacy is the knowledge and understanding of scientific concepts and processes required for personal decision making, participation in civic and cultural affairs, and economic productivity [10, p. 3]. Scientific literacy expands and deepens over a lifetime, not just during the school years. But the attitudes and values established toward science in the early years will influence a person's development of scientific literacy as an adult. DeBoer [13] described how ideally students should be introduced to science and the issues that science raises so that they like science and care enough about the impact of science on society to stay informed about scientific issues as adults (p. 598). Scientific literacy means that a person can ask, find, or determine answers to questions derived from curiosity about everyday experiences. A scientifically literate citizen is able to evaluate the quality of scientific information on the basis of its source and the methods used to generate it [10, p. 22]. Preparing students to successfully face their future must be considered one of the main responsibilities of the educational system.

According to Shamos' (as cited in [13]) proposal, scientific literacy includes the following elements: (i) having an awareness of how the science/technology enterprise works, (ii) having the public feel comfortable with knowing what science is about, (iii) having the public understand what can be expected from science, and (iv) knowing how public opinion can best be heard in respect to the enterprise (p. 229). Roberts [14] argued that, although the definition of scientific literacy has varied and lacks consensus, it is necessary to require the next generation of citizens to develop sufficient knowledge to effectively face social-science-related problems. However, there are several critics of this approach to scientific literacy. Some researchers claim that school science does not cultivate students' scientific literacy or support their success as adults during an age in which science dominates everyday life [15, 16]. Millar [17] criticized that sciences dealing with risk (such as environment or health) are not normally taught or even included in mainstream school science curricula. Lee and Roth [18] also argued the traditional point of view about scientific literacy as lacking a link between formal education and action. They advocated for what they called *community-based science* and claimed that being a scientifically literate "good citizen" entailed showing one's interest with respect to one's local community. This hybrid allows the modern citizen to make good use of scientific information in decision making and in evaluating public policy issues related to science. Osborne [19] also pointed out that the consequent failure to recognize the centrality of literacy to science education leaves the majority ill equipped to become critical consumers of science. The need to reconceptualize the priorities for science education has to be through a mix of new curricula and new strategies. To be realistic, Shamos [16] suggests "science appreciation" and "science awareness" to be the emphases of science teaching in school practice.

As for chemistry literacy, the definition consists of understanding four dimensions: chemical ideas (such as using microscopic structure of matter to explain a phenomenon), contextual aspects (such as seeing the relevance and usability of chemistry in many related contexts), cognitive aspects (such as being able to raise a question and look for related information), and affective aspects (such as showing one's interest in chemical issues [20, 21]).

Civic scientific literacy and media literacy. Miller [12] noticed that the public plays the role of the final arbiter of disputes, especially when the scientific community and political leadership are divided on a particular issue. She claims that it is essential to have a significant proportion of the electorate capable of understanding important public policy disputes involving science and technology in the twenty-first century. The operational definition for civic scientific literacy is conceptualized as the level of understanding of science and technology needed to function as citizens in a modern industrial society [12, p. 4]. As one of the research pioneers in the area of civic scientific literacy, Miller [3] identified three dimensions for measuring civic scientific literacy. These dimensions require a citizen to display the following:

1) An understanding of basic scientific concepts and constructs, such as the molecule, DNA, and the structure of the solar system;
2) An understanding of the nature and process of scientific inquiry; and
3) A pattern of regular information consumption.

On the basis of the elements listed above, Miller [12] found that only 17% of American adults are qualified as civic scientifically literate, although it was still higher than the numbers in Canada, the European Union, or Japan using similar measures. In 1988, the number of Americans said to be civic scientifically literate was around 10%, and in 1997 it was 15%, demonstrating an increasing trend [22].

Other studies also stressed the importance of media literacy. On the basis of the contents of the *Core Principles of Media Literacy Education in the United States* [23], Hobbs and Jensen [24] identified two issues that can potentially impact media literacy and science education: (i) media literacy's relationship to the integration of educational technology into the K-12 curriculum and (ii) the relationship between media literacy education and the humanities, arts, and sciences. Connecting school science content to media reports helps students appreciate science and understand how science relates to their daily lives. This then increases their levels of scientific and media literacy. As DeBoer [25] pointed out, scientific literacy should also include media literacy, which allows future citizens to be able to critically follow reports and discussions on science in the media.

Gilbert and Lin [6] stated that the education of adults is receiving little if any systematic attention. They therefore argued that there should be a focus on the use of informal educational resources for this task, which is of great cultural significance (p. 288). Miller [12] used a structural equation analysis of the 1999 U.S. dataset to identify the factors associated with civic scientific literacy. The analytic model revealed that the use of informal science education resources (such as science magazines, news magazines, science books, science museums, home computers, science Web sites, and the public libraries) was positively related to civic scientific literacy (0.30), which was second to college-level science courses. This demonstrates the positive effect informal science education resources can have on enhancing the scientific literacy of non-science members.

The purpose of this chapter is to show how terms commonly used in local chemistry textbooks relate to media reports of science topics in terms of frequencies of the selected keywords in chemistry. With the analyses of this study, we aim to highlight the emerging needs for reform in school science curricula and textbooks as well as the need to develop civic scientific literacy in the general population in order to bridge the gap between school science practice and daily life.

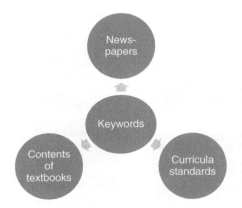

Figure 3.1 Framework of the study.

To form the foundation of our discussion of the issues described above, three main themes will be introduced, namely teaching keywords-based recommendation system (TKRS); analyses of selected chemistry keywords in the specific textbooks and the selected newspapers in Taiwan, the United Kingdom, and the United States; and implications of the use of TKRS and connections between scientific and media literacy (see Figure 3.1).

3.3
Introduction of Teaching Keywords-Based Recommendation System

The teaching keywords-based recommendation system (TKRS) searches for selected keywords in the national and international newspapers, curriculum standards, and textbooks (see Figure 3.2 for its framework). This integrated search system comprises two subsystems: an integrated search system for course outlines (e.g., education cloud) and an integrated search system for hard-copy news archives (e.g., media cloud). The education cloud includes five modules, namely the index module, the parse module, the record restore module, the data query module, and the database maintenance module. In this study, the media cloud includes media reports from *The China Times*, *The Liberty Times*, and *The United Daily News* in Taiwan, *The New York Times* in the United States, and *The Times* in England. Integration of the results of two subsystems results in a comprehensive literacy-based keyword search and analysis system for news media, curriculum standards, and textbooks. TKRS allows users to link the same concepts among different sources of data, such as newspapers, curriculum guidelines, and contents of textbooks, in intra and interdisciplinary curricula (see details in [26]). In this study, we used TKRS to investigate the frequencies of specific terms used in media reports in the local and international newspapers, curriculum guidelines, and textbooks in Taiwan.

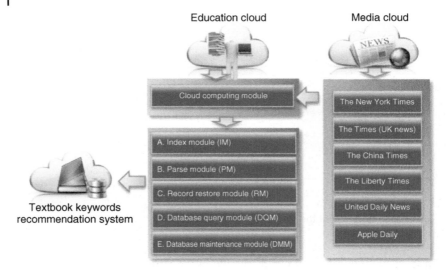

Figure 3.2 Integrated system architecture [26].

3.4
Method

Through the use of TKRS, we selected specific terms to investigate the frequencies of their use in newspapers in Taiwan, the United Kingdom, and the United States. The selected local newspapers were *The China Times*, *The Liberty Times*, and *United Daily News*, which were the major daily newspapers in Taiwan. The other comparative newspapers were *The Times* from United Kingdom and *The New York Times* from the United States.

The specific terms (i.e., keywords) were global warming (全球暖化), sustainability (永續), energy (能源), acid (酸), atomic structure (原子結構), chemical equilibrium (化學平衡), and ethylene (乙烯). The first four terms were drawn from daily life due to social scientific events and concerns and daily life experiences, while the other three were drawn from chemistry textbooks and were commonly covered in the secondary school chemistry curriculum. For our special interest, we also chose melamine (三聚氰胺) and nano (奈米) as the last two terms to be analyzed, even though there were not listed in the national curriculum guidelines in Taiwan. The Chinese characters are kept in the text to show which specific concepts they correspond to in English.

The curricula guidelines investigated in this study included two main documents: the Curriculum Guidelines for Science and Technology for Grades 1–9 (2010) and the Curriculum Guidelines for Chemistry for Grades 10–12 (2008), announced by the Ministry of Education in Taiwan.

Forty-one local textbooks on science were chosen, which included 24 textbooks by three publishers on science and technology for elementary schools (published during 2009–2012 for grades 3–6), 12 textbooks on science and technology for junior high schools (published in 2009–2013 for grades 7–9) by two publishers, and 5 textbooks on chemistry for senior high schools by one publisher (published in 2010–2012 for grades 10–12). All the textbooks were scanned to pdf files except those provided by the publishers. The recognition of Chinese characters of our scanned documents had 92% accuracy by Adobe version 11.

The following sections are the analyses of frequencies of the selected keywords used during the period 2009–2013.

3.5 Results

3.5.1 Example 1: Global Warming

As seen in Table 3.1, the term *global warming* appeared much more often in *The New York Times* compared to the others newspapers over the past 5 years. It appeared on average 3.8 times more often in *The New York Times* from 2009 to 2013 than in any other newspaper. As for the local newspapers in Taiwan, relatively little attention was paid to this term when compared with frequencies in the well-known international newspapers.

As for the local curriculum guidelines for earth science for grades 10–12, we found global warming was listed as one of the main concepts to be taught but was not included in the chemistry curriculum specifically for high schools.

As for science and technology textbooks, the term *global warming* appeared 26 times in six textbooks for grades 6–11. The term was found only three times in the high school chemistry textbooks.

3.5.2 Example 2: Sustainability

As shown in Table 3.2, the average frequency of the term *sustainability* in local and international newspapers was 961 times over the past 5 years. The frequencies showed consistent interest by the newspaper editors in Taiwan, with the average frequency being about 600 except for *United Daily News*, which had around 900

Table 3.1 Number of articles that used the term *global warming* during 2009–2013.

Newspaper Year	United Daily News	Taiwan, The China Times	The Liberty Times	USA, The New York Times	UK, The Times	Average
2009	176	249	175	3620	963	1036.6
2010	140	195	110	2080	663	637.6
2011	128	112	77	1630	504	490.2
2012	68	94	59	2240	395	571.2
2013	54	84	68	1550	409	433.0
Average	113.2	146.8	97.8	2224.0	586.8	633.7

Table 3.2 Number of articles that used the term *sustainability* during 2009–2013.

Newspaper Year	United Daily News	Taiwan, The China Times	The Liberty Times	USA, The New York Times	UK, The Times	Average
2009	897	506	556	1610	2602	1234.2
2010	1036	574	556	1040	1790	999.2
2011	806	611	639	899	1206	832.2
2012	861	666	610	1150	1165	890.4
2013	956	767	731	631	1158	848.6
Average	911.2	624.8	618.4	1066.0	1584.2	960.9

references to sustainability, a level that was closer to the frequency found in *The New York Times*. Reference to *sustainability* was relatively high in the international newspapers as well. In particular, *The Times* had a consistently high frequency of reporting on this topic in the United Kingdom over the past 5 years. The difference between the frequencies of *sustainability* in the local papers compared to the international papers was not as large as for global warming.

As for the local curriculum guidelines, we found this term across different disciplines (environmental education, ocean education, English, geography, civic education and social science, and science and technology). That means the term *sustainability* is a cross-domain concept for grades 7–9 in Taiwan. It appeared in the curriculum guidelines for grade 10 in high school general introductions to chemistry, biology, and earth science but not for grades 11 and 12.

As for textbooks, we found this term in science and technology for elementary and junior high schools (i.e., grades 4 and 6–9). Among the 41 textbooks examined, this term appeared 56 times in 13 textbooks.

3.5.3
Example 3: Energy

As shown in Table 3.3, on average, the frequency of the term *energy* appeared about 7000 times in local and international newspapers. As might be expected given the public concern about energy, the frequencies of the keyword *energy* appeared consistently high in *The New York Times* and *The Times* across the 5 years. In particular, this term appeared close to 25 900 times in *The New York Times*. This keyword had stable frequencies in Taiwan as well, even though the figure was not as high as those in the international newspapers.

Taking the curriculum guidelines into account, we found that *energy* appeared in the guidelines for different subjects from grades 3 to 12, including science and technology, applied geography, earth science, biology, life technology, and introduction to environmental science. From the analyses, it was found that the concept of energy was incorporated across different disciplines and was the focus of many curriculum guideline experts.

Table 3.3 Number of articles that used the term *energy* during 2009–2013.

Newspaper Year	United Daily News	Taiwan, The China Times	The Liberty Times	USA, The New York Times	UK, The Times	Average
2009	1365	880	1032	32800	6450	8505.4
2010	1194	807	814	23200	6312	6465.4
2011	1134	807	1141	24000	6209	6658.2
2012	1208	784	1035	28400	6078	7501.0
2013	1217	850	1120	21100	5645	5986.4
Average	1223.6	825.6	1028.4	25900.0	6138.8	7023.3

As for textbooks, we found that *energy* was introduced from grade 3 to grade 12 in the textbooks. It mainly appeared in application sections such as "energy and life." The frequency of the term *energy* was 554 times in 26 textbooks.

It must also be noted that one is likely to find the term *energy* in many places that might have potentially different meanings unrelated to chemistry or science in Chinese newspapers. Therefore, the popularity of this term can be attributed to the many other meanings of "energy" as it is used in Chinese. The incomparability of the two-language system can be seen here. This finding also shows the limitation of cross-country comparisons on this type of term analysis.

3.5.4
Example 4: Acid

Table 3.4 reveals that the local newspaper *United Daily Times* used the character for "acid" regularly compared to the other newspapers, with the average frequency of 2726 times over the past 5 years.

Table 3.4 Number of articles used the term acid during 2009 and 2013.

Newspaper Year	United Daily News	Taiwan, The China Times	The Liberty Times	USA, The New York Times	UK, The Times	Average
2009	2716	NA	NA	2280	710	1902.0
2010	2633	NA	NA	1250	739	1540.7
2011	2658	NA	NA	1090	656	1468.0
2012	2669	NA	NA	1170	617	1485.3
2013	2954	NA	NA	793	640	1462.3
Average	2726.0	NA	NA	1316.6	672.4	1571.7

The China Times and *The Liberty Times* could not be analyzed with a single character, such as "acid" while the *United Daily News* had no such problem.

As for the curriculum guidelines, the term *acid* was commonly found across different grades. Most often, "acid," "base," and "salt" appeared simultaneously in the science curriculum guidelines. In our analysis, we found the term *acid* in the science and technology curriculum for grades 3–9 as well as in chemistry curriculum guidelines for grades 10–12. This suggests that *acid* is not an interdisciplinary term.

As for the textbooks, *acid* appeared in every science and technology textbook from grades 3 to 9 and in chemistry textbooks for grades 10–12. Among the 41 textbooks, *acid* was mentioned 4186 times in 28 textbooks. This high frequency was consistent with the frequency of the term in the local newspapers.

Because of the constraint of the language structures between English and Chinese and the search engine, some characters/words could not be directly used for both language systems simultaneously. For instance, *acid* in Chinese can also refer to the adjective of acid, "acidic," in some situations. Therefore, some uncertainty of the exact frequency of appearance of *acid* might exist. The vaguer the term, the more uncertainty in finding it.

3.5.5
Example 5: Atomic Structure

Atomic structure is a fundamental concept in learning science. As seen in Table 3.5, it was rarely mentioned in newspapers in Taiwan. If the term *atomic structure* appeared in the newspaper, it was likely to be mentioned in relation to the national examination rather than something related to socio-scientific events. Even in the United States and United Kingdom, there were only 55–57 occurrences on average over the past 5 years.

The concept of atomic structure is commonly introduced during junior high school as basic knowledge for learning science. The curriculum guidelines for grades 7–9 and guidelines of chemistry and physics for grades 10–12 cover the term *atomic structure*.

As for the textbook analysis, the frequency was 68 times in seven textbooks in the area of science and technology for grade 8 and chemistry for grades 10–12.

Table 3.5 Number of articles that used the term *atomic structure* during 2009–2013.

Newspaper Year	United Daily News	Taiwan, The China Times	The Liberty Times	USA, The New York Times	UK, The Times	Average
2009	3	0	1	65	98	33.4
2010	4	1	1	38	46	18.0
2011	3	2	5	79	52	28.2
2012	1	1	2	62	57	24.6
2013	4	0	0	34	35	14.6
Average	3.0	0.8	1.8	55.6	57.6	23.8

From the analysis shown above, we concur with Linn and colleagues' [27, 28] criticism of the uses of atomic and molecular models of thermodynamics in K-12 education, which lack relevance in students' everyday lives. However, as the fundamental concept in chemistry and physics, an introduction of atomic and molecular models is essential and critical for learning advanced concepts in chemistry.

3.5.6
Example 6: Chemical Equilibrium

Chemical equilibrium is a central concept in chemistry and has received a lot of attention from teachers as well as students (e.g., [29, 30]). However, according to our analysis, this term was rarely mentioned in newspapers (close to zero for *The China Times and The Liberty Times*). *The New York Times* printed this term more than the other four newspapers but the figure was still low (28 occurrences). Compared to the other terms discussed above, on average this term appeared even less often than the keyword *atomic structure* (see Table 3.6).

As for the curriculum guidelines, *chemical equilibrium* was included for grades 7–9 as well as in the chemistry curriculum guidelines for grades 10–12. Therefore, *chemical equilibrium* was not a cross-subject term. In other words, this was a domain-specific concept that did not relate to other school subjects.

As for the textbooks, we found that the term *chemical equilibrium* appeared in science and technology textbooks for grades 8, 11, and 12. Out of 41 textbooks, there were 26 appearances of *chemical equilibrium* in only four textbooks.

In sum, similar to the term *atomic and molecular models*, *chemical equilibrium* was not commonly used in daily life but plays important roles in learning advanced concepts in chemistry (e.g., chemical reaction), which have to be taught in school science practice. If this concept is not frequently used in daily life, the misconception on chemical equilibrium that students have might not be directly influenced by their experiences. However, research suggested that students have

Table 3.6 Number of articles used the term *chemical equilibrium* during 2009 and 2013.

Newspaper Year	United Daily News	Taiwan, The China Times	The Liberty Times	USA, The New York Times	UK, The Times	Average
2009	2	0	0	39	5	9.2
2010	3	1	2	25	3	6.8
2011	3	1	0	23	1	5.6
2012	1	0	0	35	2	7.6
2013	0	0	0	20	3	4.6
Average	1.8	0.4	0.4	28.4	2.8	6.8

difficulty in understanding the concept of chemical equilibrium (e.g., [31, 32]). It might have been influenced by the concept of static balance in physics and subsequently misapplied in their understanding of the concept of equilibrium in chemistry.

3.5.7
Example 7: Ethylene

As shown in Table 3.7, *ethylene* was not as popular a term in *The New York Times* and *The Times*. On average, the frequencies for the three local newspapers were higher than for the international publications (16 occurrences) but still relatively low. We speculate that the frequencies for the local newspapers were higher because the media focused on the socio-scientific issues surrounding Taiwan, such as building petroleum refineries over the past few years. This issue was often in the headlines, and politicians used this issue in their campaigns for parliament. This is one potential reason why the local newspapers had higher frequencies than *The Times and The New York Times* (see Table 3.7). Also, ethylene and polyvinylchloride (PVC, 聚氯乙烯) are different words in English but PVC (聚氯乙烯) included ethylene (乙烯) in Chinese except for adding 聚氯, which implies its relation with ethylene and is easier to be identified as an ethylene derivatives.

In the selected curriculum guidelines between grades 3 and 12, there were no direct statements discussing the concept of *ethylene* from grades 3 to 7. At the second semester of grade 8, *ethylene* was introduced in the chapter on organic compounds, in the chapter on commonly seen polymers in daily life for grade 10, in the chapter of produced organic compounds at grade 11, and then related to chemical reaction properties of *ethylene* in the unit on chemical polymers for grade 12 as a selected course.

Table 3.7 Number of articles that used the term *ethylene* during 2009–2013.

Newspaper Year	United Daily News	Taiwan, The China Times	The Liberty Times	USA, The New York Times	UK, The Times	Average
2009	50	24	49	16	16	30.0
2010	109	70	124	13	13	65.8
2011	152	111	182	6	6	91.2
2012	75	55	113	36	36	57.4
2013	60	51	88	9	9	43.0
Average	89.2	62.2	111.2	16.0	16.0	57.5

As for the textbook analysis, the findings revealed that the frequencies of descriptions and discussions of *ethylene* totaled 211 times in eight textbooks for grades 8–12.

3.5.8
Example 8: Melamine

To show the special case of the term *melamine*, we have included 2005–2008 for reference purposes in Table 3.8. As shown in the table, *melamine* was mentioned over 200 times in *The New York Times* in 2007 because in that year pet food sold in the United States was found to be contaminated with the industrial chemical melamine, which is not permitted in food or pet food products [33]. As for Taiwan, it reported limited information about this event in 2007 (about six occurrences). In 2008, when it was reported that Chinese domestic baby formula was also contaminated with melamine, it had a greater impact on Taiwan because quite a few enterprises from Taiwan were located in China and, as a result, there were increasing numbers of reports on melamine in the Taiwanese newspapers (436 occurrences on average in 2008). However, *The Times* did not have frequent reports on melamine compared to the other newspapers. This is a good example that demonstrates how learning scientific terms is dynamic and how *melamine* needs to arouse the public's awareness of socio-scientific issues as related real-life events occur.

As for the curriculum guidelines and textbooks, we found there was no mention of melamine.

Table 3.8 Number of articles that used the term *melamine* in the past 9 years.

Newspaper Year	United Daily News	Taiwan, The China Times	The Liberty Times	USA, The New York Times	UK, The Times	Average
2005	0	0	0	10	7	3.4
2006	0	0	0	20	17	7.4
2007	6	5	7	209	14	48.2
2008	409	256	642	271	27	321.0
Average	103.8	65.3	162.3	127.5	16.3	95.0
2009	55	29	83	106	19	58.4
2010	26	19	41	46	18	30.0
2011	46	49	59	17	20	38.2
2012	33	17	48	20	14	26.4
2013	35	30	55	32	21	34.6
Average	39.0	28.8	57.2	44.2	18.4	37.5

This table included the statistics of frequencies before 2009 for the sake of comparisons.

3.5.9
Example 9: Nano

To extend our own interest in terms that show trends in science education research, we also investigated the frequencies of *nano*, which was not present in the curriculum guidelines for grades 3–12.

According to Table 3.9, *nano* was mentioned twice more often in the newspapers in Taiwan (205 occurrences on average) than in *The Times* in United Kingdom and *The New York Times* in the United States (109 occurrences on average). This might be due to the promotion of nanotechnology and nanotechnology education by the industry and the Ministry of Education in Taiwan.

As for the textbooks, even though the term was not included in curriculum guidelines for grades 3–12, the term did appear as additional information in science and technology textbooks and started formally to be introduced in junior high schools. Among the 41 textbooks, the term *nano* was mentioned 177 times in 14 textbooks. Apparently, the textbook writers recognized the importance of introducing this concept at an early grade in school science because of its essential role in the twenty-first century. The other possible reason might be the large budget for national-level projects on nanotechnology sponsored by the National Science Council since 2003 [35] and K-12 nanoscience and nanotechnology education development projects sponsored by several governmental sections (including the National Science Council) starting from 2003 (see *http://pesto.lib.nthu.edu.tw/*).

Stevens *et al.* [36] pointed out that scientific citizens need to have knowledge about nanoscience and nanotechnology. Gilbert and Lin [6] argued that, with the rapidly growing economic, personal, and social significance of nano, there is a great need for early and universal education about what this emerging science area

Table 3.9 Number of articles that used the term *nano* during 2009–2013.

Newspaper Year	United Daily News	Taiwan, The China Times	The Liberty Times	USA, The New York Times	UK, The Times	Average
2009	264	109	163	259	115	182.0
2010	258	142	193	150	50	158.6
2011	247	108	185	137	51	145.6
2012	231	169	392	162	45	199.8
2013	210	124	281	89	34	147.6
average	242.0	130.4	242.8	159.4	59.0	166.7

entails. We consider the keyword *nano* as relative to daily life and a topic worthy of public attention. Nano should be included in school science subjects and be used as an exemplar for linking scientific concepts with school learning and daily experiences.

3.6 Concluding Remarks and Discussion

According to the analyses described in this study, we have summarized the average frequencies in Table 3.10 and Figures 3.3–3.7. Six major findings are described as follows.

Table 3.10 A summary table for the average frequencies of the terms during 2009–2013.

Newspaper Year	United Daily News	Taiwan, The China Times	The Liberty Times	USA, The New York Times	UK, The Times	Average
Global warming	113.2	146.8	97.8	2224.0	586.8	633.7
Sustainability	911.2	624.8	618.4	1066.0	1584.2	960.9
Energy	1223.6	825.6	1028.4	25900.0	6138.8	7023.3
Acid	2726.0	NA	NA	1316.6	672.4	1571.7
Atomic structure	3.0	0.8	1.8	55.6	57.6	23.8
Chemical equilibrium	1.8	0.4	0.4	28.4	2.8	6.8
Ethylene	89.2	62.2	111.2	16.0	8.8	57.5
Melamine	39.0	28.8	57.2	44.2	18.4	37.5
Nano	242.0	130.4	242.8	159.4	59.0	166.7

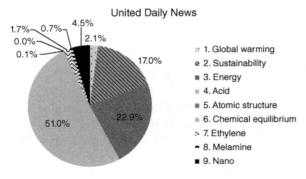

Figure 3.3 Distribution of nine terms in *The United Daily News*.

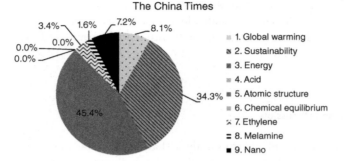

Figure 3.4 Distribution of nine terms in *The China Times*.

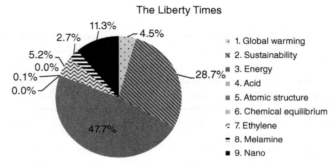

Figure 3.5 Distribution of nine terms in *The Liberty Times*.

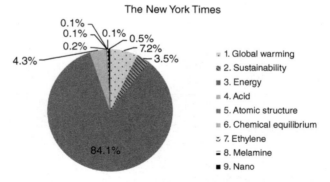

Figure 3.6 Distribution of nine terms in *The New York Times*.

First, among the four selected terms, namely *global warming, sustainability, energy,* and *acid*, we found *energy* and *acid* to have high frequencies on average. The term *energy* had the highest frequency with 25 900 occurrences of which 6138 occurrences were in *The Times* over the past 5 years.

Second, in general, the frequencies of some selected key concepts from chemistry curriculum guidelines (e.g., atomic structure, chemical equilibrium) did not

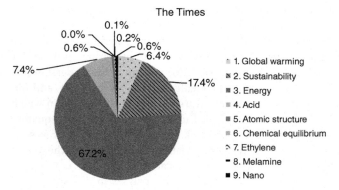

Figure 3.7 Distribution of nine terms in *The Times*.

correlate with the frequencies of the commonly used terms (e.g., energy and global warming) that appeared in the newspapers. Also, students might not benefit from these "school-learned" terms when it comes to understanding and linking their school knowledge to their daily life. We are not against teaching these fundamental concepts in chemistry, but we hope to highlight the gap between school learning and daily life experiences that needs to be remedied through educational reforms. It is important for students to read and absorb relevant scientific knowledge while in school, so that they can continue to expand their interest and interaction with socio-scientific issues.

Third, "*ethylene* (乙烯)" was more often mentioned in local newspapers in Taiwan than in the two international newspapers. Although we might not have direct evidence to show the relationship between local events and the high frequency of the term, we speculate that recent political issues about building petrochemical industries might have impacted the frequency of appearance of *ethylene* in local newspapers.

Fourth, as for the term *melamine*, since it was associated with a unique event that happened in Asia in 2008, it generally appeared less often in the international newspapers. To continue the second point addressed above, developing sustainable interest in socio-scientific issues is essential to improve the students' future quality of life.

Fifth, the frequencies of the term *nano* were higher in the local newspapers in Taiwan compared to *The New York Times* and *The Times*. Given the special concern from the government for promoting nanoscience and nanotechnology as well as nano educational projects, the high frequencies were to be expected. However, the Committee on Curriculum Guidelines is only now responding to this emerging need.

In sum, we noticed that there were four major findings: First, there were more news about *sustainability* in Taiwan than in *The New York Times* and *The Times*. Second, there was more news related to energy issues in the international newspapers than the three local newspapers. Third, the local newspapers tended to report more news related to nanotechnology compared to *The New York Times* and

The Times. Fourth, the remaining terms had limited frequencies in the five selected newspapers.

Stuckey *et al.* [37] pointed out that school curricula should consider three dimensions of relevance in teaching: (i) relevance for preparing students for potential careers in science and engineering, (ii) relevance for understanding scientific phenomena and coping with the challenges in a learner's life, and (iii) relevance for students becoming effective future citizens in the society in which they live. Miller [22] also pointed out, based upon a set of structural equation models analyses, that there exists a strong link between the learning of science during the school years and the persistence of this knowledge into the young adult years. What is taught in school should be carefully decided, as it will be carried on into students' later years as adults. The analyses in this study shed light on areas to target for curriculum reform and highlight the importance of carefully considering what should to be taught in schools.

Osborne [19] stated that, if science teachers' instruction focuses on traditional practice with no connection with daily life contexts, it will limit students' thinking and understanding of science. As Gilbert and Lin [6] claimed, school systems are reacting too slowly and in too limited a way to what students need. Several studies have recommended methods to link school science and daily life experiences, such as using technological knowledge to increase the connection between school science and students' out-of-school experiences (e.g., [28, 38]). Fensham and Harlen (1999) reviewed studies measuring public understanding of science and commented that none of them was helpful on determining how to prioritize the science students should leave school knowing, or in what form that knowledge should be tested except to point out the importance of these matters. Researchers also commented that some curriculum reforms (such as Science–Technology–Society, proposed by Solomon and Aikenhead [40]) did not have any real impact in general education in the United States because of the vague purpose associated with such projects [22, 41]. Therefore, we need to explicitly state what epistemological approach to take in order to link scientific knowledge, methods, and skills with daily experiences. We advocate that using every day experiences helps teachers to build upon the students' prior knowledge. A profound analysis by TKRS might provide a new avenue for identifying some key concepts for school learning.

3.7
Implications for Chemistry Education

Through the analysis in this study, we identified the following implications for research and education:

Keywords Analysis as Connections between Scientific and Media Literacy
1. Based upon the analyses shown above, there is an emerging need to construct a curriculum map system for cultivating civic literacy in chemistry.

2. The dynamic progression of introducing appropriate and popular concepts to laypersons should be well conveyed to media, so that appropriate concepts can be introduced.

Keywords Analysis as Guidelines for Curriculum Design
1. The integration of frequencies of keywords in textbooks and daily-life reports (including newspapers) should be emphasized when a society intends to elicit civic understanding of science.
2. From the analysis of the selected terms over the past 5 years in this study, we were able to identify (i) whether or the frequencies of specific terms changed because of a specific event or (ii) whether central concepts designed for school teaching should be learned as necessary knowledge and competence in life.
3. On the basis of the design of TKRS, it is possible to find keywords across as well as within subjects and then identify the contextualized link with the terms in the textbooks.

Keywords Analysis as a Source for Linking School Science and Daily Life
1. Media have influenced the public's understanding of science. Media reporters have to realize the responsibility and authority they hold and have to convey the idea of socio-scientific issues carefully.
2. A concept with high frequency implies that the topic is important for the public to understand.
3. As far as we found, the contents of the newspapers were easy to access for analysis via the Internet. Therefore, we have to make good use of this resource.

Keywords Analysis Informing the Need for Developing Instruments for Measuring Public Understanding of Science

Unlike many search engines, TKRS allows users to find the relationships between specific terms that occur frequently in media reports as opposed to textbooks. With this function, a term can be categorized as either interdisciplinary or domain specific. In addition, due to the capacity of the search system, we were able to identify the loci of the specific terms in media reports, make a plausible suggestion for their importance in relation to daily life, and then propose a suggestion for potential inclusion in the science curriculum. In other words, if a term was commonly mentioned in media reports, then it should be considered for inclusion in school science.

Civic Scientific Literacy Should Plant Its Seeds in Precollegiate Education and Then Extend and Enhance Its Impact on General Education in University for Both Science Majors and Non-science Majors

On the basis of our understanding that the science and technology policy will be decided by citizens through normal democratic processes in most countries, and that political issues require understanding of science and technology for problem-solving and policy design, we advocate developing an usable and functional scientific literacy precollegiate education that allows most students to become well-informed citizens capable of making appropriate political decisions for their society. Furthermore, at the college/university level, general education should also provide opportunities for non-science majors to acquaint themselves with and learn to appreciate how science and technology impacts their daily lives. One has to recognize the need to develop scientific literacy for the economic and political good of one's country.

To echo the title of this chapter, we advocate that the keywords in the science curriculum should reflect the local needs and events that are closely related to the students' daily lives. Those terms might not be related to global needs (such as melamine in Taiwan and the United States but not in the United Kingdom), but once it becomes a local issue, curriculum experts should take this into consideration so as to better prepare students for what they would possibly encounter in their daily lives. While thinking about local socio-scientific issues, we also need to broaden our view to include global economic, environmental, and political concerns. After all, we have to think globally and act locally, and then become well-informed citizens of the Earth.

Acknowledgment

The authors would like to thank the National Science Council of Taiwan for financially supporting this research under Contract No. NSC100-2511-S-241-006-MY2 and NSC 102-2515-S-003-015-MY2.

References

1. American Association for the Advancement of Science (1989) *Project 2061. Science for all Americans*, American Association for the Advancement of Science, Washington, DC.
2. UNESCO (1983) *Science for All: Report of a Regional Meeting*, UNESCO Office for Education in Asia and the Pacific, Bangkok.
3. Miller, J.D. (1998) The measurement of civic scientific literacy. *Public Understanding Sci.*, **7**, 203–223.
4. National Research Council (1996) *National Science Education Standards*, National Academy Press, Washington DC.
5. Shen, B.S.P. (1975) in *Communication of Scientific Information* (ed S. Day), Karger, Basel, pp. 44–52.
6. Gilbert, J.K. and Lin, H.S. (2013) How might adults learn about new science and technology? The case of nanoscience and nanotechnology. *Int. J. Sci. Educ.*, **3** (3), 267–292.
7. OECD (2007) OECD Programme for International Student Assessment (PISA), http://www.pisa.oecd.org (accessed 25 July 2014).
8. OECD (2007) *PISA 2006: Science Competencies for Tomorrow's World*, Vol. 1: Analysis, OECD, http://www.oecd.org/fr/education/scolaire/programme-internationalpourlesuividesacquisdeselevespisa/pisa2006results.htm (accessed 25 July 2014).
9. Lin, S.F., Lin, H.S., and Wu, Y.Y. (2013) Validation and exploration of instruments for assessing public knowledge of and attitudes toward nanotechnology. *J. Sci. Educ. Technol.*, **22**, 548–559.
10. National Research Council (1996) *National Science Education Standards*, National Academy of Science, Washington, DC.

11. OECD (2013) OECD Programme for International Student Assessment (PISA), http://www.oecd.org/pisa/pisaproducts/Draft%20PISA%202015%20Science%20Framework%20.pdf (accessed 25 July 2014).
12. Miller, J.D. (2002) Civic scientific literacy: A necessity in the 21st century. *Public Interest Rep.*, **55** (1), 3–6.
13. DeBoer, G.E. (2000) Scientific literacy: another look at its historical and contemporary meanings and its relationship to science education reform. *J. Res. Sci. Teach.*, **37** (6), 582–601.
14. Roberts, D.A. (2007) in *Handbook of Research on Science Education* (eds S. Abell and N. Ledermann), Lawrence Erlbaum Associates, Mahwah, NJ, pp. 729–780.
15. Jenkins, E.W. (2000) in *Improving Science Education: The Contribution of Research* (eds R. Millar, J. Leach, and J. Osborne), Open University Press, Buckingham, pp. 207–226.
16. Shamos, M. (1995) *The Myth of Scientific Literacy*, Rutgers University Press, New Brunswick, NJ.
17. Millar, R. (2007) Twenty first century science: insights from the design and implementation of a scientific literacy approach in school science. *Int. J. Sci. Educ.*, **28** (13), 1499–1521.
18. Lee, S. and Roth, W.M. (2003) Science and the "good citizen": community-based scientific literacy. *Sci. Technol. Hum. Values*, **28** (3), 403–424.
19. Osborne, J. (2007) Science Education for the Twenty First Century. *Eurasia J. Math., Sci. Technol. Educ.*, **3** (3), 173–184.
20. Shwartz, Y. (2004) Chemical literacy: defining it with teachers, and assessing its expression at the high-school level. Doctoral dissertation, Department of Science Teaching, The Weizmann Institute of Science, Israel.
21. Shwartz, Y., Ben-Zvi, R., and Hofstein, A. (2005) The importance of involving high-school chemistry teachers in the process of defining the operational meaning of chemical literacy. *Int. J. Sci. Educ.*, **27** (3), 323–344.
22. Miller, J.D. (2000) The development of civic scientific literacy in the United States. In D.D. Kumar and D.E. Chubin (Eds.), *Science, technology, and society: A sourcebook on research and practice*, Kluwer Academic/Plenum Press, New York, pp. 21–47.
23. National Association for Media Literacy Education (2007) Core Principles of Media Literacy Education in the United States, http://namle.net/publications/core-principles/ (accessed 14 December 2013).
24. Hobbs, R. and Jensen, A. (2009) The past, present, and future of media literacy education. *J. Media Literacy Educ.*, **1**, 1–11.
25. DeBoer, G.E. (2005) in *Science Education: Major Themes in Education* (ed. J.K. Gilbert), Routledge, New York, pp. 220–245.
26. Lin, P.J. and Chou, C.C. (2013) Keywords-based recommendation system: a case study of civic chemical literacy. *Chem. Educ. J.*, **15** (2), http://www.t.soka.ac.jp/chem/cej_temp/CEJ1502/index-e.html (accessed 26 July 2014).
27. Linn, M. and Muilenburg, L. (1996) Creating lifelong science learners: what models form a firm foundation? *Educ. Res.*, **25** (5), 18–24.
28. Linn, M. and Songer, N.B. (1991) Teaching thermodynamics to middle school students: what are appropriate cognitive demands? *J. Res. Sci. Teach.*, **28** (10), 885–918.
29. Gussarsky, E. and Gorodetsky, M. (1990) On the concept "chemical equilibrium": the associative framework. *J. Res. Sci. Teach.*, **27**, 197–204.
30. Voska, K.W. and Heikkinen, H.W. (2000) Identification and analysis of student conceptions used to solve chemical equilibrium problems. *J. Res. Sci. Teach.*, **37** (2), 160–176.
31. Chiu, M.H., Chou, C.C., and Liu, C.J. (2002) Dynamic processes of conceptual change: analysis of constructing mental models of chemical equilibrium. *J. Res. Sci. Teach.*, **39** (8), 688–712.
32. Quílez-Pardo, J. and Solaz-Portolés, J.J. (1995) Students' and teachers' misapplication of le chatelier's principle: Implications for the teaching of chemical

equilibrium. *J. Res. Sci. Teach.*, **32** (9), 939–957.

33. Martin, A. (2007) Melamine from U.S. Put in Feed. The New York Times (May 31), *http://www.nytimes.com/2007/05/31/business/31food.html?_r=1&* (accessed 1 August 2014).

34. Jenkins, E.W. (1999) School science, citizenship and the public understanding of science. *Int. J. Sci. Educ.*, **21** (7), 703–710.

35. National Science Council (2011) *Science and Technology Yearbook*, National Science Council, Taipei, *http://yearbook.stpi.org.tw/html/2011.htm* (accessed 25 July 2014).

36. Stevens, S.Y., Sutherland, L.M., and Krajcik, J.S. (2009) *The Big Ideas of Nanoscale Science and Technology*, STATE: NSTA, Arlington, TX.

37. Stuckey, M., Hofstein, A., Mamlok-Naaman, R., and Eilks, I. (2013) The meaning of relevance in science education and its implications for the science curriculum. *Stud. Sci. Educ.*, **49** (1), 1–34.

38. Jenkins, E.W. (1992) Public understanding of science and science education for action. *J. Curriculum Stud.*, **26**, 601–611.

39. Fensham, P.J. and Harlen, W. (1999) School science and public understanding of science. *Int. J. Sci. Educ.*, **21** (7), 755–763.

40. Solomon, J. and Aikenhead, G.S. (eds) (1994) *STS Education: International Perspectives on Reform*, Teachers College Press, New York.

41. Cajas, F. (1999) Public understanding of science: using technology to enhance school science in everyday life. *Int. J. Sci. Educ.*, **21** (7), 765–773.

4
Changing Perspectives on the Undergraduate Chemistry Curriculum
Martin J. Goedhart

In this chapter, I will discuss the role of universities in this changing world, and more particularly how we can train future chemists in such a way that they are adequately equipped to contribute to meet these future challenges. This chapter describes how ideas about teaching and curriculum in universities have changed over the last decades. Although some ideas have influenced the way we teach chemistry in the undergraduate curriculum, the way the curriculum is structured in courses has stayed more or less unchanged. I will present a new division of chemistry from a competency-based perspective, which can be used as the basis for the structure of a new curriculum.

4.1
The Traditional Undergraduate Curriculum

I begin this chapter with a characterization of what I call the "traditional curriculum" for undergraduate chemistry programs at colleges and universities. Hake [1] defines traditional courses as "[courses that] make little or no use of IE [interactive-engagement] methods, relying primarily on passive-student lectures, recipe labs, and algorithmic-problem exams" (p. 65). Hake's wordings make clear that the word "traditional" has a negative connotation, but I will use the word "traditional" in a more neutral sense to describe teaching methods that have a long history. In this meaning, the word "traditional" opposes "modern" teaching methods, which have become popular during the last decades, but I realize that a sharp distinction between "traditional" and "recent" cannot be made.

In this chapter, I will use the description of curriculum given by Hass [2]. Hass states that a curriculum means "all of the experiences that individual learners have in a program of education whose purpose is to achieve broad goals and related specific objectives, which is planned in terms of a framework of theory and research or past and present professional practice" (p. 5). Also, other authors define curriculum broadly, including all decisions taken in a school to enable the learning of students. This involves the attainment targets of the entire program and objectives of individual courses, the organization of the program in subjects and courses, the

Chemistry Education: Best Practices, Opportunities and Trends, First Edition.
Edited by Javier García-Martínez and Elena Serrano-Torregrosa.
© 2015 Wiley-VCH Verlag GmbH & Co. KGaA. Published 2015 by Wiley-VCH Verlag GmbH & Co. KGaA.

teaching strategies employed, the learning activities of the students, the learning materials, and the testing methods.

The most striking characteristic of the traditional curriculum in universities is the role of the lecture in the transmission of knowledge to students. The lecture has a long tradition in universities. It has been practiced since medieval times and is nowadays still a dominant teaching strategy. Although lectures offer opportunities for interaction between the students and the lecturer, many lectures are conducted in such a way that the lecturer transmits his knowledge to the students without getting feedback on the results of his teaching. This is particularly the case in lectures for large student groups. In lecture courses, textbooks or lecture notes are used to support students in their knowledge acquisition. Students' knowledge and skills are commonly tested in written exams. Problem-solving classes or tutorials are used to prepare students for the type of problems in the exam. These involve smaller groups, generally some dozens of students. Students work individually or in small groups, tutored by graduate students.

An important part of training chemistry students is done in laboratories. Originally, the aim of lab courses was to teach manipulative or technical skills, such as learning to use chemicals and instruments or to perform standard procedures [3, 4]. In the traditional curriculum, lab work may also have the function of providing learners with experiences to expand or foster their theoretical knowledge, for instance, by verification of the knowledge they have acquired from textbooks or lecture courses.

4.2
A Call for Innovation

From the 1960s, several developments have influenced the ideas about the university curriculum. These developments came from different sources. Some were caused by novel conceptions of student learning, and some underpinned by findings from educational research. Other relevant developments involved the role of computer technologies in teaching, the changing nature of chemical research, and changes in the society.

4.2.1
Constructivism and Research on Student Learning

Nowadays, constructivism is the dominant paradigm on learning. Constructivism is more than a learning theory, since it is also a philosophy about knowledge and science. Constructivism emphasizes that knowledge and science are personal constructions of the world [5]. Learning is considered as an active construction process in which students interpret new information in their existing frameworks. This notion has led to extensive research into students' pre-knowledge, revealing misconceptions or alternative conceptions they hold on many chemical topics, mostly at the secondary level [6–8]. Research has shown that students' conceptions are often difficult to influence, and that changing those conceptions may be a laborious and long-term process. As a consequence, many alternative conceptions survive the secondary school and can be found among university students [9].

A number of constructivist teaching approaches have been developed and empirically tested. These are generally summarized as conceptual change and as social constructivism. An example of the conceptual change approach, applied in college teaching, is the learning cycle [10]. The learning cycle uses a cognitive conflict, in which students meet new experiences that are incompatible with their existing conceptions. Research has shed light on the conditions and strategies needed to bring about the intended changes in students' ideas [11]. Social constructivist methods focus on the knowledge construction process in groups. Social-constructivism is rooted in the work of Lev Vygotsky [12], a Russian psychologist from the 1920s to the 1930s. Social constructivist theorists recognize the importance of language, culture, norms, and values shared by communities in the process of knowledge construction [13, 14].

Although constructivism did not permeate immediately into universities, the insight that learning is an active process has led to doubts about the effectiveness of "passive teaching" methods. A number of studies, mainly on physics courses, have investigated the learning gains of traditional lectures. Halloun and Hestenes [15] found that after a course in classical mechanics many freshmen lacked a conceptual understanding of the subject. Their conclusion about the ineffectiveness of lectures is supported by a number of other studies [1, 16, 17]. Although these examples come from physics classes, it is likely that chemistry students do not respond very differently on lecture instruction.

Constructivism has drawn attention to novel pedagogical approaches applying "active learning," such as collaborative learning [18–20], and to metacognition. Metacognition is sometimes defined as "thinking about thinking" and enables students to monitor their own thinking and action. Metacognitive skills are considered as an important aspect of an expert's competence [21]. Cooper and colleagues have developed instruments to measure chemistry students' metacognition [22, 23], but research on the development of students' metacognition in higher level chemistry courses is scarce. One of the few examples is a study by Tien *et al.* [24] on the development of college students' metacognition in inquiry-oriented laboratory work.

Another influence of constructivism is the attention for meaningful learning in authentic environments. The traditional curriculum is remote from the situations for which we prepare our students, either in research or in other professions. Constructivism has led to novel approaches in which authentic contexts play a role, such as research-based teaching, problem-based learning, project work, context-based teaching, competency-based teaching, and cognitive apprenticeship (see Section 4.3).

Different learning approaches and their effects are visualized by the cone of learning or the Dale pyramid (see Figure 4.1). Although the pyramid is not based

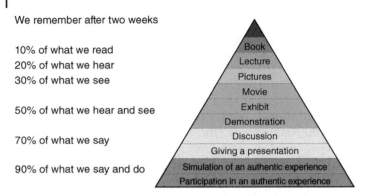

Figure 4.1 The Dale pyramid (Adapted from Biology Forums Gallery).

on research evidence, it is a powerful way to communicate the assumed effects of active learning.

4.2.2
New Technologies

The advent of modern information and communication technology (ICT) and new media has had the most drastic influence on university teaching. The use of computers has become an integral part of chemical research, in chemical measurement, molecular modeling, and communication. Nowadays, smartphones and computers are an inextricable part of students' lives. There can be no debate that computers should be used in chemistry teaching, although opinions on how and when one can use computer technology effectively in the learning process differ. I give a very brief and nonrepresentative overview of the options of computer technology in higher chemistry education. The use of computers in teaching followed the application of computers in chemical research, as in the case of the use of computers in animations, simulations, and modeling of reactions and structures. Some studies have shown that the use of animations stimulates student learning [25, 26]. A second example is the use of computers in accessing information sources: with a few mouse clicks the entire body of scientific literature and instructional resources is available to the students. This makes it easier to use chemical databases and literature for teaching purposes.

From the perspective of student learning, computers offer promising opportunities. The introduction of ICT in teaching at universities occurred more or less simultaneously with the call for active learning. Computers have been applied widely to enable or to support active and student-centered learning strategies [27]. Computers enhance communication between the students and the teachers. Teachers can give online instructions and personal feedback to students and stimulate discussion and collaboration between students. Further, ICT may be used for personalized instruction and formative assessment through quizzes or assignments with individual feedback (either computer-generated or teacher-generated) and adaptive learning [28]. Computers make learning place- and time-independent, offering opportunities for distance learning or virtual laboratories, for instance, by using video technology in online courses (see Section 4.3.1).

4.2.3
The Evolving Nature of Chemistry

After World War II, chemical research developed rapidly. Both in industry and in universities, applied and fundamental research led to new knowledge and insights. As a result of the expansion of chemical research and growth of chemical knowledge, new interdisciplinary research fields emerged, such as organometallic chemistry and bio-inorganic chemistry. Moreover, the disciplinary borders of chemistry with other research disciplines became fuzzy. Chemistry interacts with a large number of disciplines, such as pharmacy, physics, environmental science, earth science, and others. During the course of the twentieth century, the nature of chemical research has changed dramatically, partly due to the development of electronic instruments and ICT. Time-consuming and laborious operations disappeared from the labs.

These developments affected chemistry teaching at the undergraduate level. The enormous variation in knowledge, instruments, and laboratory procedures used by chemists has led to a selection problem: which topics should be introduced to undergraduate classes? Lloyd and Spencer [29] describe how general chemistry courses changed as a result of the changes in chemistry. More and more, the content of introductory courses changed from descriptive chemistry (properties of substances and reactions) to fundamental principles of bonding, structure, and thermodynamics. Because of this change, the nature of general chemistry courses became more physical and mathematical.

The selection problem led to the redefinition of goals of laboratory courses [4]. The emphasis on procedural skill-learning moved to one on cognitive skills, such as problem solving and research skills. This includes formulating hypotheses, designing experiments, interpreting data, and reporting the experimental findings. At many colleges, more open-ended or inquiry-based practical work replaced the standardized lab work. In fact, this redefinition of goals of practical work followed the developments at the secondary level [30].

4.2.4
Developments in Society and Universities

Although I do not want to give a comprehensive description of the way changes in society influenced university education, some developments are relevant for my reflections on the undergraduate curriculum. One factor that has had a huge impact on teaching in universities is the rapid growth of student numbers and student diversity, particularly after 1960s [31, 32]. This has led to an increasing gap between teaching and research. In early days, students were absorbed into the scientific community soon after they entered the university. But with hundreds of freshmen, contacts between students and staff became scarce and teaching more distant. This constrained the way chemistry was taught in colleges and resulted in the continuation of lectures and standardized lab work at many places.

Another relevant development is the increasing influence of employers on university programs. Companies have stressed the importance that students be not only trained as bench chemists but also equipped with other skills – commonly referred to as *academic, generic, transferable,* or *key skills* – such as management, collaboration, and communication skills. Nowadays, these skills are explicitly mentioned in the guidelines for undergraduate programs, for instance, by the American Chemical Society [33] and the European Eurobachelor guidelines [34].

A third factor is the globalization, resulting in increasing mobility of students and graduates. This has led to the harmonization of higher education and the tuning of undergraduate chemistry programs within and between countries by stating national standards, for instance, in the United States [33] or in the European Higher Education Area [34]. Accreditation, in which universities have to comply with agreed standards, stimulates standardization and explicates teaching aims in institutions.

4.3
Implementation of New Teaching Methods

Influenced by the developments mentioned in the last section, the traditional curriculum has been criticized by many authors. Such comments concerned the lack of effectiveness and the emphasis on the transmission of facts in lectures, the training of algorithmic procedures in problem-solving classes and in laboratories, and the use of cookbook-like manuals in the lab. Although the debate on the goals and teaching methods in undergraduate courses already started a long time ago, it is still going on today. Apparently, the traditional curriculum – or parts of it – still exists in many universities. For instance, Talanquer and Pollard [35] characterized the current situation as follows:

> *the first-year chemistry curriculum at most universities is still mostly fact-based and encyclopedic, built upon a collection of isolated topics, oriented too much towards the perceived needs of chemistry majors, focused too much on abstract concepts and algorithmic problem solving, and detached from the practices, ways of thinking, and applications of both chemistry research and chemistry education research in the 21st century.* (p. 74)

Similar comments on the traditional curriculum led to attempts to innovate the curriculum. These innovations will be presented in the sections below.

4.3.1
The Interactive Lecture

As stated in Section 4.2.1, lectures have acquired a bad reputation because of their impersonal nature and disappointing learning gains. Since small-group teaching is an expensive option for large-enrolment courses, attempts have been made to improve the effectiveness of large-group lectures [1, 36] using active learning or formative feedback. Mazur, for instance, has introduced the concept of peer instruction, in which students answer multiple-choice questions and discuss their answers with peers [37]. This method can be applied with large numbers of attendees. Research has shown that this is an effective strategy [38]. Mazur's peer instruction is also the basis for the use of clickers. Different clicker technologies have been developed and operated in classrooms. In a review study, MacArthur and Jones [39] report that the use of clickers contributes to the use of formative feedback by instructors and fosters student collaboration. Refer to Chapter 13 of this book for further information.

Other effective strategies have been reported, such as the use of prelecture activities. Seery [40] presents some options for prelecture activities from a cognitive load perspective. His assumption is that students will learn more effectively during class if they had already studied some of the contents of the lecture. He emphasizes the importance of quizzes to give information on student preparation, both to the students and to the teacher. A variation of prelecture preparation is "flipping the classroom." Students prepare for their classes by viewing teachers' instructions online, and subsequent meetings are used to give additional information and apply concepts. At the moment, some experiences have been published [41, 42], but profound studies on the effect of flip teaching are still lacking.

Video technology and the Internet have brought new opportunities for lecture-based teaching. Video lectures give more opportunities for interactive use by students than live lectures, because students have the option to repeat or skip parts of the lectures. In addition, video lectures may be supported by interactive tools such as quizzes or tutor support. In addition, the Internet enables students to view online video lectures at any place in the world. Many institutions offer MOOCs (Massive Open Online Courses). Because of their open access nature, these MOOCs attract large numbers of students. Leontyev and Baranov [43] give an overview of MOOCs in chemistry. Does the popularity of MOOCs lead to revival and revaluation of lectures? This is hard to say, because experiences with these online courses are diverse. Research has provided some clues that learning effects are similar to face-to-face courses [44], but MOOCs have large withdrawal rates (generally over 90% [43]). This supports the idea that only small groups of well-prepared students benefit from online courses [45]. Further, critics have pointed at the lack of interaction between students and lecturers and the consequences of this for the students' academic development.

4.3.2
Problem- and Inquiry-Based Teaching

Many new instructional methods have been implemented in colleges and universities, aiming at students' development of cognition and process skills (problem-solving, inquiry, communication, and collaboration skills). These methods combine ideas from constructivism, such as the use of authentic situations, collaborative work, and learning cycle approaches. Literature provides numerous examples of these strategies, and I will present only a few of them.

The focus on problem solving has been stimulated by insights from psychological research [21]. Research has provided heuristics that support students in becoming successful problem solvers. Problem-solving strategies have been widely adopted in chemistry teaching at the university level [46–49]. George Bodner [50–52] used the problem-solving perspective in organic and general chemistry courses for freshmen and more advanced students. He emphasizes the importance of representations in problem solving in organic chemistry [52].

Zoller used the concept of higher order cognitive skills (HOCS), namely "the capabilities of question asking, problem (not exercise) solving, decision making, and critical system thinking, the latter comprising rational, logical, reflective, and consequential *evaluative thinking* ... followed by a decision ... followed by accordingly responsible action" [53, p. 584]. He opposes the traditional teaching of lower order cognitive skills, focusing on "knowing, remembering, and/or application of memorized algorithms" [53] (p. 584). The basis of the HOCS-oriented classroom is the solving of complex problems, both in large classes and in small groups. Students socially interact to stimulate their critical, creative, and evaluative thinking. Zoller and Pushkin [54] give several examples – both from the classroom and from the laboratory – to stimulate the development of HOCS.

Since 1994, POGIL (project-oriented guided inquiry learning) has been used by chemistry departments in the United States for discovery-based learning [55]. The focus of POGIL is on concept development, using a learning cycle approach (exploration–concept invention–application). Students work in small groups on problems. POGIL has published student texts, classroom activities, and lab activities for a number of chemical topics. Research has shown that guided inquiry classes, such as POGIL, give lower attrition rates and higher student grades in general and in organic chemistry classes compared to traditional classes [55–58].

4.3.3
Research-Based Teaching

Several teaching strategies have been suggested to strengthen the link between research and teaching. Jenkins and Healey [59] used the term "research-based teaching" for strategies in which students are involved in research processes as participants. Research-based strategies aim at developing the students' research skills and enculturation into the research community. In addition, research-based teaching is expected to contribute positively to the students' motivation. Research-based teaching uses similar methods as inquiry-based teaching, but is more focused on research in the laboratory.

Goedhart *et al.* [60] state that many different terms are used to denote examples of research-based teaching, such as undergraduate research, (research) projects, inquiry-based (or inquiry-oriented) teaching, and problem-based learning. Borders between these are not sharp, and different terms are used to indicate the same kind of teaching. Goedhart *et al.* [60] give an overview of the examples of research-based chemistry education. A first group of examples refers to undergraduate research. This is particularly popular in the United States [61]. Many examples have been described in the literature [62] for chemistry, especially in the *Journal of Chemical Education* [63]. A second group is called inquiry-based laboratory work [64]. A third group consists of research-based projects [65–68]. Recently, the teaching of argumentation skills has drawn attention in science education practices and research. Walker and colleagues [69] developed an argument-driven instructional (ADI) model for a chemistry laboratory course.

The similarity between these kinds of research-based teaching is that students are involved in some way in research activities, but the authenticity of the research projects differs. In some cases, students fully participate in a research group. In other cases, typical instruction experiments are modified in such a way that students have to argue methodological decisions. In the latter situation, experiments are usually done in the student laboratory, which is a less authentic environment.

The implementation of research-based teaching raises many questions, for instance, on students' guidance and the assessment of students' work. The literature gives many practical examples. A good organization is essential to get positive appreciation by students. The ownership of the task and the challenges students experience may be motivating to students, but students may become frustrated when they feel having insufficient knowledge and skills to complete a research task.

4.3.4
Competency-Based Teaching

Traditionally, curricula are based on objectives in terms of student knowledge, skills, and attitudes. In the traditional undergraduate curriculum, the teaching of knowledge and skills tends to be separated. Students are expected to acquire knowledge from lectures and textbooks, and acquire their skills in the lab, in problem-solving classes, in (research) projects, and sometimes in specific skill classes. In traditional curricula, the teaching of attitudes and academic skills (e.g., interpersonal, management, and communication skills) is underrepresented. The separation of knowledge and skill acquisition leads to a situation in which students are burdened with the problem how to integrate these when they perform a complex task. Research has provided evidence that students have difficulty in transferring their knowledge and skills to new situations, for instance, when performing complex tasks at the graduate level or in professional environments [21].

Considerations on students' preparation to professional careers have drawn attention to the concept of *competence*. A competent person is able to perform a task in a successful way adhering to professional standards. Closely related to the term *competence* is the term *competency*. Eraut [70] articulates the distinction between competence and competency as follows:

> ... the term 'competence', which is given a generic or holistic meaning and refers to a person's overall capacity, and the term 'competency', which refers to specific capabilities. (p.179)

In Eraut's definition, competence is an overarching term. A competent person has certain competencies, such as a competent synthetic chemist has the competency to design a synthesis route for a given substance or perform a synthesis in the lab. Van Merriënboer and Kirschner [71] define competency as follows:

> A combination of complex cognitive and higher order skills, highly integrated knowledge structures, interpersonal and social skills, and attitudes and values. Acquired competencies can be applied in a variety of situations (transfer) and over an unlimited time span (lifelong learning). (p. 280)

In this definition, Van Merriënboer and Kirschner point at the integrated nature of competencies as well as their transferability and sustainability. It is important to note that competence differs from expertise, and that a competent professional differs from an expert [72]. An expert has specialized in a certain field, and is being recognized by the professional community because of his specific competence. So, a competent professional is "a position on a graduated scale from novice to expert" [70, p. 167].

Competencies and competence standards are derived from professional practice and professional standards. Since competencies refer to professional activities, they pertain to authentic settings. From the 1980s, vocational schools and institutions for higher education have adopted competency-based approaches [73]. Especially, in training programs for those professions with a well-defined profile, such as physicians, architects, or lawyers, competency-based education is popular.

Cultural-historical activity theory, which originates from Vygotsky [12] and Leont'ev [74], has brought interesting view points to competency-based learning, especially through the contributions by Engeström [72]. His theoretical framework describes communities of professionals with "a model of the activity system" consisting of six different components (Figure 4.2).

Transposing this model to the example of a chemical research group, we can describe the components as follows: A researcher (*subject*) investigates the mechanism of a reaction (*object*). He uses research instrumentation and knowledge (*instruments*). He discusses his results with his colleagues (*community*). In the

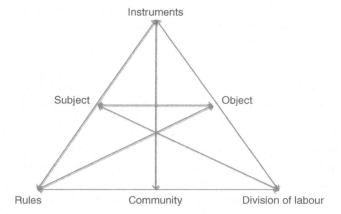

Figure 4.2 Schematic representation of an activity system (Based on Ref. [75]).

community, the persons (researchers, graduate students, professors, technicians) have different roles (*division of labor*). The work within the discipline and the work by the group follow certain explicit and implicit *rules*, for instance, to publish the findings in journals. With this model, Engeström analyzed the activities of professionals and showed how the elements from the scheme interact in professional practices [75, 76].

An interesting notion of activity theory is that professional practices are part of a collective activity system. For university teaching, this shifts the focus of teaching from the abilities of individual students to the enculturation of students as becoming a member of a professional community.

Summarizing, competency-based teaching shows following characteristics:

- achievement goals and assessment based on professional standards,
- integrated development of knowledge, skills, and student attributes,
- learning in authentic and professional contexts, and
- learning as a process of enculturation in a community.

Not many examples of competency-based education in chemistry higher education can be found in the literature. Witt and colleagues [77] describe how in an engineering school at a Spanish university ten key competencies for chemical engineers were selected and how these were implemented into a curriculum.

4.4
A Competency-Based Undergraduate Curriculum

In Section 4.3, I listed a number of teaching approaches that have become popular during the last decades. To some extent, these approaches have found their way to colleges and universities. However, we are still left with an important question, which is about the contents of the program. Which competences – or knowledge, skills, and attitudes – should our students acquire? This question concerns the goals and the structure of the curriculum.

Several organizing principles may be used to device a curriculum for undergraduate chemistry. A *discipline-based* (or *disciplinary*) *curriculum* is based on the structure of the discipline – in our case chemistry. A *competency-based curriculum* is based on the competencies chemists need in their professional environments, in industry, in governmental institutions, in research, or in other sectors. Some curriculum designers have proposed a *theme-based* (or *thematic*) *curriculum*. In this case, themes are generally chosen to integrate chemistry learning with societal issues, such as solar energy, polymers, or climate change, offering opportunities for interdisciplinary learning [35, 78]. Context-based curricula are also arranged according to themes (for a more elaborate treatise, see Chapter 2 of this book).

4.4.1
The Structure of the Curriculum

A discipline-based curriculum is based on our views as chemistry as a discipline. These views have changed over time. In Section 4.2.3, we saw how introductory chemistry courses developed from descriptive to more theoretical. More advanced courses were grouped according to the five classical disciplines (analytical, organic, inorganic, physical, and biochemistry) [3, 79, 80]. At that time, this division in subdisciplines reflected what chemists were doing and how research institutions were organized in research groups. The emergence of new interdisciplines and the increasing multidisciplinary character of chemical research have questioned this classical division of chemistry. Several chemists made proposals for other divisions of chemistry, reflecting new developments. In 1970, a committee, chaired by Hammond and Nyholm [81], published a report on the structure of chemistry as a science and its consequences for chemistry teaching at universities. The committee made several proposals for a new division of chemistry. One of their proposals was to make a distinction between *structure, dynamics,* and *synthesis*. Although the report is written from a research perspective, the authors highlight the significance of this division for education. Another proposal was put forward by the Nobel Laureate Roald Hoffmann [82], advocating a division into *analysis, synthesis,* and *mechanisms*. More recently, a proposal for a new division of chemistry was made in the report *Beyond the Molecular Frontier* [83] by the U.S. National Science Foundation (NSF). The report identifies future challenges for chemistry and chemical engineering in research and industry and mentions four main areas in which modern chemical scientists work: *analysis, synthesis, transformation, and modeling.*

Most of the proposals above take a research perspective: they follow from analyses of current chemical research and identify research challenges for the future, both in industry and in universities.

When designing a competency-based curriculum, we have to take a more educational perspective. As indicated, we start with analyzing the activities chemists perform. In part, this analysis concurs with the proposals mentioned above. All three proposals mention *synthesis* as a core activity by chemists. There is no doubt that the creation of new substances and materials is an important activity of research chemists and chemical engineers. Also, chemists seem to agree upon the role of *analysis* in research. It is explicitly mentioned by Hoffmann and by the NRC report, and it is included as structure in the Hammond and Nyholm report. More differences – at least in terminology – can be found in the third category, which is indicated as *dynamics* (Hammond and Nyholm), *mechanisms* (Hoffmann), and *transformation and modeling* (NSF). Both dynamics and mechanisms and transformation refer to the study of processes of change in which substances and materials are involved, such as chemical reactions, phase transitions, and industrial manufacturing.

Some time ago, I proposed a division of chemistry from the viewpoint of teaching research competencies to students at the undergraduate level [60, 80]. This proposal was based on observations when students designed experiments in the laboratory. In an inquiry-based laboratory course, we assigned students to use their theoretical knowledge in the design of experiments in analytical and physical chemistry [84]. One of our observations concerned the way students designed spectrometric measurements. They experienced problems in taking and underpinning decisions about the use of control measurements, the preparation of standard solutions, the need for calibration, and so on. In an analytical measurement, the concentration of a substance in a sample is determined by calibration of the absorption scale with a number of standard solutions. Students were wondering why one or two standard solutions were not sufficient. In a physical chemistry measurement, students determined the UV–vis spectrum of a compound to deduce its electronic structure, and they were expected to calibrate the wavelength scale. In an organic synthesis, the UV–vis spectrum of a synthesized compound was used for identification, and accurate calibration was not necessary. These three cases illustrate different methods of calibration, depending on the purpose of the measurement. This means that a learning task, such as the performance of a spectrometric measurement, can be learned only if the context of the measurement is explicit to the students. The context defines the purpose of the measurement, its desired accuracy, the way the instrument is handled, and the way data is analyzed. Here, I just gave one example, but it is not difficult to find other examples.

What is the consequence of this analysis for the structure of chemistry in a competency-based curriculum? I distinguished between three so-called context areas in chemistry: synthesis, analysis, theory development [80]. In this chapter, taking a competency-based curriculum perspective, I adopt the name *competency areas*, and distinguish between *synthesis, analysis,* and *modeling*. Modeling is a more concise and familiar term than theory development and is more closely connected to the divisions above. Analysis and synthesis are familiar terms to chemists, although I give these concepts a somewhat deviating meaning. In the next sections, I will define and elaborate these areas.

4.4.2
Competency Area of Analysis

Analysis is the acquisition of knowledge on the qualitative and quantitative composition of substances (in elements) and mixtures (in substances) and on the surface and inner structure of substances and mixtures, also at the molecular and atomic levels. Different from the familiar definitions of analysis, here structural analysis is included. In the division followed in the NSF report [80], structural analysis is also combined with analytical methods.

Analytical chemistry is the oldest activity in experimental chemistry. Starting from Lavoisier's theory of elements, this branch of chemistry was important in the search for elements and the determination of the chemical composition of all kinds of substances. Because these substances had to be isolated from the mixtures in which they were present, improvement in separation methods strongly influenced the effectiveness of analytical methods. During the nineteenth century, better instruments and methods for analytical work became available, particularly through the work by Justus Liebig (1803–1873) in Giessen, Germany [79]. His methods were based on chemical reactions of the substance to be analyzed and subsequent weighing of the reaction products. During the first half of the twentieth century, more chemical methods (such as gravimetric and volumetric analysis) were developed. The disadvantage of many chemical methods is that they are laborious, time-consuming, and expensive. In the second half of the twentieth century, two revolutions in analytical chemistry took place. The first one was the introduction of physical – especially, spectroscopic – methods. The second revolution was the introduction of the computer, making it possible to analyze large datasets and to automate analytical measurements. These revolutions enabled analytical chemists to determine the chemical composition of mixtures and substances more rapidly and accurately. Further, continuous monitoring of industrial processes and environmental parameters became possible. The specificity of spectroscopic methods avoided the necessity of isolating compounds. Spectroscopy enables the analysis of complex mixtures in industrial production, in biological cells, and in other complex systems. Many analytical procedures have become routine in quality control labs, clinical labs, research labs, and industrial labs. The design of complex analytical instruments is mainly done by specialized commercial companies. Many chemists and lab technicians only know the main features of their instruments and do not have detailed knowledge about its functioning. The routine nature of analytical measurement is one of the reasons why analytical chemistry research has lost its prominent position in research universities. On the other hand, analysis is one of the fundamental activities in many disciplines, and it has been absorbed by many areas such as environmental science and biochemistry. So, the significance of analysis in chemical work is still undisputed, and for this reason the teaching of analysis to university students is of

Table 4.1 Key competencies and knowledge areas in the competency area analysis.

Aim	Acquiring information about the composition and structure of substances and mixtures
Key competencies	Sampling
	Using instruments
	Data interpretation (accuracy, precision, reliability, validity, etc.)
Knowledge areas	Sampling and sample preparation procedures
	Instruments, methods, techniques (chemical methods, spectrometric methods, surface analysis methods, structural analysis methods)
	Instrument properties (sensitivity, detection limit, calibration, instrument errors)
	Knowing physical measurement principles (signal transformation, etc.)
	Methods of data analysis (statistics, computer software)

Adapted from Ref. [80].

great significance. But which competencies do we wish our students to develop, which knowledge and which lab and cognitive skills do they need?

When choosing competencies for a curriculum, we have to realize that the field of analysis has expanded enormously. The large variety of methods and instruments makes it impossible to introduce all of these to the students. So, we have to limit our attention to the key features of the field. Many students will pursue a career in this field, and an urgent question is which competencies they need. In Table 4.1, I present an overview of the competency area, although I immediately admit that it is not complete.

One of the most important issues is the choice of appropriate methods to solve analytical problems. Students should have some idea of the properties of the methods and the instruments available. When designing and conducting an analytical method, questions should be answered about the sampling procedures, the properties (physical, chemical, biological) of substances involved, the specificity and sensitivity of the measurements, the handling of instruments, the reliability and validity of the data, and the ways to process the data (automation, statistics, software tools).

To become competent in this area, students should learn to ask and answer these questions. Much of this information can be found in textbooks on analytical chemistry, but most textbooks describe the most frequently used methods and instruments and do not always discuss the selection of instruments for a specific analytical problem. Many students will consider analytical chemistry as an endless catalog of facts, without seeing the challenges of the area. Therefore, I advocate that textbooks adopt a more problem-based or competency-based perspective.

4.4.3
Competency Area of Synthesis

Synthesis is defined as the making of substances and mixtures with specific properties. This encompasses traditional organic and inorganic synthesis, and also the preparation of mixtures and the isolation and purification of substances from mixtures. Synthesis is done at different scales, ranging from nano- and micro-scale laboratory synthesis to large-scale production in industries. A major challenge of this area is to synthesize substances and materials with the desired properties. These may be mechanical, electrical, optical, chemical, or metabolic properties. Synthesis borders on materials science, and it does not make sense to draw strict borderlines between chemistry and materials science because in practice there is a strong overlap, necessitating professionals with both backgrounds to cooperate. Traditionally, the isolation and purification of substances are not denoted as synthesis. But, considering the aim of isolation and purification – namely the production of a substance with certain properties – I classify these as synthesis.

Synthesis in industries is different from synthesis in research environments. In industries, chemists feel a pressing need to synthesize substances in such a way that minimal energy is consumed and minimal waste products are generated. This perspective has led to green synthesis methods that are cost effective and ecologically sustainable. In research universities, costs of chemicals and production of waste are generally less important issues.

When analyzing what a competent synthetic chemist should know and be able to do, we come to the overview presented in Table 4.2. First, synthesis requires a

Table 4.2 Key competencies and knowledge areas in the competency area synthesis.

Aim	Making products with specified properties
Key competencies	Designing synthesis routes or industrial processes
	Using synthetic procedures, including separation and isolation
	Controlling reaction processes (temperature, pressure, solvent, catalysts)
	Determining the quality of the process (purity, yield)
	Determining the identity and quality of the product (characterization, identification)
Knowledge areas	Synthesis principles (retrosynthesis, group protection, multistep synthesis, combinatorial chemistry, etc.)
	Separation and isolation procedures
	Process engineering
	Substance properties and reactions (descriptive (in)organic chemistry, biochemistry)
	Processes and influence of variables on chemical processes (thermodynamics, kinetics)
	Characterization and identification methods

Adapted from Ref. [80].

broad knowledge of substances and reactions in inorganic, organic, or biochemistry, depending on the area in which the chemist works. Second, competent synthetic chemists know methods of synthesis, purification, and identification and are able to use them. Third, synthetic chemists know how to design synthetic routes for substances with certain properties. This requires that they know how the structure and properties are related. For chemical engineers, knowledge of industrial processes and the ability to design these processes are additional fields of competence. Fourth, synthetic chemists know how to influence the way a reaction proceeds by choosing proper conditions (temperature, solvent, catalysts). Fifth, chemists know how to identify compounds with specific – presently, mainly spectroscopic – methods.

Synthesis is certainly one of the most appealing parts of chemistry. The ability to make complex biomolecules *in vitro* and to make substances not originally present in nature is one of the most fascinating achievements of chemistry. The synthesis of drugs and artificial materials has had a big impact on our lives, and synthesis of materials has made our lives at present far more healthy and comfortable than a century ago. The teaching of descriptive chemistry (properties of substances and reactions) has almost disappeared from colleges and universities. However, this knowledge is extremely important when performing syntheses in the lab or in industries. The use of computers has made information sources more easily accessible, and finding relevant information using these sources is an important competency.

4.4.4
Competency Area of Modeling

Modeling in chemistry aims at understanding and explaining structures and transformations of substances. Modeling is about "why" questions. Modeling involves developing and validating qualitative and quantitative models and theories about chemical substances and their transformations. Traditionally, this has been the field of physical chemistry and, partly, biochemistry. Physical chemistry investigates properties of chemical substances (surface and colloid chemistry, electrochemistry), their structures (quantum chemistry), and chemical reactions (kinetics, thermodynamics, computational chemistry). Biochemistry is dedicated to the study of chemical processes in living organisms, often studied *in vitro* [85].

Modeling applies measurements to find the relations between quantities (mathematical modeling) as well as qualitative modeling as in the development of structural models (see Table 4.3). Chemical modeling is related to physical modeling but differs in the objects studied. Measurement and physical methods take a central position in this area, and this means that students have to be knowledgeable about instrumentation (not necessarily the same as in analysis) and data analysis methods (particularly statistics). Further, students have to be aware of the epistemology of science: the role of models and theories in scientific explanations and the nature of causality.

Table 4.3 Key competencies and knowledge areas in the competency area modeling.

Aim	Development and validation of scientific models and explanations on properties of substances and reactions
Key competencies	Using experimental methods in testing hypotheses and models
	Knowing the nature and structure of scientific models and explanations
	Using instruments to measure physical quantities
	Using methods of data analysis
Knowledge areas	Models and theories at the microscopic and macroscopic level (bonding theory, thermodynamics, etc.)
	Instruments, methods, and techniques
	Methods of data analysis

Adapted from Ref. [80].

The area of modeling interacts in various ways with the other two areas. First, the areas of analysis and synthesis are informed by models and theories. For instance, synthetic chemists use theories about bonding and structure, reaction mechanisms, thermodynamics, and kinetics when they synthesize compounds. Analytical chemists use theories on the structure, for instance, when using spectroscopic methods. But the scope in synthesis and analysis differs from modeling, because these areas *use* theory and models and their first concern is not to *develop* models, although the work in synthesis and analysis may contribute to modeling.

4.4.5
The Road to a Competency-Based Curriculum

The importance of this distinction into three competency areas is that, when we consider the structure of the chemistry undergraduate curriculum, it offers three clear lines for the development of competencies. The use of professional contexts gives students an orientation on the role of chemistry in research, industry, and society and on their future profession, and this might enhance students' motivation. The professional contexts also give coherence to the knowledge and skills the students learn. Instead of incoherent and noncontextual knowledge, students learn how knowledge and skills are integrated in professional activities. These knowledge and skills go beyond the chemical knowledge and skills that I presented in Tables 4.1–4.3. The learning of generic skills, such as communication, collaboration, and management skills, should be an integral part of the activities, enabling students to develop these skills in the direction of professional level. Another benefit of competency-based education is that, through the use of professional contexts, students get acquainted with the role of chemistry in multidisciplinary issues, in which multiple perspectives contribute to the solution of problems in society, research, or industry.

I presented only a blueprint for a curriculum for chemistry undergraduates. Curriculum designers can use this as a start, but there is still a lot of work to do. For a more detailed elaboration, we can follow the approach published by Van Merriënboer and Kirschner [71]. Their 4C/ID (four-component instructional design) approach consists of 10 steps. We will illustrate these steps with an example from chemistry: the performance of a synthesis. The first three steps of 4C/ID involve the selection of tasks and setting standards. For step 1, we select authentic synthesis tasks from research or industrial practice. In step 2 tasks are ordered according to their complexity. For instance, single-step synthesis precedes multistep synthesis. Step 3 includes setting standards for each task. This may refer to the quality of the synthetic products (yield, purity), the processes (organization, efficiency), and communication (report, oral presentation). Different from regular curriculum design practice, we do not start with objectives but with activities chemists perform in their profession. When we design the next steps 4–7, we reflect on the selection of supportive information such as heuristics for problem solving (for designing a synthetic route, for choosing reaction conditions), organization of relevant knowledge (e.g., reaction mechanisms, the use of catalysts), and procedures by providing manuals. Synthetic chemists use recurrent procedures, such as purification and identification procedures. Students have to gain experience in using these procedures by training in steps 8–10. These steps involve identification and analysis of these procedures and the knowledge needed to perform them. Specific tasks are used to offer students additional training on these routine procedures, such as distillation. Students work on these professional tasks, from simple to complex, and with increasing autonomy. Initially, teachers give a good deal of support (scaffolding), but eventually their support decreases (fading). Although the 4C/ID method gives some direction to the structure of a curriculum, tasks have to be selected and the way students are coached have to be filled in yet. This requires much effort from teachers and educators, but the perspective is interesting.

More and more, entrepreneurship is considered as an important attribute students should acquire during their university training. Entrepreneurship is defined as "a positive frame of mind that is aimed at the identification and realization of value" [86, p. 261]. Especially, governments and companies have stressed the importance of Entrepreneurship Education and Training (EET). According to Al-Atabi and DeBoer [86], EET should include the development of collaborative, management, and financial skills, and of positive thinking. Martin and colleagues [87] reviewed the literature on the effects of EET and presented evidence of a correlation between EET and entrepreneurship. Some examples of the teaching of entrepreneurship in chemistry and engineering courses can be found in the literature [88, 89].

In Figure 4.3, the way chemical competencies and professional competencies interact in the curriculum is visualized.

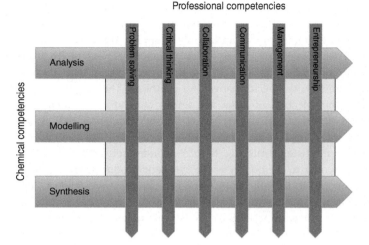

Figure 4.3 Schematic representation of a competency-based chemistry curriculum.

4.5
Conclusions and Outlook

In this chapter, I sketched how curricula in universities changed as a result of a changing environment. In many universities, new teaching methods have penetrated the curriculum. These methods combine pedagogical ideas, partly derived from constructivism, such as collaborative learning, emphasis on process skills, and the use of authentic learning environments. A variety of new methods for concept and skill learning have been developed and implemented in classrooms. However, our knowledge of effects of these pedagogical approaches is still limited. Effect studies, especially in specific domains such as chemistry, are relatively rare [90].

It should be noted that the traditional curriculum – or its remnants – is still present in many institutions. How is it possible that lecturing is still the most dominant teaching method, although we know that it is inefficient and can put off students? Practical constraints, such as the student numbers, may be the reason for the eternal popularity of lectures, but I think that the narrow pedagogical repertoire of lecturers also plays a significant role. For most lecturers in colleges and universities, the most obvious thing is to lecture, because they have never experienced other ways of teaching. Professional development of faculty staff to broaden their pedagogical repertoire has yet to begin in many institutions. Sunal and colleagues [91] have given the necessary conditions for successful faculty development. They mention that teachers have to be dissatisfied with their way of teaching. Dissatisfaction seems to be the initiator for the implementation of new teaching methods. Teachers can be dissatisfied by the lack of interaction during their classes or by disappointing learning results. Educators have a role in making lecturers aware of the limited yields of their teaching. Dissatisfied teachers are more open to use new pedagogies. Experiments with new pedagogies and reflecting on their outcomes will lead to an increase in personal efficacy. Support by management and administration is essential to initiate these processes and make innovations sustainable.

In this chapter, I emphasized the importance of designing a new structure of the undergraduate program. The structure of the program is the way the program is divided into learning activities for students. A competency-based perspective, based on the professional activities of research chemists, offers clear guidance to develop such a structure. This perspective leads to new views on the division of chemistry as a science.

I proposed a division based on analysis, synthesis, and modeling. In my opinion, this division offers well-defined tracks in a chemistry undergraduate program. Different approaches are possible in this program, depending on the emphases institutions would like to put. Studies that systematically design and evaluate teaching and learning activities are necessary to develop, ultimately, a research-informed curriculum.

References

1. Hake, R.H. (1998) Interactive-engagement versus traditional methods: a six-thousand-student survey of mechanics test data for introductory physics courses. *Am. J. Phys*, **66**, 64–74.
2. Hass, G. (1987) *Curriculum Planning: A New Approach*, Allyn and Bacon, Boston, MA.
3. Lloyd, B.W. (1992) The 20th century general chemistry laboratory. *J. Chem. Educ.*, **69**, 866–869.
4. Reid, N. and Shah, I. (2007) The role of laboratory work in university chemistry. *Chem. Educ. Res. Pract.*, **8**, 172–185.
5. Bodner, G.M. (1986) Constructivism: a theory of knowledge. *J. Chem. Educ.*, **63**, 873–878.
6. Nakhleh, M.B. (1992) Why some students don't learn chemistry: chemical misconceptions. *J. Chem. Educ.*, **69**, 191–196.
7. Taber, K. (2002) *Chemical Misconceptions: Prevention, Diagnosis, and Cure: Theoretical Background*, vol. 1, Royal Society of Chemistry, London.
8. Kind, V. (2004) *Beyond Appearances: Students' Misconceptions about Basic Chemical Ideas*, School of Education, Durham University, Durham.
9. Taber, K. (2000) Chemistry lessons for universities? A review of constructivist ideas. *Univ. Chem. Educ.*, **4**, 63–72.
10. Karplus, R. (1977) Science teaching and the development of reasoning. *J. Res. Sci. Teach.*, **14**, 33–46.
11. Posner, G.J., Strike, K.A., Hewson, P.W., and Gertzog, W.A. (1982) Accommodation of a scientific conception: toward a theory of conceptual change. *Sci. Educ.*, **66**, 211–227.
12. Vygotsky, L.S. (1978) *Mind in Society: The Development of Higher Psychological Processes*, Harvard University Press, Cambridge, MA.
13. Lave, J. and Wenger, E. (1991) *Situated Learning: Legitimate Peripheral Participation*, Cambridge University Press, Cambridge.
14. Brown, J.S., Collins, A., and Duguid, P. (1989) Situated cognition and the culture of learning. *Educ. Res.*, **18**, 32–42.
15. Halloun, I.A. and Hestenes, D. (1985) The initial knowledge state of college physics students. *Am. J. Phys*, **53**, 1043–1055.
16. Redish, E.F. and Steinberg, R.N. (1999) Teaching physics: figuring out what works. *Phys. Today*, **52**, 24–31.
17. Wieman, C. and Perkins, K. (2005) Transforming physics education. *Phys. Today*, **58**, 36–41.
18. Johnson, D.W., Johnson, R.T., and Smith, K.A. (1991) *Active Learning: Cooperation*

in the College Classroom, Interaction Book Company, Edina, MN.

19. Cooper, M.M. (1995) Cooperative learning. An approach for large enrollment courses. *J. Chem. Educ.*, **72**, 162–164.

20. Eilks, I., Markic, S., Bäumer, M., and Schanze, S. (2009) in *Innovative Methods of Teaching and Learning Chemistry in Higher Education* (eds I. Eilks and B. Byers), RSC Publishing, Cambridge, pp. 103–122.

21. Bransford, J.D., Brown, A.L., and Cocking, R.R. (eds) (2000) *How People Learn. Brain, Mind, Experience, and School*, National Academy Press, Washington, DC.

22. Cooper, M.M. and Sandi-Urena, S. (2009) Design and validation of an instrument to assess metacognitive skillfulness in chemistry problem solving. *J. Chem. Educ.*, **86**, 240–245.

23. Cooper, M.M., Sandi-Urena, S., and Stevens, R. (2008) Reliable multimethod assessment of metacognition use in chemistry problem solving. *Chem. Educ. Res. Pract.*, **9**, 18–24.

24. Tien, L.T., Rickey, D., and Stacy, A.M. (1999) The M.O.R.E. thinking frame: guiding students' thinking in the laboratory. *J. Coll. Sci. Teach.*, **28**, 318–324.

25. Williamson, V.M. and Abraham, M.R. (1995) The effects of computer animation on the particulate mental models of college chemistry students. *J. Res. Sci. Teach.*, **32**, 521–534.

26. Abraham, M., Varghese, V., and Tang, H. (2010) Using molecular representations to aid student understanding of stereochemical concepts. *J. Chem. Educ.*, **87**, 1425–1429.

27. Brouwer, N. and McDonnell, C. (2009) in *Innovative Methods of Teaching and Learning Chemistry in Higher Education* (eds I. Eilks and B. Byers), RSC Publishing, Cambridge, pp. 123–152.

28. Van Seters, J.R., Wellink, J., Tramper, J., Goedhart, M.J., and Ossevoort, M.A. (2012) A web-based adaptive tutor to teach PCR primer design. *Biochem. Mol. Biol. Educ.*, **40**, 8–13.

29. Lloyd, B.W. and Spencer, J.N. (1994) New directions for general chemistry. *J. Chem. Educ.*, **71**, 206–209.

30. Hofstein, A. (2004) The laboratory in chemistry education: thirty years of experience with developments, implementation, and research. *Chem. Educ. Res. Pract.*, **5**, 247–264.

31. Scott, P. (1995) *The Meanings of Mass Higher Education*, Open University Press, Bristol.

32. Teichler, U. (1998) Massification: a challenge for institutions of higher education. *Tertiary Educ. Manag.*, **4**, 17–27.

33. American Chemical Society (2003) *Undergraduate Professional Education in Chemistry. Guidelines and Evaluation Procedures*, American Chemical Society, Washington, DC.

34. ECTN (n.d.) The Chemistry "Eurobachelor", http://ectn-assoc.cpe.fr/chemistry-eurolabels/doc/officials/Off_EBL070131_Eurobachelor_Framework_2007V1.pdf (accessed 19 November 2013).

35. Talanquer, V. and Pollard, J. (2010) Let's teach how we think instead of what we know. *Chem. Educ. Res. Pract.*, **11**, 74–83.

36. DeHaan, R.L. (2005) The impending revolution in undergraduate science education. *J. Sci. Educ. Technol.*, **14**, 253–269.

37. Mazur, E. (1997) *Peer Instruction: A User's Manual*, Prentice-Hall, San Francisco, CA.

38. Crouch, C.H. and Mazur, E. (2001) Peer instruction: ten years of experience and results. *Am. J. Phys*, **69**, 970–977.

39. MacArthur, J.R. and Jones, L.L. (2008) A review of literature reports of clickers applicable to college chemistry classrooms. *Chem. Educ. Res. Pract.*, **9**, 187–195.

40. Seery, M. (2012) Jump-starting lectures. *Educ. Chem.*, **49**, 22–25.

41. Smith, J.D. (2013) Student attitudes towards flipping the general chemistry classroom. *Chem. Educ. Res. Pract.*, **14**, 607–614.

42. Teo, T.W., Tan, K.C.D., Yan, Y.K., Teo, Y.C., and Yeo, L.W. (2014) How flip teaching supports undergraduate chemistry laboratory learning. *Chem. Educ. Res. Pract.*, **15**, 550–567.

43. Leontyev, A. and Baranov, D. (2013) Massive open online courses in chemistry: a comparative overview of platforms and features. *J. Chem. Educ.*, **90**, 1533–1539.
44. Means, B., Toyama, Y., Murphy, R., Bakia, M., and Jones, K. (2009) *Evaluation of Evidence-Based Practices in Online Learning: A Meta-Analysis and Review of Online Learning Studies*, US Department of Education, Washington, DC.
45. Smith Jaggars, S. and Bailey, T. (2010) *Effectiveness of Fully Online Courses for College Students: Response to a Department of Education Meta-Analysis*, CCRC Columbia University, New York.
46. Mettes, C., Pilot, A., Roossink, H., and Kramers-Pals, H. (1980) Teaching and learning problem solving in science. Part I: a general strategy. *J. Chem. Educ.*, **57**, 882–885.
47. Mettes, C., Pilot, A., Roossink, H., and Kramers-Pals, H. (1981) Teaching and learning problem solving in science: part II: learning problem solving in a thermodynamics course. *J. Chem. Educ.*, **58**, 51–55.
48. Wilson, H. (1987) Problem-solving laboratory exercises. *J. Chem. Educ.*, **64**, 895–896.
49. Browne, L.M. and Blackburn, E.V. (1999) Teaching introductory organic chemistry: a problem-solving and collaborative-learning approach. *J. Chem. Educ.*, **76**, 1104–1107.
50. Bowen, C.W. and Bodner, G.M. (1991) Problem-solving processes used by students in organic synthesis. *Int. J. Sci. Educ.*, **13**, 143–158.
51. Bodner, G.M. (1991) in *Toward a Unified Theory of Problem Solving: Views From the Content Domain* (ed. M.U. Smith), Lawrence Erlbaum, Hillesdale, NJ, pp. 21–34.
52. Bodner, G.M. and Domin, D.S. (2000) Mental models: the role of representations in problem solving in chemistry. *Univ. Chem. Educ.*, **4**, 24–30.
53. Zoller, U. (1999) Scaling-up of higher-order cognitive skills-oriented college chemistry teaching: an action-oriented research. *J. Res. Sci. Teach.*, **36**, 583–596.
54. Zoller, U. and Pushkin, D. (2007) Matching higher-order cognitive skills (HOCS) promotion goals with problem-based laboratory practice in a freshman organic chemistry course. *Chem. Educ. Res. Pract.*, **8**, 153–171.
55. POGIL (n.d.) *www.pogil.org* (accessed 19 November 2013).
56. Farrell, J.J., Moog, R.S., and Spencer, J.N. (1999) A guided inquiry general chemistry course. *J. Chem. Educ.*, **76**, 570–574.
57. Lewis, J.E. and Lewis, S.E. (2005) Departing from lectures: an evaluation of a peer-led guided inquiry alternative. *J. Chem. Educ.*, **82**, 135–139.
58. Hanson, D. and Wolfskill, T. (2000) Process workshops – A new model for instruction. *J. Chem. Educ.*, **77**, 120–130.
59. Jenkins, A. and Healey, A. (2010) Undergraduate research and international initiatives to link teaching and research. *CUR Q.*, **30**, 36–42.
60. Goedhart, M.J., Finlayson, O.E., and Lindblom-Ylänne, S. (2009) in *Innovative Methods of Teaching and Learning Chemistry in Higher Education* (eds I. Eilks and B. Byers), RSC Publishing, Cambridge, pp. 61–84.
61. The Council on Undergraduate Research – Learning through Research (n.d.) *www.cur.org* (accessed 19 November 2013).
62. Doyle, M.P. (ed.) (2000) *Academic Excellence: The Sourcebook. A Study on the Role of Research in the Physical Sciences at Undergraduate Institutions*, Research Corporation, Tucson, AZ.
63. Special issue on undergraduate research (1984) *J. Chem. Educ.*, **61** (6), 477–566.
64. Pavelich, M.J. and Abraham, M.R. (1977) Guided inquiry laboratories for general chemistry students. *J. Coll. Sci. Teach.*, **7**, 23–26.
65. Read, J.R. and Kable, S.H. (2007) Educational analysis of the first year chemistry experiment 'thermodynamics think-in': an ACELL experiment. *Chem. Educ. Res. Pract.*, **8**, 255–273.
66. Buntine, M.A., Read, J.R., Barrie, S.C., Bucat, R.B., Crips, G.T., George, A.V. *et al.* (2007) Advancing chemistry by enhancing learning in the laboratory

(ACELL): a model for providing professional and personal development and facilitating improved student laboratory learning outcomes. *Chem. Educ. Res. Pract.*, **8**, 232–254.
67. Kelly, O.C. and Finlayson, O.E. (2007) Providing solutions through problem-based learning for the undergraduate 1st year chemistry laboratory. *Chem. Educ. Res. Pract.*, **8**, 347–361.
68. McDonnell, C., O'Connor, C., and Seery, M.K. (2007) Developing practical chemistry skills by means of student-driven problem based learning mini-projects. *Chem. Educ. Res. Pract.*, **8**, 130–139.
69. Walker, J.P., Sampson, V., and Zimmerman, C.O. (2011) Argument-driven inquiry: an introduction to a new instructional model for use in undergraduate chemistry labs. *J. Chem. Educ.*, **88**, 1048–1056.
70. Eraut, M. (1994) *Developing Professional Knowledge and Competence*, Routledge, London.
71. Van Merriënboer, J.J.G. and Kirschner, P.A. (2007) *Ten Steps to Complex Learning. A Systematic Approach to Four-Component Instructional Design*, Lawrence Erlbaum, Mahwah, NJ.
72. Engeström, Y. (1987) *Learning by Expanding: An Activity-Theoretical Approach to Developmental Research*, Orienta-Konsultit, Helsinki.
73. Bowden, J.A. and Masters, G.N. (1993) *Implications for Higher Education of a Competency-Based Approach to Education and Training*, Australian Government Publishing Service, Canberra.
74. Leont'ev, A.N. (1978) *Activity, Consciousness, and Personality*, Prentice-Hall, Englewood Cliffs, NJ.
75. Engeström, Y. (2000) Activity theory as a framework for analyzing and redesigning work. *Ergonomics*, **43**, 960–974.
76. Engeström, Y. (1996) in *Cognition and Communication at Work* (eds Y. Engeström and D. Middleton), Cambridge University Press, Cambridge, pp. 199–232.
77. Witt, H., Alabart, J.R., Giralt, F., Herrero, J., Vernis, L., and Medir, M. (2006) A competency-based educational model in a chemical engineering school. *Int. J. Eng. Educ.*, **22**, 218–235.
78. Hopkins, T.A. and Samide, M. (2013) Using a thematic laboratory-centered curriculum to teach general chemistry. *J. Chem. Educ.*, **90**, 1162–1166.
79. Brock, W.H. (1992) *The Fontana History of Chemistry*, Fontana Press, London.
80. Goedhart, M.J. (2007) A new perspective on the structure of chemistry as a basis for the undergraduate curriculum. *J. Chem. Educ.*, **84**, 971–976.
81. Hammond, G. and Nyholm, R. (1971) The structure of chemistry. *J. Chem. Educ.*, **48**, 6–13.
82. Hoffmann, R. (1995) *The Same and Not the Same*, Columbia University Press, New York.
83. National Science Foundation (2003) *Beyond the Molecular Frontier. Challenges for Chemistry and Chemical Engineering*, The National Academies Press, Washington, DC.
84. Goedhart, M.J. and Verdonk, A.H. (1991) The development of statistical concepts in a design-oriented laboratory course in scientific measuring. *J. Chem. Educ.*, **68**, 1005–1009.
85. Jacob, C. (2002) Philosophy and biochemistry: research at the interface between chemistry and biology. *Found. Chem.*, **4**, 97–125.
86. Al-Atabi, M. and DeBoer, J. (2014) Teaching entrepreneurship using massive open online course (MOOC). *Technovation*, **34**, 261–264.
87. Martin, B.C., McNally, J.J., and Kay, M.J. (2013) Examining the formation of human capital in entrepreneurship: a meta-analysis of entrepreneurship education outcomes. *J. Bus. Ventur.*, **28**, 211–224.
88. Creed, C.J., Suuberg, E.M., and Crawford, G.P. (2002) Engineering entrepreneurship: an example of a paradigm shift in engineering education. *J. Eng. Educ.*, **91**, 185–195.
89. Runge, W. and Bräse, S. (2009) Education in Chemical Entrepreneurship: Towards Technology Entrepreneurship for and in Chemistry-Related Enterprises, http://ce.ioc.kit.edu/download/Chem_Entrepreneurship_Paper.pdf (accessed 16 May 2014).

90. Singer, S., Nielsen, N.R., and Schweingruber, H.A. (2012) *Discipline-Based Education Research: Understanding and Improving Learning in Undergraduate Science and Engineering*, The National Academies Press, Washington, DC.

91. Sunal, D.W., Hodges, J., Sunal, C.S., Whitaker, K.W., Freeman, L.M., Edwards, L. *et al.* (2001) Teaching science in higher education: faculty professional development and barriers to change. *Sch. Sci. Math.*, **101**, 246–257.

5
Empowering Chemistry Teachers' Learning: Practices and New Challenges

Jan H. van Driel and Onno de Jong

5.1
Introduction

For chemistry education to be effective in terms of promoting students' understanding of, and interest in, chemistry, highly qualified chemistry teachers are necessary. This chapter focuses on the initial and continued education of chemistry teachers. It is based on the idea that chemistry teachers need to develop their professional knowledge and practice throughout their entire career. Obviously, programs for chemistry teacher preparation play an important role in supporting the development of the competencies that are needed as a beginner teacher of chemistry. However, such programs can contribute to this development only to some extent [1]. Therefore, continued professional development for chemistry teachers, both formal and informal, is needed to support them in facing the dynamic and challenging issues that are inherent in the teaching of chemistry. These issues include dealing with the diversity among learners in terms of their abilities and interest to learn chemistry, integrating recent findings of chemistry research in the already overloaded curricula, and applying new instructional methods and techniques (e.g., educational software). In other words, for high-quality chemistry education, empowering chemistry teachers' professional learning is vital.

In general, professional learning of teachers, both preservice and in-service, is a complex process. Only recently has research begun to demonstrate what teachers learn from teacher education and professional development programs, and how this impacts on their professional knowledge and practice [2]. Several review studies have revealed that, for strategies aimed at promoting professional learning of teachers to be successful, the following elements are important: (i) an explicit focus on teachers' initial knowledge, beliefs, and concerns; (ii) opportunities for teachers to experiment in and investigate their own practice; (iii) collegial cooperation or exchange among teachers; and (iv) sufficient time for changes to occur [3, 4]. Many professional development programs, however, have been found lacking with respect to stimulating teacher learning [5, 6] because they tend to neglect the knowledge, beliefs, and attitudes these teachers bring into the program [7] and

Chemistry Education: Best Practices, Opportunities and Trends, First Edition.
Edited by Javier García-Martínez and Elena Serrano-Torregrosa.
© 2015 Wiley-VCH Verlag GmbH & Co. KGaA. Published 2015 by Wiley-VCH Verlag GmbH & Co. KGaA.

ignore the context in which teachers work [8]. Furthermore, many professional learning programs also fail to take into account existing knowledge about how teachers learn [9].

There is general agreement in the educational research community about the importance of teachers' professional learning as one of the ways to improve education. However, there is no consensus on how such a process occurs and how it can be analyzed and promoted. This may be because it was only in the last decades that the nature and development of teachers' knowledge started to be understood by educational researchers [10]. A major question in teacher learning literature relates to the issue of how changes in knowledge, beliefs, and attitudes relate to changes in teacher practice, and vice versa [11]. For a long time, it had been widely assumed that when teachers change their knowledge, beliefs, and attitudes on, for example, new instructional methods, their teaching practice will improve and accordingly result in better student outcomes. A linear model of teacher change has been proposed [12], assuming that a professional development program causes changes in teachers' practice, which in turn will lead to changes in students' learning and consequently result in changes in teachers' knowledge, beliefs, and attitudes (see Figure 5.1).

Other researchers, however, have cautioned that teacher learning is not a linear process but consists of more complex and reciprocal processes. According to Sprinthall *et al.* [13], there are three main models that explain teachers' professional development: the craft model, the expert model, and the interactive model.

- The *craft* model advocates the view that teachers develop professionally as a result of becoming experienced teachers. In this case, professional knowledge is seen to emerge from classroom experiences. However, this model does not make clear how teachers produce new meanings from their experiences, or why some teachers seem to reproduce the same experience many times without learning from it.
- The *expert* model is based on the idea that teachers learn through what is taught to them by experts. This model is reflected in the term "teacher training," that is, of something that is done to teachers and in which they are relatively passive participants. Typically, research on this approach has demonstrated that whatever the teacher learned, either barely or only temporarily, impacted their practice.

Figure 5.1 A model of teacher change [12] (With permission from Elsevier, 2002).

- In order to be more effective, many researchers have recognized that programs of professional development of teachers should involve meaningful learning activities. This is the basis of what was characterized as the *interactive* model. In this model, reciprocal relationships between teachers' knowledge and their practice are assumed; that is, teachers may develop their knowledge by reflecting on practical experiences, and, in turn, they may implement changes in their practice, inspired by newly acquired knowledge.

An example of an interactive model is the Interconnected Model of Teacher Professional Growth [14], which is constituted of four different domains: (i) the personal domain which is concerned with teachers' knowledge, beliefs, and attitudes; (ii) the external domain which is associated with external sources of information or stimuli; (iii) the domain of practice which involves professional experimentation; and (iv) the domain of consequence which is comprised of salient outcomes related to classroom practice (see Figure 5.2).

This model can be used to understand the way chemistry teachers develop their professional knowledge and practice, and how this development can be supported. For instance, chemistry teachers may participate in workshops on teaching and learning of chemistry with information and communications technology (ICT) (external domain). Such workshops may provide the teachers with ideas they could use in their practice. Next, teachers may try out some of these ideas in their classrooms (domain of practice), and collect data among their students to find out how they responded to their new teaching approach (domain of consequence). Such activities may help teachers to understand to what extent their teaching approach is successful for their students (personal domain). This new knowledge may then

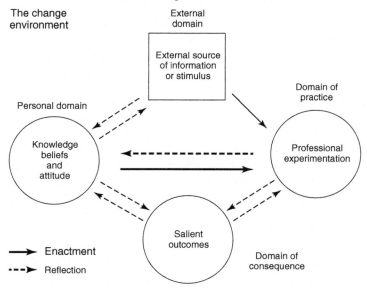

Figure 5.2 The interconnected model of teacher professional growth (IMTPG) [15] (With permission from Elsevier, 2002).

inspire them to adapt their approach for a next teaching cycle (domain of practice). In other words, this sequence of enactment and reflection steps results in teacher learning, in terms of new knowledge and changed practice.

5.2
Chemistry Teachers' Professional Knowledge Base

In several of studies on chemistry teachers, the focus is on their professional knowledge and how they develop and extend their professional knowledge base. We view a teacher's professional knowledge as the total knowledge that he or she has at his or her disposal at a particular moment which underlies his or her actions [16]. This does not imply that all the knowledge a teacher has does actually play a role in his or her actions. Teachers can, either consciously or not, refrain from using certain insights during their teaching. The basic idea, however, is that a reciprocity exists between the whole of a teacher's cognition (in the broad sense) and his or her activities and that, consequently, it makes sense to investigate teachers' knowledge [17].

5.2.1
The Knowledge Base for Teaching

In two highly influential publications, Lee Shulman [18, 19] outlined a model for what he termed *the knowledge base for teaching*. Central in this model was a new category of teacher knowledge, called *pedagogical content knowledge* (PCK), described as "that special amalgam of content and pedagogy that is uniquely the province of teachers, their own special form of professional understanding" [19, p. 8]. In his model, PCK is shaped by teachers' knowledge of subject matter on one hand, and by their general pedagogical knowledge (e.g., about classroom management, instructional principles, learning theories) on the other. Moreover, the knowledge of their own professional context (school, students, community) impacts on teachers' PCK.

At the heart of PCK lies what teachers know about how their students learn specific subject matter, or topics, and the difficulties or misconceptions students may have regarding this topic, related to the variety of representations (e.g., models, metaphors) and activities (e.g., explications, experiments) teachers know to teach this specific topic. These components are mutually related: the better teachers understand their students' learning difficulties with respect to a certain topic, and the more representations and activities they have at their disposal, the more effectively they can teach about this topic. From this perspective, Loughran and colleagues defined PCK as "the knowledge that a teacher uses to provide teaching situations that help learners make sense of particular science content" [20, p. 289].

Several authors have pointed out that it is not always possible to make a clear distinction between PCK and subject matter knowledge (SMK) [21, 22]. However,

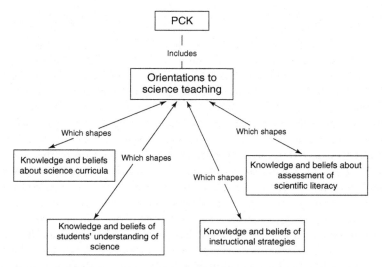

Figure 5.3 The Magnusson model of pedagogical content knowledge (PCK) [23] (With permission from Springer, 1999).

studies on the teaching of unfamiliar topics and in the context of preservice teacher education suggest that a thorough and coherent understanding of subject matter acts as a prerequisite to the development of PCK [1]. At the same time, studies on preservice teachers have demonstrated how teaching experiences may stimulate the (further) development of teachers' knowledge [23]. Magnusson and colleagues presented a strong case for the existence of PCK as a separate and unique domain of knowledge related to teaching of specific topics [24]. These authors conceptualized PCK as consisting of five components: (i) orientations toward science teaching, (ii) knowledge of the curriculum, (iii) knowledge of science assessment, (iv) knowledge of science learners, and (v) knowledge of instructional strategies (see Figure 5.3). This model of PCK has been particularly influential in research on science teacher knowledge since 2000 [25].

In the next sections, we will review several recent studies on preservice and in-service chemistry teachers' professional knowledge base. In the following section, we will concentrate on studies that investigated chemistry teachers' SMK and/or PCK at one moment in time. Next, we will discuss studies focused on the development of chemistry teachers' SMK and/or PCK. Since these studies concern a huge variety of chemical topics, we have chosen not to organize these sections around the chemical content.

5.2.2
Chemistry Teachers' Professional Knowledge

Research on science teachers' SMK has demonstrated that teachers often have very similar weaknesses and misconceptions of specific science topics as their students [26]. Whereas this may not be surprising for elementary or beginner teachers, studies of experienced secondary teachers of chemistry (or physics) often led to the same conclusion. Needless to say, this outcome is worrying [27]. An example of a study in the domain of chemistry focused on secondary teachers' understanding of chemical equilibrium [28]. Interviews were conducted with 12 chemistry teachers in Hong Kong with 3–18 years of teaching experience. They were asked to predict what would happen to the equilibrium system $N_2(g) + 3H_2(g) \rightleftharpoons 2NH_3(g)$ if a small amount of nitrogen gas is added to the system at constant pressure and temperature. Analyses of the interview protocols revealed that all these teachers had misconceptions and made wrong predictions, among others, because of the way they misunderstood Le Châtelier's principle. As another example, a study reported on 30 Turkish beginner chemistry teachers, focusing on their SMK and their beliefs about science teaching on the topic of chemical reactions [29]. Deficits were reported in the SMK of the teacher, who also appeared to hold very traditional and teacher-centered beliefs about chemistry teaching at the secondary level. The authors suggest that the limited SMK, plus the traditional beliefs, hindered the emerging development of these teachers' PCK.

Studies on the PCK of in-service chemistry teachers have found that these teachers may differ substantially in the extent to which they are knowledgeable about, and focused on, the learning of chemistry of their students. For example, a study was carried out on experienced Swedish chemistry teachers' knowledge of the learning difficulties of students in the area of acid–base chemistry [30]. Authentic student work (i.e., student responses to test questions) was used in an interview setting to probe the teachers' understanding. All nine teachers recognized at least some of the preconceptions and learning problems of the students, in particular, confusion among students between different models of acids and bases, and confusion between the phenomenological and the particle levels. Interestingly, within their sample, the authors distinguished two groups of teachers: one group showing a much greater awareness of student learning difficulties, and planning their teaching explicitly to address these difficulties using different models of acids and bases for this purpose, and the other group less focused on student learning and assuming it was sufficient to distinguish in their teaching between the phenomenological and the particle level.

Another study of preservice chemistry teachers' PCK focused on the topic of ozone layer depletion [31]. Three groups of 25 Turkish preservice teachers were interviewed, who had been previously identified from a sample of 219 by an SMK test as having high, average, and low SMK in the domain, respectively. The interviews were structured according to the PCK components from the Magnusson model given in Figure 5.3. The authors found medium to strong correlations between SMK and the various PCK components. There were also significant intra-relationships among the components of the preservice teachers' PCK, except that their knowledge of assessment was not correlated with the other PCK components.

In conclusion, these studies suggest that chemistry teachers' SMK and PCK are mutually related. Clearly, when teachers' SMK is limited, this places restrictions on their PCK. In the next section, we will focus on studies that explored the development of chemistry teachers' SMK and PCK.

5.2.3
Development of Chemistry Teachers' Professional Knowledge

Studies on the development of chemistry teachers' professional knowledge mostly focus on changes in SMK or PCK, in the context of either preservice teacher education programs or in-service teacher professional development programs. The latter are often connected with curriculum reforms in chemistry. In a project aimed at improving the SMK of Greek primary school teachers about chemical change, a specific intervention was conducted and the development SMK of 130 teachers was studied [32]. It was found that the teachers' SMK, which was limited initially, had significantly improved. Among others, a relationship between teachers' ideas about their particles and their explanations that incorporated particle ideas was found. However, the authors concluded that even though "the present training course had a significant effectiveness on teachers' understanding of chemical changes [..] one 'shot' is not enough" [32, p. 23].

Another successful intervention was reported [33] about a series of modules on pharmacology (i.e., drugs) for secondary schools in the United States plus an accompanying distance learning program of professional development workshops. In total, 121 chemistry and biology teachers participated in the workshops to learn how to incorporate the modules into their teaching. A quantitative assessment of their SMK at several moments in time revealed that the participants' SMK of chemistry and biology increased significantly after the workshops and was maintained for at least a year. Moreover, their students demonstrated a significant increase in the knowledge of chemistry and biology concepts, with higher scores as the number of modules used increased. The authors suggest that the long-term SMK retention was due to teachers' use of the modules in their classes because those who actually used the modules had additional knowledge gains, whereas those who did not use the modules lost some of their initial knowledge gains.

A New Zealand study focused on the development of PCK of four preservice chemistry teachers [34]. These teachers designed the so-called content representations (CoRes, see [20]) in consultation with their school-based mentor teachers on topics they were teaching on practicum. Preservice teachers refined their CoRes based on their teaching experiences and through ongoing discussions with their mentors. The findings indicated that the CoRe task improved preservice teachers' awareness of the different PCK components and provided a useful framework for focused conversations about chemistry teaching between preservice teachers and their mentor teachers.

Another study focused on the PCK development of experienced chemistry teachers [35]. In a case study, they observed two Turkish chemistry teachers teaching redox reactions and electrochemical cells. The authors used the

Magnusson model of PCK (see Figure 5.3) and studied its development by focusing on the process of integration of the PCK components. The results revealed that knowledge of chemistry learners and instructional strategies were the central components in the integration process. The other components, that is, knowledge of assessment and of curriculum, appeared less prominent in shaping these teachers' PCK. Additionally, more coherent integration was observed in the way these two teachers taught the topic of electrochemical cells. This confirms the topic-specific nature of PCK. In another study of experienced chemistry teachers [36], the Magnusson model (see Figure 5.3) was also used to analyze PCK development. Three in-service Dutch chemistry teachers were studied in the context of a curriculum innovation, focusing on their design and subsequent use of curricular materials. The researchers reported that the teachers cooperating in a network under supervision of an expert were able to develop innovative learning materials, which appeared to be a powerful way to prepare them for the innovation. Moreover, the PCK increased in all five components of the Magnusson model for each teacher during the cycle of developing the new materials and implementing these in classroom teaching.

As an alternative to curriculum innovation, teacher certification may also act as a stimulus to PCK development. For instance, PCK development of three chemistry teachers working in the same U.S. high school was investigated in the context of the National Board Certification (NBC) process [37]. The authors found that the NBC process "pushed the teachers to not only add to but also integrate the five components of this [Magnusson] model in ways they had not done previously" [37, p. 828]. An important factor in stimulating the integration of different knowledge components was the NBC requirement that teachers analyze and articulate their pedagogical decisions and actions.

To conclude this section, we like to mention a comprehensive and ambitious project aimed at improving chemistry teaching in Manitoba, Canada. The project was in response to a new chemistry curriculum that advocated a tetrahedral orientation to the teaching of chemistry [38]. The outcomes at the end of the fourth year of a 5-year research and professional development program which involved three cohorts of in-service chemistry teachers (74 in total) were reported [39]. The program incorporated the following features by design: the program was sustained over time, had a strong focus on both SMK and PCK, was in line with specific curriculum and professional standards, and was supported at both the school and the district level. Chemistry teachers were actively involved in the program. The authors used a Chemistry Teacher Inventory to collect teacher responses over time, and concluded that the teachers showed progress in their classroom toward implementing the tetrahedral orientation, in a manner consistent with the curriculum. However, they also found that this development tended to be limited. Projects like this show that a sustained effort is needed to promote changes in chemistry teachers' PCK and SMK, but these changes do not always easily transfer into innovative teaching practices.

5.3 Empowering Chemistry Teachers to Teach Challenging Issues

This section deals with empowering chemistry teachers for teaching three challenging issues in modern chemistry education. First, the issue of context-based teaching is addressed. This issue is interesting because the use of contexts for teaching chemistry topics aims at fostering among students a more positive attitude and a better understanding of chemistry. Second, the issue of teaching the nature and functions of models and modeling is presented. This issue fits very well with the growing interest in teaching of "nature-of-chemistry" issues, and therefore teaching *about* chemistry models is added to teaching *of* specific models in chemistry. Third, the issue of using computer-based technologies for teaching chemistry is discussed. This issue is important because the use of modern technologies is a new challenge for supporting current chemistry teaching practices.

5.3.1 Empowering Chemistry Teachers for Context-Based Teaching

Context-based teaching became a core part of many chemistry curricula in the 1980s, and the interest in this teaching, as well as empowering teachers for this teaching, is still going on (see Table 5.1). In the 1980s, several curriculum projects that included contexts were launched such as the U.S. project "Chemistry in the Community" [40] and the U.K. project "Chemistry: The Salters Approach" [41]. Teachers were asked to relate chemistry concepts and processes to situations from the personal or society domain that were believed to be of interest for students. To support teachers for implementing this teaching, extended in-service courses were launched. However, in general, the results of this innovation were rather disappointing [42, 43]. For instance, students did hardly see the relevance of the given

Table 5.1 Context-based chemistry curricula and ways of empowering teachers.

Period	Example of context-based curriculum project	Way of empowering teachers
1980s	USA: "Chemistry in the Community" UK: "Chemistry: The Salters Approach"	In-service courses presenting new topics and teaching strategies
1990s	USA: "Chemistry in Context" UK: "Salters Advanced Chemistry"	In-service courses combining workshops with classroom teaching
2000s	The Netherlands: "Chemistry between Context and Concept" (CCC) Germany: "Chemie im Kontext" (ChiK)	Learning communities of experts and teachers with strong focus on common development of teaching and learning modules

contexts for understanding related concepts and processes, and teachers encountered difficulties in guiding students for context-based learning. These problems can be explained by several factors such as a lack of sufficient understanding of students' real interests, and a lack of attempts to actively involve teachers in the process of developing innovative teaching.

About a decade later, a new generation of curriculum projects was launched such as the U.S. project "Chemistry in Context" [44], and the U.K. project "Salters Advanced Chemistry" [45]. These projects of the 1990s often combined workshops with context-based teaching in the classroom. Soon, several projects followed in the 2000s, such as the Dutch project "Chemistry between Context and Concept" (CCC) [46] and the German project "Chemie im Kontext" (ChiK) [47]. The latter two projects are the focus of this subsection, not only because they are the most recent projects but also because they include the interesting feature of involving experienced chemistry teachers at an early stage of curriculum reform. The teachers cooperated with experts (teacher educators, curriculum specialists) at the stage of developing and field-testing context-based modules. Studies reported that the involvement of experienced teachers in these learning communities had a positive impact on their teaching. For instance, German chemistry teachers involved in designing ChiK modules became empowered for more context-based and student-oriented teaching than before their involvement [48]. Another comparative study indicated that Dutch science teachers, among them chemistry teachers, who were involved in teams for designing context-based materials showed more context-based competence than their non-designing colleagues [49].

Several studies on projects, including the development of PCK of context-based teaching, reported on a specific difficulty in using context-based materials. Regarding the German ChiK project, it was found that, although students became aware of the relevance of chemistry in everyday life and societal issues, they sometimes experienced a sense of getting lost in the context [48]. In line with this outcome, it was reported that ChiK teachers encountered difficulties in using students' questions that were evoked by the introductory context as an orientation event for the subsequent lessons [50]. The contexts given were too general and broad to be effectively applicable as a setting in which such activities as students developing their ideas and exploring them systematically could take place.

Regarding the Dutch CCC project, a similar hindering factor was found [51]. When preparing interested teachers in how to implement the first version of a CCC teaching unit in their classroom, university experts asked these teachers to design strategies for connecting the introductory context with related chemistry concepts. Some teachers wanted to use a "look for unknown words" teaching strategy. That is, after the context, students should read the background text in the unit and, when they stumble upon a word they do not understand, they can look it up in their textbook. Other teachers wanted to use a "carefully guiding" approach focused on helping students in shifting focus from the context to related concepts. After applying these strategies, it became clear to the teachers that their approaches were not sufficient to evoke the students' need-to-know sufficiently for

Table 5.2 Strategy for context-based chemistry teaching.

Phase of context-based teaching	Aim of the phase
Offering introductory contexts	Evoking students' need-to-know, that is, questions
Collecting and, if necessary, adapting questions in the classroom	Preparing students for finding answers by learning about relevant chemistry concepts
Restructuring textbook content or selecting appropriate Web site info	Enhancing the links between the students' questions and information in textbooks or Web site
Offering follow-up inquiry contexts	Evoking students' need to apply their knowledge

connecting context and concepts. As a follow-up, a revised strategy for context-based chemistry teaching was implemented [51] (see Table 5.2). It was reported that the teachers wanted to select and reformulate students' questions about the given context in such a way that students were encouraged to find answers to their own questions by using appropriate chemistry concepts [52]. The teachers appreciated the application of this strategy, and afterwards they designed a set of "do's and don'ts" for handling students' questions. Most teachers considered this set useful for their teaching practice, although its contribution to stimulating students to connect contexts with concepts remained unclear.

In conclusion, the outcomes of these projects suggest that a crucial aspect of context-based teaching; that is, connecting an introductory context with underlying chemistry concepts in a meaningful way for students is difficult to implement for teachers. This difficulty can explain why the student outcomes of context-based teaching have been somewhat disappointing from a cognitive development point of view, despite the improvement of student learning motivation and the promotion of positive attitudes of students toward chemistry in general [53]. Revisions of the existing projects are needed to improve relevant curriculum materials and to further empower teachers for context-based chemistry teaching.

5.3.2
Empowering Chemistry Teachers to Teach about Models and Modeling

Another important issue in modern chemistry education is the teaching *about* models and modeling, especially their nature and functions. This implies that teachers should not only have a well-developed knowledge of specific models used in chemistry but also a sound knowledge about models in general and their use (modeling) in the process of developing new scientific insights of nature. Moreover, they should have adequate knowledge of how to teach specific models and modeling and how to teach general aspects of models and modeling. However, several studies have shown that, in general, teachers' knowledge about models and modeling is quite limited and often inadequate [54]. For instance, although teachers present the idea that models are simplified representations of specific parts of reality, they do not generally acknowledge the important function

> In a post-graduate teacher education workshop at Utrecht University, three pre-service teachers (PST's) discuss the issue of particle models. Part of their discussion is given below.
>
> PST#1: We talk now about a model for an atom and it seems to me that we say that atoms exist, but is the atom concept itself not a model that we have designed to explain reactions?
>
> PST#2: You can simply explain things with it, but whether atoms exist? I don't know.
>
> PST#3: According to me, atoms exist.
>
> PST#2: In lower secondary classes, I would say, they exist for sure. There are those little needles and you can use them for scanning a substance's surface [PST refers to scanning tunnelling microscopy (STM)]. Then, you can say, look, they exist. In upper secondary classes, I would say, they don't exist at all, of course not. Atoms cannot be seen.
>
> PST#3: But that does not mean that they don't exist.
>
> PST#1: I would say to my students: you cannot see or prove atoms, for that reason, chemistry has designed a model that could represent reality. If observations do not fit anymore with the model, then, you have to revise the model. I would talk in this way.
>
> Although all these pre service teachers already had obtained their Master's degree in chemistry, they show important differences between their SMK of the concepts of atom and model. PST#1 and PST#2 also show interesting differences between their PCK of teaching these concepts.

Figure 5.4 A preservice teachers' discussion about particle models [56].

of models for making predictions of phenomena, and they consider modeling as a straightforward, rational process [55]. Another example is given in Figure 5.4.

In the realm of empowering prospective teachers, a post-graduate teacher education course on developing SMK and PCK about models and modeling was examined [57]. The participants were five Dutch prospective science teachers (four of them of chemistry). They designed a lesson series on models and modeling, and conducted an inquiry project about this theme in their high school classes (as part of the course). The results showed that the teachers developed personal knowledge about models and modeling, especially about the role of models in the development of scientific knowledge, the nature and role of modeling, and the use of both teaching models and modeling activities in teaching. It was found that, in particular, reflective activities (such as writing reports and sharing experiences in collective meetings) stimulated the development of SMK and PCK of these preservice teachers. Another study reported on a course about the use of particle models in teaching chemistry [58]. The participants were 12 Dutch prospective chemistry teachers. The course emphasized their PCK development through learning *from* teaching by connecting authentic school teaching experiences with follow-up institutional workshops. The outcomes of the study revealed that, after teaching, all prospective teachers demonstrated a deeper understanding of their students' problems with the use of particle models and were able to describe specific learning difficulties of students in relating properties of substances to their constituent particles. In addition, about half of the participants had become more aware of the possibilities and limitations of using particle models in specific teaching situations.

In relation to empowering experienced teachers, the development of SMK of models in general and PCK of teaching about models and modeling was investigated in the context of the implementation of a new syllabus, which emphasized models and modeling [59]. The study followed nine Dutch science teachers (three of them of chemistry) during the first year of its implementation. The results showed three related types of knowledge development. First, the teacher's learning of model content was combined with a critical reflection on the nature and role of models. Second, the teacher's learning about modeling as an activity undertaken by students was combined with the learning of specific model content. Third, the teacher's learning of model content involved both the production and revision of models and a critical examination of the nature of models in general.

Some studies have reported on teacher courses that included the use of dynamic computer-based molecular modeling. For instance, a course on developing SMK and PCK about models and modeling was studied, which included a software package that enabled three-dimensional molecular visualizations and the availability of several types of representation [60]. The participants were 34 Israeli pre- and in-service chemistry teachers who chose molecules from a database, measured their bond lengths and angles, rotated them, and watched them in various styles. They discussed the nature and function of models and the possibilities and limitations computerized molecular modeling offers as a teaching tool. The results showed that most of the teachers thought of a model as a way to describe phenomena that could not be seen. They all agreed that models help explain and understand phenomena through simplification and visualization. In an interesting follow-up stage of the study, four participants of the course, all experienced chemistry teachers, became involved in teaching about models and modeling. Two of them taught three (experimental) classes of students by presenting various models and computerized molecular modeling. Two other participants taught three (control) classes of students in the traditional way, without computer software. The findings showed a noticeable difference between the student groups. Experimental group students scored significant higher than those of the control group on a model perception questionnaire, especially regarding the meanings of chemistry models.

In conclusion, the reported studies show the importance of courses for chemistry teachers focused on improving their knowledge about models and modeling and how to teach these issues [61]. However, studies of the design and outcomes of relevant courses are quite scarce. Seeking to improve this situation will be one of the important challenges for chemistry teacher education research in the near future.

5.3.3
Empowering Chemistry Teachers to Use Computer-Based Technologies for Teaching

In the last decades, a number of new educational technologies have been introduced in modern chemistry curricula for supporting or transforming current educational practices: for instance, interactive computer software for

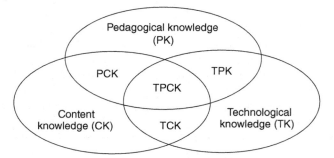

Figure 5.5 A model of technological pedagogical content knowledge (TPCK) [63]. Reproduced by permission of the publisher, © 2012 by tpack.org.

students, computer-assisted instruction, and Web-based interactive learning environments. Many chemistry teachers, at least in several parts of the world, possess personal computers, and by connecting them to Internet at home they get a positive attitude toward Internet use [62]. However, it is well known that many of them, especially older teachers, are not very familiar with relevant educational software and other ICT tools because they have not grown up with them. Their technological pedagogical content knowledge (TPCK; see Figure 5.5) is not well developed [63]. Therefore, simply dumping computer software in the classroom is, in itself, unlikely to transform chemistry teaching practices. Chemistry teachers, especially teachers with many years of teaching, need empowerment to become familiar with innovative computer-based tools and to use them.

Most of the courses on ICT software tools and their use in chemistry teaching are concerned with computer-based inquiry laboratories and computer-based molecular modeling. A course for experienced Israeli teachers focusing on the development of their PCK of using modules that include computer-based inquiry laboratories was investigated [64]. These modules implied that students read authentic problems, carry out inquiry-based lab experiments, process data collected by sensors, and then interpret the resulting graphs that appeared on their desktop screen. The results of the study showed that the teachers initially did not feel very confident to integrate computer applications into their chemistry teaching. However, at the end of the course, about half of the teachers were willing to implement the modules. The most important reasons for adopting them were making chemistry appealing and demonstrating up-to-date chemistry in action. The reasons for not willing to implement the modules were not reported.

More information about factors that influence chemistry teachers in deciding to use ICT software tools were provided by a study in which 19 experienced Swedish chemistry teachers were interviewed who had previously participated in a course on computer-based molecular modeling [65]. The teachers' answers showed the following main reasons for using this kind of modeling: (i) it helps the teachers to illustrate and explain various phenomena in chemistry and (ii) it supports the students to develop visualization skills and understanding by providing models of difficult concepts. The main reasons for not using this kind of modeling were poor local conditions: no programs or opportunities to use computers. The main reasons for hardly using this kind of modeling were not only school-related, namely, tight lesson timetables and large class sizes, but also personal-related, namely, feelings of insufficient personal skills. The researchers also asked the teachers what need for empowerment they felt using computer-based molecular modeling in teaching. The findings showed that the teachers wanted to get more support, particularly on where and how modeling can be best used. They would also appreciate (i) the sharing of experiences with colleagues, (ii) personal mentoring or independent study via the Internet, and (iii) access to examples on how computer-based molecular modeling can be applied in the teaching of concepts in chemistry.

Web 2.0 provides educational technological environments including a range of tools for use in the classroom. A rather new tool for teaching is YouTube. A teacher course focusing on using and editing YouTube videos was investigated [66]. The course participants were 16 experienced Israeli chemistry teachers who were first taught the fundamental skills of searching, using, and downloading YouTube videos. Then, they were taught how to use "Movi Maker," a freeware used to edit videos. Finally, each video that was produced by the teachers was tried out in their classrooms and was also presented during the course meetings. The findings showed that the teachers improved their knowledge of video editing. They also developed their knowledge of using videos in their classroom lessons. Most of the teachers trusted in their ability to integrate videos in their teaching but not all of them were confident in their video editing skills.

In conclusion, new technological tools can play an important role in modern chemistry teaching. However, the implementation of these tools can evoke resistance among teachers because of local or personal constraints. Teacher courses should take this problem into account and empower teachers in a careful way, for instance, by improving their knowledge and skills through creating opportunities for sharing experiences with colleagues.

5.4
New Challenges and Opportunities to Empower Chemistry Teachers' Learning

In this section, two interesting challenges to modern chemistry teacher education are presented. The first challenge concerns the preparation of preservice chemistry teachers for research-based teaching practices. The second challenge deals with the professional development of experienced teachers by implementing learning communities of practice.

5.4.1
Becoming a Lifelong Research-Oriented Chemistry Teacher

Modern chemistry education is more and more aimed at implementing research-based teaching practices [67]. This development can be successful only when appropriate chemistry teacher programs are developed, that is, programs that prepare chemistry teachers to use existing research as input for their practice, and to investigate their practice, for instance, in terms of the impact it has on their students.

An interesting example of such a program in the field of preservice teacher education has been studied [68]. The reported 5-year program aimed at educating future Swedish chemistry teachers to become lifelong research-oriented teachers. These teachers should be able to follow developments both in chemistry and chemistry teaching, to implement up-to-date research findings in their work as teachers, and to engage in research on chemistry teaching. The program consisted of eight courses which dealt with chemistry teaching and learning from four different perspectives incorporating high-quality evidence: "(i) The nature of chemistry and scientific literacy, (ii) concepts and phenomena in chemistry and their learning, (iii) supporting concept building and interest in chemistry through a variety of learning strategies (i.e., inquiry, molecular modeling, informal learning), and (iv) applied chemistry (e.g., renewable resources) in teaching" [68, p. 87]. Each course provided a range of activities and various meetings such as topical seminars. The course also used modern ICT tools such as virtual learning environments that offer materials on research and a discussion platform.

A case study on 18 Swedish preservice chemistry teachers collected their views about the phenomenon of the "teacher as researcher" at the end of a relevant course in the fourth year of the program [68]. The results showed the following most commonly identified characteristics of the "teacher as researcher": (i) continuous development of own expertise of chemistry and chemistry education, (ii) use of various teaching approaches, (iii) skills in the field of creativity and collaboration (networking), and (iv) interest in chemistry and how it is learned. The participants of the case study were also asked to report on the influence of the course on these views. The most commonly given reasons for the change of views were (i) reading various research articles, for example, identifying different learning difficulties and (ii) using different sources of information, not only literature but also own experiences with investigations.

In conclusion, the case study suggested that future chemistry teachers could appreciate and incorporate the idea of becoming lifelong research-oriented teachers as an important condition for improving the quality of teaching chemistry. The aims of the program reported above can function as a realistic and useful challenge to chemistry teacher education elsewhere.

5.4.2
Learning Communities as a Tool to Empower Chemistry Teachers' Learning

Many modern courses for the professional development of teachers are influenced by socio-constructivist views on learning. They conceive teachers as active and reflective practitioners and create conditions for collaborative and self-directed learning. Discussing and sharing experiences with colleagues similarly engaged can contribute to providing a context that builds professional learning. This has stimulated the interest in establishing teacher courses that are designed as

5.4 New Challenges and Opportunities to Empower Chemistry Teachers' Learning | 115

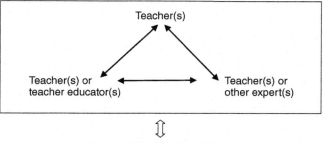

Figure 5.6 Participants of a learning community of practice.

interactive networks: learning communities of practice. These networks consist of groups of teachers but other groups can also be involved, such as teacher educators, curriculum specialists, and, in case of use of modern technologies, ICT experts (Figure 5.6).

Learning communities of practice can play an important role in involving teachers in the development of small-scale and evidence-based chemistry curriculum innovations. For instance, a study reported on a community focused on new context-based materials and teaching ([52]; see also Section 5.3.1). A group of seven Dutch chemistry teachers came together with teacher educators and curriculum specialists to discuss a first draft of an innovative context-based module. The context under consideration was superabsorbent materials used in disposable diapers. This context was related to the chemistry concepts of polymers, monomers, and their properties. The discussions focused on the related curricular framework, the specific module content, and the relevant teaching strategies. As a result, the draft was adapted for use in the classroom. In a series of lessons, the teachers taught the module, and followed it up by sharing their experiences with the community participants. The outcomes were used for revising the module and designing outlines for other context-based modules. The results of the study showed that teachers became empowered for modern context-based teaching. The results also indicated that they became empowered in designing context-based modules, provided they have sufficient time and resources.

In another study on a chemistry teacher learning community, the focus was on the research-based development of a plan for a series of context-based lessons on shower gels and added musk fragrances [69]. The lesson plan not only focused on the chemistry of these gels but also on artificial musk fragrances as potentially harmful ingredients. The community consisted of 15 German chemistry teachers, together with teacher educators and researchers in chemistry education. A first-draft lesson plan was discussed and adapted within the community until consensus was reached. Through subsequent cycles of field-testing, evaluation, and reflection/revision, the community improved the lesson plan. The findings of the study indicated that the teachers considered the lesson plan to be highly feasible, very motivating for students, and a fruitful initiator of evidence-based discussions about the controversial issue of adding potentially harmful ingredients to consumer products. The teachers became empowered for using a socio-critical and problem-oriented approach to chemistry teaching.

Learning communities of practice can include the use of computer-mediated technologies. A study reported on a professional development program that used Wiki [15]. This Web 2.0 software application allowed the users to create and edit Web page content, and to add, delete, change, and comment on content created by others. The study focused on a learning community consisting of 20 Israeli chemistry teachers and program facilitators. The main aim was to develop a community of experienced teachers who would be able to support other chemistry teachers in their school or district. The 2-year program included face-to-face meetings combined with online collaboration in the Wiki environment. The collection of Wiki pages that the teachers developed was mainly related to the difficult topic of chemical bonding. The results of the study showed two successful methods for promoting collaboration and interaction. The first was "fading scaffolding," that is, the level and means of interactions increased over time, whereas the explicit instruction provided decreased. The second was creating small groups and changing the group for every task to allow teachers to meet and work with more peers. All teachers indicated that they use or intend to use the materials developed in the Wiki context in their lessons. However, nearly half of the teachers did not appreciate the collaborative online platform very well. The findings suggested that their disappointment with the Wiki was not correlated with technical difficulties of the chemistry content. The researchers believed that it had to do with the teachers' perception of the relevance of the Wiki environment itself (as a model of other virtual environments) to classroom teaching.

In conclusion, learning communities of practice can empower chemistry teachers to accept innovative ideas and practices, especially when the implementation is supported by materials that engage teachers in teaching and foster a sense of experimentation ("learning by doing"). Moreover, by involving learning communities of chemistry teachers in the development of small-scale chemistry curriculum innovations, these teachers can become "co-owners" of the innovations. This is an important condition for success in bringing new teaching materials and approaches into classroom practice. Finally, organizers of learning communities, including online collaborators, should be aware of possible aversion toward this mode of collaboration. To cope with resistance among teachers, they should create appropriate conditions for promoting successful online collaboration. A concise description of these conditions is incorporated, in addition to other issues, in the section below.

5.5
Final Conclusions and Future Trends

In this chapter we argued that high-quality chemistry education requires teachers with a high level of expertise in both chemistry content and pedagogy. Moreover, because chemistry and chemistry teaching are constantly changing, chemistry teachers need to keep expanding and developing their professional knowledge and competencies throughout their career. On the basis of international development and research in chemistry education, we described several ways to empower to chemistry teachers and identified challenges that teachers may encounter. In our opinion, several interesting new trends in empowering chemistry teachers' learning can be indicated. Three of these are concisely addressed below.

The first trend regards empowering teachers for teaching practices that fit context-based chemistry curriculum reform. The past 40 years have seen a fundamental change of teachers' role: from "consumers" of innovations to "co-producers" of new topics and teaching approaches (see also Section 5.3.1). The replacement of a "top down" implementation of curriculum innovations by a "bottom up" approach contributes to an early involvement of chemistry teachers. As participants of pilot projects, they play an active role in developing, field-testing, and evaluating new curriculum materials. In these projects, teacher learning is facilitated by creating communities of practice. In our opinion, these networks in the pilot stage can function as sources of inspiration for other teacher networks at the stage of the implementation of the definite version of a new chemistry curriculum in the classroom.

The second trend concerns the launching of Internet-based collaboration of chemistry teachers. In our opinion, these online networks can be successful when they fulfill the following three specific conditions. First of all, the school management should give support, for instance, by providing chemistry teachers with sufficient time for online participation and access to Internet by high-quality tools. Second, the structure of networks should reduce chemistry teachers' relative isolation, despite online chats, by combining online activities with face-to-face meetings of special interest groups. Third, the culture of networks should contribute to evoke feelings of personal trust and psychological safety among the participating chemistry teachers.

The third trend regards empowering chemistry teachers to develop research-based teaching practices. This empowerment can grow in the context of specific teacher courses (see also Section 5.4.1), but also through consulting research on chemistry education, for instance, published by the monthly e-journal *Chemistry Education: Research and Practice*. This free-access site provides a range of useful findings and suggestions that teachers can use as input for research-based teaching practices.

Finally, we see a need for more in-depth and longitudinal studies of empowering chemistry teachers' learning. For instance, there is a lack of these studies in the field of promoting teachers' knowledge of modern issues such as "green" chemistry with a strong focus on environmental issues and chemistry for sustainability, nano-technological chemistry, and socio-critical issues including chemistry aspects. These studies should be combined with investigations of chemistry teachers' knowledge of how to teach these issues and studies of related classroom practices. In the field of chemistry teachers' e-learning, there is a lack of empirical research on the role and impact of new technologies and online professional networks on chemistry teachers' knowledge, beliefs, and classroom activities. A focus on chemistry teaches' learning acknowledges that teachers, as professionals, hold the key to teaching chemistry in ways that make sense to and motivate twenty-first century students.

References

1. Van Driel, J.H., Verloop, N., and De Vos, W. (1998) Developing science teachers' pedagogical content knowledge. *J. Res. Sci. Teach.*, **35**, 673–695.
2. Fishman, B.J., Marx, R.W., Best, S., and Tal, R.T. (2003) Linking teacher and student learning to improve professional development in systematic reform. *Teach. Teach. Educ.*, **19**, 643–658.
3. Garet, M., Porter, A., Desimone, L., Birman, B., and Yoon, K.S. (2001) What makes professional development effective? Results from a national sample of teachers. *Am. Educ. Res. J.*, **38**, 915–945.
4. Hewson, P.W. (2007) in *Handbook of Research on Science Education* (eds S.K. Abell and N.G. Lederman), Lawrence Erlbaum, Mahwah, NJ, pp. 1179–1203.
5. Ball, D.M. and Cohen, D. (1999) in *Teaching as the Learning Profession: Handbook of Policy and Practice* (eds L. Darling-Hammond and G. Sykes), Jossey-Bass, San Francisco, CA, pp. 3–32.
6. Little, J.W. (2001) in *Teachers Caught in the Action: Professional Development that Matters* (eds A. Lieberman and L. Miller), Teachers College Press, New York, pp. 28–44.
7. Van Driel, J.H., Beijaard, D., and Verloop, N. (2001) Professional development and reform in science education: the role of teachers' practical knowledge. *J. Res. Sci. Teach.*, **38**, 137–158.
8. Kennedy, M.M. (2010) Attribution error and the quest for teacher quality. *Educ. Res.*, **39**, 591–598.
9. Borko, H. (2004) Professional development and teacher learning: mapping the terrain. *Educ. Res.*, **33** (8), 3–15.
10. Munby, H., Russell, T., and Martin, A.K. (2001) in *Handbook of Research on Teaching*, 4th edn (ed. V. Richardson), American Educational Research Association, Washington, DC, pp. 877–904.
11. Richardson, V. and Placier, P. (2002) in *Handbook of Research on Teaching*, 4th edn (ed. V. Richardson), AERA, Washington, DC, pp. 905–947.
12. Guskey, T.R. (1986) Staff development and the process of teacher change. *Educ. Res.*, **15**, 5–12.
13. Sprinthall, N.A., Reiman, A.J., and Thies-Sprinthall, L. (1996) in *Handbook of Research on Teacher Education*, 2nd edn (eds J. Sikula, T.J. Buttery, and E. Guyton), Macmillan, New York, pp. 666–703.
14. Clarke, D. and Hollingsworth, H. (2002) Elaborating a model of teacher professional growth. *Teach. Teach. Educ.*, **18**, 947–967.
15. Shwartz, Y. and Katchevitch, D. (2013) Using wiki to create a learning community for chemistry teacher leaders. *Chem. Educ. Res. Pract.*, **14**, 312–323.
16. Carter, K. (1990) in *Handbook of Research on Teacher Education* (ed. W.R. Houston), Macmillan, New York, pp. 291–310.
17. Verloop, N., Van Driel, J., and Meijer, P. (2001) Teacher knowledge and the knowledge base of teaching. *Int. J. Educ. Res.*, **35** (5), 441–461.
18. Shulman, L.S. (1986) Those who understand: knowledge growth in teaching. *Educ. Res.*, **15**, 4–14.
19. Shulman, L.S. (1987) Knowledge and teaching: foundations of the new reform. *Harv. Educ. Rev.*, **57**, 1–22.
20. Loughran, J., Milroy, P., Berry, A., Mulhall, P., and Gunstone, R. (2001) Science cases in action: documenting science teachers' pedagogical content knowledge through PaP-eRs. *Res. Sci. Educ.*, **31**, 289–307.
21. Tobin, K., Tippins, D.J., and Gallard, A.J. (1994) in *Handbook of Research on Science Teaching and Learning* (ed. D.L. Gabel), Macmillan, New York, pp. 45–93.

22. Kind, V. (2009) Pedagogical content knowledge in science education: perspectives and potential for progress. *Stud. Sci. Educ.*, **45**, 169–204.
23. Lederman, N.G., Gess-Newsome, J., and Latz, M.S. (1994) The nature and development of preservice science teachers' conceptions of subject matter and pedagogy. *J. Res. Sci. Teach.*, **31**, 129–146.
24. Magnusson, S., Krajcik, J., and Borko, H. (1999) in *Examining Pedagogical Content Knowledge* (eds J. Gess-Newsome and N.G. Lederman), Kluwer Academic Publishers, Dordrecht, pp. 95–132.
25. Friedrichsen, P., Van Driel, J., and Abell, S. (2011) Taking a closer look at science teaching orientations. *Sci. Educ.*, **95**, 358–376.
26. Wandersee, J.H., Mintzes, J.J., and Novak, J.D. (1994) in *Handbook of Research on Science Teaching and Learning* (ed. D.L. Gabel), Macmillan, New York, pp. 177–210.
27. Abell, S.K. (2007) in *Handbook of Research on Science Education* (eds S.K. Abell and N.G. Lederman), Lawrence Erlbaum, Mahwah, NJ, pp. 1105–1149.
28. Cheung, D. (2009) Using think-aloud protocols to investigate secondary school chemistry teachers' misconceptions about chemical equilibrium. *Chem. Educ. Res. Pract.*, **10**, 97–108.
29. Usak, M., Özden, M., and Eilks, I. (2011) A case study of beginning science teachers' subject matter (SMK) and pedagogical content knowledge (PCK) of teaching chemical reaction in Turkey. *Eur. J. Teach. Educ.*, **34**, 407–429.
30. Dreschler, M. and Van Driel, J. (2008) Experienced teachers' pedagogical content knowledge of teaching acid–base chemistry. *Res. Sci. Educ.*, **38**, 611–631.
31. Kaya, O.N. (2009) The nature of relationships among the components of pedagogical content knowledge of pre-service science teachers: "ozone layer depletion" as an example. *Int. J. Sci. Educ.*, **31**, 961–988.
32. Papageorgiou, G., Stamovlasis, D., and Johnson, P. (2012) Primary teachers' understanding of four chemical phenomena: effect of an in-service training course. *J. Sci. Teach. Educ.* doi: 10.1007/s10972-012-9295-y
33. Schwartz-Bloom, M.D., Halpin, M.J., and Reiter, J.P. (2011) Teaching high school chemistry in the context of pharmacology helps both teachers and students learn. *J. Chem. Educ.*, **88**, 744–750.
34. Hume, A. and Berry, A. (2013) Enhancing the practicum experience for pre-service chemistry teachers through collaborative CoRe design with mentor teachers. *Res. Sci. Educ.*, **43**, 2107–2136.
35. Aydin, S. and Boz, Y. (2013) The nature of integration among PCK components: a case study of two experienced chemistry teachers. *Chem. Educ. Res. Pract.*, **14**, 615–624.
36. Coenders, F., Terlouw, C., Dijkstra, S., and Pieters, J. (2010) The effects of the design and development of a chemistry curriculum reform on teachers' professional growth: a case study. *J. Sci. Teach. Educ.*, **21**, 535–557.
37. Park, S. and Oliver, J.S. (2008) National board certification (NBC) as a catalyst for teachers' learning about teaching: the effects of the NBC process on candidate teachers' PCK development. *J. Res. Sci. Teach.*, **45**, 812–834.
38. Mahaffy, P. (2006) Moving chemistry education into the 3D: a tetrahedral metaphor for understanding chemistry. *J. Chem. Educ.*, **83**, 49–55.
39. Lewthwaite, B. and Wiebe, R. (2011) Fostering teacher development to tetrahedral orientation in the teaching of chemistry. *Res. Sci. Educ.*, **41**, 667–689.
40. ACS (American Chemical Society) (1988) *ChemCom: Chemistry in the Community*, Kendall-Hunt, Dubuque, IA.
41. UYSEG (University of York Science Education Group) (1989) *Salters' GCSE Chemistry: 16 Unit Guides*, UYSEG, York.
42. Postlethwaite, T. and Wiley, D. (1991) *Science Achievement in Twenty-Three Countries*, Pergamon Press, Oxford.
43. Wahlberg, H.J. (1991) Improving school science in advanced and developing countries. *Rev. Educ. Res.*, **61**, 25–69.
44. ACS (American Chemical Society) (1994) *Chemistry in Context*, Kendall-Hunt, Dubuque, IA.

45. SAC Project (Salters Advanced Chemistry Project) (1994) *Chemical Storylines; Chemical Ideas; Activities and Assessment Pack*, Heinemann, Oxford.
46. Driessen, H.P.W. and Meinema, H.A. (2003) *Chemistry between Context and Concept: Designing for Renewal*, SLO, Enschede.
47. Gräsel, D., Nentwig, P., and Parchmann, I. (2005) in *Evaluation as a Tool for Improving Science Education* (eds J. Bennett, J. Holman, R. Millar, and D. Waddington), Waxmann, Münster, pp. 53–66.
48. Parchmann, I., Gräsel, D., Baer, A., Nentwig, P., Demuth, R., Ralle, B., and the ChiK Project Group (eds) (2006) "Chemie im Kontext": a symbiotic implementation of a context-based teaching and learning approach. *Int. J. Sci. Educ.*, **28**, 1041–1062.
49. De Putter-Smits, L., Taconis, R., Jochems, W., and Van Driel, J. (2012) An analysis of teaching competence in science teachers involved in the design of context-based curriculum materials. *Int. J. Sci. Educ.*, **34**, 701–721.
50. Vos, M.A., Taconis, R., Jochems, W.M., and Pilot, A. (2011) Classroom implementation of context-based chemistry education by teachers: the relation between experiences of teachers and the design of materials. *Int. J. Sci. Educ.*, **33**, 1407–1432.
51. Stolk, M., De Jong, O., Bulte, A., and Pilot, A. (2011) Exploring a framework for professional development in curriculum innovation: empowering teachers for designing context-based chemistry education. *Res. Sci. Educ.*, **41**, 369–388.
52. Stolk, M., Bulte, A., De Jong, O., and Pilot, A. (2012) Evaluating a professional development framework to empower chemistry teachers to design context-based education. *Int. J. Sci. Educ.*, **34**, 1487–1508.
53. Bennett, J., Lubben, F., and Hogarth, S. (2007) Bringing science to life: a synthesis of the research evidence on the effects of context-based and STS approaches to science teaching. *Sci. Educ.*, **91**, 347–370.
54. Justi, R. and Gilbert, J.K. (2002) Science teachers' knowledge about and attitudes towards the use of models and modelling in learning science. *Int. J. Sci. Educ.*, **24**, 1273–1292.
55. Van Driel, J.H. and Verloop, N. (1999) Teachers' knowledge of models and modelling in science. *Int. J. Sci. Educ.*, **21**, 1141–1154.
56. De Jong, O. (1998) *Development of Pre-service Teashers' Knnowledge Base*, Utrecht University, Utrecht (internal research report, May 1998).
57. Justi, R. and Van Driel, J.H. (2005) The development of science teachers' knowledge on models and modelling: promoting, characterizing, and understanding the process. *Int. J. Sci. Educ.*, **27**, 549–573.
58. De Jong, O., Van Driel, J.H., and Verloop, N. (2005) Preservice teachers' pedagogical content knowledge of using particle models in teaching chemistry. *J. Res. Sci. Teach.*, **42**, 947–964.
59. Henze, I., Van Driel, J., and Verloop, N. (2007) The change of science teachers' personal knowledge about teaching models and modelling in the context of science education reform. *Int. J. Sci. Educ.*, **29**, 1819–1846.
60. Barnea, N. and Dori, Y. (2000) Computerized molecular modelling – the new technology for enhancing model perception among chemistry educators and learners. *Chem. Educ. Res. Pract.*, **1**, 109–120.
61. De Jong, O., Blonder, R., and Oversby, J. (2013) in *Teaching Chemistry – A Studybook* (eds I. Eilks and A. Hofstein), Sense Publishers, Rotterdam, Taipei, pp. 97–126.
62. Tekerek, M. and Ercan, O. (2012) Analysis of teachers' attitude towards internet use: example of chemistry teachers. *Creat. Educ.*, **3**, 296–303.
63. Mishra, P. and Koehler, M.J. (2006) Technological pedagogical content knowledge: a framework for teacher knowledge. *Teach. Coll. Rec.*, **108**, 1017–1054.
64. Barnea, N., Dori, Y., and Hofstein, A. (2010) Development and implementation of inquiry-based and computerized-based laboratories: reforming high school chemistry in Israel. *Chem. Educ. Res. Pract.*, **11**, 218–228.

65. Aksela, M. and Lundell, J. (2008) Computer-based molecular modelling: Finnish school teachers' experiences and views. *Chem. Educ. Res. Pract.*, **9**, 301–308.
66. Blonder, R., Jonathan, M., Bar-Dov, Z., Benny, N., Rap, S., and Sakhnini, S. (2013) Can you tube it? Providing chemistry teachers with technological tools and enhancing their self-efficacy beliefs. *Chem. Educ. Res. Pract.*, **14**, 269–285.
67. Gilbert, J.K., De Jong, O., Justi, R., Treagust, D.F., and Van Driel, J.H. (eds) (2002) *Chemical Education: Research-Based Practice*, Kluwer Academic Publishers, Dordrecht.
68. Aksela, M. (2010) Evidence-based teacher education: becoming a life-long research-oriented chemistry teacher? *Chem. Educ. Res. Pract.*, **11**, 84–91.
69. Marks, R. and Eilks, I. (2010) Research-based development of a lesson plan on shower gels and musk fragrances following a socio-critical and problem-oriented approach to chemistry teaching. *Chem. Educ. Res. Pract.*, **11**, 129–141.

6
Lifelong Learning: Approaches to Increasing the Understanding of Chemistry by Everybody
John K. Gilbert and Ana Sofia Afonso

6.1
The Permanent Significance of Chemistry

Those who have studied chemistry extensively know that the key ideas of the subject and the products of chemical technologies (which we will collectively abbreviate to "chemistry") have impacts on all four aspects of the lives of everybody. At the personal level, chemistry underpins the nature of all the things we put on or in our bodies, for example, clothes, cosmetics, food, medicines. At the social level, chemistry determines the nature of many decisions that affect us all, for example, the purification of water. At the economic level, chemistry provides the foundation for many of the industries that are economically important and which therefore produce jobs, for example, the production of microprocessors, of fabrics, of building materials. Last, and not least, chemistry plays a key role, if under-recognized, in cultural life by defining to a substantial extent how we see ourselves in relation to our epistemological and physical surroundings.

These impacts began to be felt in Western Europe from about 1800 CE and have gradually accelerated in scope, in intensity, and in geographical spread. There is every reason to believe that this trend will continue, not least because chemistry continues to produce new themes, most recently nanoscience and nanotechnology. The implication of this trend is that everybody will need access to chemical education throughout their lives. The issues then become what specifically is to be done, and how, if general engagement is to take place.

6.2
Providing Opportunities for the Lifelong Learning of Chemistry

6.2.1
Improving School-Level Formal Chemistry Education

Many, if not most, countries now require students to study a "national curriculum" in science until the end of the statutory period of school attendance. Such curricula normally contain some chemistry. This has too often been found to be fragmented and uninteresting by students and therefore a disincentive to the further study of the subject [1]. Efforts are being made to "slim down" the chemistry curriculum by focusing on "key explanatory stories" [2], concepts that have a broad explanatory value, and by teaching those stories though the study of "contexts," situations, problems, and events that are thought to be of interest to students [3]. The engagement of students with these contexts does seem to increase if the socio-cultural issues to which they give rise – the social aspects of chemical research and the impact of chemistry on people's lives – are emphasized [4]. If generally successful, these innovations ought to improve the likelihood of individuals being more willing to engage in lifelong learning of the subject.

However, the current norms of formal education, including that in chemistry, have particular characteristics that govern the scope and limitations of what can ever be achieved within its framework. The curriculum of school chemistry is fixed and changes only very slowly [5]. New ideas – especially those that represent radical change – conceived on behalf of a central curriculum authority become progressively diluted as they are included in the prescribed curriculum, disseminated to teachers via in-service training courses, enshrined in textbooks, taught in classrooms, and learnt by students. Those involved in each link in the chain interpret the message transmitted in terms of what they already know and believe to be valuable. Effective change can therefore take place only if all the successive links in the chain are overseen by the central curriculum authority: this implies a substantial increase in both organization and funding. Additionally, and most importantly, the key aspects of any desired change must be reflected in any national scheme of student assessment. Students, faced with heavy workloads, will give preference to what they know they have to learn.

Curricular conservatism is to a considerable degree inevitable, given that the major ideas of the fundamentals of chemistry now seem to be firmly established. Additionally, this conservatism does ensure that those major established ideas are introduced to everybody. However, it also has a series of unintended consequences. There is often a failure to address new ideas arising from current research or industrial practice. Moreover, ideas are often presented in an abstract manner, an approach that seems not to lead to transfer of learning to new situations and problems. As the structuring of the content to be learnt is done by the teacher, this means that the sequence followed does not necessarily constitute a "need-to-know" order for many students. The "coverage" of the curriculum is paced to the needs of "average achievers," an approach that often leaves both high and low achievers frustrated. The learning that takes place (or does not take place!) is frequently assessed, invariably for the benefit of the school and not for that of the individual students. The control by the teacher does emphasize those aspects of the curriculum that are deemed vital by her/him, but this it can result in an overlooking of themes, such as the social implications of ideas and technologies that are motivating for individual students [6, p. 9].

Overcoming all these limitations within the confined of a "one-size-fits-all" approach to the chemical curriculum would, almost certainly, result in an increase

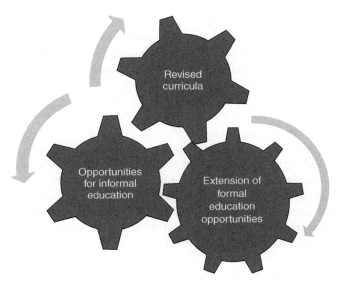

Figure 6.1 Key aspects that facilitate lifelong chemical education.

in the already substantial curriculum load. This might lead to a reduction in the number of other subjects that students could study, which would be culturally undesirable. One approach would be to provide a range of curriculum options in chemistry, depending on the likely future commitment of students to the study of the subject. However, this alternative also has its problems, for it implies decisions that may not be readily altered within the age framework of formal schooling.

There are two other measures that can be taken. The formal curriculum can be made available to adults throughout their lives, so the pressure to learn "everything" in the school years is reduced, and more use can be made of informal educational resources by students of any age. The three measures are summarized in Figure 6.1.

6.2.2
Formal Lifelong Chemical Education

The extension of access of the formal chemistry curriculum to adults falls within an orthodox meaning of "lifelong learning," in that it:

> ... includes people of all ages learning in a variety of contexts – in educational institutions, at work, at home, and through leisure activities. It focuses mainly on adults returning to organised learning than on the initial period of education or on incidental learning [7, p. 10].

The focus on "returning to organized learning" implies attendance at/participation in structured "courses" having a predetermined content.

Much of such provision is made by employers and is therefore focused on the immediate concerns of that employer, and is not accessible to outside scrutiny. The various national "Open Universities" have also made an invaluable contribution to adult chemical education for many years through the provision of student-selected modules that can be accumulated toward the award of a degree, but involvement in such programs does imply a high degree of sustained commitment.

The introduction and rapid uptake since 2011 of "Massive Open Online Courses" (MOOCs) [8] – essentially a dramatic expansion and development of the "open university" idea – by commercial companies, individual universities, and syndicates of universities, both nationally and internationally, is very interesting and seems likely to have a major impact on adult learning including that of chemistry. The genre, the nature of which is evolving as the provision of such courses expands worldwide and very rapidly, cuts across the the distinction between formal and informal types. Some MOOCs can be viewed as "formal" courses as they consist of video recordings of the whole or part of conventional lectures. These video recordings are sometimes interspersed within conventional classroom teaching, to produce the so-called blended course, or are available outside the classroom, to produce the so-called flipped classroom [9]. Inevitably, such materials have a fixed structure, and relate to established themes in orthodox chemical education, for example, "thermodynamics" [10] Thus course requires 8–10 h of study per week for 12 weeks, does not involve a final examination, and is free unless a certificate of completion is required.

6.2.3
Informal Chemical Education

The provision of formal school-level and adult-level chemical education can be augmented by access to informal resources. Informal resources, on whatever subject, have a number of distinct characteristics. Their content spread can be wide and optional, in that learners can choose what they specifically want to know about. Such materials are not provided in extended structures, so that learners can choose when they want to study a topic. It follows that they can be studied at any pace, thus accommodating people whose learning skills are underdeveloped or which have been forgotten. If they include assessment activities, they are for the immediate benefit of the learner, that is, to facilitate further learning. Lastly, and perhaps most importantly, they can focus attention on aspects of a subject that are not commonly addressed in formal educational curricula.

Many MOOC courses [8] fall under the heading of "informal chemical education" in that they often address themes not usually found in the formal chemistry curriculum. They are usually provided in unorthodox ways, for example, "Kitchen Chemistry: Chemistry without a laboratory" [11], which requires 2 h study per week for 6 weeks, and which is free. MOOCs have a number of distinctive characteristics [12]:

- Their uptake is global in distribution and very large, indeed huge, and growing rapidly;
- Their basic requirement is access to a computer and the Internet, plus some computer literacy;
- They can be used flexibly, to fit in with the individual student's requirements, but within a fixed timeframe;
- They involve quizzes and computer-managed assessments;
- All the basic materials needed for learning are provided or cheap and readily accessible sources cited;
- Discussion forums, where interactions between students are facilitated, are a key feature. "Google + Hangout" and "Twitter" feature prominently.

Being a new educational phenomenon, the use MOOCs has not been evaluated yet. The results of small scale studies, for example [13], show what one would expect: students value well-designed teaching materials, the quizzes included, and appropriate assessment questions. Interestingly, although the discussion forums were also valued, some students found their demands overwhelming.

Once all the available slots in the formal chemistry curriculum are available in MOOC form, it seems likely that this type of course will grow rapidly in diversity and number. However, as the current completion rate of all MOOCs, both "formal" and "informal," seems anecdotally to be about 9%, there will be a growth in the use of prior "learning to the study of chemistry" courses. In the meantime, the snags in their management and use are becoming evident, not least in their implicit demands on libraries. A list of the MOOCs in chemistry available in 2013 is given in Ref. [12].

An overview of the elements that contribute to lifelong chemical education is given in Figure 6.2.

6.2.4
Emphases in the Provision of Lifelong Chemical Education

If lifelong chemical education can be facilitated by a combination of revised school curricula, the extension of formal education opportunities to adults, and by the expansion of opportunities for informal education, how might these three approaches be harmonized?

The four-stage model for lifelong learning proposed by Schuller and Watson [7, p. 1] can be adapted to accommodate both a broader age range of learners and a focus on chemical education. This suggests, as a first approximation, the following structure for lifelong chemical education:

1) *Compulsory school age range and up to 25 years*
   The emphasis here would be on ensuring a thorough introduction to the key concepts that underlie chemistry, to the way that they are established, and to the recognition that these have implications for all citizens. This provision can be made primarily by formal educational systems, provided that these emphases are recognized and implemented: a reorientation of chemistry

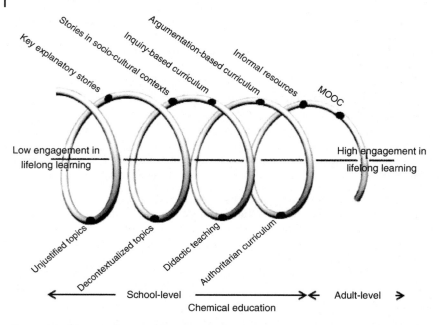

Figure 6.2 The contribution of formal school-level and adult-level provision to lifelong chemical education.

teacher education is called for. This provision is likely to require extensive access to informal resources where necessary.

2) *Age range 25–50 years*

The emphasis here would be on those aspects of chemistry that have particular significance for industry and for the creation and retention of employment opportunities, as these are of major signification for this age group. This emphasis can be provided by the use of tailor-made formal courses and through MOOC provision. The implications of chemistry for the personal, social, and cultural development of the individual would also form a central focus of provision here. These latter needs can best be met by access to a multiplicity of informal resources, as they will be manifest in the very diverse life trajectories of individuals.

3) *Age range 50–75 years*

This age group, which is rapidly rising in demographical significance throughout the world, will want to remain intellectually, socially, and economically active. This implies that they might be interested in chemical education that emphasize recent developments in the field and their social and economic implications. Informal resources will play a dominant role in chemical education for this age group, for whom the major issue is how to "keep in touch" with the concerns of younger people and a rapidly changing world.

4) *Age 75 and over* Although this age group is becoming more numerous, it is very hard to identify those aspects of chemical education in which they would

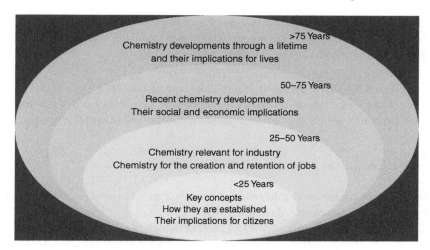

Figure 6.3 Age-related succession of themes.

be most interested. We suggest that they might wish to look back over the developments in chemistry that have taken place during their lifetimes and the changing implications for their lives.

The cumulative relation of the contribution to lifelong chemical education at the different age cohorts is emphasized in Figure 6.3.

This pattern of changing emphases and access carries with it a need to focus of what will be learnt and how that can be facilitated.

6.3
The Content and Presentation of Ideas for Lifelong Chemical Education

6.3.1
The Content of Lifelong Chemical Education

A general overview of chemistry identifies the content of the subject that must be addressed both in school-level chemistry education and in the later provision of "catch-up" learning. Thus

- chemistry has a specific vocabulary that is needed when discussing phenomena with a chemical dimension;
- chemical phenomena can be explained in terms of the structures of the entities (atoms, ions, free radicals, molecules) of which they are composed;
- chemical reactions proceed with characteristic dynamics that depend on the energy changes involved;
- chemical ideas have a role in explaining many everyday phenomena, including those that involve personal, social, and economic decision taking;

- the understanding of the world-as-experienced provided by chemistry advances in ways similar to those of the other sciences, into which the history of the subject can provide considerable insights.

Although it is beyond the remit of this chapter to discuss that process, chemical educators with a sound and extensive knowledge of chemistry and of learning should be able to design accurate and attractive courses and informal resources.

6.3.2
The Presentation of Chemistry to Diverse Populations

The design of courses and informal resources has to address three challenges which arise from the nature of chemistry itself.

The first challenge arises from the fact that the models of phenomena that make chemistry a separate and distinctive science are abstract in nature, for example, those of atomic structure, bonding, chemical energetics, and chemical kinetics. The meaning of these words, and the way that they relate to the world as experience, are difficult to understand by many, if not most, people. The obvious solution to this problem is to make extensive use of representations to assist in the mental visualization of these ideas, for such visualizations are an essential ingredient in all thinking [14]. Representations can be created in a wide variety of modes: the material mode, for example, the ball-and-stick representations of crystal structures; the visual mode, generally known as *diagrams*, for example, of the Krebs cycle; the gestural mode, the use of body posture to emphasize aspects of an explanation, for example, the relative distance between a series of entities; and the symbolic mode, of which the "chemical equation" is an exemplar [15]. It is no overemphasis to say that the selection of suitable types of representations and their use in relation to text are essential ingredients of chemical explanations provided to any audience [16].

The second challenge arises because most of the chemicals used in experiments are toxic, corrosive, or inflammable, and almost invariably expensive. This means that direct experience of chemical phenomena must inevitably be limited in extent and confined to simple examples, for example, the testing of acidity with indicator paper. As experience of a phenomenon is regarded as a highly desirable phase of the learning of science [17], indirect experience must be provided instead. Video recordings of demonstration experiments are the most convenient way of doing so.

The third challenge arises because many of the contexts within which chemical ideas were historically developed are often either not of interest to people today, for example, the electrolysis of a dilute acid, or are now regarded as unethical, for example, the demonstration of the composition of air by the exposure of birds to a partial vacuum. There can be no simple recipe for the identification of "contexts" that would meet the two criteria of relevance and ethics. Students must already know something about candidate contexts, a prerequisite for being interested in

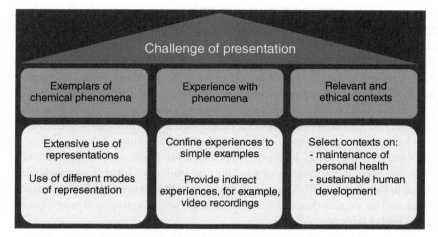

Figure 6.4 Challenge of representing chemical ideas.

them, and they must be such that the concepts being taught are an essential ingredient in understanding those contexts. There are two themes that seem likely to figure highly in the selection of contexts for study over the course of a lifetime. The first of these is the maintenance of personal health [7, p. 4]. Although many countries now provide medical services that are either free or subsidized at the point of access, an increasing general emphasis is being placed on "preventative medicine," which implies continued personal self-education. The second of these is "sustainable human development", defined as involving learning about

> the role of technological and economic tools in shifting individuals, group, and industry activities towards a more sustainable path of economic development [18, p. 3].

Chemical education has a clear role to play in lifelong education for sustainability, the need for which will become increasingly evident as the twenty-first century progresses [19].

The challenges to the presentation of ideas if lifelong chemical education is to be effectively provided are summarized in Figure 6.4.

However, these three aspects of lifelong learning will be realized in practice only if individuals meet ideas in ways that are conducive to their being learnt.

6.4 Pedagogy to Support Lifelong Learning

The established view amongst the adult education community is that adults learn most effectively in particular circumstances. First is when they are able to concentrate on what they think they need to learn. Second is when they have had direct practical experience of a phenomenon or problem, for this gives them a realistic focus for their learning. Third is when they are able to use those elements of their extensive fund of experience that are relevant to the task. Fourth is when they are able to learn how and when they wish. Known as *andragogy* [20], this view has validity when compared to most learning by school-age students who, as we have outlined earlier, are required to do so in ways dictated by the formal educational system. While accepting that young people have to learn the basic ideas of science and that the formal system has been established for them to do just that, they could also learn a great deal by engaging in informal learning if they are allowed and encouraged to do so. In short, the principles of informal learning and andragogy are very similar: both young people and adults can learn to a considerable extent in the same way. This view suggests that the foundations of lifelong learning could, and we suggest should, be laid down in the school system and progressively developed in later life.

A meta-analysis of research [21, pp. 113–125] provides evidence that all learning – of whatever age and experience of the learner – best takes place in particular social circumstances. That is, if

- enough time is allowed for it to take place. Most people can only learn a particular, personal, rate;
- the learner can be motivated to make an active mental effort to learn. That is, the person must want to know more about the topic under consideration;
- it is recognized that an individual's attention spans are short. Most people lose their concentration on a particular task quite soon;
- it provides for practice in recalling and applying new ideas to be spread throughout the period of study. "Taking stock" of what has been learn is a necessary part of memorizing something;
- it is recognized that, having elected to learn something, the learner very probably have some prior knowledge of the subject that will influence the learning that then takes place. The phenomenon of "alternative conceptions" has been established for many years [22];
- strategies to make the material to be learned memorable (e.g., summaries, analogies) are included in its presentation. These enable a person to mentally process material in a number of complementary ways, and are especially important for the theme of informal lifelong learning of chemistry;
- the material to be learnt can be experienced through a range of media, for this enables words and visualizations to be interrelated. We deal with this issue in the next section.

It then follows that all the media used in the presentation of chemical ideas to anybody must meet all the above conditions.

6.5
Criteria for the Selection of Media for Lifelong Chemical Education

There are a finite number of types of media, either singly or in combination, that can serve as possible substantial contributors of lifelong learning. The selection of any one of them does depend on its inherent characteristic: what it can do well. So, while all of them must meet the above criteria for the support of good learning, there are a number of other considerations. These are that any medium chosen must be

- readily available to all those who might want to use it;
- able to meet the needs of any specific interest sector of the population;
- usable either by individuals, small groups, or large groups;
- capable of presenting complex ideas, if necessary by introducing simplifications using visualizations and analogies;
- capable of relating possible prior knowledge to what is to be learnt;
- capable of relating the science and technology to any relevant personal, social, or economic interests of learners;
- able to present a range of exemplars of relevant contexts; and
- able to show the relationship between science and any associated technology.

In the following seven sections, we present outlines of some media that might meet these criteria and therefore have a substantial role in conveying the ideas of chemistry. We also provide a preliminary evaluation of their scope and limitations for this purpose. The treatments are not of equal weight, for the general natures of some of them (museums, science centers) have been extensively reviewed elsewhere [23]. We have concentrated on those that are less widely known at the moment: the Internet and other modern electronic media, popular books, graphic novels, and citizen science.

6.6
Science Museums and Science Centers

Museums, together with planetariums, zoos, aquariums, botanical gardens, and science centers, constitute what [23, p. 127] we call *designed settings* for informal learning. Of these, only science museums and science centers have any major relevance to the lifelong learning of chemistry.

6.6.1
Museums

Historically, science museums were created to house collections of objects of scientific interest. They still strongly maintain that tradition, but have added an educational dimension that has steadily grown in importance over the last century or so. They can be defined as consisting of a collection of static exhibits, often grouped together around historical or taxonomic themes. As museums are often in city centers, they can usually be reached by means of public transport and are therefore accessible to groups of a wide range of sizes. Although every effort is made to provide textual panels that explain exhibits, it is not usual for them to have a specific declared educational purpose and they, therefore, do not explicitly relate to visitor's prior knowledge.

The exhibits about chemistry in science museums usually consist of samples of substances of interest, for example, slices of silicon used in semiconductors, or of animated displays of chemical processes, for example, of the industrial plant used for the fractional distillation of oil. Issues of cost, safety, and feasibility limit the range of exhibits to this narrow band, which means that science museums can, at present, have only a limited contribution to make to the lifelong chemical education.

6.6.2
Science Centers

Science centers consist, like museums, of a series of individual exhibits. However, unlike those in conventional museums, the exhibits in science centers are interactive: they require the visitor to take a specific action, and the exhibit then produces an automatic response to what has been done. Science centers are intended both to provide the visitor entertainment and also to be educational. Although originally established as separate institutions, science centers are increasingly attached to conventional museums. Indeed, interactive exhibits are often placed with conventional exhibits on the same theme, increasing the educational value of both if the link between them is explicit or self-evident.

The range of chemical phenomena that can form the basis of interactive exhibits is small. Chemical reactions tend to be very fast or very slow, thus taxing the patience of the visitor: it is only the use of those that produce a reaction in a few seconds that seem likely to hold the visitors. Relatively few chemical phenomena – typically those involving the transition-group metals – produce visually dramatic effects. Lastly, usable exhibits have to be both physically robust and cheap to run, not characteristics that one associates with chemical equipment.

6.7
Print Media: Newspapers and Magazines

Print media play diverse roles in lifelong chemical education. For newspapers and magazines, this role is small but important. The proportion of the content of those daily newspapers that is devoted to scientific topics, in general, is small in many countries but is slowly increasing [24, 25]. Although inclined toward headline-grabbing, sensational stories, articles concerned with specific matters likely to be of interest to adults, for example, global warming [26], appear from time to time. Collectively, newspaper articles alert the public to issues of personal, social, and economic importance that in any way concern chemistry, for example, the nature and impact of new medicines.

Few magazines carry any articles that are overtly related to chemistry, the exception being specialist periodicals such as *New Scientist* and *Scientific American*, although these are only really accessible to readers who are already "scientifically literate" [27].

6.8
Print Media: Popular Books

"Popular" books are those written for the science education of the general public [28]. They are sold through a wide variety of outlets, for example, supermarkets. They provide an extended treatment of topics thought by the authors to be of general interest, while their very nature enables the reader to concentrate on a particularly interesting or difficult passage over a sustained period by re-reading and reconsidering them. This potential has been tapped for books on a wide variety of themes, of which we have identified eight [29]:

- *Chemical cookbook/hands-on chemistry.* These books present chemical reactions and phenomena that can safely be explored at home.
- *Science fiction with an emphasis on chemistry.* These are books of fiction, including those about crime detection, using real or imagined chemical ideas.
- *Foundation ideas in chemistry.* These books deal with the core ideas of chemistry, presented in such a way as to be understood by a person without a background in the subject.
- *History of developments in chemistry.* Such books deal with how the subject of chemistry has emerged and developed over the years.
- *The use of chemistry in other disciplinary fields.* These books discuss how particular chemical ideas make an important, often central, contribution to research and development in other disciplinary fields such as genetics, pharmacy, materials science, nanotechnology, and to the chemical technologies in general.
- *Chemistry in everyday life.* Such books deal with the way that chemistry contributes to an understanding of everyday personal life (e.g., cooking) and social life (e.g., forensic science).
- *Particular chemicals.* This broad category of books addresses the chemistry, applications, and implication of particular substances and chemical species, including those of contemporary importance in both basic research and in chemical technologies.
- *Biographies of chemists.* The books in this category recount the contributions to science made by individual chemists, often setting these against the background of the person's life trajectory.

Our recent analysis of the current catalogs of publishers in English is summarized in Table 6.1.

Table 6.1 Informal chemistry books available in June 2013.

Category	Number of titles ($N = 217$)
Chemical cookbook/hands-on chemistry	5
Science fiction with an emphasis on chemistry	13
Foundation ideas in chemistry	14
History of developments in chemistry	27
The use of chemistry in other disciplinary fields	29
Particular chemicals	33
Biographies of chemists	44
Chemistry in everyday life	52

This analysis illustrates the great versatility of the genre, in particular the capability, to address the history and philosophy of chemistry and the implications of the subject for personal, social, and economic life. The category "Foundation ideas in chemistry" contains items that must be of great importance in the provision of lifelong learning for individuals who need access to basic ideas.

6.9
Printed Media: Cartoons, Comics, and Graphic Novels

6.9.1
Three Allied Genre

While all formal educational systems aspire to lead to universal literacy, there is always a proportion of any population who either cannot read, are disinclined to read, or would welcome a change from the normal form of texts. In addition to fully exploiting the inherent pedagogic value of visualizations, this genre as a whole, consisting of the overlapping categories of cartoons, comics, and graphic novels, does provide a step on the road to chemical education for some people.

The basic entity of the genre consists of a simple image of a place, people, or an event in framed boxes and includes the concise use of a few words, the meanings of which are made clear. *Cartoons* usually consist of only one image, which is free-standing, while *comics* consist of a series of images, called *panels*, typically arranged in horizontal strips, are sequential and

> ... have as one of their main aims to communicate science or to educate the reader about some non-fictional, scientific concept or theme, even if this means using fictional narratives to convey the non-fictional information [30].

Comic books are book-length or magazine-length narratives based on comic strips [31]. The term *graphic novel* describes an extension of this form, as discussed below. Although the literature tends to conflate the use of the three terms, it is possible to identify the nature and value of the "graphic novel," for this would perhaps be the most valuable form with which to in address either a series of themes or one theme in some depth.

6.9.2
The Graphic Novel

The term *graphic novel* emerged in the 1970s with the wish of publishers to try out new styles, sophisticated formats, or new storylines, so as to create comic books for adults covering themes such as terrorism [31]. The term is now used to describe any book-length publication that makes use of drawn images along with a narrative [32]. Graphic novels make use of all the different modes of visual external representation, namely gestural, verbal, visual, and symbolic (e.g., short broken lines indicating jumping). In a graphic novel, the relation between visualizations and the accompanying text is complex: visualizations often carry information that extend the text and that are not present in the text (e.g., facial and body language, thoughts appearing in "clouds"), and the text often conveys information not presented in the visualizations. When reading a graphic novel, readers need to infer synchronically meanings that emerge from both the visualizations and text and to follow the flow of panels as they carry the narrative [33].

Few empirical studies have analyzed what constitutes a good graphic novel or what shortcomings in other print media might be met by a graphic novel. A famous writer of science graphic novels [34] reported, when interviewed, that in a good graphic novel there is an effective interplay of images, text, and narrative. In this interplay, words and pictures have different roles; the text is embedded in the image, forcing the reading of the visual material; some anthropomorphism is used, so that animals, plants, cells, or molecules become the protagonists of the stories; and images are part of the story so that, in order to comprehend the narrative, the images cannot be skipped.

The form of a graphic novel rather than its content is what makes it capable of overcoming the limitations of other print media. Sabeti [35] carried out a study of five students (16-year-old) attending a graphic novel reading group as an extracurricular school activity. The data showed that, when compared with image-free books, graphic novels

- are less detailed than plain text, but in a theme it is easier to identify the information on which the reader should focus attention. Among the strategies used by creators of comics to highlight the relevant information are repetition of a motive, or detailing of one thing while another remains in sketch only;
- allow reading to be slowed down. While in an image-free text the writer carries the reader though a story by means of sentences, in a graphic novel the reader can stop at each panel and take his/her time to engage in an analysis of a particular image;

Table 6.2 Graphic novels about chemistry in English available in June 2013.

Categories	N = 24
Chemical cookbook/hands-on chemistry	0
Science fiction with an emphasis on chemistry	4
Foundation ideas in chemistry	6
History of developments in chemistry	1
The use of chemistry in other disciplines	4
Particular chemicals	0
Biographies of chemists	4
Chemistry in everyday life	5

- produce a better memorization of a scene and a better recall of the thoughts that occurred during the processing of reading;
- are easy to discuss since it seems easier to talk about images than about a written text;
- require a shorter attention span for the major message to be conveyed.

There are several sub-genres of graphic novels including both fictional and non-fiction types. Non-fiction graphic novels on science and technology themes have often addressed pressing social issues. For example, graphic novels specifically concerned with chemistry, such as atomic energy, rocket propulsion, and their exploitation in Earth-orbiting satellites, inspired the publication in the 1950s of the special series "Adventures in Science" [36]. More recently, the need to increase public awareness of illness such as AIDS, cancer, or diabetes has led to the publication of graphic novels on these themes [37].

While the range of straightforward chemical themes in graphic novel format is fairly wide, the coverage is fairly shallow. When the categories of chemical content outlined earlier are applied to the graphic novels that are currently available in English in publishers' lists, the result is Table 6.2.

This analysis confirms that relatively few graphic novels on a chemistry theme exist. The form has educational potential, perhaps most significantly on themes of high social salience, such as health issues and environmental protection, while also being capable of addressing fundamental chemical ideas.

6.9.3
The Educational Use of Graphic Novels in Science Education

Graphic novels have been used for educational and training purpose in science and technology contexts for many years. For example, during World War II, the Pentagon used comic book artists to design a variety of military training manuals [38]. They have since been seen as a potential medium for school science education because they are popular among children and teenagers. They motivate reading, their content is easily memorable, they make the whole process of learning fun [39], they can target students having a low literacy level, they can promote social cohesion (e.g., by engaging students in creating their own comics), they can be useful in conveying complex political and moral issues to young readers, and they can be linked to the content of the curriculum [35]. The fictional sub-genre of graphic novels does often contain bits of scientific information [36] but also a large amount of misleading pseudoscience. Nevertheless, this sub-genre can be of pedagogical use, for example, by encouraging students to consult reference books to critically evaluate them [40].

The educational effectiveness of graphic novels has been evaluated only occasionally and then for their use with university students. Hosler and Boomer [39] analyzed how they can affect student learning and attitudes about biology. A graphic novel produced by the above authors covers the theme of evolution, which was introduced into a science course for a mixture of science major and non-science major students. During the course, the students were required to read specific passages of the graphic novel and to discuss the story and the science beyond it. Data collected through pre and post-instruments showed an increase in content knowledge for both groups and a statistically significant improvement in the non-science major students' attitudes toward biology. The authors of the paper were cautious about the results, believing that it was unclear whether the attitude improvement was specifically the result of using a graphic novel or merely a result of the "Hawthorne Effect": any innovation in teaching produces a short-term improvement in attitudes toward the subject.

In medical courses, graphic pathographies (i.e., graphic novels that tell stories of the impact of an illness on patients, families, and caregivers) have been used in order to expand medical students' perspectives on a disease from the simple biological to include other perspectives [37, 41]. Medical students also became aware of patients' concerns and misconceptions about disease and their treatment, which might not be mentioned in the usual clinical setting. Some authors [37, 41] took advantage of graphic pathographies in a medical course by asking fourth-year medical students to reflect on how they might convey the experience of illness, to analyze how patients and their families communicate with the medical system, and to develop their own story in a graphic novel format.

Science graphic novels have also been considered an efficient medium for science communication in a broader sense. In particular, graphic pathographies, by reporting personal narratives on how someone struggled with a disease, could help patients to learn more about an illness and its treatment, to trigger questions to be discussed with the doctor; and to identify a community of similarly affected people [37]. They can be a valuable means of communication for the less text-literate patients or to those that are more attached to the visual genre [42].

A brief case study will illustrate the use of such graphic novels in the education of adults.

6.9.4
Case Study: A Graphic Novel Concerned with Cancer Chemotherapy

The graphic novel *Cancer Vixen* [43] is based on the real-life experience of the author (Marisa Aconcella Marchetto) and recounts her personal experience of contacting and surviving treatment for cancer of the breast. The book reports how the author receives surgery, chemotherapy, and radiotherapy. During her trajectory, she meets several care providers, such as a radiologist, an oncologist, and a nurse, and experience several emotions from anxiety (while waiting for the diagnosis), through struggle with the side effects of the treatment, to grief (when she found out that she became infertile). Chemistry is embedded in the story, although at a very superficial level.

A graphic novel on this theme has potentially wide application because 1.45 million new cases of cancer currently occur yearly in developed countries (records elsewhere being uncertain). Chemistry plays a vital role in the design and development of antitumor drugs to slow down the growth of, or kill, malignant tumor cells. Information about cancer chemotherapy is important for patients because it contributes to a reduction in anxiety, allows informed decision taking, and supports a feeling of satisfaction with the choice of treatment made. However, not all patients search for information beyond that provided by the health systems. Those who do exhibit seeking behavior are often high educated, younger in age, and married, but their desire for information decreases 3–6 months after diagnosis [44]. Another group of people that also benefit from such information are carers and family members, for they need to monitor and manage the side effects of any treatment being received [45].

While little is known about the knowledge needs of patients, family, and carers about cancer chemotherapy [44], there are widespread myths and misconceptions surrounding cancer treatments, such as follows: chemotherapy is poisonous; there is a cure for cancer, but pharmaceutical companies are suppressing it; cancer can be cured by ingesting apricot seeds, baking soda, or combinations of products from herbs to vitamins [46]. Graphic novels have a potentially large educational role in dispelling such misconceptions and misinformation. However, there are no studies on the types of patient experiences for which graphic novels could achieve particular educational goals [42]. The expansion of the genre into chemistry more generally seems both possible and called for.

6.10
Radio and Television

Radio broadcasting can include chemistry-related material in a wide variety of formats: news programs, science "magazines," dramas, and documentaries. The strength of the medium is that it allows the contribution of experts to be included in "interactive" formats with a presenter, or even with "phone-ins" from members of the public. The weakness of the medium is the absence of visual images, but these are often not needed for the presentation of the applications and implications of chemistry.

Radio is very widely accessible and can be either of a "broadcasting" type, aimed at very large populations, for example, of mainland China, or of a "narrowcasting" format, where the needs of a specific or geographically dispersed audience can be met, for example, of those living in the "outback" of Australia. The educational efficacy of radio has been little researched, although some works [47] do provide a sympathetic overview of its potential for science education generally.

The capability of television to project images at a human scale of very small things (e.g., , viruses) and very large objects (e.g., chemical production complexes), that are both nearby (e.g., down a microscope) and very far away (e.g., on the sun), all in full color, means that the medium should make a major contribution to lifelong chemical education. The contribution of television to the education of adults has not been systematically researched [6, p. 20], but there are some pointers to "good practice." A very large scale study in the United States showed that newscasts can make a valuable contribution, especially when concerned with health issues [48]. Documentaries about wildlife always attract very high viewing figures [49]. Popular television dramas that feature scientific ideas also have an impact, for example, the internationally popular series *Crime Scene Investigations* (CSIs) [23, pp. 259–260]. In all these cases, the secret of success is the provision of a narrative built around scientific ideas. As Dhingra [50] points out,

> The function of television is in telling stories. Its intersection with science as a collection of stories about people, their collaborations, controversies, disputes, and ideas, and with television practitioners as institutions and people with their own constraints and preferences, merit continued attention (p. 118).

As has already been made clear, the greatest problem in facilitating informal chemical education is in providing access to chemical phenomena. While the direct experience of chemical phenomena is extremely difficult to provide in any circumstance, indirect provision through television offers a partial but acceptable substitute.

6.11
Digital Environments

The explosive growth in the private ownership of computers throughout the world along with their use both to gain access to and to create knowledge has been one of the major social events in the last decade. The pace of change shows no signs of slowing down, making it impossible to provide, with conviction, an overview of its implications for the lifelong learning of chemistry. Of one thing we can be certain: those implications will be profound for science education in general [23, pp. 261–264].

The Internet was the first digital environment to become available. It is capable of meeting all the criteria for a successful source of informal learning of chemistry set out in Section 6.5. However, as there are no general quality controls applied to material that is to be placed on a Web site, the information provided on sites within a common Web theme may be of varying quality. The most reliable information will be provided by established sources, that is, where the staff are known to be experts in some aspect of science/technology education and where statements made have been subjected to validation, for example, those provided by the Science Museum in London (*http://www.sciencemuseum.org.uk/online*). This implies that opportunities for "Internet education" will be needed so that adults will be able to better judge the relative worth of what they find in the offerings of very diverse providers.

Most of the research that has been done on the use of the Internet in respect of science education has concentrated so far on its use with and by school-age children. A recent review [51] of the 65 papers on the theme that have appeared in journals entered in the Social Science Citation Index between 1995 and 2008 produced evidence that use of the Internet was associated with improved attitudes toward science and better conceptual learning and problem solving abilities, especially amongst higher achieving students.

The use of other digital media is growing apace, their common characteristic being the interactivity that they make possible with the game format, particularly when involving other players. Education-oriented computer games are a popular sub-genre, the learning they facilitate being greatest when they (i) require extensive mental activity by the gamer, (ii) provide surrogate experience of people, places, or events, (iii) are placed in a definite physical or social context, and (iv) are based on an address to a definite problem, with the program providing immediate feedback to the gamers [52]. Role play is an allied sub-genre, in which several players contribute interactively to the evolution of environments and activities [53].

The recent evolution of the hand-held microcomputer seems very likely to lead to a very rapid expansion of the range of games and role play available. These new ideas are making their mark rapidly, for the main characteristic of anything to do with microelectronics is the narrow time gap between conception and mass production [54].

One of the major forces in the expanded use of digital environments lies in the recent advent and rapid expansion of MOOCs, including those in chemistry, as was discussed earlier in this chapter. A rationale for this approach to learning has been put forward as the theory known as *connectivism* [55]. Resting on the ideas of Vygotsky [56], connectivism asserts that knowledge is produced in an individual by interactions with other people and with physical resources, mediated by the use of communication technologies. MOOCs, particularly those concerned with issues that lie outside the orthodox canon of science, for example, those concerned with the social implications of new chemical technologies, seem likely to make considerable us of these technologies. The particular techniques used can include computer- and phone-mediated interactive forums with academics/other students, the use of the video conferencing facility "Skype," group address to both preset problems and to creative projects, and engagement in podcast activities.

While the adoption of new digital technologies by older people certainly lags behind that of younger people, there seems every indication to the authors of this chapter that the gap is closing rapidly, if unevenly.

6.12 Citizen Science

There is evidence that the best way for science teachers to learn about "nature of science" is by a combination of direct instruction and reflection on the processes of their involvement in scientific inquiries [57]. It seems possible that this combination would be suitable for anybody, including school-age children or members of the general public, in the context of life-long learning. Where it is desired to learn about the nature of chemistry, the use of the Internet would probably provide the most convenient access to the knowledge component. The problem then arises in how to facilitate actual practical enquiry in which chemical ideas played a central role and skills were deployed, given the risks associated in handling laboratory-grade chemicals. "Citizen science with a chemistry focus" does provide a limited alternative, albeit one that has not been explored extensively so far.

Citizen science involves volunteers carrying out scientific investigations, often in collaboration with professional scientists [23, pp. 189–192]. The idea has gained considerable momentum, such that a central database, associated with a Citizen Science Association, exists for existing activities in the United States [58]. Environmental monitoring, whether of the soil, water, or air, provides the theme of the majority of projects, which suggests that chemistry-themed projects, for example, of ground water quality, could readily be mounted, for example, as recorded by Overdevest et al. [59]. Little research has been done into the impact of citizen science projects on the knowledge and attitudes of those involved, but the evidence that is available suggests that the outcomes are positive [60]. The latter concluded, for the two programs studied in detail, that

> … one (was) explicitly linked to environmental stewardship and the other more closely associated with a basic scientific mission. These differences in primary goal may impact on who chooses to participate----as well as on the particular knowledge and skills they develop through participation [60, p. 192].

There is extensive data about citizen science researchers available from a large-scale and long-duration project to measure precipitation (rain, hail, snow) in the United States, a topic that has a direct link to environmental chemistry [61]. It seems that sustained and successful engagement in this project occurred as a result of the intersection of several factors. Many of the participants were educated to degree level and were retirees. The daily measurements required

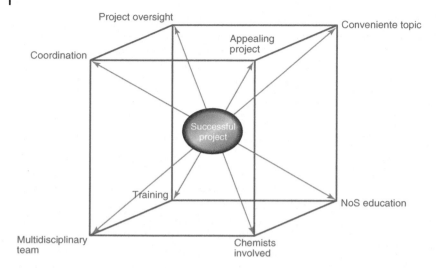

Figure 6.5 Aspects of successful citizen science.

could readily be taken using the equipment provided and about which they had received training. Perhaps most importantly, all participants were kept in touch with regularly by a podcast that included all recently acquired data.

General approaches to the organization and conduct of successful citizen science projects that endorse the above case study have emerged [62]. The new hand-held mobile phones should provide further impetus to the approach, for they facilitate the collection and aggregation of data from a range of individuals "in the field." The major challenges that have to be faced if the approach is to be used more widely in lifelong chemical education are threefold: increasing the willingness of professional scientists – here, chemists – to play a leading role in projects; persuading those chemists to work in multidisciplinary teams with other scientists, for "environmental chemistry" is likely to be a major focus; designing projects that both appeal to younger adults and which can be readily accommodated in their existing lifestyles. A summary of the factors that contribute to successful citizen science projects is given in Figure 6.5.

6.13
An Overview: Bringing About Better Opportunities for Lifelong Chemical Education

There is much to be done if lifelong chemical education is to be made available to and participated in by people of all ages and backgrounds. For example, in the United Kingdom, the proportion of Social Class A/B who currently engage in any form of adult education (53%) is far greater than that for Social Class D/E (24%) [7, p. 66]. Also for the United Kingdom, while the overall participation for men and women is roughly equal, the ratio varies between different ethnic groups

[7, pp. 70–71]. These sorts of disparities, which we suspect are roughly applicable elsewhere, are not acceptable, most certainly in chemical education.

There are many opportunities that, if taken singly or collectively, should lead to a drastic improvement in these situations. The social and technological trends of the twenty-first century, as currently manifest in everyday life, point to an increased need to know about chemistry. An emphasis on environmental education for sustainability, together with health issues, in public life is the foremost of these. The notion of "lifelong education" is, in itself, now widely accepted, so the scene is set for a steady expansion of chemical education. This acceptance is being aided by the slow – perhaps unduly slow – evolution of the school chemistry curriculum.

A wide range of media is now available for everybody, on all topics, whenever and wherever wanted. Figure 6.6 represents a rough estimate of the particular value of each medium for different purposes.

Realizing these opportunities does mean, inevitably, that a series of challenges have to be met. Much more needs to be known about the type and effectiveness of chemical education provided by some of the new media, for example, graphic novels, popular books, and citizen science. The advent of MOOCs has, by analogy, all the characteristics of an epidemic: they are spreading rapidly, yet we currently know little about their significance or how to manage them. A great

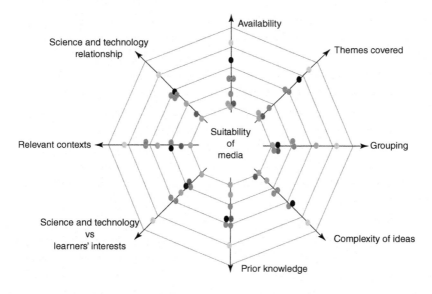

- Cartoons, comics and graphic novels
- Digital environments
- Magazies
- Newspapers
- Popular science books
- Radio
- Science museum and science centers
- Television

Figure 6.6 Value of different media for varied aspects of lifelong chemical education.

deal is known about what is successful in science education in general: the extensive use of representations, the framing of topics within contexts, the adoption of suitable pedagogies, and the matching of the demands of materials to the needs and wants of learners. The major challenge must be to get this knowledge validated in respect of chemistry, widely available, and consistently acted upon by chemical educators.

References

1. Osborne, J. and Collins, S. (2001) Pupils' views of the role and value of the science curriculum: a focus group study. *Int. J. Sci. Educ.*, **23** (5), 441–468.
2. Millar, R. and Osborne, J. (1998) *Beyond 2000*, King's College London, London.
3. Gilbert, J.K. (2006) On the nature of 'context' in chemical education. *Int. J. Sci. Educ.*, **28** (9), 957–976.
4. Sadler, T. (2009) Situated learning in science education: socioscientific issues as contexts for practice. *Stud. Sci. Educ.*, **45**, 1–42.
5. Van den Akker, J. (1998) in *International Handbook of Science Education* (eds B.F. Fraser and K.G. Tobin), Kluwer Academic Publishers, Dordrecht, pp. 421–448.
6. Stocklmayer, S., Rennie, L., and Gilbert, J. (2010) The roles of the formal and informal sectors in the provision of effective science education. *Stud. Sci. Educ.*, **46** (1), 1–44.
7. Schuller, T. and Watson, D. (2009) *Learning Through Life: Inquiry into the Future of Lifelong Learning*, National Institute of Adult Continung Education, Leicester.
8. MOOC (2013) MOOC List, http://www.mooc-list.com (accessed 23 November 2013).
9. Pienta, N. (2013) Online courses in chemistry: salvation or downfall? *J. Chem. Educ.*, **80** (11), 1244–1246.
10. Mooc List (2014) edx. Thermodynamics, http://www.mooc-list.com/course/me209x-thermodynamics-edx.
11. FutureLearn (2014) Kitchen Chemistry, https://www.class-central.com/mooc/1757/futurelearn-kitchen-chemistry-without-a-laboratory.
12. Leontyev, A. and Baranov, D. (2013) Massive open online courses in chemistry: a comparative overview of platforms and features. *J. Chem. Educ.*, **90**, 1533–1539.
13. Zutshi, S., O'Hare, S., and Rodafinos, A. (2013) Experiences in MOOCs: the perspective of students. *Am. J. Distance Educ.*, **27**, 218–227.
14. Gilbert, J.K., Boulter, C.J., and Rutherford, M. (2000) in *Developing Models in Science Education* (eds J.K. Gilbert and C.J. Boulter), Kluwer Academic Publishers, Dordrecht, pp. 193–208.
15. Gilbert, J.K. and Treagust, D.F. (2009) *Multiple Representations in Chemical Education*, Springer, Dordrecht.
16. Eilam, B. and Gilbert, J.K. (2014) *Science Teachers' Use of Visual Representations*, Springer, Dordrecht.
17. Hofstein, A. and Lunetta, V. (1982) The role of the laboratory in science teaching: neglected aspects of research. *Rev. Educ. Res.*, **52**, 201–217.
18. Fien, J. and Tilbury, D. (2002) The global challenge of sustainability, in *Education nd Sustainability: Responding to the Global Challenge* (eds D. Tilbury et al.), Commission on Education and Communication, IUCN, Gland.
19. Burmeister, M. and Eilks, I. (2012) An example of learning about plastics and their evaluation as a contribution to education for sustainability in secondary school chemistry teaching. *Chem. Educ. Res. Pract.*, **13** (2), 93–102.
20. Knowles, M., Holton, E., and Swanson, K. (2011) *The Adult Learner*, Elsevier, Oxford.
21. Hattie, J. and Yates, G. (2013) *Visible Learning and the Science of How We Learn*, Routledge, London.

22. Gilbert, J.K. and Watts, D.M. (1983) Conceptions, misconceptions, and alternative conceptions: changing perspectives in science education. *Stud. Sci. Educ.*, **10**, 61–98.
23. Bell, B. *et al.* (2009) *Learning Science in Informal Environments: People, Places, and Pursuits*, The National Academies Press, Washington, DC.
24. Metcalfe, J. and Gascoigne, T. (1995) Science journalism in Australia. *Public Underst. Sci.*, **4**, 411–428.
25. Pellenchia, M. (1997) Trends in science coverage: a content analysis of three US newspapers. *Public Underst. Sci.*, **6**, 49–68.
26. Stamm, R., Clark, F., and Eblacas, P. (2000) Mass communication and public understanding of environmental problems: the case of global warming. *Public Underst. Sci.*, **9**, 219–237.
27. Roberts, D. (2007) in *Handbook of Research on Science Education* (eds S. Abell and N. Lederman), Erlbaum, Mahwah, NJ, pp. 729–780.
28. MacPherson, S. and Della Sala, S. (2008) *Reviews of Popular Science Books*, vol. 44, Cortex, p. 763.
29. Afonso, A. and Gilbert, J.K. (2013) The role of 'popular' books in informal chemical education. *Int. J. Sci. Educ.*, **3** (1), 77–99.
30. Tatalovic, M. (2009) Science comics as tools for science education and communication: a brief, exploratory, study. *J. Sci. Commun.*, **8** (4), 1–16.
31. Denesi, M. (2103) in *Encyclopedia of Media and Communication* (eds M. Danesi *et al.*), University of Toronto Press, Toronto, pp. 165–167.
32. Masuchika, G. and Boldt, G. (2010) Japanese manga in translation and American graphic novels: a preliminary examination of the collections in 44 academic libraries. *J. Acad. Librariansh.*, **36** (6), 511–517.
33. Goldsmith, E. (1984) *Research into illustration: An approach and a review*, Cambridge University Press, Cambridge.
34. Meier, J. (2012) Science graphic novels for academic libraries: collections and collaborations. *College Research Library News*, **73**, 662–665, in press.
35. Sabeti, S. (2011) The irony of the 'cool club': the place of comic book reading in schools. *J. Graphic Novels Comics*, **2** (2), 137–149.
36. Carter, H. (1989) Chemistry in comics, part 2: classical chemistry. *J. Chem. Educ.*, **66** (2), 118–127.
37. Green, M. and Myers, K. (2010) Graphic medicine: use of comics in medical education and patient care. *Br. Med. J.*, **340**, 574–577.
38. Lavin, M. (1998) Comic books and graphic novels for libraries: what to buy. *Serials Rev.*, **24** (2), 31–45.
39. Hosler, J. and Boomer, K. (2011) Are comic books an effective way to engage non-majors in learning and appreciating science? *Life Sci. Educ.*, **10**, 309–317.
40. Kauffman, G. (1992) Chemistry in comics. *J. Chem. Educ.*, **62** (1), 83.
41. Green, M. (2013) Teaching with comics: a course for fourth-year medical students. *J. Med. Humanit.*, **34**, 471–476.
42. Lo-Fo-Wong, D. *et al.* (2013) Cancer in full colour: use of a graphic novel to identify distress in women with breast cancer. *J. Health Psychol.*, Dio: 10.1177/1359105313495905.
43. Marchetto, M.A. (2007) *Cancer Vixen*, Harper-Collins, London, p. 217.
44. Eheman, C.R. (2009) Information-seeking styles amongst cancer patients before and after treatment, by demographics and use of informational sources. *J. Heath Commun.*, **14**, 487–502.
45. Leeming, C. (2013) Carers walk the cancer journey with patients but need more support. http://www.theguardian.com/society/2013/june/26/carers-cancer-patients-support.
46. Grimes, D. (2013) Six Stubborn Myths about Cancer, http://www.theguardian.com/science/2013/aug/30/six-stubborn-myths-cancer (accessed 30 August 2013).
47. Mazzonetto, M., Merzagora, M., and Tola, E. (2005) *Science in Radio Broadcasting*, Polimetrica, Milan.
48. Miller, J. *et al.* (2006) Adult learning from local television newscasts. *Sci. Commun.*, **28**, 216–242.
49. Dingwall, R. and Aldridge, M. (2006) Television wildlife programming as a source of popular scientific information:

a case of evolution. *Public Underst. Sci.*, **13**, 131–152.
50. Dhingra, K. (2006) Science on television: storytelling, learning and citizenship. *Stud. Sci. Educ.*, **42**, 89–123.
51. Lee, S. *et al.* (2011) Internet-based science education: a review of journal publications. *Int. J. Sci. Educ.*, **33** (14), 1893–1925.
52. Boyle, E., Connolly, T., and Hainey, T. (2011) The role of psychology in understanding the impact of computer games. *Entertain. Comput.*, **2**, 69–74.
53. Gee, J.P. (2007) *What Video Games Have to Teach Us About Learning and Literacy*, 2nd edn, Palgrave Macmillan, New York.
54. Falloon, G. (2013) Young students using iPads: App design and content influences on their learning pathways. *Comput. Educ.*, **68**, 505–521.
55. Siemens, G. (2005) Connectivism: a learning theory for the digital age. *Int. J. Instr. Technol. Distance Learn.*, **2** (1), 1–8.
56. Vygotsky, L. (1978) *Mind in Society: The Development of Higher Psychological Processes*, Harvard University Press, Cambridge, MA.
57. Abd-El-Khalick, F. (2005) Developing deeper understanding of nature of science: the impact of philosophy of science courses on pre-service teachers' views and instructional planning. *Int. J. Sci. Educ.*, **27** (1), 15–42.
58. Cornell (2013) Citizen Science Central, http://www.citizenscience.org (accessed December 2013).
59. Overdevest, C., Orr, C.H., and Stupenuck, K. (2004) Volunteer steam monitoring and local participation in natural resource issues. *Res. Hum. Ecol.*, **11** (2), 177–185.
60. Brossard, D., Lewenstein, B., and Bonney, R. (2005) Scientific knowledge and attitude change: the impact of a citizen science project. *Int. J. Sci. Educ.*, **27** (9), 1099–1121.
61. Holzer, M. *et al.* (2011) Lessons learned from a participation survey of a citizen science project, in *Citizen Science Symposium*, University of Maine.
62. Dickinson, J. and Bonney, R. (eds) (2012) *Citizen Science: Public Participation in Environmental Research*, Cornell University Press, New York.

Part II
Best Practices and Innovative Strategies

7
Using Chemistry Education Research to Inform Teaching Strategies and Design of Instructional Materials
Renée Cole

7.1
Introduction

The results of chemistry education research provide a significant resource to inform teaching strategies and the design of instructional materials. However, there are gaps in the bridge between chemistry education research and chemistry teaching. A recent National Research Council (NRC) report [1] concluded that discipline-based education research, which includes chemistry education research, has not yet resulted in widespread changes in practice. Dick Zare [2] has pointed out that the work done by chemical education researchers "will have little impact unless chemists and chemical education researchers can communicate clearly to one another." Henderson and Dancy [3] have explored a similar disconnect that exists in the physics community. It is critical to identify the specific gaps in this bridge and ways to promote communication and understanding among chemists, chemistry educators, and chemistry education researchers.

Each instructor starts with some knowledge of teaching and learning. However, Martin *et al.* [4] state that "the critical issue is not how much teachers know or what their level of teaching skill is, but what it is they intend their students to know and how they see teaching helping them to know" (pp. 387–388). They argue that a key factor in teaching practice is the instructor's intentions concerning what students should learn. The analysis of strategies used in teaching and the intentions associated with their use led to the characterization of five approaches to teaching [5], as shown in Table 7.1.

The use of the research literature to improve teaching and learning in chemistry starts with reflection on the part of the instructor. As a first step, instructors should identify which approach to teaching is most consistent with their intentions for student learning. This is important because adaptations to a teaching innovation that are inconsistent with the strategy may result in a significant loss of effectiveness [5]. Instructors must also identify which aspects of their teaching or student learning they wish to improve. The aspect of instruction to be improved will also

Chemistry Education: Best Practices, Opportunities and Trends, First Edition.
Edited by Javier García-Martínez and Elena Serrano-Torregrosa.
© 2015 Wiley-VCH Verlag GmbH & Co. KGaA. Published 2015 by Wiley-VCH Verlag GmbH & Co. KGaA.

Table 7.1 Categorization of five approaches to teaching [5].

Approach	Focus of strategy	Intention
A	Teacher	Information transmission
B	Teacher	Concept acquisition
C	Teacher/student	Concept acquisition
D	Student	Conceptual development
E	Student	Conceptual change

depend on their stage of teaching, which determines how they think about teaching and where they focus their attention. Kugel [6] described the development of college professors as teachers as a progression through two phases and five stages, illustrated in Figure 7.1.

Instructors who are looking to the literature to improve their teaching are likely in stage 3 or beyond. At stage 3, instructors are typically looking to make their teaching better and increase the engagement of students in the course material. However, Elton [7] warned of the "dangers of doing the wrong thing righter," and recommended less focus on "doing things better" and more on "doing better things." This chapter will discuss some areas of research that should inform chemistry instruction and describe how research has been used to develop innovative strategies for teaching. The goal is to provide a resource for instructors wishing to learn about, adapt, and incorporate research-based materials and teaching strategies into their courses, hopefully with the goal of "doing better things" to create more effective learning environments to promote students achieving a deeper understanding of chemistry and developing appropriate science process skills.

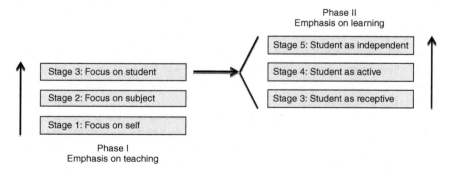

Figure 7.1 Stages and phases of college professor development as teachers. From Ref. [6]. Reprinted by permission of the publisher (Taylor & Francis Ltd, http://www.tandf.co.uk/journals).

7.2
Research into Student Learning

Instructors can draw from a significant body of research into student learning in chemistry. Much of this work has been done on student misconceptions and concepts or tasks that students find challenging, but more recent efforts have focused on learning trajectories and the impact of active learning strategies. Much research into teaching and learning in chemistry can be found in chemistry education journals such as the *Journal of Chemical Education, Chemistry Education Research and Practice, The Chemical Educator,* and the *Australian Journal of Education in Chemistry,* but there are many other journals that publish chemistry education research. Additionally, there is a wealth of research in cognitive science, learning sciences, and other discipline-based research areas that can inform teaching and learning in chemistry. As a body, this research should inform future research, the development of research-based teaching materials and pedagogies, and the teaching practice of reflective practitioners.

Several active areas of research in chemistry education were identified in the 2013 National Research Council [1] report. They include students' understanding of concepts, particularly of the particulate nature of matter, the use of technology to support student learning, analysis of student discourse and argumentation patterns, the use of heuristics in student reasoning, and the development of assessment tools to measure student thinking about chemistry. A review of the peer-reviewed literature in chemistry education research between 2000 and 2010 was conducted by Towns and Kraft [8]. The studies included in the review represent work in a variety of focus areas (including pedagogy, misconceptions, particulate nature of matter, instrument development, and student achievement) in a range of instructional levels and courses, and used a variety of research designs (qualitative, quantitative, and mixed methods). The findings from the reviewed studies support the use of socially mediated forms of learning that employ some form of student interaction as a method to improve student learning. There is significant evidence from studies in the review that indicates the challenges in creating learning environments that result in meaningful learning. The review also includes a number of instruments that could be used by instructors to inform many different aspects of their teaching, including measuring students' conceptual understanding, their cognitive expectations, metacognition, or formal reasoning ability, and their beliefs, attitudes, or perceptions.

Complementing research into aspects of student learning specifically related to chemistry, research into how people learn more generally is helpful in identifying approaches to creating learning environments or materials that more fully support student learning. The book *How People Learn* [9], first published in 1999 and expanded in 2000, has had a significant influence in many areas of educational research and development as evidenced by the thousands of articles and books that cite it. The authors highlight three key findings that have strong implications for creating effective learning materials and teaching strategies: instruction must acknowledge and engage students' preconceptions; instruction should enable students to develop factual knowledge, understand facts, and ideas in context, and organize knowledge in a way that supports retrieval and application; and instruction should use a metacognitive approach that helps students take control of their learning and monitor their progress in achieving defined learning goals. They go on to discuss how research into learning and learning theory can be used to design more effective learning environments.

There is an increasing amount of work in chemistry education research that explicitly draws on the work done in understanding how people learn. One example is the learning cycle approach, which is a generalized teaching model used to design curriculum materials and instructional strategies, and is partially derived from Piaget's theory of development. A general description of a learning cycle is a three-phase process of exploration, concept invention, and application. Traditional instruction typically follows a sequence of inform, verify, and practice. Studies have been conducted exploring the sequence of instruction and its impact on student achievement and attitudes [10, 11]. Abraham [12] provides an overview of the inquiry and the learning cycle approach to teaching chemistry in the *Chemists' Guide to Effective Teaching* [13]. Even a cursory search of the literature gives numerous examples of the use of inquiry and the learning cycle in both lecture and laboratory materials for a range of chemistry courses.

7.3
Connecting Research to Practice

The process of connecting research to practice can range from making minor modifications to practice, to adopting or adapting existing research-based teaching practices, to developing new instructional materials and strategies based on this research. A few examples of more fully developed research-based teaching practices are described in the following section, but this section will focus on examples of how a few particular areas of research can be used to inform practice.

7.3.1
Misconceptions

A significant body of work has been devoted to elucidating students' conceptual understanding (or lack thereof) of a wide variety of topics in chemistry. The study of identifying misconceptions (sometimes called *alternative conceptions* or *naïve conceptions*, among other names) is one of the most active areas of research in chemistry education [1]. Barke *et al.* [14] provide an overview of many of the common misconceptions in chemistry. Duit has provided a bibliography of work on students' alternative frameworks and science education [15] until 2009. The topic of misconceptions is explored in more detail in this volume in the chapter by Barke. Having knowledge of commonly held misconceptions can provide instructors with insights into designing lectures, course materials, and exam questions.

However, not all misconceptions need equal attention. Instructors should place a priority on identifying and addressing misconceptions that are directly related to the fundamental ideas of chemistry.

Becoming familiar with the literature on misconceptions is certainly one step forward in creating an instructional environment that is consistent with the finding from *How People Learn* [9] that instruction must acknowledge and engage students' preconceptions. The ways in which instructors use the literature on misconceptions will depend on their approach to teaching and their stage of development. Instructors at stages 2 and 3 using more teacher-focused strategies will likely focus on how to improve their presentation of material to make it more likely that students will acquire a correct understanding of concepts. As instructors move to stage 4 or 5 and adopt more student-centered teaching approaches, there are more strategies that can be used to both assess the degree to which students hold misconceptions and provide opportunities for them to actively confront their ideas and build more scientifically accepted understandings.

The topic of stoichiometry provides one example of how research on misconceptions can be used to inform teaching strategies and materials design. Stoichiometry is a basic concept for chemistry, but it is clear that students struggle with many ideas associated with the topic. It has been reported [16] that students confuse molar mass, the amount of material given, and the reacting mass. This confusion is likely related to students' difficulty with understanding the concept of limiting reagents [17–19]. There is also evidence that students do not understand the role of coefficients in chemical reaction equations, frequently including this information when determining molar mass [20, 21], or struggling with the fact that atoms must be conserved in chemical reactions while molecules need not be [22]. Many additional studies have addressed students' understanding of stoichiometry, and it is clear that traditional approaches to the topic leave many students with an incomplete understanding.

Instructor responses to this information could be limited to updating presentations or following the recommendation to avoid particular types of questions that students can answer correctly using an incorrect approach [16]. A next level of approach would be to incorporate questions into the lecture that provide feedback about student learning to both students and instructors in real time. Structuring these questions such that the distractors are based on common misconceptions provides a mechanism for instructors to assess how many students appear to hold these misconceptions. The use of electronic student response systems (SRSs), such as "clickers," is a relatively easy way to collect and display this information. (More on SRSs will be presented in the next section.) Of course, the key is to do something with the information – either adapting the presentation of material, prompting students to discuss the material further, or engaging students in an activity that requires them to confront these misconceptions.

A more extensive response would be to design instructional practices that more directly address the common conceptual challenges highlighted in the literature. For example, an instructor could choose to use a guided-inquiry approach using a learning cycle that allows students explore and analyze the appropriate relationships to gain a conceptual understanding rather than simply memorizing an algorithmic approach to solving problems. This approach was taken in designing a unit for AP chemistry teachers [23]. The design of the activity was informed by research into students' understanding of stoichiometry and research regarding inquiry and the learning cycle.

The structure of the activity uses a learning cycle approach to help students develop both a qualitative and quantitative understanding of the connections among the number of atoms, molecules, and mass in a balanced chemical reaction. The activity begins by having students explore a series of cartoons depicting "snapshots" of a chemical reaction in progress (the sequence is shown in Figure 7.2). The exploration requires students to translate between particulate-level representations and symbolic representations. Students then analyze the changes that occur during this progression. Students follow this by developing an understanding of the notation used to represent chemical reactions. At the end of the activity, students are asked to use the law of conservation of mass to explain why a chemical reaction equation must be "balanced."

The framework for understanding limiting reaction problems is established by having students reflect on what information is provided by a chemical reaction equation. Unlike most instructional materials that introduce reaction equations, the cartoons do not show a series where all reactants are converted to products. This was done intentionally to help students see that reactions represent a process, rather than a depiction of the contents of a container before and after a reaction takes place. Additionally, the activity includes explicit prompts for students to work as a group to generate explanations. Research has shown that cooperative learning has a significant and positive effect on student learning [24]. However, the experience of the author was that students often work on materials in parallel rather than collaboratively unless explicitly prompted to do so by either the instructor or the materials themselves. Providing a few explicit prompts for the students to work as a group makes it more likely that students will complete the activity in the manner intended.

The use of learning teams and inquiry-oriented instructional materials that help students address and avoid misconceptions is consistent with student-centered approaches to teaching that focus on conceptual development and change. Instructors using these strategies are in Phase II and create learning

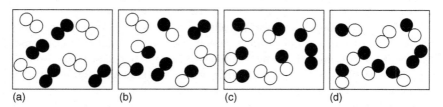

Figure 7.2 (a–d) Cartoon of a series of "snapshots" depicting a chemical reaction in progress.

environments where the students are active and somewhat independent in developing their knowledge of concepts. However, the trade-off to these approaches is that they can take more class time to implement as compared to a didactic presentation of the same concepts. The use of these types of teaching and learning strategies takes more skill on the part of the instructor in identifying the desired learning outcomes for the course and then determining how to best achieve them through classroom activities, course readings, homework, laboratory experiments, and so on.

7.3.2
Student Response Systems

The use of SRSs, or "clickers," is frequently cited as a way to make lectures more interactive and encourage greater student participation, particularly in large-enrollment courses. While "SRSs" can range from very low tech (having students use their fingers to indicate responses [25]) to very high tech (the Learning Catalytics program [26]), the basic premise is to provide a way for students to respond to questions during class in a manner that allows the instructor to assess the current level of understanding. The term "clicker" was popularized when instructors began using stand-alone devices like those found for audience polling in game shows. Students clicked a button to indicate their responses – hence, the moniker clicker. Even as platforms that use students' mobile phones or other Web-based technologies have been developed to collect student responses, the name clicker has stuck, likely because it is simply easier to refer to clickers and clicker questions rather than SRSs and electronic voting questions.

A search for "clickers" and "chemistry" resulted in more than 2000 citations, indicating that there is a significant number of people working on incorporating clickers into chemistry classrooms. In 2008, MacArthur and Jones published a review of the literature regarding the use of clickers in college chemistry classrooms [27]; the reader is referred to this study for more details. The overarching view was that the use of clickers had many benefits and few drawbacks if used appropriately. The biggest drawbacks are related to time for implementation and issues related to helping students learn the technology and how to respond if the system is not working properly. Best practices for using SRSs have been summarized in multiple publications [28–31] and new studies continue to appear in the literature, particularly as new technologies are developed and clicker questions are integrated into other teaching and learning strategies. Effective use has centered on both the types of questions asked and the nature of implementation.

Questions that are more conceptual in nature and are written at a level that requires discussion and explanation rather than recall are more effective in developing student understanding. Instructors can write their own questions, but other sources of questions include concept inventories (see following section), books [30, 32], and various published articles and Web sites. Most textbook publishers now include slides with "clicker questions" as part of the instructor resource materials, but the quality of these questions is frequently inconsistent with research into best practices.

The nature of the implementation of SRSs in the classroom is also key to having a positive impact on student learning. Using the technology as a mechanism for taking attendance is generally perceived by students as a means to control behavior rather than as a support for learning. This tends to result in more negative student attitudes and no improvement in learning gains. The use of SRSs for formative assessment and to promote student argumentation (see Section 7.3.4) is more consistent with the research into how people learn. Studies that have shown increased learning gains through the incorporation of clickers into the lecture have all included some aspect of student collaboration and discussion into their implementation [27, 33].

7.3.3
Concept Inventories

The development of concept inventories to help identify and measure changes in student understanding of particular topics has also gained significant attention in recent years [34, 35]. Some of the strengths and weaknesses of concept inventories are summarized in the NRC report [1]. While not used in chemistry to the extent they have been in physics, particularly the force concept inventory [36], concept inventories have the potential to drive reform in teaching practice in chemistry as well. Treagust has done significant work in developing two-tiered diagnostic tests to identify student misconceptions [37–40]. The items in these multiple-choice tests include a question that focuses on content knowledge paired with a question where students must choose the best explanation for their answer to the first question. Mulford and Robinson developed a concept inventory to assess alternate conceptions about topics typically covered in first-semester general chemistry [41]. More recently, concept inventories have been developed focusing on foundational topics addressed in introductory chemistry courses which then reappear in subsequent courses such as biochemistry [42] or engineering [43]. Work has also been done to develop concept inventories on much more specific topics such as acid strength in organic chemistry [44], enzyme–substrate interactions [45], and chemical equilibrium [46].

Concept inventories and diagnostic tests can be used in a variety of ways to inform teaching strategies. The use of a comprehensive concept inventory as a pre-test followed by a comparable post-test at the end of the course can provide evidence of the degree to which instruction improved students' understanding of those concepts. Of more use to most instructors is determining preexisting ideas held by students prior to instruction. The information provided by students' performance on one of these instruments can assist instructors in designing a learning environment that explicitly addresses misconceptions and builds on prior knowledge. Concept inventories that focus on foundational knowledge can also be used to help students assess their own mastery of prerequisite knowledge to better prepare for further coursework. This can be particularly useful if resources (or links to resources) are provided that can be used for remediation to help students achieve the level of understanding required.

However, concept inventories are not without criticism. Smith and Tanner [47] explore the promise of concept inventories but raise questions about what inventories actually measure. They also challenge the community to think more deeply about other ways to obtain similar information about student thinking. Questions have been raised about the construction of the instruments [48] and how well they reflect actual student thinking. Concept inventories are also limited to probing predetermined concepts and a limited set of responses – those included on the inventory. Despite these concerns, they may provide a data source to help document the effect of changes to instruction for those instructors who take a more scholarly approach to teaching and want to measure how changes to the learning environment impact student learning.

It is unclear to what extent concept inventories are useful to most instructors in everyday use. Probably the most accessible use of concept inventories and diagnostic tests is as good resources for questions that can be used (or modified) to engage students with SRSs (clicker questions), Peer Instruction [49], or other active learning strategies in the classroom. One of the primary challenges in using SRSs is creating effective questions [28]. As indicated in the previous section, clicker questions are more likely to be effective in promoting interactive engagement if they address specific learning goals, probe conceptual understanding, and provide an opportunity to identify misconceptions and confusions [27–30]. The questions in most concept inventories have been specifically designed to have these very characteristics.

7.3.4
Student Discourse and Argumentation

There has been significant work on student discourse and argumentation that can be used by instructors to inform teaching practice. Having students participate in a process in which they construct understanding and defend their ideas has been suggested as one way to support students' conceptual understanding of science [50–53]. However, the process of asking students to make claims and justify these claims with a reasoned argument does not appear to be normative practice for most chemistry instructors. Research suggests that students struggle in learning how to generate effective arguments and in how to engage in productive scientific argumentation as they propose and critique ideas [53]. The research on argumentation provides a wealth of resources that instructors can use to create learning environments that promote student interactions and guide how students think about chemistry concepts. Classroom and laboratory materials can be developed that prompt students to generate arguments or parts of arguments. Engaging students in classroom discourse provides opportunities for students to construct ideas, to negotiate the relationships among different types of evidence or representations, and to practice making scientific arguments about science content [52, 54]. The nature of classroom facilitation can also shape students' reasoning about chemical and physical properties, including characteristics of what counts as acceptable types of justifications in student argumentation [55].

Strategies for establishing a learning environment that supports student argumentation can range from simply prompting students to articulate the reasoning behind their answers to questions to developing new curricular materials and instructional models for the laboratory. Instructors can use a process of questioning, revoicing, and elaboration to model appropriate reasoning for students, which will help establish classroom norms for argumentation and can even promote the use of reasoning using the particulate nature of matter [55]. Students can be asked to develop written arguments and should be provided support in how to back up claims with credible evidence, establish linkages among ideas, and substantiate assertions with a variety of reasoning strategies [56]. A combination of short questions and probing questions has been shown to elicit components of scientific arguments from students [57]. Laboratories can be structured such that students engage in the identification of the research question, develop and justify a method for data collection, develop claims, and justify those claims with evidence. Two models for this approach are the Science Writing Heuristic [58] and Argument-Driven Inquiry [59, 60]. The structure of guided inquiry activities can also promote student argumentation [55, 61, 62]. Questions with prompts such as "explain," "why," "predict," and "show" elicit the production of arguments, as students provide evidence and reasoning to support their answers.

While the benefits for establishing learning environments that promote student discourse and argumentation are clear to most instructors, many find it challenging to find time in their courses to provide students with this opportunity. This is particularly true for instructors who have a more teacher-focused approach to teaching. It can also be challenging to provide feedback to the students on the quality of their reasoning, particularly in large classes. A simple way to promote student reasoning is to simply have students explain the choices for their answers. This can even be done in large lectures, particularly if SRSs are used. While only a small number of students will have the opportunity to express their reasoning and receive feedback, it does provide the expectation that students are prepared to justify their answers. The manner in which the instructor responds to explanations and models reasoning will also have an impact on student learning. The use of active learning strategies such as POGIL (process-oriented guided inquiry learning), peer instruction, or inquiry-based laboratory experiments provides more support for students to engage in discourse and argumentation, but these generally require a commitment by instructors to invest the time to learn how to implement these strategies and to make changes to their course structures. Additionally, simply using these materials is not always effective if appropriate classroom facilitation strategies are not used.

7.3.5
Problem Solving

Problem solving is a key skill in chemistry and is often cited by instructors as an area in which they would like to see students improve. To create a learning environment that supports student development of problem-solving skills requires an understanding of the challenges students encounter in solving problems and knowledge of approaches that are more likely to result in improved performance. Many instructors do not explicitly address problem solving in their course, but they hope students improve simply through working chemistry problems or conducting laboratory experiments.

The challenge for most instructors is how to explicitly help students develop stronger problem-solving skills. Many instructors often conflate working example "problems" for students with teaching students how to solve problems [63]. This creates an unreasonable expectation on the part of students that they simply need to memorize a set of steps to work out a particular type of problem and promotes an algorithmic approach to learning material rather than a deeper understanding of the concepts. It also creates the very common situation in which students and instructors have very different perceptions of how closely related exams questions are to assigned homework questions. Instructors focus on the concepts addressed by the question, while students focus on the surface features for which they have memorized a procedure that can be used to find an answer.

The findings summarized in the NRC report [1] suggest that instructors need to emphasize the role of good representations and good solution methods in successfully solving problems. Instructors should also encourage students to take a deeper approach to problem solving rather than focusing on surface features. There is a need for a systemic approach with multiple forms of support and scaffolding to help students learn to identify the important structural features of problems. Research in chemistry education has shown that students tend to stabilize on a problem-solving strategy after a relatively limited number of attempts [64]. However, having students work on problems in collaborative groups improves their ability to describe their problem-solving strategies and they retain those strategies when they work individually [64, 65]. Students may also benefit from instruction and activities that engage them in both imagistic reasoning and algorithmic strategies for problem solving [66]. Bodner presents a more extensive review of the research on problem solving in chemistry in this volume, which will provide more resources for instructors.

7.3.6
Representations

Research demonstrates that students often struggle with translating among the different ways in which chemical concepts are represented [67]. As was noted in the NRC report [1], " ... students have difficulty translating among alternative representations that describe the same set of relationships, such as videos, graphs, animations, equations, and verbal descriptions" (pp. 5–24). Experts can easily translate among the different representations, but students often do not focus on the relevant aspects of these models; they get distracted by irrelevant surface features. Kozma and Russell [68] suggest that being able to use different representations and their features as evidence to support claims about chemical and physical properties is a key competence necessary for the practice of chemistry. However, research has shown that students often struggle with these skills. For example, Lewis structures are ubiquitous in chemistry, providing structural information that can be used to predict the properties of chemical substances. In contrast, while students indicate that structural information can be obtained from Lewis structures, a significant number do not make the connections between these structures and chemical properties, even though their utility in predicting properties is what makes these representations useful [69]. Students' limited interpretations of diagrams also impede their abilities to solve problems or make connections among representations [70, 71]. Even when instructors provide diagrams with the intention of facilitating qualitative reasoning to answer a question, many students ignore the representation and use a quantitative approach [72]. The struggle to connect representations is not limited to pictorial or diagrammatic representations. Students' difficulty in interpreting mathematical representations, including relating the mathematical descriptions to macroscopic or particle-level descriptions of chemical concepts, has also been documented [73–75].

One approach to help students visualize the particulate nature of matter is through the use of animations and simulations. The incorporation of animations during lectures appears to have a positive effect in helping students explicitly link molecular, macroscopic, and symbolic representations for a variety of topics in chemistry, particularly when implemented in a manner supported by research on student learning. In one study, the animations were shown at least twice, with the first presentation having no explanation, followed by a second presentation with the instructor pointing out important features of the animation and relating the molecular representation to symbolic and laboratory representations [76]. The impact of the order in which representations are presented was also seen in a study, that suggested that having students view a phenomenon at a macroscopic level followed by a particulate-level animation results in improved understanding of a particulate-level explanation for the phenomenon [77]. Additionally, research suggests that three-dimensional representations can be more effective than two-dimensional representations in learning stereochemistry [78]. The use of these images is even more effective if students are provided with opportunities to link the two representations.

Jaber and BouJaoude [79] suggest that explicit discussions of the nature of chemical knowledge and the relationships between representations may be necessary to improve students' ability to translate between representations to construct chemical understandings. While the use of multiple representations can help learning, explicit prompts should be provided to help students make connections among representations. Instructors should provide students with detailed introductions to different representations, including highlighting the relationships between complementary representations and explaining how different representations are optimized for different tasks. Explanations of the limitations of representations are also needed to help prevent the development of misconceptions. Students need extensive practice interpreting and producing different representations, including determining which one is the most appropriate for a particular application. In addition to classroom presentations, interactive lecture demonstrations or guided inquiry activities provide mechanisms to engage students in actively working with different representations and articulating the connections among them. Using a learning cycle approach, instructors could use directed questions to focus attention on salient details as students explore different representations. These could be followed by questions that support students in developing the concepts associated with the representations and connecting the representations with important aspects of the concept. Finally, students could apply this knowledge to solve both conceptual and more quantitative questions.

7.3.7
Instruments

In addition to the concept inventories discussed previously, a variety of instruments have been developed that can be used both to inform teaching practice and to measure the impact of changes to practice. A quick search results in the identification of dozens of instruments, many of which have been described in relation to chemistry education research [8, 80]. Bretz [81] has also recently published a chronology of assessment in chemistry education. The intent here is not to summarize all the options available to instructors, but rather to describe a few examples of how different types of instruments could be used to inform or assess the selection and implementation of instructional strategies.

One class of instruments relates to measuring students' beliefs about chemistry and their expectations for learning chemistry. Two instruments are CHEMX (Chemistry Expectations Survey) [82] and CLASS-Chem (Colorado Learning Attitudes about Science Survey) [83]. These instruments are designed to compare novice and expert beliefs about chemistry and about learning chemistry. Monitoring students' beliefs can provide instructors with information about how the learning environment they have created has influenced students' views about chemistry and what it means to learn chemistry. While it is reasonable to assume that most instructors would hope that students would shift toward more expert beliefs about chemistry as they progress through the curriculum, a shift toward more novice-like beliefs after completing introductory chemistry courses was noted with both instruments [82, 83]. This inconsistency suggests a mismatch between instructors' beliefs about learning chemistry and the learning environments experienced by students. Administration of these surveys at the beginning of the semester allows instructors to assess students' current beliefs about learning chemistry. This information can then be used to inform the way different topics are presented and assessed in the course to improve the alignment between students' expectations and those of the instructor. The instructor may want to explicitly address some of these expectations and model the behavior they hope to observe in their students.

Students' beliefs about a discipline can also have a significant impact on how they learn new information and how they react to different learning environments, so this information may provide an instructor with additional insight into how to introduce different learning activities into a course. An instrument such as the ASCI (Attitude toward the Subject of Chemistry Inventory) [84] can give an instructor a sense of students' emotional response to learning chemistry, including how interesting and useful students find chemistry, how anxious the subject makes them feel, and how intellectually accessible they find the subject. One would expect that different approaches would be more successful with a class full of students with high levels of anxiety who find chemistry very inaccessible than would work best for a class full of students who enjoy chemistry and are confident in their ability to master the material.

Piagetian tests of logical thinking such as the GALT (Group Assessment of Logical Thinking) and the TOLT (Test of Logical Thinking) provide a measure of students' formal reasoning abilities [85]. These tests can also be used to assign students to one of three levels of intellectual developments: formal, pre-formal, or concrete. Both instruments have been shown to be good predictors of students' success in chemistry [85–87], so they may be useful in identifying students who may need additional support to be successful in chemistry courses, particularly introductory courses. Students with lower TOLT or GALT scores will likely benefit from having concepts introduced in a more concrete way or by using activities that support the development of formal reasoning skills [85]. Students with different formal reasoning abilities may also respond differently to changes to teaching practices [88]. These differential responses can sometimes mask effects if not taken into account when analyzing assessments of student outcomes.

Scores on the GALT or TOLT can be used as one factor in assigning students to collaborative learning groups or laboratory partners. Studies comparing the effect of group composition on problem-solving ability suggest that some combinations of partners lead to better learning gains than others [64]. Most pairings of students resulted in gains in their problem-solving ability, with the exception of concrete students paired together. The other notable difference in learning gains was for pre-formal students paired with a concrete student, who were observed to have learning gains significantly higher than the average. These results would suggest that instructors should avoid pairing concrete students together, and that an effective strategy for assigning students to partners (at least for collaborative learning activities) would be to pair concrete students with pre-formal students. This approach has proven to be useful in the experience of the author, particularly in assigning laboratory partners in introductory courses.

Chemistry education research also provides resources for instructors who are developing materials. Rubrics can provide a guide to faculty developing materials to ensure consistency with the expectations of the community and serve as a mechanism for reviewers to provide consistent feedback [89]. For example, both the NRC [1] and the President's Council of Advisors on Science and Technology (PCAST) [90] recommend greater use of inquiry in undergraduate laboratories. However, an analysis of the levels of inquiry in published laboratory manuals found that most experiments were either confirmation experiments or structured inquiry [91]. Even self-described inquiry laboratories showed significant variations in the level of inquiry [92]. Departments and instructors can use the levels-of-inquiry rubric [91, 92] to characterize their current practice and to guide the transition from a curriculum comprised of primarily confirmatory laboratory activities to one that also provides opportunities for students to engage in more authentic science practice. Developers can then use the rubric to characterize the level of inquiry in newly developed experiments when disseminating laboratory activities.

In contrast to much of the research described previously, most of these instruments are not directly related to strategies to be used in the classroom. Instead, they provide a means for instructors to inform and assess changes to instruction. However, any instrument should be used with caution, and it is still the responsibility of the user to obtain psychometric measures which ensure that the instrument is functioning as intended for the specific population and purpose being investigated. One step is conducting appropriate statistical analyses to determine whether the instrument is working the same for the population under study as was reported in the literature [93]. Arjoon *et al.* [94] examine a number of instruments and make recommendations to the community regarding gathering and reporting additional data for published instruments. They also point out the need for instructors (and researchers) to have a more complete understanding of the psychometric evidence associated with the instruments they want to use. As the area of chemistry education research continues to develop, it will be important for both researchers and practitioners to continue to contribute to the further development and refinement of instruments.

7.4
Research-Based Teaching Practice

In addition to research into understanding how students learn (or do not learn) chemistry, there are many teaching practices and materials that have been developed based on this research. The majority of these research-based teaching practices involve some aspect of active learning. The use of student-centered pedagogies is consistent with one of the conclusions from the 2013 NRC report: "... cognitive science research and science education research have shown that students are more likely to change their conceptions when they interact more with the content and the learning process" (1, pp. 4–11). The advantage of research related to more fully developed pedagogies is that it often provides more resources to instructors looking for "better" approaches to teaching. Three examples are discussed below, but the intention is not to provide an extensive list of what is available or endorse a particular approach. Several additional examples such as peer instruction are discussed in more depth in other chapters in this volume.

Some of the analyses include discussion of the features of learning materials and strategies that impact the likelihood that they will be adopted by faculty other than the developers [95]. The greater the degree of change required by adopters, the higher the degree of support that is likely to be needed to support the change. This is described in more detail in the section on implementation.

7.4.1
Interactive Lecture Demonstrations

There is a strong tradition of using classroom demonstrations to illustrate concepts in chemistry. However, the impact of these demonstrations on learning is limited when students are only observers [96]. Zimrot and Ashkenazi [97] developed a set of interactive lecture demonstrations for introductory chemistry courses to facilitate conceptual learning by students. The structure of the implementation of the demonstrations was informed by work in conceptual change [98] and modeled after work done in physics [99]. A set of demonstrations was created for each topic addressed by the course. The development of the demonstrations themselves was informed by research into misconceptions. While some of the demonstrations validated students' prior conceptions, others were intentionally designed to confront common misconceptions through the use of a "discrepant event." Additional demonstrations that lead to refinement of concepts were also developed [100].

These approaches are consistent with research showing that creating a need to know, which is likely to happen when students make an incorrect prediction and then want to know why the event happened as it did, promotes conceptual change. However, simply watching the demonstrations was not sufficient to achieve the desired outcomes. Consistent with the work on how people learn, students needed to actively engage in the learning process through making predictions about what they thought would happen, observing the demonstration, and then trying to explain why they observed what they did. This predict – observe – explain model was demonstrated in this work (among others) to result in much more meaningful learning by a larger proportion of students [97].

Of the three examples of research-based teaching practices discussed in this section, this approach is likely the easiest to be adopted by the greatest range of instructors. The demonstrations themselves can be categorized as "flexible instructional tools to promote engagement in the lecture" [95]. While demonstrations were developed for each topic addressed in a typical general chemistry course, an instructor could certainly choose which demonstrations to use in a particular class. This would allow individual instructors to implement the demonstrations in a course with minimal change to the overall course structure and approach to teaching. This would make the materials more accessible to instructors who are in stage 3 and making a transition from teacher-centered to student-centered approaches to teaching. They could also be used as a component of the learning environment for instructors using more student-centered approaches to teaching that require students to be more active participants in developing their understanding of key concepts and skills in chemistry. Indeed, in order for the demonstrations to have a significant impact on their learning, students had to have a more active role in making sense of the demonstrations. The use of lecture demonstrations is also conducive for use in institutions with coordinated sections of general chemistry. One instructor could carry out the demonstrations without requiring that all instructors teaching the course use them. This makes it easier for individual instructors to make changes to their instruction.

As is the case with many changes to instruction that shift from a learning environment focused on presenting materials to students to one where students are actively engaged in developing their knowledge and skills, an extensive use of interactive lecture demonstrations requires some trade-offs in terms of time spent presenting particular topics. The research supports the claim that this trade-off ultimately benefits students, but it does require instructors to prioritize learning outcomes and evaluate what material actually needs to be presented to students. The use of lecture demonstrations also requires access to a stockroom and a classroom that is conducive to conducting the demonstrations.

7.4.2
ANAPOGIL: Process-Oriented Guided Inquiry Learning in Analytical Chemistry

In response to a National Science Foundation report that called for an overhauling of the analytical chemistry curriculum to include active learning approaches, a group of analytical chemistry faculty and chemistry education researchers developed a comprehensive analytical chemistry curriculum that involves active learning and fosters the development of process skills such as critical thinking, problem solving, and communication skills [101]. The decision was made to employ a POGIL approach due to its research-based foundation and evidence of effectiveness [102–106]. In the POGIL method, groups of students engage in worksheet activities that are based on the learning cycle [107–109]. Research on process skill areas such problem solving, critical thinking, and communication was used to inform the structure of the activities and the nature of tasks students are asked to complete. The activities were also designed to promote collaborative learning and argumentation. This was done by regularly including questions that prompt students to analyze data and explain observations and then decide on a consensus answer.

An iterative approach, as shown in Figure 7.3, was used in the development of the activities. Instruments developed by the chemistry education community were used as part of this process. A rubric was used to ensure consistency in the design of activities and feedback process [89]. The American Chemical Society Division of Chemical Education Examinations Institute 2007 Analytical Chemistry Exam [110] was used to measure students' content knowledge. This

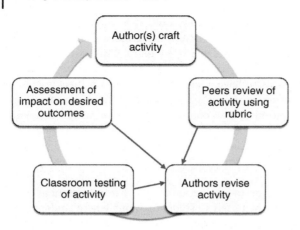

Figure 7.3 Iterative process used in developing ANAPOGIL activities.

was done because one barrier to the adoption of active learning strategies is the fear that content coverage will suffer. Comparison of students' achievement in courses using these strategies to nationally normed data provided evidence that this was not the case. The materials were developed and tested by faculty at a variety of institutions, for which the CHEMX [82] was used as one of the measures to compare student expectations for learning chemistry at the different institutions. A range of approaches were used to assess the impact on student process skills, and work is going on to develop strategies and instruments that support instructors in assessing these skills and providing feedback to students. Another area for future research is more in-depth analysis of implementation of the materials and facilitation strategies at different institutions and how those different implementations impact student outcomes.

The structure of the ANAPOGIL materials is consistent with the work on how people learn, particularly with the emphasis on helping students develop and apply knowledge in a way that supports meaningful learning and the incorporation of prompts for metacognitive and independent approaches to learning. However, these materials are not amenable to be incorporated into lecture sessions without making substantial changes to instruction. The adoption of ANAPOGIL materials (or any other POGIL materials) for a course requires a greater degree of change than was necessary for interactive lecture demonstrations. The use of ANAPOGIL activities in a course would include the use of a new pedagogy that requires an investment of time by the instructor to learn about effective strategies for classroom facilitation and implementation. It would also require instructors to spend some time revising their syllabi and thinking about how to best integrate the materials into their courses. While it is not necessary to completely replace traditional lecture presentation of course material (although one could), the use of ANAPOGIL activities does require replacing some presentations with time for learning teams to work through the materials during class. These materials are now commercially available [111], which makes it easier to obtain them but also increases the cost to students.

The manner in which the ANAPOGIL materials were developed also resulted in the creation of an extensive instructor's guide with substantial tips and insights from both the development team and initial classroom testers of the activities. This provides an additional resource that can help instructors implement particular activities in their own courses. Instructors can also learn more about the creation and facilitation of materials through workshops offered by the POGIL project. Access to an extensive network of other practitioners provides a way for instructors to obtain feedback and support from other faculty, which makes sustained implementation more likely. Of course, the drawback is that instructors have to have resources (both in terms of time and finances) to attend a workshop to learn more about the design and implementation principles that support successful implementation. It is unlikely that instructors who have not adopted student-centered approaches to teaching would find these materials of interest. It is most likely that instructors at stage 4 or 5 who are looking for materials and strategies that will support students as independent, active learners would be most likely to find value in this instructional approach.

7.4.3
CLUE: Chemistry, Life, the Universe, and Everything

Cooper and Klymkowsky have developed a new approach to general chemistry based on research on teaching and learning at the college level [112, 113]. The work was informed by a survey of the history of general chemistry and the issues that have been raised over the years. Rather than focusing on simply creating a new pedagogical approach, they also changed the content and reported a model that could be used to redesign other courses [112].

They began by identifying a set of learning outcomes for the course, explicitly identifying what students should know and be able to do at the end of the course. They used research indicating the need for foundational knowledge and the processes used by other groups to identify big ideas that should be addressed. These were used to develop a set of knowledge statements and science practices, which were then used to produce a set of performance expectations. To determine the optimum order of topics in the curriculum, they used work on learning progressions that described "conceptual and logical pathways that are more likely to lead to improved understanding of core ideas in science" [112]. In determining how to teach and assess the course, they turned to the misconceptions literature. A variety of strategies were developed to elicit student thinking about different topics. One strategy was the use of "clicker questions" during class, some of which were based on questions found in concept inventories.

To fully support student learning, they used research to determine which materials and experiences would be most appropriate for the different types of knowledge and skills they were trying to develop. As a result, course materials included not only a textbook but also carefully selected (or developed) videos, simulations, and resources to promote meaningful learning. They also used research to guide the nature and types of assessments used in the curriculum, particularly the need to use formative assessment and assessments that are consistent with promoting conceptual mastery. The development of the curriculum is part of an ongoing research-based design process where assessment leads to modifications of the curriculum. The project has also resulted in work to identify additional learning progressions and areas of conceptual difficulty that can inform instructors who may not be using the CLUE curriculum as well as informing continuing improvements to the curriculum itself [69].

The CLUE curriculum provides an exciting opportunity for a stage 4 or 5 instructor who wants to adopt a research-based, student-centered approach to teaching that focuses on conceptual development and change. The text, worksheets, activities, and notes versions of slide sets are provided on the website, although there is not a great deal of supporting information such as instructor notes or suggestions for classroom implementation. The curriculum is consistent with the key findings in *How People Learn* [9], particularly in engaging students' preconceptions and enabling students to develop factual knowledge, understand facts and ideas in context, and organize knowledge in a way that supports retrieval and application of knowledge. It also supports students in taking ownership of their learning and monitoring their own progress in mastering the material. The research into the impact of the curriculum on student learning gains and depth of knowledge certainly supports the value of the approach in meeting desired outcomes for first-year chemistry courses.

However, many of the aspects that make adopting the curriculum an exciting opportunity will make it more challenging for many instructors to adopt (at least at the moment). The text and student learning materials are still works in progress as the research to optimize the structure and learning activities is still ongoing. For this reason, some of the tools are only available upon request from the author, including the BeSocratic tool and instructor slide sets. The text is available only as a series of Web pages, and the tools are not well integrated into the text – they are accessed from different sections of the website. As the project matures, it is reasonable to expect that this will be put together in a way that is easier to navigate. Teaching with the CLUE curriculum also requires instructors to make substantial changes to the way they think about chemistry content and creating an effective learning environment. It is not an approach that lends itself to integration into a course with very little modification to "traditional" teaching practices. It would certainly require some time beyond usual lesson planning to implement initially. For instructors who teach at institutions with multiple sections or instructors where faculty agree to use a common text and exams, it would also require getting all of the faculty on board.

7.5
Implementation

Instructors generally look to the chemistry education research literature when they have identified an issue in their course they would like to improve. Most reflective practitioners regularly look for ways to improve student understanding in their courses. However, an additional component of using research to inform teaching practice and design instructional materials is in the actual implementation of those materials and strategies.

Bridging the gap between chemistry education research and the practice of teaching is not trivial. Research in physics has shown that a significant number of instructors who are aware of research-based instructional practices never actually try using the strategies or discontinue their use after briefly trying new instructional strategies [114]. Potential explanations for the high levels of discontinuance are that the instructors either lack the knowledge needed to customize a research-based practice to their local situation or underestimate the factors that tend to work against the use of innovative instructional practices [114, 115]. This is illustrated in Figure 7.4. This means that, in addition to learning about a particular practice, instructors must also understand the principles behind the practice so that modifications made in adoption and implementation do not lead to "inappropriate assimilation" [116]. A primary reason for developers to avoid this phenomenon is that instructors who think they have adopted a particular instructional strategy but have inappropriately assimilated it into their instructional practice may erroneously conclude that the strategy is ineffective rather than their implementation.

To improve the likelihood that changes in instructional practice result in the desired outcomes, instructors may have to engage in reflection and conceptual change in order to effectively incorporate research-based changes into their instruction [117]. Because of conflicts with existing beliefs and values about teaching (as well as conflicts with situational factors), instructors may keep surface features of an innovation but fail to make the intended changes to instruction [116, 118]. For example, it has been shown that self-described student-centered instruction is often not confirmed by observations [119]. Instructors incorporating new instructional practices should also realize that it frequently takes

Figure 7.4 Gap in understanding the principles behind a new teaching strategy, often leading to inappropriate implementations.

multiple semesters before they feel they can implement the practices effectively [120]. Both instructor inexperience and student resistance to new instructional strategies can contribute to this. Instructors are more likely to achieve successful changes to practice if they endeavor to understand the principles behind the innovation they are implementing and ensure that the strategies are aligned with their intentions [4, 5]. Becoming part of a community of practice where they can receive feedback and support also enhances the likelihood that instructors will move toward sustained practice [121]. Participation in scientific teaching learning communities has been shown to be a successful mechanism for sustaining and expanding the use of active learning strategies [122, 123].

7.6 Continuing the Cycle

The process of using chemistry education research to inform teaching strategies and the design of instructional materials begins with instructors reflecting on current practices and outcomes. What are their goals for teaching? In what areas are they satisfied with current practice and outcomes? In what areas are they frustrated with the learning environment or lack of achievement? Once these have been identified, a search of the literature can help instructors find options that may help them address their concerns. There are many published articles, activities, and experiments in the chemistry education literature that provide a range of options that can support improved student learning by improving information transmission by instructors or promoting concept acquisition, conceptual development, and conceptual change on the part of students. However, to be consistent with the work on how people learn, instructors should focus on approaches that engage students' preconceptions, enable students to develop an organized knowledge structure that supports deeper learning and application, and support students in becoming independent learners who can take control of their learning.

The research reported here (and elsewhere) supports the use of more student-centered instructional approaches that require students to be active participants in developing both content knowledge and transferable skills, such as communication and problem solving. These can be as simple as the use of the misconceptions literature to develop high-quality clicker questions to obtain formative assessment data of student knowledge during class, or as extensive as the adoption of an entirely new curriculum and pedagogy, such as the CLUE curriculum. The selection of aspects of a course to change or materials and strategies to adopt will largely depend on an instructor's approach to teaching as well as his or her phase and stage of professional development as a teacher. Using clickers or lecture demonstrations, providing small opportunities for students to engage in discourse, or adopting a few inquiry laboratory experiments is often seen as more accessible by instructors who are still focused on concept acquisition and teacher-oriented approaches to teaching and learning. As instructors transition to more student-centered approaches to teaching and learning and focus more on conceptual development with students being active participants in instruction, the research can be used to make more significant changes to instructional practice, including the adoption of research-based curricula, which may require substantial changes for implementation.

The process of improving teaching and learning starts with reflection, but it should not end with implementation. If instructors use the research literature to change their teaching practice, they should also assess the success of those changes. This cycle of reflection, modifications to teaching practice, and assessment is at the heart of scholarly teaching. If done systematically and documented appropriately (including obtaining permission to share student data), the results of this process can add to the body of work on teaching and learning, as shown in Figure 7.5. The process of expanding the audience from oneself to the broader community by sharing the results of changes to instruction moves scholarly teaching into the realm of pedagogical research. An advantage of engaging in pedagogical research is that it can help instructors to become more effective teachers as they become more aware of classroom practices, make measurements of student learning outcomes, and work to create more systemic change [124].

Pedagogical research encompasses a broad range of activities, including scholarship of teaching and learning, action research, and discipline-based educational research [124]. Particularly, given the complexities of conducting research in real classrooms, collecting evidence of how changes to teaching strategies and

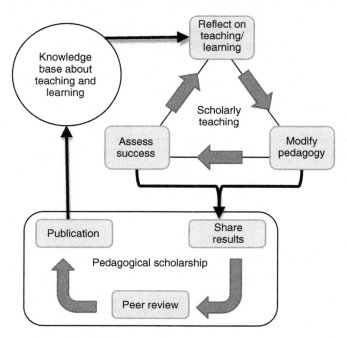

Figure 7.5 Cycle of pedagogical research.

instructional materials change student learning in a variety of contexts and institutions contributes to having a more robust and generalizable body of work about teaching and learning in chemistry. Each implementation, when placed in the context of relevant published literature, adds to our understanding of which strategies work best for different populations and leads to new research questions and improvements in those strategies and materials. To quote Healey [125],

> The scholarship of teaching involves engagement with research into teaching and learning, critical reflection of practice, and communication and dissemination about the practice of one's subject. This provides a challenging agenda for the development of subject-based teaching. Implementing this agenda includes applying the principles of good practice in the disciplines; developing the status of teaching; developing the complementary nature of teaching and research; and undertaking discipline-based pedagogic research.

The process of bridging the gap between research and practice, to focus on "doing better things," requires a culture change where all instructors become more scholarly teachers. Greater use of chemistry education research to inform approaches to teaching chemistry, which include a diversity of styles, materials, and strategies that are consistent with how people learn, should result in more students achieving a deeper understanding of chemistry and developing appropriate science process skills. While significant strides have been made, continued progress requires greater communication and collaboration among students, chemistry instructors, and chemistry education researchers with the common goal of improving teaching and learning chemistry. One goal would be to obtain feedback on how the results of chemistry education research can be made more accessible to a greater range of instructors. Identifying the most appropriate channels to reach the largest audience and developing support systems to help instructors sustain and continue to improve the implementation of research-based teaching strategies will require continued efforts on the parts of all members of these communities. The ultimate goal is for this process to lead to the transformation of chemistry teaching to include best practices and strategies as part of normative practice rather than being classified as innovative practice.

References

1. National Research Council (2012) *Discipline-Based Education Research: Understanding and Improving Learning in Undergraduate Science and Engineering*, The National Academies Press.
2. Zare, R.N. (2007) in *Nuts and Bolts of Chemical Education Research* (eds D.M. Bunce and R.S. Cole), Oxford University Press, pp. 11–18.
3. Henderson, C. and Dancy, M.H. (2008) Physics faculty and educational researchers: divergent expectations as barriers to the diffusion of innovations. *Am. J. Phys*, **76** (1), 79–91.
4. Martin, E., Prosser, M., Trigwell, K., Ramsden, P., and Benjamin, J. (2000)

What university teachers teach and how they teach it. *Instrum. Sci.*, **28** (5), 387–412.
5. Trigwell, K., Prosser, M., and Taylor, P. (1994) Qualitative differences in approaches to teaching first year university science. *High Educ.*, **27** (1), 75–84.
6. Kugel, P. (1993) How professors develop as teachers. *Stud. High. Educ.*, **18** (3), 315–328.
7. Elton, L. (2000) *Evaluate and Improve: Teaching in the Arts and Humanities*, Humanities and Arts Higher Education Network (HAN), Milton Keynes, pp. 7–9.
8. Towns, M. and Kraft, A. (2011) Review and synthesis of research in chemical education from 2000–2010. Paper Presented at the Second Committee Meeting on the Status, Contributions, and Future Directions of Discipline-Based Education Research.
9. Bransford, J.D., Brown, A.L., and Cocking, R.R. (2000) *How People Learn: Brain, Mind, Experience, and School*, National Academy Press, Washington, DC.
10. Abraham, M.R. (1989) Research & teaching: research on instructional strategies. *J. Colloid Sci. Teach.*, **18** (3), 185–87,200.
11. Abraham, M.R. and Renner, J.W. (1986) The sequence of learning cycle activities in high school chemistry. *J. Res. Sci. Teach.*, **23** (2), 121–143.
12. Abraham, M.R. (2005) *Chemists' Guide to Effective Teaching*, vol. 1, Prentice Hall, Upper Saddle River, NJ, pp. 41–52.
13. Pienta, N.J., Cooper, M.M., and Greenbowe, T.J. (2005) *Chemists' Guide to Effective Teaching*, Pearson Prentice Hall.
14. Barke, H.-D., Hazari, A., and Yitbarek, S. (2009) *Misconceptions in Chemistry: Addressing Perceptions in Chemical Education*, Springer.
15. Duit, R. Bibliography – STCSE, http://www.ipn.uni-kiel.de/aktuell/stcse/stcse.html (accessed December 2013).
16. Schmidt, H.J. (1990) Secondary school students' strategies in stoichiometry. *Int. J. Sci. Educ.*, **12** (4), 457–471.
17. Huddle, P.A. and Pillay, A.E. (1996) An in-depth study of misconceptions in stoichiometry and chemical equilibrium at a South African university. *J. Res. Sci. Teach.*, **33** (1), 65–77.
18. Gauchon, L. and Meheut, M. (2007) Learning about stoichiometry: from students' preconceptions to the concept of limiting reactant. *Chem. Educ. Res. Pract.*, **8** (4), 362–375.
19. Chandrasegaran, A.L., Treagust, D.F., Waldrip, B.G., and Chandrasegaran, A. (2009) Students' dilemmas in reaction stoichiometry problem solving: deducing the limiting reagent in chemical reactions. *Chem. Educ. Res. Pract.*, **10** (1), 14–23.
20. BouJaoude, S. and Barakat, H. (2000) Secondary school students' difficulties with stoichiometry. *Sch. Sci. Rev.*, **81** (296), 91–98.
21. BouJaoude, S. and Barakat, H. (2003) Students' problem solving strategies in stoichiometry and their relationships to conceptual understanding and learning approaches. *Electron. J. Sci. Educ.*, **7** (3).
22. Mitchell, I. and Gunstone, R. (1984) Some student conceptions brought to the study of stoichiometry. *Res. Sci. Educ.*, **14** (1), 78–88.
23. Cole, R. (2013) in *AP Chemistry: Guided Inquiry Activities for the Classroom* (eds T.J. Greenbowe and M. DeWane), The College Board, New York, pp. 9–16.
24. Bowen, C.W. (2000) A quantitative literature review of cooperative learning effects on high school and college chemistry achievement. *J. Chem. Educ.*, **77** (1), 116.
25. Shaver, M.P. (2010) Using low-tech interactions in the chemistry classroom to engage students in active learning. *J. Chem. Educ.*, **87** (12), 1320–1323.
26. Schell, J., Lukoff, B., and Mazur, E. (2013) Catalyzing learner engagement using cutting-edge classroom response systems in higher education. *Cutting-edge Technol. High. Educ.*, **6**, 233–261.

27. MacArthur, J.R. and Jones, L.L. (2008) A review of literature reports of clickers applicable to college chemistry classrooms. *Chem. Educ. Res. Pract.*, **9** (3), 187–195.
28. Kay, R.H. and LeSage, A. (2009) Examining the benefits and challenges of using audience response systems: a review of the literature. *Comput. Educ.*, **53** (3), 819–827.
29. Keller, C., Finkelstein, N., Perkins, K., Pollock, S., Turpen, C., and Dubson, M. (2007) Research-based practices for effective clicker use. *AIP Conf. Proc.*, **951** (1), 128–131.
30. Asirvatham, M.R. and Bierbaum, V.M. (2009) *Clickers in Action: Increasing Student Participation in General Chemistry*, WW Norton & Company.
31. Lantz, M.E. and Stawiski, A. (2014) Effectiveness of clickers: effect of feedback and the timing of questions on learning. *Comput. Hum. Behav.*, **31**, 280–286.
32. Landis, C.R. (2001) *Chemistry ConcepTests: A Pathway to Interactive Classrooms*, Prentice-Hall.
33. Turpen, C. and Finkelstein, N.D. (2010) The construction of different classroom norms during Peer instruction: students perceive differences. *Phys. Rev. ST Phys. Educ. Res.*, **6** (2), 020123.
34. Libarkin, J. (2008) Concept inventories in higher education science. BOSE Conference, 2008.
35. Evans, D., Gray, G.L., Krause, S., Martin, J., Midkiff, C., Notaros, B. M., Pavelich, M., Rancour, D., Reed-Rhoads, T., and Steif, P. (2003) Progress on concept inventory assessment tools. 33rd Annual Frontiers in Education, 2003. FIE 2003, IEEE, 2003, Vol. 1, p. T4G-1-8.
36. Hestenes, D., Wells, M., and Swackhamer, G. (1992) Force concept inventory. *Phys. Teach.*, **30**, 141.
37. Peterson, R.F., Treagust, D.F., and Garnett, P. (1989) Development and application of a diagnostic instrument to evaluate grade 11 and 12 students' concepts of covalent bonding and structure following a course of instruction. *J. Res. Sci. Teach.*, **26** (4), 301–314.
38. Treagust, D., Chandrasegaran, A.L., Crowley, J., Yung, B.W., Cheong, I.A., and Othman, J. (2010) Evaluating students' understanding of kinetic particle theory concepts relating to the states of matter, changes of state and diffusion: a cross-national study. *Int. J. Sci. Math. Educ.*, **8** (1), 141–164.
39. Treagust, D.F. (1988) Development and use of diagnostic tests to evaluate students' misconceptions in science. *Int. J. Sci. Educ.*, **10**, 159–169.
40. Treagust, D.F. (1995) in *Leaning Science in the Schools: Research Reforming Practice* (eds S. Glynn and R. Duit), Erlbaum, Hillsdale, NJ, pp. 327–346.
41. Mulford, D.R. and Robinson, W.R. (2002) An inventory for alternate conceptions among first-semester general chemistry students. *J. Chem. Educ.*, **79** (6), 739.
42. Villafañe, S.M., Bailey, C.P., Loertscher, J., Minderhout, V., and Lewis, J.E. (2011) Development and analysis of an instrument to assess student understanding of foundational concepts before biochemistry coursework. *Biochem. Mol. Biol. Educ.*, **39** (2), 102–109.
43. Krause, S., Birk, J., Bauer, R., Jenkins, B., and Pavelich, M.J. (2004) Development, testing, and application of a chemistry concept inventory. Frontiers in Education, 2004. FIE 2004, 34th Annual, IEEE, Vol. 1, p. T1G-1-5.
44. McClary, L.M. and Bretz, S.L. (2012) Development and assessment of a diagnostic tool to identify organic chemistry students' alternative conceptions related to acid strength. *Int. J. Sci. Educ.*, **34** (15), 2317–2341.
45. Bretz, S.L. and Linenberger, K.J. (2012) Development of the enzyme–substrate interactions concept inventory. *Biochem. Mol. Biol. Educ.*, **40** (4), 229–233.
46. Banerjee, A.C. (1991) Misconceptions of students and teachers in chemical equilibrium. *Int. J. Sci. Educ.*, **13** (4), 487–494.
47. Smith, J.I. and Tanner, K. (2010) The problem of revealing how students think: concept inventories and beyond. *CBE-Life Sci. Educ.*, **9** (1), 1–5.

48. Adams, W.K. and Wieman, C.E. (2010) Development and validation of instruments to measure learning of expert-like thinking. *Int. J. Sci. Educ.*, **33** (9), 1289–1312.
49. Crouch, C.H. and Mazur, E. (2001) Peer instruction: ten years of experience and results. *Am. J. Phys*, **69**, 970.
50. Asterhan, C.S. and Schwarz, B.B. (2007) The effects of monological and dialogical argumentation on concept learning in evolutionary theory. *J. Educ. Psychol.*, **99** (3), 626.
51. Sampson, V. and Clark, D. (2009) The impact of collaboration on the outcomes of scientific argumentation. *Sci. Educ.*, **93** (3), 448–484.
52. Osborne, J. (2010) Arguing to learn in science: the role of collaborative, critical discourse. *Science*, **328** (5977), 463–466.
53. Sampson, V. and Clark, D.B. (2008) Assessment of the ways students generate arguments in science education: current perspectives and recommendations for future directions. *Sci. Educ.*, **92** (3), 447–472.
54. Duschl, R.A. and Osborne, J. (2002) Supporting and promoting argumentation discourse in science education. *Stud. Sci. Educ.*, **38** (1), 39–72.
55. Becker, N., Rasmussen, C., Sweeney, G., Wawro, M., Towns, M., and Cole, R. (2013) Reasoning using particulate nature of matter: an example of a sociochemical norm in a university-level physical chemistry class. *Chem. Educ. Res. Pract.*, **14** (1), 81.
56. Aydeniz, M., Pabuccu, A., Cetin, P., and Kaya, E. (2012) Argumentation and students' conceptual understanding of properties and behaviors of gases. *Int. J. Sci. Math. Educ.*, **10** (6), 1303–1324.
57. Kulatunga, U. and Lewis, J.E. (2013) Exploration of peer leader verbal behaviors as they intervene with small groups in college general chemistry. *Chem. Educ. Res. Pract.*, **14**, 576–588.
58. Burke, K., Greenbowe, T.J., and Hand, B.M. (2006) Implementing the science writing heuristic in the chemistry laboratory. *J. Chem. Educ.*, **83** (7), 1032.
59. Walker, J.P. and Sampson, V. (2013) Learning to argue and arguing to learn: argument-driven inquiry as a way to help undergraduate chemistry students learn how to construct Arguments and engage in argumentation during a laboratory course. *J. Res. Sci. Teach.*, **50** (5), 561–596.
60. Walker, J.P., Sampson, V., and Zimmerman, C.O. (2011) Argument-driven inquiry: an introduction to a new instructional model for use in undergraduate chemistry labs. *J. Chem. Educ.*, **88** (8), 1048–1056.
61. Kulatunga, U., Moog, R.S., and Lewis, J.E. (2014) Use of Toulmin's argumentation scheme for student discourse to gain insight about guided inquiry activities in college chemistry. *J. Coll. Sci. Teach.*, **43** (5), 78–86.
62. Cole, R., Becker, N., Towns, M., Sweeney, G., Wawro, M., and Rasmussen, C. (2012) Adapting a methodology from mathematics education research to chemistry education research: documenting collective activity. *Int. J. Sci. Math. Educ.*, **10** (1), 193–211.
63. Bodner, G.M. (1987) The role of algorithms in teaching problem solving. *J. Chem. Educ.*, **64** (6), 513.
64. Cooper, M.M., Cox, C.T. Jr., Nammouz, M., Case, E., and Stevens, R. (2008) An assessment of the effect of collaborative groups on students' problem-solving strategies and abilities. *J. Chem. Educ.*, **85** (6), 866.
65. Sandi-Urena, S., Cooper, M., and Stevens, R. (2012) Effect of cooperative problem-based lab instruction on metacognition and problem-solving skills. *J. Chem. Educ.*, **89** (6), 700–706.
66. Stieff, M. and Raje, S. (2010) Expert algorithmic and imagistic problem solving strategies in advanced chemistry. *Spatial. Cognit. Comput.*, **10** (1), 53–81.
67. Gilbert, J.K. and Treagust, D.F. (2009) *Multiple Representations in Chemical Education*, Springer.
68. Kozma, R. and Russell, J. (2005) in *Visualization in Science Education* (ed. J.K. Gilbert), Springer, pp. 121–126.

69. Cooper, M.M., Grove, N., Underwood, S.M., and Klymkowsky, M.W. (2010) Lost in Lewis structures: an investigation of student difficulties in developing representational competence. *J. Chem. Educ.*, **87** (8), 869–874.
70. Kraft, A., Strickland, A.M., and Bhattacharyya, G. (2010) Reasonable reasoning: multi-variate problem-solving in organic chemistry. *Chem. Educ. Res. Pract.*, **11** (4), 281–292.
71. Linenberger, K.J. and Bretz, S.L. (2012) Generating cognitive dissonance in student interviews through multiple representations. *Chem. Educ. Res. Pract.*, **13** (3), 172–178.
72. Orgill, M. and Crippen, K. (2010) Teaching with external representations: the case of a common energy-level diagram in chemistry. *J. Coll. Sci. Teach.*, **40** (1), 78–84.
73. Towns, M. and Grant, E.R. (1997) I believe I will go out of this class actually knowing something, Cooperative learning activities in physical chemistry. *J. Res. Sci. Teach.*, **34**, 819–835.
74. Hahn, K.E. and Polik, W.F. (2004) Factors influencing success in physical chemistry. *J. Chem. Educ.*, **81**, 567–572.
75. Hadfield, L.C. and Wieman, C.E. (2010) Sudent interpretations of equations related to the first law of thermodynamics. *J. Chem. Educ.*, **87**, 750–755.
76. Dalton, R., Tasker, R., and Sleet, R. (2012) In research into practice: using molecular representations as a learning strategy in chemistry. Proceedings of The Australian Conference on Science and Mathematics Education (Formerly UniServe Science Conference), 2012.
77. Williamson, V.M., Lane, S.M., Gilbreath, T., Tasker, R., Ashkenazi, G., Williamson, K.C., and Macfarlane, R.D. (2012) The effect of viewing order of macroscopic and particulate visualizations on students' particulate explanations. *J. Chem. Educ.*, **89** (8), 979–987.
78. Abraham, M., Varghese, V., and Tang, H. (2010) Using molecular representations to aid student understanding of stereochemical concepts. *J. Chem. Educ.*, **87** (12), 1425–1429.
79. Jaber, L.Z. and BouJaoude, S. (2012) A macro–micro–symbolic teaching to promote relational understanding of chemical reactions. *Int. J. Sci. Educ.*, **34** (7), 973–998.
80. Bauer Christopher, F., Cole Renée, S., and Walter Mark, F. (2008) *Nuts and Bolts of Chemical Education Research*, vol. 976, American Chemical Society, pp. 183–201.
81. Bretz, S.L. (2013) *Trajectories of Chemistry Education Innovation and Reform*, vol. 1145, American Chemical Society, pp. 145–153.
82. Grove, N. and Bretz, S.L. (2007) CHEMX: an instrument to assess students' cognitive expectations for learning chemistry. *J. Chem. Educ.*, **84** (9), 1524.
83. Adams, W.K., Wieman, C.E., Perkins, K.K., and Barbera, J. (2008) Modifying and validating the colorado learning attitudes about science survey for use in chemistry. *J. Chem. Educ.*, **85** (10), 1435.
84. Bauer, C.F. (2008) Attitude toward chemistry: a semantic differential instrument for assessing curriculum impacts. *J. Chem. Educ.*, **85** (10), 1440.
85. Jiang, B., Xu, X., Garcia, A., and Lewis, J.E. (2010) Comparing two tests of formal reasoning in a college chemistry context. *J. Chem. Educ.*, **87** (12), 1430–1437.
86. Bunce, D.M. and Hutchinson, K.D. (1993) The use of the GALT (Group Assessment of Logical Thinking) as a predictor of academic success in college chemistry. *J. Chem. Educ.*, **70** (3), 183–187.
87. BouJaoude, S., Salloum, S., and Abd-El-Khalick, F. (2004) Relationships between selective cognitive variables and students' ability to solve chemistry problems. *Int. J. Sci. Educ.*, **26** (1), 63–84.
88. Cole, R.S. and Todd, J.B. (2003) Effects of web-based multimedia homework with immediate rich feedback on student learning in general chemistry. *J. Chem. Educ.*, **80** (11), 1338.
89. Bauer, C.F. and Cole, R. (2012) Validation of an assessment rubric via

controlled modification of a classroom activity. *J. Chem. Educ.*, **89** (9), 1104–1108.
90. Olson, S. and Riordan, D.G. (2012) Engage to Excel: Producing One Million Additional College Graduates with Degrees in Science, Technology, Engineering, and Mathematics. Report to the President. Executive Office of the President.
91. Bruck, L.B., Bretz, S.L., and Towns, M.H. (2008) Characterizing the level of inquiry in the undergraduate laboratory. *J. Coll. Sci. Teach.*, **38** (1), 52–58.
92. Fay, M.E., Grove, N.P., Towns, M.H., and Bretz, S.L. (2007) A rubric to characterize inquiry in the undergraduate chemistry laboratory. *Chem. Educ. Res. Pract.*, **8** (2), 212–219.
93. Heredia, K. and Lewis, J.E. (2012) A psychometric evaluation of the colorado learning attitudes about science survey for use in chemistry. *J. Chem. Educ.*, **89** (4), 436–441.
94. Arjoon, J.A., Xu, X., and Lewis, J.E. (2013) Understanding the state of the art for measurement in chemistry education research: examining the psychometric evidence. *J. Chem. Educ.*, **90** (5), 536–545.
95. Henderson, C., Cole, R., Froyd, J., Khatri, R., and Stanford, C. (2014) Analysis of Propagation Strategies Rubric, http://www.increasetheimpact.com/resources (accessed June 2014).
96. Crouch, C., Fagen, A.P., Callan, J.P., and Mazur, E. (2004) Classroom demonstrations: learning tools or entertainment? *Am. J. Phys*, **72**, 835.
97. Zimrot, R. and Ashkenazi, G. (2007) Interactive lecture demonstrations: a tool for exploring and enhancing conceptual change. *Chem. Educ. Res. Pract.*, **8** (2), 197–211.
98. Posner, G.J., Strike, K.A., Hewson, P.W., and Gertzog, W.A. (1982) Accommodation of a scientific conception: toward a theory of conceptual change. *Sci. Educ.*, **66** (2), 211–227.
99. Sokoloff, D.R. and Thornton, R.K. (1997) Using interactive lecture demonstrations to create an active learning environment. *Phys. Teach.*, **35** (6), 340–347.
100. Ashkenazi, G. and Weaver, G.C. (2007) Using lecture demonstrations to promote the refinement of concepts: the case of teaching solvent miscibility. *Chem. Educ. Res. Pract.*, **8** (2), 186–196.
101. Lantz, J. and Cole, R. (2007) *A New Approach to Analytical Chemistry: The Development of Process Oriented Guided Inquiry Learning Materials*, National Science Foundation: Drew University, University of Iowa.
102. Farrell, J.J., Moog, R.S., and Spencer, J.N. (1999) A guided inquiry general chemistry course. *J. Chem. Educ.*, **76** (4), 570–74.
103. Hanson, D.M. and Wolfskill, T. (1998) Improving the teaching/learning process in general chemistry. *J. Chem. Educ.*, **75**, 143–147.
104. Lewis, S.E. and Lewis, J.E. (2005) Departing from lectures: an evaluation of a Peer-led guided inquiry alternative. *J. Chem. Educ.*, **82** (1), 135–139.
105. Moog, R.S., Creegan, F.J., Hanson, D.M., Spencer, J.N., and Straumanis, A.R. (2006) Process-oriented guided inquiry learning: POGIL and the POGIL project. *Metropol. Univ.*, **17** (4), 41–52.
106. Straumanis, A. and Simons, E.A. (2008) in *Process Oriented Guided Inquiry Learning* (eds R.S. Moog and J.N. Spencer), Oxford University Press, New York, pp. 226–239.
107. Karplus, R. and Butts, D.P. (1977) Science teaching and the development of reasoning. *J. Res. Sci. Teach.*, **14** (2), 169–175.
108. Bodner, G.M. (1986) Constructivism: a theory of knowledge. *J. Chem. Educ.*, **63** (10), 873–878.
109. Musheno, B.V. and Lawson, A.E. (1999) Effects of learning cycle and traditional text on comprehension of science concepts by students at differing reasoning levels. *J. Res. Sci. Teach.*, **36** (1), 23–37.
110. American Chemical Society Division of Chemical Education Examinations Institute http://chemexams.chem.iastate.edu/ (accessed December 2013).

111. Lantz, J., Cole, R., Bauer, C., Dalton, C., Falke, A., Fischer-Drowos, S., Fish, C., Langhus, D., Riter, R., Salter, C., and Walczak, M. (2014) *Analytical Chemistry: A Guided Inquiry Approach: Quantitative Analysis Collection*, John Wiley & Sons, Inc., Hoboken, NJ.

112. Cooper, M. and Klymkowsky, M. (2013) Chemistry, life, the universe, and everything: a new approach to general chemistry, and a model for curriculum reform. *J. Chem. Educ.*, **90** (9), 1116–1122.

113. Cooper, M.M. and Klymkowsky, M.W. CLUE: Chemistry, Life, the Universe, and Everything, http://besocratic.colorado.edu/CLUE-Chemistry/ (accessed December 2013).

114. Henderson, C., Dancy, M., and Niewiadomska-Bugaj, M. (2012) Use of research-based instructional strategies in introductory physics: where do faculty leave the innovation-decision process? *Phys. Rev. Spec. Top. Phys. Educ. Res.*, **8** (2), 020104.

115. Henderson, C. and Dancy, M.H. (2009) Impact of physics education research on the teaching of introductory quantitative physics in the United States. *Phys. Rev. Spec. Top. Phys. Educ. Res.*, **5** (2), 020107.

116. Henderson, C. and Dancy, M. (2005) When one instructor's interactive classroom activity is another's lecture: communication difficulties between faculty and educational researchers. American Association of Physics Teachers Winter Meeting, Albuquerque, NM, 2005.

117. Weiss, T.H., Feldman, A., Pedevillano, D.E., and Capobianco, B. (2004) The implications of culture and identity: a professor's engagement with a reform collaborative. *Int. J. Sci. Math. Educ.*, **1** (3), 333–356.

118. Henderson, C. (2005) The challenges of instructional change under the best of circumstances: a case study of one college physics instructor. *Am. J. Phys*, **73**, 778.

119. Boice, R. (1991) New faculty as teachers. *J. Higher Educ.*, **62**, 150–173.

120. Pfund, C., Miller, S., Brenner, K., Bruns, P., Chang, A., Ebert-May, D., Fagen, A.P., Gentile, J., Gossens, S., Khan, I.M., Labov, J.B., Pribbenow, C.M., Susman, M., Tong, L., Wright, R., Yuan, R.T., Wood, W.B., and Handelsman, J. (2009) Summer institute to improve university science teaching. *Science*, **324** (5926), 470–471.

121. Fixsen, D.L., Naoom, S.F., Blase, K.A., and Friedman, R.M. (2005) *Implementation Research: A Synthesis of the Literature*, University of South Florida.

122. Sirum, K.L. and Madigan, D. (2010) Assessing how science faculty learning communities promote scientific teaching. *Biochem. Mol. Biol. Educ.*, **38** (3), 197–206.

123. Gallos, M.R., van den Berg, E., and Treagust, D.F. (2005) The effect of integrated course and faculty development: experiences of a university chemistry department in the Philippines. *Int. J. Sci. Educ.*, **27** (8), 985–1006.

124. Gurung, R.A.R. and Schwartz, B.M. (2009) *Optimizing Teaching and Learning*, Wiley-Blackwell, New York, pp. 1–17.

125. Healey, M. (2000) Developing the scholarship of teaching in higher education: a discipline-based approach. *High. Educ. Res. Dev.*, **19** (2), 169–189.

8
Research on Problem Solving in Chemistry
George M. Bodner

8.1
Why Do Research on Problem Solving?

It is slightly more than 40 years since the author began his academic career by teaching a large lecture section of a general chemistry course for science and engineering majors in the School of Chemical Sciences at the University of Illinois at Urbana-Champaign. Several things soon became apparent. First, there was the shear enjoyment he found in teaching. And then there was the pleasure he felt, at the end of the semester, when he looked at his course evaluations and found that the students reacted favorably to his course. But he clearly remembers the feeling of surprise during the course of that first semester when he looked at the results of the exams he gave in this class. In spite of what he thought were clear explanations of topics such as molarity, he found that it was difficult to create a "simple" problem which more than about 65% of the bright, hard-working students in this class could answer correctly. Eventually, the author decided to shut down his NMR research and pursue a career dedicated to trying to understand why so many "good" students struggle to understand topics (such as molarity) that have been called *threshold concepts* [1], fundamental ideas that serve as the foundation upon which an understanding of chemistry is based.

Having eventually accepted a faculty position at Purdue in chemical education, he soon encountered the constructivist theory, and began a paper summarizing some of the tenets of this theory in a language that chemists might understand. In the first paragraph of this paper, he proposed the following hypothesis: *Teaching and learning are not synonymous; we can teach, and teach well, without having the students learn* [2]. When one considers the experiences that had drawn him to a career in chemical education, it is not surprising that, as he learned how to do discipline-based educational research, he started by looking into the teaching and learning of problem solving in the large-enrollment general chemistry course for science and engineering majors he was asked to teach.

Some might ask: *Why do research on problem solving*? First of all, the author would respond, problem solving is what chemists do. The goal of his research as a graduate student and young faculty member, for example, was not to collect

Chemistry Education: Best Practices, Opportunities and Trends, First Edition.
Edited by Javier García-Martínez and Elena Serrano-Torregrosa.
© 2015 Wiley-VCH Verlag GmbH & Co. KGaA. Published 2015 by Wiley-VCH Verlag GmbH & Co. KGaA.

Fourier transform nuclear magnetic resonance (FT NMR) spectra; it was to use FT NMR as a tool to get a better understanding of the structure and bonding of the systems being studied [3].

Another reason for doing research on problem solving can be found by asking: *What do we expect chemistry students to do?* The answer is simple: Problem-solving of one form or another is a characteristic skill expected of chemistry students across the undergraduate curriculum. Consider the following end-of-chapter problem from one of the author's textbooks [4], for example:

> A lecture hall has 50 rows of seats. If laughing gas (N_2O) is released from the front of the room at the same time hydrogen cyanide (HCN) is released from the back of the room, in which row (counting from the front) will students first begin to die laughing?

Interest in problem solving is not something that is a characteristic of only chemistry or other STEM domain courses. Consider a classic question from research in cognitive science [5] that starts with a drawing of four cards, as shown in Figure 8.1.

Figure 8.1 The four-card problem.

Each of these cards is described as having a letter on one side and a number on the other side. Which card (or cards) would you have to turn over in order to find out whether the following rule is true or false?

> If a card has a vowel on one side, it has an even number on the other side.

The optimum way to test this rule might be to consider each card, one at a time. You need to turn over the card labeled "A," but not the card with the "D" because it is not a vowel. Nor do you have to turn over the card labeled "4" because the rule doesn't say that even numbers can only occur when there is a vowel on the other side. But you have to turn over the card with the "7" to make sure there is no vowel on the other side. So the answer is: Two cards, the one labeled "A" and the one labeled "7."

A third reason for doing research on problem solving reflects the importance placed on problem-solving skills by potential employers of the students who take our courses. Consider the following question from a book that describes the process by which the most competitive companies select potential employees [6]:

> Suppose you have eight balls. One of them is defective — it weighs more than the others. How do you tell, using a balance, which ball is defective in two weighings?

The author has given this problem to practicing chemists and asked them to solve it using a "think-aloud" protocol [7]. They almost always use a "trial and error" strategy, which frequently starts with a drawing or representation of a two-pan balance in which the eight balls are initially divided into two groups, as shown in Step 1 in Figure 8.2. They then divide the four balls on the heavier pan into two groups of two.

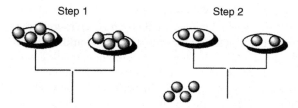

Figure 8.2 Representation of the most common approach to this problem.

This obviously doesn't work. So after a significant pause, they try something else. They often ask: What would happen if we started with three balls on each pan? If the two pans are in balance, as shown in Figure 8.3, the defective ball must be one of the two that weren't included in this step; that is, the problem is solved.

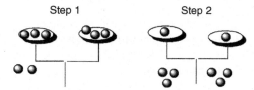

Figure 8.3 Representation of what might happen if two sets of three balls are compared in the first step.

But, they then ask: What happens if the pans are not balanced, as shown in Figure 8.4.

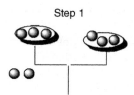

Figure 8.4 Representation of an alternative result of comparing two sets of three balls in the first step.

The defective ball must be on the pan that appears to be heavier. But our practicing chemists soon realize that they only need one more step: Compare the weight of any two of the three balls on that pan. If they are the same, the defective ball was the one left out of the second weighing. Of course, if they are not the same, the defective ball must be on the pan that seems "heavier."

The billiard-ball problem provides us with a glimpse of several ideas that will be discussed in more detail in subsequent sections. For now, let's concentrate on the idea that successful problem solvers – such as the group of practicing chemists to whom this problem has been given – often rely on a problem-solving strategy (or heuristic) that involves trial and error. Furthermore, they often rely on a simple drawing or representation that captures and organizes information when they encounter a novel problem.

Ten years ago, we tried to write a chapter covering the literature on studies of problem solving in chemistry [8]. It was a difficult task then, and even more daunting today. And there is no evidence that a chapter of that nature would be useful, other than as a critical bibliography for individuals interested in extending work of this nature. The author has therefore chosen to focus on the development of research in problem solving within his own group as the basis for readers to think about the implications of this research for their own classroom. And as a way for readers who might be interested in doing discipline-based educational research to think about how research studies build upon each other.

8.2
Results of Early Research on Problem Solving in General Chemistry

Having a strong quantitative background, it is not surprising that the first project that came to mind, when a graduate student asked the author if she could do her M.S. thesis in "chemical education" with him as her major professor, was a quantitative study of differences in problem-solving ability of the students in general chemistry; the course to which he had the easiest access.

The paper based on this study [9] concluded that problem solving "involves more than learning how to do gram-gram calculations or titration problems." It is at this point in the evolution of our model of problem solving that we argued that a particular task cannot inherently be a problem; it can simultaneously be a *routine exercise* for those with many years of experience and yet a *novel problem* for individuals encountering this kind of task for the first time. We also differentiated between an algorithm (which can be used to get an answer to a routine exercise) and a heuristic (a general strategy that can be applied to helping one extract meaning from the statement of a novel problem).

Our first paper on problem solving in chemistry featured a battery of four tests of spatial ability, including one that is known as the "Purdue Visualization of Rotations Test" or "ROT" [10]. One item from this test is shown in Figure 8.5.

In our first study, we found a small, but statistically significant correlation ($r = 0.3$, $p < 0.001$) between students' performance on the spatial tests and their performance on sub-scores from exams that grouped highly spatial concepts in the general chemistry course, such as questions on crystal structures. But correlations of similar magnitude were also seen for a sub-score that grouped questions on stoichiometry from the first exam and for the students' total score on a comprehensive final exam ($r = 0.3$, $p < 0.001$).

We had created a methodology that we hoped would differentiate between questions on concepts that were highly "spatial" and those that didn't seem to have any

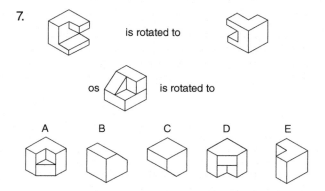

Figure 8.5 Item 7 from the Purdue Visualization of Rotation (ROT) Test.

"spatial" component, but no difference was found. What we saw was a correlation between spatial skills and students' ability to solve *novel problems* when they encountered them for the first time in an exam. We explained this by arguing that there is an early holistic stage in problem solving during which students have to "disembed" relevant information from a question and restructure the problem into one they can solve, and that this skill is similar to what happens when one has to solve problems students encounter on tests of spatial ability.

What would you expect someone to do when the results obtained in their first study seemed to be simultaneously interesting, statistically significant, and yet unexpected? The answer is obvious: Try a similar study, under more diverse conditions, that focused on the interesting but unexpected preliminary results [11]. Once again, the total sample size was large ($n = 2498$). Roughly two-thirds of these students were in an introductory course for students majoring in STEM disciplines, and the other one-third were enrolled in a similar but lower level course for students from nursing and agriculture.

An interesting result was seen when students in the two courses were compared. The average ROT exam score in the course for STEM majors ($\bar{x} = 13.96$, $\sigma = 3.96$) was significantly larger than the average score of the students in the lower level course ($\bar{x} = 11.66$, $\sigma = 3.79$): a result that was then extended across a variety of levels of first- and second-year courses [10]. But the key result can be summarized as follows: The correlations between students' scores on the spatial ability exam and their performance on sub-scores that grouped similar items taken from either mid-term exams or comprehensive final exams were largest for sub-scores that grouped questions that probed the students' problem-solving ability rather than either rote memory or the application of simple algorithms. Correlations were also large for verbally complex questions that required students to disembed and restructure the relevant information in the question.

We argued at this point in our work that the essence of problem solving can be found in the early stage(s) of students' response to a task or question, while they are building a representation that helps them "understand" the problem. We then built on that foundation in a paper that outlined the difference between what beginning instructors (e.g., teaching assistants) did when they tried to explain to their students how a problem is worked and the process these individuals had used to solve the problem themselves [12]. The teaching assistants "taught" the students how to answer a question as if it was a routine exercise when they inevitably admitted that it had been a novel problem when they encountered it for the first time.

8.3
What About Organic Chemistry

When we first thought about the possible relationship between spatial ability and performance in undergraduate chemistry courses, we thought that the most significant results would be obtained when we looked at students taking organic chemistry. It was easier to start with the students in general chemistry however, because they were a "sample of convenience" [13] – we had access to the exam data and could collect student scores on the spatial ability exams by administering them during one of the lab sessions at the beginning of the course.

The results we obtained with students in general chemistry reinforced our interest in the organic chemistry course, so we designed a study that examined the relationship between spatial ability and performance in four organic chemistry courses designed for students from a variety of majors, including agriculture, biology, health sciences, pre-med, pre-vet, pharmacy, medicinal chemistry, chemistry, and chemical engineering [14]. Three of these courses were taught in the Department of Chemistry ($n = 69$, 127, and 158), the fourth was offered to students in the Medicinal Chemistry program ($n = 68$).

It was not surprising that we found a small, but statistically significant correlation ($p < 0.01$) between students' spatial ability scores and "spatial" tasks in organic chemistry. What we found interesting was that the correlations between spatial scores and performance on exam sub-scores were statistically significant for questions that required problem-solving skills even when the questions – such as predicting the product of a reaction or outlining a multistep synthesis – did not explicitly involve "spatial" tasks. But no correlations were observed for questions that could be answered by either rote memory or the application of simple algorithms learned in class.

Because we soon recognized the limits of quantitative research methods for understanding what students do when they solve problems, we took the first step toward qualitative research methods [15, 16] in our study of organic chemistry students by analyzing their exam answers. We noted that there was a difference in the way high- versus low-spatial ability students worked exam questions. High-spatial-ability students were more likely to draw preliminary line drawings, even for questions that did not explicitly require these drawings. When questions required preliminary or extra line drawings, low-spatial-ability students were more likely to draw figures that were incorrect. They were also more likely to draw structures that were lopsided, ill-proportioned, and nonsymmetric.

Consider the differences between the way students in these two categories answered a question that asked them to predict the products of the following reaction:

PhCOOH + SOCl$_2$ →

The high-spatial-ability students were more likely to draw preliminary structures in which the "Ph" or phenyl group was represented by a six-membered ring and the "COOH" carboxylic acid group was represented by an −OH group attached to a C=O function, and they were significantly more likely to get the correct answer. Low-spatial ability students were more likely to write equations such as

PhCOOH + SOCl$_2$ → PhCl + SO$_2$ + HCl

or

PhCOOH + SOCl$_2$ → PhCOOCl + SO$_2$ + HCl

On the surface, these answers can be viewed as incorrect because the students did not get the "correct" answer (PhCOCl) and, furthermore, the equations were not balanced. On reflection, however, the most significant limitation to these answers is that one cannot get these products from the starting materials by the process by which chemical reactions occur, i.e., by the making and breaking of bonds.

All of our evidence so far had suggested that the correlation between spatial ability and student performance in both general and organic chemistry was statistically significant on tasks that measured students' problem-solving ability. We therefore changed the focus of our research to studies of problem solving that used qualitative research methods, and chose to examine the problem-solving process of students enrolled in a graduate-level course on organic synthesis [17]. The sample population for this study consisted of seven graduate students working toward a Ph.D. in organic chemistry and three graduate students from other departments whose research involved organic synthesis. The approach we took to generating qualitative data in this study was to ask the participants to use the "think aloud" approach [7] while working on a series of tasks we provided them. The students were given traditional questions such as proposing a sequence of reactions that would lead to a given target molecule, or identifying the reagents needed to transform a given starting material into the indicated product in a multistep synthesis. They were also asked to evaluate two reported syntheses of *cis*-2-methyl-5-hexanolide to determine which was "better." Because we were interested in different stages in the problem-solving process, we also asked questions such as, What would be the most difficult portion of molecules, such as those shown in Figure 8.6, to synthesize?

The results obtained in this study were analyzed within the context of three stages of the problem-solving process: construction of a representation of the problem, reformulation of knowledge to produce a portion of a solution, and a check on the usefulness and viability of the solution that was constructed.

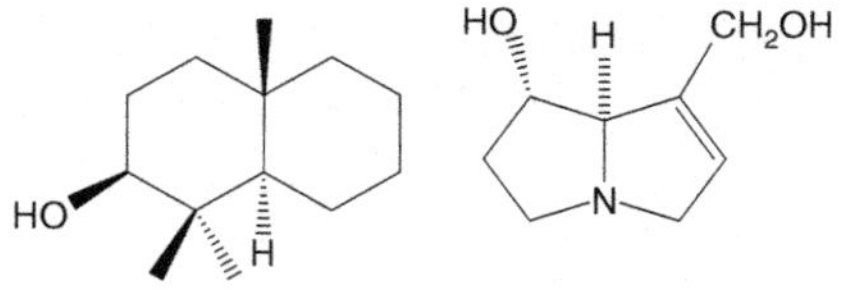

Figure 8.6 Typical items for questions that asked participants to focus on identifying the most difficult portion of a molecule to synthesize.

A subsequent study of problem solving by organic chemistry graduate students provided insight into the two characteristic phases of organic synthesis: (i) the generation of a hypothetical sequence of reactions and (ii) the process by which these reactions are carried out in the lab [18]. We concluded: "Among practicing chemists, these components are tightly linked ... Interviews with ... students involved in solving synthesis problems, however, suggest that there is little (if any) connection between the synthetic schemes the students propose on paper and the procedures that would have to be carried out in the lab ... "

The focus of our work with graduate students [19] turned to the curved-arrow/electron-pushing formalism because it is one of the most important tools in the practicing organic chemist's repertoire. A prototypical example of this formalism from the introductory organic chemistry course is shown in Figure 8.7. By convention, curved arrows show the flow of electrons from the source to the sink.

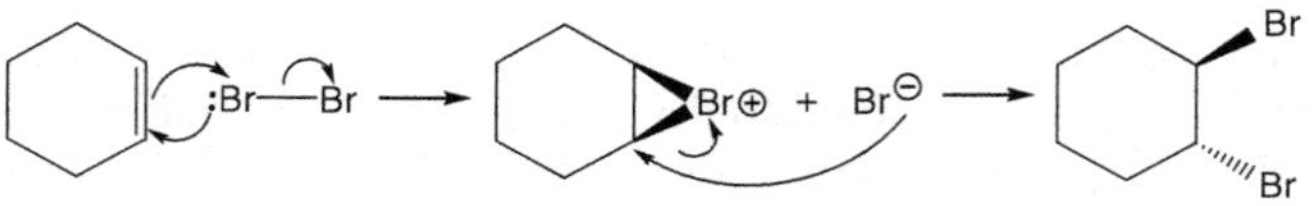

Figure 8.7 Electron-pushing formalism for the addition of bromine across the C=C double bond in cyclohexene.

The participants in our study of how graduate students use this formalism were enrolled in a first-semester graduate-level course in organic chemistry. We noted that the results of this study could be distilled into a single sentence: *The curved arrows used in the electron-pushing formalism held no physical meaning for the graduate students involved in this study.* We went on to argue: "Like the arrows drawn when equations are first used to represent simple chemical reactions, they were nothing more than a vehicle for getting from the reactants to the products." The students' comments focused almost exclusively on the *products*, not the *process*, involved in transforming starting materials into products. It is important to emphasize that "the participants in this study were the kind of students others would classify as 'good.' They were bright, conscientious, and working hard to master the elements of organic chemistry." Like so much of the research on problem solving in introductory courses, the graduate students produced correct answers to the mechanism tasks they were given without an understanding of the chemical concepts behind their responses.

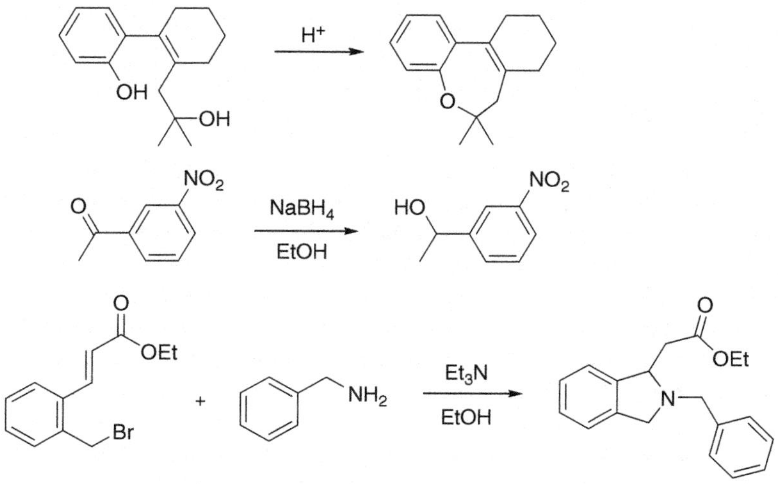

Figure 8.8 Three tasks used to probe the use of the curved-arrow symbolism by second-year organic chemistry students.

The results of work with graduate students on the use of the curved-arrow formalism led us to look at how students in the undergraduate organic chemistry course made sense of this formalism [20]. The sample population consisted of 16 chemistry or chemical engineering majors, and the interviews on which this study was based were conducted toward the end of the second semester. Three of the seven questions used in this study are shown in Figure 8.8.

As defined in a leading textbook [21], the curved-arrow/electron-pushing formalism is supposed to be a "symbolic device for keeping track of electron pairs in chemical reactions." In the paper based on our study of the undergraduate organic chemistry students, we referred to this formalism using the term "arrow-pushing." As might be expected, we received criticism from reviewers who argued: "We don't push arrows, we push electrons." We deliberately used the term *arrow-pushing* however, because our results suggest that students were not "pushing electrons," they were pushing "arrows."

Four different categories of barriers to sense-making emerged from our data. The first was labeled *inability to recall*, and surfaced when " ... participants relied on memory for an answer, as opposed to predicting the answer on the basis of a conceptual understanding, and their memory failed." A second category, *inability to apply or understand* occurred when students " ... either misapplied information they recalled from memory or did not understand the information that was remembered." The third category was used to describe situations in which the student used *poorly understood content*, which occurred in situations such as students' poorly differentiating between the concepts of nucleophile and Brønsted base, or when a word was used without a clear understanding of what it meant. Consider the following excerpt from an interview with "Barb" that occurred when she used the term *reduction* and was asked: *What does this term mean to you?*

> A reducing agent is going to take electrons, accept electrons? No, it's going to reduce the species … [pause] … OK, a reducing agent reduces this [the carbon]. So, it's going to take away a pair of electrons, I mean … I just confused myself.

We noted that Barb's use of the terms *reduction* and *reducing agent* might be considered examples of what Vygotsky [22] called a verbalism: " … a parrot-like repetition of words …, simulating a knowledge of the corresponding concepts but actually covering up a vacuum" (p. 150). We concluded that Barb knew the term *reduction* and correctly associated it with $NaBH_4$, but could not relate it to either the mechanism of the reaction, an explicit definition of the term, or to the lab experiment she had done in the course where it was used as a reducing agent. We concluded that

> On the surface, there was a resemblance between what the students drew on paper as they worked on these mechanism problems and what a practicing organic chemist would draw. In spite of this resemblance, there was little, if any, connection between what the students did and what chemists think about when they visualize a chemical reaction. For many of the students, the process of drawing curved arrows was purely mechanical; it had little (if any) intrinsic meaning. … The participants viewed arrow-pushing as an academic exercise, producing a mechanism on paper because they were asked to do this [20].

The results of a study that followed seven students through the same organic chemistry course for chemistry and chemical engineering was summarized in an article entitled "What can we do about 'Parker'?" [23]. "Parker" was a pseudonym given to a student who had been one of the top students in the general chemistry course for chemistry majors. But his experiences in the first semester of the organic chemistry course eventually led him to change his major. Parker's case study was " … extracted from an extended study of second-year chemistry majors because discussions with our colleagues who teach organic chemistry have led us to believe he is far from unique; that, for better or worse, there are a lot more 'Parkers' out there."

Parker was a bright, dedicated student, who demonstrated an excellent understanding of general chemistry concepts and principles, and began the year "enthusiastic about chemistry." But Parker never felt comfortable with his understanding of organic chemistry, and expressed the opinion that he was unable to learn the material being presented to him. So, for us, the obvious question became: *Why was Parker so unsuccessful in organic chemistry when he had been so successful in the first-year course?*

Elsewhere [24], we have argued that:

> … students who do poorly in organic chemistry … tend to handle chemical formulas and equations that involve these formulas in terms of letters

and lines and numbers that cannot correctly be called symbols because they do not represent or symbolize anything that has physical reality.

Within the context of Parker's case study, we argued that:

> ... chemical symbols — including Lewis structures, condensed or skeleton structures, and reaction mechanisms — mean very different things to students and practicing organic chemists. ... this inability to attribute useful meaning to chemical symbols can present a significant barrier to student learning.

Parker was uncomfortable with what he called "diagrams," which he encountered in the context of (i) deriving the structure of a compound from a line structure, (ii) translating 2D structures on exam questions into 3D mental images, (iii) visualizing the structure and properties of an entire molecule rather than focusing on individual atoms, and (iv) interpreting the curved-arrow, electron-pushing formalism. Parker reported no problems learning organic chemistry until he reached the point at which mechanistic reasoning became critical: " ... by the time I was at chapter eight I was so lost that I was reading it and I'd go back and read the paragraph I had just read, and I'd be like, I don't completely even remember reading any of this."

Parker provided valuable insight into the difference between students' and their instructor's perception of what is being taught. Parker argued that his primary goal was to understand *why* chemicals behaved the way they did, and he believed his instructor did not share this goal. Having attended the same lectures, we believe that the instructor did, in fact, have exactly the same goal, and that the textbook reinforced this goal of understanding the *whys* of organic chemistry. Parker was a "good student" in the sense that he attended class, read the textbook, and tried to do well in the course. But we concluded that Parker didn't believe that his instructor or the author of his textbook were interested in the *whys* of organic chemistry because of his inability to view the letters, lines, dots, and arrows with which organic chemists communicate as true symbols; because the *whys* were being communicated to him through letters, lines, numbers, and drawings that were not symbols for Parker.

Parker could be described as an instrumental learner, " ... recognizing a task as one of a particular class for which one already knows a rule" [25]. He approached organic by focusing on the rules for particular reactions. But he soon found himself encumbered with so many rules that he had difficulty remembering them. "So, the problem I guess I'm having is there are so many rules I'm struggling to associate which rules go with which ... " He also had difficulty identifying situations in which he could apply a given rule. " ... my current strategy is just: memorize the rules, and that's obviously not working. ... I need a way, I guess, to match up why the rules are with what, because just memorizing it though that way doesn't really tell me why, you know. ... "

Although qualitative results are frequently said to be incapable of being generalized, Parker provides an interesting idea for researchers interested in qualitative methods [16]. We argue that Parker is not unique to our study. Once we understood his perspective and the barriers he encountered to success in organic chemistry, we concluded that there are undoubtedly many students like him, and the probability is high that there are students like him in every organic chemistry class.

As noted above, we began our work on problem solving by looking at quantitative tasks such as those encountered by beginning students in the chapters on stoichiometry and gas law problems. But an important take-home message can be found in a study of the problem-solving ability of organic chemistry graduate students and faculty within the domain of problems that involved the determination of the structure of a molecule from the molecular formula of the compound and a combination of both IR and ^{1}H NMR spectra [26]. Whereas the problem-solving literature has often focused on the difference between *experts* and *novices*, our work has historically focused on the difference between *successful* and *unsuccessful* problem solvers at any point in their epistemic development. Thirteen of the 15 organic chemistry graduate students and faculty who participated in this study could be classified as either "more successful" or "less successful." (Two of the students were characterized by intermediate levels of success.) The "more successful" participants adopted a consistent approach to solving these problems, were better at mining the spectral data, and were more likely to check their final answer against the spectra upon which their answer was based. But there was one important characteristic of the "more successful" participants that we would like to emphasize, because it has been seen in so many of the domains of problem solving we have studied: The "more successful" problem solvers were more likely to write down fragments of the structure of the molecule as they were deduced during the intermediate stages in the problem-solving process. The "less successful" problem solvers tried to solve the problem in their minds, without writing anything down.

This result is consistent with the results of a study of how sophomore organic chemistry students solve "synthesis" problems [27]. Successful students within this nonmathematical problem-solving domain were more likely to "write something down." Unsuccessful students tended to exhibit a phenomenon that might be described as a "dead start." The idea of a "dead end" is familiar to anyone who has graded students work; it is exhibited when the student makes some progress toward the solution but then seems to be unable to further cross the gap between the initial information and the solution to the problem being solved. A "dead start," on the other hand, is characterized by the student returning a blank sheet of paper, which could be given to the next student being interviewed because there is nothing written on the page.

8.4
The "Problem-Solving Mindset"

A qualitative study of both chemistry and physics majors enrolled in courses on quantum mechanics at the undergraduate level provided insight into the "problem-solving mindset" these students brought to their quantum mechanics

Figure 8.9 An example of a task second-year organic chemistry students were asked that ask them to propose a synthesis of limonene.

courses [28], which can be described as follows: Propose a synthesis of linonene whose structure is shown in Figure 8.9:

> The approach that the majority of the students in this study took to learning quantum mechanics was based on a single, common, unstated assumption: their goal was to solve problems. This assumption was so pervasive it can be best termed a "problem-solving mindset." Students with this mindset organize their behavior and thinking around the idea that the main objective in this class, as in so many other courses they have taken, was to solve problems. Moreover, students with this mindset perceive that the rationale for taking the course was to learn additional ways of solving new and more comprehensive types of problems.

The "problem-solving mindset" had an effect on three aspects of the student's experience in class: (i) the expectations the students had of what the class *should* be like, (ii) the students' beliefs about what material in the course would be useful, and (iii) the behaviors the students exhibited to get through the course.

The students expressed dissatisfaction with the course because they expected to see numerical examples, and believed that numerical examples would overcome all of the troubles they were having with the course. They expressed the belief that examples are valid only when they are numerical; non-numerical examples were viewed as more "theory." As might be expected from the "problem-solving mindset," the students also expected answers to be "correct," when the appropriate equation was correctly applied to a problem, for example, it should give the correct, real-world value.

Another consequence of the "problem-solving mindset" could be seen in the students' desire to learn material that was "useful." For many students in this study, usefulness was equated with the ability to solve problems.

The effect of the problem-solving mindset on the students' behavior was clear: The students seldom focused on the conceptual aspects of the material. This could best be seen in the way they studied for exams. They tended to believe that there were many equations they had to learn, and adopted the technique of brute force memorization in the sense of rote or nonmeaningful learning [29].

8.5
An Anarchistic Model of Problem Solving

The long-term goal of our work has been to develop a model of problem solving that has three fundamental characteristics. First, it has to be consistent with what we have observed when *successful* problem solvers solve *novel* problems. Second, it has to be *teachable*; we don't want another model that explains why students are not successful. We want a model we can give to students such that, when they use it, they become better problem solvers. Finally, it has to be *transferable*; we want a model that can be used in other chemistry courses as well as in the courses for which chemistry is a prerequisite.

An important step toward this model involves providing, at last, explicit definitions of the terms *problem* and *problem solving*. Hayes [30] defined a problem as follows: "Whenever there is a gap between where you are now and where you want to be, and you don't know how to find a way to cross that gap, you have a problem." Wheatley [31] captured the essence of problem solving when he argued that problem solving is: "What you do when you don't know what to do." As we have noted elsewhere [32], these definitions provide the basis for our distinction between tasks that are *routine exercises* and those that are *novel problems*. Consider the following questions the author has used on exams, for example, taken from Bodner and Herron [8]:

> What weight of oxygen is required to burn 10.0 g of magnesium?
>
> $2Mg(s) + O_2(g) \rightarrow 2MgO(s)$

And:

> Robinson annulation reactions involve two steps: Michael addition and aldol condensation. Assume that Michael addition leads to the following intermediate. What would be produced when this intermediate undergoes aldol condensation?

The first question was once a problem for the readers of this chapter, but is not likely to be more than a routine exercise for them at this point in their career. We have found that the second question is far more likely to be a novel problem for most people when it is used in a seminar, but it is a routine exercise for organic chemists because the task is so familiar to them.

The distinction between the concepts of *exercise* and *problem* can be understood in terms of a common observation of what happens when a student asks a teaching assistant for help with a problem [12]. Over and over again, I have sat in the back of a recitation session, observing a TA. When the class is over, the students often come over to me and say: "I understood what the TA told us when he described how to 'solve' the problem we asked about. What I don't understand is how he knew how to do this?"

Models of problem solving go back at least 100 years, to the work of John Dewey [33]. But the model upon which so much of the early problem-solving literature was based appeared in a book published by George Polya in 1945 [34]. Polya proposed a stepwise model of problem solving that begins by trying to understand the problem. One then devises a sequence of steps; a plan for solving the problem. Then one carries out the plan, and, finally, one looks back to evaluate the answer and to consolidate any gains in terms of an understanding of related problems one might encounter in the future. It is easy to challenge practicing chemists' confidence in the legitimacy of this model by asking them to think about what they would do while trying to solve the following problem taken from a general chemistry textbook [35]:

> A sample of a compound of xenon and fluorine was confined in a bulb with a pressure of 24 torr. Hydrogen was added to the bulb until the pressure was 96 torr. Passage of an electric spark through the mixture produced Xe and HF. After the HF was removed by reaction with solid KOH, the final pressure of xenon and unreacted hydrogen in the bulb was 48 torr. What is the empirical formula of the xenon fluoride in the original sample?

We have given this question to Ph.D. level organic chemists because it is not likely to be a routine exercise for them. When we ask them to solve it while talking aloud, the audiotape of the interview clearly shows that they are not using Polya's model. So far, they have all gotten the "correct" answer, but the path along which they proceed is anything but a straight line that would characterize the steps of "devising a plan" and then "carrying out the plan." It is interesting to note that, when they eventually get an answer, they often comment: "Oh, it's an empirical formula problem." Our experience using the think-aloud technique with this and a variety of other tasks suggests that virtually all of the "problem solving" occurs during the stage in the problem-solving process when one is trying to "understand the problem."

We can summarize the results of our studies of successful problem solvers working on novel problems across a wide variety of tasks in terms of a model that is *anarchistic* in the sense of "lacking order or control." It begins with the following steps:

1) Read the problem
2) Read the problem again
3) Draw a picture that represents the structure of the system being modeled
4) Write down what you *hope* is the relevant information
5) Try something
6) Try something else
7) See where this gets you.

It is worth noting that one of the features of a task that makes it a novel problem might be some uncertainty on what is the relevant information. Successful problem solvers often miss what will turn out to be important information in this phase of the problem-solving process, or write down information that is not going to be useful. But their goal is to organize the information in the problem. It is also worth noting that students who struggle with chemistry problems have often told us that "trial and error" is not a legitimate problem-solving strategy. This scares the author because it seems to be one of the most powerful strategies the successful participants in our problem-solving interviews possess. It is also important to emphasize the role that drawing a picture can play in the process successful problem solvers use when they encounter a novel problem.

Whereas routine exercises are often worked in a logical, forward-chaining manner, novel problems are worked in a cyclic, reflective approach that may appear irrational to subject-matter experts watching us struggle with the task. The next stage in the problem-solving process therefore often requires that we return to the process that has gotten us this far.

8) Read the problem again
9) Try something else
10) See where this gets you.

Something often happens while successful problem solvers are working on a problem. They seem to reach a point where an important step takes place:

11) Test intermediate results to see whether you are making any progress toward an answer
12) Read the problem again.

With the tongue firmly in cheek, we now invoke a 13th step:

13) When appropriate, strike your forehead and say, "son of a gun."

The next steps are easy to imagine:

14) Write down "an" answer (not necessarily "the" answer)
15) Test the answer to see if it makes sense
16) Start over if you have to, celebrate if you don't.

Students often write absurd answers to questions because they either can't or won't test them to see if they make sense. This is something that successful problem solvers do when working on a novel problem, but not when they are working on a routine exercise.

Although the anarchistic model of problem solving was first derived from data based on the process successful problem solvers applied to mathematically oriented problems encountered toward the beginning of the first-year college or university chemistry course, we have since found that it works equally well when applied to nonmathematical domains, such as proposing a synthetic pathway. The model of problem solving that most closely fit the results of the study of sophomore organic chemistry students solving problems such as creating a multistep synthesis of compounds such as limonene, for example, was the anarchistic model of problem solving we had proposed many years earlier [27].

The importance of a drawing as a means of organizing information during the problem-solving process can be illustrated by considering the following problem from a typical introductory college-level chemistry course: *What is the solubility*

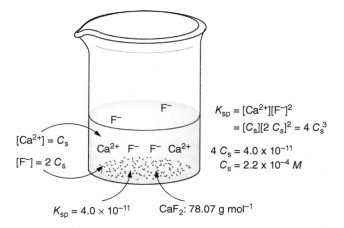

Figure 8.10 Drawing that can be used to organize students' work when they are asked to predict the solubility of CaF_2 in grams per liter.

in grams per liter of calcium fluoride? While doing this in class, we start with a simple drawing that emphasizes that there are twice as many F^- ions as Ca^{2+} ions in this solution and note the magnitude of the equilibrium constant for this system. We can then use this diagram to organize the calculation of the solubility of the salt, in moles per liter, as shown on the right side of the drawing in Figure 8.10.

Once we calculate the number of grams per mole of CaF_2, we are ready to calculate the solubility of the salt in the specified units: 0.017 g/l. In other words, the solubility of CaF_2 was too small when it was first considered as a potential source of the fluoride ion for use in toothpaste. As a result, a more soluble salt (SnF_2) was used.

The impact of a drawing like this on students' understanding of solubility calculations can be demonstrated by asking students: *Which of the following equations correctly describes the relationship between the concentration of the Ca^{2+} and F^- ions in a saturated solution of CaF_2 in water? $[Ca^{2+}] = 2[F^-]$ or $[F^-] = 2[Ca^{2+}]$.* For years, the author has watched significant fractions of the students in an introductory course for science and engineering majors vote for one of these equations while another significant fraction of the students voted for the other. And, as you might expect, a third significant fraction raised their hand when asked: *How many of you are uncertain about which is the correct equation?* When he then creates a drawing such as the one in the center of Figure 8.10, which illustrates the fact that starting with a handful of CaF_2 formula units we get twice as many F^- ions as Ca^{2+}, he notes that he can almost always hear laughter as students who were confused realize how to get the right answer.

The author's experience over the course of many years, in various courses, has suggested that students who pick up this approach to problem solving do better than those who do not. To illustrate the power of this approach, let's consider a question from the homework assignment in the author's physical chemistry course for students from the life sciences as adapted from a well-known biophysical chemistry textbook [36].

> Myoglobin is a skeletal muscle protein that binds oxygen. The equilibrium constant for the following reaction is 1.85×10^5 at 25 °C and pH 7.
>
> $$\text{Myoglobin} + O_2 \rightleftharpoons \text{Oxymyoglobin}$$
>
> Assume ideal gas behavior for O_2, that Henry's law is valid for O_2 dissolved in water, and that the standard state for O_2 dissolved in water is its concentration in molarity. What is the ratio of the oxygenated to nonoxygenated forms of myoglobin at equilibrium when the partial pressure of oxygen is 30 mmHg if the Henry's law constant for oxygen dissolved in water is 43×10^3 atm?

It should come as no surprise that students have difficulty with this problem when they first encounter it. But an interesting observation has been made when students who come for advice about this problem are given help on creating a drawing, such as Figure 8.11, that captures the relevant information in the statement of the problem.

This illustration tries to capture the fact that the system consists of three components at equilibrium (myoglobin, oxygen, and oxygenated myoglobin), and that it is a Henry's Law problem that probes the implications of the fact that there is a partial pressure of oxygen gas pushing down on the solution.

We can now organize our work around this figure. We might start by asking: *What can we do based on the fact that Henry's law is invoked in this problem*? There is no guarantee that this would be useful, but a "trial and error" strategy might suggest that we substitute what we know into the equation for Henry's law and solve for the only unknown: the mole fraction of O_2. We can then ask: *What do we know about the concept of mole fraction*? And recognize that the number of moles of oxygen in this solution is a very small fraction of the total number of moles of both water and oxygen. This means that we can calculate the concentration of dissolved oxygen in moles per liter by assuming that we have 1 l of water, and therefore 55.5 mol of water as shown in Figure 8.12.

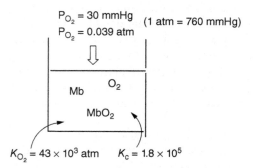

Figure 8.11 The first step in organizing information.

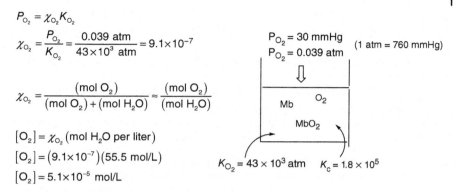

Figure 8.12 "Try something" applied to the information in Figure 8.11.

We can now turn to the equilibrium constant expression Figure 8.13

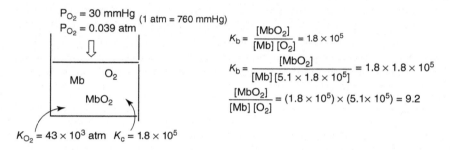

Figure 8.13 The result of "trying something else."

Nothing we have done so far makes the problem "easy," but it gives us a basis for thinking about the problem.

8.6 Conclusion

Over a career in discipline-based educational research that spans roughly 30 years, our group has tried to study the problems associated with the teaching and learning of chemistry, and more recently engineering, in a variety of different context, and at levels ranging from the introductory college-level course to the end of graduate school. Along the way, we have found certain characteristics that make someone a good problem solver in chemistry. Successful problem solvers

- are able to "disembed" the relevant information from the statement of the problem;
- often draw a figure that helps them construct a representation (or, at times, multiple representations) that organizes this information so it can be used in the problem-solving process [37];

- are willing to "try something," even if they don't know where this will lead, as a step in the problem-solving process. Unlike beginners, they seldom suffer from a "dead start";
- often step back from the problem-solving process, to see if they are getting anywhere;
- are likely to keep track of what they have achieved, so that they can combine information, as needed, to get to an answer;
- when working on a novel problem outside their immediate area of expertise, they tend to check their answer to see if it makes sense.

On the basis of more than 30 years of discipline-based research and more than 40 years of experience teaching undergraduate chemistry courses at all levels of the curriculum, the author has been able to boil advice given to students down to two simple ideas. First, organize the problem-solving process around a drawing that captures the physical reality of the system and the relevant information extracted both from the statement of the problem and the thought process applied to working the problem. Second, when in when in doubt: Try something, and see where it gets you. When discussing this work with other chemistry instructors, he encourages them to think carefully about how these ideas are communicated to students. He recommends that you never tell the students that this is something *they* should do; tell them this is something that *you* (as their instructor) have to do in order to solve novel problems.

References

1. Meyer, J.H.F. and Land, R. (2006) *Overcoming Barriers to Student Understanding: Threshold Concepts and Troublesome Knowledge*, Routledge, Oxon.
2. Bodner, G.M. (1986) Constructivism: a theory of knowledge. *J. Chem. Edu.*, **63**, 873–878.
3. Bodner, G.M. and Todd, L.J. (1974) A Fourier transform carbon-13 nuclear magnetic resonance study of arenetricarbonylchromium complexes. *Inorg. Chem.*, **13**, 360–363.
4. Bodner, G.M. and Pardue, H.L. (1994) *Chemistry: An Experimental Science*, 2nd edn, John Wiley & Sons, Inc., New York.
5. Johnson-Laird, P.N. and Wason, P.C. (1977) *Thinking: Readings in Cognitive Science*, Cambridge University Press, Cambridge, pp. 143–157.
6. Poundstone, W. (2003) *How Would You Move Mount Fuji? Microsoft's Cult of the Puzzle: How the World's Smartest Companies Select the Most Creative Thinkers*, Little, Brown and Company, New York.
7. Ericsson, K.A. and Simon, H.A. (1993) *Protocol Analysis: Verbal Reports as Data*, Revised edn, MIT Press, Boston, MA.
8. Bodner, G.M. and Herron, J.D. (2002) in *Chemical Education: Research-based Practice* (eds J.K. Gilbert, O. DeJong, R. Justi, D.F. Treagust, and J.H. van Driel), Kluwer Academic Publishers, Dordrecht, pp. 235–266.
9. Bodner, G.M. and McMillen, T.L.B. (1986) Cognitive restructuring as an early stage in problem solving. *J. Res. Sci. Teach.*, **23** (8), 727–737.
10. Bodner, G.M. and Guay, R.B. (1997) The purdue visualization of rotations test. *Chem. Educ.*, **2** (4), 1–18.
11. Carter, C.S., LaRussa, M.A., and Bodner, G.M. (1987) A study of two measures of spatial ability as predictors of success in different levels of general chemistry. *J. Res. Sci. Teach.*, **24**, 645–657.

12. Bodner, G.M. (1987) The role of algorithms in teaching problem solving. *J. Chem. Ed.*, **64** (6), 513–514.
13. Patton, M.Q. (2002) *Qualitative Research and Evaluation Methods*, 2nd edn, Sage Publications, Thousand Oaks, CA.
14. Pribyl, J.R. and Bodner, G.M. (1987) Spatial ability and its role in organic chemistry: a study of four organic courses. *J. Res. Sci. Teach.*, **24**, 229–240.
15. Bodner, G.M. (2004) Twenty years of learning how to do research in chemical education. *J. Chem. Educ.*, **81** (5), 618–628.
16. Bodner, G.M. and Orgill, M. (2007) *Theoretical Frameworks for Research in Chemistry/Science Education*, Prentice-Hall, Englewood Cliffs, NJ.
17. Bowen, C.W. and Bodner, G.M. (1991) Problem-solving processes used by graduate students while solving tasks in organic synthesis. *Int. J. Sci. Educ.*, **13**, 143–158.
18. Bhattacharyya, G., Calimisiz, S., and Bodner, G.M. (2004) Strange bedfellows: organic synthesis and essay writing. *IEEE Trans. Prof. Commun.*, **46** (4), 320–326.
19. Bhattacharyya, G. and Bodner, G.M. (2005) It gets me to the product: how students propose organic mechanisms. *J. Chem. Educ.*, **82**, 1402–1407.
20. Ferguson, R.L. and Bodner, G.M. (2008) Making sense of arrow-pushing formalism by chemistry majors enrolled in organic chemistry. *Chem. Educ. Res. Pract.*, **9**, 102–113.
21. Loudon, G.M. (1995) *Organic Chemistry*, 3rd edn, Benjamin/Cummings Publishing Company, Menlo Park, CA.
22. Vygotsky, L. (1986) *Thought and Language*, The MIT Press, Cambridge, MA.
23. Anderson, T.L. and Bodner, G.M. (2008) What can we do about "Parker"?: a case study of a good student who didn't "get" organic chemistry. *Chem. Educ. Res. Pract.*, **9**, 93–101.
24. Bodner, G.M. and Domin, D.S. (2000) Mental models: the role of representations in problem solving in chemistry. *Univ. Chem. Educ.*, **4**, 24–29.
25. Skemp, R.R. (1979) *Intelligence, Learning and Action: A Foundation for Theory and Practice in Education*, John Wiley & Sons, Inc., New York.
26. Cartrette, D.P. and Bodner, G.M. (2010) Non-mathematical problem solving in organic chemistry. *J. Res. Sci. Teach.*, **47** (6), 643–660.
27. Calimsiz, S. (2003). How undergraduates solve organic synthesis problems: a problem solving model approach. Unpublished M.S. thesis, Purdue University, West Lafayette, IN.
28. Gardner, D.E. and Bodner, G.M. (2007) *Advances in Teaching Physical Chemistry*, ACS Symposium Series, American Chemical Society, Washington, DC, pp. 155–173.
29. Ausubel, D., Novak, J., and Hanesian, H. (1978) *Educational Psychology: A Cognitive View*, 2nd edn, Holt, Rinehart, and Winston, New York.
30. Hayes, J.R. (1989) *The Complete Problem Solver*, 2nd edn, Lawrence Erlbaum, Hillsdale, NJ.
31. Wheatley, G.H. (1984) Problem Solving in School Mathematics. MEPS Technical Report 84.01, School Mathematics and Science Center, Purdue University, West Lafayette, IN.
32. Bodner, G.M. (1991) in *Toward a Unified Theory of Problem Solving: Views from the Content Domain* (ed. M.U. Smith), Lawrence Erlbaum Associates, Hillsdale, NJ, pp. 21–34.
33. Dewey, J. (1910) *How We Think*, Heath and Company, Boston, MA.
34. Holtzlaw, H.F., Robinson, W., and Nebergall, W.H. (1984) *General Chemistry*, 7th edn, D. C. Heath, Lexington, MA.
35. Polya, G. (1945) *How to Solve It; A New Aspect of Mathematical Method*, Princeton University Press, Princeton, NJ.
36. Tinoco, I., Sauer, K., and Wang, J.C. (1995) *Physical Chemistry: Principles and Applications in Biological Sciences*, 3rd edn, Prentice Hall, Upper Saddle River, NJ.
37. Domin, D. and Bodner, G.M. (2012) Using students' representations constructed during problem solving to infer conceptual understanding. *J. Chem. Educ.*, **89**, 837–843.

9
Do Real Work, Not Homework
Brian P. Coppola

9.1
Thinking About Real Work

9.1.1
Defining Real Work: Authentic Learning Experiences

In his 1997 essay, "Situated Cognition and How to Overcome It," Carl Bereiter [1] uses the fictionalized experiences of two students, Flora and Dora, to make an important and often-repeated point about learning. Both of these students pass the same Algebra I course with flying colors. But in Algebra II, while Flora continues her success, Dora barely passes. Bereiter's point is that in Flora's case, she actually learned the mathematics of algebra in her Algebra I class, and so she could transfer (transport) this learning to Algebra II. Dora, it is supposed, did not *learn algebra*, but *learned about doing algebra problems*, and succeeded in Algebra I through rote repetition and recognition.

Many versions of the Flora/Dora story have featured in discussions of science learning, where contrast is made between *meaningful learning* and *rote learning* [2, 3], or the distinction between *learning about science, learning science*, and *learning to be a scientist* is emphasized [4]. Unlike a traditional apprenticeship, where one learns specific artisanship in a materially relevant setting under the direction of a master craftsperson, school learning is distilled, abstracted, and idealized, with the validation of correct answers standing in as the nearly universal indicator of learning. Even at the introductory college level, there is compelling evidence that students can produce or select the proper answer in a way that is disconnected from a deep understanding of the underlying subject matter [5–10].

As one proceeds in higher education, it becomes increasingly more likely that science instructors are skilled in the art of their specific discipline-centered subject, and so the chances increase that the limitations and basis for this knowledge will feature more prominently than it might have at the introductory level, and that science-related values, such as skepticism, are appended to the lessons [11]. Yet, the main experience students have, even in upper level science classes, is the canonized textbook, its in-text homework exercises, and the sort of testing that allows both the Floras and the Doras to do well.

Over in the School of Art and Design, things are different. Here, the work is automatically more authentic: drawing, painting, and sculpting classes result in artifacts that can be scrutinized by practitioners for their direct, real-world merit, as does public performance of dance, music, or theater. While there is no guarantee that each student taking classes in the arts is investing the same level of purpose, meaning, emotion, and experience in conveying a story that others can or will reflect upon, *learning to be an artist* in a standard art class, through the production of and conversation about art [12], is a more likely outcome than *learning to be a scientist* is in a standard science class.

Stein *et al.* have reviewed the use of the authentic learning experiences in the context of university-level instruction [13]. They have synthesized a definition of authentic classroom practice (p. 241) as "that which reflects, for the students, a combination of personal meaning and purposefulness within an appropriate social and disciplinary framework. The learning experience is authentic for the learner while simultaneously being authentic to a community of practice." The key features embedded here are (i) work that matters outside the context of an assignment whose sole outcome is allocating course points (i.e., a painting is hung in a gallery for contemplation and reflection by observers) and (ii) work that is new, and not derivative or replicative, and which can be treated as any other work created by one who is practiced in the art (i.e., spectroscopic data collected on a purified product from a reaction that has never been carried out previously).

Herrington has contributed significantly to the development and identification of useful design elements for authentic tasks [14] as well as the instructional design to support authentic learning [15], in which "learners must be engaged in an inventive and realistic task that provides opportunities for complex collaborative activities" [16, p. 1]. Herrington's catalog of design elements derives from a variety of researchers and theorists (Table 9.1), and is neutral to the mode of delivery, instead focusing on concrete analytical features that can be easily assessed.

Lombardi has also reviewed the use of the term "authentic learning" [40], and contributes to placing it into the larger context of real classrooms, and how changing one element (e.g., student roles) will necessarily cascade throughout the entire learning environment (e.g., learning goals, instructor roles, content, assessment). Intentional and deliberate change in one part of a complex, interlocked system with lots of moving parts requires corresponding changes in expectations, performance, training, and support in the other components. The rapid emergence and proliferation of new in-class and online instructional methods has disrupted, in many cases, what Elmore calls the "instructional core" [41] and can result in fragmentation and incoherence in the learning environment [42–44].

In his seminal review, "Academic Work," Doyle [45] covers the research in which the nature of an academic task dictates or even predicts the corresponding academic performance (p. 165).

"A comparison of memory and comprehension tasks suggests that preparation suitable for one type may not necessarily be suitable for the other. Accomplishing

Table 9.1 Design elements for authentic learning tasks.

Real-world relevance	Authentic activities match the real-world tasks of professionals in practice as nearly as possible. Learning rises to the level of authenticity when it asks students to work actively with abstract concepts, facts, and formulae inside a realistic – and highly social – context mimicking "the ordinary practices of the (disciplinary) culture" [17–25]
Ill-defined problem	Challenges cannot be solved easily by the application of an existing algorithm; instead, authentic activities are relatively undefined and open to multiple interpretations, requiring students to identify for themselves the tasks and subtasks needed to complete the major task [25–27]
Sustained investigation	Problems cannot be solved in a matter of minutes or even hours. Instead, authentic activities comprise complex tasks to be investigated by students over a sustained period of time, requiring significant investment of time and intellectual resources [18, 26–28]
Multiple sources and perspectives	Learners are not given a list of resources. Authentic activities provide the opportunity for students to examine the task from a variety of theoretical and practical perspectives, using a variety of resources, and requires students to distinguish relevant from irrelevant information in the process [22, 27–29]
Collaboration	Success is not achievable by an individual learner working alone. Authentic activities make collaboration integral to the task, both within the course and in the real world [22, 26, 30]
Reflection (metacognition)	Authentic activities enable learners to make choices and reflect on their learning, both individually and as a team or community [22, 30, 31]
Interdisciplinary perspective	Relevance is not confined to a single domain or subject matter specialization. Instead, authentic activities have consequences that extend beyond a particular discipline, encouraging students to adopt diverse roles, and think in interdisciplinary terms [18, 32]
Integrated assessment	Assessment is not merely summative in authentic activities but is woven seamlessly into the major task in a manner that reflects real-world evaluation processes [33–35]
Polished products	Conclusions are not merely exercises or substeps in preparation for something else. Authentic activities culminate in the creation of a whole product, valuable in its own right [30, 36, 37]
Multiple interpretations and outcomes	Rather than yielding a single correct answer obtained by the application of rules and procedures, authentic activities allow for diverse interpretations and competing solutions [27, 32, 37–39]

Adapted from Ref. [14]. by Jan Herrington, with permission from Jan Herrington.

a comprehension task can, because of the effects of semantic integration, interfere with the ability to reproduce specific facts or the surface features of the original text. On the other hand, accomplishing a memory task can produce knowledge in a form that is not easily applied to recognizing new instances or making inferences to new situations. Thus, reading for comprehension may be inappropriate for a recall task. It is probably for this reason that students typically adjust study strategies to fit the nature of the test they expect to take. A parallel argument can be made for procedural and comprehension tasks. Learning to use an algorithm does not necessarily enable one to understand why the algorithm works or when to use it. Similarly, learning to understand why an algorithm works or when it should be used does not necessarily lead to computational proficiency."

Bereiter supposes that Dora has approached her learning of Algebra I as an example of what Doyle calls "school work" [1] (memory, recall, algorithm), while Flora, in the same classroom situation, made the connection to "real life" needs (building on prior knowledge, application, evaluation). In Algebra I, both of these strategies worked; in Algebra II, they did not. To be clear: Doyle would argue that Flora is also using memorization and recall because these "school work" strategies accomplish different things than the "real life" strategies, and what she may have is a better metacognitive sense of how to manage and decide what to do and when. An underlying challenge in teaching is how to design instructional tasks to elicit and support the decisions Flora made about her learning, and how to keep those made by Dora from succeeding, even in the short term, and even if and when they have been successful in the past.

In their 1990 book *Teaching Writing that Works*, Rabkin and Smith [46] contrast explicitly different instructional strategies and associated student behaviors they have observed in English composition classes, and provide strong and practical recommendations that reflect nearly all of Herrington's design elements for authentic learning. The key underpinning of their approach is to put every aspect of writing, as is true for any of us who write for any reason whatsoever, into a larger social and intellectual context. Writing results from a negotiated understanding, within student groups, of purpose, meaning, and audience, and it improves through evaluation, editing, and feedback that are as critical for students to give, as they are to receive. Rabkin and Smith summarize their approach in the clear and concisely stated principle "real work is better than home work" (p. 159), which Rabkin has further elaborated [47], and which has inspired the title for this chapter.

9.1.2
Doing Real Work: Situated Learning

From a science education perspective, the Dora/Flora problem often appears when thinking about laboratory instruction. There is a core belief that students are not *doing science*, and are only *learning about science*, if they do not participate in experimental design, data collection, data analysis, and other components of experimental laboratory science. The influential arguments for this position come from Gabel [48], Lunetta [49], and Hofstein [50].

From a theoretical perspective, situated cognition provides a convincing framework for understanding how *thinking like a scientist* is coupled with the social settings in which science is done, or situated [237]. Sweeney and Paradis [51, p. 195] assert that "students are unable to fully appreciate the scientific method and the essence of scientific inquiry unless they have the opportunity to acquire and analyze data first-hand." They contend that the fundamental postulate of situated cognition theory [17], that knowing and doing cannot be separated, is key to the design of science learning environments.

Critics insist that the locus of control of cognition is in the mind-brain [52] and not the physical setting where the brain happens to be located. They argue for context as a necessary but insufficient feature to understanding learning. Bereiter [1] appeals to the community to overcome its obsession with situated cognition because, after all, both Flora and Dora were in exactly the same situation but their learning outcomes were divergent. A more moderate view integrates these positions by pointing out that a setting, such as a research laboratory, reflects an environment in which situated cognition theory is fruitful for understanding scientific practices and the production of knowledge (i.e., you are less likely to learn it in the kitchen), and "follows the literature in terms of communities of practice, cognitive apprenticeship, scaffolded learning, affordances, constraints, and the production of valued products, through a social epistemology" [53] and where the "agency and intention of people" matter (p. 310). In fact, Bereiter agrees with the importance of teaching intentionality [54].

Historically, the rhetoric of situated learning followed on the heels of situated cognition. In reflecting on the development of her ideas on situated learning [55], Lave also "take[s] issue with ... work characterized [using either/or characterizations of learning], for it either maintains overly simple boundaries between the individual (and this is the "cognitive") and some version of a world "out there," or turns into a radical constructivist view ... Learning, it seems to me, is neither wholly subjective not fully encompassed in social interaction, and it is not constituted separately from the social world (with its own structures and meanings) of which it is part" [56, p. 64].

Situated learning theory was the foundation for the development of authentic learning pedagogy. Prior to her development of the design elements for authentic learning environments [14], Herrington was interested in the critical characteristics of situated learning [57], particularly regarding electronic media: "computer-based applications are a further step removed from real life work situations, and criticisms have been leveled at computer-based materials that claim to use a situated learning framework in their design." Herrington's list of design elements for authentic learning [57] aligns with features of McLellan's elements of situated learning [237], and with the later work on authentic tasks [14]. In addition, there is substantial correspondence with Rabkin's notion of "Real Work" in designing writing assignments (Table 9.2).

Table 9.2 Alignment of design elements for situated learning (1997) versus authentic learning (2000), authetnic tasks (2004) and real work (1990).

Situated learning [237]	Authentic learning [57]	Authentic tasks [14]	Real work [46]
	Authentic context	Real-world relevance	Purpose
	Authentic activities	Ill-defined problem	
		Sustained investigation	
Multiple perspectives	Multiple perspectives	Multiple sources and perspectives	Audience
Collaboration	Collaborative construction of knowledge	Collaboration	Collaboration
Reflection	Reflection	Reflection (metacognition)	Editing and feedback
	Authentic assessment	Integrated assessment	Integrated assessment
Cognitive apprenticeship	Access to expert performance and models		
Coaching	Coaching and scaffolding		
Articulation of learning skills, Stories	Articulation		Making meaning
		Interdisciplinary perspective	
		Polished products	Polished products
		Multiple interpretations and outcomes	
Technology			

Since the early 1990s, my collaborators and I have been interested in designing chemistry learning environments that feature many of the elements present in both authentic and situated learning, and at looking at others' work through this lens. We adopted Rabkin's *Real Work* rhetoric because of its intuitive appeal.

Our inspiration for thinking about *Real Work* in undergraduate chemistry instruction derives from the studio instruction model that dominates the visual and performance arts, where young painters learn to paint and young dancers learn to dance. Our focus, however, was not in the laboratory, where this translation is most easily made (*young researchers learn to research*), but rather in the classroom setting, where learning canonical facts contained in canonical textbooks (or videos) still dominates.

When art students get an assignment, they generate and create, bringing about new objects and interpretations of the world; when science students get an assignment, it is usually a problem set with a fixed set of prescribed answers. There are benefits and strengths resulting from both kinds of assignments, and the translation from science to art was obvious: reproducing a faithful copy of a master's existing work is the classic apprenticeship model. There are things to be learned by getting the right answer, mainly gauging one's own performance against a clear set of standards.

Translating from art to science, on the other hand, is reminiscent of C. P. Snow's [58] famous phrase "The Two Cultures," in which he describes the inability of scientists and non-scientists to communicate, and that to do so required bridging "a gulf of mutual incomprehension" (p. 4). Beginning in the 1990s, engineering programs began integrating didactic instruction with design and fabrication activities in the same space, and adopted the term "studio instruction" to characterize this change [59]. In 1993, Wilson adapted this to physics instruction [60], and by1999 a few chemistry departments, particularly those located at the polytechnic universities where the studio engineering classrooms had been invented, had begun to experiment with the integrated lab–lecture format [61, 62].

Studio instruction is a powerful metaphor for education because it not only changes how students carry out their assignments but also changes what the assignments are. For decades, reform in science education has placed its strongest emphasis on *how* students do their assignments rather than any fundamental re-imagining of *what* assignments they are doing. The rhetoric surrounding these reforms emphasizes adjectives about the type of learning that is intended, which is in turn used to describe the pedagogical methods: conceptual, student-centered, algorithmic, problem-based, team-based, creative, peer-led [63]. An examination of the assignments given to students reveals a uniform culture of "problems and exercises" with "solutions" and pedagogies that seek to move the students more efficiently and productively toward the correct answer.

In developing our ideas about *Real Work*, we were not interested in criticizing or abandoning goals such as "getting to the right answer" as a part of learning science. Instead, we wanted to develop a way of broadening the landscape of student assignments in the sciences, to examine the value-added benefits, and to give others a useful way of thinking about their own work. In doing so, we have identified six attributes that derive or align strongly from the areas of authentic and situated learning, and which have provided particularly good entrees for work in introductory undergraduate chemistry education (Table 9.3).

9.2
Attributes of Real Work

9.2.1
Balance Convergent and Divergent Tasks

There is a historical tendency to see classroom teaching and learning in the sciences and the arts as fundamentally different because of Snow's cultural divide. Assignments and tasks in the arts are creative, generating new artifacts that express individual values, background, experience, and perspective. Students

Table 9.3 Attributes for *Real Work* assignments.

Balance convergent and divergent tasks	Convergent tasks are evaluated against a given standard (or "the right answer") and so learners can assess how successful a given pathway is in achieving the prescribed goal. Divergent tasks focus on the construction of individual outcomes within a set of common guidelines, and so individual learners come to a defensible position
Peer presentations, review, and critique	Developing explanatory knowledge allows learners to deepen their understanding and to anticipate arguments, revealing strengths and weaknesses in understanding
Balance teamwork and individual work	Successful communities of practice rely on individual members with a diverse base on knowledge and experiences, and where, in the aggregate, a common understanding encompasses more than any individual might have achieved
Students use the instructional technologies	As a matter of principle, learners should be trained in at least the instructional technologies deemed useful by a teacher. It is often more informative to see what learners construct in representing their understanding using multiple modes than for the learners to see only the practiced, expert view
Use authentic texts and evidence	Understanding and interpreting empirical evidence can be accomplished by direct experimentation, and also by using the original primary literature. Parts of these resources will be inaccessible to new learners, but other parts will not be (alternatively, significant historical papers can be used)
As important to the class as the teacher's work	The work generated by students, resulting from divergent tasks, in particular, can be returned to the class as student-generated instructional materials. Subsequent assignments and/or testing based on student-generated materials explicitly distributes the role of "teacher" in the instructional setting

in science classes learn the facts, and how to be encultured by them, against a background of "right answers" and "wrong answers." There is also a historical tendency, in science education [64], to create false dichotomies when an advocate for one position wants to highlight his or her point of view: content *versus* process, lecture *versus* discussion, facts versus concepts. Lave's commentary (see above) on not treating situated cognition as an either/or proposition is an appeal to reject a false dichotomy.

Comparing assignments and tasks in the arts and sciences is more a matter of traditional emphasis than an intrinsic difference in the disciplines. Assignments in the arts can certainly be intended to get a student to compare his or her work against a fixed standard (e.g., making a faithful copy of another work), as clearly as assignments in the sciences can generate new artifacts (e.g., research carried out on an undergraduate thesis project). A useful way to understand this comparison is found in the classic attributes of creativity: convergent and divergent thinking.

Over 60 years ago, Guilford introduced the concepts of convergent and divergent thinking [65, 66], and these ideas have proved to be robust, as they continue to be used and studied, including by neuroscientists who observe differences in brain activity based on differences in tasks, and who contend that these ways of thinking may indeed be separate and distinct [67–70]. Convergent thinking, as the name implies, relies on taking prior knowledge and an understanding of existing standards to generate the single best ("accurate") answer in a speedy and logical manner. Divergent thinking, which also relies of one's existing skills and experience, generates as many possible contextually meaningful answers as possible, relying on fluidly lateral rather than linearly vertical reasoning [71]. Straker and Rawlinson [72] provide a balanced view on these different parts of a creative process that resonates strongly with scientific practices: divergent thinking produces multiple and varied interpretations (or hypotheses) which can then be subjected to a convergent analysis in order to narrow the field of options to a few, or perhaps the single best defensible one, in order to carry a plan forward. The principles of divergence and convergence are widespread. In statistics, for instance, data modeling can randomly produce observable data over a distribution of outcomes (called a *generative model*), or narrowly define an outcome according to constraints or conditions (called a *discriminative model*), combining which, according to Bishop and Lasserre, results in "the best of both worlds" [73].

Both convergent and divergent design strategies can be used for student assignments, each to a different end. Convergent assignments produce work that is measured against an externally prescribed outcome and against a predefined standard of practice. These assignments are designed for learners to encounter the most commonly anticipated errors made by novices, and provide ways for learners to self-monitor and self-regulate their progress by understanding how their work compares with the prescribed standard, not only in its final form but also in the process, or pathway, used to get there. The benefit derived from well-designed convergent assignments is in making explicit as many of the implicit pitfalls typically associated with learning the lessons.

Divergent assignments produce work that is measured against the community of practice into which the work, thus produced, can be evaluated for its fit. Divergent assignments are open enough for learners to encounter errors that, in general, will require the use of standards such as comparison with others' work, debate, justification, and defensibility in order to identify, refine, and understand the choices made in the production of the work. The benefit derived from well-designed divergent assignments is that contributions made to the community of understanding broadens the community itself, and through comparison and debate of these contributions promotes a deep and reflective understanding of the lessons.

Poorly designed convergent assignments have been a target for criticism. If the single correct answer can be produced, or selected, by simple decoding or pattern recognition, without needing to follow a pathway in which the learner engages the underlying ideas, then two things happen: (i) getting the right answer for the wrong reason creates a sense of false confidence in the learner that productive learning is taking place [74] and (ii) the learning that does occur is indistinguishable from nonsense [75]. The benefit derived from any of the active learning classroom strategies [76] is to focus explicit attention on the steps in the path in order to promote better understanding of the basic ideas, and upon which improved test performance (converging on the single, right answer), the ubiquitous metric for evaluating these strategies, is the outcome [77]. The challenge for researchers in fixed-response methods of assessment is that the pathway is inferred: there is no direct evidence to differentiate Flora's deeper understanding of the subject from Dora's improved test-taking skills [78].

9.2.1.1 Convergent Assignments

Convergent tasks can be both generative and creative. In organic chemistry, an excellent example of a well-designed convergent assignment can be found in spectroscopic identification of molecular structure. Although there is a single best answer to the question of "what is the structure of this compound," everyone, from a seasoned experimentalist to a first-year undergraduate, has to deal with the same sources of data in order to answer the question. Structure determination, by its nature, is a convergent task: there is a structure; the work is to figure out what it is. Figure 9.1 illustrates the sort of experimental data that one might commonly encounter in such a problem. There is no single, prescribed path from these data to the structural solution that the character in the figure is thinking about. Teaching structural determination is challenging precisely because it defies prescribed algorithmic thinking.

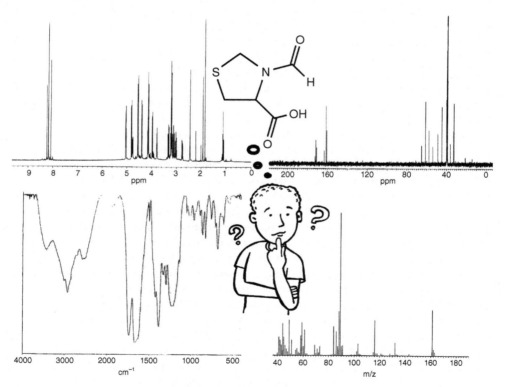

Figure 9.1 A typically good convergent assignment: a structure determination problem.

9.2.1.2 Divergent Assignments

First-year organic chemistry students also encounter divergent assignments. These are equally challenging to both teach and learn. Perhaps the most common example of a well-designed divergent assignment in organic chemistry is retrosynthetic analysis [79]. The starting point for retrosynthetic analysis is a molecular structure. The task is to take your existing knowledge of chemical reactions, including constraints derived from experimental design, and to construct as many different rational pathways as you can think of that will result in this molecular structure. Experimental constraints might include the availability of certain chemical substances you require, the compatibility of different components in the system, the likelihood for alternate reaction pathways that will reduce the production of this target substance, and the ease with which those by-products might be separated from the desired product. Figure 9.2 gives a simple example of a retrosynthetic analysis for an example target molecule. Unless the person who proposes these pathways has specific knowledge of these exact experiments having been performed, the only way to truly differentiate among them is by carrying out the experiments. However, an important decision made by an experimentalist is a convergent one: which pathway is deemed to be the most likely to succeed, and why? This decision is often a topic of heated debate between students and their advisors, between faculty members whose research groups are competing to get to the same finish line, and between individuals on a National Science Foundation review panel, who might debate who deserves a better score on their research proposal.

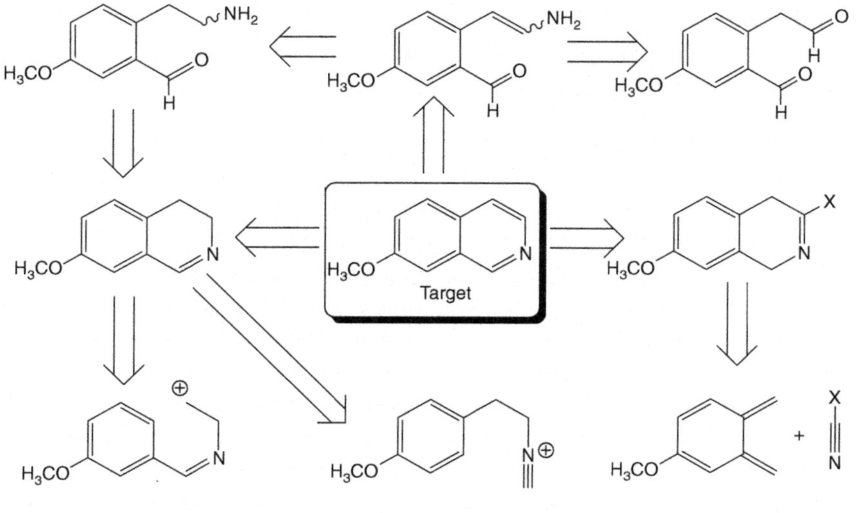

Figure 9.2 A typically good divergent assignment: retrosynthetic analysis.

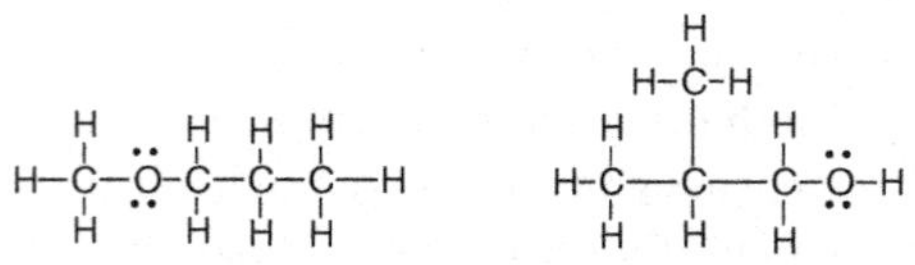

> H H H H
> H-C-O-C-C-C-H
> H   H H H
>
> CH₃OCH₂CH₂CH₃
>
> H-C-H
> H | H
> H-C—C—C-O-H
> H   H   H
>
> (CH₃)₂CHCH₂OH
>
> Besides the two compounds shown above, there are five more ways in which four carbon atoms, one oxygen atom, and ten hydrogen atoms can be connected. Find them. Molecular models will be very helpful in this task. Write Lewis structures and condensed formulas for each compound. Describe the connectivity of each compound in a way that makes clear how it differs from the other compounds with the same molecular formula.

> Draw three-dimensional representations of the following compounds.
> (a) CCl_4 (b) $CHCl_3$ (c) CH_3Br (d) CH_3CH_2Cl (e) CH_3OH
> (f) CH_3NH_2 (g) CH_3OCH_3 (h) CH_3SCH_3
>
> For the cases for which you have the necessary information, show the approximate bond lengths and bond angles you expect to find in these compounds

Figure 9.3 Examples of convergent textbook problems.

9.2.1.3 Balancing Convergent and Divergent Assignments

During the first few weeks of an organic chemistry class, students encounter many convergent textbook exercises related to the basics of molecular structure and the relationship between structure and reactivity (Figure 9.3) [80, pp. 11, 25].

In 1994, in our supplemental instruction program, called Structured Study Groups, we used divergent task design as the core of what we called a *studio format*, which was intended to be more like the classroom design for a drawing class than for fabricating artifacts in an engineering program [81–83]. For their first (divergent) studio assignment, students select from a list of recent chemistry journals, each one getting a unique combination of title and year. After thumbing through the journals to find a molecule with between 10 and 13 carbon atoms, the students are asked to (i) construct (draw) five new rational molecular structures with the same molecular formula as the one they found, (ii) rank the molecules based on selected properties (e.g., magnitude of dipole moment, boiling point, and solubility), and (iii) write out the rationales for their rankings.

The likelihood that two students will find molecules with the same molecular formula is, effectively, zero, and the number of possible drawings (rational plus irrational) they might make is, effectively, infinite. All the students are working in the same task, where collaboration and conversation about strategy is useful and encouraged, and almost nothing short of someone else doing your work for you is going to be the way to shortcut the process. Note that task (ii) has no known reliable solution. Although students learn some general trends about these properties using simple examples, the sorts of molecules being drawn in this assignment still defy the best predictive theories.

There is no solution to converge to. Unless the experimental results are known over this entirely imagined series of molecules, the best answer is a combination of (i) extrapolating a set of simple principles to these complex cases and (ii) arriving at the most defensible answer. Students bring their written-out assignments to the 2-h studio session and, under the guidance of their leader (an upper level undergraduate), participate in a structured round of peer review and critique (see below). After the review process, the students get back their work (along with the comments – it is not a grading process) and the experience of having participated in this hour-long discussion, and they then decide whether or not there is something about their work they want to revise; they always do.

Each week, there is a divergent assignment that becomes the main topic of discussion and debate during the subsequent session. Some of these assignments include the following, over the course of the year:

1) Finding an example of a given type of reaction from a journal article and reformulating it as a quiz question [84];
2) Writing essays in response to the need for introducing the formal study of research ethics in science classes, and eventually writing their own case studies [85];
3) In a term-long project during the second semester, working in teams of three to four to transform an individual segment of a journal article into a fairly detailed print and Web-based resource [86].

In all these cases, the design follows the generalization of Straker and Rawlinson [72] for creative work: an initial divergent (or generative) phase to sample the diverse options derived from individual character, background, and experience, followed by a convergent (or discriminative) phase, in which some sense of community and/or disciplinary standards are applied to sort through and understand how (or even whether) the different options might be evaluated.

9.2.1.4 Convergent Assignments in Team Learning

Convergent assignments, idealized by a "problem set," can be done in isolation, referencing only an authority ("the answer key"), and reveal information about the pathway (decision-making process) taken to solve the problem only by inference. Discussion or recitation periods can help uncover some of the implicit decisions that arise during problem solving, as can modeling help the thinking process by an instructor's thinking out loud. An excellent contribution of the Peer-Led Team Learning (PLTL) [87] model to teaching has been to structure small group discussion around the problem-solving process, making it an explicit topic of discussion. The PLTL strategy, in working with challenging convergent problems, sets a six- to eight-person threshold for the group size, finding that more than this makes a productive discussion more difficult.

Although it is tempting to have access to the authoritative answer, either in the form of an answer key or an individual – study group leader or faculty member – who acts as a living answer key, there is robust literature on the value of error making in learning [88, 89], on the value of difficult and challenging problems [90], and an emergent understanding that easy access to authoritative answers can lead to a psychological state of self-deception [91], wherein a learner confuses understanding someone else's answer with having understood the process by which it was constructed.

9.2.1.5 Divergent Assignments in Team Learning

Divergent assignments, by their nature, require larger groups to produce an array of different outcomes, with a structure that allows time for analysis, debate, and defense of those outcomes. In our divergent studio assignments, we have found that between 18 and 22 students is ideal, perhaps because it allows for enough diversity in the collective work of the group, and enough examples to begin to see recurring themes in the analysis of the outcomes.

Using divergent assignments is a useful teaching tool for training peer facilitators. Each term, the Science Learning Center at the University of Michigan runs the peer-led study group (PLSG) program, where small groups of 8–10 undergraduates taking various introductory science classes are matched with an upper level undergraduate student who serves as the group facilitator. Upwards of 60–70% of any given class participate in these groups. As a part of their training, the nearly 80–90 facilitators for the organic chemistry classes meet once a week, in their own small groups, with a graduate student instructor. Each week, as a way for them to review and think about the variation in organic chemistry, they bring their solutions to a set of divergent problems to their session with the graduate instructor. The facilitators, after having to write out or post their solutions to the tasks, are guided by the graduate instructor in a discussion about the relevant subject matter, the "party-line" for how certain topics are treated by the department's faculty, and the pit falls and recurring problems that students have with the given topics. A sampling of these tasks is provided in Table 9.4.

There is no argument that supports pitting convergent and divergent design against each other. Both designs have unique strengths, and should be used together. The challenge for convergent design is to create problems that avoid the "unfortunate coincidence," where the correct answer can easily result from the wrong pathway [92, 93]. A supportive pedagogical design (e.g., PLTL) can intervene by giving students a chance to make thinking about their pathways more explicit. The challenge for divergent design is to create a group context where there is enough pedagogical confidence and expertise to deal with the diversity of potential solutions while at the same time finding the commonalities among them, particularly where breadth and depth of subject matter knowledge are needed to compare and evaluate a large collection of student work.

Real Work uses a balance of convergent and divergent tasks because creativity is a core value which combines the ability to think openly and laterally of options, without constraint, with the ability to evaluate, rank, and defend those options according to rational criteria.

Table 9.4 Divergent tasks for training organic chemistry peer facilitators.

Week 1 Draw a rational molecular structure with the molecular formula $C_{15}H_{16}N_2O_3$	*Discussion topics* Line abbreviation representations Formal charges Hybridization, electronic, and observable geometries Predicting and evaluating resonance contributors 3D orbital drawings for substructures
Week 2 Using a pK_a table, construct a diprotic acid that would have different structures in solutions of pH 1, 7, and 12	*Discussion topics* Identify and estimate the ratio of species in solution Identify a base that could completely mono-deprotonate the most acidic form, but not fully deprotonate anything else, even in excess Identify a base that could completely mono-deprotonate the most acidic form and fully deprotonate any other acidic functions if used in excess pH versus pK_a Estimating pK_a values from representative examples
Week 4 Create and name a molecule with the molecular formula $C_7H_{13}BrO_2$. Select a nontrivial bond and draw a Newman projection for its most stable conformation	*Discussion topics* Identifying gauche, syn, and anti relationships Translating between 3D drawings and Newman projections Evaluating attractive and repulsive forces (steric, electrostatic, H-bonds) Identifying equivalent sets of hydrogen and carbons atoms for NMR spectroscopy Nomenclature
Week 7 Prepare responses to a student who asks	*Discussion topics* Reach consensus on the responses In class: Given $C_8H_{13}Br$. Draw a molecule that meets these criteria; show the mechanism

(continued overleaf)

Table 9.4 (Continued)

a) What are structural isomers? b) What is the relationship between these two structures? c) How to predict the expected reaction product(s) in substitution and elimination reactions? d) How to identify mechanistic pathway from the product(s) of a reaction?	a) Can undergo one S_N2 reaction, demonstrates inversion of configuration using lithium azide b) Can undergo an E2 reaction with sodium ethoxide and gives exactly three different alkene products c) Can form a resonance-stabilized carbocation when heated in water, resulting in a racemic mixture of substitution products along with some elimination products d) Is a trans di-substituted cyclohexane that can undergo an SN2 reaction with sodium cyanide to give a cis isomer e) Cannot undergo an S_N2 reaction with anything, but can undergo an E2 reaction with a strong base in order to produce an allene ($R_2C=C=CR_2$) f) Cannot undergo S_N or E reactions g) For part (a), what happens if you treat your molecular with potassium *tert*-butoxide? h) For part (b), what happens is you heat your molecule with ethanol? i) For part (c), what happens if you treat your molecule with sodium acetate? j) For part (d), what happens if you treat your molecule with sodium ethoxide?

9.2.2
Peer Presentations, Review, and Critique

Peer-led teaching, learning, and assessment are currently well-established strategies in higher education [94–101], although they are processes that have also fallen in and out of favor for 2000 years [102]. In the first major review of this area, Goldschmid and Goldschmid [96] hypothesized that the most recent interest in teacher-directed peer-to-peer work, which started in the 1960s, was a practical response to the growth of university class sizes and the concomitant concern that students were becoming more passive in their role as learners. As Topping observes [100], early programs emulated the traditional classroom structure, where the tutor was seen as a surrogate or extension of the primary instructor. In their meta-study on the skills of peers relative to faculty instructors [99], Falchikov and Goldfish found that peer and teacher assessment generally resembled each other, particularly when the criteria being used were clear and explicit, and, consistent with this, that upper level undergraduates were no better than beginners.

Over the past 60 years, the interest in moving away from replicating didactic instruction to promoting greater engagement of the learners has grown, where students bring their own ideas, experiences, and work into multiple low-stakes environments. Students consistently report an overwhelming sense of a social contract with structured peer-to-peer interactions in which they are more comfortable ("safe") making errors in front of one another [82]. Goldschmid and Goldschmid [96] also outlined the psychosocial benefits to members of peer groups who continually alternate between "teacher" and "learner" roles, including the ability to form a more open and intimate relationship with a peer. There is a fundamental difference in audience, as Wagner [102] points out, between thinking you are providing assistance to a person who genuinely needs to your help in understanding what you know (a peer who does not know) and a professor (whom the student does not perceive as someone who lacks the knowledge).

Not surprisingly, the literature on the use of peers in writing is extensive. Rabkin and Smith [46] built their entire program around building a strong and interconnected social context for improving writing in a large first-year university setting. The clearer the sense of audience and purpose, the easier it is to reflect on what you want to say and how you want to say it. By making the decision making, and then the work, public and subject to negotiated understanding, the writer gets direct and relevant feedback. Moving up a level, this sort of classroom setting is metaphorically related to any community of scholarly discourse that relies on peer review and publication. As Shulman [103] points out, a key feature of scholarly discourse is the sense of community property, as scholars contribute, borrow, improve, and advance understanding through presentation, review, and critique of their ideas.

Two mechanisms seem to operate together in identifying the origin of the value to learners in peer presentation, review, and critique: having recent expertise and the need to explain it. In a deep study of college students' comments, reactions, revisions and results, Herrington and Cadman [104] conclude that the "process of active reciprocal decision-making represents the primary value of peer review" (p. 184). This conclusion is aligned with Palincsar and Brown's notion of reciprocal teaching [105, 106], wherein studying and understanding the practices of new learners who come easily to their success in a complex task is broken down into sensible instructional design. Although developed in the context of young readers, reciprocal teaching neatly overlaps with what Schwenk and Whitman [107] observed in teaching skills needed by physicians in their medical residencies. Newly minted residents were likely to be more "consciously competent" in their understanding of a medical procedure for having just learned it, relative to the senior physician, in still needing to think more explicitly through the steps, and so end up more likely to share better and more relevant information. The most meaningful questions arising from reciprocal teaching were what and where were the benefits, to learners, from teaching? Was it is from the preparation, the enactment, the discussion, or some combination of these?

The second mechanism is an aspect of teaching that is still significantly overlooked in higher education, namely explanatory knowledge. A common refrain among new teachers at all levels is one version of another of "I never knew the topic so well until I had to teach it," a statement which reflects the questions that followed from reciprocal teaching, above. There is a body of research suggesting that this knowing/teaching relationship is profound: when you are consciously aware of the future need to teach what you are learning, you learn it better (more deeply) than you would if you are learning something for your own needs.

The potential benefits from peer-to-peer instruction were first examined with children. In 1973, Allen and Feldman [108] observed that low-achieving fifth grade students scored better when tutoring third graders, on a given topic, after only 2 weeks, than their peers who spent an equivalent amount of time studying alone. In subsequent experiments, Bargh and Schul [109] observed higher retention rates in college-level students who were told they were preparing to teach compared with their peers who did not have this condition as a part of their learning. In 1984, Benware and Deci [110] used two levels of questions (rote and conceptual understanding) to look at two groups of students under conditions similar to Bargh and Schul. After 2 weeks, the performance of the two groups on the rote learning questions was the same, but the students who had been told that they were preparing to teach (they did not) outscored their peers on the conceptual questions. Coleman *et al.* [111] used three levels of questions (recall, near, and far transfer) questions in examining a complex scientific topic in university-level biochemistry, and observed the same pattern, where students who were studying with the idea that they needed to teach what they were learning outperformed their peers in greater proportion as the level of question grew more sophisticated. Coleman calls this difference "explanatory knowledge," and concludes that the "preparation to teach the contents of a text versus to understand it personally may influence the mental representations that are created from text" (p. 348). In 2007, Roscoe and Chi reviewed the area of "tutor learning" [112]. They conclude that, while there is strong evidence to support the idea that teaching promotes deeper learning, not every peer instructor breaks out of the information-telling mode when they actually teach, and so not everyone who participates ends up realizing the benefits.

What is missing from this "tutor learning" discussion, universally, is that all students are tutors, and this is why this topic is an interesting attribute of *Real Work*. During any traditional class, all students need to express and explain their understanding at least a few times; these events are called *examinations*, *term papers*, *projects*, or some other form of performance on which grading is based. Unless the examination does not involve high-level concepts and relies on multiple-choice testing of recall-based items, all students are responsible for teaching what they know in response to extemporaneous prompts (exam questions). The results above suggest that students who are not actively aware of their future need to teach, and do not understand that test-taking is a form of teaching, are missing an opportunity to move away from being Dora and toward Flora.

In the area of peer review and critique, Chang and Chang [113] have observed that having structures in place to support the review process is important to getting quality critiques. In a study involving students who critiqued teacher and peer-generated molecular representations for a combustion phenomenon, they observed that the review process, as an act of reflection and feedback, resulted in the student participants developing a more sophisticated understanding of the underlying science than students who did not participate in a review process.

9.2.2.1 Calibrated Peer Review

Instructional technology has provided a solution to the problem of scale when it comes to peer review and feedback. Coursera's Massive Open Online Courses (MOOCs) include an online calibrated peer review (CPR) system that involves human participation as opposed to computer-based feedback [114]. The CPR^{TM} system [115, 116] is an online tool that provides a great deal of flexibility for users to facilitate large-scale peer-to-peer feedback on writing assignments in their courses. Users submit written work into the system, and then calibrate their abilities to review writing on that assignment by applying an instructor's rubric to three sample essays that range from poor to excellent. Information about a reviewer's competence is logged, and students then provide anonymous reviews for some specified number of dis-identified writing assignments from their peers. An instructor may also choose to have the students then review their own work. The program has been used in many chemistry and chemical sciences classes, including lower and upper level chemistry laboratory courses [117, 118], environmental chemistry [119], biochemistry [120], and neuroscience [121].

9.2.2.2 Guided Peer Review and Revision

In the Structured Study Group program [82], we have integrated a structured peer review process into the face-to-face meeting time that the students have. When they come to the 2-h session with their work on the divergent tasks, they exchange their papers and examine a peer's work. The examination is guided by a review worksheet, on which the name of the original owner's paper and the reviewer's are entered. The worksheet is a series of questions with simple yes/no prompts [83] (Is the citation formatted correctly? Does the selected molecule meet the three criteria?). The original papers are not graded; only the yes/no questions are answered. This time is generally an open, freewheeling discussion, where the students are facing a solution to the divergent task that is different from theirs, and is yet an answer to the exact same charge. The prompt used by the upper class facilitators is also simple: as you think about whether the paper you are looking at satisfies the criteria of the assignment, think about your own work, too. On average, it takes 45 min to complete a review that does not involve marking the papers but only involves deciding whether the answer to the written prompt is "yes" or "no." During this time, however, there has been a lot of discussion and presentation of thorny issues, ranging from one-to-one questioning to whole group talk.

A second round of review is carried out the same way, except that now it only takes about 20 min to come to resolution. At this point, the original owner of the paper gets his or her paper returned, along with the review sheet of circled yes/no answers. Each student has now reviewed the work of two others, and so they are now charged with deciding, on their own work, whether there are any changes needed before they turn them in for evaluation. In studying what happens next, we evaluate an unmodified copy of the original paper, from the start of the session, and compare this with the modified post-review paper. We use a simply three-point scale (outstanding, satisfactory, unsatisfactory) to evaluate the students' work. On average, the work they bring to the start of the session can be as high as 75% unsatisfactory/25% satisfactory/0% outstanding; after the review process, when they decide whether something needs to be improved in their work, the resulting papers have moved almost completely into the satisfactory/outstanding range.

9.2.2.3 Argumentation and Evidence

Evidence and theory-based argumentation is an important form of discourse characteristic of scholarly communities. Recently, the literature around how to bring students into this discourse community as a part of their education is growing, especially in science education [122]. In chemistry, this extends to the use of appropriate representational forms while constructing an argument [123].

Students in the organic Structured Study Groups participate in a 4-week *Real Work* assignment designed to promote discussion about the characteristics of scientific knowledge. For the first week, students prepare short essays answering the question: "What is the nature of scientific knowledge and how does it compare with other forms of knowledge?" After peer review and discussion that is geared toward building consensus on the answer to this question, the students receive a copy of a published journal article that is intentionally selected for its potential controversy. For example, Mehmet ("Dr. Oz") Oz is a coauthor on this report, which bears all the surface features of a standard research article, on the effects of healing energy on tumor cell proliferation [124]. The students are asked to read and react to this paper in the context of the prior discussion and be prepared to discuss whether or not, and why or why not, this paper satisfies their definition of scientific knowledge. Over a 2-week period, the students have time to discuss this in their individual group meetings. Finally, all the students participating in the program are brought together for a 1-h plenary session that features two guest faculty members as discussion facilitators (usually one from the bio-related sciences and one from the humanities or social sciences).

9.2.3
Balance Teamwork and Individual Work

9.2.3.1 Team-Based Learning: Face-to-Face Teams

Team-based learning [125] is an area that is distinct from group learning [126], although it relies on it. Varma-Nelson and Coppola have reviewed the use of teams in chemistry education [83], and proposed a model for team learning as a second-generation instructional strategy, in that it intentionally integrates a collection of existing pedagogical practices and theories: cooperative and collaborative group learning [127], reciprocal teaching [105], Vygotsky's learning theory [128, 129], and studio instruction [130]. Ideally, students are tasked with something that is complex enough to require its work to be divided up into sub-tasks and brought back together.

The following is an example of a convergent problem from the PLTL group [131]. Conceptually, the problem is complex enough for introductory students that they tend to not be able to balance all the parts, that is, to suspend judgment, in a way that allows a successful solution. Instead, the aggregate experience of a team working on the problem and sharing their thinking moves them into full consideration of the issues involved.

*Two isomeric compounds **A** and **B** are known to each have a mono-substituted benzene ring (C_6H_5-). Both have the formula $C_6H_5C_3H_5O_2$ and both are insoluble in water. However, when they are treated with dilute NaOH, A dissolves but B does not. Give structures for A and B consistent with this information.* **Explain your reasoning.**

9.2.3.2 Team-Based Learning: Virtual Teams

In a noteworthy modification of PLTL, the design and implementation of cyber-peer-led team learning (cPLTL) has been accomplished [132]. Now it is possible to create a virtual team space that maintains the features of the face-to-face meetings with the use of small document cameras, standard webcams, and networking software. The cPLTL group reports that the gains in student performance are the same in both formats. In addition, preliminary research results indicate that students in the cPLTL groups are more task-oriented than the face-to-face groups, and use more explicit explanatory language, but at the cost of less of a sense of the social benefits realized by their face-to-face counterparts [132].

9.2.3.3 Team-Based Learning: Laboratory Projects

Laboratory work is an obvious target for team-based learning and *Real Work*. In their studio General Chemistry course, Gottfried *et al.* had four-student teams, each of who selected and designed experiments for investigating the status of a local watershed that had been potentially exposed to an underground contamination [133]. The class cooperated with local and state authorities, and the student teams needed to think deeply about real-world compliance and communication as they divided their various tasks (sampling, monitoring, data recording, data storage, etc.).

9.2.3.4 Team-Based Learning: Collaborative Identification

Even in a traditional laboratory setting, some of the mundane and skill-based tasks, such as learning how to record a melting point or obtaining an infrared spectrum, can be turned into *Real Work* tasks in which individual and team-based efforts are both critical components. The following example works in settings where they might be well over a thousand students in a given course [134]. First, select 8–12 colorless organic solids that can be pulverized into identical looking powders. Add small amounts to vials, and code them all uniquely. For laboratory rooms with 22 students, create random mixtures of 30 vials, with one, two, or three vials of a given substance. At the start of the laboratory session, each student takes a vial. Their goal is as simple as it is real: by the end of the laboratory period, find out who else in the room has the same substance as you. In order to solve this problem, students now need to gather and share experimental data (e.g., melting points, solubility, thin-layer chromatographic behavior, infrared spectra) and think about shared standards (what is a solubility test and how uniform does the testing need to be), and even the student who may not have anyone in class with the same substance needs to communicate and cooperate as fully as anyone else. This blueprint can be extrapolated to any type of laboratories where comparative data can be collected on identical-looking samples (e.g., colorless liquids, solutions of acids or bases). If the coding scheme remains hidden from the instructors, which is recommended, then the solution to the problem is genuinely driven by the experimental data and the experiments that must be designed in order to reconcile ambiguities and outlying points.

9.2.3.5 Team-Based Learning: Experimental Optimization

Another *Real Work* example that takes advantage of team learning is to start with a traditional laboratory exercise and to pose a simple question for which there is no known answer but which a laboratory-sized group of students can answer. For example, in her 1000-plus student second term organic chemistry course, Shultz [135] starts with a standard textbook Wittig reaction, Fisher esterification, or aldol condensation, and poses a driving question. In the case of the Wittig reaction: using the principles of green chemistry, how could one modify this experiment? As a group, the laboratory section might decide to examine different solvents and design a series of experiments to systematically investigate different proportions of binary mixtures. Although each student is responsible for individual work, the compilation and analysis of the collective information is required in order to see trends and make generalizations. In subsequent terms, the driving question might be to start with the solvent information and decide how to optimize the yield by carrying out systematic investigations of temperature, time, concentration, and other experimental variables.

9.2.4
Students Use the Instructional Technologies

9.2.4.1 Learning by Design

Whatever technologies enhance the ability of teachers to design instruction must also enable the ability of students to convey their emergent understanding [136]. According to Coleman's ideas about explanatory knowledge in the sciences [111], as well as Kozma's work in representational competence [137, 138], arming students with multiple and varied explanatory or representational tools will promote deeper learning of the baseline subject matter as they think about and organize their understanding in anticipation of conveying it to others [139]. Exactly the same caveat about effective and ineffective choices in faculty use of presentational tools applies to students, as any technological tool "will not turn a bad presentation into a good one, and will not convert an ineffective presenter into an effective one" [140].

As a cornerstone of work in the arts, using student-generated work in order to promote learning has been adopted in science education under the "learning by design" framework, which emphasizes, as the name implies, the value of actively designing and constructing something [141]. In a study of students who generated videos in K-12 settlings, Kearney and Schuck [142] observed *Real Work* outcomes. Open-ended tasks with a high level of control over task development resulted in learners who took a high degree of ownership and autonomy, with a strong sense of audience (peer instruction). The tasks, which ranged in context from language classes to topics in the sciences, were complex enough to elicit high levels of meaningful team learning. Comparable results were observed with university undergraduates in the case of a virtual solar system project that resulted from a highly peer-based learning design [143].

9.2.4.2 Electronic Homework System: In the Classroom

Moore has merged the use of a commercial homework system for organic chemistry with a modified classroom design [144]. Students in his organic chemistry class learn to operate within a software platform that allows drawing organic molecular structures, mechanisms, and energy diagrams. Moore directs in-class activities, all of which are mediated by students generating drawings in response to these prompts. With the level of comfort for using this system that is developed this way, the students then take their examinations in the same platform, which also handles multiple chances and the subsequent grading.

9.2.4.3 Student-Generated Videos

In collaboration with the campus's writing center, Shultz's (G.V. Shultz, private communication) students, in General Chemistry Structured Study Groups, generate video-based lessons for introductory chemistry topics after having identified existing video lessons that they critique as less effective than they could be. Over a term-long period, these students use principles of rhetorical analysis to craft the framework for their lessons: introduction, sources of authority, ethos, logos, pathos, metaphor, coherence, and conclusion. They are required to integrate and explain the use of at least three of these principles into their video. Students almost always include "pathos" (an emotional appeal) in their video and recognize it as an essential element that is often missing in their course lectures or videos they find online. Also, pathos is an element that is almost always present in the videos that they rated highly.

9.2.4.4 Student-Generated Animations

High school students who generate visual representations for chemical phenomena are more successful in understanding the underlying concepts than their counterparts who select between pre-existing representations, depending on whether the selections are simple or complex, suggesting that the combination of student-generated work with identification may be a powerful combination [145]. Student-generated dynamic representations (movies) for these arguably dynamic phenomena are rarer. University-level preservice primary school teachers learned to create "Slowmation" (low frame rate) animations [146], which contributed positively to their learning and helped clear up errors in their thinking. Organic chemistry students have constructed animated gifs of complex mechanisms for

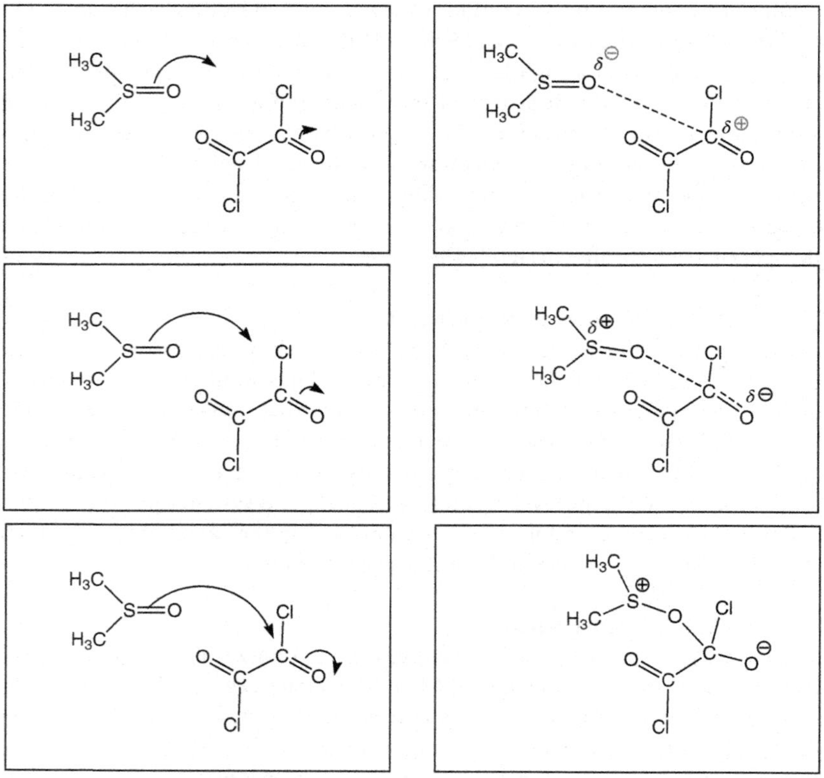

Figure 9.4 Six selected slides (out of 50) from a student-generated animated gif for the Swern oxidation mechanism.

literature-based reactions [86] (Figure 9.4). They report that the fine grain size required for creating a hundred or so frames of these easily constructed animations causes them to think more deeply about, and in some cases self-correct, the impression they got from the traditional, static, single-panel, curved-arrow representation. ChemSense [147] and Chemation [148, 149] are two software environments that were built, based on the earlier work with animated gifs, to provide greater scaffolding for creating animations and linking these together with molecular properties.

9.2.4.5 Student-Generated Video Blogs
Lawrie and Bartle [150] asked students to create 2–3-min video blogs (vlogs) that targeted explanations for structure–property relationships, and found a positive correlation between the level of sophistication of the representations, the depth of the explanation, and the standard achievement level of the students who created the vlog. Shultz [151] has integrated a video blog platform (VoiceThread) into the laboratory experience of a large-scale organic chemistry course, with a particular emphasis on students making self-explanations [97, 112] for their process of

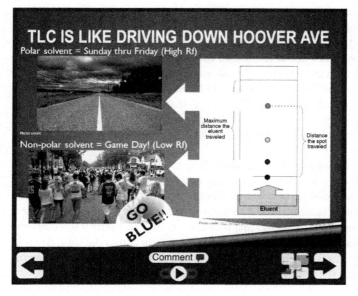

Figure 9.5 Still frame extracted from a video blog describing an analogy for thin-layer chromatography. (Original presented in Ref. [151].)

analyzing experimental data when carrying out structural identification. Her students have reported that it was helpful to see how others had perceived the same content and that explaining helped them to better understand the material being presented. A frame from one of Shultz's students' explanations, which used an analogy for thin-layer chromatography, is shown in Figure 9.5.

9.2.4.6 Wikipedia Editing
Student authorship becomes more meaningful when the purpose becomes more real [46]. Moy *et al.* integrated learning to edit and author Wikipedia articles into a graduate-level polymer chemistry course [152], and others in this department followed suit at both the graduate and undergraduate level. Undergraduate students improved articles that involved standard named reactions (e.g., Jones Oxidation, Appel Reaction, Ritter Reaction).[1]

9.2.4.7 Wiki Environment
Moore asked students to master a collection of six molecular representational tools and then to construct and peer-review, in a Wiki environment, details of the mechanisms of action of a series of biologically active molecules [153].

9.2.4.8 Student-Generated Metaphors
Inherently, technologies are things that enable other things. In this attribute of *Real Work*, we are gathering those enabling strategies that an instructor might

---
1) Wikipedia.

use to teach effectively, and looking at how encouraging students to use those selfsame technologies, according to a larger precept of reciprocal teaching, provides an important lens through which student understanding can be viewed. These technologies are not at all limited to computer-based strategies. And so, to the degree that an instructor might use metaphor in order to activate the prior knowledge of students in order to build a bridge to a new concept, instructors who ask students to provide metaphors is, in effect, sharing and encouraging an effective teaching strategy. Lancor [154] reported an analysis of metaphors generated by students from multiple disciplines in order to examine their understanding of energy concepts.

Visual metaphors can be useful guideposts for navigating complex ideas [155]. In chemistry, where visualization of unobservable species is central to the science, there is substantial overlap between a scientific models and metaphors [156]. In the "HTML Project" (see Section 9.2.6.3), students use visual metaphors to contextualize the instructional materials that they are creating. In Figure 9.6, a synthetic sequence in organic chemistry is displayed as a road trip across the United States, with different features of the work (e.g., the animation of a mechanism) showing up under links for the analogical site in a city (e.g., the site of a theater).

9.2.5
Use Authentic Texts and Evidence

Reading and making meaning from text is a core skill in learning, and something that Bereiter supposes that Flora does better than Dora, because she can transfer what she has learned from one situation to another. Understanding how successful learners do this, and turning this into ways to help others, has been of long-standing interest to educational psychologists. In a 1996 review of this area, Mayer [157] concludes "the most effective method for teaching students how to make sense out of expository text is for students to participate in selecting, organizing, and integrating information within the context of authentic academic tasks" (p. 357).

Tasks involving learners using the primary research literature have been unusual in undergraduate science education, particularly at the introductory level [82, 158, 159]. In the laboratory, authentic evidence can arise from research activities [160, 161] that are observed to develop a deeper understanding about the nature of science, or might arise from students being provided with empirically derived data upon which new questions are asked [162]. Perhaps because organic chemistry is the first discipline-centered chemistry class students take [11], using the primary literature as a source for assignments has become increasingly represented in chemistry education as access to and availability of electronic resources has grown [163].

9.2.5.1 Literature Summaries
Students in Gallagher and Adams's [164] Honors organic chemistry class used journal articles to connect their classroom learning to contemporary scientific

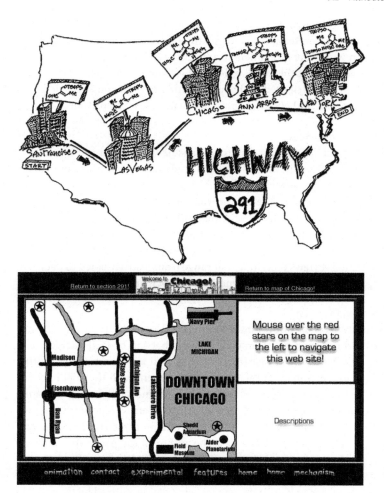

Figure 9.6 Navigating "The HTML Project." The students in Section 291 used the visual metaphor of a road trip across the United States to contextualize their five-step synthetic sequence.

writing, and through a multi-week assignment, they needed to collaborate with both peers and professors in order to make meaning from these articles. A similar approach involving first-year students and two brief summary assignments of accessible literature sources was reported by Forest and Rayne [165] to lower the intimidation barrier for subsequent use of primary sources, particularly searching for what constitutes relevant subject matter.

9.2.5.2 Literature Seminars

Almeida and Liotta [166] carried out a literature-based seminar for students taking both organic chemistry and cell biology in an effort to better integrate the learning in these two related classes. After one round of reading and discussion led by a faculty instructor, teams of students then select and lead the discussion of publications in which the fundamental subject matter from both classes is relevant. Students report that this work reduced their anxiety and hesitation about research, which likely contributed to the high fraction of the seminar participants who actively sought out and obtained undergraduate research positions.

9.2.5.3 Public Science Sources

Critical reading of public science sources, such as newspapers or the Internet, is another use of authentic sources. In a detailed study of using newspaper sources with high school chemistry students, Oliveras [167] found that topics such as identifying the purpose or intent of the author was a major challenge for students, as was understanding what constituted evidence and how it was used to warrant an argument. Glaser's [168] taxonomy of "Chemistry is in the News" includes the combination of reading relevant news sources, connecting these with course content, and conducting reviews of writing that range from instructor-led to peer-led, and including, at the highest level, international participants.

9.2.5.4 Generating Questions

Generating questions is an important skill of critical reading [169–171]. Deciding what constitutes a good question is an activity that can be facilitated by peer review. Structured Study Group participants [82] read a research paper written by one of the faculty members in their home department with the goal of creating questions for the author. They find that generating questions is easy, but some questions ("What does DMSO stand for?") are better answered own your own than taking up someone's time. In a 3-week assignment, these students generate questions that would be worthwhile to ask the author of the paper, discuss what constitutes a good question, and share and peer-review their questions. Finally, in a 1-h plenary session, the group meets with the author, confident with their peer-reviewed questions in hand, who carries out a dialog spurred by those questions [172]. Invariably, in addition to the subject matter of the article, the questions include many topics for which one cannot simply look up answers: how research is funded, what motivated the research, why did you get into science, what practical applications are there, do you have children (asked nearly exclusively of women), what about environmental concerns, what if the research done with an industrial partnership is contributing to activities with which the researcher has moral objections?

9.2.5.5 Course-Based Undergraduate Research Experiences (CURE)

Using authentic texts in a laboratory in order to propose, design, and carry out actual research is a growing area of activity. Access to instrumentation combined with the ability to get meaningful information from small-scale preparations has contributed to this increase. In an early example of this in organic chemistry [84], students in a large laboratory course were given a journal article in which a zinc-mediated allylation was carried out on a reported set of 10 or so aldehyde substrates. During the first laboratory period, students repeat one of the reported examples (a convergent task), to test their skills against the standards reported in the paper. During the subsequent two laboratory periods, the students (i) apply their learned skills to a substrate or reagent that was not reported in the paper and (ii) examine whether their result (positive or negative) is reproducible by a peer. The Center for Authentic Science Practice in Education (CASPiE, [173]) has developed research-based undergraduate curriculum materials, as well as guidelines for how individuals can develop research-based projects. At the end of a 2-year study in biology education, a group has reviewed this type of work and proposed that the following criteria might define course-based undergraduate research experiences (CURE): use of scientific practices, discovery, broadly relevant or important work, collaboration, and iteration [174].

9.2.5.6 Interdisciplinary Research-Based Projects

Using interdisciplinary research as a design for *Real Work* is a compelling target, given the increasing need for graduates to communicate (read, write, collaborate) across traditional disciplinary lines. Chemists and biologists at Seattle University [175] have created an environmental research project on pyrethroid pesticides that is carried out across three laboratory classes. Approximately 100 students, who are enrolled in Ecology Laboratory, Instrumental Analysis Laboratory, or Organic Laboratory III, begin the term in a plenary session to view and discussion a video documentary on water pollution. Students in the Ecology class collect water samples and carry out a series of abiotic and biotic analyses. The samples are passed to the organic chemistry students, who carry out reverse-phase solid-phase extraction. The resulting organic materials are then passed to the analytic chemistry students for LC-MS/MS (liquid chromatography/tandem mass spectrometry) analysis.

Kosinski-Collins and Pontrello [176] report a chemistry-to-biology hand-off where over 80% of the students are co-enrolled in the two involved laboratory courses. Students in the organic course synthesize polypeptides that are mutants with respect to a known anti-aggregation factor that is a potential drug candidate for Huntington's Disease. About 10 new inhibitors are synthesized each time the course is offered. The molecules are then tested by the biology students in Drosophila, which express the polymer target as potential inhibitors, via fluorescence spectroscopy and lifespan analysis.

The Distributed Drug Discovery (D^3) project [177, 178] is an international collaboration, and argues "that if simple, inexpensive equipment and procedures are developed for research in each of the core drug-discovery stages, computational chemistry, synthetic chemistry, and biochemical screening, this large research challenge [drug discovery for diseases in the developing world] can be divided into manageable smaller units and carried out, in parallel, at multiple academic and industrial sites" (p. 3). To date, a project involving undergraduate students in four countries has demonstrated proof of concept for the synthetic chemistry piece, generating a catalog of over 24 000 acylated unnatural amino acids.

9.2.6
As Important to the Class as the Teacher's Work

9.2.6.1 Student-Generated Instructional Materials

The attributes of *Real Work* include using authentic sources (text and data), using the tools that professionals also use, collaboration and peer review to bring intentional reflection into the generation and refinement of artifacts, and encouraging creativity through the balance of convergent and divergent tasks. In the previous section, a high sense of purpose accompanies *Real Work* tasks that target the creation of new knowledge when students are carrying out research. Our last attribute is a highly underdeveloped area: student-generated instructional materials. Another way to instill immediate purpose for student work is to design tasks that return to the classroom, and for which all members of the class are responsible for, in the deepest sense; they are responsible for generating instructional materials, and they are responsible to these, collectively, as a part of their learning in the class. Lee and McLoughlin have reviewed the emergent area of using "learner-generated content" in instruction [179], and Perez-Mateo *et al.* have proposed an extensive set of quality criteria for those assessing this kind of work [180].

Computer Science faculty have been particularly active in this area, which they call "contributing student pedagogy" [181, 182]. There are a variety of proposals on why this pedagogy is so effective, including introducing high value (purpose) in a peer-to-peer context [183], requiring formative peer review [184], developing intrinsic motivation [185], building upon social constructivism and community-based learning [186], and facilitating constructive evaluation [187].

9.2.6.2 Wiki Textbooks

New technologies have enabled authorship for individuals, and can also enable students to author their own textbooks, or supplementary text materials [188, 189]. Wheeler *et al.* [190] posit the growth of "user-created knowledge" through Wiki technologies as an "architecture of participation" (p. 989). Kidd *et al.* [191] have reviewed the brief history of Wiki textbooks, and the corresponding challenges [192], in the context of preservice teacher education, with benefits accrued to future teachers that derive from their thinking about teaching (writing a text) in such a concrete way. The same group has also provided perspective on a peer-review system that is appropriate to a term-long project with multiple contributors, such as a Wiki textbook [193]. Subject areas such as reading and second language acquisition have benefitted from using books derived from student-generated instructional materials [179]. In chemistry, Pazicni *et al.* [194] have described how student-generated materials were used to supplement the traditional text in an introductory physical chemistry class through a term-long, on-going Wiki project. Bruce *et al.* found that a student-generated E-glossary was particularly useful for nontraditional students and non-native English speakers because of the direct relevance to the course and its one-point access [195].

9.2.6.3 Print and Web-Based Textbooks

Since 1994, in the second term of the organic chemistry Structured Study Group program [82], students carry out a project that integrates all of the *Real Work* attributes. Over the duration of the 13-week term, a class of about 125, in five teams of 25 students, each led by an experienced upper-level undergraduate facilitator, carries out "The HTML Project" [86]. As a part of their weekly group work, each 25-student team is divided into subgroups with three to four members. Each of the five large teams is responsible for learning the chemistry in a specific synthetic organic chemistry reaction sequence located in a journal article selected by the instructor. Each team has a different article and a different sequence for which they are responsible. The chemistry in these sequences is more or less relevant to a second-term course in organic chemistry, and new papers are used every year. In each team, the reaction sequence is divided up between the subgroups. In other words, subgroup 1 might have responsibility for learning the transformation of compound **13** to compound **14** (in some sequence), subgroup 2 is then responsible for the transformation of **14–15**, and so on (Figure 9.6).

Every week, there is work that needs to be done on pieces of the project, which are (i) to research what is known about the mechanism of the reaction, (ii) to display the mechanism in its traditional one-panel, curved-arrow format and to construct an animated gif (e.g., Figure 9.4) using the freeware program GIFBuilder [196], (iii) to provide a detailed interpretation of the relevant spectral data and to create both text and graphical correlations (e.g., Figure 9.7), (iv) to answer some leading questions on the chemistry, and (v) to annotate the experimental procedure. All of the intermediate work, for example, the storyboard for the animated mechanism, is peer-reviewed, critiqued, and revised, accordingly. A few weeks before the end of the term, the work of individual subgroups is compiled, and the work of all five teams is gathered together, resulting in a 200–250 page printed text and its accompanying Web site, where members of this class of 125 students have all contributed to the analysis and interpretation of (i.e., "unpacked") the chemistry behind these syntheses and presented them as instructional materials for their peers.

In addition to its obvious purpose in having these students dig into the detail of these journal articles for deep understanding, the HTML serves one last relevant purpose as instructional material: the final examination in this course is based on questions posed by the instructor about the work in the student-generated text, in the context of their second-term organic chemistry course. Moreover, the students are told that the questions are only based on whatever lingering errors might exist in their work. The examination is open-book, that is, they need to bring the book they wrote, which can be annotated to whatever degree they want. The instructor stands aside and does not engage in the resulting discussion about possible errors, as the students have the name of the people who wrote every word and drew every structure as classmates. From an instructor's perspective, this examination is a

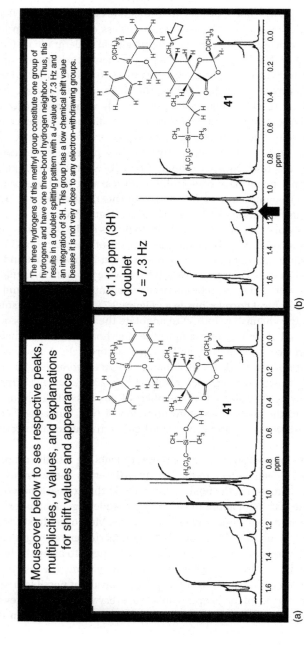

Figure 9.7 (a,b) A typical mouse-over spectral correlation and analysis generated by students as instructional material for their peers (arrows have been substituted for the usual technique, where the relevant portion of the spectrum and the group of atoms in the structure are indicated in color). The cursor can be placed over the signal or over a set of atoms, and the corresponding information and explanation are then provided to the user.

critical culmination to the project, giving the work the highest possible value in the context of a course, and teaching an overarching lesson about scientific skepticism: when you open a resource and begin to read, you should always be asking yourself a few questions: do I believe this? Does the evidence warrant the claim? A typical examination question might be "On page 65 of your book, the authors showed dichloromethane as a proton source for the second step of the transformation. Why is that incorrect, and provide the more reasonable mechanism for this reaction."

One of the recurring assignments in the organic chemistry Structured Study Group program [82] is for students to transform an example from the primary literature into an appropriate examination question. The organic chemistry program at the University of Michigan uses literature-based examination questions in both of its introductory courses [84, 197, 198], and so this task is highly relevant (e.g., Figures 9.8 and 9.9). Students are divided up among various organic chemistry journals with different publication years, and given the divergent task of locating an appropriate reaction (e.g., a bimolecular nucleophilic substitution reaction, drawing resonance contributors, an electrophilic addition reaction) and transforming it into an examination problem, complete with a brief contextualizing statement, a citation, and an answer key. During the subsequent weekly session, the question and all of its associated parts are peer-reviewed, and the author makes whatever corrections might be needed on the spot. The reviewed questions are then exchanged once again, without the answer key, and the students solve the peer-generated problem as practice.

9.2.6.4 Electronic Homework Systems

Fergus and Kirton [199] report training 237 students on the PeerWise platform [200], which enables students to generate multiple-choice problems that can be accessed by their peers. Here, the undergraduate chemistry students were required to generate two problems, answer five, comment on three, and provide

When treated with a Bronsted acid catalyst, compound **P** was transformed into compound **Q** (isocumene) through a mechanism involving two reactive intermediates (*JACS* **1981**, *103*, 82). Provide the structure of the first intermediate (resulting from protonation), the mechanism for its transformation to the second intermediate, the structure of the second intermediate (which is then deprotonated to give compound **Q**). The protonation and deprotonation mechanisms do not need to be shown.

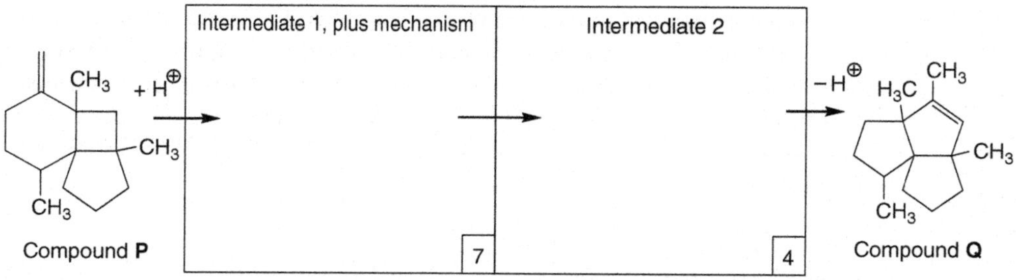

Figure 9.8 Typical example of a faculty-generated, literature-based examination problem.

Squaric acid (Compound **D**) has been used to treat warts in children who have a skin infection caused by the human papillomavirus (*J Am Acad Dermatol* **2000**, *42*, 803). The doubly deprotonated form of squaric acid (Compound **E**) is a di-anion that has a set of 3 other significant resonance contributors that (i) have all closed shell atoms, (ii) maintain charges of −1/0/+1 on the atoms, and (iii) keep the negative charges located on the oxygen atoms.

Draw these three other resonance contributors. Be sure to show all electron pairs and formal charges.

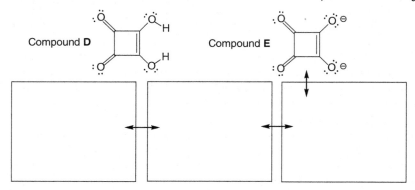

Figure 9.9 Typical example of a student-generated, literature-based examination problem.

ratings. Only three students did not produce any questions, and over 75% of the students answered more than the required number. Burkhalter and Wilson [201], teaching in online chemistry and biology courses, have used student-generated multiple-choice questions as an assignment which then increases the number and diversity of the problems in their test banks.

During the Fall 2013 semester, Coppola and McNeil (Unpublished results) trained 142 first-term first-year undergraduate students in the organic chemistry Structured Study Group program on how to be authors for the Sapling Learning e-homework system [202]. At the end of the term, out of a total of 167 completed problems, which were all targeted at skill-building exercises for learning the curved-arrow representational system, there were 61 that were rejected for content problems severe enough to not warrant revision, 46 that needed major revision, 33 that needed minor revision, and 27 that were publication quality (Figure 9.10). The major issue was that, once an undetected conceptual error entered the system, it was propagated through the numerous incorrect answers to the point where revision was simply too laborious. The solution: a more robust review process during the development stage. During the Spring 2014 term, 31 of the already-trained students generated 637 problems in 10 different skill areas, and nearly all of them were publication-ready without the need for revision. These student-generated instructional materials will be available to all students in subsequent offerings of this class.

9.2.6.5 Podcasts

Student-generated podcasts, including visual enhancement, are another growing area of student-generated instructional materials [203]. Even in creating brief

Citric acid exists in a variety of fruits and vegetables, but it is most concentrated in lemons and limes, where it can comprise as much as 8% of the dry weight of the fruit. Based on the data given in the table below, provide the major species present in a pH 4 buffer solution.

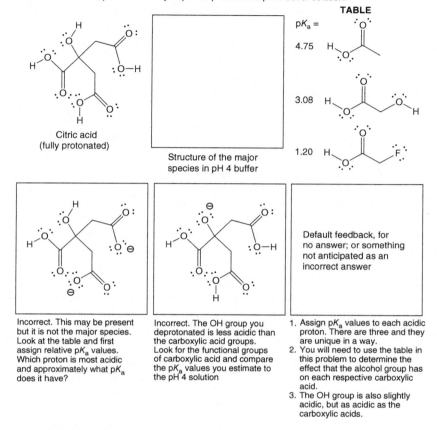

Figure 9.10 A student-generated problem for the Sapling Learning environment, including two anticipated incorrect answers, feedback for them, and the default feedback.

segments, students need to consider a broad array of issues in teaching and learning as they prepare their products for broadcast to their peers: scriptwriting and editing, presentation, audio recording and editing, publishing, and distribution [204]. Between 2006 and 2012, teams of two to three students in the first-term Structured Study Group program [82] selected a test problem from the annual collection of old examinations, which is one of the learning resources in the organic chemistry courses. Although these problems intentionally do not have a readily available answer key as a part of the strategy for getting students to work together and share their ideas, podcasts provided a way to create a new resource. Thus, over a 4-week period, each student team was tasked with creating a 3–5-min visually enhanced podcast that was aimed at providing a step-by-step tutoring lesson on how one would solve the problem. Over the 4-week period, the team would bring to the session their draft script, their storyboard, their draft podcast, and the revised podcast, and the rest of the group would provide feedback. Each student participated in the construction of one podcast during the term, thus only a subset of students generated these podcasts from problems relevant to each of the exams. The block of problems, and the podcasts generated by this group of about 160 students, then became available to the class as a whole during subsequent years. Over a 6-year period, almost 400 of these visually enhanced problems were created. About 15% of the students in the large 1500-student class report regularly using this resource and perceive them to be useful. The students have not shied away from creative contexts for building their explanations, with scripts that might be done in the form of a newscast (Figure 9.11) or a rap performance (Figure 9.12).

9.2.6.6 Classroom: Active-Learning Assignments

Providing structure and guidance for students is an important feature in the design of this *Real Work* attribute. By breaking down the task into manageable pieces and providing ample opportunity for feedback, the quality of the work produced can be acceptable for release to other students. In a two-term Computer Science course, Gehringer and Miller [205] assigned students to create curricular activities which then constituted part of the course syllabus. Starting with 30–40 min of class time immediately following a lesson, their students drafted an active-learning exercise based on the course material. Over the next week, the activities were (i) posted on a Wiki page, (ii) reviewed and commented upon by peers, (iii) revised, and (iv) reviewed a second time before incorporation into the course.

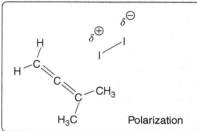

Figure 9.11 A student-generated, visually enhanced podcast explaining the mechanism of an electrophilic addition reaction (excerpt from script).

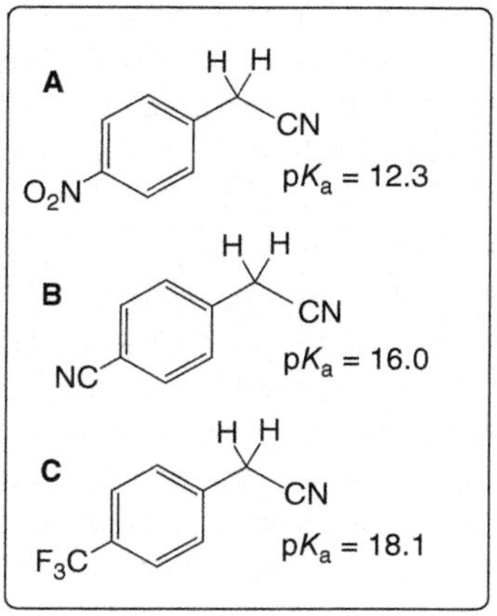

A: 4-nitro-benzyl cyanide, $pK_a = 12.3$

The pK_a values in DMSO,
For Molecules ABC, given below,
This question's testing you to see if you know,

B: 4-cyano-benzyl cyanide, $pK_a = 16.0$

The Reasons for the pK_a's that they show.

So look at **A**, right? thinking about the NO_2 group you might say.

So, nitro why you make the pK_a low,
it's cuz you pull electrons across the aromatic ring, yo

C: 4-trifluoromethyl-benzyl cyanide, $pK_a = 18.1$

I'm like, oh, induction, I see

Are there any other reasons for the reactivity?

I'm looking at this think wondering what else you got...

Figure 9.12 A student-generated, visually enhanced podcast explaining the difference in pK_a values for a series of compounds (excerpt from script).

9.2.6.7 Laboratory: Safety Teams

Turning over some of the actual face-to-face instruction in a course is another way to think about student-generated instructional materials. Alaimo *et al.* [206] created 2–3 student safety teams from each general and organic chemistry laboratory section. Each team was responsible for (i) preparing a pre-lab discussion on some prescribed safety topic(s), including handouts, (ii) monitoring the laboratory during their assigned period for safety issues, and (iii) carrying out a post-lab inspection. Students who took responsibility for this work outperformed a group of their peers, in whose course these safety teams were not formed, on a set of basic safety questions.

9.3
Learning from Real Work

Integrating *Real Work* into instructional design requires greater organization and effort than standard didactic teaching and testing methods. The attributes of *Real Work* described in these examples all attempt to elevate Dora's penchant for schoolwork [1] ("homework" that emphasizes memory, recall, and algorithmic heuristics) to Flora's real-life approach ("real work" that builds on prior knowledge, application, and evaluation). Unfortunately, it is far easier for research to evaluate Dora's skills than Flora's, and so they both do well on their Algebra I tests and on other evaluation instruments that focus solely on providing answers to problems. Compelling evidence for Flora's deep understanding, the transformation in her learning because of her instructional environment, is contextualized in her ability to have learned Algebra I in a way that allows her to succeed in Algebra II.

In a meta-analysis of 40 years worth of innovation and reform in higher education, Slavich and Zimbardo [207] have proposed a compelling framework for the basic principles that derive from transformational teaching and learning. In order for Flora to have been considered educated, she should exhibit evidence three things (p. 582): (i) acquisition and mastery of key course concepts, (ii) enhanced strategies and skills for learning and discovery, and (iii) positive learning-related attitudes, values, and beliefs. The authors propose educational psychological gains in motivation and self-efficacy as exemplars for their interesting third category. But this category could easily be expanded, in discipline-centered science education [208], as the "ways of thinking, feeling, and behaving," that is, scientific dispositions: those things that accompany learning the subject matter [209] but are not easily quantified and do not appear on the syllabus, such as improving one's evidence-based skepticism.

9.3.1
Evidence of Creativity through the Production of Divergent Explanations

Trivic *et al.* [210] studied the explanations given by two groups of students learning stoichiometry, one of whom experienced a highly convergent instructional design (demonstrations and calculations), while the other group was participating in experimental and textbook work that included the chance to generate, share, and review stoichiometry problems with one another. The second group not only showed a larger gain on standard pre/post comparisons, but they also generated a more diverse, multirepresentational set of replies, integrating more conceptual categories, such as solution chemistry, the periodic table, and a historical approach to chemistry.

9.3.2
Peer Review and Critique Reveal Conceptual Weaknesses

The CPR™ platform [211] has been an enabling technology because it allows the aggregation of information about student writing on a large scale. In using CPR, which is a Web-based tool, students deposit written assignments. They then read examples of written assignments provided by the instructor, which range in quality, and use rubrics to train themselves (calibrate) against these examples. Finally, they are provided access to other student work to read, review, and leave feedback, guided by the rubrics and by their training. Pelaez [212] compared two physiology classes, one of which used problem-based writing with peer review in place of didactic lectures, and observed a sharp difference on exam performances. She was also able to use the original writing, and its reviews, as evidence from which to generalize problem areas in student learning which, in turn, allowed subsequent classroom discussions to be more targeted at specifically those areas. In addition to recurring difficulties with specific topics (e.g., vesicles) and higher level organizational concepts (relationship between intra and intercellular features), the student writing also revealed conceptual difficulties with broader ideas about scientific thinking (cause and effect; overgeneralization of principles).

The majority of users of this online peer review system report high levels of satisfaction with student engagement and the ability to provide feedback to students in both classroom and laboratory sessions [116, 121]. Margerum *et al.* used CPR to integrate environment chemistry subject matter into a laboratory course, and noted (i) the ease with which this allowed them to bring these aligned topics into the course with minimal effort and (ii) that working through a written description of the laboratory work improved the quantitative analysis of the experimental data. Berry and Fawkes also report an increase in the quality of student writing, by focusing sequential assignments on different portions of the laboratory report, with minimal added effort [117].

Using a longitudinal series of assignments, Walvoord *et al.* [213] carried out a careful study on whether using CPR^{TM} assignments improved the general quality of student writing and their ability to convey scientific knowledge. Using comparisons between student and faculty scores, the researchers looked at the overall quality of the student reviews. Two external writing experts reviewed the student work for its technical communication skill. Based on this study, the overall quality of the writing did not improve over the term, and neither did the technical communication skills, which is different from what Gunersel *et al.* observed using a repeated measures analysis in an upper level biology course [214]. The most important conclusion here [215] is that the use of any instructional methodology is not one-size-fits-all, and is highly conditional on the setting, the specific assignments, the degree of support, the level of student experience in the subject, and the willingness of the faculty member to invest appropriate time aligning the pieces into a coherent learning environment.

9.3.3
Team Learning Produces Consistent Gains in Student Achievement

The most typical data collected in studies on team learning are examination performance, overall grade distributions, and retention. The results across many studies, including longitudinal ones, are robust: integrating structured, facilitated, peer-to-peer time into classes improves all these factors [216]. In general, the observed gains do not favor gender, race, or ethnicity. The degree to which gains result from deep, conceptual understanding versus targeted, time-on-task preparation is still an open question; that is, %Flora versus %Dora [78]. Some of the most impressive results reported suffer from a combination of self-selection bias and comparison groups with no equivalent time on task, so it is difficult to make confident claims on the mechanism of these observed changes [217].

Quitadamo *et al.* [218] used existing survey instruments to try and gauge whether critical thinking skills were being affected, but the results were modest and not triangulated with other data. Deeper insights into what is happening in these teams, using multiple methods including discourse analysis, have been obtained [219]. The results are consistent with the conclusion above: there are no bulletproof designs [220]. The groups that were studied ranged from productive and interactive to didactic and answer-telling, depending on the teacher–student dynamic [112]. Differences in the face-to-face and cyber settings have also been studied using both quantitative methods and discourse analysis, and compelling differences based on the degree of intellective commitment and social engagement was observed (P. Varma-Nelson, private communication January 17, 2014).

Facilitators for peer-led groups might be predicted to experience the "tutor learning" effect [112]. Gafney and Varma-Nelson studied in detail almost 120 peer leaders at multiple institutions [221]. They affirm that these students, who are not tutors but rather discussion leaders, also have the knowledge of the subject matter reinforced, both in breadth and in depth, and that they realized personal benefits such as increased confidence, perseverance, public speaking, and organizational skills.

In the Structured Study Group program [82, 83], through the peer review and critique process learners start with student-generated work at the start of their sessions and invariably end up needing to reject and correct parts of their assignments which they thought were correct. In order to see whether we could observe this as a transferable skill, we looked the ability of students in this program to work their way out of a discrepant, counterintuitive observation after they have made an incorrect prediction about a familiar phenomenon, and compared their performance with high-achieving students who were all in the same course but who did not elect the Structured Study Group option [197]. While none of the students in either of the groups made the correct prediction, thus creating a level playing field for the study, 80% of the first-year university students who had spent 2 h per week engaging in this structured review process, including rejecting and correcting work they thought was originally acceptable, resolved the discrepant observation in a way that was similar to a comparison group of first- and second-year graduate students. In contrast, only 10% of high-scoring students who did not participate in this group work demonstrated skills that were comparable to the graduate students.

9.3.4
Students Use Instructional Technologies

Lazzari [222] looked at multiple dimensions of student learning for a group of students who generated, as well as used, short podcast lessons, relative to another group of students who only used them but did not generate them. Using both quantitative and qualitative methods, the students who generated podcasts placed this activity as a significant contributor to exam scores, course satisfaction, and as a dominant talking point during interviews. He concludes that it is likely "that podcasting design, recording, and editing spurred the development of reflective learning skills, stimulated students to go deep into the questions they had to face, and fostered positive collaborative behaviors, promoting growth of students' collaborative learning skills" (p. 32).

9.3.5
Using Authentic Materials Result in Disciplinary Identification and Socialization

Russell and Weaver [160] have compared the effects of traditional, inquiry-based, and research-based laboratories in chemistry, and the results are consistent with the prediction that the more authentic context results in greater gains in the more authentic conceptions about science and scientific practices. Their study covers large and small settings over a range of institutional types, and involved a combination of surveys and randomly selected students for in-depth interviews where they were asked questions about scientific theory and practice. Although the surface descriptions of scientific experiments and theories was equivalent across the different types of laboratory designs, the students in the research-based design "described experiments with more scientific purpose in mind, described theories from a more informed scientific perspective, and incorporated their own experiences into their descriptions of creativity in science" (p. 66). This personal identification with science, and the culture of science, is likely part of the socialization [209] process that brings new people into the discipline [11]. Weaver et al. [223] also report that research-based curricula result in increased interest in science and a greater sense of having "done science." In their summary of the existing research in the biological and life sciences, Auchincloss et al. [174] noted a strong but inferred relationship between CURE and regular undergraduate research experiences, themselves, and in addition they identified large gaps in the knowledge base for understanding how these experiences affect student learning.

With these sorts of potential gains at stake, changing the laboratory instruction program is a difficult and challenging task. In a national survey of faculty from 279 institutions, representing a broad array of instructional settings, Spell et al. examined why faculty members define authentic research experiences for classroom settings (experimental design, data collection and analysis), the degree to which these had been integrated into these programs (low), and the perceived barriers to their implementation (faculty time and resources) [224]. Brownell and Tanner [225] also hypothesize that the professionalization of academic scientists might well create tensions in their identities, putting efforts in teaching at odds with the norms they intuited during their training.

In shifting to a first-year organic chemistry course that is built explicitly on literature-based testing, students in a parallel, traditional general chemistry course showed differences, over time, with respect to the organic chemistry students on motivation and learning strategies measures [84]. Using the Motivated Strategies for Learning Questionnaire (MSLQ [226]), students in the organic chemistry course perceived a higher task value and intrinsic motivational orientation than their peers in the general chemistry course, while at the same time a lower sense of self-efficacy and higher test anxiety. Similar to the results in the research-based versus traditional laboratories, the organic and general chemistry students showed the same level of use of surface strategies in their learning, while the organic chemistry students also responded with using higher levels of deeper learning strategies. All of these trends are consistent with students who are in a more challenging environment, but who also recognize the higher value of the work.

Reading and understanding the primary literature is another enculturation feature for bringing new students into science. In studying a group of first-year university students in a life science program, Lacum *et al.* [159] observed that the design of the assignments was critical for student success. First-year students had a difficult time, relative to more experienced scientific readers, in identifying conclusions and the salient grounds for a given argument. This result mirrors that of Oliveras *et al.* [167], where students who read newspaper articles also had a more difficult time identifying both purpose and evidence than more sophisticated readers did. In both these cases, the assignments would benefit from breaking down the reading into explicit points of peer discussion, such as a milestone assignment where the students simply address "what is the writer's purpose?" or "what is the writer claiming?" and then another one, after these discussions, that asks them to consider "what is the evidence that is used to support the claim?" (and then: "does the evidence support the claim?").

9.3.6
Student-Generated Instructional Materials Promotes Metacognition and Self-Regulation

At the present time, metacognition represents our best understanding about why Flora is an effective learner [227]. She understands her strengths and weaknesses as an individual and has developed ways to exploit the former while compensating for the latter; she understands the tasks in front of her, understands what the underlying purposes are, and understands how to construct a reasonable pathway to the destination; she is aware of multiple learning strategies, how to use them effectively, and how they intersect with her own skill set; and she knows how to monitor what she is doing and how to evaluate where she is along the way. In designing an ideal learning environment for Flora, an instructor should provide both the necessary access to the knowledge that is needed and plan a set of experiences that allows her to deploy these skills in a meaningful way [228].

Student-generated instructional materials are an ideal way to encourage the development of metacognitive skills precisely because the learner is given a strong and familiar purpose: how to design effective instruction for one's peers that causes a learner to examine deeply her own learning in a reflective and responsible way. The construction of simple multiple-choice questions with no embedded feedback, which might be the least complex version of this practice, results in higher student engagement and increased academic performance [229].

In a study where students designed electrical circuit problems for peers, whose artifacts were examined along with transcripts from semistructured think-aloud sessions [230], subjects not only drew upon their prior knowledge in appropriate ways, but "when thinking about alternatives (to provide feedback to peers about errors), subjects reflect on the procedure for the solution and think about possible mistakes that can be made … when thinking about mistakes, subjects seem to realize better what they themselves used to do wrong." In addition, when addressing more complex computational problems, the subjects "did not just write down what they knew, but tried to organize it in such a way that it would be understandable" (p. 870).

Using content analysis of student-generated podcasts, in addition to examining transcripts of focus-group interviews, McLoughlin et al. [231] studied the levels of reflection and metacognition being used by five students who created supplementary audio (podcast) materials for beginning students in an information technology course. The metacognitive features they found in the student discourse included incidences of self-knowledge, task knowledge, strategic knowledge, and self-monitoring. As one of their subjects put it: "You've got to learn how to communicate with other people and understand them and you've got to learn how to get that across … " (p. 7).

9.4 Conclusions

Metaphorically, Bereiter's caricature of Dora, who sees education as a game of information rely and test-taking, also represents the challenge of school for school's sake. Dora does what is necessary to pass the test, and moves on to the next hurdle of school work that she perceives is being put in her way. Flora, on the other hand, is building toward a time when she is not in school, and when she will be called upon to learn, and to apply what is learned, in the real world of her career choice. The distinctiveness of a university education [208], to build upon prior knowledge and develop discipline-centered conceptual understanding, to enhance and diversify the ability to learn, and to promote the attitudes, values, and beliefs of a future professional [207], includes the self-conscious need to move students from school-based homework to a more highly contextualized kind of work we are calling *Real Work*. *Real Work* is not only more characteristic of real-world tasks but also aims at developing the skills and habits of mind that are needed in the real world.

Unfortunately, the momentum in higher education has been drifting toward a "homework" mentality. Twenty-five years ago, Charles Sykes' polemic, "ProfScam," [232] which included the claim that academic standards were fraying, was easily ignored. Today, well-considered criticism is mounting from the inside. In 2010, Hacker and Dreifus [233], backed by considerable investigate prowess, returned to many of the themes raised by Sykes, particularly the lack of a strong, intellectually driven agenda for advancing the education of students at many of the institutions they examined. About the same time, Arum and Roksa [234] advanced an even stronger indictment, backed with considerable data. Nearly half the students in their sample did not develop the higher order skills that one expects in college-educated students, with a bit more than a third failing to do so even by the end of their university experience. Derek Bok, in a more recent and significant analysis of higher education in America [235], provides a balanced and optimistic view that change can take place, but he is clear that improvements in the quality of undergraduate education need to occur, and that there has been a slow decline characterized by incoherence and a reduction of standards.

In this chapter, we started with two related frameworks, authentic learning and situated learning. These frameworks are compelling precisely because they presume education derives from a focus on real-world skills, needs, and tasks, and is governed by the dispositions of the disciplines and the mastery of faculty instructors who derive intellectual strength from the depth of their understanding [11]. We have anchored the attributes of designing *Real Work* on solid practices that faculty instructors have demonstrated result in high quality values-based outcomes. Every example, for instance, of integrating structured, peer-led work, where students must explicitly deal with learning and then teaching what they know, results in higher achievement [216]. While the evidence does not always support deeper understanding and higher order learning, there is a preponderance of evidence for the general benefit of peer-led work, and the examples allow us to understand what might be necessary (e.g., careful training and consistent monitoring of peer leaders) to avoid the nonproductive results (e.g., some leaders still default to telling rather than facilitating discussion) [112].

Every instructional setting is different. Students range from teenagers, leaving home for the first time and living in a university residence, to mid-career parents with full-time jobs. Faculty roles are equally broad. The application of the *Real Work* attributes is going to be governed by creativity, adaptability, and compromise on the part of both students and instructors. Varma-Nelson has demonstrated that virtual group interactions for cPLTL retain many of the critical features of the face-to-face version (P. Varma-Nelson, private communication January 17, 2014) [132]. Classroom undergraduate research experiences can be full-fledged, multidisciplinary projects [175] or they can be the combinatorial optimization of an existing procedure [135]. Other changes, such as using student-generated instructional materials, probably rely on the willingness of an instructor to relinquish direct control over every learning resource in favor of directing and supervising the construction of materials that might not be as polished as those of an experienced instructor, but for which the construction might provide students with a uniquely valuable learning situation.

As a discipline, chemistry offers a few advantages for developing *Real Work* compared with the other sciences. A great deal of the primary literature, including experimental work, is accessible to beginning students. The more accessible the material is, the more deeply engaged students can be when participating in peer review and critique. Similarly, transforming information from original sources into student-generated instructional materials relies on the depth and degree to which students understand those sources. *Real Work* also generates artifacts that can be studied, and puts students into settings where they can be observed doing activities that reveal their internalization of disciplinary dispositions and display their learning strategies [197]. Thus, as illustrated throughout Section 9.3, those who are interested in evidence to support a claim for higher level learning and/or conceptual understanding can do so [236].

Real Work can restore and invigorate a post-secondary education in ways that Slavich and Zimabardo would define as transformational [207]. *Real Work* makes it less likely for Dora to slip under the radar as she relies on the rote, retention, and recall methods that have worked for her in the past. Flora is anticipated to thrive in a *Real Work* context, and she needs to be engaged to help Dora change her approach: because designing *Real Work* draws from the disciplinary expertise of an instructor, and one of its greatest promises is to energize the link between research excellence and teaching excellence, on which the historical strength of our system of higher education rests.

Acknowledgments

The author thanks his organic chemistry colleagues in the Department of Chemistry, University of Michigan, for their continued support and enthusiasm that makes being part of that faculty teaching team such a rewarding experience: Prof. Masato Koreeda, Prof. Anne K. Mapp, Prof. Anne J. McNeil, Prof. Melanie S. Sanford, Prof. John Montgomery, Prof. Pavel Nagorny, and Prof. John P. Wolfe. He also thanks the the 100 student leaders who have driven the Structured Study Group program since its inception in 1994, and for all of their real work. Finally, he thanks his friend and long-time colleague, Eric S. Rabkin, for, among a million other things, understanding and inspiring the value of *Real Work*.

References

1. Bereiter, C. (1997) in *Situated Cognition: Social, Semiotic, and Psychological Perspectives* (eds D. Kirshner and J.A. Whitson), Erlbaum, Hillsdale, NJ, pp. 281–300.
2. Mayer, R.E. (2002) Rote versus meaningful learning. *Theor. Pract.*, **41** (4), 226–232.
3. Bretz, S.L., Fay, M., Bruck, L.B., and Towns, M.H. (2013) What faculty interviews reveal about meaningful learning in the undergraduate chemistry laboratory. *J. Chem. Educ.*, **90** (3), 281–288.
4. Van Oers, B. and Wardekker, K. (1999) On becoming an authentic learner: semiotic activity in the early grades. *J. Curriculum Stud.*, **32** (2), 229–249.
5. Nurrenburn, S. and Pickering, M. (1987) Concept learning versus problem solving: is there a difference? *J. Chem. Educ.*, **64**, 508–510.
6. Pickering, M. (1990) Further studies on concept learning versus problem solving. *J. Chem. Educ.*, **67**, 254–255.
7. Sawrey, B.A. (1990) Concept learning versus problem solving: revisited. *J. Chem. Educ.*, **67**, 253–254.
8. Nakhleh, M.B. and Mitchell, R.C. (1993) Concept learning versus problem solving. *J. Chem. Educ.*, **70**, 190–192.
9. Beall, H. and Prescott, S. (1994) Concepts and calculations in chemistry teaching and learning. *J. Chem. Educ.*, **71**, 111–112.
10. Francisco, J.S., Nakhleh, M.B., Nurrenburn, S., and Miller, M.L.

(2002) Assessing student understanding of general chemistry with concept mapping. *J. Chem. Educ.*, **79**, 248–257.
11. Coppola, B.P. and Krajcik, J.S. (2013) Discipline-centered post-secondary science education research: understanding university level science learning. *J. Res. Sci. Teach.*, **50** (6), 627–638.
12. Barton, G. (2013) The arts and literacy: what does it mean to be arts literate? *Int. J. Educ. Arts*, **14** (18), http://www.ijea.org/v14n18 (accessed 25 July 2014).
13. Stein, S.J., Isaacs, G., and Andrews, T. (2004) Incorporating authentic learning experiences within a university course. *Stud. Higher Educ.*, **29** (2), 241–258.
14. Reeves, T.C., Herrington, J., and Oliver, R. (2002) Authentic activities and online learning. HERDSA 2002 Quality Conversations, 7–10 July 2002, Perth, Western Australia, http://researchrepository.murdoch.edu.au/7034/1/authentic_activities_online_HERDSA_2002.pdf (accessed 25 July 2014); subsequently published in J. Herrington, T.C. Reeves, R. Oliver, and Y. Woo (2004) Designing authentic activities in web-based courses, *Journal of Computing in Higher Education*, **16** (1), 3–29.
15. Herrington, J., Reeves, T.C., and Oliver, R. (2014) in *Handbook of Research on Educational Communications and Technology*, 4th edn (eds J.M. Spector et al.), Springer, pp. 401–412.
16. Herrington, J., Reeves, T.C., and Oliver, R. (2012) *A Guide to Authentic e-learning*, Routledge, London, New York.
17. Brown, J.S., Collins, A., and Duguid, P. (1989) Situated cognition and the culture of learning. *Educ. Res.*, **18** (1), 32–42.
18. Jonassen, D. (1991) Evaluating constructivistic learning. *Educ. Technol.*, **31** (9), 28–33.
19. Lebow, D. (1993) Constructivist values for instructional systems design: five principles toward a new mindset. *Educ. Technol. Res. Dev.*, **41** (3), 4–16.
20. Oliver, R. and Omari, A. (1999) Using online technologies to support problem based learning: learners responses and perceptions. *Aust. J. Educ. Technol.*, **15**, 158–179.
21. Cronin, J.C. (1993) Four misconceptions about authentic learning. *Educ. Leadersh.*, **50** (7), 78–80.
22. Young, M.F. (1993) Instructional design for situated learning. *Educ. Technol. Res. Dev.*, **41** (1), 43–58.
23. Winn, W. (1993) Instructional design and situated learning: paradox or partnership. *Educ. Technol.*, **33** (3), 16–21.
24. Resnick, L. (1987) Learning in school and out. *Educ. Res.*, **16** (9), 13–20.
25. Cognition and Technology Group at Vanderbilt (1990) Anchored instruction and its relationship to situated cognition. *Educ. Res.*, **19** (6), 2–10.
26. Lebow, D. and Wager, W.W. (1994) Authentic activity as a model for appropriate learning activity: implications for emerging instructional technologies. *Can. J. Educ. Commun.*, **23** (3), 231–244.
27. Bransford, J.D., Vye, N., Kinzer, C., and Risko, V. (1990) in *Dimensions of Thinking and Cognitive Instruction* (eds B.F. Jones and L. Idol), Lawrence Erlbaum, Hillsdale, NJ, pp. 381–413.
28. Cognition and Technology Group at Vanderbilt (1990) Technology and the design of generative learning environments. *Educ. Technol.*, **31** (5), 34–40.
29. Spiro, R.J., Vispoel, W.P., Schmitz, J.G., Samarapungavan, A., and Boeger, A.E. (1987) *Executive Control Processes in Reading*, vol. 31, Lawrence Erlbaum Associates, Hillsdale, NJ, pp. 177–199.
30. Gordon, R. (1998) Balancing real-world problems with real-world results. *Phi Delta Kappan*, **79**, 390–393.
31. Myers, S. (1993) A trial for Dmitri Karamazov. *Educ. Leadersh.*, **50** (7), 71–72.
32. Bransford, J.D., Sherwood, R.D., Hasselbring, T.S., Kinzer, C.K., and Williams, S.M. (1990) in *Cognition, Education and Multimedia: Exploring Ideas in High Technology* (eds D. Nix and R. Spiro), Lawrence Erlbaum, Hillsdale, NJ, pp. 115–141.
33. Reeves, T.C. and Okey, J.R. (1996) in *Constructivist Learning Environments: Case Studies in Instructional Design* (ed

B.G. Wilson), Educational Technology, Englewood Cliffs, NJ, pp. 191–202.
34. Young, M.F. (1995) Assessment of situated learning using computer environments. *J. Sci. Educ. Technol.*, **4** (1), 89–96.
35. Herrington, J. and Herrington, A. (1998) Authentic assessment and multimedia: how university students respond to a model of authentic assessment. *Higher Educ. Res. Dev.*, **17** (3), 305–322.
36. Barab, S.A., Squire, K.D., and Dueber, W. (2000) A co-evolutionary model for supporting the emergence of authenticity. *Educ. Technol. Res. Dev.*, **48** (2), 37–62.
37. Duchastel, P.C. (1997) A web-based model for university instruction. *J. Educ. Technol. Syst.*, **25** (3), 221–228.
38. Bottge, B.A. and Hasselbring, T.S. (1993) Taking word problems off the page. *Educ. Leadersh.*, **50** (7), 36–38.
39. Young, M.F. and McNeese, M. (1993) A situated cognition approach to problem solving with implications for computer-based learning and assessment, in *Human-Computer Interaction: Software and Hardware Interfaces* (eds G. Salvendy and M.J. Smith), Elsevier Science Publishers, New York.
40. Marilyn, M.L. (2007) Educause Learning Initiative (ELI Paper 1:2007) May 2007, *http://net.educause.edu/ir/library/pdf/eli3009.pdf* (accessed 26 July 2014).
41. City, E.A., Elmore, R.F., Fiarman, S.E., and Teitel, L. (2009) *Instructional Rounds in Education*, Harvard Education Press, Cambridge, MA.
42. Strayer, J.F. (2012) How learning in an inverted classroom influences cooperation, innovation and task orientation. *Lear. Environ. Res.*, **15**, 171–193.
43. Turpen, C. and Noah, D.F. (2009) Not all interactive engagement is the same: variations in physics professors' implementation of Peer Instruction. *Phys. Rev. Spec. Top. Phys. Educ. Res.*, **5** (2), 1–16.
44. Judson, E. and Sawada, D. (2002) Learning from past and present: electronic response systems in college lecture halls. *J. Comput. Math. Sci. Teach.*, **21** (2), 167–181.
45. Doyle, W. (1983) Academic work. *Rev. Educ. Res.*, **53** (2), 159–199.
46. Rabkin, E.S. and Smith, M. (1990) *Teaching Writing that Works: A Group Approach to Practical English*, University of Michigan Press, Ann Arbor, MI.
47. Rabkin, E.S. Real Work is Better than Homework: a Principle to Teach By, *http://www-personal.umich.edu/~esrabkin/realwork/index.html* (accessed 25 July 2014).
48. Gabel, D. (1998) in *International Handbook of Science Education* (eds B.J. Fraser and K.G. Tobin), Kluwer Academic Publishers, Dordrecht, pp. 233–248.
49. Lunetta, V.N. (2003) The school science laboratory: historical perspectives and contexts for contemporary teaching, in *International Handbook of Science Education* (eds B.J. Fraser and K.G. Tobin), Kluwer Academic Publishers, Dordrecht, pp. 249–262.
50. Hofstein, A. (2004) The laboratory in chemistry education: thirty years of experience with developments, implementations, and research. *Chem. Educ.: Res. Pract.*, **5** (3), 247–264.
51. Sweeney, A.E. and Paradis, J.A. (2004) Developing a laboratory model for the professional preparation of future science teachers: a situated cognition perspective. *Res. Sci. Educ.*, **34**, 195–219.
52. Bechtel, W. (2009) in *Cambridge Handbook of Situated Cognition* (eds P. Robbins and M. Aydede), Cambridge University Press, Cambridge, pp. 155–170.
53. Janet Bond-Robinson and Amy Preece Stucky (2005) Grounding Scientific Inquiry and Knowledge in Situated Cognition, *http://csjarchive.cogsci.rpi.edu/proceedings/2005/docs/p310.pdf* (accessed 25 July 2015).
54. Bereiter, C. and Scardamalia, M. (1989) *Knowing, Learning, and Instruction: Essays in Honor of Robert Glaser*, Lawrence Erlbaum Associates, Hillsdale, NJ, pp. 361–392.

55. Lave, J. and Wenger, E. (1991) *Situated Learning: Legitimate Peripheral Participation*, Cambridge University Press, Cambridge.
56. Lave, J. (1991) in *Perspectives on Social Shared Cognition* (eds L.B. Resnick, J.M. Levine, and S.D. Teasley), American Psychological Association, Washington, DC, pp. 63–82.
57. Herrington, J., and Oliver, R., (2000) An instructional design framework for authentic learning environments *Educational Technology Research and Development*. **48**, 3, 23–48.
58. Snow, C.P. (1959) *The Two Cultures*, Cambridge University Press, Cambridge.
59. Thompson, B.E. (2002) Studio pedagogy for engineering design. *Int. J. Eng. Educ.*, **18** (1), 39–49.
60. Wilson, J. (1994) The CUPLE physics studio. *Phys. Teach.*, **32**, 518–523.
61. Apple, T. and Cutler, A. (1999) The rensselaer studio general chemistry course. *J. Chem. Educ.*, **76**, 462–463.
62. Bailey, C.A., Kingsbury, K., Kulinowski, K., Paradis, J., and Schoonover, R. (2000) An integrated lecture-laboratory environment for general chemistry. *J. Chem. Educ.*, **77**, 195–199.
63. Eberlein, T. *et al.* (2008) Pedagogies of engagement in science: a comparison of PBL, POGIL, and PLTL. *Biochem. Mol. Biol. Educ.*, **36**, 262–273.
64. Wandersee, J.H. (1991) Mantras, false dichotomies, and science education research. *J. Res. Sci. Teach.*, **28** (3), 211–212.
65. Guilford, J.P. (1950) Creativity. *Am. Psychol.*, **5**, 444–454.
66. Guilford, J.P. (1967) *The Nature of Human Intelligence*, McGraw-Hill, New York.
67. Razoumnikova, O.M. (2000) Functional organization of different brain areas during convergent and divergent thinking: an EEG investigation. *Cogn. Brain Res.*, **10** (12), 11–18.
68. Akbari Chermahini, S. and Hommel, B. (2010) The (b)link between creativity and dopamine: Spontaneous eye blink rates predict and dissociate divergent and convergent thinking. *Cognition*, **115**, 458–465.
69. Akbari Chermahini, S. and Hommel, B. (2012) Creative mood swings: divergent and convergent thinking affect mood in opposite ways. *Psychol. Res.*, **76** (5), 634–640.
70. Colzato, L.S., Szapora, A., Pannekoek, J.N., and Hommel, B. (2013) The impact of physical exercise on convergent and divergent thinking. *Front. Hum. Neurosci.*, **7**, 824. doi: 10.3389/fnhum.2013.00824
71. De Bono, E. (1967) *The Use of Lateral Thinking*, Jonathan Cape, London.
72. Straker, D. and Rawlinson, G. (2003) *How to Invent (Almost) Anything*, Spiro Press, London.
73. Bishop, C.M. and Lasserre, J. (2007) in *Bayesian Statistics 8* (eds J.M. Bernado *et al.*), Oxford University Press, Cambridge, pp. 3–24.
74. Baldwin, B.A. (1984) The role of difficulty and discrimination in constructing multiple-choice examinations: with guidelines for practical application. *J. Account. Educ.*, **2** (1), 19–28.
75. Gross-Glenn, K., Jallad, B., Novoa, L., Helgren-Lempesis, V., and Lubs, H.A. (1990) Nonsense passage reading as a diagnostic aid in the study of familial dyslexia. *Reading Writing*, **2** (2), 161–173.
76. M. Svinicki and W. J. McKeachie (2011) *McKeachie's Teaching Tips*, 13th edn Wadsworth: Belmont, CA.
77. Hake, R. (1998) A six-thousand-student survey. *Am. J. Phys*, **66**, 64–74.
78. Johnstone, A.H. (2003) *Effective Practice in Objective Assessment*, Physical Science Centre, Hull.
79. Corey, E.J. (1988) Retrosynthetic thinking – essentials and examples. *Chem. Soc. Rev.*, **17**, 111–133.
80. Seyhan, N. (2003) *Ege, Organic Chemistry*, 5th edn, Houghton Mifflin, Boston, MA.
81. Coppola, B.P. and Daniels, D.S. (1996) The role of written and verbal expression in learning. *Lang. Learn. Across Discip.*, **1** (3), 67–86.
82. Coppola, B.P., Daniels, D.S., and Pontrello, J.P. (2001) in *Student Assisted Teaching and Learning* (eds J. Miller, J.E. Groccia, and D. DiBiasio), Anker, New York, pp. 116–122.

83. Varma-Nelson, P. and Coppola, B.P. (2005) in *Chemist's Guide to Effective Teaching* (eds N. Pienta, M.M. Cooper, and T. Greenbowe), Pearson, Saddle River, NJ, pp. 155–169.
84. Coppola, B.P., Ege, S.N., and Lawton, R.G. (1997) The University of Michigan undergraduate chemistry curriculum 2. Instructional strategies and assessment. *J. Chem. Educ.*, **74**, 84–94.
85. Coppola, B.P. (2000) Targeting entry points for ethics in chemistry teaching and learning. *J. Chem. Educ.*, **77**, 1506–1511.
86. Hayward, L.M. and Coppola, B.P. (2005) Teaching and technology: making the invisible explicit and progressive through reflection. *J. Phys. Ther. Educ.*, **19** (3), 83–97.
87. Gosser, D.K. Jr., Kampmeier, J.A., and Varma-Nelson, P. (2010) Peer-Led learning: 2008 James Flack Norris award address. *J. Chem. Educ.*, **87** (4), 374–380.
88. Stellan, O. (1996) Learning from performance errors. *Psychol. Rev.*, **103** (2), 241–262.
89. Priest, A. and Roach, P. (1991) Learning from errors. *Cogn. Syst.*, **3** (1), 79–102.
90. Tanner, K. and Allen, D. (2005) Approaches to biology teaching and learning: understanding the wrong answers - teaching toward conceptual change. *Cell Biol. Educ.*, **4** (2), 112–117.
91. Chance, Z., Norton, M.I., Gino, F., and Ariely, D. (2011) Temporal view of the costs and benefits of self-deception. *Proc. Natl. Acad. Sci. U.S.A.*, **108** (Suppl. 3), 15655–15659.
92. Davidson, S., Stickney, C.P., and Weil, R.L. (1980) *Intermediate Accounting Concepts:Methods and Uses*, Dryden Press, Fort Worth, TX.
93. Hoffmann, R. and Coppola, B.P. (1996) Some heretical thoughts on what our students are telling us. *J. Coll. Sci. Teach.*, **25**, 390–394.
94. Boud, D., Cohen, R., and Sampson, J. (eds) (2001) *Peer Learning in Higher Education: Learning with and from Each Other*, Routledge, London.
95. Whitman, N.A. (1988) Peer Teaching: To Teach is to Learn Twice, ASHE-ERIC Higher Education Report No. 4, Association for the Study of Higher Education, Washington, DC.
96. Goldschmid, B. and Goldschmid, M.L. (1976) Peer teaching in higher education: a review. *High. Educ.*, **5**, 9–33.
97. Topping, K. (1998) Peer assessment between students in colleges and universities. *Rev. Educ. Res.*, **68** (3), 249–276.
98. Freeman, M. (1995) Peer assessment by groups of group work. *Assess. Eval. High. Educ.*, **20** (3), 289–300.
99. Falchikov, N. and Goldfinch, J. (2000) Student peer assessment in higher education: a meta-analysis comparing peer and teacher marks. *Rev. Educ. Res.*, **70** (3), 287–322.
100. Topping, K. (1996) The effectiveness of peer tutoring in further and higher education: a typology and review of the literature. *High. Educ.*, **32**, 321–345.
101. Marcoulides, G.A. and Simkin, M.G. (1991) Evaluating student papers: the case for peer review. *J. Educ. Bus.*, **67**, 80–83.
102. Wagner, L. (1982) *Peer Teaching: Historical Perspectives*, Greenwood Press, Westport, CT.
103. Shulman, L.S. (1993) Teaching as community property: putting an end to pedagogical solitude. *Change*, **25** (6), 6–7.
104. Herrington, A.J. and Cadman, D. (1991) Peer review and revising in an anthropology course: lessons for learning. *Coll. Compos. Commun.*, **42** (2), 184–199.
105. Palincsar, A.S. and Brown, A.L. (1984) Reciprocal teaching of comprehension-fostering and comprehension-monitoring activities. *Cogn. Instr.*, **1**, 117–175.
106. Brown, A.L. and Palincsar, A.S. (1989) in *Knowing, Learning, and Instruction: Essays in Honor of Robert Glaser* (ed. L.B. Resnick), Lawrence Erlbaum Associates, Hillsdale, NJ, pp. 393–451.
107. Schwenk, T.L. and Whitman, N. (1984) *Residents as Teachers*, University of

Utah School of Medicine, Salt Lake City, UT.

108. Allen, V.L. and Feldman, R.S. (1973) Learning through tutoring: low-achieving children as tutors. *J. Exp. Educ.*, **42** (1), 1–5.

109. Bargh, J.A. and Schul, Y. (1980) On the cognitive benefits of teaching. *J. Educ. Psychol.*, **72** (5), 593–604.

110. Benware, C.A. and Deci, E.L. (1984) Quality of learning with an active versus passive motivational set. *Am. Educ. Res. J.*, **21** (4), 755–765.

111. Coleman, E., Brown, A., and Rivkin, I. (1997) The effect of instructional explanations on formal learning from scientific texts. *J. Learn. Sci.*, **6** (4), 347–365.

112. Roscoe, R.D. and Michelene, T.H. Chi (2007) Understanding tutor learning: knowledge-building and knowledge telling in peer tutors' explanations and questions. *Rev. Educ. Res.*, **77** (4), 534–574.

113. Chang, H.-Y. and Chang, H.-C. (2013) Scaffolding students' online critiquing of expert- and peer-generated molecular models of chemical reactions. *Int. J. Sci. Educ.*, **35** (12), 2028–2056.

114. Balfour, S.P. (2013) Assessing writing in MOOCs: automated essay scoring and calibrated peer review. *Res. Pract. Assess.*, **8**, 40–48.

115. Russell, A.A., Chapman, O.L., and Wegner, P.A. (1998) Molecular science: network-deliverable curricula. *J. Chem. Educ.*, **75** (5), 578–579.

116. Robinson, R. (2001) An application to increase student reading and writing skills. *Am. Biol. Teach.*, **63** (7), 474–480.

117. Berry, D.E. and Fawkes, K.L. (2010) Constructing the components of a lab report using Peer review. *J. Chem. Educ.*, **87** (1), 57–61.

118. Gragson, D.E. and Hgen, J.P. (2010) Developing technical writing skills in the physical chemistry laboratory: a progressive approach employing peer review. *J. Chem. Educ.*, **87** (1), 62–65.

119. Margerum, L.D., Gulsrud, M., Manlapez, R., Rebong, R., and Love, A. (2007) Application of Calibrated Peer Review (CPR) writing assignments to enhance experiments with an environmental chemistry focus. *J. Chem. Educ.*, **84** (2), 292–295.

120. Hartberg, Y., Gunersel, A.B., Simpson, N.J., and Balester, V. (2008) Development of student writing in biochemistry using calibrated Peer review. *J. Scholarsh. Teach. Learn.*, **2** (1), 29–44.

121. Roxanne Prichard, J. (2005) Writing to learn: an evaluation of the calibrated Peer review program in two neuroscience courses. *J. Undergrad. Neurosci. Educ.*, **4** (1), A34–A39.

122. Berland, L.K. and Victor, R.L. (2012) In pursuit of consensus: disagreement and legitimization during small-group argumentation. *Int. J. Sci. Educ.*, **34** (12), 1857–1882.

123. Hand, B. and Choi, A. (2010) Examining the impact of student use of multiple model representations in constructing arguments in organic chemistry laboratory classes. *Res. Sci. Educ.*, **40**, 29–44.

124. Shah, S. *et al.* (1999) A study of the effect of healing energy on in vitro tumor cell proliferation. *J. Altern. Complement. Med.*, **5** (4), 359–365.

125. Michaelson, L.K., Knight, A.B., and Fink, D.L. (eds) (2004) *Team-Based Learning: A Transformative Use of Small Groups in College Teaching*, Stylus Publishing, Sterling, VA.

126. Hills, H. (2001) *Team-Based Learning*, Gower, Hampshire.

127. Cooper, M.M. (2005) in *Chemist's Guide to Effective Teaching* (eds N. Pienta, M.M. Cooper, and T.J. Greenbowe), Pearson, Saddle River, NJ, pp. 117–128.

128. Vygotsky, L. (1978) *Mind in Society: The Development of Higher Psychological Processes*, Harvard University Press, Cambridge, MA.

129. Vygotsky, L. (1985) *Thought and Language*, The MIT Press, Cambridge, MA.

130. Rieber, L.P. (2000) The studio experience: educational reform in instructional technology, in *Best Practices in Computer Enhanced Teaching and Learning*, Wake Forest Press, Winston-Salem, NC.

131. Kampmeier, J.A., Varma-Nelson, P., and Wedegaertner, D. (2001) *Peer-Led Team Learning: Organic Chemistry*, Prentice-Hall, Upper Saddle River, NJ.
132. Smith, J., Wilson, S.B., Banks, J., Zhu, L., and Varma-Nelson, P. (2014) Replicating Peer-Led Team Learning in cyberspace: Research, opportunities, and challenges, *J. Res. Sci. Teach.*, **51** (6), 714–740.
133. Gottfried, A.C. et al. (2007) Deisgn and implementation of a studio-based general chemistry course. *J. Chem. Educ.*, **84** (2), 265–270.
134. Coppola, B.P. and Lawton, R.G. (1995) "Who has the same substance that I have?" A blueprint for collaborative learning activities. *J. Chem. Educ.*, **72**, 1120–1122.
135. Shultz, G.V. (2012) *CHEM 216 Online Laboratory Manual*, Hayden-McNeil, Plymouth, MI.
136. Vermaat, H., Kramers-Pals, H., and Schank, P. (2003) The use of animations in chemical education. Proceedings of the International Convention of the Association for Educational Communications and Technology, Anaheim, CA, pp. 430–441.
137. Kozma, R. (2003) Material and social affordances of multiple representations for science understanding. *Learn. Instr.*, **13** (2), 205–226.
138. Kozma, R. and Russell, J. (2005) in *Students Becoming Chemists: Developing Representational Competence* (ed. J. Gilbert), Springer, Dordrecht.
139. Michalchik, V. et al. (2008) in *Visualization: Theory and Practice in Science Education* (eds J. Gilbert, M. Nakhlah, and M. Reiner), Springer, New York, pp. 233–282.
140. Dunmire, R.E. (2010) The Use of Instructional Technology in the Classroom: Selection and Effectiveness, http://www.usma.edu/cfe/Literature/Dunmire_10.pdf (accessed 26 July 2014).
141. Schank, P. and Kozma, R. (2002) Learning chemistry through the use of a representation-based knowledge building environment. *J. Comput. Math. Sci. Teach.*, **21** (3), 253–279.
142. Kearney, M. and Schuck, S. (2006) Spotlight on authentic learning: student developed digital video projects. *Australas. J. Educ. Technol.*, **22** (2), 189–208.
143. Barab, S.A., Hay, K.E., Barnett, M., and Keating, T. (2000) Virtual solar system project: building understanding through model building. *J. Res. Sci. Teach.*, **37** (7), 719–756.
144. Janowicz, P.A. and Moore, J.S. (2009) Chemistry goes global in the virtual world. *Nat. Chem.*, **1**, 2–4.
145. Zhihui, H.Z. and Linn, M.C. (2013) Learning from chemical visualizations: comparing generation and selection. *Int. J. Sci. Educ.*, **35** (13), 2174–2197.
146. Hoban, G. and Nielsen, W. (2013) Learning science through creating a "Slowmation": a case study of preservice primary teachers. *Int. J. Sci. Educ.*, **35**, 119–146.
147. International SRI ChemSense, http://chemsense.sri.com (accessed 26 July 2014).
148. ACM http://dl.acm.org/citation.cfm?id=1017862 (accessed 26 July 2014).
149. Chang, H.-Y. and Quintana, C. (2006) Student-generated animations: supporting middle school students' visualization, interpretation and reasoning of chemical phenomena. ICLS '06 Proceedings of the 7th International Conference on Learning Sciences, 2006, pp. 71–77.
150. Lawrie, G. and Bartle, E. (2013) Chemistry vlogs: a vehicle for student-generated representations and explanations to scaffold their understanding of structure-property relationships. *Int. J. Innov. Sci. Math. Educ.*, **21** (4), 27–45.
151. Shultz, G.V., Winschel, G.A., Inglehart, R., and Coppola, B.P. (2014) Eliciting student explanations of experimental results. *J. Chem. Educ.*, **91**, 684–686.
152. Moy, C., Locke, J.R., Coppola, B.P., and McNeil, A.J. (2010) Improving science education and understanding through editing wikipedia. *J. Chem. Educ.*, **87** (11), 1159–1162.
153. Evans, M.J. and Moore, J.S. (2011) A collaborative, wiki-based organic chemistry project incorporating free

154. Lancor, R.A. (2014) Using student-generated analogies to investigate conceptions of energy: a multidisciplinary study. *Int. J. Sci. Educ.*, **36** (1), 1–23.
155. Williams, V. and Dwyer, F. Jr., (1999) The effects of metaphoric (visual/verbal) strategies in facilitating student achievement of different educational objectives. *Int. J. Instructional Media*, **26** (2), 205–211.
156. Kretzenbacher, H.L. (2003) The aesthetics and heuristics of analogy. *HYLE-Int. J. Philos. Chem.*, **9** (2), 191–218.
157. Mayer, R.E. (1996) Learning strategies for making sense out of expository text: the soi model for guiding three cognitive processes in knowledge construction. *Educ. Psychol. Rev.*, **8** (4), 357–371.
158. Houde, A. (2000) Student symposia on primary research articles. *J. Coll. Sci. Teach.*, **30** (3), 184–187.
159. van Lacum, E., Ossevoort, M., Buikema, H., and Goedhart, M. (2012) First experiences with reading primary literature by undergraduate life science students. *Int. J. Sci. Educ.*, **34** (12), 1795–1821.
160. Russell, C.B. and Weaver, G.C. (2011) A comparative study of traditional, inquiry-based, and research-based laboratory curricula: impacts on understanding of the nature of science. *Chem. Educ.: Res. Pract.*, **12**, 57–67.
161. Alaimo, P.J., Langenham, J.M., and Suydam, I.T. (2014) Aligning the undergraduate laboratory experience with professional work: the centrality of reliable and meaningful data. *J. Chem. Educ.* Article ASAP (October 10, 2014) DOI:10.1021/ed400510b.
162. Bondeson, S.R., Brummer, J.G., and Wright, S.M. (2001) The data-driven classroom. *J. Chem. Educ.*, **71**, 56–57.
163. Camill, P. (2000) Using journal articles in an environmental biology course. *J. Coll. Sci. Teach.*, **30** (1), 38–43.
164. Gallagher, G.J. and Adams, D.L. (2002) Introduction to the use of primary organic chemistry literature in an honors sophomore-level organic chemistry course. *J. Chem. Educ.*, **79** (11), 1368–1371.
165. Forest, K. and Rayne, S. (2009) Incorporating primary literature summary projects into a first-year chemistry curriculum. *J. Chem. Educ.*, **86** (5), 592–594.
166. Almeida, C.A. and Liotta, L.J. (2005) Organic chemistry of the cell: an interdisciplinary approach to learning with a focus on reading, analyzing, and critiquing primary literature. *J. Chem. Educ.*, **82** (12), 1794–1799.
167. Oliveras, B., Marquez, C., and Sanmarti, N. (2013) The use of newspaper articles as a tool to develop critical thinking in science classes. *Int. J. Sci. Educ.*, **35** (6), 885–905.
168. Glaser, R.E. and Carson, K.M. (2005) Chemistry is in the news: taxonomy of authentic news media-based activities. *Int. J. Sci. Educ.*, **27** (9), 1083–1098.
169. King, A. (1994) Autonomy and question asking: the role of personal control in guided student-generated questioning. *Learn. Individ. Differ.*, **6** (2), 163–185.
170. Middlecamp, C.H. and Nickel, A.-M.L. (2005) Doing science and asking questions II: an exercise the generates questions. *J. Chem. Educ.*, **82** (8), 1181–1186.
171. Colbert, J.T., Olsen, J.K., and Clough, M.P. (2007) Using the web to encourage student-generated questions in large-format introductory biology classes. *CBE-Life Sci. Educ.*, **6**, 42–48.
172. Delbanco, N. and Cheuse, A. (2012) *Literature: Craft and Voice*, 2nd edn, McGraw-Hill Higher Education, New York.
173. The Center for Authentic Science Practice in Education http://www.caspie.org (accessed 26 July 2014).
174. Auchincloss, L.C. et al. (2014) Assessment of course-based undergraduate research experiences: a meeting report. *CBE - Life Sci. Educ.*, **13**, 29–40.
175. Latch, D.E., Whitlow, W.L., and Alaimo, P.J. (2012) in *Science Education and Civic Engagement: The Next Level*, ACS Symposium Series, Vol. 1121 (eds R.D. Sheardy and W.D. Burns),

American Chemical Society, Washington, DC, pp. 17–30.
176. Kosinski-Collins, M. Responding to the Age–Old Question: "Why do I Have to Take Orgo?" (2012) http://educationgroup.mit.edu/HHMIEducationGroup/wp-content/uploads/2012/02/KC-Presentation-slides.pdf (accessed 2 August 2014).
177. Scott, W.L. and O'Donnell, M.J. (2009) Distributed drug discovery, part 1: linking academic and combinatorial chemistry to find drug leads for developing world diseases. *J. Comb. Chem.*, **11**, 3–13.
178. Scott, W.L. et al. (2009) Distributed drug discovery, part 2: global rehearsal of alkylating agents for the synthesis of resin-bound unnatural amino acids and virtual D catalog construction. *J. Comb. Chem.*, **11** (1), 14–33.
179. Lee, M.J.W. and McLoughlin, C. (2007) Teaching and learning in the Web 2.0 Era: empowering students through learner-generated content. *Int. J. Instr. Technol. Distance Learn.*, **4** (10), 21–34.
180. Pérez-Mateo, M., Maina, M.F., Guitert, M., and Romero, M. (2011) Learner generated content: quality criteria in online collaborative learning. *Eur. J. Open, Distance E-Learn, 2011 Special Issue on Creativity and Open Educational Resources (OER)*, http://www.eurodl.org/materials/special/2011/Perez-Mateo_et_al.pdf.
181. Hamer, J., Sheard, J., Purchase, H., and Luxton-Reilly, A. (2012) Contributed student pedagogy. *Comput. Sci. Educ.*, **22** (4), 315–318.
182. Hamer, J. et al. (2008) Contributing student pedagogy. *ACM SIGCSE Bull.*, **40** (4), 194–212.
183. Cajander, A., Daniels, M., and McDermott, R. (2012) On valuing Peers: theories of learning and intercultural competence. *Comput. Sci. Educ.*, **22** (4), 319–342.
184. Sondergaard, H. and Mulder, R.A. (2010) Collaborative learning through formative Peer review: pedagogy, programs and potential. *Comput. Sci. Educ.*, **22** (4), 343–367.
185. Herman, G.L. (2010) Designing contributing student pedagogies to promote students' intrinsic motivation. *Comput. Sci. Educ.*, **22** (4), 369–388.
186. Katrina, F. and Falkner, N.J.G. (2012) Supporting and structuring "contributing student pedagogy" in computer science curricula. *Comput. Sci. Educ.*, **22** (4), 413–443.
187. Luxton-Reilly, A. and Denny, P. (2010) Constructive evaluation: a pedagogy of student-contributed assessment. *Comput. Sci. Educ.*, **20** (2), 145–167.
188. The Horzon Report (2007) http://www.nmc.org/pdf 2007_Horizon_Report.pdf.
189. Morrison, T.G., Bryan, G., and Chilcoat, G.W. (2002) Using student-generated comic books in the classroom. *J. Adolescent Adult Lit.*, **45** (8), 758–767.
190. Wheeler, S., Yeomans, P., and Wheeler, D. (2008) The good, the bad and the wiki: evaluating student-generated content for collaborative learning. *Br. J. Educ. Technol.*, **39** (6), 987–995.
191. Kidd, J., O'Shea, P., Allen, D., and Tamashiro, R. (2008) Student-authored textbooks: the future or futile. Society for Information Technology and Teaching Education International Conference 2008, Chesapeake, VA, pp. 3274–3279.
192. O'Shea, P., Chappell, S., Allen, D., and Baker, P (2007) Issues confronted while designing a student-developed online textbook. Society for Information Technology & Teacher Education International Conference 2007, Chesapeake, VA, Vol. 2007, pp. 2074–2079.
193. Gehringer, E.F., Kadanjoth, R., and Kidd, J. (2010) Software support for peer-reviewing wiki textbooks and other large projects. Proceedings of the Workshop on Computer-Supported Peer Review in Education, 2010.
194. Vasquez, A.V. et al. (2012) Writing-to-teach: a new pedagogical approach to elicit explanative writing from undergraduate chemistry students. *J. Chem. Educ.*, **89**, 1025–1031.
195. Rees, S., Bruce, M., and Nolan, S. (2013) Can i have a word please – strategies to enhance understanding of subject specific language

in chemistry by international and non-traditional students. *New Dir.*, **9** (1), 8–13.
196. GIFBuilder *https://www.macupdate.com/app/mac/235/gifbuilder*.
197. Coppola, B.P. (2010) in *Assessment in the Disciplines*, Vol. 5: Assessment in Chemistry (eds J. Ryan, T. Clark, and A. Collier), Association for Institutional Research, Tallahassee, FL, pp. 175–199.
198. Coppola, B.P. (2001) in *College Pathways to the Science Education Standards* (eds E.D. Siebert and W.J. McIntosh), NSTA Press, Arlington, VA, pp. 84–86.
199. Fergus, S. and Kirton, S. (2013) N106 Peerwise Presentation June 2013, *http://www.studynet1.herts.ac.uk/intranet/lti.nsf//Teaching+Documents/3E20D3CA17259DDA80257B94005A94B9/$FILE/N106%20Peerwise%20presentation%20June%202013.pptx* (accessed 2 August 2014).
200. PeerWise *http://peerwise.cs.auckland.ac.nz/* (accessed 26 July 2014).
201. Kelly, R. (2012) Have Students Generate Content to Improve Learning, 2012 October, *http://www.magnapubs.com/newsletter/online-classroom/issue/1405/* (accessed 26 July 2014).
202. Sapling Learning *http://www2.saplinglearning.com/* (accessed 26 July 2014).
203. Forbes, D., Khoo, E., and Johnson, E.M. (2012) It gave me a much more personal connection: student-generated podcasting and assessment in teacher education. in Future Challenges, Sustainable Futures. Proceedings Ascilite Wellington, 2012, pp. 326–330.
204. Lee, M.J.W., McLoughlin, C., and Chan, A. (2008) Talk the talk: learner-generated podcasts as catalysts for knowledge creation. *Br. J. Educ. Technol.*, **39** (3), 501–521.
205. Gehringer, E.F. and Miller, C.S. (2009) Student-generated active-learning exercises. ACM SIGCSE Bulletin – SIGCSE '09, New York, Vol. 41 (1), pp. 81–85.
206. Alaimo, P.J., Langenham, J.M., and Tanner, M.J. (2010) Safety teams: an approach to engage students in laboratory safety. *J. Chem. Educ.*, **87** (8), 856–861.
207. Slavich, G.M. and Zimbardo, P.G. (2012) Transformational teaching: theoretical underpinnings, basic principles, and core methods. *Educ. Psychol. Rev.*, **24** (4), 569–608.
208. Coppola, B.P. (2013) The distinctiveness of a higher education. *J. Chem. Educ.*, **90** (8), 955–956.
209. Smart, J., Feldman, K.A., and Ethington, C.A. (2000) *Academic Disciplines: Holland's Theory and the Study of College Students and Faculty*, 1st edn, Vanderbilt University Press, Nashville, TN.
210. Trivic, D., Tomasevic, B., and Vukovic, I. (2012) Student creativity in chemistry classes. CnS-La Chimica nella Scuola XXXIV-3 Proceedings ICCE-ERICE, 2012, pp. 393–398.
211. Calibrated Peer Review Website *http://cpr.molsci.ucla.edu* (accessed 26 July 2014)
212. Pelaez, N.J. (2002) Problem-based writing with peer review improves academic performance in physiology. *Adv. Physiol. Educ.*, **26**, 174–184.
213. Walvoord, M.E., Hoefnagels, M.H., Gaffin, D.D., Chumchal, M.M., and Long, D.A. (2008) An analysis of Calibrated Peer Review (CPR) in a science lecture classroom. *J. Coll. Sci. Teach.*, **37** (4), 66–73.
214. Gunersel, A.B., Simpson, N.J., Aufderheide, K.J., and Wang, L. (2008) Effectiveness of calibrated peer review[TM] for improving writing and critical thinking skills in biology undergraduate students. *J. Scholarsh. Teach. Learn.*, **8** (2), 25–37.
215. Reynolds, J. and Moskovitz, C. (2008) Calibrated peer review assignments in science courses: are they designed to promote critical thinking and writing skills. *J. Coll. Sci. Teach.*, **38** (2), 60–66.
216. Tien, L.T., Roth, V., and Kampmeier, J.A. (2002) Implementation of a Peer-Led team learning instructional approach in an undergraduate organic chemistry course. *J. Res. Sci. Teach.*, **39** (7), 606–632.

217. Wamser, C.C. (2006) Peer-led team learning in organic chemistry: effects on student performance, success, and persistence in the course. *J. Chem. Educ.*, **83** (10), 1562–1566.
218. Quitadamo, I.J., Jayne Brahler, C., and Crouch, G.J. (2009) Peer-led team learning: a prospective method for increasing critical thinking in undergraduate science courses. *Sci. Educ.*, **18** (1), 29–39.
219. Sawyer, K., Frey, R., and Brown, P.L. (2013) in *Productive Multivocality in the Analysis of Group Interactions*, Computer-Supported Collaborative Learning Series, vol. 16 (eds D.D. Suthers *et al.*), Springer, New York, pp. 191–204.
220. Feldon, D.F. (2010) Why magic bullets don't work. *Change*, **42** (2), 15–21.
221. Gafney, L. and Varma-Nelson, P. (2007) Evaluating Peer-led team learning: a study of long-term effects on former workshop leaders. *J. Chem. Educ.*, **84** (3), 535–539.
222. Lazzari, M. (2009) Creative use of podcasting in higher education and its effect on competitive agency. *Comput. Educ.*, **52**, 27–34.
223. Weaver, G.C., Russell, C.B., and Wink, D.J. (2008) Inquiry-based and research-based laboratory pedagogies in undergraduate science. *Nat. Chem. Biol.*, **4** (10), 577–780.
224. Spell, R.M., Guinan, J.A., Miller, K.R., and Beck, C.W. (2014) Redefining authentic research experiences in introductory biology laboratories and barriers to their implementation. *CBE - Life Sci. Educ.*, **13**, 102–110.
225. Brownell, S.E. and Tanner, K.D. (2012) Barriers to faculty pedagogical change: lack of training, time, incentives, and. tensions with professional identity? *CBE - Life Sci. Educ.*, **11**, 339–346.
226. Pintrich, P.R., Smith, D.A., Garcia, T., and McKeachie, W.J. (1993) Reliability and predictive validity of the Motivated Strategies for Learning Questionnaire (Mslq). *Educ.Psycholog. Meas.*, **53**, 801–813.
227. Boekaerts, M., Pintrich, P.R., and Zeidner, M. (2000) *Handbook of Self-Regulation*, Academic Press, San Diego, CA.
228. Volet, S.E. (1991) Modelling and coaching relevant metacognitive strategies for enhancing university students' learning. *Learn. Instruction*, **1**, 319–336.
229. Bates, S. *et al.* (2012) Assessment and Feedback Programme SGC4L: Final Evaluation Report, Final Evaluation Report, 2012 October, The University of Edinburgh, http://repository.jisc.ac.uk/4994/1/SGC4L-final-evaluation-report.pdf (accessed 26 July 2014).
230. Cornelise Vreman-de, O. and de Jong, T. (2004) Student-generated assignments about electrical circuits in a computer simulation. *Int. J. Sci. Educ.*, **26** (7), 859–873.
231. McLoughlin, C., Lee, M., and Chan, A. (2006) Using student generated podcasts to foster reflection and metacognition. *Aust. Educ. Comput.*, **21** (2), 34–40.
232. Sykes, C.J. (1988) *Profscam: Professors and the Demise of Higher Education*, Regnery Publishing, Washington, DC.
233. Hacker, A. and Dreifus, C. (2010) *Higher Education?: How Colleges are Wasting Our Money and Failing Our Kids — and What We Can Do About It*, Times Books, New York.
234. Arum, R. and Roksa, J. (2010) *Academically Adrift: Limited Learning on College Campuses*, University of Chicago Press, Chicago, IL.
235. Bok, D. (2013) *Higher Education in America*, Princeton University Press, Princeton, NJ.
236. Hill, R. and Plantenberg, K. (2014) Assessing a conceptual approach to undergraduate dynamics instruction. Proceedings of the 2014 ASEE North Central Conference-14, 2014.
237. McLellan, H. (1996) Situated learning: Multiple perspectives. (ed H. McLellan), *In Situated learning perspectives*, Educational Technology Publications, Englewood Cliffs, NJ, pp. 5–18.

10
Context-Based Teaching and Learning on School and University Level

Ilka Parchmann, Karolina Broman, Maike Busker, and Julian Rudnik

10.1
Introduction

Research often reports that students are not interested in learning chemistry, even though they see the relevance of chemistry for the development of our societies [1]. Students might enjoy chemistry lessons, especially when they contain the demonstration or even experiments which they can carry out themselves, but they often do far less enjoy the effort of interpreting what they have seen or carried out. Moreover, only a small number of students actually consider science as an area they would like to study or work with in the future [1]. Why do we find those discrepancies? One reason often given is the lack of the students' perception of relevance for themselves, either in their daily lives or in the future. Approaches of context-based learning (CBL; for an overview see [2, 3]) have been developed and rather widely introduced at the school level as a promising way to overcome such challenges. CBL also aims at making chemistry knowledge more applicable, as the students have to use their knowledge and conceptual understanding to raise, investigate, discuss, and evaluate problems that can occur in real life, either in personal, societal, or professional situations. This closely refers to the goal of developing "scientific literacy" for all members of a modern society, to be able to participate in societal issues in a reflected way [4, 5]. Last but not least, it aims at giving students insights into contemporary science, both looking at research and development, and not only into the history of chemistry as it is often presented in textbooks. Research on CBL has shown effects regarding students' interest and perception of relevance with either the same or sometimes even better results in understanding as well (e.g., [6–8]). However, the findings on cognitive learning are diverse and need to be investigated further.

Currently, comparable observations of problems, unsatisfactory conditions, and results are reported for university studies, especially at the introductory level, which become apparent in the large dropout rates in the science programs [9]. Here, similar reasons can be given: a mismatch between the students' interests and their programs of study; the level of expectations and exams with regard to the students' abilities and self-efficacy; and their perception of relevance with regard

Chemistry Education: Best Practices, Opportunities and Trends, First Edition.
Edited by Javier García-Martínez and Elena Serrano-Torregrosa.
© 2015 Wiley-VCH Verlag GmbH & Co. KGaA. Published 2015 by Wiley-VCH Verlag GmbH & Co. KGaA.

to their choice of career [10, 11]. CBL can and has again built bridges here [12, 13], taking into consideration the different experiences and expectations that students at school and university show today.

This chapter describes and discusses examples of CBL at the school and university level, based on a short exploration of the theoretical background and learning assumptions for CBL. It also gives insights into an on-going empirical study investigating the still-not-well-understood problem-solving approaches and the resulting effects of CBL on the development of the students' conceptual understanding.

10.2
Theoretical and Empirical Background for Context-Based Learning

Gilbert *et al.* [14] highlight five problems context-based approaches try to address: (i) to reduce curriculum overload, (ii) to make content knowledge less fragmented, (iii) to enhance students' possibilities to transfer knowledge, (iv) to show the relevance of science in everyday life events, and, finally, (v) to point out to the students why they should study science. To provide students with meaningful and relevant learning situations that allow the building up an interconnected knowledge base and stimulating students' interests [3, 6, 15–17] is high expectation. Reasons for those expectations can be found in theoretical frameworks and empirical findings on teaching and learning in general (summarized in [18]). Most important is the assumption that all learning is situated and linked to the situation where it is first grounded. Further on, CBL takes into consideration that learning has to be undertaken by each individual, starting from already existing knowledge and beliefs, as pointed out by constructivistic learning theories and conceptual change frameworks. Last but not least, it has been shown that personal interest and motivation are fostered by the perception of relevance, competence, and autonomy. This leads to different methodological approaches next to the contextualization of knowledge.

However, challenges are also related to CBL approaches: "The greater the social or cultural relevance associated with canonical content, the greater the student *motivation*, but the greater the complexity to learn it meaningfully" [19, p. 85]. This quote describes an important benefit as well as an important challenge of CBL at the same time. Context-based tasks often represent less prestructured and more complex conceptual problems to the students [20]. In general, the findings of research on the impact of CBL indicate positive effects on students' interest and motivation [6, 8, 21], while findings with regard to the impact on students' cognitive learning outcomes are less coherent [17, 22, 23]. Beyond methodological difficulties [24], the incoherent findings of test results for CBL approaches might be explained by at least two reasons: (i) the different levels of complexity of the context-based tasks and (ii) the insufficient strategies students apply when mainly trained for algorithmic problem-solving approaches. The difficulties of transferring knowledge and the use of varying contexts for the application of knowledge set high demands on students, even though, as King and Ritchie [21] stress, these concerns are not specific to context-based approaches. Hence, frameworks are needed that can be used to analyze both the demands of different tasks and of different strategies that students apply to solve problem-based tasks in general, or context-based problems in particular. This aspect will be discussed further on as part of the on-going research.

In the following, we will explore how goals and challenges have been taken into consideration for the development of CBL frameworks at the school and university level.

10.3
Context-Based Learning in School: A Long Tradition with Still Long Ways to Go

As most research in science education around the world has focused at the school level for a long time, most findings also result from this field. The already mentioned and frequently pointed out decrease of interest and motivation, the lack of applicable knowledge, and in many countries the low numbers of students and workforce have led to the development of CBL approaches in different countries (i.e., [25, 26]). Among the first, the Dutch "PLON" project for physics, the American "Chemistry in the Community," and the UK projects "Salters Science" and "Salters Chemistry" can be named as starting points for other CBL developments in other countries, such as the German "Chemie, Physik, and Biologie im Kontext." The framework of "Chemie im Kontext" connects contexts of assumed interest and relevance for students with chemical content and the underlying basic concepts to build up a transferable structure of knowledge [7, 18] (see Figure 10.1).

The "Chemie im Kontext" units are structured according to four phases (see Figure 10.2). The students' and teachers' activities and roles change during these four phases. In the beginning, the teacher prepares an environment that allows

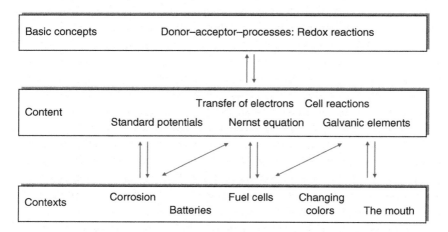

Figure 10.1 Framework of connecting contexts, content, and basic concepts.

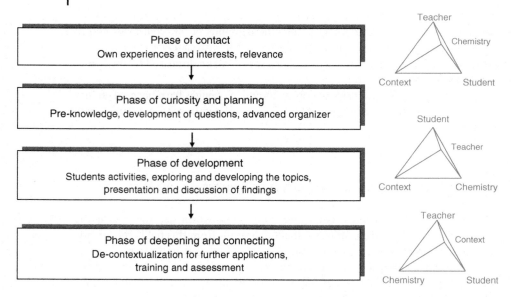

Figure 10.2 Phases of "Chemie im Kontext" units, connected to different roles of the context, the content, the students, and the teachers (The model of the tetrahedron is due to Jan Apotheker).

interaction between the context and the students, initiating questions of interest to the students. The chemical content is "waiting in the background" to be explored and applied on demand later during the unit. The next phase incorporates the students' ideas and interests into the planning of any further exploration, applying previous knowledge and identifying suitable methods of investigation. After the main phase of exploration, in which the students should be highly active and the teacher steps back, it is again for the teacher to point out knowledge that should be transferred to following contexts (see Figures 10.2 and 10.3).

On the basis of these framework guidelines, many units for lower and upper secondary education have been developed by teachers and chemistry education researchers in "symbiotic learning communities" [7]. They were tested in schools, some of them being implemented into statewide curricula and some transferred into teacher training courses [27]. Empirical research accompanied that process, funded by the German Ministry of Education (BMBF). In line with other research findings, the learning outcomes had been pleasant overall, however diverse [27]. In some studies, the students had outperformed others, while in other studies we saw diverse effects for classes even though they had all been taught the same unit of "Chemie im Kontext." Therefore, we still investigate the effects and especially the processes of CBL. The following study gives an insight into this on-going research.

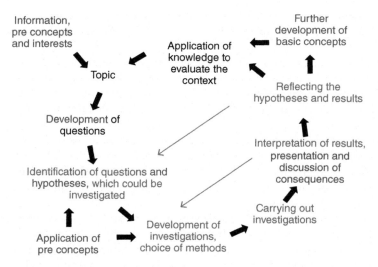

Figure 10.3 Idealized cycle of exploring a "Chemie im Kontext" unit. The colors in the cycle: Blue stands for the application and development of basic concepts, green for methods of investigation, red for communication and grey for judgement.

10.4
Further Insights Needed: An On-Going Empirical Study on the Design and Effects of Learning from Context-Based Tasks

Context-based tasks differ from typical school textbook exercises, which rather ask for the application of routines or the replication of factual knowledge than for thinking skills leading to the development of problem-solving approaches for real problems. A problem is, according to John Hayes, construed as a "gap between where you are now and where you want to be but you don't know how to cross that gap" [28, p. xii]. Problem solving is a fundamental part of chemistry education but, even though investigated for many years, is still found problematic for students [28, 29]. Problem solving is often studied by comparing experts' and novices' different behavior when approaching a problem or by relating more successful problem solvers to less successful [30]. When studying problem-solving patterns, four behaviors are common among more successful problem solvers: they approach the problem in a consistent way; they use given information efficiently; they use fragments and put them into a comprehensive picture; and they go back and check the accuracy of their result [30]. Ngu and Yeung conclude that "experts categorize problems on the basis of problem structures that guide them to solution procedures whereas novices are readily influenced by surface features that may be unrelated to the solution" [31, p. 16]. Similarly, novices often tend to use algorithmic approaches to handle problems [30, 32] and struggle in transferring algorithmic and non-algorithmic/conceptual problem-solving strategies to unfamiliar problems [20]. Transfer of content knowledge from one context to another is considered difficult for students [31]. One reason for this difficulty is that transfer, in the sense of improved performance in problem-solving situations, does not occur spontaneously but needs previous knowledge and specific learning opportunities.

For the design and investigation of CBL tasks, better knowledge is needed about the approaches that learners and experts undertake at different stages to solve chemical problems in authentic contexts. To be able to support students in their learning process in an effective way, difficulties occurring in the process need to be reflected on.

The study described in the following lays a ground for investigating such questions by exploring different approaches and challenges for an exemplary CBL task. The comparison of experts and different levels of student abilities shows the variety of approaches that can be used to reflect on different perspectives and expectations by learners and teachers.

In this explorative study, a context-based problem on medical drugs was chosen to investigate students' and experts' problem-solving strategies. Knowing that the relevance of context is important to students [6], "medical drugs in the environment" was selected as topic and context, that is, health in a societal context. This specific context-based task dealt with swine flu and Tamiflu, very much reported in media at the time of the empirical study, in the school year 2010–2011. The task is presented below (see Figure 10.4). A newspaper article was mentioned in the introductory text connected to the task. This article presented in a general way that medical drugs are spread into the environment but the article had no specific chemical focus.

In a first step, 236 upper secondary chemistry students of schools in Sweden solved the task in writing; however, to get a more in-depth insight into problem-solving strategies, an interview study was conducted, which will be presented in this chapter. One week after the written examination, students were selected for semistructured interviews [33]. The students represented high achievers and low achievers, boys and girls, and more or less interested and talkative students to be able to characterize approaches from a broad spectrum. The number of interviews was limited according to the findings until no basic differences had been observed anymore, and as a consequence nine student interviews were conducted. The selection of students was done by one of the authors together with the students' teacher, and this was enabled by previously accomplished classroom observations. The interviewer had also read the students' written exam answers in advance to validate the findings based of 236 exam papers.

A medical drug against swine flu frequently discussed in media this year, Tamiflu® (an antiviral agent), is unfortunately spreading in the environment. Explain how this can occur. What chemical properties does the drug have that will make it spread in the environment? See the structural formula for Tamiflu® on the right.

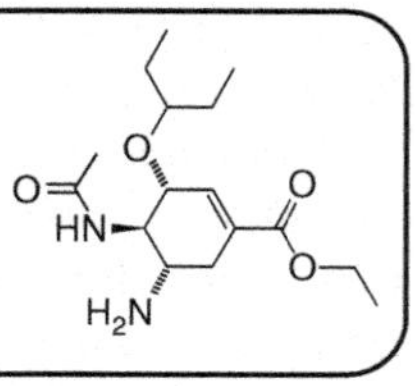

Figure 10.4 The exemplary context-based task.

In a second step, the study was expanded to investigate how chemistry experts, in this case university chemistry professors, approach and process the same context-based task. Semistructured interviews were conducted with five chemistry professors having a broad and deep chemistry content knowledge, for example, professors in organic chemistry, biochemistry, and medicinal chemistry. The chemistry experts had experience from both teaching university chemistry students and performing chemistry research. After reading the task and the newspaper article, the experts were asked how they themselves would solve the problem and what problem-solving strategies they use when they approach a new chemistry task. Thereafter, they were given the opportunity to present ideas about students' problem-solving strategies, in other words, how the students would solve it and what difficulties students might have with the task.

The following leading questions had been guiding the analyses:

> Which strategies can be characterized for novices and experts approaching context-based tasks?
> Which chemical concepts are applied, and which problems can be identified?

10.4.1
Strategies to Approach Context-Based Tasks

All students started with the newspaper article, trying to find relevant factual information to solve the problem. They realized that the article could not give this support, as one student said: *the article is not giving any "real" chemistry, only useless daily-life knowledge*. The important demand of context-based tasks, namely to identify chemical concepts that can be applied to solve the problem, was too difficult for the students at this stage. The need for initiating and training strategies for approaching problems that do not directly ask for the content of importance became obvious in all interviews and might be equivalent in education in other countries as well.

A second strategy was to try to remember what the teacher had said during the lessons or what they had read in their chemistry textbook. One student explained: *I always try to remember something our teacher has said or something I remember from the book, for instance, facts that medical drugs often are organic molecules*.

A third strategy was to use the structural formula, as a student said: *since it is a chemistry test, we're supposed to look at the formula; this is the chemistry part of the task*. The difficulty with the structural formula was that the students did not know how to approach and apply it to solve the problem; students' modeling skills might be problematic which results in transfer difficulties between different molecular representations [34]. Here again, all students began with trying to remember the teacher's words from the chemistry lessons and to remember something helpful from the textbook, using facts. It was also apparent that they focused on the small parts of the molecules, namely the individual functional groups (see Figure 10.5). This again raises the question about how students are educated to practice their "chemical thinking": do they learn and apply a series of facts, or do they learn to

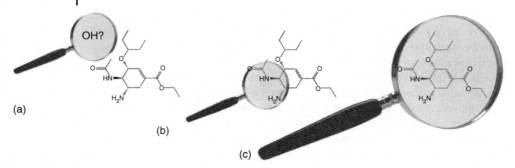

Figure 10.5 Strategies of three interviewees in comparison. (a) The student was searching for an OH group to explain polarity but could not find it in the molecule. (b) The student identified the relevant groups to explain polarity. (c) The experts investigated the whole molecule before focusing on functional groups.

look at the whole molecule and situation and then ask for the important facts and concepts?

In summary, students are primarily used to lower order cognitive strategies such as recall of facts and rote memorization. Moreover, students said they were not accustomed to tasks where they were supposed to further elaborate and reason about chemical properties and transfer knowledge from one area to another. They indicated mostly getting questions where they had to answer with short, factual answers. This, of course, might be different in different countries. The German standards, for example, ask explicitly for context-based tasks as part of the final exams. However, first (unpublished) results of tests show that the more complex demands are still difficult for many students, as expected.

The interplay between concepts and context is important to highlight to make students aware of the relevance of chemistry [35]. On the other side, the context seems to make the problem more complex, but the obstacle might be that students really do not know how to solve a problem posed in an unfamiliar way. Many students said that reasoning and exhaustive expositions were associated with school subjects such as social science and history. This again is a barrier to be approached in future research and development.

The results of the students had been contrasted with approaches of experts to be able to describe the spectrum of strategies. Even though the chemistry experts were from varying research fields within chemistry, they all applied the same strategy when approaching the task. Everyone used the structural formula to consider the Tamiflu molecule's polarity. The experts discussed the different parts of the molecule, the functional groups, but still perceived the molecule as one entity; in other words, they could combine the details of the molecule into a whole (see Figure 10.5). They got an overview by looking at the whole molecule and at the same time scrutinizing the important functional groups. One of the experts expressed that: *this small molecule is obviously polar because of electron attracting substituents and thereby water soluble, I look for hydrolyzable functionalities and acidic protons, pK_a is very important ... the molecule is regarded as a system,* *even though looking at the parts, the functional groups.* Experts also connected the molecule with metabolism and what could happen to the molecule in the body or the environment. Because of the experts' broad and deep chemistry content knowledge, they had no problem in applying or transferring knowledge from the area of chemical bonding to organic chemistry. The experts' answers may rather be used to develop supporting tools for student material in the future [36], like "expert lifts," an analogy to the "history lifts" described by Jansen and Matuschek for historical CBL [37].

10.4.2
Application of Chemical Knowledge

The task expects the students to apply knowledge about the structure of the molecule, especially aspects of polarity, chemical bonding, and interaction. However, this was not evident to the students. Even when being asked directly about the Tamiflu molecule and hydrogen bonding, they did not see the connection. In their textbooks and chemistry lessons, examples of hydrogen bonding were described in relation to water and/or alcohols, containing oxygen–hydrogen bonding. The Tamiflu molecule consists of an amino and amide group, which had only been presented as bases, not as possible participators in hydrogen bonding. Students were also not aware of the possibility for hydrogen bonding through hydrogen acceptors (e.g., carbonyl groups, C=O). The focus on the example learned in class – OH groups for hydrogen bonding – has been described in the literature as well. School students often reduce rules to examples instead of exploring the general concept behind the rule, which might cause difficulties in understanding and application later on.

Another main conceptual problem was that the students did not know how to apply electronegativity, one central concept to understand chemical bonding [38]. All students agreed that the teacher frequently discussed relevant chemical properties during teaching (e.g., hydrogen bonding and polarity), but they could not connect these to the task. One student said: *Yes, our teacher often writes $\delta+$ and $\delta-$ (to describe the bond polarizability due to the difference in electronegativity) in structural formulas on the whiteboard, but I'm not sure how to use that information. I don't ask her since I'm supposed to remember that from reading about chemical bonding.* These problems with intra and intermolecular bonding were also apparent in the 236 written answers collected before, but the interviews distinctly revealed electronegativity as the specific concept in demand. After scaffolding from the interviewer, all students finally came to the conclusion on how to use and apply the concept. The concept of electronegativity is essential [39] and could be a concept teachers have to emphasize more thoroughly to remind students of its importance. It is therefore striking that Linus Pauling, the introducer of the electronegativity scale, also was famous for his exceptional problem-solving skills [24].

Furthermore, it was apparent that students tried to solve the problem starting from content areas recently studied in the course. Students who answered the task after reading organic chemistry gave other responses than those who had studied biochemistry at the time of the exam. An example of a student's reply after studying organic chemistry is: *Medical drugs are organic molecules with carbon since humans are built of organic molecules*. One example of a student's reply after studying biochemistry is: *The drug contains a protein that for sure is denatured when it is connected to other compounds. A denatured protein can change its properties and the changed structure will affect the environment negatively*. Chemistry concepts linked to biochemistry (e.g., denaturation of proteins and a molecule's isoelectric point) were not mentioned by students who had their examination test on organic chemistry because biochemistry is studied after organic chemistry in the chemistry course.

All chemistry experts agreed on the importance of the chemistry concept of electronegativity to understand chemical bonding, polarity, and solubility. They were surprised about the interviewed students' negative reply to the direct question of whether the Tamiflu molecule could develop hydrogen bonding. The experts were not aware of the restrictions that are often made in school, such as introducing and discussing hydrogen bonds with connection to hydroxyl groups primarily or only.

Another problem occurs from the structure of courses at school and university. Students studying organic chemistry gave answers different from those studying biochemistry. Also, in university chemistry courses concepts are not always transferred explicitly from one course to another. Sometimes they might even be introduced differently by different lecturers.

In summary, our results are in line with the findings of Cartrette and Bodner [30], who show that less successful problem solvers (i.e., students) show declarative knowledge only, whereas more successful (i.e., experts) break down the problem into subproblems that are easier to handle. The experts have a more holistic way of problem solving, something also highlighted by Cartrette and Bodner [30]. Experienced chemists are able to "see" entities; they can transform and deconstruct content knowledge. Consequently, "an important goal of chemical educators is to make students […] 'see' chemistry as chemists do" [34, p. 71]. The setting in this context-based problem is not familiar to the students but to the experts who are habituated to broader and more complex problems because the research process in itself emanate from open and complex issues. Context-based tasks give opportunities even for students to challenge their problem-solving ability and improve their higher order thinking by seeing entities and not only recall facts about limited parts of a molecule or a problem. Of course, this development needs several steps, support, and training, but without challenging students to think as a chemist, they will probably stay with their belief that chemistry is only about learning facts and rules without being able to understand the background.

10.4.3
Outlook on the Design of Tasks and Research Studies

One conclusion from this study is that the design of context-based conceptual problems has to be more carefully elaborated, regarding both the contextualization and the content. We suggest further distinction between the "context," the "topic," and the "content" of a task. For the follow-up research, 15 context-based problems were systematically varied through a structured design [40]. Another important conclusion was to more thoroughly investigate the impact of the interviewer to investigate how the process of scaffolding through hints or prompts can help the students to higher order thinking. To analyze the complexity of students' chemistry knowledge, the Model of Hierarchical Complexity in Chemistry, MHC-C [41, 42], has been applied. For school chemistry, five levels of hierarchical complexity have been considered: everyday experiences, facts, processes, linear causality, and multivariate interdependencies [41]. These levels reflect the central aims of science education, for example, to make students able to describe and explain scientific phenomena, and are operationalized as specific performance expectations. The MHC-C focuses less on general cognitive activities, for example, problem solving, but on the combination of cognitive operations and the content structure of the domain, for example, making causal inferences to explain a context-based phenomenon. The five levels of the MHC-C thus provide a system of categories to describe task demands more fine-grained (as a matter of course because of the larger number of categories, but also because of the more focused description of the categories) and content-bound (as facts, processes, and causal elements are inevitably part of the domain in question). Besides analyzing the tasks in themselves, the MHC-C also shows promising qualities to promote ideas on how to scaffold students in their problem-solving process and will therefore be applied as a supporting scaffolding analytical tool in the follow-up research [40].

10.5
Context-Based Learning on University Level: Goals and Approaches

The situation for university students differs from that of school students. While many students at school have to do chemistry, students at university have chosen to study chemistry at least to some degree, depending on the study course. Nevertheless, problems of motivation and interest, as well as of applying suitable knowledge and competencies, are also relevant at that level of study [9]. One reason is that, especially at the introductory year, students from several different programs visit the same courses, for example, general chemistry. Some would like to become researchers and are interested in the development of a deeper understanding of both the theoretical background and the experimental and model-based approaches. Others might like to become teachers and are searching for insights into best practice explanations. Again, others will become biochemists,

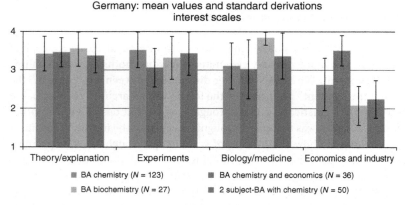

Figure 10.6 Differences between fields of interest for different study courses [44].

medical doctors, environmental chemists, industrial chemists, and so on, and will have different interests and fields of application.

The expectable differences regarding interest and other variables have been shown in research studies. In one of our own projects, we investigated the students' interests for different contexts as well as for different activities, applying the "IPN-model of interest" (summarized in [43]). The results show significant differences between different study courses (see Figure 10.6).

CBL offers a framework to link chemical phenomena, theories, and representations to different contexts of study programs, following Mahaffy's "tetrahedron" as an enlargement of Johnstone's "chemical triangle" [45] (see Figure 10.7).

Therefore, the adaptation of CBL promises to build bridges for those challenges also at the university level, as the same basic topics can be applied in different fields. Especially in tutorials, students can work on applications relevant for their

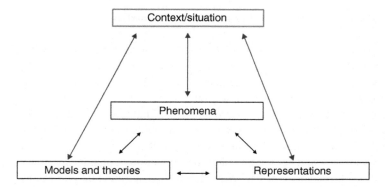

Figure 10.7 Connecting phenomena, theories, and representations to different contexts (Adapted from Mahaffy's "chemical tetrahedron").

area of study, fostering motivation and showing the importance of understanding basic concepts, according to the theories of CBL. Mahaffy and others [13] have given impressive examples of such tasks. In our projects, we have focused on the design of CBL task for tutorials accompanying joint lectures for different study courses in science. The tutorials aim at supporting the students' basic knowledge necessary for the exam. They also aim at highlighting the importance of basic knowledge for the different fields of study.

10.5.1
Design of Differentiated CBL-Tasks

In order to respond to the students' different interests and expectations, tasks were grouped into different categories: basic tasks, context-based tasks with relevance to different fields of study, and (in the new project "PerLe" at the University of Kiel) tasks giving an insight into authentic research, aimed at motivating also the best students already in the beginning. This design builds on the tetrahedron shown in Figure 10.7.

For the CBL tasks, contexts have been defined as current and interdisciplinary problems and questions, within which the meaningful contributions of the scientific discipline for the different degree programs become apparent and with which the structures of the subject of chemistry can be grasped. The goal was to show all groups of students the rationale behind dealing with chemical aspects and to empower students to develop their professional competence. As in the framework of "Chemie im Kontext," connections between the content, the basic concepts, and the contexts were made. Each task framed the chemical content in a certain context. For example, redox reactions with iron compounds were related to biochemical systems such as the human body, industrial systems such as the production of iron, or environmental systems. During the tutorial, the basic concept (e.g., donor–acceptor reactions) and the specific content (redox reactions with iron and other compounds) were pointed out, also referring back to the accompanying lecture.

The variation of the tasks also took the heterogeneity of the group of students into account, in regard to both students' interests and their prior knowledge. Students with little prior knowledge could choose the easiest tasks as a first introduction to the topic. Those tasks mostly took up contents already covered in school (such as examples and definitions of acids and bases). Tasks with a moderate level of difficulty corresponded to the minimum standards for a successful completion of the course exam applying the basic concept (e.g., for predicting and explaining the behavior of an acidic substance in water). In order to allow students to compare their present knowledge with the expected learning outcomes, tasks corresponding to the final exam's level of difficulty and design were marked as such. For students with an interest in topics beyond the moderate understanding required for the course exam, tasks with a high level of difficulty were now added. These tasks dealt with problems that might appear later in the students' study programs or were meant as an in-depth puzzle over the content (e.g., donor–acceptor reactions in complex molecules manipulated in research projects).

The following examples of context-based university tasks show the general outline of the different levels and CBL applications of the tasks.

10.5.2
Example 1 Physical and Chemical Equilibria of Carbon Dioxide – Important in Many Different Contexts

Carbon dioxide is a popular substance. We find it in the air, in drinks, and in our blood. It is also blamed for fostering the greenhouse effect and changes in our climate system. Indeed, CO_2 molecules are able to absorb infrared radiation and, therefore, have an impact on the global balance of radiation, absorption, and emission. It is therefore highly important to follow the emission balances and the resulting processes in the atmosphere. However, the atmosphere is not independent from other spheres. The solubility of carbon dioxide in water influences the uptake in the oceans, for example, and hence also the concentration in the atmosphere.

Basic level Describe the reactions between CO_2 and H_2O and give the reaction schemes for the resulting equilibria.

Context-based relevance Find examples of the given equilibria in different contexts. Look into the environment (hint: oceans.), into your own health (hint: blood buffer.), and into societal products (hint: still water or water with gas.), to name some areas of relevance.

Explain the ability of the reaction system to act as a buffer.

Task for application and training The Lake Nyos is a natural phenomenon. On Wikipedia, you can find the following description for the lake:

> "A pocket of magma lies beneath the lake and leaks carbon dioxide (CO_2) into the water, changing it into "carbonic acid." Nyos is one of only three known "exploding lakes" to be saturated with carbon dioxide in this way ..... On August 21, 1986, possibly as the result of a landslide, Lake Nyos suddenly emitted a large cloud of CO_2, which "suffocated" 1,700 people and 3,500 livestock in nearby towns and villages."
> (http://en.wikipedia.org/wiki/Lake_Nyos)

Describe the processes occurring at Lake Nyos from a chemical perspective, beginning with the enrichment of carbon dioxide until the emergence of the gas cloud. Take into consideration the solubility of carbon dioxide in water and, connected to it, chemical equilibrium reactions. Explain all the factors influencing the solubility of carbon dioxide during the different steps of the process.

Today, there is a vertical pipe in the middle of the lake which reaches all the way to the bottom. From it, a fountain of water with a height of 40 m emerges continuously on its own after initial pumping. Explain how this procedure helps secure the population in the area.

10.5.3
Example 2 Chemical Switches – Understanding Properties like Color and Magnetism

At school, you certainly dealt with phenomena of color and magnetism, which you also know from your daily life. How would you explain them based on the knowledge that you have?

Both phenomena are also of interest for researchers. In a research group for coordination chemistry, two scientists carried out work on the synthesis and characterization of iron(II) complexes. These can change their structure if they were "switched" by a stimulus, such as temperature or light. Why should that be of interest? the change of the structure corresponds to an emerging change of properties, such as color and magnetism.

Copying switches from nature, such as hemoglobin, could offer new ways for biochemistry and medicine. For a chemist working in industry, those molecular switches are also interesting, for example, to develop new memory devices with smaller compounds and faster circuit times.

How can we understand what is actually causing these switches of properties?

Let us have a look at an example: In the left picture (Figure 10.8) you see the color change of the iron compound from red-violet to violet when it is cooled down with liquid nitrogen (N_2 (l)). In the right picture (Figure 10.8) you see the change of magnetism from paramagnetic at room temperature to diamagnetic at lower temperature if the compound is placed in front of a neodymium magnet.

How would you explain the phenomena of color change and magnetism change of this iron(II) complex? At school level, those two properties are not explained in combination. For the explanation of color, you might have learned the concept of "alternating double bonds," this is not really helpful here. For magnetism, you might have learned something about poles and magnetic field lines (Figure 10.9

Figure 10.8 Change of color and magnetism of the iron(II) complex. (Copyright by H. Naggert, CAU Kiel.)

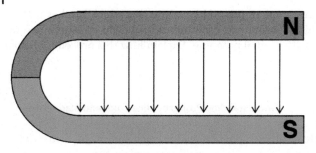

Figure 10.9 Poles and magnetic field lines. (Graphic by J. Rudnik, IPN Kiel.)

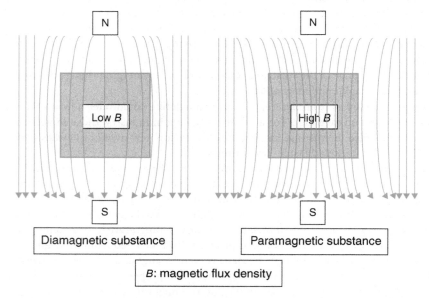

Figure 10.10 Diamagnetic and paramagnetic substances within a magnetic field. (Graphic by J. Rudnik, IPN Kiel.)

and 10.10), which are also not related to the change occurring here. Therefore, models have to be enlarged and developed, and you will learn more about this during your studies.

Chemists explain many phenomena with models about the structure of electrons and orbitals. What do you know about the electronic structure of the complex compound? Think about the different "spin states" you see in Figure 10.11. How is the change of color and magnetism represented by the electronic structure? *Hints*: Check for paired/unpaired electrons and the gap between the different energy levels (e_g and t_{2g}).

When the iron compound is cooled down to −108 °C, the metal–ligand bond distance gets smaller (smaller bond length). This results in a change of the ligand field energy. The unpaired electrons couple to a low-spin system. The change of

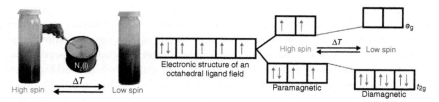

Figure 10.11 Electronic structure of the complex compound. (Copyright by H. Naggert, CAU Kiel.)

electronic structure causes a change in color (this is exactly caused by a different metal–ligand charge transfer) and, due the paired electrons, diamagnetism at low temperatures.

10.5.4
Feedback and Implications

The first feedback from working with these tasks was positive. A lot of students responded positively to the support of basic knowledge for preparing the exam. They were interested in exercises but also preferred applications of basic knowledge for their field of study. There were also students who declared that they were interested in CBL tasks dealing with topics beyond the basic understanding, such as research topics, even though they do not see the relevance yet. However, we also found that students with weaker prior knowledge often did not choose context-based tasks. Similar to the findings described for CBL at the school level, it would be helpful to investigate the students' problem-solving approaches and aspects stimulating motivation for CBL tasks at the university level as well.

In addition to the feedback students gave about the tasks, they named other challenges as well. One topic was the preparation of lab journals and support with practical work. The amount of parallel work, for example, for lectures, seminars, and lab courses, hindered some of them from participating in the tutorials which they would have liked to do for further training. The influence of external and personal factors for the improvement of CBL at the university, aiming at building bridges between the necessary knowledge and skill foundation and the individual fields of study should be discussed furthermore.

In conclusion, context-based tasks can be attractive for tutorials for first-semester courses but there is also a need for open discussions and reflection of approaches and challenges in linking the basic concepts to authentic applications.

10.6
Conclusions and Outlook

CBL has shown significant positive effects on students' interest and motivation at the school level. The outcomes on learning and conceptual understanding are more diverse. Further research is needed to investigate how students approach context-based tasks, how they use contextual information to identify important chemical aspects, and how they apply and deepen their chemical knowledge to work on the given problem. An explorative interview study described in this chapter gives first insights into students' difficulties while working with context-based tasks. Novices do not approach a contextual problem by looking for chemical starting points; they instead look for direct answers in their memory from classroom instruction. The contrast to experts' problem-solving strategies offers a foundation for the development of supporting tools and training measures to foster students' chemical thinking already at the school level. Further research is under way for the design and investigation of CBL.

At the university level, CBL seems to be promising especially for introductory courses where students with different backgrounds and often different study courses visit the same lectures, seminars, and lab courses. Context-based tasks give them an option to work on similar chemical content with different applications to their fields of interest. Initial trials have shown that students appreciate this, but they still find context-based tasks more difficult and also wish to have a "straightforward preparation" for their end-of-term exams in tutorials.

One goal for future CBL studies is to design and implement context-based tasks in a way that would challenge students to develop chemical thinking but also offer them stepped supporting tools to be successful with their problem-solving process. Context-based tasks point out the variety of different applications of chemistry in daily life and fields of research and development; but to raise and keep students' interest, they must perceive themselves as being competent to solve context-based problems by studying and applying chemical knowledge. The first results of our on-going studies indicate that it is possible to combine such positive effects on interest and learning.

References

1. Sjøberg, S. and Schreiner, C. (2010) *The ROSE Project: An Overview and Key Findings*, University of Oslo, Oslo.
2. Pilot, A. and Bulte, A.M.W. (2006) The use of "Contexts" as a challenge for the chemistry curriculum: its successes and the need for further development and understanding. *Int. J. Sci. Educ.*, **28** (9), 1087–1112.
3. Nentwig, P. and Waddington, D. (eds) (2005) *Making it Relevant: Context Based Learning of Science*, Waxmann, Münster.
4. Bybee, R. (1997) *Achieving Scientific Literacy*, Heinemann, Portsmouth.
5. Bøe, M.V., Henrikson, E.K., Lyons, T., and Schreiner, C. (2011) Participation in science and technology: young people's achievement-related choices in late-modern society. *Stud. Sci. Educ.*, **47** (1), 37–72.
6. Bennett, J., Lubben, F., and Hogarth, S. (2007) Bringing science to life: a synthesis of the research evidence on the effects of context-based and STS approaches to science teaching. *Sci. Educ.*, **91**, 347–370.
7. Parchmann, I., Gräsel, C., Baer, A., Nentwig, P., Demuth, R., and Ralle, B. (2006) "Chemie im Kontext": a symbiotic implementation of a context-based teaching and learning approach. *Int. J. Sci. Educ.*, **28** (9), 1041–1062.

8. Fechner, S. (2009) *Effects of Context-Oriented Learning on Student Interest and Achievement in Chemistry Education*, Logos Verlag Berlin GmbH, Berlin.
9. Ulriksen, L., Madsen, L.M., and Holmegaard, H.T. (2010) What do we know about explanations for drop out/opt out among young people from STM higher education programmes? *Stud. Sci. Educ.*, **46** (2), 209–244.
10. Dalgety, J. and Coll, R.K. (2006) Exploring first-year science students' chemistry self-efficacy. *Int. J. Sci. Math. Educ.*, **4**, 97–116.
11. Nagy, G., Trautwein, U., and Lüdtke, O. (2010) The structure of vocational interests in Germany: different methodologies, different conclusions. *J. Vocat. Behav.*, **76** (2), 153–169.
12. Mahaffy, P. (1992) Chemistry in context. How is chemistry portrayed in the introductory curriculum? *J. Chem. Educ.*, **69** (1), 52.
13. Mahaffy, P., Tasker, R., Bucat, B., Kotz, J., Weaver, G., Treichel, P., and McMurry, J. (2011) *Chemistry: Human Activity, Chemical Reactivity*, Cengage Learning, Stamford, CT.
14. Gilbert, J.K., Bulte, A.M.W., and Pilot, A. (2011) Concept development and transfer in context-based science education. *Int. J. Sci. Educ.*, **33** (6), 817–837.
15. Fensham, P.J. (2009) Real world contexts in PISA science: implications for context-based science education. *J. Res. Sci. Teach.*, **46** (8), 884–896.
16. King, D. (2012) New perspectives on context-based chemistry education: using a dialectical sociocultural approach to view teaching and learning. *Stud. Sci. Educ.*, **48** (1), 51–87.
17. Ramsden, J.M. (1997) How does a context-based approach influence understanding of key chemical ideas at 16? *Int. J. Sci. Educ.*, **19** (6), 697–710.
18. Nentwig, P., Parchmann, I., Gräsel, C., Ralle, B., and Demuth, R. (2007) Chemie im kontext – a new approach to teaching chemistry, its principles and first evaluation data. *J. Chem. Educ.*, **84** (9), 1439–1444.
19. Aikenhead, G.S. (2006) *Science Education for Everyday Life: Evidence-Based Practice*, Teachers College Press, New York.
20. Salta, K. and Tzougraki, C. (2011) Conceptual versus algorithmic problem-solving: focusing on problems dealing with conservation of matter in chemistry. *Res. Sci. Educ.*, **41**, 587–609.
21. King, D. and Ritchie, S.M. (2012) in *Second International Handbook of Science Education* (eds B.J. Fraser, K.G. Tobin, and C.J. McRobbie), Springer, pp. 69–80.
22. Bennett, J., Gräsel, C., Parchmann, I., and Waddington, D. (2005) Context-based and conventional approaches to teaching chemistry: comparing teachers' views. *Int. J. Sci. Educ.*, **27** (13), 1521–1547.
23. Taasoobshirazi, G. and Carr, M. (2008) A review and critique of context-based physics instruction and assessment. *Educ. Res. Rev.*, **3** (2), 155–165.
24. Taasoobshirazi, G. and Glynn, S.M. (2009) College students solving chemistry problems: a theoretical model of expertise. *J. Res. Sci. Teach.*, **46** (10), 1070–1089.
25. Bennett, J. and Lubben, F. (2006) Context-based chemistry: the salters approach. *Int. J. Sci. Educ.*, **28** (9), 999–1015.
26. Schwartz, A.T. (2006) Contextualized chemistry education: the American experience. *Int. J. Sci. Educ.*, **28** (9), 977–998.
27. Demuth, R., Gräsel, C., Parchmann, I., and Ralle, B. (eds) (Hrsg.) (2008) *Chemie im Kontext – Von der Innovation zur nachhaltigen Verbreitung eines Unterrichtskonzepts. (Chemistry in Context – From an Innovation towards a Sustainable Distribution of a Conceptual Approach)*, Waxmann, Münster, New York, München, Berlin.
28. Hayes, J.R. (1989) *The Complete Problem Solver*, 2nd edn, Lawrence Erlbaum Associates, Hillsdale, NJ.
29. She, H.-C., Cheng, M.-T., Li, T.-W., Wang, C.-Y., Chiu, H.-T., Lee, P.-Z., and Chuang, M.-H. (2012) Web-based undergraduate chemistry problem-solving: the interplay of task performance, domain knowledge and web-searching strategies. *Comput. Educ.*, **59**, 750–761.

30. Cartrette, D.P. and Bodner, G.M. (2010) Non-mathematical problem solving in organic chemistry. *J. Res. Sci. Teach.*, **47** (6), 643–660.
31. Ngu, B.H. and Yeung, A.S. (2012) Fostering analogical transfer: the multiple components approach to algebra word problem solving in a chemistry context. *Contemp. Educ. Psychol.*, **37**, 14–32.
32. Nakhleh, M.B. and Mitchell, R.C. (1993) Concept learning versus problem solving. *J. Chem. Educ.*, **3**, 190–192.
33. Denscombe, M. (2010) *The Good Research Guide for Small-Scale Social Research Projects*, 4th edn, Open University Press, Maidenhead.
34. Dori, Y.J. and Kaberman, Z. (2012) Assessing high school chemistry students' modeling sub-skills in a computerized molecular modeling learning environment. *Instruct. Sci.*, **40**, 69–91.
35. King, D. (2009) Context-based chemistry: creating opportunities for fluid transitions between concepts and context. *Teach. Sci.*, **55** (4), 13–20.
36. Fach, M., de Boer, T., and Parchmann, I. (2007) Results of an interview study as basis for the development of stepped supporting tools for stochiometric problems. *Chem. Educ. Res. Pract.*, **8** (1), 13–31.
37. Jansen, W., Matuschek, C., Fickenfrerichs, H., and Peper, R. (1992) in *Grundlagen deutscher Chemiedidaktik* (eds N. Just and H.-J. Schmidt), Westarp, Essen, pp. 207–228.
38. Nahum, T.L., Mamlok-Naaman, R., Hofstein, A., and Krajcik, J. (2007) Das historisch-problemorientierte Unterrichtsverfahren-Geschichte der Chemie im Chemieunterricht [The historical-poblemoriented teaching method-history of chemistry in chemistry teaching]. In N. Just & H.-J. Schmidt (eds.), Grundlinien deutscher Chemiedidaktik (pp. , 207-228). Essen: Westarp.
39. Gilbert, J.K. (2006) On the nature of "Context" in chemical education. *Int. J. Sci. Educ.*, **28** 9, 957–976.
40. Broman, K. and Parchmann, I. (2014) Students' application of chemical concepts when solving chemistry problems in different contexts. *Chemistry Education Research and Practice*, **15** 4, 516-529.
41. Bernholt, S. and Parchmann, I. (2011) Assessing the complexity of students' knowledge in chemistry. *Chem. Educ. Res. Pract.*, **12**, 167–173.
42. Commons, M.L., Trudeau, E.J., Stein, S.A., Richards, F.A., and Krause, S.R. (1998) Hierarchical complexity of tasks shows the existence of developmental stages. *Dev. Rev.*, **18** (3), 237–278.
43. Menthe, J. and Parchmann, I. (2015) Getting Involved. Context-Based Learning in Chemistry Education. In: Kahveci, M., & Orgill, M. (Eds.):. Affective dimensions in chemistry education. Berlin Heidelberg: Springer-Verlag.
44. Klostermann, M., Höffler, T., Bernholt, A., Busker, M., and Parchmann, I. (2014) Erfassung und Charakterisierung kognitiver und affektiver Merkmale von Studienanfängerinnen und Studienanfängern im Fach Chemie. (Characteristics of cognitive and affective attributes of chemistry students at the beginning of their university studies). ZfDN. doi:10.1007/s40573-014-0011-7.
45. Mahaffy, P. (2006) Moving chemistry education into 3D – a tetrahedral metaphor for understanding chemistry. *J. Chem. Educ.*, **83** (1), 49–55.

11
Active Learning Pedagogies for the Future of Global Chemistry Education
Judith C. Poë

- *Rewriting Art History*: Raman Reveals All
- *Perfect Recycling*: From Rubbish to Resource
- *Food Special*: Cooking up a Chemical Feast
- *Eco-friendly Paint*: Organic Solvents Brushed Away
- *Mass Spectrometry*: In the Extremes

all were cover stories in *Chemistry World* [1], which deal with real-world issues but rarely find their way into university general chemistry curricula. However, since the majority of students in North American introductory chemistry courses have no intention of pursuing the subject professionally, it is especially important that, throughout those courses, such students are made to appreciate just why anyone would want to know the material included in the curriculum, both in the context of development of the subject and in terms of existing or potential applications. They need to see the relationship between material in the course and data upon which they will draw for making personal, social, and political decisions in the future. They should come to see chemistry, like music and art, as an integral part of our culture.

It is common for such beginning students to be set routine tasks that impart content knowledge. One must learn to conjugate the irregular verbs in French before one can read the poetry of Beaudelaire. So too, in chemistry, one must study the many reactions of organic chemistry before being able to recognize the beauty of a natural product synthesis. But, because such drill may dull interest and imagination, we must guard against thwarting in an early stage of a student's career those qualities that will be essential for success at a later stage.

Our traditional introductory university classes, held in large lecture halls, often rely heavily on the hydraulic model of teaching which is based on the assumption that the transfer of knowledge from professor to student is analogous to a liquid being poured from one flask to another. This works remarkably well for the students in the front rows of our classes because they are usually the best students, the ones who, without provocation, actively participate in their education. But often it fails to penetrate to the middle rows where sit the invisible majority. Sometimes, weeks may pass without their contributing a word, with no communication and not much learning going on. As Lee Shulman, President Emeritus of the Carnegie Foundation for the Advancement of Teaching, pointed out, "The greatest enemies of learning are anonymity and invisibility. People who are invisible don't learn. In no sense are they accountable, in no sense are they responsible, and therefore they can simply turn off [2]." As well, a contributing reason why many of our front-row students do not pursue a program in the physical sciences is the perceived lack of relevance of the courses to real-world human experiences. While a chemist might ask, incredulously, "What in the world isn't chemistry?," students can easily be blinded by the theories, formulae, and equations and fail to see their relevance and application to societal matters.

Thus, we need to employ teaching techniques that engage the front rows and penetrate to those who silently lurk behind them. In our efforts to actively involve these students in their learning, we need to consider how we can place our course content in the context of real-world, scientific, or societal problems. We need to see how we can integrate academic instruction with community engagement. This chapter will advocate the use of problem-based learning (PBL) and service learning as pedagogies by which to stimulate and engage our students, to link them to their community and, with the use of technology, to expand the scope of that community.

11.1
Problem-Based Learning

PBL is a process by which the content and methods of a discipline are learned in an environment in which they are to be used to address a real-world problem (originally called *learning in a functional context* [3]). It is a strategy in keeping with the recommendation of the American Association for the Advancement of Science (AAAS) to teach science as it is practiced at its best [4]. Students are presented with problems whose solutions require certain information and skills that they do not yet possess. They must identify the required material that is relevant to the problem, locate its sources, and develop strategies for its use. This then defines the curriculum, and students immediately realize, on a need-to-know basis, the reasons for including certain material in the curriculum. Skillful supplementation with more conventional, prescriptive methods of teaching can maximize the contribution of the instructor who acts both as a content authority and as a facilitator of student activities. By this process, the desired outcomes can be achieved: enhancing and integrating knowledge, developing problem-solving strategies, generating the skills and motivation for continued learning, and gaining confidence in the assessment of one's own work. The key feature is that the learning process is initiated by the problem. The problem, in effect, directs the learning. This is in contrast to the more usual instruction in which the problem is used to illustrate how to use the knowledge after it has been learned or to test what knowledge has been learned.

11.1.1
History

Learning in a functional context was first described by Shoemaker in 1960 [3]. In teaching potential radio technicians about electronics, he decided to provide the students with radios that did not work and set students the task of repairing them. Simultaneously he directed the students to books on transformers and condensers, and so on, and offered to discuss them with interested students. Subsequently, he reported his amazement at how much more quickly the students became effective radio technicians than did those taught by more conventional methods.

The late Howard Barrows of Southern Illinois University School of Medicine was for many years the leading American authority in the use of PBL. A memorial to his enormous contribution to the use of innovative educational methods based on self-directed learning in medical education was recently published [5]. He very broadly defined PBL as "the learning which results from the process of working toward the understanding of, or resolution of, a problem [6]." Barrows himself credited the late James Anderson of McMaster University Faculty of Medicine as being the originator of the use of PBL in medical education (H. Barrows, personal communication). The pedagogy was subsequently adopted and adapted to the field of engineering by the late Donald Woods of McMaster University's Department of Chemical Engineering. His book, *Problem-based Learning: How to Gain the Most from PBL*, has eased and guided generations of students into effective use of PBL [7]. For these individuals, all from professional faculties, the practical problems were defined by their profession. However, increasingly, the pedagogy of PBL is being extended to arts and sciences disciplines [8] where it is used to emphasize the relevance of the course content while helping students to develop skills of communication, research, analysis, and critical assessment.

11.1.2
The Process

In a sense, PBL is a universal learning process, practiced from the time an infant learns to satisfy its physical and emotional needs. But it is a technique that becomes submerged by experiencing years of prescriptive learning, and the transition back to a less instructor-dependent form of learning is not an easy one.

Once presented with the problem, students, normally working in groups of four or five, proceed with the following tasks:

1) analyze the problem and identify precisely what is being asked;
2) compile a list of what knowledge and skills are required in order to address the problem;
3) assess what knowledge that is needed is already present within the group and what will have to be sought out;
4) determine where and how to seek the additional information that is required;
5) divide up the search amongst members of the group;

6) search;
7) regroup and synthesize, applying the new knowledge and skills acquired back to the problem (often this leads to additional questions that need to be researched);
8) reiterate, researching at a deeper level or redefining the problem until a satisfactory conclusion is reached; and
9) construct and report a solution to the problem.

It is worth noting here that this process is, in keeping with the AAAS recommendations [4], compatible with the nature of scientific inquiry as applied in research settings, with these comparable steps:

1) recognize a good (interesting, important, and tractable) problem;
2) learn more about the problem;
3) decide which experiments/observations/calculations would contribute to a resolution of the problem;
4) perform the experiments;
5) decide whether the results really do contribute to a better understanding of the problem;
6) go back to step 2 or 3 if they don't;
7) go on to step 8 if they do; and
8) communicate your results.

That PBL is excellent preparation for undergraduate research activities is highlighted in the recent article by Hal White, 2011 winner of McMaster University's Howard Barrow Award for promoting student engagement through PBL [9].

The range of options in the application of PBL go from making PBL the primary mode of curriculum delivery, to using it for projects within a conventionally structured course, to merely using it to introduce topics which will then be dealt with in the lecture format (the latter being valuable in large introductory courses). Thus, for faculty members the key tasks (after defining the course objectives in terms of content, methods, and skills to be learned) include the following:

1) generating appropriate problems so as to ensure that the desired content is discovered (with a well-designed PBL problem, when the students decide what they need to know in order to address the problem, that coincides with what the professor intended the learning outcomes to be);
2) integrating lectures with problems while maintaining a coherent lecture pattern and, at the same time, being prepared to interrupt the lecture pattern to address areas needed to assist with the problem;
3) modeling the PBL strategies in lectures, demonstrating questions that should be asked in order to deal with the problem;
4) challenging and questioning the students in order to test their understanding, to clarify points, and to help students remain focused;
5) serving as a cognitive coach;
6) facilitating interaction between the PBL group members;

7) making the course evaluation procedure consistent with the PBL process so that not only content learned but also skills gained are assessed; and
8) integrating the PBL curriculum with the broader departmental curriculum.

11.1.3
Virtual Problem-Based Learning

When using PBL as the primary mode of content delivery or for projects, students normally work in groups of four or five. In upper division courses, this works quite well (see Section 11.1.4.2). But it can be problematic in large introductory courses, as the requisite cooperative aspect of the process may not work well due to the heterogeneity of the students in terms of both background and motivation. When surveyed about their experience with group work, one of the more philosophical students expressed it as follows: "Each person has a role that they choose. Leader lead, workers work, loafers loaf-around, and sponge off humanity. That's life." While a valuable lesson to learn, it was not the one I sought to teach. And it was echoed by approximately 75% of the class who claimed that they carried their group and did all of the work. The benefits inherent in PBL were outweighed by the resentments that developed among some students.

To address this problem, virtual problem-based learning (VPBL) was developed [10]. PBL problems were mounted on a Web site along with some text-book-level background information relevant to the problems. Within the background information were links to other topic-related Web sites, only some of which were actually relevant to the problems. Also linked to the Web site were chat rooms, one for each problem, which students used to communicate with each other about the problem to which they were assigned. Through use of the chat rooms, groups of workers emerged and worked cooperatively on their problem, dividing up the work and sharing their findings. Each student was, however, responsible for submitting his or her own report on the solution to the problem (unlike PBL in which a single group report is usual). And although the loafers could also read the material posted on the discussion boards, they were not resented by the workers because the two groups were not forced together. Moreover, by the end of the projects, over 80% of the students had actively participated, that is, posted questions and information. The professor's role here was to monitor the chat rooms, acting as a facilitator of the discourse, reining it in when it veered in unproductive directions, and helping to refocus it. Figure 11.1 shows the distribution of chat room postings by day and Figure 11.2 by time of day, clearly showing that this on-line asynchronous communication accommodates the schedules of all involved.

VPBL opened the learning process to groups of students who had been disenfranchised by the requirement for face-to-face, small group meetings that are a part of PBL, for example, students with long commutes and those with family and/or employment obligations. It was particularly notable that some students from cultures that discourage their public assertiveness emerged as leaders in the chat rooms. And, of course, VPBL opens up the opportunity to form global

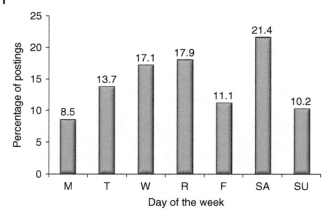

Figure 11.1 Postings on VPBL chat room by day.

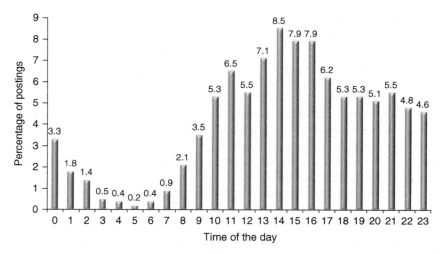

Figure 11.2 Postings on VPBL chat room by time of day.

learning communities involving students from anywhere in the world with internet access.

There are some who believe that any form of distance or online education is simply a profit-driven plot concerning the commodification of higher education. Others believe that the Web is effective for the transfer of information but that it will stifle meaningful communication. In the fifth century BC, Plato, who favored reasoned discourse via the spoken word over the written transmission of knowledge, predicted that writing would weaken the tradition of poetry. But 24 centuries later, poetry remains alive. Similarly, over 550 years ago many believed that the printing press would end the tradition of oral debate. But few would seriously argue today that the intellectual enterprise of teaching and learning has suffered as a result of the use of writing and the invention of the printing press.

Rather, new opportunities have emerged. So too will they emerge as we enjoy the active, collaborative, and self-directed approaches to learning enabled by the online experience. As predicted by Marshall McLuhan, the new technologies are shrinking the world and enhancing society [11]. Others simply believe that there is no substitute for face-to-face communication. This is a belief with very high face validity (i.e., it seems so reasonable and rational that it needs no verification by data). But growing amounts of data now contradict this. The Web supports student conversations at least as well as the traditional small group discussions [12] and retains the social constructivism essential to the PBL process [13]. Surprisingly perhaps, rather than engendering a feeling of disenfranchisement due to distance, it seems to result in increased camaraderie and support.

At the University of Toronto Mississauga (UTM), VPBL projects have been included within a conventionally structured (lecture based) introductory chemistry courses of 300–500 students. The course, which has Grade 12 chemistry and mathematics as prerequisites and university calculus as a corequisite, includes students from all disciplines that require basic chemistry knowledge (astronomy, biology, chemistry, earth science, environmental studies, and forensic science) and students with a wide range of ability and motivation. PBL projects have also been included in a bioinorganic chemistry course with enrollment of 50–75 students and taken by third- and fourth-year students in biological chemistry, chemistry, and molecular biology programs. These projects have taken the place of the more conventional term-paper assignments in upper level courses.

11.1.4
The Problems

PBL and VPBL problems should be ones that contain content that is authentic, which build upon the experience of the students but require knowledge that the students do not yet possess. The problems posed should allow students to use the same cognitive processes as are required in real life. And, as in life, they may be ill-structured problems, involving information that is ambiguous, conflicting, and confusing. They may be open-ended without having a "right" solution, and may even be redefined as new knowledge is gained. The students must learn to be comfortable dealing with probability rather than certainty; they must learn to tolerate doubt. And they should come to appreciate that learning is an ongoing process and that educated, albeit imperfect, judgments can be made throughout the stages of this process. A library of problems for chemistry and a variety of other disciplines are available at the University of Delaware PBL Clearing House [8].

11.1.4.1 Selected Problems for Introductory Chemistry at the UTM

Descriptive Chemistry of Metals

- You are an aircraft engineer examining the materials used in the construction of the exterior of a plane. The principal metal used is aluminum. In the aerospace industry, primer paints containing dichromate pigments are used to inhibit corrosion on aluminum. Despite the high toxicity of these primers, the aerospace industry still prefers to use them rather than alternative nontoxic substitutes. How would you explain to representatives of the Ministry of Labour why the industry is so strenuously resisting the move away from the use of these dichromate-based paints?

This problem was used to teach the descriptive chemistry of metals and their binary oxides. In its application, cost effectiveness, chemical toxicity, environmental concerns, and political factors must be considered. The report requires providing an explanation of scientific data to nonscientists.

- You are a biomedical engineer working on a new design for an artificial hip. Through your research, you have narrowed down the choice of materials for this prosthesis to aluminum, titanium, chromium, or an alloy thereof. Define the factors that need to be considered in making your final choice of material. Then, based on your knowledge of these factors, choose the most suitable material for the task and prepare an argument in defense of your choice.

The learning outcomes of this problem are also focused on the descriptive chemistry of metals, here with a biomedical application. It could be used equally well as a PBL project or in the assessment of learning in a PBL course.

Thermochemistry

- The two most common household stoves are powered by electricity or natural gas. For speed of cooking, the use of electrically powered microwave ovens has been substituted for both. Design an experiment to compare the efficiency of using either an electric stove or a gas stove to heat water to that of using a microwave oven to heat water. Are there factors other than efficiency which should inform one's choice? (Assume that the natural gas used in a gas stove is 100% methane, CH_4). The efficiency of the each type of stove is calculated as follows:

$$\text{efficiency of electric stove} = \frac{\text{amount of energy absorbed by water in a pot}}{\text{amount of energy supplied by electrical power}}$$

$$\text{efficiency of gas stove} = \frac{\text{amount of energy absorbed by water in a pot}}{\text{amount of energy supplied by burning methane}}$$

Assume that standard pieces of laboratory and/or kitchen equipment are available. Your experimental design should include a procedure for carrying out the experiment and a list of data to be collected.

This problem from which basic thermochemistry is learned also requires students to think about how to design a properly controlled experiment from which meaningful data could be collected. It also serves as one of a series of problems assigned to students who, due to a health or disability, are unable to perform physical laboratory work. As with much of PBL, the process mimics that of investigative research.

- In economies with limited supplies of petroleum and natural gas (which is primarily methane), coal (both bituminous and anthracite) has been a major fuel for power plants. Compare the burning of methane with the burning of *either* bituminous *or* anthracite coal in terms of the thermodynamic efficiency, the practicality, and the effects on the environment of these processes.

  In 2000, the Citizens Electric Corporation of Ste. Genevieve, Missouri, sold 1.055×10^9 kWh of power from their gas powered plant. How much natural gas was consumed to generate that much power? How much coal would have been required to generate the same amount of power?

- If the central heating in your home goes off, you could warm your kitchen by turning on the stove and keeping the oven door open. If your air conditioning goes off, could you cool your kitchen by leaving the refrigerator door open?

A short problem such as this is appropriate for use in discussion/tutorial classes and for use in the assessment of thermochemistry learning in a PBL course.

Organic Chemistry

- Devise a reaction scheme by which 3-ethylpent-2-ene could be synthesized from ethanol as the only source of carbon atoms.

This type of problem, presented after students have studied atomic and molecular structure but before any study of organic chemistry, makes a good introduction to the topic. The class, working as a group of the whole, can compile a list of what knowledge they would need in order to address the problem (one of the key cognitive steps in the process):

1) Organic nomenclature, in order to know what 3-ethylpent-2-ene is;
2) How to make carbon–carbon bonds, in order to build up a seven-carbon compound from a two-carbon compound;
3) How to break carbon–carbon bonds, in order to make a compound with an odd number of carbons from one with an even number;
4) How to eliminate the OH group;
5) How to generate the carbon–carbon double bond (related, of course, to 3 above).

Introduced in this way, the *raison d'être* for the lecture material that follows becomes much more apparent to the students. Students also learn a strategy for synthesis based upon making and breaking carbon–carbon bonds and interconverting functional groups, which can be applied to other syntheses.

Introductory chemistry students, when initially faced with PBL, view it as a daunting task. For most of their past experience, they were first taught content and then asked to do problems applying that content knowledge. By contrast, with PBL, content is learned, largely independently, on a need-to-know basis in an effort to address a problem. It takes a few problems before students become comfortable with this new responsibility. VPBL participation data is shown in Figure 11.3. The descriptive chemistry problems were considered by the students

Year	1		2		3	
VPBL topic	Descriptive chemistry of metals	Thermo-chemistry	Descriptive chemistry of metals	Thermo-chemistry	Descriptive chemistry of metals	Thermo-chemistry
Total no. of students	~175 ~125	~195 ~195	~175 ~175	~250 ~250	~230 ~230	
No. of students posting	110 81	153 137	149 154	222 203	209 200	
Percentage of students posting	63 65	78 70	85 87	89 81	91 87	
Total no. of posts	352 315	562 479	849 890	1130 1286	1227 1304	
Total no. of chat visits	4311	4394	4697	9136	10 143	

Figure 11.3 VPBL participation.

to be more difficult than the thermochemistry problems. Thus, in year 1, when descriptive chemistry was used to introduce PBL, the active participation rate was only ~65%. In years 2 and 3, when PBL was introduced through the thermochemistry problems and then followed by the descriptive chemistry problems, active participation was significantly improved. Notable in all years, however, was the fact that the number of visits to the chat rooms was far in excess of the number of messages and questions posted. This, supported by anecdotal information provided by students on their course evaluation surveys, indicated that students valued and felt the benefit from this virtual collaboration with their classmates.

11.1.4.2 Project for a UTM Upper Level Bioinorganic Chemistry Course

This 13-week bioinorganic chemistry course taken by third- and fourth-year students was designed around a PBL project. Students experiencing PBL for the first time and working in groups of four were given a problem with the following components.

- Design a compound that you think may behave as a catalytically active model of one of the following metalloenzymes.
- Design a synthesis for your model compound.
- Write a research grant proposal, using the NSERC (National Science and Engineering Research Council) format, for funds to support the synthesis and testing of your model compound.

In the first meeting, the class generated a list of the kinds of background information that they would require in order to address the problem. For the remaining weeks 1–3, lectures were given to survey that required background material. PBL can be most effective when reinforced by other learning experiences, each selected and timed to help toward the attainment of specific objectives of the problem. At the end of week 3, the groups were assigned the particular metalloenzyme that they were to model in their project (including, for example, Cu–Zn superoxide dismutase and polyphenoloxidase both of which are implicated in mammalian diseases). Lectures in weeks 4 through 12 dealt with a variety of bioinorganic systems, focusing in all cases on the relationship between biological functions and structures, acidities and reduction potentials, and on the strategies for the synthesis of structural and functional models. In addition to imparting new knowledge about particular systems that would form the core material in any bioinorganic chemistry course, lectures modeled the investigative strategy that might be used in the projects. Throughout the course, weekly meeting with each group were held in order to facilitate the work of the groups to ensure that students understood the information that they had gathered for their project and that they realized what was yet unknown.

The objectives of this PBL project were for the student to

- learn to apply the integration of knowledge of inorganic and organic chemistry to the understanding of biochemical systems;
- develop a problem-solving strategy that depends on background knowledge, research, analysis, and creativity;
- learn how to read the chemical literature in the context of using it;
- develop an appreciation of research and become more comfortable in dealing with a constantly changing body of knowledge; and
- increase their retention of factual material by learning it in the context of a problem-solving approach.

The final examination was consistent with the general pedagogy and objectives of the course; 1 of the 3 h of the examination was devoted to reading and interpreting the relevance of a communication from the *Journal of the American Chemical Society* which the students had not previously seen. The high quality of the project reports and the final examination scripts suggested that the first four of the project objectives were realized. The middle rows of students had clearly been reached as reflected by their increased participation in all aspects of the course and by an average increase of 7% in their final marks as compared to their achievements in their more didactically taught courses. As for the fifth objective, research by both medical educators and education psychologists suggests that this approach results in enhanced retention of new knowledge as well as increased capacity for critical thinking and continued independent learning. While conventional pedagogies impart more knowledge in the short term, knowledge gained through PBL is retained over a longer period [14–16].

11.1.5
PBL – Must Content Be Sacrificed?

A final question merits consideration. Is content sacrificed by using a pedagogy that focuses so much on teaching students how to learn? Are the very best students being cheated out of content in the effort to penetrate beyond the front rows? On the contrary, as with other examples of universal design, all students benefit. While it cannot be denied that most PBL courses trade some content for teaching critical thinking skills, this is in line with the underlying principle of Project 2061 [17], an

AAAS initiative to reform science education. It advises us to "teach less in order to teach it better."

Students need to understand that the content of their chemistry courses and the material in the literature are not immutable truth. It rather represents only our current understanding of the structure and dynamics of matter as it is embodied today in a discipline that is constantly changing and being enlarged. They should come to appreciate that chemistry is an exciting discipline in which discoveries, sometimes fundamental discoveries that challenge textbook content wisdom, are being made.

11.2
Service-Learning

Service learning is an active learning pedagogy that combines credit-bearing coursework with community service and critical reflection. In response to needs identified by community partners, students provide service related to their discipline. In the process, they acquire the three types of knowledge described by Altman: content knowledge, process knowledge (i.e., skills), and socially relevant knowledge [18]. Through it, they gain an enhanced appreciation of the role of their discipline and advance their sense of civic responsibility and their role as citizens. The process depends on active involvement of a triumvirate: determination of the academic content is contributed by the professor; contribution of the service opportunity relevant to that content is made available by the community partner; and connection between the academic material and the service experience is made through critical reflection by the student.

Furco placed service-learning in the center of an experiential education continuum consisting of volunteerism, community service, service-learning, field education, and internships [19]. He argued that the position in this continuum is determined by (i) who the primary intended beneficiary of the activity is and (ii) what the primary focus of the activity is. While there is much to learn from volunteerism, the learning goals of the volunteer are not paramount; the primary focus is the service, and the intended beneficiary is the recipient of the service. At the other end, internships normally focus on learning the content and skills of a particular profession, with the intended beneficiary being the student. Service-learning falls in between these two extremes, focusing equally on the student and the interests of the community partner.

A broad scope of service outreach has been reported, which varies from service within a given academic institution, for example, senior students tutoring more junior students, to that focused on local community agencies, to international service as exemplified by the University of British Columbia's International Service-Learning program [20]. The delivery varies from dedicated service-learning courses such as Indiana University's Hoosier Riverwatch Water Quality Testing course [21], in which the content is designed to suit the service experience, to conventional courses with service-learning projects incorporated in order to apply the content. In the latter case, after establishing clear learning objectives as one would do for any course, professors must articulate these objectives to the community partners to ensure that the expectations of those partners are compatible with the objectives, bearing in mind that the academic credit is for the learning and not for the service.

11.2.1
The Projects

General resources to support incorporating service-learning projects into chemistry courses are provided on the American Chemical Society Web site [22]. The efforts required to use this pedagogy can be simplified by involving other parts of the university as the community partners [23]. Shumer has provided a self-assessment tool for service-learning projects which can be used for formative assessment in the planning process and monitoring, and ultimately for summative assessment of the project. Designed for use in the K-12 environment, it is easily modified for use in higher education [24]. The vast majority of chemistry service-learning projects described in the literature originate in the United States and fall in one of two areas: chemistry teaching in its broadest sense, and analytical/environmental chemistry, both done largely by second- and third-year students. But the international interest and applicability is illustrated by Seddeqi's list of projects from Umm Al Qura University, which includes assisting with science projects for schools, analyzing ground water quality and its impact on health, and monitoring air quality in homes where people smoke [25].

11.2.1.1 Selected Analytical/Environmental Chemistry Projects

Students at Westminster College in Pennsylvania tested the efficacy of a passive water treatment site that was established in an attempt to improve the water quality at an abandoned mine drainage site. They found this treatment regime to be effective in reducing the acidity and the metal content of the water and reported their findings at a recent meeting of the Slippery Rock Water Coalition [26]. Similarly, partnered with the city of Bloomington, students at Indiana University used gas chromatography-mass spectrometry and atomic absorption spectroscopy to carry out environmental testing on the Clear Creek Watershed near the site of an abandoned railroad and creosol plant [27].

Gardella's students at the State University of New York-Buffalo have worked cooperatively with local agencies and the United States Army Corps of Engineers to analyze soil in the Lewiston school district and generate geographic maps of soil contamination [28]. Similar projects are of interest in Bangladesh where soil concentrations of arsenic are high.

Thompson of the University of Akron reported an interesting project in collaboration with the local zoo. The anesthesia used in surgery on fish affects the pH of the aquarium water. First-year students monitored the acidity of aquarium water before and after surgeries, and prepared pH trend charts which informed the veterinarians with regard to maintaining the required constant pH environment for the fish [29].

At the University of Utah, students surveyed a neighborhood for the presence of lead in house paint. After distributing information about the health hazards of lead-containing paints, they were allowed to sample the paint in some of the older homes. Samples were subjected to atomic absorption spectroscopic analysis [30]. On reflection, it was apparent that when older neighborhoods are gentrified and the older homes are renovated, infants and small children are put at risk because of the possible inhalation of lead-containing particulates. Bucknell University students analyzed for potential health threats due to the use of pressure-treated lumber, which were treated with chromated copper arsenate, in playground equipment [31]. This preservative is also commonly used on wood that is destined for other outdoor structures including poles, decks, and garden furniture.

In all these examples, increased student engagement with their subject has been reported. Realizing that decisions could be made based in part upon their work, students' attention to the accuracy of their experimental work and the sense of responsibility to the community were enhanced.

11.2.1.2 Selected Projects in Chemistry Education

From large research-intensive universities to smaller liberal arts colleges, from North American universities to Tamkang University in Taiwan, service-learning projects which bring post-secondary students in contact with K-12 students abound. Projects include mentoring, tutoring, carrying out demonstrations, guiding science fair projects, supervising experimental work, and preparation of lesson plans.

Saitta reports university students tutoring high school students via Web conferencing [32], one of a number of technologies that can facilitate the formation of international service-learning collaborations. After learning about the synthesis of azo dyes, organic chemistry students at Rhodes University in South Africa taught this material to high school students from schools with limited resources and then used the products of their syntheses to dye clothing [33].

Chemistry education projects with, undoubtedly, the potential for the most widespread effects are those involving contributions to Wikipedia. Bioinorganic chemistry students at the UTM identified Wikipedia stubs (articles identified by the Wikipedia editors as requiring editing and/or enhancement) and topics related to their course material that had not yet been addressed on Wikipedia. So Wikipedia became the community partner in this service-learning project. The project description was as follows.

- There are two components to this project. For the service-learning component, you will edit or create a Wikipedia entry on your assigned topic. The entry should be accessible to the public but aimed more particularly at the level of students who are studying chemistry or biochemistry.

- Instructions on how to edit and create material on Wikipedia are provided on the course Blackboard site. Also, the technology librarian is available to provide technical advice.
- Note that all material must be original, all material must be properly referenced, and any copyrighted material can be included only if the permission of the copyright holder has been secured.
- The second component of the project is to give a 15-min oral presentation to the class on the information contained on your Wikipedia page. This presentation should focus on the bioinorganic chemistry of the topic and should be aimed specifically at the level of your classmates. The presentation should be accompanied by a hard copy of your Wikipedia page which should be provided to each student in the class at the beginning of the class in which you are to make your presentation.
- Marks for this group project will be determined 50% by peer assessment using forms provided to you and 50% by your instructor. All students in a group will receive the same mark for the project (excepting in unusual and extraordinary circumstances) [34].

Some of the topics included were zinc fingers and the treatment of HIV/AIDS, effect and detection of blood doping, drug treatment for thalassemia, chromium treatment of diabetes, gold treatment for rheumatoid arthritis, drug treatment of Wilson's disease, and zinc fingers and schizophrenia (protein 804A).

Recognizing their responsibility in providing information to be freely available to the global community on the World Wide Web, the UTM students were highly motivated to be clear and accurate in their presentation and to make it simultaneously accessible to the lay reader while useful to the technical reader. This experience was echoed in the work of McNeil at the University of Michigan [35]. In a Wikipedia assignment to graduate students, intended to improve the research skills of the students while simultaneously enhancing the information available to the public, her students also exhibited an increased motivation to be accurate and reported an increased ability to identify appropriate resources to develop an argument. Although all of the UTM students had previously used Wikipedia as a source of information, few knew how material came to be on Wikipedia nor how easily or with what motivation Wikipedia content could be altered. As a result of this project, while developing a respect for the efforts of the Wikipedia editors, they vowed to be more cautious in verifying from primary sources the content that they found on Wikipedia before assuming it to be accurate.

11.2.1.3 Project for an Upper Level Bioinorganic Chemistry Course at UTM

Students working in groups of four or five were assigned a metallic element. The description of the project was as follows:

- Design an information pamphlet, of the type that one sees in a doctor's office, explaining an illness or disease whose cause is somehow related to the metabolism of your assigned element. The intended audience for the pamphlet is the student body that frequents the UTM Health Service.

- Prepare a 20-min presentation on the same topic. Here the audience will be your fellow bioinorganic chemistry students. The presentation should be accompanied by a hard copy of your pamphlet, which should be provided to each student in the class at the beginning of the class in which you are to make your presentation.

In this project, the community partner was the University Health Service, and a challenge was to identify topics that would be relevant to the types of issues brought to the Health Service by undergraduate students, topics that would be of interest to the recipients of the service. Topics included the role of zinc in alcohol metabolism, inorganic toxins in cigarette smoke, the involvement of zinc in eating disorders, the relationship between zinc and the immune system, the facts and myths about mercury, dietary sodium requirements and overload, the health effects of lead in commercial products, iron and physical endurance, and radon – the second leading cause of lung cancer. As with the Wikipedia projects, because students knew that their work would be distributed for public use and scrutiny, they paid careful attention to the accuracy and relevance of their information. Figure 11.4 shows an example of the product of one student group.

11.2.2
Benefits of Service-Learning

The benefits of service-learning in chemistry accrue to all members of the triumvirate. Faculty members have the opportunity to liaise with community agencies. This opens the possibility of new research partnerships as well as providing a new source of applications to enhance their teaching. It also strengthens the connection between the university and the community, increasing the public's perception of the value of the university and, in particular, increasing the public understanding of the role of chemistry as an integral part of our culture.

Community partners benefit, of course, from the support of volunteer workers. This may allow them to deal with back-burner issues and ideas for which they do not have sufficient regular staff to address. It raises their awareness of the expertise that rests within the university while simultaneously educating the university partners with regard to the expertise that resides within the community agency, resulting in enhanced mutual respect. Involving students in the work of public agencies lays a foundation for support of those agencies in the future.

Most importantly, the use of this pedagogy is of benefit to the students. Experience is a powerful teacher. And the reflective component of service-learning is what links this experience with the new knowledge gained [36]. In addition to improved academic achievement [37], students become more engaged with course material as it is applied in real-world settings. They develop an awareness of the importance of honest and ethical behavior as chemists, and increase their level of civic responsibility. As well, Bringle has documented the positive effect of service-learning on retention of first-year university students to the second year [38].

Figure 11.4 Service-Learning Medical Pamphlet Project (open views of trifold pamphlet).

11.3
Active Learning Pedagogies

Active learning pedagogies represent alternatives to the hydraulic model of course content delivery. They may involve a wide variety of strategies from the use of clickers to PBL, from studio courses which blend the laboratory and lecture environment to service-learning in the community, and from peer-led instruction to the inverted or flipped classroom. The common element is that they are all student-centered and rely upon active rather than passive participation of students in their learning.

The surge in interest in active learning pedagogies in the late twentieth century stemmed from a growing literature showing them to result in improved learning outcomes for students in terms of cognitive achievements and retention. This was expressed by the 1990 World Declaration on Education for All, which states that "active and participatory instructional approaches are particularly valuable in assuring learning acquisition and allowing learners to reach their fullest potential [39]." There was also the perception that use of these pedagogies could indirectly enhance economic development and prepare learners for their roles as responsible global citizens.

In 2009, Ginsberg reported his analysis of five case studies on the introduction of active learning pedagogies as educational reform initiatives [40]. These studies were sponsored by the United States Agency for International Development (USAID) and produced by the American Institutes for Research under the Educational Quality Improvement Program (EQUIP 1). The studies focused on teacher training for the introduction of active learning pedagogies in schools in five countries: Cambodia, Egypt, Jordan, Kyrgyzstan, and Malawi [41–45]. Many of the findings, however, are readily applicable to higher education environments. Ginsberg concluded that effective teacher training in the methods of the pedagogies was the core element necessary to facilitate the reforms. Some progress was observed in those schools that received direct support from agencies dedicated to professional development of the teachers. But the rhetoric of administrative policy documents advocating the introduction of active learning pedagogies was ineffective beyond those schools. Key challenges to the reforms that were identified included the fear of using active learning pedagogies in situations where examinations that focus on memorization are the norm, the limiting physical design of the classrooms, the number of students per class, the tendency of teachers to teach as they were taught, the heavy investment of time required to implement active learning pedagogies in courses, and the lack of incentives (financial, promotion, and recognition). Key, then, to facilitating educational reform and increasing the use of active learning pedagogies is to provide instructors with accessible training and to engender the feeling that there is value in their investment of time. While the immediate value may be expressed in terms of personal incentives, long-term value to the students, to the institution, and to the greater society should be apparent.

11.4
Conclusions and Outlook

Through active learning pedagogies such as PBL and service-learning, the front rows can be penetrated, with all students emerging as participants in their learning. Such students who become more engaged as they see their academic course content being applied to real-world situations invigorate the classroom environment and enrich the institution. Moreover, with technology available to establish e-learning environments, collaborative projects involving students from different institutions and in different countries can be mounted. And, through this group work, students can emerge with increased skills for communication and critical thinking as well as with an enhanced sense of ethical behavior and their role as community and global citizens. With the development of massive open online courses, we have the possibility of delivering the highest quality chemistry content to every remote corner of our globe. Examinations and accreditation remain as obstacles with this new technology. However, the potential to incorporate VPBL into these courses would allow students from around the world to collaborate on problems of both general relevance, for example, effects of global warming, and those of especial relevance to particular communities, for example, recycling in urban communities or water purification in remote communities. Such PBL projects could bring not only chemistry content but greater global awareness to our students. Independent of the medium, the more students actively participate in their learning, the greater the learning will be.

> Education is an admirable thing, but it is well to remember that nothing that is worth learning can be taught.
> Oscar Wilde (1891) *The Critic as Artist* [46]

References

1. Chemistry World *www.rsc.org/chemistryworld* (accessed 27 December 2013).
2. The New York Times *www.nytimes.com/2005/04/24/education/edlife/merrow24.html?pagewanted=5* (accessed 27 December 2013).
3. Shoemaker, H.A. (1960) The functional context method of instruction. *IRE Trans. Med. Educ.*, **3**, 52–57.
4. American Association for the Advancement of Science (1990) *The Liberal Art of Science: Agenda for Action*, AAAS, Washington, DC.
5. Dorsey, K.J. and Rangachari, P.K. (2012) Students Matter: The Rewards of University Teaching, Southern Illinois University School of Medicine, Springfield, IL.
6. Barrows, H.S. and Tamblyn, R.M. (1980) *Problem-Based Learning: An Approach to Medical Education*, Springer, New York.
7. Woods, D.R. (1994) *Problem-Based Learning: How to Gain the Most from PBL*, Woods, Waterdown.
8. University of Delaware PBL Clearinghouse The Motivation to Learn Begins with a Problem, *www.udel.edu/pbl* (accessed 27 December 2013).
9. White, H.B. III, (2007) in *Designing, Implementing, and Sustaining a Research Supportive Undergraduate Curriculum*

(eds K.K. Karukatis and T.E. Elgren), Council on Undergraduate Research, Washington, DC, pp. 9–19.
10. Poë, J. (2011) Problem Based Learning, www.utm.utoronto.ca/~w3chm140/pbl (accessed 27 December 2013).
11. McLuhan, M. (1962) *The Gutenberg Galaxy*, University of Toronto Press, Toronto.
12. Harasim, L., Hiltz, S.R., Teles, L., and Turoff, M. (1995) *Learning Networks*, MIT Press, Cambridge, MA.
13. Bodner, G., Klobuchar, M., and Geelan, D. (2001) The many forms of constructivism. *J. Chem. Educ.*, **78** (8), 1107, http://pubs.acs.org/doi/pdf/10.1021/ed078p1107.4 (accessed 27 December 2013).
14. Eisenstaedt, R. (1990) Problem-based learning: cognitive retention and cohort traits of randomly selected participants and decliners. *Acad. Med.*, **65**, 511–512.
15. Douchy, F., Segers, M., den Bossche, V., and Gijbels, D. (2003) Effects of problem-based learning: a meta-analysis. *Learn. Instr.*, **13**, 533–568.
16. Norman, C.R. and Schmidt, H.G. (1992) The psychological basis of PBL: a review of the evidence. *Acad. Med.*, **67**, 557–565.
17. AAAS Project 2061, www.project2061.org (accessed 27 December 2013).
18. Altman, I. (1996) Higher education and psychology in the millenium. *Am. Psychol.*, **51**, 371–398.
19. Furco, A. (1996) *Expanding Boundaries: Serving and Learning*, Corporation for National Service, Washington, DC, pp. 2–6.
20. University of British Columbia (2013) International Service Learning Program, www.student.ubc.ca/global/learning-abroad/international-service-learning (accessed 27 December 2013).
21. Tait, S.L. (2012) Tait Research Group, www.indiana.edu/~taitlab/?p=courses/waterq (accessed 27 December 2013).
22. American Chemical Society www.acs.org/content/acs/en/education/resources/undergraduate/service-learning-resources-for-chemistry-faculty.html (accessed 27 December 2013).
23. Sutheimer, S. (2008) Strategies to simplify service-learning efforts. *J. Chem. Educ.*, **85**, 231–233.
24. Shumer, R.D. (2000) Shumer's Self Assessment for Service-Learning, www.talloiresnetwork.tufts.edu/wp-content/uploads/Shumer_sservice-learningself-assessment.pdf (accessed 27 December 2013)
25. Seddeqi, Z. (2013) Some Examples for Chemistry Service Learning by Dr. Zaki Seddeqi, www.dirasat.org/1/post/2013/05/some-examples-for-chemistry-service-learning-by-dr-zaki-seddeqi.html (accessed 27 December 2013).
26. Boylan, H. (2013) Westminster College Chemistry/Biochemistry Majors present Results of Service-Learning Study, www.westminster.edu/news/releases/release.cfm?id=4420 (accessed 27 December 2013).
27. Reck, C., Lindberg, K., and Tait, S. (2012) Service partnership with local city government and Hoosier River Watch, www.bcceprogram2012.haydenmcneil.com/p714-service-partnership-local-city-government-hoosier-river-watch (accessed 27 December 2013).
28. Gardella, J., Milillo, T.M., Sinha, G., Manns, D.C., and Coffey, E. (2007) Linking community service learning and environmental analytical chemistry. *Anal. Chem.*, **79**, 811–818.
29. Thompson, J. (2011) University of Akron Chemistry with the Zoo, www.uakron.edu/service-learning/service-learning-highlight.dot (accessed 27 December 2013).
30. Kesner, L. and Eyring, E.M. (1999) Service-learning general chemistry: lead paint analysis. *J. Chem. Educ.*, **76**, 920–923.
31. Draper, A.J. (2004) Integrating project based service-learning into an advanced environmental chemistry course. *J. Chem. Educ.*, **81**, 221–224.
32. Saitta, E.K.H., Bowdon, M.A., and Geiger, C.L. (2011) Incorporating service-learning technology and reserch supportive teaching techniques into the university chemistry classroom. *J. Sci. Educ. Technol.*, **20**, 790–795.

33. Glover, S.R., Sewry, J.D., Bromley, C.L., Davies-Coleman, M.T., and Hlengwa, A. (2013) The implementation of a service-learning component in an organic chemistry laboratory course. *J. Chem. Educ.*, **90**, 578–583.
34. Poe, J. (2013) Assignments, *www.utm.utoronto.ca/~w3chm333/assign.html#group* (accessed 27 December 2013).
35. May, C.L., Locke, J.R., Coppola, B.P., and McNeil, A.J. (2010) Improving science education and understanding through editing Wikipedia. *J. Chem. Educ.*, **87**, 1159–1162.
36. Bringle, R. and Hatcher, J. (1999) Reflection in service-learning: making meaning of experience. *Educ. Horiz.*, summer, 179–185.
37. Gray, M.J., Ondaatje, E.H., Fricker, R.D., and Geschwind, S.A. (2000) Assessing service-learning: results from a survey of "learn and serve America, higher education." *Change*, **32**, 30–39.
38. Bringle, R.G., Hatcher, J.A., and Muthiah, R.N. (2010) The role of service-learning in the retention of first-year students to second year. *Mich. J. Community Serv. Learn.*, spring, 38–47.
39. Inter-Agency Commission (1990) World declaration on education for allDocument adopted by the, in *World Conference on Education for All: Meeting Basic Learning Needs, Jomtien, Thailand, March 5–9, 1990*, Inter-Agency Commission, New York.
40. Ginsberg, M. (2009) Active-learning pedagogies as a reform Initiative: synthesis of case studies, in *EQUIP1 Research Report*, American Institutes for Research, Washington, DC, *www.equip123.net/docs/e1-ActiveLearningSynthesis.pdf* (accessed 27 December 2013).
41. Bunlay, N., Wright, W., Sophea, H., Bredenburg, K., and Singh, M. (2009) Active-learning pedagogies as a reform initiative: the case of Cambodia, in *EQUIP1 Research Report*, American Institutes for Research, Washington, DC.
42. Megahed, N., Ginsburg, M., Abdellah, A., and Zohry, A. (2008) *Active-Learning Pedagogies as a Reform Initiative: The Case of Egypt EQUIP1 Research Report*, American Institute for Research, Washington, DC.
43. Roggemann, K. and Shukri, M. (2009) *Active-Learning Pedagogies as a Reform Initiative: The Case of Jordan*, American Institute for Research, Washington, DC.
44. Price-Rom, A. and Sainazarov, K. (2009) *Active-Learning Pedagogies as a Reform Initiative: The Case of Kyrgyzstan*, American Institute for Research, Washington, DC.
45. Mizrachi, A., Padilla, O., and Susuwele-Banda, W. (2008) *Active-Learning Pedagogies as a Reform Initiative: The Case of Malawi*, American Institute for Research, Washington, DC.
46. Oscar Wilde Quotes (2014) *www.brainyquote.com/quotes/authors/o/oscarwilde.html* (accessed 24 October 2014).

12
Inquiry-Based Student-Centered Instruction
Ram S. Lamba

12.1
Introduction

> Here we use the Socratic Method: I call on you; I ask you a question; you answer it. Why don't I just give you a lecture? Because through my questions you learn to teach yourself. By this method of questioning-answering, questioning-answering, we seek to develop in you the ability to analyze that vast complex of facts that constitutes the relationship of members within a given society.
>
> *Prof. Kingsfield in "Paper Chase" [1].*

Lecturing is a fairly common method of presenting new material in most classrooms of universities and colleges. Students familiar with lecturing since their school days accept what the teacher says in the classroom whether or not they understand the material; therefore, they are not accustomed to asking questions even when they do not grasp the essence of the material. It is common to observe instructors dictating lectures and students taking notes. These lecture notes become the study notes of the students, which they memorize for exams and quizzes. The interesting thing is that, if one looks at the lecture notes of all students in the class, each of them would have emphasized different aspects of the same concept delivered by the instructor. In fact, many lecture notes are likely to be completely different from what the instructor was trying to convey! In the students' minds, learning science – in particular chemistry – is to memorize the lecture notes taken while listening to the instructor [2–4] whether or not they depict correctly the concept taught. Therefore, the majority of the students in introductory courses study for grades rather than for learning. The whole process is basically teacher-centered.

Students acquire and confirm lifelong beliefs and attitudes about science and in particular chemistry in their introductory courses. This leads them to make the decision whether to major in these fields, whether to take further courses, and whether it is important to be literate on science issues. It is common to hear from students that most chemistry classes are dull and contain little relevant and interesting material. These dull courses, consisting mainly of lectures and *canned* labs, keep students passive and detached from interaction, while the instructor presses on at a fast pace for the sake of "coverage." Needless to say, this teaching model slams the door on positive attitudes toward chemistry. We should not be surprised that students find that chemistry courses consist of a series of abstractions which appear to them to have no connection with their daily lives or even any application outside of the classroom. The first experience of learning chemistry is often one of frustration and failure. Courses labeled *introductory* turn out to be *terminal.* Students encounter barriers which engender conflicting perceptions of chemistry, and other scientific disciplines. Consequently, early in their college years they are lost from the human resource pipeline that leads to scientific careers. Ironically, science instruction is not consistent with the way scientists study the natural world.

The retention and graduation rates of chemistry majors are much lower than in other subjects. Barely 4% of the chemistry majors entering introductory chemistry courses in the United States actually graduate with a major in chemistry [5].

In a nutshell, the introductory courses are taught in colleges and universities in a teacher-centered instructional strategy as follows: The teacher lectures and students are expected to attend, listen passively while taking lecture notes, and are given assignments based on the material presented during the classroom. Usually, the instructor lectures on the concepts covered in the textbook, which seem to cover "almost everything" the students need. This is followed by a laboratory experiment, often taught by a teaching assistant, that either verifies the concepts that the students were lectured on, or is completely unrelated to (detached from) the lecture.

Throughout history, team work has been important for the progress of society. However, the teaching methodology utilized in chemistry courses does not foster team work and deters the full realization of students' future work as professionals [6]. The same holds true in the laboratory. Experiments are relegated to the function of verifying what is taught in the lectures; hence they become mere adjuncts to them, and quite frequently lectures and laboratory exercises have no relationship whatsoever. Furthermore, "*canned*" experiments performed in the laboratory are intended to yield results that are already known and are assigned to the incoming students every year. Students go through these experiments perfunctorily to achieve these preordained results and are rewarded with a good grade. Students are simply procedurally bound, going through the motions without reflecting on their meaning. The heuristic helps the learner not to become caught up in procedures without understanding them. The grades assigned, therefore, do not take into consideration the students' skills and depth of observations. In essence, students are like cooks, who learn and follow instructions from recipes.

12.2 Inquiry-Based Instruction

The key features of our approach in introductory courses include an emphasis on processes and general concepts as well as real problem solving in laboratory experiments that do not have a *"right answer."* The primary goal is to involve students in the discovery of chemical principles from observations and data and to engage them in scientific ways of thinking. In order to do so, we seek to expose students to the thinking and qualitative reasoning processes, by which chemists make observations, organize data, develop principles, make predictions, and design experiments. By emphasizing qualitative reasoning rather than memorizing the mathematical algorithm, we encourage students to visualize that chemistry is different from what they perceive [7].

With the *inquiry* method of instruction, students are led to understanding concepts for themselves, and the responsibility for learning rests with them. The teacher acts as the catalyst, directing students' interaction, activities, and discussions rather than serving as the source of all information. *Inquiry* does not necessarily lead students to discover information unknown to anyone before; it simply allows students to discover information new to them.

In our inquiry approach, the laboratory is used to discover concepts instead of verifying them; it is also known as the *inductive approach*. Students learn not only manipulative skills but also how to work the way scientists do, reasoning from a specific example to a generalization. This is in contrast to the more common deductive instructional laboratory which requires the student to reason from a known concept to a specific example to verify the principle. It is more productive to use the inductive laboratory approach as a discovery experience in which students develop understanding of key chemical concepts from the experiment. This approach encourages scientific thinking in which the emphasis is on *what does the data mean?* rather than on *chemically correct* deductions. This process reflects scientific research. The important aspect is how students reached these reasoning processes to get to the final result. Here lies the difference between a scientist and a technician. As science educators, it is more important to prepare "scientists" than "technicians."

> ... a "scientist" is a seeker of new knowledge about reality. A scientist will check old accepted theories to see if someone might have missed something or actually have been wrong. A "technician" is more interested in restating the old knowledge and being an authority on it. Technicians are reluctant to question existing theories for fear they might be wrong [8].

In inquiry-based teaching methodology, the professor acts like a choreographer, to enliven facts, not merely as the source of information. Emphasis is placed on questions raised by the professor, to guide and promote reasoning and interaction in the classroom, thus promoting the skills to become a "scientist."

Scientific inquiry refers to the diverse ways in which scientists study the natural world and propose explanations based on the evidence derived from their work. Inquiry also refers to the activities of students in which they develop knowledge and understanding of scientific ideas, as well as an understanding of how scientists study the natural world [9].

The following section discusses and explains why the learning cycle approach [10, 11], that is, inquiry-based, student-centered instruction, is the most effective model for learning in the introductory chemistry courses.

12.3
The Learning Cycle and the Inquiry-Based Model for Teaching and Learning

Any model of teaching and learning must consider ways in which students acquire knowledge. *Constructivism* is an epistemological theory that enjoys widespread acceptance among behavioral and natural scientists [12]. It deals with the nature and origin of knowledge, recognizing that students actually construct their own knowledge as they interpret new information and reconstruct what they already know. Cognitive interests provide a framework to consider the purposes behind curricular processes. The inquiry-based instruction model integrates both of these aspects: active involvement of students in building new knowledge (constructivism), and demonstration of the use of the scientific concepts and relationships they have constructed based on the framework (cognitive interests). The two together not only promise more effective modes of instruction, they are also a powerful medium for students to monitor their learning relative to content and process knowledge.

Inquiry-based instructional strategy is a student-centered approach which uses the *learning cycle* [11, 13]. The learning cycle is a three-stages methodology: exploration (E), invention (I), and application (A). In the exploration stage, students collect data to invent (derive) the concept, and finally, in the application stage they apply the concept to other situations. In contrast to the traditional teaching strategy, inquiry-based approach involves active participation of students in all phases (E, I, A). Laboratory activities play a fundamental role by exploring and introducing concepts rather than verifying them [11]. Classroom discussions are focused on using data to derive concepts rather than informing students of the concepts. In addition, textual materials are used to apply, reinforce, review, and extend concepts rather than introduce them. This approach encourages more active, deeper and constructive learning by students [14].

Jean Piaget's theory suggests that human beings have mental structures and sensorial mechanisms that interact with the environment [15]. The interaction in the exploration and the concept invention stages help students to derive a concept from the classroom data, which allows them to accommodate that information into their existing mental structures. Finally, during the application stage, students organize and reinforce the concepts they have just derived with concepts already existing in their mental structures. Activities that allow the student to understand the relationship between that which is already known and newly acquired information reinforce preexisting knowledge. At times, due to an existing misconception, the new influx of information leads to a state of confusion. Equilibrium is once again achieved when the preexisting misconception is revised once the discrepancy between what we already know and the newly received information is resolved.

There is a direct relationship between the learning cycle, the inquiry-based strategy, and Piaget's theory of cognitive development, where the E component (exploration) is assimilation, I (invention) is accommodation, and A (application) is organization. It works because the sequence of instruction in inquiry is consistent with how students learn according to a constructivist model like Piaget's cognitive model [14].

According to Alex Johnstone [4], humans "... *have a filtration system that enables us to ignore a large part of sensory information and focus upon what we consider to matter.*" Filtering information is positive for humans. One would be overwhelmed, and possibly go mad, if one recalls all the experiences lived. Our filter helps us recall that which is more important to us, makes better sense, and is vital.

New situations or pieces of information go through our perception filter and interconnect with the knowledge we already possess in our long-term memory. The previously existing knowledge allows us to recognize and categorize meaning to any instruction, event, or observation. When the newly obtained knowledge bonds with the already stored information and builds upon it, it is easily recalled, thus allowing the process of learning to be more efficient and to be applied subsequently to new situations. Johnstone [4] and Baddeley [16] show that in our brain we have a working space, where interpretation, rearrangement, comparisons of information take place, and a long-term storage system (memory) which stores the previously acquired knowledge. Each event or piece of information passes through the filter which first goes to the working space and then goes on further to the long-term memory. Constant interaction occurs among the working space, the long-term memory, and the filter. Whether the information is interesting, boring, useful, or mind-numbing is determined by the filter and the information stored in the long-term memory.

According to *constructivism*, our responses are based on our prior experiences and what we already know [15, 17]. We usually cannot interconnect and thus reject information, whether new or already known, if it is presented in a manner different from the previously stored information [4]. As an example, our first glimpse in Figure 12.1a,b show an old woman and a young princess with a crown, respectively. However, it is the same figure obtained by turning it around. We can recognize them by recalling the information found in our long-term memory. Said differently, our cognitive system connects with our existing knowledge.

If our long-term memory were not to have previous information about the appearance of a young princess or an old woman, these images would not mean

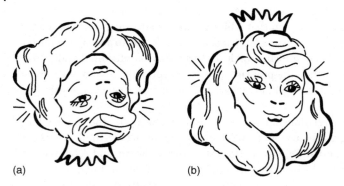

Figure 12.1 (a) An old woman. (b) A princess with a crown.

anything to us. These examples illustrate important aspects of the teaching learning process. As stated by Lamba [18]

> Similar situations happen in our classrooms when information (for example, a chemical formula) is presented to students in a manner different than previously seen or learned. In many introductory texts the formula of acetic acid is presented in different forms: $HC_2H_3O_2$, $C_2H_4O_2$, CH_3CO_2, H_3CCO_2H, among others. For a novice it is not easy to recognize that all of these represent the same species. The meaning of these representations is supplemented by what has already been learned. However, it is assumed that since the information had been provided earlier, the students must have retained it and are able to build upon it.

An analogous situation occurs when students who have never seen or used basic laboratory glassware and equipment take a chemistry course for the first time. Invariably, the first "experiment" in an introductory course is *Basic Laboratory Techniques,* which introduces the use and handling of common glassware, basic equipment, lab techniques, guidelines for writing a lab report, significant figures, and safety rules, among others. The amount of information provided by the instructor in this experiment is overwhelming. The use and handling of the basic glassware and equipment component in itself is breathtaking. For example, the names and usage of Florence flask, Erlenmeyer flask, pneumatic trough, collecting bottle, evaporating dish, short-stem and long-stem funnels, beakers (of different sizes), test tubes, test tube rack, forceps, spatula, pipet, burette, burette clamp, graduated cylinders, utility clamps, wire gauze, metal ring, lab balance, and burner, among others, are only about 60% of the basic glassware the students are supposed to be familiar with in the introductory courses. When the instructor presents more complex tasks during the same lab period, such as the use of the burner to get maximum heat (the amount of air–gas mixture) for the hottest flame, decanting and filtering techniques, folding of the filter paper, use of the glass rod during filtering, use of balance-taring, working with the glass tubing and cutting glass tubing using a file, polishing the edges, inserting the tubing into a rubber stopper, and so on, the result is overload. There are about five pages of diagrams for the glassware and basic equipment and over 25 different instructions on the laboratory techniques. There is a limit to what a human brain is able to process and retain in order to store and then later assimilate new information in a certain period of time. When we try to conceptualize or memorize too much information, invariably one gets a headache and is able to retain only a part of it [4].

Most freshmen have little or no previous knowledge of the names and the specific use of most of this glassware and equipment. Introducing safety rules, writing lab reports using significant figures, and performing other complex tasks at the same time as the instructor explains the use of equipment add to the confusion. Over time, as the students use and manipulate different types of glassware and equipment frequently, they gradually acquire deeper understanding of their effective use and become more familiar with their names. Once the names and the usage of the glassware and basic laboratory equipment are stored in their long-term memory, they are able to recall while trying to follow a new lab procedure. Each new experiment brings its own information barrage and, since experiments are seldom repeated, experience plays little part in reducing this load.

This is comparable to the experience that we encounter in our daily lives when we attempt to follow instructions for using different features of a new mobile phone. If we repeat the process many times, the operation may become easy for that feature. But if one is confronted with several new mobile operations at once, one feels frustrated, like a student, a first time learner confronting all the difficulties involved. This is very much the situation in which students find themselves in laboratories.

The overload of information in introductory courses, lectures, and labs is a common problem [19]. A typical situation in a 3 hour laboratory experiment from a traditional lab manual proves the point. For example, the amount of the information to be processed by a student for complete understanding in a common experiment amounts to over 15 ideas from only 11 lines of font 8 out of a 2-page experiment:

> Select two regular test tubes; when filled with distilled water, they should appear to have identical color when you view them down the tubes against a white background. Draw 50 mL of each of the following solutions into clean, dry, 100-mL beakers, one solution to a beaker: $4M$ acetone, $1M$ HCl, $0.005M$ I_2. Cover each beaker with a watch glass. With your graduated cylinder, measure out 10.0 mL of the $4M$ acetone solution and pour it into a clean 125-mL Erlenmeyer flask. Then measure out 10.0 mL $1M$ HCl and add that to the acetone in the flask. Add 20.0 mL distilled H_2O to the flask. Drain the graduated cylinder, shaking out any excess water, and then use the cylinder to measure out 10.0 mL 0.005 M I_2 solution. Be careful not to spill the Iodine solution on your hand or clothes.

Under these circumstances, it is not surprising that students understandably determine to use the manual as a recipe book.

12.4
Information Processing Model

Johnstone's information processing model [4] can be summarized in layman's terms in three steps: encoding, storage, and retrieval. Let's recall what happened when we saw Figure 12.1a,b. Our brain perceived it as a unique sensory stimulus, and thus kept it as something worthy of being retained. Since we see so many faces daily, our filtration system processes and omits most common faces which are not important or do not mean anything to us and only recalls (retrieves) a princess or an old lady. When we see someone often or someone unique, our brain retains that face and recalls it from our long-term memory.

If you concentrate on something and pay more attention (for example, while watching an interesting thriller movie), you may stop paying attention to people sitting around you as they become less noticeable. If a particular scene in this movie is really impacting, and you are paying more attention to it, you will probably remember it for the rest of your life, while the rest of the movie gets filtered out as details. It may be likened to experiencing the *aurora borealis*: once seen, you never forget its initial thrill!

The more novel an object and the greater your attention, the stronger and more specifically it is imprinted in your memory. You move from simply encoding that you are watching this movie, to the scene that had the greatest impact on you, and then to the specific details in this scene. That is why some performers sing with dark glasses and noise-canceling headsets in order to be able to concentrate and recall. Similarly, some people close their eyes while expressing their ideas as they try to concentrate and retrieve specific information from their long-term memory. As indicated earlier, constant interaction among the encoding (working space), the long-term memory (storage), and the filter. The more our brain can focus on a particular stimulus, the greater will be its encoding and eventual retrieval.

12.5
Possible Solution

We as chemistry instructors firmly believe that the role of the laboratory is very important in the formation of scientists and of chemists in particular. It is common to hear statements like, *Chemistry is an experimental science, and therefore students must have hands-on experience.* On the other hand, as indicated earlier, instructors in most colleges and universities have students follow recipes. Students can gain manipulative skills through these *canned* labs, but a valuable opportunity to stimulate creative thought is lost when these skills are acquired by following mindlessly the laboratory instructions in a manual. The predominantly rote-mode teaching practices and algorithmic calisthenics practiced in the traditional learning have put "braces on the brains" of many students. Even excellent teaching does not guarantee learning of concepts, especially if the learner is not actively involved at the appropriate cognitive level.

Based on the author's own experiences and with the aid of a National Science Foundation Division of Undergraduate Education grant (DUE 9354432) and the U.S. Department of Education-Minority Science and Engineering Improvement Program (P120A30018) [7], we identified the shortcomings of the traditional teaching of chemistry and decided to make radical changes and develop a new and innovative approach. We decided that the curriculum would be inquiry-based and laboratory-centered, in which the students' experiences would reflect the way that the scientific process actually occurs. In other words, instead of traditional recipes, concepts should be introduced in the laboratory and then discussed after the laboratory activity is completed.

In this new approach, rote memorization and algorithmic manipulation of the concepts are replaced, to a great extent, by conceptual development. The nature of this instructional paradigm requires more varied and sophisticated tools to promote the learning process.

Student learning in the inquiry-based laboratory-driven curricula differs significantly from that of traditional lecture-driven courses. Rote memorization and algorithmic manipulation are replaced, to a significant extent, by concept development. Students express their understanding of chemical phenomena in multiple valid ways. The nature of the instructional model requires more varied and more sophisticated experimental paradigms.

To illustrate the innovative student-centered laboratory-driven program, it is necessary to briefly describe the nature of these introductory courses and how they differ from traditional ones. The essential elements of our approach, which have been described before [7], are as follows:

1) The laboratory is the central focus of the course, where concepts are introduced inductively.
2) The laboratory activities have an investigative approach in which students develop scientific methods of inquiry. This is the opposite of the traditional recipe-type approach, in which students are given instructions on how to execute each procedure to obtain a predetermined result.
3) The laboratory-based activities promote the development of creative critical reasoning skills. Students learn to pose questions and seek answers rather than blindly accepting explanations.
4) The laboratory provides an effective environment to foster cooperative learning. Students pool and share data using computer technology (if available) to create a sufficient body of information from which they can develop their own explanations that are consistent with their observations.
5) The students are actively engaged and empowered to participate in their own learning, which encourages creativity.

In this approach, the curriculum is not driven by a horizontal arrangement of the concepts, textbook style, but by data produced by the students themselves in the labs. Instead of using textbook materials to introduce concepts, they are used to apply, emphasize, and extend concepts. This approach empowers and stimulates the students in their own learning process.

In general, in the traditional introductory courses the instructors are assigned large numbers of students depending upon the size of the department and the student faculty ratio. For large classes, the students are usually divided into sections of 20–25 for laboratory experiments working in groups of two or more. In our approach, during the laboratory sessions, the students generate and pool data as explained in the next section, while in the post-lab session (classroom) the instructor gathers this combined data to make graphs and tables from several lab groups. Classroom discussions are focused between the peers and the instructor to develop concepts from students' data. The derived concept is consistent with students' observations in the laboratory experiences. The role of the instructor is one of a facilitator of learning, asking probing questions to help guide the students to develop deeper understanding on their own. Textual materials are used to apply, reinforce, review, and extend concepts to explore new situations. In this manner, the inquiry-based student-centered instruction can be effectively implemented in the classroom as well as in the laboratory independent of the size of the group.

12.6
Guided Inquiry Experiments for General Chemistry: Practical Problems and Applications Manual

A guided inquiry-based lab manual for general chemistry [20] has been developed by Kerner and Lamba to use the laboratory as a process experience in which students develop understanding of key chemical concepts from experiments. The manual is designed for two semesters and is divided into 12 units. Each unit has several experiments focusing on different concepts based on the topics for general chemistry. It provides an exciting and meaningful laboratory experience for students by challenging them to solve a practical problem or answer a real-life question (*How Long Will It Take?*, *Why Is Harley Chrome Plated?*) by drawing conclusions from experimental data. Each experiment ends with questions requiring feedback on the solution to the problem in the form of a report. A closing "extension and applications" section emphasizes the practical applications of chemistry by asking questions that require students to apply and extend experimental results to untested systems and real-world situations. The approach uses Piaget's theories and the learning cycle approach. In addition, it utilizes Johnstone's information processing model as described earlier.

This approach encourages scientific thinking in which the emphasis is on asking *what do the data mean?* rather than on making *chemically consistent* deductions. This approach also more closely reflects what occurs in scientific research. After conducting the labs, students improve their communication and inquiry skills, including forming and testing hypotheses, analyzing data, and even designing experiments.

Several examples will illustrate how the laboratory-based materials bridge the students' common experience (as part of a process or as part of content) with the objectives of their investigation. Learning theory indicates the need for students to restructure their prior knowledge via active learning, and thus this process of restructuration/reorganization of knowledge takes place within individual laboratory exercises and between various laboratory sessions.

A typical laboratory session begins with a question being posed to a group of students to which they offer a variety of answers based on their previous experiences. They are stimulated to help generate the answer to the question or to side with one of the answers (hypotheses) which will be put to test. All the students then carry out the same general procedures, though each one is required to run his or her own experiment under a slightly different set of conditions. A computer interface collects the pooled data, which is utilized by the instructor to discuss trends and help the students to discover the concepts. A post-lab session (usually a classroom discussion) assembles the students after the activity. Group results are discussed, which allows them to discover a wide range of new concepts, as well as to answer the question originally posed. This discussion can lead to additional class meetings [7].

The following sections are contained in each experiment: Introduction, Specific Objectives, Procedures, Results and Conclusions, Extensions, and Applications. Furthermore, all experiments contain section(s) on safety precautions, as required by the experiment under study.

Every experiment is designed to provide enough information and guided inquiry directives to achieve successful learning outcomes. Every experiment is featured in the accompanying Teacher's Guide, in which the instructor can find information on the skills used, the basic concepts the experiment intends to develop, the anticipated data, the chemicals and materials needed, ideas to implement in the classroom, and answers to sample questions. It is important to mention that most of the chemicals and materials used are quite inexpensive and many are available in hardware stores, pharmacies, or supermarkets as can be seen in the examples that follow. Overall, the cost involved in conducting these experiments is relatively low compared to those found in common lab manuals.

The first example is based on a guided discovery experiment designed at Holy Cross College [21]. In this experiment, students make a first-hand observation that the compositions of pre- and post-1982 pennies are different. Students make a few notches with a file on both pennies. Then, they drop each penny in a separate beaker containing dilute hydrochloric acid. Most are surprised to observe that the beaker containing the post-1982 penny shows a vigorous evolution of a gas, while the pre-1982 penny does not, and that after a few hours the post-1982 penny is floating while the other is left unreacted at the bottom of the beaker. Through this experiment, students are empowered to discover for themselves that under the same conditions Zn reacts with dilute HCl producing H_2 while Cu does not.

The experiment becomes open-ended as students explore the reaction of HCl with some other metals [7]. Through these experiments, the students discover the need to understand the mole concept, the chemical equations, stoichiometry, and reactions rates. Students work in groups of two and measure quantitatively the amount of hydrogen gas evolved at room temperature when different amounts of Zn (mossy), Mg ribbon, and Al foil react with dilute HCl (1 M for Mg and 4 M for

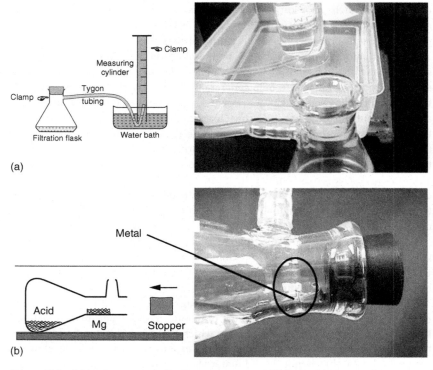

Figure 12.2 (a) Hydrogen gas collection apparatus and (b) flask with the metal.

Zn and Al). Mg ribbon and Al foils are first cleaned with a synthetic scrub sponge to remove any coating. They assemble the apparatus in Figure 12.2a for collecting hydrogen gas using inverted graduated cylinders completely filled with water into the water bath. Tygon tube connected to the filtration flask is carefully inserted into the mouth of the inverted graduated cylinder and held in position with a clamp. Different graduated cylinders of capacity 100–500 ml are used depending upon the volume of gas expected to be produced.

With a graduated cylinder, they pour 25 ml of HCl to the filtration (Erlenmeyer) flask. Then they weigh a known amount of the metal accurate to ±0.01 g and, carefully inclining the flask, place the folded metal into the neck of the flask as shown below Figure 12.2b, and then immediately seal the top of the flask with the rubber stopper before the metal comes in contact with the acid so that no gas is lost. They carefully straighten the flask (Figure 12.2a) and clamp it loosely. The flask is gently swirled – as needed – until all of the metal has reacted. The volume of hydrogen gas collected in the inverted graduated cylinder is recorded in the table. The observations are repeated and recorded using a total of six to eight samples of different known metal mass (0.02–0.20 ± 0.01 g).

As shown in Table 12.1, with pooled data they go on to measure quantitatively the amount of hydrogen gas evolved when different amounts of Zn, Mg, and Al

Table 12.1 Compiled class data: mass (g) of Al, Mg, and Zn versus gas volume (ml).[a]

Mass of metal (g)		Volume of gas (ml)		
		Mg	Zn	Al
1	0	0	0	0
2	0.02	18	8	30
3	0.05	60	20	65
4	0.08	88	35	115
5	0.11	112	45	160
6	0.14	150	54	200
7	0.17	180	70	245
8	0.20	228	79	282

a) The precision of the gas volumes depend upon the capacity of the cylinder.

react with dilute HCl and discover that the ratio of the mass of Zn to Mg required to produce the same amount of hydrogen gas is the same as that of their molar masses. They also observe that it takes a greater mass of Zn than Mg to produce the same amount of hydrogen. The Al results, which require both accounting for the differences in relative molar mass of the metal and the differences in the balancing coefficients in the reaction equation, are more challenging. By working in groups and plotting experimental data in various ways, students figure out how to account for the differences among the metals.

When data for the entire class are combined, it is possible to construct a single graph which enables the students to predict the mass of each metal required to prepare a given volume of gas. The graph consists of three lines which pass through the origin, as shown in Figure 12.3.

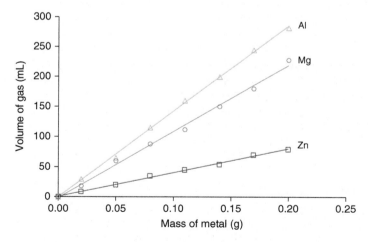

Figure 12.3 Mass of metal (g) versus gas volume (ml).

When the results obtained with zinc and magnesium are compared, the students find that the ratio by mass of the samples of these metals required to prepare a given volume of gas is virtually the same as the ratio of the molar mass of these elements.

$$\frac{\text{Mass Zn}}{\text{Mass Mg}} = \frac{\text{molar mass Zn}}{\text{molar mass Mg}}$$

The implications of this relationship can be probed by rearranging this equation as

$$\frac{\text{Mass Zn}}{\text{molar mass Zn}} = \frac{\text{Mass Mg}}{\text{molar mass Mg}}$$

The results of this experiment therefore suggest that it takes an equivalent number of moles of either zinc or magnesium to prepare a given amount of hydrogen gas. We may therefore assume that the stoichiometry of the two reactions is the same.

$$Zn(s) + 2H^+(g) \rightarrow Zn^{2+}(aq) + H_2(g), \quad Mg(s) + 2H^+(g) \rightarrow Mg^{2+}(aq) + H_2(g)$$

The ratio by mass of magnesium and aluminum needed to prepare a given volume of gas differs from the ratio of their molar mass by a factor of about 1.5 or 3/2. In other words, it takes 1.5 times as many moles of magnesium as moles of aluminum to prepare a given volume of hydrogen gas.

$$\frac{\text{Mass Mg}}{\text{molar mass Mg}} = \frac{3}{2} \times \frac{\text{Mass Al}}{\text{molar mass Al}}$$

This suggests that the stoichiometry of the two reactions must be different. The experimental results can be explained by noting that the balanced equation for the reaction between aluminum and hydrochloric acid consumes a different amount of acid per mole of metal.

$$Mg(s) + 2H^+(g) \rightarrow Mg^{2+}(aq) + H_2(g)$$

$$2Al(s) + 6H^+(g) \rightarrow 2Al^{3+}(aq) + 3H_2(g)$$

By analyzing a plot of the average volume of gas given off by fixed amounts of the different metals, the students can predict the mass of each metal that would be required to produce a given volume of gas.

The experimental results serve as a natural introduction to reaction yields, limiting reaction, and solution stoichiometry [20].

12.7
Assessment of the Guided-Inquiry-Based Laboratories

The "Guided-Inquiry-Based *Laboratories*" approach we have designed has been successfully used by us since 1994. External evaluators provided evidence of this success the first time around by (i) measuring improved student attitudes toward the study of chemistry and (ii) comparing the distribution of grades in the traditional courses with those obtained in inquiry-based laboratories, both of which were linked to the development of greater confidence in the students' ability to perform adequately in such courses [22]. The external evaluators also collected qualitative data found in *Reflective Diaries* completed by the students on a regular basis during the course.

Because the guided-inquiry-based laboratory method integrates lectures and laboratory experiences, both were conducted in the laboratory. It is important to mention that the laboratory component of the course in the guided-inquiry model drives the instruction; students basically earned the same grade for both components of the course: the lectures and the laboratory. In most instances, in the traditional courses different faculty members gave the laboratory component and the lectures, so there is little, if any, integration of both components. Typically, in the traditional format larger numbers of students earn better grades in the laboratory component, as opposed to the lecture component.

For the purposes of comparison, quantitative measures were taken into consideration. The Chi-square analysis of the traditional versus guided-inquiry course showed that there was a significant difference in the grades between the guided-inquiry-based modalities and the traditional course and laboratories. The percentages of A's and B's are much higher in guided inquiry as compared to the traditional course. Consequently, the percentage of failures (D, F, and W) was significantly lower than in the traditional course. Thus, a group of students (D, F, and W) moved toward C, whereas students obtaining grade C moved to A and B in the guided-inquiry course [23].

The use of *Reflective Diaries*, which provided qualitative data related to the course and the laboratory, became an important assessment tool. Designed by the external evaluators, they were analyzed on a monthly basis. Feedback was given to the instructor anonymously. Examples of topics included in the diaries are skills in which students need further practice, concepts easy for the students to learn, difficulties in the laboratories, the more difficult concepts, areas of achievement, recommendations to the professor instructor, and their overall perception of the course. The anonymous nature of the feedback provided the opportunity for instructors to include further discussion of some concepts or provide additional time to emphasize some skills. The following are some quotes written by students in various diaries [23]:

> *The guided inquiry course is wonderful, you learn a lot and you know how things happen and why do they happen. The class is based on experiments and you learn a lot more by observing what you are doing than by listening to a professor.*

> *The system allows for you to obtain knowledge by yourself through the experiments, you do not have to rely only on textbooks.*

> *Allow me to make you an invitation to the world of chemistry. This class is like a window to a world that you could not think exists. It is a way of showing you that chemistry is something that has to do with you in your daily life, even if you had not realized it before. And it is in this course that you realize this. It is a totally progressive method, and you learn the concepts by doing, that are almost impossible to master well unless you take the guided inquiry course.*

Informal interviews held with students taking the traditional course as well as those taking the guided-inquiry approach, as well as with general chemistry faculty members, indicate clearly that students in the traditional chemistry course find it difficult to integrate chemical concepts discussed in the classroom vis-à-vis the phenomenon observed in the laboratory. Students felt overwhelmed when attempting to effectively integrate the information from the lecture to the laboratory, and vice versa. In the guided-inquiry courses, however, the student found the transferring of knowledge and information from course to the laboratory, and vice versa, could be done with relative ease. The basic focus of the course was to be hands-on and minds-on, delegating more responsibility to the students for their own learning. This aspect played an important role when a follow-up study was prepared of their grades in subsequent chemistry courses, such as organic chemistry. A successful guided-inquiry full-year organic chemistry course has been developed and implemented at Washington College [23].

Assessing innovative methods of teaching and learning is always time consuming and thought provoking. On the other hand, it is not easy to quantify the outcomes on a short-term basis. One needs to collect assessment data in each semester in order to determine the gains. The fact that novel methods take a long time to assess should not be an excuse for not pursuing them. The important thing is to develop an assessment component akin to the teaching and learning strategy [24].

12.8
Conclusions

Research and experience demonstrate that we can replace rote memorization and the cook-book type features of the introductory chemistry courses by engaging students as part of the process in discovering concepts. When students have active participation in the classroom and the laboratory component using hands-on and minds-on strategies, the outcomes are better assimilation of knowledge from the lecture and the laboratory, and vice versa. The learning cycle approach is an inquiry-based instructional strategy that has great promise for chemistry instruction. This approach gives an instructor the opportunity to expose students to different modes of learning. In the inquiry-based instructional strategy, the concepts are usually introduced in a laboratory setting or some other source of data from which the concept can be developed through discussion, usually in a classroom. Students' attitudes toward chemistry through this approach are positive and they develop scientific methods of analysis. The inquiry-based laboratory activities help the students to acquire creative and critical reasoning skills. There is greater emphasis on cooperative learning and team approach. Furthermore, our follow-up study of students in the organic chemistry course shows that students achieve greater accomplishment and better retention of concepts, more self-confidence, improved attitudes toward chemistry, improved reasoning ability, and superior process and research skills than with traditional instructional approaches. They also obtain higher grades and better retention rates.

Using such strategies, more students are retained in the human resource pipeline which leads to scientific careers. Science instruction becomes consistent with the way scientists study the natural world. The whole process becomes "student-centered" rather than "teacher-centered." As faculty members, it is our responsibility to look for teaching models that can be more effective in achieving student learning. Every new model involves a lot more work in the beginning. However, in the long run, students' achievements and personal satisfaction should be the key factors motivating change.

References

1. Wirth, K.R. and Perkins, D. Curricular Materials, *http://www.macalester.edu/geology/wirth/CourseMaterials.html* (accessed 15 October 2013).
2. Ward, R.J. and Bodner, G.M. (1993) How can lectures undermine the motivation of our students? *J. Chem. Educ.*, **70**, 198–199.
3. Zoller, U. (1993) Are lectures and learning compatible may be: for LOC's unlikely for HOC's. *J. Chem. Educ.*, **70**, 195–197.
4. Johnstone, A.H. (1997) Chemistry teaching – science or alchemy. *J. Chem. Educ.*, **74**, 262–268.
5. Bopegadera, A.M.R.P. (2011) Putting laboratory at the center of teaching chemistry. *J. Chem. Educ.*, **88**, 443–448.
6. Halpern, D.F. (ed.) (1994) *In Changing College Classrooms: New Teaching and Learning Strategies for an Increasingly Complex World*, Jossey Bass, San Francisco, CA.
7. Lamba, R.S. (1994) Laboratory-driven instruction in chemistry. *J. Chem. Educ.*, **71**, 1073–1074.
8. Ryokan *http://www.physicsforums.com* (accessed 8 November 2013).
9. National Academic Press (1995) National Science Education Standards, p. 23.
10. Lawson, A.E., Abraham, M.R., and Renner, L.W. (1989) *A Theory of Instruction: Using the Learning Cycle to Teach Science Concepts and Thinking Skills*, Monograph Number One, National Association for Research in Science Teaching, Kansas State University, Manhattan, KS.
11. Lawson, A.E. (1995) *Science Teaching and the Development of Thinking*, Wadsworth Publishing Company, Belmont, CA.
12. Pienta, N.J., Cooper, M.M., and Greenbowe, T.J. (eds) (2005) *Chemists' Guide to Effective Teaching*, Prentice Hall, Upper Saddle River, NJ, pp. 41–52.
13. Marek, E.A. and Cavallo, A.M.L. (1997) *The Learning Cycle: Elementary School Science and Beyond*, Heinemann, Portsmouth, NH.
14. Abraham, M.R. (2011) What can be learned from laboratory activities? Revisiting 32 years of research. *J. Chem. Educ.*, **88**, 1020–1025.
15. Piaget, J. (1952) *The Origins of Intelligence in Children*, International Universities Press, Inc., New York.

16. Baddeley, A. (1986) *Working Memory*, Oxford Psychology Series, Number II, Oxford University Press, Oxford.
17. Bodner, G.M. (1986) Constructivism: a theory of knowledge. *J. Chem. Educ.*, **63**, 873–878.
18. Lamba, R.S. (2008) in *Process Oriented Guided Inquiry Learning (POGIL)* (eds R. Moog et al.), American Chemical Society, Washington, DC, pp. 26–39.
19. Johnstone, A.H. and Letton, K.M. (1991) Practical measures for practical work. *Educ. Chem.*, **28**, 81–83.
20. Kerner, N.K. and Lamba, R.S. (2008) *Inquiry Experiments for General Chemistry: Practical Problems and Applications*, John Wiley & Sons, Inc., Hoboken.
21. Ricci, R.W. and Ditzler, M.A. (1991) Discovery chemistry: a laboratory-centered approach to testing general chemistry. *J. Chem. Educ.*, **68**, 228–232.
22. Spencer, J.N. (2006) New approaches to chemistry teaching: 2005 George C. Pimentel Award. *J. Chem. Educ.*, **83**, 528–533.
23. Lamba, R.S. and Creegan, F.J. (2008) in *Process Oriented Guided Inquiry Learning (POGIL)* (eds R. Moog et al.), American Chemical Society, Washington, DC, pp. 186–199.
24. Holme, T. (2011) Assessment data and decision making in teaching. *J. Chem. Educ.*, **88**, 1017.

13
Flipping the Chemistry Classroom with Peer Instruction
Julie Schell and Eric Mazur

13.1
Introduction

Universities are hotbeds of innovation. Since the advent of our first institutions of higher learning, millions of inventions borne on campuses have radically transformed our lives. However, the innovators at the core of our institutions of higher learning have done too little to transform the most fundamental activities of the university – teaching and learning. Until recently, you could safely assume that behind most doors of any university classroom anywhere in the world, a professor would be standing alone, lecturing to rows and rows of students. The classroom you would observe would be structurally identical to those constructed in Bologna. This amphitheater-inspired structure symbolizes the vestiges of the antiquated view that formal education means listening to an expert for a set period of time, writing down his words of wisdom, and reviewing them in the evening. In the new millennium, this assumption would be riskier: indeed, experts in educational change suggest that we are on the precipice of a golden age in higher education [1]. The resounding calls for something different and better – among students, governmental leaders, educational researchers, and especially scientists – that have gone unanswered for more than a century are finally starting to gain traction. For the first time in modern history, pedagogy in higher education is experiencing widespread disruption with the simplicity of putting learners first.

The flipped classroom is one such educational disruption. In the first section of this chapter, we introduce the concept of the flipped classroom, offer three big ideas about flipped classrooms, connect the flipped classroom method to the larger taxonomy of blended learning, briefly review the history of the concept, confront common myths about the flipped classroom based on our conversations with instructors across the globe, and provide a protocol for flipping any classroom. Where appropriate, we have inserted case examples from chemical education.

In the second section, we introduce Peer Instruction as one research-based pedagogy chemistry educators have used to successfully flip their classrooms. We define Peer Instruction and its key features, offer examples of Peer Instruction

Chemistry Education: Best Practices, Opportunities and Trends, First Edition.
Edited by Javier García-Martínez and Elena Serrano-Torregrosa.
© 2015 Wiley-VCH Verlag GmbH & Co. KGaA. Published 2015 by Wiley-VCH Verlag GmbH & Co. KGaA.

in chemistry, and conclude with strategies for avoiding common Peer Instruction pitfalls. The purpose of this chapter is twofold: We aim to provide chemistry educators (i) a basic framework for flipping the classroom and (ii) a more specific framework for flipping the chemistry classroom with Peer Instruction.

13.2
What Is the Flipped Classroom?

Discussions of flipped classrooms among educators, researchers, and the media are often based on the assumption that everyone agrees on what they are. However, the concept of a flipped classroom suffers from a lack of definitional clarity, even among experienced flippers. Flipped classroom experts and chemistry educators Jonathan Bergmann and Aaron Sams [2] offer the most universal definition: "Basically the concept of a flipped class is this: that which is traditionally done in class is now done at home, and that which is traditionally done as homework is now completed in class" (p. 12). They go further in their definition to add the caveat that viewing the method as one where students watch lectures out of class and do homework in class is oversimplified. Bergmann and Sams are considered the "most important populizers" [3] of the modern flipped classroom movement. They define flipping as more of a mindset than a method: "Flipping the classroom is more about a mindset: redirecting attention away from the teacher and putting attention on the learner and the learning" [2].

When instructors embrace this mindset, flipped classrooms are examples of student-centered pedagogies. A student-centered pedagogy is one where students take an active role in their own learning process, versus serving as receptacles of knowledge who receive direct instruction from a teacher. However, not all flipped classrooms are equally student-centered. If an instructor simply flips her class by asking students to watch her delivery of a 60-min lecture before class and then observes them completing homework problems from the back of the book during class time, she is not shifting responsibility to the student; she is simply inverted what typically happens in and out of class.

The instructional design inherent in a flipped classroom is often described as one that leverages advances in education technology to deliver lectures online for asynchronous viewing outside of class. During synchronous class time, students and instructors engage in more interactive activities such as problem solving or discussion [4, 5]. Even more specific, Derek Bruff [6], director of the Vanderbilt University Center for Teaching, defines the flipped classroom as "a teaching approach in which students get a first exposure to course content before class through readings or videos then spend class time deepening their understand[ing] of that content through active learning exercises" (para 4). However, when social science and humanities educators hear such definitions of the flipped classroom, they often wonder, "How is this different than reading Hamlet before you come to class and then discussing the theme of melancholy in depth?"

For the purposes of this chapter, we define an efficacious flipped classroom as a three-step process that shifts the traditional model from an instructor-centered to a student-centered learning environment. Operationally, in this flipped classroom process there are three phases, although the third phase is almost always underemphasized in published materials:

1) *Before class*: students gain first exposure to content via text, video, or other transmissionist (i.e., direct instruction) activity. Before-class activities can take many shapes, for example, a reading from a text book, a video the instructor creates, or a video the instructor finds online. The essential design component for this phase is to select activities that will direct students to enact the kind of thinking about the subject-matter that they need in order to complete more complex cognitive tasks and to develop deeper subject-matter knowledge. This phase should not be considered homework.

2) *During class*: students spend class-time learning in community through experiential activities and frequent feedback. The essential design component for this phase is to select activities that will require students to engage in higher order cognitive tasks such as application, analysis, and transfer of their knowledge to new contexts.

3) *After class*: students engage in self-directed learning based on feedback garnered from peers and instructors; instructors use feedback to plan additional learning activities. This phase is often ignored in flipped classroom literature. The essential design component for this phase is to help direct students' practice to the areas they need the most help and to solidify their understanding. Homework is an example of an after-class activity that can help students continue learning in a flipped-class process.

To see a video definition of the flipped classroom we adopt, see *What is a Flipped Classroom*, in 60 s (*http://www.youtube.com/watch?v=r2b7GeuqkPc*).

13.2.1
Three Big Ideas about Flipped Classrooms

In our view, there are three big ideas, tied to the phases above, educators need to embrace when adopting a flipped classroom.

> *Big Idea 1: Prior knowledge is required to scaffold deeper learning.* A main rationale for flipped classrooms is that they help strengthen students' prior knowledge before deeper learning activities. Successful flipped classrooms require facilitating students' development of surface-level prior knowledge *before* they attend class so that they are prepared for the more rigorous work of experiential and deep learning during class time. The most promising strategy is to design your course so that students complete the easier cognitive work involved in surface learning through transmissionist[1]

---

1) Referring to the transmission of information from instructor to student, as in lecture, or the transmission of information from a book to a student, as in reading.

activities, such as listening and note-taking, individually, at their own pace. The prior knowledge they develop will serve as a scaffold for the more demanding cognitive work of deep learning [7].

Big Idea 2: People learn best when they are engaged. The second big idea is that most people learn best when they are provided with opportunities for social and/or experiential learning [8, 9]. In his seminal theoretical work on college student's academic development, Astin [10] emphasized that the "greater the student's involvement in college, the great will be the amount of student learning" (p. 529). Such involvement includes intellectual interactions with peers and faculty. In a flipped classroom, in-class interactions serve a number of purposes, including activating prior learning and capitalizing on how people learn best.

Big Idea 3: Flipped classrooms enable a sustained learning path. The third big idea is that the learning path does not cease after the class ends; it is extended outside of class, as students engage in self-regulated learning activities. Instead of studying all material *en masse*, students follow the signals they receive through in-class feedback to direct their practice in the areas where they need the most work. Thus, in each phase of the flipped class process, the learning shifts from instructor-centered teaching to student-centered learning. In this way, the flipped classroom transforms definitions of what education is into what education can be.

13.2.2
Blended Learning and Flipped Classrooms

Pedagogically, flipped classrooms fall under the larger taxonomy of blended learning [11]. Leading blended learning researchers, Staker and Horn [11] define blended learning as follows:

> A formal education program in which a student learns at least part through online delivery of content and instruction with some element of student control over time, place, path, and/or pace and at least in part at a supervised brick-and-mortar location away from home. (p. 3)

The word *formal* is important to this definition because it differentiates blended learning from learning activities that students might do on their own, such as running the online PhET molecular model simulation [12]. Blended learning is a hybrid innovation in education: a merging of traditional approaches with innovative ones, rather than simply replacing the traditional approach altogether [13]. For example, just as a hybrid vehicle offers options for gas (traditional approach) and electric charging (innovative approach), a blended learning model too attempts to pool the benefits of traditional classrooms with those of more innovative methods.

13.2.3
A Brief History of the Flipped Classroom

The concept of the flipped classroom is not new, even though the term is. In 2000, Lage, Platt, and Treglia published "the inverted classroom," which described a group of economists' efforts to improve student learning by providing videotaped lectures for viewing outside of class; in-class, the instructors facilitated discussion of the preclass lectures, followed by an economics experiment or lab. Lage et al. [14] defined the inverted classroom as one where "events...traditionally taken place inside the classroom now take place outside the classroom and vice versa...[where] students view lectures either in computer labs or at home, whereas homework assignments can be done in class, in groups" (p. 32).

Ten years earlier, the second author of this chapter began experimenting with the idea of pushing transmission activities outside the class to open up time for more interactivity in his undergraduate physics course at Harvard University. Contrasted with most science courses at the time, Mazur's course design required students to engage in computer-assisted learning before class and, subsequently, incorporated Just-in-Time Teaching (JiTT) [15]. Using JiTT, Mazur required students to read the text and answer an online assignment before class, with the intention to help students build prior knowledge and to set the stage for more in-depth conceptual thinking during class time (see [16]). This combined method, JiTT with Peer Instruction, will be detailed in Sections 13.3.1 and 13.3.3.

The basic idea of using a flipped classroom as a way to scaffold deeper learning dates even earlier, to the 1890s. At that time, the dean of the Harvard Law School, Christopher Columbus Langdell, developed the case study method, whereby law students prepared for class by reading cases and then engaged in Socratic dialog during class. The case study method has been universally adopted by law schools and business schools; we consider it to be the first and last sustained pedagogical innovation to reach such scale in higher education.

13.2.4
Traditional versus a Flipped Chemistry Classroom

A flipped classroom is an example of a blended learning model that forms a hybrid between traditional and innovative teaching approaches. In the traditional approach in chemistry, students usually get a first exposure to a concept during class time, not before class. For example, if the concept for the day is thermal expansion, in a traditional classroom, students will come to class having no recent prior exposure to that concept (although they may have learned about it in high school) and thus buried or no prior knowledge. In this scenario, the instructor provides a lecture during class time on thermal expansion; students listen, take notes, and the bravest might ask questions. At home and on their own, students continue the learning path by recovering the material transmitted during class, completing problem sets, and receiving feedback on problem sets several days later. Students might revisit the concept of thermal expansion several weeks later in preparation for an exam.

Alternatively, in a flipped chemistry classroom, students gain first exposure to the basic concept of thermal expansion before they come to class. This first exposure can occur in a number of forms; the most common, but not only form, is a video lecture or a screencast recorded by the instructor or by a third party. A screencast is a recording of all activities that occur on a computer screen, including audio, once the user presses record. Instructors can use screencasting in a variety of ways, such as using the computer's built-in camera to record a lecture, or from an audio walkthrough of a PowerPoint presentation, or by using additional software and hardware to record handwork and problem solving. Kahn Academy videos are examples of screencasts, with additional software added to record handwork, see the following as an example from Chemistry: *http://www.khanacademy.org/science/chemistry*.

During a flipped chemistry class, because students have already developed some prior knowledge of the underlying concepts, the instructor can execute deeper learning activities, such as group problem solving, in an effort to engage students in more intense cognitive work. This guided application of prior knowledge provides opportunities for students to gain immediate, expert feedback in the more difficult stages of learning. For example, instead of doing problem sets about thermal expansion alone and receiving delayed feedback on their understanding after grading, students might work on those problem sets with each other and their instructor during class, thus increasing the opportunity for frequent immediate signals about their learning states, that is, a snapshot of students' knowledge and the distance they need to travel to gain mastery, in that moment in time.

After a flipped chemistry class, students can use the immediate feedback gained in class to continue their learning with further reading, additional videos, or additional problem solving on their own. This basic implementation of a flipped classroom is outlined in Figure 13.1.

13.2.5
Flipped Classrooms and Dependency on Technology

While the above discussions of blended learning and flipped classroom examples emphasize online learning, flipped classrooms do not require online components. A common phrase in the flipped learning community is that flipping is about pedagogy not about technology. Indeed, in the most basic implementation of a flipped classroom, which Bergmann and Sams [2] describe as Flipped 101, teachers would assign students a book chapter on thermal expansion before coming to class and then ask the students to complete the problems at the end of the chapter during class time using pen and paper, and then assign further problems for out-of-class work. Designing ways to flip chemistry classrooms without technology similar to the Flipped 101 model is critical for students for whom technology requirements present a barrier to access high-quality learning.

	1. Before class	2. In-class	3. After class
Flipped Chemistry classroom	Students gain first exposure to concept of thermal expansion through a reading or a video providing direct instruction or content coverage.	Instructor delivers mini-lecture on thermal expansion recovering the biggest ideas and then provides students with problems that help them apply the concept. Students work in groups, and instructor and course staff rotate throughout the room to provide immediate feedback.	Students use feedback from in-class time to point to areas for further study about thermal expansion and complete an online quiz with isomorphic questions to receive another round of feedback.
Traditional Chemistry classroom	-	Students gain first exposure to the concept of thermal expansion through a 50-minute lecture.	Students complete problem sets about thermal expansion at home.

Figure 13.1 Example of basic implementation of a flipped classroom lesson on thermal expansion.

Technology enhances a flipped chemistry classroom best when it is used to facilitate pedagogy in ways that would not otherwise be possible. For example, experienced flipped classroom instructors do not make lecture videos and post them online simply because the technology exists to do so. Done well, online lecture videos can serve multiple pedagogical purposes. First and foremost, online lecture videos provide a channel for instructors to contribute interesting and motiving lectures for students around the most difficult concepts, *and* to move them outside the class so that students can do the easier cognitive work of listening and note-taking on their own, asynchronously, and spend class time on the more difficult cognitive work of application with their peers and instructor.

13.3
How to Flip the Chemistry Classroom

While there are differing opinions on what a flipped classroom is in educational communities, there is wide consensus that there is no one right way to flip a classroom [2]. In this section, we begin by addressing the most common myths about how to teach within a flipped classroom based on conversations we have had with teachers across the globe. We will then offer a general protocol for flipping any concept in chemistry, and conclude with a brief discussions of students' reactions to flipped classrooms.

13.3.1
Common Myths about Flipped Classrooms

13.3.1.1 Myth 1: Flipped Classrooms are Just Video Lectures

The most pervasive myth about flipped classrooms we hear in our conversations with instructors is that flipped classrooms are primarily about creating lecture videos. As noted previously, lecture videos are not a defining characteristic of flipped classrooms; again, the most important feature of a flipped classroom is to provide students opportunities to develop prior knowledge so that they can take full advantage of opportunities for deeper learning during class. Many teachers flip their classroom without any lecture videos at all. See Figure 13.2 for a case example of out-of-class work for a chemistry classroom that does not rely on lecture videos. That said, lecture videos are a popular way to help build students' prior knowledge. See Figure 13.3 for a case study example of how instructors at The University of Texas at Austin flip their chemistry classroom with videos.

13.3.1.2 Myth 2: Flipped Classrooms Have No Lectures

We speak with many instructors who fear that flipping their classroom means they can never deliver a live lecture again. This is not true; in fact, because students are so socialized to believe they learn through lectures in chemistry, we suggest

---

Case example

Just-in-Time Teaching (JiTT) is a research-based method for motivating students to prepare for class by doing warm-up activities online [15]. The most common implementation of a warm-up is to have students complete a reading assignment and then answer at least one question to test students' conceptual understanding and at least one open-ended question where students have the opportunity to provide qualitative feedback to their instructor on points of confusion or difficulty.
Warm-ups are scored but are low-stakes, in that they do not count for a significant portion of students' grades. The main feature of a JiTT course design is the feedback cycle: instructors use students' responses to warm-ups to focus in-class activities on the subject matter students indicate they find most confusing or difficult.

According to JiTT co-developer Andrew Gavrin ([17]), warm-ups can be used in a variety of ways in chemistry, for example, to frame a big idea, to "promote the understanding of visual representations" (p. 123), and to help students connect chemistry with their lives. Below is an example of JiTT questions for use in an introductory chemistry course related to the topic of chemical change [17].

A sidewalk is formed when concrete sets (hardens). Is this an example of a chemical change? A physical change? Explain your answer.

What aspects of the reading on chemical change did you find most difficult or confusing?

Figure 13.2 Case example of Just-in-Time Teaching as a technology-independent flipped teaching method.

> **Case example**
>
> Dr. Cynthia LaBrake teaches a 400 student chemistry course at The University of Texas at Austin for students pursuing degrees in science or medicine (but not chemistry). Dr. LaBrake and her colleague Dr. David Vanden Bout have transformed their lecture-based course into a blended learning experience that incorporates features of the flipped classroom. On their website, they provide lecture videos and screencasts students can watch to prepare for in-class meetings [18]. There are several videos for each unit: Gases, Atomic, IMFs, and Thermodynamics. For example, for The Ideal Gas Law chapter, there are 10 videos, some which introduce content such as Boyle's Law, Charles' Law, and Avogardo's Law. There are additional videos, which cover problem solving and a Gas Law Simulator, with which students engage before coming to class. At the beginning of each class period, students complete a low-stakes quiz using a classroom response system, or clicker, which covers their out-of-class activities.

Figure 13.3 Case example of flipping a large chemistry classroom at a major research university.

instructors lecture a little every class period. One caveat is to tailor the lectures to the key points that students find difficult or confusing. Another option is for instructors to spend time before the semester begins documenting the typical most challenging points for students through years of teaching.

13.3.1.3 Myth 3: Students Won't Be Prepared for Class

Another common concern teachers across the disciplines report to us is that their students will not do out-of-class work in a flipped classroom. Giving students frequent low-stakes incentives in the form of points is the most important thing you can do to motivate them to do their preclass work. Frequent low-stakes quizzes during the first 15 min of class is the first author's favorite strategy because it provides students with the opportunity for frequent retrieval practice, the practice of retrieving and using information they have already learned [19]. Research in the cognitive sciences stresses that retrieval practice, especially when it is distributed or spaced after a learning event, is the most effective strategy for helping students retain knowledge over time [20].

13.3.1.4 Myth 4: Flipping Your Classroom Means Changing Everything You Do

Despite the wording, you do not need to actually flip your entire classroom.

To some, the phrase *flip your classroom* implies that instructors need to transform their entire course. This is false. You can flip your entire department, a full course, one unit, one concept, or even just the laboratory. If you are new to flipped teaching in chemistry, we suggest an agile approach: begin with one or two concepts, solicit feedback from your students, and adjust. How do you choose which concepts to flip? Start with a concept students always have difficulty learning, and flip that one concept. See Figure 13.4 for an example of an agile approach to flipping in chemistry education.

> **Case example**
>
> Dr. Vikul Rajpara teaches introductory chemistry at Union Community College in New Jersey. New to the concept of the flipped classroom, he decided to start by flipping his chemistry laboratory sections before trying to flip his lecture-based course. He started planning to flip his lab by identifying the big ideas for each experiment and then developing learning outcomes based on those big ideas. For example, his first lab related to Freezing Point Depression. After completing the lab, he expected students to be able to (i) observe the change in freezing points of a pure solvent and a solution; (ii) predict the freezing point of a solution based on molar mass and number of moles; and (iii) identify practical applications of freezing point depression based on molar mass. To help his students achieve these learning outcomes, he selected a set of pre-lab videos and created a pre-lab quiz for students to complete before coming to class. After the completion of the first lab, he noticed that most of the students did well on the quiz. Some students reported that watching the videos before classed helped them understand the concept, and some resisted completing extra work.

Figure 13.4 Example of an agile approach to flipping the chemistry laboratory.

13.3.1.5 Myth 5: Flipped Classrooms Solve All Students' Problems Immediately

Another common myth we notice in our work with science instructors is that flipped classrooms are a cure-all. Many instructors try flipping their classrooms and do not see the radical change in results they expect after the first implementation. Like any good teaching effort, creating an efficacious flipped classroom takes time and multiple iterations. This is why we suggest taking an agile approach to flipping your classroom before you go full scale. You will immediately notice that students are more engaged, that their discussions are richer, and that they are having a more positive learning experience.

In summary, when flipping the chemistry classroom, start with these five basic principles:

1) Create opportunities for students to build prior knowledge before class and apply that knowledge through experiential or social learning in class.
2) Lecture a little; mini lectures can help ease students into the flipped classroom process.
3) Provide incentives for students to complete out-of-class work through frequent low-stakes assessments that reward their doing that out-of-class work.
4) Take an agile approach to flipping your classroom; start with one concept or lesson, pilot it with your students, integrate their feedback, and make adjustments before going all out.
5) Do expect to feel more engaged and excited about your work; however, do not expect huge changes in student achievement in your first implementation. Do expect to see evidence of increased student engagement, which is a precursor to deeper learning.

F	Find a concept or lesson to flip
L	Lift some of the direct instruction of concept out of the class to provide students with opportunity to build prior knowledge
I	Incentivize students by awarding points for the completion of out-of-class work through pre-class and in-class quizzes
P	Prioritize in-class activities that allow students to engage in more intensive cognitive work, practice applying that concept, and engaging with their peers and instructor for feedback

Figure 13.5 Four-step protocol for flipping any classroom[2)].

13.3.2
FLIP

Figure 13.5 provides a simple protocol instructors can use to flip any chemistry concept [21].

13.3.3
Student Attitudes toward Flipping General Chemistry

In a recent article, J. D. Smith [22] reported on a survey of research of students ($n = 235$) in a flipped general chemistry (I and II) classroom at Lipscomb University. The course design for this study involved 200 online mini lectures for student viewing before class with a number of learning activities including problem solving and discussion, facilitated during class time. J. D. Smith reported "students overwhelmingly [agreed] the pre-recorded lectures [were] generally useful (97%)" (np). Even though some of J. D. Smith's students found that the flipped class required more work, 65% of the students agreed that the flipped classroom approach made in-class time "less boring and/or more engaging" (np). Students also used the preclass lectures for continued study – not just for preparing before class – helping them during homeworking, clarifying concepts, and preparing for exams.

It is our experience that some students initially resist the flipped classroom approach because they believe they learn better through lectures. Student perceptions of how much they are learning, however, may not always be accurate when compared with their performance [23].

13.4
Flipping Your Classroom with Peer Instruction

Flipped classrooms are flexible enough to accommodate a large range of variations in implementation. Chemistry instructors should choose from an array of

---

2) Protocol inspired by Josh Walker, Center for Teaching and Learning, University of Texas at Austin.

methods one that they are most comfortable with and will promote the development of prior knowledge before class and various activities that will promote further learning in and after class. In this section, we will propose one low-threshold research-based method for flipping your classroom, called *Peer Instruction*. We will provide an overview of Peer Instruction, outline the key components of an efficacious implementation, summarize Peer Instruction research, and conclude with two brief case studies of how an instructor could use Peer Instruction to flip their chemistry classroom.

13.4.1
What Is Peer Instruction?

In classifications of pedagogy, Peer Instruction falls under "interactive teaching" or "interactive engagement" [24–26]. Like other flipped classroom methods, Peer Instruction involves using a design that includes before-class work to accomplish four goals: (i) help students develop prior knowledge before class; (ii) facilitate students' engagement in frequent self-monitoring [8, 9] of their learning states; (iii) solicit feedback from students on areas of difficulty, misconception, or misunderstanding; and (iv) help the instructor make data-driven decisions to advance learning in the classroom.

The second author, Eric Mazur, developed Peer Instruction in the 1990s at Harvard University, initially for use in large, introductory physics classrooms. However, the method is now used in classrooms all over the world, in both small and large classes, at a range of institutional types, and in a large variety of disciplines including the humanities and social sciences [27].

At a basic level, in a Peer Instruction course, class time is organized by a sequence of questioning, interactive discussion, and explanation. Instructors pose prepared questions, called *ConcepTests*, and students think and then formulate an answer to the ConcepTest individually, called *round one*. Next, students discuss their response with their peers; in round two, students formulate a final response to the question. Finally, instructors bring the sequence to a close through explanation of the concept.

> The basic goals of Peer Instruction are to exploit student interaction during lectures and focus students' attention on underlying concepts. Instead of presenting the level of detail covered in the textbook or lecture notes, lectures consist of a number of short presentations on key points, each followed by a ConcepTest – short conceptual questions on the subject being discussed. [16, p. 10]

Peer Instruction is a low-threshold pedagogy because it can be implemented with very little effort and without the purchase of any new equipment or technology. Like flipping the classroom in general, Peer Instruction can be used in any discipline, with small or large class sizes, and with students at any level of preparation. We recommend instructors take an agile approach to using Peer

Instruction by first identifying one concept or lesson to flip following a Peer Instruction course design.

13.4.2
What Is a ConcepTest?

What makes a ConcepTest different from a question? ConcepTests try to elicit, confront, and resolve (ECR) student misconceptions, rather than test simple recall [28]. As such, writing effective ConcepTests requires instructors to have a sense of common student misconceptions. For example, the ConcepTest in Figure 13.6 aims to test students' conceptions of a spontaneous reaction based on the knowledge that students often misunderstand that spontaneous processes can happen slowly as well as quickly [29]. The ConcepTest provided in Figure 13.6 is relatively easy; many instructors scaffold their questions during class time from easier to harder in a sequence. Therefore, an instructor might start with a mini lecture on spontaneous reactions followed by a series of ConcepTests of graduated difficulty. Figure 13.6 provides an example ConcepTest from Chemistry.

One of the best ways to gain an understanding of student misconceptions is to prepare open-ended questions for students. For example, if an instructor was not aware of student misconceptions about spontaneous reactions, the night before class she might ask students to write an open-ended response to the question: *Explain spontaneous reactions, and provide two distinct examples.* The instructor can then quickly analyze the responses and prepare a set of ConcepTests the next day.

The format of ConcepTests can vary; they can be multiple-choice or free-response. When using multiple-choice ConcepTests, it is important to include distractors that represent authentic student misconceptions. In addition, ConcepTests do not have to have one right answer – they can be used to direct thinking and encourage debate around controversial points. Finally, ConcepTests should be at a level of desirable difficulty [26] for students; that is they should not be too difficult or too easy. For ConcepTests, instructors should aim for a range of 30–70% of students getting the question correct during round one of responding.

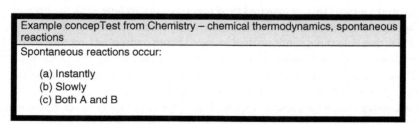

Figure 13.6 Chemistry ConcepTest example.

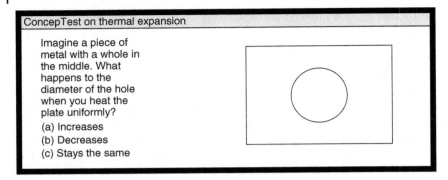

Figure 13.7 Chemistry ConcepTest example.

The best ConcepTests are those that test students' ability to transfer their understanding to new contexts. For example, to evaluate students' conceptual understanding of thermal expansion, an instructor could start with a mini lecture about the behavior of atoms in a metal that is heated and in a metal that is cold. She could begin the mini lecture with a statement such as, "When you heat metals up, they expand." Instead of asking students a simple recall question, such as, "What happens when you heat metals up?," a better choice is to provide students with an opportunity to transfer their understanding of the concept to a new context. The ConcepTest in Figure 13.7 illustrates a knowledge transfer question that could follow a mini lecture on thermal expansion.

Writing good ConcepTests is not easy; this skill also becomes easier with practice. However, a simple Google search for chemistry ConcepTests yields a number of Web sites and a book [30] with chemistry ConcepTests available. In addition, Learning Catalytics (*www.lcatalytics.com*), a classroom response system developed by the second author, has over 8000 ConcepTests from across the disciplines, including chemistry.

13.4.3
Workflow in a Peer Instruction Course

The general workflow for learning in a typical Peer Instruction course is depicted in Figure 13.8 (see [16]). Instructors plan preclass work to help students prepare for application activities during class. After class, students complete homework activities that help them continue their learning based on feedback on their learning they receive from their peers and their instructors. The most time-intensive part is developing and finding ConcepTests, so we recommend instructors find sample ConcepTests when they are first starting.

For ConcepTests with correct answers, instructors can expect to see a positive shift in the percentage of wrong answers to the correct ones immediately after peer discussion [31].

Peer Instruction workflow	
Before class	1. Students prepare for class by completing an assignment meant to help them develop prior knowledge and pinpoint misconceptions. 2. Instructors analyze student responses and other feedback to develop a series of ConcepTests with the aim of eliciting, confronting, and resolving student misconceptions.
During class	1. Instructors provide a mini lecture on select concepts. 2. Instructors pose ConcepTests. For each ConcepTest: 3. Students are given time to think. 4. Students record individual answers to their questions and submit the first round of responses. 5. Instructors review responses. 6. Students discuss their answers in groups. 7. Students record revised answers and submit the second round of responses. 8. Instructors review feedback. 9. Instructors give explanation of the correct answer.
After class	1. Students complete problem sets to further solidify knowledge. We recommend students also review the ConcepTests posed during class and ensure they understand the explanations provided. In addition, students may go back and re-read or re-watch the pre-class assignment to further solidify their knowledge.

Figure 13.8 Peer Instruction course workflow.

13.4.4
ConcepTest Workflow

While Figure 13.8 details the general workflow for a Peer Instruction class, there is also a specific workflow for ConcepTests. This is outlined in Figure 13.9. The ConcepTest is the centerpiece of Peer Instruction. ConcepTests provide extensive opportunities for students to engage in high-order thinking skills, such as application of knowledge. When you begin class, we recommend using feedback from students' JiTT responses to plan the brief presentation. It is important to have enough students with different answers for the peer discussion components, which is why we recommend you aim for having between 30 and 70% of student who answer correctly after the first response. We recommend that you never skip giving students the opportunity to submit a first response, to discuss with their peers, to submit a second response, and to come to resolution at the end through an explanation.

13.4.5
Peer Instruction and Classroom Response Systems

While Peer Instruction does not require technology, it does require the use of a classroom response system to facilitate the two question response rounds.

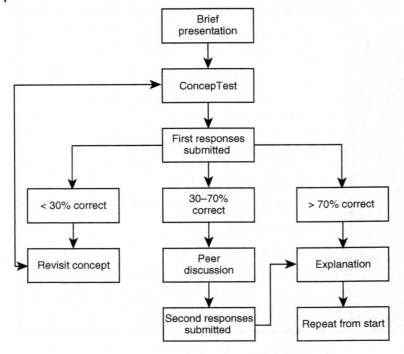

Figure 13.9 ConcepTest workflow in a Peer Instruction course.

Instructors can chose from low-tech or high-tech versions.[3] Flashcards are examples of low-tech versions: instructors can print a version of Figure 13.10 on an 8.5 × 11 piece of paper and hand it out to students to display their response. With white boards, student use small, individual dry erase boards and colored erasable markers to write their responses and display them.

For higher technology versions, chemistry instructors who use Peer Instruction often use clickers to facilitate the ConcepTest work. Clickers provide instructors with real-time visualizations, primarily in the form of histograms, of student response. More recently, instructors have adopted cloud-based response systems, such as Learning Catalytics, which allow students to respond to ConcepTests using their own devices, including phones and tablets.

13.4.6
The Instructional Design of a Peer Instruction Course

A variety of principles from the cognitive science literature underlie the instructional design of a Peer Instruction course. In particular, *retrieval practice* demonstrates that students both perform better on subsequent tests (the testing effect)

---

3) See [32] for a thorough review of classroom response systems.

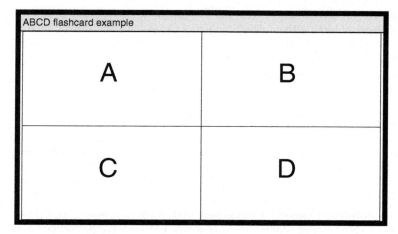

Figure 13.10 Example of a low-tech classroom response system.

and retain what they learn most effectively when they engage in a variety of learning activities that require them to retrieve information [19, 20]. During a single Peer Instruction session, ConcepTests require students to repeatedly retrieve both the prior knowledge they began developing before class and the deeper knowledge they are developing during class time.

Cognitive science research also recommends that retrieval practice is enhanced when it is spaced over time, called *distributed* or *spaced practice* [33–35]. In a Peer Instruction course, spaced practice can be facilitated in a number of ways: through JiTT warm-up questions that students complete before class, ConcepTests in class, and problem sets assigned after class. This distributed practice contrasts with mass studying (or cramming) that generally occurs during one long study session before an exam.

In order for material to be retained over a long period in our minds, we must spend a significant amount of time thinking about that material [36]. Peer Instruction works to maximize the amount of time that students are both thinking about the concepts key to the discipline *and* engaging in metacognition, or self-monitoring, of their understanding (or lack thereof) of those concepts in and out of class. In a traditional course, students are rarely given time during class to pause and think about the content, let alone think about their understanding of the material. Throughout a semester-long Peer Instruction course, instructors provide students with hundreds of opportunities to think about the subject matter on their own and to gain awareness of their knowledge or learning states.

Finally, Peer Instruction furthers self-monitoring of learning through the peer discussion component of the sequence. As students explain their responses to ConcepTests to one another, they often have "Aha!" moments that take them further than their individual thinking processes. One explanation of the power of peer discussion may be related to the social-learning concept of the zone of proximal development. The zone of proximal development, an idea first developed by

Vygotsky, refers to the difference between what students can learn on their own and what they can learn with guidance from more knowledgeable others, such as instructors or peers (as cited in [9]). In a Peer Instruction course, students have frequent opportunities to work individually and then be pulled into deeper learning through interaction with their peers.

13.4.7
Research on Peer Instruction

More than 15 years of research primarily supports the use of interactive teaching methods, such as Peer Instruction, to improve student learning [37]. The body of research on Peer Instruction, primarily from physics education researchers, indicates that, when compared to the traditional lecture and when implemented effectively, Peer Instruction significantly improves student learning outcomes, such as conceptual understanding, problem-solving ability, and academic performance on exams [16, 25, 38–40]. Lasry *et al.* [41] observed learning gains in Peer Instruction courses across institutional types, such as elite universities and 2-year institutions. Improvements in student learning and engagement when Peer Instruction is used have been reported in other science, technology, engineering, and math (STEM) disciplines, including engineering [42], biology [43, 44], chemistry [30], and mathematics [45].

In addition to learning outcomes, research suggests that interactive teaching approaches that incorporate Peer Instruction in physics may also have positive effects on the most persistent problems in STEM education. Researchers have observed that Peer Instruction, in combination with other interactive teaching techniques, may have an effect on reducing the racial and gender gap in physics and drop-out rates in STEM courses and majors [46–48]. In a study on retention specifically, Watkins and Mazur [48] reported "that a single course [taught with Peer Instruction] can have a significant long-term impact on the retention of students in STEM majors" (p. 39). And Landis *et al.* [30] reported that in a study of general chemistry courses at the University of Wisconsin-Madison, when ConcepTests were used, "fewer than 2% of the students dropped the course, whereas 25% of the students in the traditional lecture dropped the course" (p. 31). However, Landis *et al.* also emphasize that because the courses studied incorporated other interactive approaches it is unclear how strong of a role ConcepTests played.

13.4.8
Strategies for Avoiding Common Pitfalls of Flipping the Classroom with Peer Instruction

Similar to flipped teaching, Peer Instruction is a flexible method that can accommodate customization. Turpen and Finkelstein [49, 50] document that instructors implement Peer Instruction in a variety of ways. Because individual implementation choices can significantly influence students' learning experiences [48], it is important to identify strategies to avoid common pitfalls in Peer Instruction implementation. In this section, we will review four such strategies: (i) effecting grouping; (ii) response opportunities; (iii) discussion opportunities; and (iv) response sharing.

13.4.8.1 Effective Grouping

One of the most common questions among Peer Instruction beginners is, "How do I ensure students are engaged in discussion of the ConcepTest?" The first step is to attempt to pose ConcepTests that are at a level of desirable difficulty for students, again, so that you can observe a range or 30–70% correct answers after the first round of responses. The second step is to cue students to discuss their responses with peers who have different answers. Grouping students in this manner can be done verbally by stating, "find someone who has a different answer" or by using an intelligent grouping tool if it is available in your classroom response system. Pairing by different response may put students in a state of cognitive dissonance or disequilibrium [51] that they may aim to resolve; when two students have the same answer, they may simply agree and stop discussing the subject matter. After the first round of responding, students do not need to be grouped in a way that ensures one student has the right answer in order for effective discussion, measured by a shift to the correct answer in round two, to result. Students may converge on the right answer even if both have incorrect responses during round one [26, 44].

13.4.8.2 Response Opportunities

The ConcepTest workflow in Figure 13.9 identifies two rounds in which students submit responses to ConcepTests. During a ConcepTest sequence, some instructors skip the first round of responding and cue students to discuss their answers. Providing students time to think about the question, retrieve, and commit to an answer is important because it increases their opportunities for self-monitoring; students must gain awareness of whether they know how to respond to the question or not. This awareness can help regulate what students do next, for example, how engaged they are in their peer discussions as well as how and what they study after class [26]. In addition, in a survey of undergraduate engineering students ($n = 114$), Nicol and Boyle [42] reported that students prefer the opportunity for a first response and view it as important to their learning (p. 468).

13.4.8.3 Peer Discussion Opportunities

Some instructors skip the peer discussion portion in a ConcepTest sequence and hold only one round of voting. While this is a valid strategy if a large percentage of students get the answer correct after the first response, if you are hoping to see significant learning gains, it is important to allow for peer discussion between responses [31]. Since peer discussion is the defining feature of Peer Instruction, this is an essential activity for any implementation regardless of customization.

13.4.8.4 Response Sharing

Another variation in Peer Instruction implementation among some instructors is to display or describe the results of the first round of responses, for example, by projecting the histogram or by describing the votes qualitatively: "most students responded with B." This is a valid choice if there is an even spread among answer choices or when there is no one correct answer (such as in an opinion question). However, if there is a correct answer, displaying the answer can introduce unnecessary noise in the form of "the most common response bias" between the first and second voice. Perez *et al.* [52] report that seeing "the most common response can bias a student's second vote on a question and may be misinterpreted as an increase in performance due to student discussion alone" (para 1). Therefore, when there is a correct answer, we recommend instructors do not to share the results of the first round of responses with students before they engage in peer discussion and a second round of responding. It can be useful to display for students both rounds of responses after the ConcepTest sequence has come to a close.

In summary, while Peer Instruction can be customized in a number of ways, we advise using these four strategies to avoid common pitfalls: (i) cue students for effective conversations by grouping by different answers; (ii) provide students opportunities to engage in a first round of responses; (iii) include opportunities for peer discussion; and (iv) wait until the end of the ConcepTest sequence to display the first round responses if there is a correct answer.

13.4.9
Flipping the Chemistry Classroom with Peer Instruction

Flipping the chemistry classroom with Peer Instruction offers a research-based opportunity to improve student engagement and learning of chemistry regardless of the level of preparation of students or the types of institutions in which they are learning. There are a number of resources available for instructors who wish to flip their chemistry classroom using Peer Instruction, including the Peer Instruction network. The Peer Instruction network (*www.peerinstruction.net*) is a free database of Peer Instruction users worldwide. Chemistry instructors interested in sharing Peer Instruction resources from several continents are registered on the network. Just as student learning is facilitated through peer learning, so is instructional development. We recommend beginning, intermediate, and expert Peer Instruction users use the network to share ideas, resources, and ConcepTests. In addition, *Turn to Your Neighbor: The Official Peer Instruction Blog* (*blog.peerinstruction.net*) holds over 50 articles on Peer Instruction written by the first author to address questions posed by Peer Instruction users.

In Figures 13.11 and 13.12, we have provided two hypothetical case studies, based on our experience with numerous faculty, that incorporate innovative ideas for flipping the chemistry classroom with Peer Instruction.

Both these examples illustrate how a chemistry instructor might flip their classroom with Peer Instruction using a highly customized approach. Just as

> **Case example – organic chemistry**
>
> Dr. Michael Smith is a professor at a community college in the Northeastern United States. He teaches a small organic chemistry course for 30 students preparing to take the Medical College Admissions Test.
>
> **Before class:** Dr. Smith requires students to read an electronic textbook before coming to class. The night before each class meeting, students are required to complete a 10-item online, multiple choice quiz on their readings that is graded on effort only; students only get credit if they complete all the questions, and students receive immediate feedback on the correctness of their responses. Dr. Smith receives a report on the questions students are having the most difficulty with. He also asks the following open-ended question: "What do you think is the most important thing we should discuss in tomorrow's class?" He uses the data from the quiz and the open-ended responses to select ConcepTests for delivery in class.
>
> **During class:** Dr. Smith spends the first 15 min of class sharing a few interesting, anonymized selections from students' responses to the open-ended question with the class. He spends the next 30 min engaging students with ConcepTests (following the sequence outlined in Figure 9) related to the students' areas of confusion as well as the areas he identifies as the most important to future learning. He uses ABCD flashcards (as in Figure) to collect students' responses, help facilitate decisions in class, and record decisions for later review.
>
> **After class:** For additional post-class work, Dr. Smith assigns traditional problem sets and provides access to a database of helpful videos he has located online and curated for each unit in the textbook.

Figure 13.11 Case example of flipping a classroom with Peer Instruction, small class size.

there are many ways to flip a class, there are many ways to flip a classroom using Peer Instruction and no one right away. Adherence to the strategies of grouping, response sequencing, peer discussion, and response sharing will increase the likelihood of greater gains in learning, especially in deep conceptual understanding.

13.5
Responding to Criticisms of the Flipped Classroom

There are numerous criticisms of flipped classrooms including that the approach deemphasizes the role of the subject-matter expert, the teacher. Other key, and important, criticisms of flipped classrooms include that they are simply replicating bad instruction, such as lecture, in an online environment. Finally, vocal critics suggest that flipped classrooms simply do not work and represent only hype or the latest educational trend.

The first criticism misinterprets the directive of flipped classrooms, indeed, when implemented properly the teacher plays a much more active role by guiding and coaching students to deeper understanding, rather than simply telling them what they need to do. And while many instructors do simply replicate bad teaching online by requiring students to watch 90-min videos of their classroom

> **Case example – general chemistry**
>
> Dr. Anne Bennett teaches an 800 student general chemistry course at a large public research university in Arizona. Dr. Bennett is new to Peer Instruction, so in her first year of implementation, she decided to use Peer Instruction for lecture demonstrations only.
>
> **Before class:** On days when Dr. Bennett is going to use a lecture demonstration, she assigns students to watch a 2–3 min pre-class video she made describing the concept or process they will observe in the lecture demonstration. She created the videos using screencasting software and published them online in the course learning management system.
>
> **During class:** During each demonstration, Dr. Bennett revisits the concept or process she described in the video and then explains the demonstration. She asks students to predict what they will observe using a multiple choice question and collects their responses using clickers. Once students submit their responses, she cues them to discuss their prediction with a neighbor with a different response. After 3 min of discussion, she solicits a second round of voting. Finally, she conducts the lecture demonstration and then facilitates a large group discussion and explanation of the concept being demonstrated.
>
> **After class:** After class, Dr. Bennett asks students to respond to an open-ended reflection question on their prediction of the demonstration's outcome, their observations of the demonstration, and how their prediction changed or did not change after discussion and observation. Students receive 10% of their grade for answering their after-class reflection questions.

Figure 13.12 Case example of flipping a classroom with Peer Instruction, large class size.

lectures, most employ a more thoughtful instructional design by chunking their direct instruction. However, another criticism of the chunked flipped classroom approach documented in the media is that these mini lectures strip away the nuances of content and what is left are abstractions that do not make sense and are disconnected from a larger narrative, which lectures can provide. While this may be the case, a successful flipped classroom will ensure that those nuances are discovered during class time where students have the support of their peers and teachers and solidified with more extensive practice after class.

Those who suggest that the flipped classroom is not a new approach are absolutely right. Having students come to class prepared is an essential aspect of literature instruction and is not a new idea. In the United States, the approach is as at least as old as the case study method, developed at the Harvard Law School in the 1800s. What is new is the way flipped classrooms are finally taking hold to transform how people teach and learn on a global scale.

Finally, the jury on flipped classrooms is still out. There is a tremendous opportunity for educational researchers to contribute greater understanding on flipped learning as we currently know it. However, when flipping a classroom with Peer Instruction, you can be sure that you are using a research-based methodology with extensive study supporting its use.

13.6
Conclusion: The Future of Education

While flipped classrooms can introduce instructional challenges, such as increased time and effort and student resistance to non-lecture to one's pedagogical practice, the approach also presents remarkable opportunities to radically transform our classrooms in ways that improve student learning and success. As we enter the golden age of higher education, the future promises to bring many such pedagogical innovations designed to help promote student learning in chemistry. Never before in history has there been so much global excitement about improving teaching and learning in higher education. Current innovations include massive open online courses (MOOCs) which are scaling high-quality content at unprecedented levels and Big Data processes, which are helping to personalize learning in ways that were heretofore impossible. At present, the flipped classroom represents a method that has the potential to finally disrupt obsolete methods of instruction, such as pure lecture, by supplementing past approaches aligned with how people learn best.

Peer Instruction can, but need not be, the only method a chemistry instructor uses to flip their classroom to improve chemical education. Indeed, instructors often combine Peer Instruction with many other interactive teaching methods, including POGIL (Process oriented guided inquiry learning), project-based learning, team-based learning, and discovery learning. Regardless of the method selected, flipped classrooms are bridging the gap between what education has always been and what education can be on the university campus. With their emphasis on putting students and their learning first, flipped classrooms represent the most innovative disruption in higher education in the last century. The success of this disruption in chemical education will be evident not only through the deep learning and understanding our students have about content, but in the widespread transformation of the chemistry classroom from a quiet lecture hall to a room alive, bustling, and full of activity.

Acknowledgments

We would like to thank Cassandre Giguere Alvarado, PhD, Emily A. Johnson, and the Mazur Group for their assistance in the development of this chapter. We would also like to thank Joshua Walker, PhD for inspiring ideas about the phases and sequencing of a flipped class.

References

1. Barber, M., Donnelly, K., and Rizvi, S. (2013) *An avalanche is Coming: Higher Education and the Revolution Ahead*, Institute for Public Policy Research, London.
2. Bergmann, J. and Sams, A. (2012) *Flip Your Classroom: Reach Every Student in Every Class Every Day*, International Society for Technology in Education, Washington, DC.
3. Rosenberg, T. (2013) Turning Education Upside Down. The New York Times (Oct. 09).

4. Ebbeler, J. (2013) Introduction to Ancient Rome,' the Flipped Version. The Chronicle of Higher Education (Jul. 22).
5. Ojalvo, H. and Doyne, S. (2011) Five Ways to Flip Your Classroom with the New York Times. The Learning Network: Teaching and Learning with the New York Times (Dec. 08).
6. Bruff, D. (2012) The Flipped Classroom FAQ. CIRTL Network: Center for the Integration of Research, Teaching, and Learning (Sep. 15).
7. Heath, C. and Heath, D. (2007) *Made to Stick*, Random House, New York.
8. Ambrose, S., Bridges, M.W., Dipietro, M., Lovett, M.C., Norman, M.K., and Mayer, R.E. (2010) *How Learning Works: Seven Research-based Principles for Smart Teaching*, Jossey-Bass, San Francisco, CA.
9. Bransford, J.D., Brown, A.L., and Cocking, R.R. (eds) (2000) *How People Learn: Brain, Mind, Experience, and School*, Expanded edn, National Academy Press, Washington, DC.
10. Astin, A. (1999) Student involvement: a developmental theory. *J. Coll. Stud. Develop.*, **40** (5), 518–529.
11. Staker, H. and Horn, M. (2012) *Classifying K-12 Blended Learning*, Innosight Insitutute.
12. PhET Interactive Simulations (2013) Build a Molecule, http://phet.colorado.edu/en/simulation/build-a-molecule (accessed 15 December 2013).
13. Christensen, C., Horn, M., and Staker, H. (2013) *Is K–12 Blended Learning Disruptive? An Introduction of the Theory of Hybrids*, Clayton Christensen Institutue.
14. Lage, M., Platt, G., and Treglia, M. (2000) Inverting the classroom: a gateway to creating an inclusive learning environment. *Econ. Instruct.*, **31** (1), 30–43.
15. Novak, G., Patterson, E., Gavrin, A., and Christian, W. (1999) *Just-in-Time Teaching: Blending Active Learning with Web Technology*, Prentice Hall, Upper Saddle River, NJ.
16. Mazur, E. (1997) *Peer Instruction: A User's Manual*, Prentice Hall, Upper Saddle River, NJ.
17. Simkins, S. and Maier, M. (eds) (2009) *Just in Time Teaching: Across the Disciplines, and Across the Academy*, Stylus Publishing.
18. LaBrake, C. and Vanden Bout, D. *Chemistry 301 Website*, The Univeristy of Texas at Austin, https://ch301.cm.utexas.edu (accessed 20 November 2014).
19. Karpicke, J.D. and Blunt, J.R. (2011) Retrieval practice produces more learning than elaborative studying with concept mapping. *Science*, **331** (6018), 772–775.
20. Roediger, H.L. and Karpicke, J.D. (2006) Test-enhanced learning taking memory tests improves long-term retention. *Psychol. Sci.*, **17** (3), 249–255.
21. Schell, J. (2013) How to FLIP Your Class … in 4 Basic Steps. Turn to Your Neighbor: The Official Peer Instructoin Blog (Mar. 04).
22. Smith, J.D. (2013) Student attitudes toward flipping the general chemistry classroom. *Chem. Educ. Res. Pract.*, **14**, 607–614.
23. Carpenter, S., Wilford, M., Kornell, N., and Mullaney, K. (2013) Appearances can be deceiving: instructor fluency increases perceptions of learning without increasing actual learning. *Psychonomic Bull. Rev.*, **20** (6), 1350–1356.
24. Henderson, C. and Dancy, M.H. (2009) Impact of physics education research on the teaching of introductory quantitative physics in the United States. *Phys. Rev. Spec. Top. Phys. Educ. Res.*, **5**, 1–9.
25. Mazur, E. and Watkins, J. (2009) in *Just in Time Teaching Across the Disciplines* (eds S. Simkins and M. Maier), Stylus Publishing, Sterling, VA, pp. 39–62.
26. Schell, J., Lukoff, B., and Mazur, E. (2013) in *In Increasing Student Engagement and Retention Using Classroom Technologies Classroom Response Systems and Mediated Discourse Technologies*, vol. 6E (eds C. Wankel and P. Blessinger), Emerald Group Publishing Limited, pp. 233–261.
27. Schell, J. and Mazur, E. (2012) Peer Instruction Network, 08 July 2012.
28. Heron, P., Shaffer, P.S., and McDermott, L.C. (2008) *Identifying and Addressing Student Conceptual Difficulties: An Example from Introductory Physics*, National Academy Press.

29. Baker, D. (2013) Chapter 19. Chemical Thermodynamics, Baker Chemistry, http://users.bergen.org/danbak/APChem-Resources.html (accessed 14 December 2013).
30. Landis, C., Ellis, A., Lisensky, G., Lorenz, J., Meeker, K., and Wamser, C. (2001) *Chemistry ConcepTests: A Pathway to Interactive Classrooms*, Prentice Hall, Upper Saddle River, NJ.
31. Lasry, N., Charles, E., Whittaker, C., and Lautman, M. (2009) When talking is better than staying quiet. *AIP Conf. Proc.*, **1179**, 181–184.
32. Bruff, D. (2009) *Teaching With Classroom Response Systems: Creating Active Learning Enviornments*, Jossey-Bass, San Francisco, CA.
33. Agarwal, P. (2012) *Research-Based Cognitive Strategies for Enhancing College Readiness*, University of Texas, Austin, TX, 25 June 2012.
34. Cepeda, N., Pashler, H., Vul, E., Wixted, J., and Rohrer, D. (2006) Distrubuted practice in verbal recall tasks: a review and quantiative synthesis. *Psychol. Bull.*, **132** (3), 354–380.
35. Rohrer, D. and Taylor, K. (2006) The effects of overlearning and distrubted practise on the retention of mathematics knowledge. *Appl. Cognit. Psychol.*, **20**, 1209–1224.
36. Willingham, D. (2010) *Why Don't Students Like School?*, Jossey-Bass, San Francisco, CA.
37. Hake, R.R. (1998) Interactive-engagement versus traditional methods: a six-thousand-student survey of mechanics test data for introductory physics courses. *Am. J. Phys*, **66**, 64–74.
38. Crouch, C. and Fagen, A.P. (2004) Classroom demonstrations: learning tools or entertainment. *Am. J. Phys*, **72** (6), 835–838.
39. Crouch, C.H. and Mazur, E. (2001) Peer Instruction: ten years of experience and results. *Am. J. Phys*, **69** (9), 970–977.
40. Fagen, A.P., Crouch, C.H., and Mazur, E. (2002) Peer Instruction: results from a range of classrooms. *Phys. Teach.*, **40**, 206.
41. Lasry, N., Mazur, E., and Watkins, J. (2008) Peer Instruction: from Harvard to the two-year college. (Author abstract) (Report). *Am. J. Phys*, **76**, 1066–1069.
42. Nicol, D.J. and Boyle, J.T. (2003) Peer Instruction versus class-wide discussion in large classes: a comparison of two interaction methods in the wired classroom. *Stud. High. Educ.*, **28**, 457.
43. Knight, J.K. and Wood, W.B. (2005) Teaching more by lecturing less. *Life Sci. Educ.*, **4** (4), 298.
44. Smith, M.K., Wood, W.B., Adams, W.K., Wieman, C., Knight, J.K., Guild, N., and Su, T.T. (2009) Why Peer discussion improves student performance on in-class concept questions. *Science*, **323** (5910), 122–124.
45. Miller, R., Santana-Vegas, E., and Terrell, M. (2006) Can good questions and peer discussion improve calculus instruction? *PRIMUS*, **16** (3), 193–203.
46. Lorenzo, M., Crouch, C., and Mazur, E. (2006) Reducing the gender gap in the physics classroom. *Am. J. Phys*, **74** (2), 118–122.
47. Watkins, J.E. (2010) *Examining Issues of Underrepresented Minority Students in Introductory Physics*, Harvard University, Cambridge, MA.
48. Watkins, J. and Mazur, E. (2013) Retaining students in science, technology, engineering, and mathematics majors. *J. Coll. Sci. Teach.*, **42** (5), 36–41.
49. Turpen, C. and Finkelstein, N. (2009) Not all interactive engagement is the same: Variations in physics professors' implementation of Peer Instruction. *Phys. Rev. Spec. Top. Phys. Educ. Res.*, **5** (2).
50. Turpen, C. and Finkelstein, N.D. (2010) The construction of different classroom norms during Peer Instruction: students perceive differences. *Phys. Rev. Spec. Top. Phys. Educ. Res.*, **6** (2), 020123.
51. Piaget, J. (1985) *The Equilibration of Cognitive Structures: The Central Problem of Intellectual Development*, University of Chicago Press, Chicago, IL.
52. Perez, K.E., Strauss, E.A., Downey, N., Galbraith, A., Jeanne, R., and Cooper, S. (2010) Does displaying the class results affect student discussion during Peer Instruction? *Life Sci. Educ.*, **9** (2), 133.

14
Innovative Community-Engaged Learning Projects: From Chemical Reactions to Community Interactions
Claire McDonnell

A critical review of developments in community-based learning (CBL) (also called service-learning) and community-based research (CBR) in chemistry education at the second and third levels is provided in this chapter. The focus is on innovative and interesting projects implemented over the last 5 years, but relevant initiatives prior to this are also incorporated. Evidence for the effects on student learning, motivation, and personal and professional development that arise from community engagement is examined, as are the consequences of this pedagogical approach for academic staff, educational institutions, and community partners. Other aspects that are considered include facilitating the effective and sustainable implementation of activities and projects, minimizing obstacles, maximizing potential synergies, and the impact of the local institutional culture and support framework. Current and potential future developments are then reviewed, and the evaluation made is that a critical mass of chemistry-community-engaged learning activity appears to be developing.

> The aim [of education] must be the training of independently acting and thinking individuals who, however, see in the service to the community their highest life problem.
>
> *Albert Einstein [1, p. 60]*

14.1
The Vocabulary of Community-Engaged Learning Projects

Community-engaged learning projects involve student-community engagement. Several terms may be used to describe them, and an awareness of these ensures effective communication and the ability to access information on initiatives already in place. As well as a familiarity with the terminology, it is essential to establish a shared understanding of what is meant when these terms are applied at an institutional level, and also on a regional or national basis if collaboration at that level takes place.

Chemistry Education: Best Practices, Opportunities and Trends, First Edition.
Edited by Javier García-Martínez and Elena Serrano-Torregrosa.
© 2015 Wiley-VCH Verlag GmbH & Co. KGaA. Published 2015 by Wiley-VCH Verlag GmbH & Co. KGaA.

14.1.1
Community-Based Learning

"CBL" is a term applied outside of the United States, particularly in Europe, and a number of other terms are used at a local level (for example, "pedagogies for civic engagement" [2]). "Service-learning" is the name applied extensively to this pedagogical approach in the United States, where it is very well established. It has also been used in Australia, Asia, and Africa. Other terms that have gained traction, particularly in the United States and Canada, are "community-engaged learning" and "community engagement pedagogies." These have recently been applied to encompass service-learning and other activities in higher education institutions (HEIs) such as Brandeis University, Stanford University, and Vanderbilt University [3], as well as in as a recent publication on the Irish context [4]. Also, there has been a focus on institutional change over the last couple of years, and this has resulted in the use of the term "university–community engagement" [5, 6].

Many definitions have been proposed in the literature, and the one developed for service-learning in 1996 by Bringle and Hatcher [7] is often used; "a credit-bearing educational experience in which students participate in an organized service activity that meets identified community needs and reflect on the service activity in such a way as to gain further understanding of the course content, a broader appreciation of the discipline and an enhanced sense of civic responsibility." As an institution or staff member develops a clearer understanding in their own context of the application of CBL, this statement can be used as a good basis to establish a local definition. In this regard, it has been suggested that the description above focuses exclusively on student learning and that a reference to mutually beneficial relationships between learners and community partners could be added [5].

14.1.2
Community-Based Research

CBR can also be called *community-engaged scholarship*. It involves a methodology in which a community partner determines the research question and is actively involved in and contributing to the research process [8, 9]. The research methods employed should be respectful of ways of understanding that apply for the community partner, and the main priority for the research results should be to meet the needs of the community partner or the wider community [8, 9]. It is also preferable that the projects will give learners meaningful research experience and that the outcomes have the potential to be published by the academic partner(s) [8, 9]. An important distinction between CBR and traditional academic research is that the former is undertaken *with* instead of *on* the community and the community is not treated as a laboratory [10].

A related initiative that is well established in some parts of the world is "Science Shops" [11]. They were first developed in the Netherlands in the 1970s. They were established for the purpose of providing a forum or exchange to foster a working relationship between knowledge-producing organizations, such as universities, and community groups seeking answers to relevant questions. The aim is to achieve active collaboration in research and to provide members of the public with access to the university and its facilities. Other characteristics are that there is a demand-driven and bottom-up approach [11]. The name "Science Shop" is applied for historical reasons and is a direct translation to English. It can be misleading, however, as the projects extend across all disciplines and are not restricted to the natural sciences and there are usually no fees charged [12–15].

A more recent development, particularly in the health sciences, has been community-based participatory research. This differs from CBR in that the partnership is between health and social service agencies as well as community-based groups, community members, and HEIs [16], but the two approaches can be complementary. Gardella *et al.* provide useful advice on facilitating public participation in the chemistry CBL projects they have been involved in [17], and they refer to the International Association for Public Participation guidelines. Another related approach is public participation in scientific research (PPSR), sometimes referred to as *citizen science* projects. This involves lay people (who may be school or university students) interacting with scientists in order to take part in a scientific research effort. Projects in which large quantities of data are required to be collected over a geographic area are well suited to this model, and local water quality, ornithology, and astronomy are among the areas that have been studied [18, 19]. An example of a project using a similar approach that is limited to university student participants takes place in the University of Sydney where 1000 first-year students collect samples of fungal spores in their own back garden on an agar plate using the settling method. The students identify the fungal colonies once they have grown and also provide some local information (post code and local vegetation). The data are used to inform research programs on allergens and asthma [20].

14.1.3
Developing a Shared Understanding of CBL and CBR

In the Dublin Institute of Technology (DIT), the author's own institution, the terms CBL and CBR are generally applied. However, the small team that facilitates the integration of this pedagogical approach has developed the name Students Learning With Communities (SLWC) for their program. The rationale for this name is to facilitate a more ready recognition by prospective community partners, students, and academic staff of what is involved [21]. The focus on names and terms is necessary because misunderstandings can and do arise due to confusion between the CBR and CBL projects described in this chapter and volunteering activities or student work placements. Bringle and Hatcher have observed that "creating a common understanding of what constitutes service-learning at a particular institution will pay dividends later" [7].

Boland [4] has recently identified six elements that define community-engaged learning, as follows:

- It involves a credit-bearing element of an academic module, course, or program.

- Students engage with the community, commonly providing a "service" to the not-for-profit/community sector, in response to a need identified by this community partner.
- Citizenship and engagement feature as core values and organizing principles.
- It requires the application of discipline-specific knowledge and skills as well as the integration of theory and practice.
- The pedagogy is based on the principles of experiential learning, and reflection is a key element in the learning and assessment process.
- Reciprocity and partnership characterize the relationships between parties to the engagement (students, community partners, and college staff).

She notes that, in practice, the extent to which these elements are present can vary [4]. This description can be applied to encompass CBL and CBR, and the focus on characteristics instead of names is very useful. It can be applied to establish whether the interested parties have the same interpretation of what is involved and to avoid misunderstandings and unrealistic expectations developing.

The potential to confuse CBL and CBR with other experiential learning activities such as volunteering or work placement/internships can be addressed by focusing on the balance between community and learner benefits. In volunteering, the main beneficiary is the community, while, in an internship/placement, the emphasis is on what the student will gain. The aim in CBL and CBR is to strike an appropriate balance so that both benefit to a similar extent. Furco originally summarized this relationship very effectively in a diagram [22], an adaptation of which that also draws from more recently developed models [10, 23] is shown in

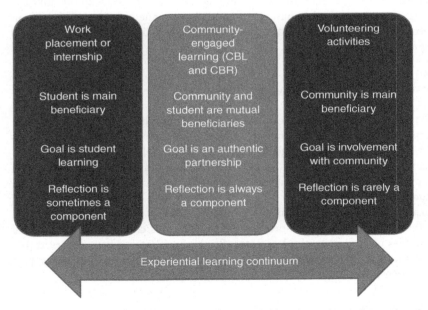

Figure 14.1 Relationship between community-engaged learning, volunteering, and work placement/internships. See Ref. [22] for original model. Developed from Refs. [10, 23].

Figure 14.1. This visual representation is often very helpful in clarifying the nature of the process to participants.

14.2
CBL and CBR in Chemistry

The way in which CBL and CBR are applied to chemistry education has been evolving over the past decade. In the last 3 years particularly, there has been a marked increase in the number of publications and in their geographic spread. It is difficult to determine a cause for this, but the focus of modern higher education in many countries at policy level on engagement as the third core responsibility in addition to research and teaching may be having an effect [6, 10, 24]. The global economic situation may also have contributed to the renewed focus on university–community engagement [25, 26].

A search strategy was applied that used a range of keywords to look for texts that dealt with the application of CBL and/or CBR in chemistry education using the Web of Knowledge^{SM} database and Google Scholar. Table 14.1 shows the 18 chemistry projects found that have been undertaken over a 6 year period from 2008 to 2013.

There is also a recent article that does not describe the application of CBL or CBR to chemistry education but is of relevance. It provides a comprehensive review of existing resources that can be applied to improve chemistry education in developing countries and would be very useful to anyone working on CBL or CBR projects to support chemistry education in these areas [46]. In addition, a conference paper presented by Stevens-Truss at the Biennial Conference on Chemical Education (BCCE) in 2012 [47, 48] discussed the development of a chemistry badge for Girl Scouts which they were then assisted in working toward. This initiative shows that there are avenues other than schools that can be explored when HEIs seek to become involved in outreach projects. Several references to science fairs and science nights are made in the projects described in Table 14.1. Other opportunities that could be exploited for CBL activities are national Chemistry Weeks and Science Weeks. Although no journal articles were found that referred to the application of CBL in this context, it is likely that projects are being undertaken.

Prior to 2008, 14 peer-reviewed articles and book chapters were found [49–62], and the earliest of these was a paper by Fitch *et al.* from 1996 [62]. Many of the projects described in these articles are also discussed in the publications in Table 14.1. Several of the early articles did not associate their approach with either service-learning, CBL, or CBR, which provides an indication of the extent to which awareness of the pedagogy of community-engaged learning has improved since they were written. In 1999, Ram used problem-based learning pedagogy to frame a project in which students worked with a community group and local government agency to perform water quality analysis [61], while, in 2000, Hope and Johnston described students addressing local urban environmental problems in terms of the application of project-based learning [58].

Table 14.1 A summary of the peer-reviewed accounts of community-based learning (CBL) and/or community-based research (CBR) in chemistry education from 2008 to 2013.

Year published	Project description	Location	Authors and reference number
2013	First-year undergraduates measure the lead content of soil samples from local residences and write a letter communicating the results to homeowners	Oregon, United States	Burand and Ogba [27]
2013	First-year college students undertake a project involving the local and/or campus community in groups of four	Connecticut, United States	Webb [28]
2013	Second-year undergraduates work with resource-limited school students and teachers to prepare azo dyes and use them to dye T-shirts	Grahamstown, South Africa	Glover et al. [29]
2013	First-year college students in groups of four prepare for and lead activities at a Halloween Science Night for school children aged 5–12	Missouri, United States	Theall and Bond [30]
2013	17 final-year college students prepare information pamphlets for a homeless shelter on topics such as hypertension, diabetes, and allergies	Pennsylvania and Iowa, United States	Harrison et al. [31]
2012	An optional project taken by first-year undergraduates who work in groups with a range of partners such as museums, a fish hatchery, and schools. Past participants act as teaching assistants	New York, United States	Donaghy and Saxton [32]
2012	Water quality in a nature reserve is analyzed for trace metals, anions, pH, dissolved oxygen, and coliform bacteria by groups of chemistry and science students from two colleges	Ohio, United States	Kammler et al. [33]
2011	College students prepare and present hands-on science experiments, with a focus on chemistry, to students in an elementary school	North Dakota, United States	Knutson-Person [34]
2011	Students select from a range of projects such as analysis of samples from the local community for lead and determination of fat levels in fast food	Washington DC, United States	Hosten et al. [35]
2011	College students prepare lessons to clarify misconceptions for secondary school students and interact using Web-conferencing tools. They also act as virtual lab partners for an experiment	Florida, United States	Saitta et al. [36]

Table 14.1 (Continued)

Year published	Project description	Location	Authors and reference number
2011	Undergraduates prepare interactive science demonstrations and work on curriculum experiments with secondary school students, give careers presentations in schools, perform safety audits in small organizations, test soil in a community garden, and raise awareness of alcohol limits for road users	Dublin, Ireland	McDonnell et al. [37]
2010	College students act as judges at science fairs in primary schools and demonstrate hands-on experiments at family science nights and science fairs	Texas, United States	Cartwright [38]
2010	Secondary school students from Australia and Tanzania participate in a Bilateral Youth Forum in Tanzania where they learn from each other about sustainable environments and develop their scientific literacy	Western Australia and Tanzania	Murcia et al. [39]
2009	College students act as activity presenters for hands on nanoscience experiments in schools and provide information about studying science in college	Pennsylvania, United States	Furlan [40]
2009	Undergraduates work in groups to perform environmental analysis for a range of organic and inorganic contaminants in partnership with six local communities and government agencies	New York, United States	Gardella et al. [17, 41]
2009	First-year college students work on community-based STEM (Science, Technology, Engineering, and Maths) projects including water contamination and coral reef degradation	Hawaii, United States	Franco [42]
2008	Undergraduates work in teams to design, perform, and explain chemistry discovery activities for primary (elementary) school children	Idaho, United States	Kalivas [43]
2008	"Simple short projects" include college students collecting water samples for a local government agency, extracting dyes from natural sources to use on baby clothes to be donated, developing a local historical exhibition on a medical kit, and using their academic institution as a community partner	Vermont, United States	Sutheimer [44] (also Sutheimer and Pyles [45])

14.2.1
Chemistry CBL at Secondary School (High School) Level

All publications that have been examined except for one [39] deal with the implementation of chemistry CBL and CBR projects in HEIs. The implementation of CBL (service-learning) is also promoted extensively at the secondary school (high school) level in the United States. There is not likely to be the same incentive to disseminate secondary school community-engaged learning projects in peer-reviewed publications, and this may account for the lack of articles found. Esson and Johnston presented a conference paper on an innovative project that involved secondary school students engaging in CBL at the BCCE in 2010 [63]. Undergraduates from their college worked with secondary school students to teach them the principles and methods relevant to a particular hands-on chemistry experience. These younger students then acted in turn as peer mentors to teach others in their own schools. The program has also been implemented as a pilot project in which college students worked with younger middle school students, who then became peer mentors in primary (elementary) school outreach activities. This model represents a resourceful way of maximizing the impact of the projects undertaken as the number of schools that benefit is substantial.

14.2.2
Chemistry Projects Not Categorized as CBL or CBR

There are some projects which have been discussed in the literature that bear most of the hallmarks of CBL or CBR but they have not been identified or categorized as such. Examples from 1999 and 2000 were discussed in Section 14.2 [58, 61]. There are also some more recent incidences such as academic staff and undergraduates working with a local museum to identify the contents of an exhibit from a Victorian pharmacy [64], as well as the Undergraduate Ambassadors Scheme which involves undergraduate students in science, engineering, and mathematics working closely with a teacher in a school to develop additional teaching materials [65, 66]. It may be that the reason why these projects are not described as examples of CBL or CBR is due to a local culture that is not aware of this pedagogical approach. This potential lack of awareness of CBL and CBR reinforces the importance of disseminating work undertaken. The opportunity to allow for learner reflection is a key feature of CBL or CBR, and this is often the aspect that is not incorporated in related projects [4]. In some cases, those involved are motivated to take the extra steps to align their work with CBL or CBR, and in others they may decide not to.

14.2.3
Guidelines and Resources for Getting Started

For anyone who is interested in initiating a pilot project that applies CBL and/or CBR to chemistry education, there are a number of resources and guidelines, some discipline-specific and some general, that can be consulted. As community-engaged learning is well established in the United States, many of the resources originate there. The National Service-Learning Clearinghouse Web site and Campus Contact Web site are very comprehensive, and the Talloires Network Web site is also useful. Resources that are specific to CBR can be found on the Living Knowledge Web site, which is the international science shop network [67]. Information that is specific to science education can be accessed on the Science Education for New Civic Engagements and Responsibilities (SENCER) Web site, and the American Chemical Society have developed their Service-learning Resources for Chemistry Faculty Web site.

Several models for planning and implementing CBL and CBR projects have been developed. The PARE (preparation, action, reflection, and evaluation) model for course planning is one such example [68]. Welch developed a rubric to facilitate conceptualization, implementation, and assessment. The mnemonic OPERA is applied, and it refers to five stages: enumerating objectives (O), exploring community partnerships (P), identifying the type of community-engaged learning students will be engaged (E) in, facilitating reflection (R), and assessing (A) to what extent the learning objectives were met [69]. Another useful model is based on the 6 R's (roles, relevance, reciprocity, reflection, risk management, and reporting), which is based on previous similar frameworks [70]. McEwen and O'Connor provide recommendations for building capacity among staff in HEIs so that effective community engagement can take place [71], and Tryon and Ross give an account of the implementation of the Science Shop model in the United States which includes a useful section on the lessons learned [15].

The logistics of incorporating community-engaged learning into a curriculum have been studied by Boland, who has identified five curriculum design options that can be applied [4]. The five possibilities are: adapting an existing stand-alone module, adding a new module to a range of electives, developing a new stand-alone module, developing a new generic module that is available across the institution, and integrating a module across several programs for a multidisciplinary project. When initiating a pilot CBL or CBR project, the approach recommended is the first one described by Boland – to identify suitable activities already in place that could be modified. Possible examples include assignments in which case studies are used or existing outreach activities that do not receive academic credit [37]. Sutheimer [44] has provided very effective guidelines for "simple short projects" in chemistry for those starting out with community-engaged learning (see Table 14.1).

14.3
Benefits Associated with the Adoption of Community-Engaged Learning

The effects of the implementation of CBL and CBR will be now examined from the perspective of the three partners involved: the students, the HEI or school, and the community. The established and potential benefits will be examined first, and the potential issues and obstacles that can arise will then be considered in Section 14.4. Table 14.2 summarizes the benefits of community-engaged learning to academic staff, students, and community partners, as well as the relationships between them.

14.3.1
How Do Learners Gain from CBL and CBR?

When CBL and/or CBR are implemented effectively, students can gain by being engaged in active learning [56, 72], applying their discipline knowledge in an authentic real-world context [52, 73], and developing key skills (teamwork, time-management, digital literacy, problem solving, critical thinking, self-efficacy, interpersonal, and communication), as well as improving their academic performance [74, 75], confidence, sense of civic engagement, and understanding of professional roles they can enter when they qualify [37, 73, 75–77]. Also, improved retention from first to second year in higher education has been observed [78], and potential employers rate what they would usually classify as extracurricular activities such as those achieved through CBL and CBR projects highly [79]. A longitudinal study has shown that, 13 years after graduating, involvement in community engagement in college had positive and indirect effects on aspects of well-being such as personal growth, purpose in life, and life satisfaction [80]. Matusovich *et al.* have observed that the CBL program at Purdue University attracts higher rates of female students in science, technology, engineering, and mathematics (STEM). They interviewed female engineering students and found that they value the opportunity to apply theory in authentic situations and to learn from their peers [81]. Fitch has made the related observation that a higher proportion of female academic staff appeared to be involved in community-engaged learning in chemistry, based on a review of the literature in 2007 [50].

14.3.1.1 Personal Development and Graduate Attributes
In previous work by the author, personal development of students engaged in CBL was considered and the critical contribution of effective reflection was recognized. Learners described feeling a sense of responsibility to ensure that they did not let anyone down. They also enjoyed working toward an identified common goal in a group and felt that they had a clearer idea of the professional roles their qualification in chemistry would lead to [37]. Community-engaged learning has been examined from the perspective of graduate attributes and has been found to provide an opportunity to develop citizenship, employability, problem solving, resilience, and self-motivation [82]. The link between student-community engagement and the concept of viewing learners as active coproducers of knowledge instead of passive recipients [83] is also explored [82].

14.3.1.2 High-Impact Educational Practices
CBL and CBR are recognized as "high-impact" practices, which means that they increase the rates of student retention and student engagement [72]. The main

Table 14.2 A summary of the potential benefits from community-engaged learning (community-based learning, CBL and/or community-based research, CBR) for academic staff, students, and community partners.

Benefits for academic staff	Benefits for students	Benefits for community partners
Increased awareness of community issues related to a discipline and opportunities to connect teaching and research	Application of academic knowledge and skills to the complexity of a real-world situation	Mutually beneficial relationship with an educational institution fostered as well as their understanding of the community partner's mission and goals
Opportunities for scholarship and publication and new pathways for research	Exploration of the role they will have in the future as a professional and how they can contribute to the community	Overcome budget and time constraints to pursue objectives (having factored in their required time commitment)
When multidisciplinary projects are implemented, networking with colleagues in other subject areas is facilitated	Development of key skills (collaboration, critical-thinking, digital literacy, organization, and communication)	Application of specialized expertize and equipment in the partner institution to community projects
New perspectives and understanding of how learning takes place	Improved self-confidence and self-efficacy	Potential for greater impact on decision-making as a result of evidence from CBR projects
No need to try to find time outside of work for civic engagement, and meaningful involvement with communities	No need to try to find time outside of academic studies for civic engagement, and meaningful involvement with communities	Contribute to the education process and create student awareness of civic engagement and of their goals and needs
Opportunities for learning (from community partners, peers, and mentoring of students), sharing knowledge and ideas and personal growth	Opportunities for learning (from community partners, peers academic staff, and own reflection), sharing knowledge and ideas and personal growth	Opportunities for learning (from academic partners and mentoring of students), sharing knowledge and ideas and personal growth

(continued overleaf)

Table 14.2 (Continued)

Benefits for academic staff	Benefits for students	Benefits for community partners
The positive effects on student learning and retention associated with CBL and CBR contribute to greater satisfaction among staff and students Involvement in a more engaging and interesting learning experience for staff and students Working toward a common goal means that interactions between students and staff (and among students) are generally positive and an effective rapport develops Better understanding and awareness of social and civic responsibility among academic staff and students		—

source of data used is the National Survey of Student Engagement in the United States, and it is shown that the benefits are greater for students from communities that are traditionally underrepresented as well as those who enter higher education with lower academic performances at second level. In addition, the case is made that the 10 "high-impact" activities identified prepare learners for twenty-first century challenges by developing the necessary key skills. It is shown that CBL and CBR have a strong positive effect on engagement in deep approaches to learning as well as self-reported practical, personal, and general gains [72]. Kuh asserts that these practices need to become the norm instead of the exception and that the aim should be to include at least two "high-impact" practices in every degree [72]. He weighs up the short-term costs of implementing these approaches against the long-term costs of not providing graduates with the necessary skills and not effectively addressing engagement and retention issues. Figures 14.2–14.5 show photographs of community-engaged learning in action in the author's institution. The projects represented in Figures 14.2 and 14.3 are referred to in Table 14.1, and more detail is provided in the relevant journal article [37]. Figure 14.4 shows a project recently initiated in which chemistry students worked with the charity Wells for Zoe in Malawi for six weeks as part of a work experience placement. Figure 14.5 relates to interactive "Slice of Science" demonstrations devised and implemented by students for a CBL project in collaboration with an after-school initiative, the Aisling Project.

14.3.2
How Do HEIs and Schools Gain from CBL and CBR?

The positive impacts that CBL and CBR can have on HEIs and schools can be considered in terms of individual staff and the college or school as a whole. One of the benefits to staff is that they can work with the community as part of their working day and do not have to try to find extra time to do so [52]. They also gain an

Figure 14.2 Demonstration of fingerprinting techniques by a DIT student to a local school group. This formed part of a case study devised for a Chemistry at Work event supported by the Royal Society of Chemistry [37].

Figure 14.3 A survey and breathalyzer testing being performed by a DIT student to raise awareness of levels of alcohol in the system the morning after consuming alcoholic drinks. This work was part of the College Awareness of Road Safety project in conjunction with the Garda Road Safety Unit [37].

increased awareness of community issues that are related to their discipline and find opportunities to connect teaching and research. Involvement in real-world projects is a more engaging and interesting experience for staff as well as students and provides opportunities for learning and personal growth and new pathways for applied research [5]. In addition, the positive effects on student learning

Figure 14.4 A DIT student helping to teach science in a secondary school in Malawi. The charity Wells for Zoe is the community partner [84].

Figure 14.5 A model of vanillin prepared during a "Slice of Science" activity devised by DIT students to get children to build models of compounds with distinctive fragrances at an after-school initiative.

contribute to greater satisfaction among staff [73]. Academic staff find it rewarding to work with more engaged and motivated learners toward a common goal, and interactions are more positive [37]. There are opportunities for scholarship and publication, and, when multidisciplinary projects are implemented, networking with colleagues in other subject areas is facilitated.

Benefits to a HEI or school as an organization include the associated good publicity, positive community relations, greater access to community resources, and improved student retention [73], as well as students having more positive attitudes toward them. Also, CBL and CBR provide a means for colleges and schools to demonstrate that they are outward-looking and are fostering graduate attributes such as civic engagement and social responsibility [79] and therefore to attract new students.

14.3.3
How Do Communities Gain from CBL and CBR?

The underpinning philosophy of community-engaged learning should be that the activities undertaken will be beneficial to the community partner and that the needs of the community are given equal consideration to those of the students (see Figure 14.1). Achieving this balance requires open, honest, and respectful communication between the academic institution and community partners as well, as revisiting the project to ensure that this balance is being maintained [85]. A clear explanation of the terms used and the underlying rationale is a fundamental initial step. Expectations of all stakeholders then need to be managed to ensure that they are realistic. The restrictions of academic semesters and the requirement for confirmation by an accredited laboratory of students' analytical data are important to communicate. Students and academic staff need to recognize that they can learn a great deal from their community partners who will have particular expertise and skills and that a collaborative approach will lead to useful outcomes. If CBL is applied in a way that emphasizes the role of students as providers of knowledge and members of the community as reliant on this assistance, stereotypes and attitudes can be reinforced by the associated "hidden" curriculum and an important opportunity for mutual learning and development will have been ignored [86].

14.3.3.1 Reciprocity

The principle of reciprocity in the community–academic institution partnership is very important to establish. Holland [87, p. 2] has described it as "respect for different sources of knowledge, different contributions of each participant, a fair exchange of value, and the assurance of benefits to all participants." She conducted research with Sandy to investigate the perspectives of community partners and found that they had a substantial depth of understanding and commitment to student learning and proposed that a partnership approach may result in HEI and community partners becoming more committed to the mission, values, and goals of the other [85]. A recent study has shown that the conceptualization of reciprocity in CBL requires further attention [88].

14.3.3.2 Maximizing Impact for Community Partners

Chupp and Joseph examined the impact of CBL on students, the HEI, and the community and compiled the lessons learned from their work on developing university–community partnerships [5]. Three of these focus on the community perspective:

- A multiyear commitment by the HEI is necessary to develop an authentic relationship.
- An awareness of the demands being made on community members is important, as is coordination, to ensure that requests for input are not duplicated and designated points of contact with the HEI are established.
- Clear communication about the constraints of the academic calendar and student assignments is required.

They also emphasize the need for an explicit focus on goals and benefits to the community partner to maximize impact. An unanticipated benefit in their case was that CBL catalyzed an increased engagement of community members in decision making with local government and other organizations [5].

14.4
Barriers and Potential Issues When Implementing Community-Engaged Learning

The implementation of CBL and/or CBR is unlikely to run perfectly on the first iteration. It is important to review and evaluate carefully at this stage and to consult with colleagues and peers who have relevant experience. There are several common issues that may arise, as identified in the literature, which will now be discussed. In some cases, they have already been alluded to earlier in this chapter.

14.4.1
Clarity of Purpose

It is important to establish a clear understanding of what CBL and CBR are, as well as the related aims at a departmental and, if possible, an institutional level before implementing them and before interacting with a potential community partner. This means that the terminology, characteristics, and the underlying rationale need to be communicated unambiguously [37]. The induction period, when the concept of community-engaged learning and the project involved is introduced to students, is a critical time. If possible, an opportunity for them to meet other students who have been involved in a CBL or CBR project or anyone responsible for promoting and facilitating community-engaged learning in the HEI should be arranged.

14.4.2
Regulatory and Ethical Issues

Chemistry practical work requires careful consideration of health and safety issues. Projects also often have legal and ethical issues [89, 90]. When an analysis is carried out by a student in a college laboratory, a disclaimer should be provided to say that results would need to be checked by an accredited laboratory [35, 37, 41]. It is also important to establish whether any screening or vetting procedures established under Child Protection Legislation need to be followed if students and staff will be working with children. Outreach and fund-raising events are likely to require that a risk assessment be completed and, if students are traveling off campus, insurance cover requirements will have to be checked. These issues and the time that they require need to be anticipated. In general, when they have been dealt with once, experience is gained on how to process these requirements efficiently. In HEIs where community-engaged learning is well developed, guidance is often available and processes have been streamlined.

14.4.3
Developing Authentic Community Partnerships

14.4.3.1 Useful Frameworks

The importance of reciprocity in academic institution–community relationships has been discussed in Section 14.3.3.1. Several frameworks have been proposed to facilitate the development of effective community partnerships. These include the Community Impact Framework designed to maximize the impact of CBR [8] and the Strengthening Communities Initiative developed to institutionalize CBL to revitalize urban neighborhoods [91]. Stoeker *et al.* have developed an adaptation of the CBR model called *project-based research* [9]. The Science Shop model described in Section 14.1.2 is another example of a method that establishes a genuine community need and develops an authentic partnership [11–15].

14.4.3.2 Case Studies on Developing Authentic Community Partnerships

Gardella *et al.* describe 12 years of experience in community-engaged learning in the area of environmental chemical analysis during which long-term partnerships have been established with six communities in Western New York [41]. This program involved a switch from traditional field studies to large class-based projects that required the validation of the methods and data collected. Key changes to the curriculum included a shift of focus to standard analytical methods instead of newer ones, a greater emphasis on statistical analysis and data quality, and a change to a model of implementing projects that responded to community needs and concerns. Essential requirements for success were community participation in the design and implementation of the study, preparation of students to interact with the community, and sustaining collaborations beyond the end of the semester by recruiting some students into longer term undergraduate and postgraduate research projects [41]. The article includes some very useful "Rules of Engagement" for academic staff and students interacting with community members.

A case study from Trent University in Canada shows how several challenges were dealt with when developing a CBL project in environmental science [92]. The partnership was with a nonprofit organization in a rural community located 125 km from the college and involved a large first-year class (104 students). The focus was to establish a mutually beneficial partnership, and the two main factors that influenced this were that the collaboration was initiated and facilitated by an independent broker located in the community and the structure of the CBL experience was co-designed in meetings between the community partner, broker, and college instructors [92]. Although the involvement of a third-party broker may not be possible in many HEIs, it is a useful model to be aware of, and details provided on how logistical issues related to this project were dealt with are very helpful.

14.4.4
Sustainability

While involvement in community-engaged learning is a rewarding experience, it often necessitates a significant time commitment from college staff and community members, particularly during initial implementation. Sustainability is therefore an important consideration, and it is important to consider the measures to achieve it as well as synergies that can be maximized. Examples include requesting that academic colleagues contribute some time on a rotational basis, particularly if there is a school outreach element involved, requiring students to find their own community partner in some instances [37], building in a separate CBL project which allows past participants to act as teaching assistants, referring to CBL projects as examples when reviewing chemistry topics learned for the semester [32], and developing CBR projects from initial CBL activities. In some cases, with large class groups, it may be prudent to provide the CBL or CBR project as an elective module [32, 92], while for smaller cohorts it may be necessary to make the community-engaged learning activity mandatory, as providing an alternative is too resource-intensive [37]. Sutheimer's proposals for simple and short projects are presented in Table 14.1 and provide an excellent basis for the development of sustainable projects, particularly if institutional resources and support for community-engaged learning are not well established [44]. The project described in Section 14.2 on working with the Girl Scouts to develop a chemistry badge allows greater flexibility when interacting with community partners than working with schools, as evening and weekend sessions are possible [47, 48]. Technology can contribute to sustainability by facilitating interactions with schools and other communities that are located a considerable distance away [36]. The embeddedness of CBL or CBR in the curriculum and an academic's orientation toward civic engagement have been found to be indicators of sustainability [4]. Concerns about workload and institutional recognition have a negative effect [4]. The institutional culture and structures have a very significant influence on sustainability [6, 9, 10].

If institutional support is in place, the most effective means to develop an authentic community partnership is to plan for the long-term, up to 5 years [93]. In this way, work performed in one academic year is built on the following year [50]. It may also be possible to have undergraduate research students work on a project over the summer [41]. Multidisciplinary projects are the most complex in organizational terms but are the most likely to achieve reciprocity [4, 93]. Although more difficult to establish, they may represent a more sustainable option for academic staff once the relationship with the community partner has been developed. Also, the synergies associated with working as part of a multidisciplinary team are valuable. If the HEI has assigned staff to support CBL and CBR, they will often focus efforts on longer term multidisciplinary projects. Examples of such projects in the author's institution are the College Awareness of Road Safety program (see Figure 14.3), the Lifeline urban renewal project [37], and a partnership with the charity, Wells for Zoe (see Figure 14.4) [84].

14.4.5
Institutional Commitment and Support

As pointed out in the previous section, commitment at the institutional level has a strong influence on sustainability. Community-engaged learning can either be facilitated or hindered by the institutional culture [10]. It is recommended that a philosophy of community engagement is in place at an institutional level in order for CBL to be effective [79, 94]. At a policy level, there has been a marked shift toward an engagement agenda internationally. This is a positive development but changing practices and the implementation of university–community engagement policies has often proved to be difficult [10]. Some recent research has examined staff reward policies in place in HEIs where community engagement is well established [95, 96].

14.4.6
An Authentic Learning Environment

The real-world context of CBL and CBR projects results in academic staff having less control than when a project is running in-house. For example, external factors may result in cancelations or a requirement to reschedule. This can result in last minute changes of plan and rearrangements. A benefit is that students gain an appreciation of the "messiness" that is involved in real-life projects where there is no one correct answer as well as the importance of resourcefulness, innovation, and pragmatism in these circumstances [97]. The challenge of developing assessments for the range and complexity of skills required of twenty-first century graduates can often be effectively addressed by the authentic nature of a CBL or CBR project [98].

14.4.7
Reflection

The requirement for critical reflection by students on their CBL or CBR experience can prove to be a significant issue [4, 37]. Unlike some other disciplines, chemistry undergraduates have not usually been asked to reflect on their feelings, attitudes, and personal experiences before. Their lecturers also often do not have much experience of guiding students through this type of reflective process. The provision to students of prompts and questions to reflect on is recommended, and academic staff may be able to seek training from academic development staff or their community-engagement support office or consult an experienced peer or colleague. The "What, So what? Now what?" framework to guide reflection developed by Rolfe et al. is very useful [99] and related rubrics have been developed for assessment [100]. Glover et al. provide a detailed account of the application of this framework to a chemistry CBL project [29]. It is not easy to ensure that critical reflection takes place, but it is very worthwhile to provide this opportunity to do so on the basis of the value of reflective learning in dealing with real-world experiences [5, 10, 37, 101]. An added benefit is that learners will have had the opportunity to consider the resulting skills and knowledge they have developed before they enter the employment market.

14.5 Current and Future Trends

14.5.1 Geographic Spread

There has been an increase in the number of publications on community-engaged learning in chemistry over the past 5 years, and their geographic spread has widened. This mirrors the situation across all disciplines. For example, some recent articles describe the implementation of CBL and/or CBR in Hong Kong [102] and Australia [79], and organizations that provide the infrastructure to promote and support civic engagement are now established in most regions [67, 103]. International service-learning (ISL) and international civic engagement (ICE) involve staff and students in a college or school working with a community partner in another country and are a means to extend community-engaged learning geographically with regard to the community involved [104–106]. Examples include a project in which staff and students at an American college worked with a not-for-profit organization that operated schools in Bolivia [107] and one in which an Australian college worked with a community in Kenya [79]. A chemistry project described in Table 14.1 involved students and staff at schools in Western Australia working with community partners in schools in Tanzania [39]. Another approach is to establish a binational partnership such as the one Austin describes [108]. This program was developed to support a community in a town on the Mexico–United States border and involves students and staff in colleges and schools in both countries working with local government, business, and not-for-profit organizations to address community problems.

A driver behind this increase in prominence and geographic spread of CBL and CBR is that the engagement agenda has come to the forefront internationally and there have been a series of calls for HEIs to develop links and partnerships with the community over the last 10 years [6, 10]. This bodes well for those implementing CBL or CBR or about to. This emphasis can be seen in recent Science in Society projects such as those in the European Union seventh Framework for Research and Innovation and public participation in research programs [19].

14.5.2 Economic Uncertainty

The economic uncertainty in the recent past has led to increased pressure on resources in community organizations and schools and colleges. This has added to the difficulty of implementing CBL and CBR [25]. However, an opposite effect has also resulted, as there is a renewed emphasis on the role of HEIs in addressing public issues and having a greater public purpose [109]. A related trend has been a focus on the role of higher education in economic development [26, 110] and the development of graduate employability skills. Employability considerations are important, but they should not be the sole focus of undergraduate study. Society needs graduates who can make a valuable contribution in a number of ways and employment is a significant facet but not the only one [82]. Achieving the appropriate balance is a challenging task, but it is important that the need to incorporate community engagement into courses is recognized. In addition, as discussed in Section 14.3.1, a range of key skills are developed by students who undertake CBL and CBR projects, thus enhancing their employment prospects.

14.5.3
The Scholarship of Community-Engaged Learning

The development of the scholarship of community-engaged learning is associated with the increased awareness and visibility [96, 111]. There are greater opportunities now for academic staff to publish in peer-reviewed journals and to present their findings at relevant conferences. The higher workload associated with CBR and CBL is more acceptable when their outcomes can be disseminated in this way [97]. A related topic is the potential of community-engaged learning as a foundation for research. Opportunities can sometimes arise that lead to research questions suitable as areas of study for postgraduate research, particularly in the case of CBR. Although this should not be the primary focus of academic partners, it does provide an additional potential benefit.

14.5.4
Online Learning

e-Learning will undoubtedly impact on community-engaged learning in the future. Table 14.1 included a chemistry project that utilized technology to facilitate communication between college and second-level students, including the opportunity for them to act as virtual lab partners [36]. Waldner, McGorry, and Widener assert that the application of online learning in part or whole to CBL and CBR has two main benefits: it extends the geographical reach of projects once the partners accept the use of online interaction instead of some or all face-to-face interaction, and it provides a means to make e-learning more engaging [112]. They also provide guidance for good practice in relation to technology, communication, and course design. These principles include providing technology training for staff, students, and community partners; providing an opportunity early on for a face-to-face meeting with the community partner(s) or a real-time video conference session; using technology tools that facilitate group collaboration such as discussion boards or wikis; and establishing a back-up method of communication. Reflection by students can be accommodated by means of discussion boards, blogs, or electronic journals with appropriate privacy settings. The author has employed technology using a blended learning approach to support CBL projects by using discussion boards to facilitate communication within a group and with academic staff. In one project, a discussion board was used for interaction with students in the school we had partnered with to establish what questions they had on the chemistry topics they were studying [37]. Several online videos have also been prepared by our students to be used as learning resources for second-level students and these include interactive science demonstrations as well as mandatory experiments on the curriculum.

14.5.5
Developments in Chemistry Community-Engaged Learning

As well as an increase in the number of chemistry CBL and CBR projects, a greater diversity in the type of projects is also apparent. In general, the type of projects implemented show considerable ingenuity and resourcefulness but they can usually be classified as involving environmental analytical chemistry or outreach to schools. These are important fields to which the discipline can contribute and represent two categories of activity that are well suited to community-engaged learning in chemistry. However, projects that relate to some other areas have been reported recently (see Table 14.1). These include students preparing information leaflets on health issues for a local homeless shelter [31], developing exhibitions in museums [32, 44], dyeing baby clothes with naturally extracted dyes [44], and measuring fat levels in fast foods [35]. The opportunity to get involved in long-term multidisciplinary projects may also provide the means to consider less obvious ways in which chemistry can be applied to the benefit of a community partner. Aspects of recent projects that are particularly innovative include offering past participants the opportunity to gain credit by acting as teaching assistants to current students [32], the extension of a schools outreach project from a college to a secondary school by mentoring the secondary school students to perform outreach to their peers [63], the application of technology to facilitate communication with students in schools [36], and the development of a chemistry badge in collaboration with local Girl Scouts so that outreach could be extended to this community group [47, 48].

14.6
Conclusion

There is much to recommend community-engaged learning as a pedagogy. As discussed in Section 14.3, evidence shows that there are benefits to the learner (development of key skills and civic engagement and applying knowledge to an authentic problem or issue), educator (learner engagement, new pathways for scholarship, and research), and their educational institutions (student retention and recruitment, positive community relations). The underlying principles dictate that there should always be a benefit to the community partner and that the activities undertaken should meet particular needs of theirs. This can be a challenging requirement, but it is an essential tenet of this pedagogical approach; measures to achieve it have been discussed in Section 14.4. The projects and activities that are reviewed in this chapter showcase creativity and commitment as well as the critical contribution that the chemical sciences can make to society.

Although not yet widespread, indications are that a critical mass of educational institutions and academic staff who practice chemistry community-engaged learning is developing and the next couple of years promise to be both exciting and challenging. In the words of Kenworthy-U'ren, "we have challenges to address, communities to partner with, students to interact with, and an unlimited world of learning to explore" [97, p. 819]. The hope is that those who have already engaged will continue to do so and that the steady increase in activity will continue as dissemination of existing practice encourages more chemistry educators and learners and community partners to experiment with community-engaged learning.

References

1. Einstein, A. (1995) *Ideas and Opinions*, Broadway Books, New York, from address, October 15, 1936, reprinted in Einstein A.
2. Boland, J.A. and McIlrath, L. (2007) The process of localising pedagogies for civic engagement in Ireland: the significance of conceptions, culture and context, in *Higher Education and Civic Engagement: International Perspectives* (eds L. McIlrath and I. Mac Labhrain), Ashgate, Hampshire.
3. Bandy, J. (2013) A Word on Nomenclature (Service Learning and Community Engagement), http://cft.vanderbilt.edu/teaching-guides/teaching-through-community-engagement/a-word-on-nomenclature/ (accessed 22 October 2013).
4. Boland, J. (2013) in *Emerging Issues in Higher Education III: From Capacity Building to Sustainability* (eds C. O'Farrell and A. Farrell), EDIN, Athlone, pp. 210–224.
5. Chupp, M.G. and Joseph, M.L. (2010) Getting the most out of service learning: maximizing student, university and community impact. *J. Community Pract.*, **18**, 190–212.
6. Holland, B. and Ramaley, J. (2008) Creating a supportive environment for community-university engagement: conceptual frameworks. Proceedings of the 31st HERDSA Annual Conference, Rotorua, New Zealand, July 1–4, 2008, HERDSA, New South Wales, pp. 11-25.
7. Bringle, R.G. and Hatcher, J.A. (1996) Implementing service learning in higher education. *J. Higher Educ.*, **67** (2), 221–239.
8. Beckman, M., Penney, N., and Cockburn, B. (2011) Maximizing the impact of community-based research. *J. Higher Educ. Outreach Engagement*, **15** (2), 83–104.
9. Stoecker, R., Loving, K., Reddy, M., and Bollig, N. (2010) Can community-based research guide service learning? *J. Community Pract.*, **18**, 280–296.
10. O'Connor, K.M., McEwen, L.J., Owen, D., Lynch, K., and Hill, S. (2011) Literature Review: Embedding Public/Community Engagement in the Curriculum: An Example of University-Public Engagement. Report for the National Co-ordinating Centre for Public Engagement, Bristol.
11. DeBok, C. and Steinhaus, N. (2008) Breaking out of the local: international dimensions of science shops. *Gateways: Int. J. Community Res. Engagement*, **1**, 165–178.

12. Leydesdorff, L. and Ward, J. (2005) Science shops: a kaleidoscope of science-society collaborations in Europe. *Public Underst. Sci.*, **14**, 353–372.
13. Mulder, H.A., Jorgensen, M.S., Pricope, L., Steinhaus, N., and Valentin, A. (2006) Science shops as science society interfaces. *Interfaces Sci. Soc.*, **1** (48), 278–296.
14. Rodríguez, F. (2011) Degree of public participation in science shops: an exploratory study. Masters dissertation. Bielefeld University.
15. Tryon, E. and Ross, J.A. (2012) A community-university exchange project modeled after Europe's science shops. *J. Higher Educ. Outreach Engagement*, **16** (2), 197–212.
16. Israel, B.A., Krieger, J., Vlahov, D., Ciske, S., Foley, M., Fortin, P., Guzman, J.R., Lichtenstein, R., McGranaghan, R., Palermo, A., and Tang, G. (2006) Challenges and facilitating factors in sustaining community-based participatory research partnerships: lessons learned from the Detroit, New York City and Seattle Urban Research Centers. *J. Urban Health*, **83** (6), 1022–1040.
17. Gardella, J.A. Jr., Milillo, T.M., Sinha, G., Oh, G., Manns, D.C., and Coffey, E. (2009) in *Civic Service: Service-Learning with State and Local Government Partners* (eds T. Rice and D. Relawask), Jossey Bass Publishers, San Francisco, CA, pp. 98–122.
18. Shirk, J.L., Ballard, H.L., Wilderman, C.C., Phillips, T., Wiggins, A., Jordan, R., McCallie, E., Minarchek, M., Lewenstein, B.V., Krasny, M.E., and Bonney, R. (2012) Public participation in scientific research: a framework for deliberate design. *Ecol. Soc.*, **17** (2), 29.
19. Cornell Lab of Ornithology Citizen Science Central Website http://www.birds.cornell.edu/citscitoolkit (accessed 23 October 2013).
20. Taylor, C. (2009) Incorporating Students into a Research Program as Part of the First Year Biology Curriculum, Enhancing Assessment in the Biological Sciences, Example of Practice, http://bioassess.edu.au/sites/default/files/Taylor%204.pdf (accessed 22 May 2014).
21. Gamble, E. and Bates, C. (2011) Dublin Institute of Technology's programme for students learning with communities: a critical account of practice. *Education + Training*, **53** (2/3), 116–128.
22. Furco, A. (1996) *Expanding Boundaries: Serving and Learning*, Corporation for National Service, Washington, DC, pp. 2–6.
23. Kenworthy-U'Ren, A., Taylor, M.L., and Petri, A. (2006) Components of successful service-learning programs: notes from Barbara Holland, Director of the US National Service-Learning Clearinghouse. *Int. J. Case Method Res. Appl.*, **18** (2), 120–129.
24. AUCEA Inc. (2008) Australian Universities Community Engagement Alliance Position Paper 2008-2010, http://admin.sun.ac.za/ci/resources/AUCEA_universities_CE.pdf (accessed 18 June 2013).
25. Holland, B.A. (2010) Preface to special issue on service learning: community engagement and partnership for integrating teaching, research, and service. *J. Community Pract.*, **18** (2-3), 135–138.
26. Trani, E.P. and Holsworth, R. (2010) *The Indispensable University: Higher Education, Economic Development, and the Knowledge Economy*, Rowman & Littlefield Publishers, Plymouth.
27. Burand, M.W. and Ogba, O.M. (2013) Letter writing as a service-learning project: an alternative to the traditional laboratory report. *J. Chem. Educ.*, **90** (12), 1701–1702.
28. Webb, J.A. (2013) Integrating SENCER into a large lecture general education chemistry course. *Sci. Educ. Civ. Engagement*, **5** (2), 32–37.
29. Glover, S.R., Sewry, J.D., Bromley, C.L., Davies-Coleman, M.T., and Hlengwa, A. (2013) The implementation of a service-learning component in an organic chemistry laboratory course. *J. Chem. Educ.*, **90**, 578–583.
30. Theall, R.A.M. and Bond, M.R. (2013) Incorporating professional service as a component of general chemistry laboratory by demonstrating chemistry to

elementary students. *J. Chem. Educ.*, **90**, 332–337.

31. Harrison, M.A., Dunbar, D., and Lopatto, D. (2013) Using pamphlets to teach biochemistry: a service-learning project. *J. Chem. Educ.*, **90**, 210–214.

32. Donaghy, K.J. and Saxton, K.J. (2012) Service learning track in general chemistry: giving students a choice. *J. Chem. Educ.*, **89**, 1378–1383.

33. Kammler, D.C., Truong, T.M., VanNess, G., and McGowin, A.E. (2012) A service-learning project in chemistry: environmental monitoring of a nature preserve. *J. Chem. Educ.*, **89**, 1384–1389.

34. Knutson-Person, J.L. (2011) What are the effects of science outreach by college students with elementary school children? Masters dissertation. Montana State University.

35. Hosten, C.M., Talanova, G., and Lipkowitz, K.B. (2011) Introducing undergraduates to the role of science in public policy and in the service of the community. *Chem. Educ. Res. Pract.*, **12**, 388–394.

36. Saitta, E.K.H., Bowdon, M.A., and Geiger, C.L. (2011) Incorporating service-learning, technology, and research supportive teaching techniques into the university chemistry classroom. *J. Sci. Educ. Technol.*, **20**, 790–795.

37. McDonnell, C., Ennis, P., and Shoemaker, L. (2011) Now for the science bit: implementing community-based learning in chemistry. *Education + Training*, **53** (2/3), 218–236.

38. Cartwright, A. (2010) Science service learning. *J. Chem. Educ.*, **87**, 1009–1010.

39. Murcia, K., Haigh, Y., and Norris, L. (2010) Learning from community service: engaging Australia Tanzania Young Ambassadors with sustainability. *Issues Educ. Res.*, **20** (3), 294–313.

40. Furlan, P.Y. (2009) Engaging students in early exploration of nanoscience topics using hands-on activities and scanning tunneling microscopy. *J. Chem. Educ.*, **86**, 705–711.

41. Gardella, J.A., Milillo, T.M., Sinha, G., Oh, G., Manns, D.C., and Coffey, E. (2007) Linking community service, learning, and environmental analytical chemistry. *Anal. Chem.*, **79** (3), 810–818.

42. Franco, R.W. (2009) Service-Learning: Reconciling Research and Teaching, Tackling Capacious Issues, SENCER (Science Education for New Civic Engagements and Responsibilities).

43. Kalivas, J.H. (2008) A service-learning project based on a research supportive curriculum format in the general chemistry laboratory. *J. Chem. Educ.*, **85**, 1410–1415.

44. Sutheimer, S. (2008) Strategies to simplify service-learning efforts in chemistry. *J. Chem. Educ.*, **85**, 231–233.

45. Sutheimer, S. and Pyles, J. (2012) in *Social Responsibility and Sustainability: Multidisciplinary Perspectives Through Service Learning* (eds T. McDonald and G.S. Eisman), Stylus Publishing, Virginia, pp. 21–34.

46. Jansen-van Vuuren, R.D., Buchanan, M.S., and McKenzie, R.H. (2013) Connecting resources for tertiary chemical education with scientists and students in developing countries. *J. Chem. Educ.*, **90**, 1325–1332.

47. Stevens-Truss, R. (2012) Starting small and going big: civic engagement at Kalamazoo College and the chemistry classroom. Proceedings of the 22nd Biennial Conference on Chemical Education, Pennsylvania State University, July 29–August 2, 2012.

48. Kalamazoo College (2013) Kalamazoo College Chemistry Department Outreach Programs Website, https://reason.kzoo.edu/chem/faculty/regina/op/ (accessed 2 August 2013).

49. Cavinato, A.G. (2007) in *Active Learning Models from the Analytical Sciences*, ACS Symposium Series (ed. P.A. Mabrouk), American Chemical Society, Washington DC, pp. 109–122.

50. Fitch, A. (2007) in *Active Learning Models from the Analytical Sciences*, ACS Symposium Series (ed. P.A. Mabrouk), American Chemical Society, Washington DC, pp. 100–108.

51. LaRiviere, F.J., Miller, L.M., and Millard, J.T. (2007) Showing the true face of chemistry in a service-learning

Outreach course. *J. Chem. Educ.*, **84**, 1636–1639.

52. Esson, J.M., Stevens-Truss, R., and Thomas, A. (2005) Service-learning in introductory chemistry: supplementing chemistry curriculum in elementary schools. *J. Chem. Educ.*, **82**, 1168–1173.

53. Karukstis, K.K. (2005) Community-based research. A new paradigm for undergraduate research in the sciences. *J. Chem. Educ.*, **82** (1), 15–16.

54. Beckman, M. and Caponigro, J. (2005) The creation of a university-community alliance to address lead hazards: three keys to success. *J. Higher Educ. Outreach Engagement*, **10** (3), 95–108.

55. Draper, A.J. (2004) Integrating project-based service-learning into an advanced environmental chemistry course. *J. Chem. Educ.*, **81** (2), 221–224.

56. Hatcher-Skeers, M.E. and Aragon, E.P. (2002) Combining active learning with service learning: a student-driven demonstration project. *J. Chem. Educ.*, **79** (4), 462–464.

57. Werner, T., Tobiessen, P., and Lou, K. (2001) The water project. *Anal. Chem.*, **73**, 84A–87A.

58. Hope, W.W. and Johnson, L.P. (2000) Urban air: real samples for undergraduate analytical chemistry. *Anal. Chem.*, **72**, 460A–467A.

59. Wiegand, D. and Strait, M. (2000) What is service learning? *J. Chem. Educ.*, **77** (12), 1538–1539.

60. Kesner, L. and Eyring, E.M. (1999) Service-learning general chemistry: lead paint analyses. *J. Chem. Educ.*, **76** (7), 920–923.

61. Ram, P. (1999) Problem-based learning in undergraduate education - A sophomore chemistry laboratory. *J. Chem. Educ.*, **76**, 1122–1126.

62. Fitch, A., Wang, Y., Mellican, S., and Macha, S. (1996) Lead lab: teaching instrumentation with one analyte. *Anal. Chem.*, **68** (23), 727A–731A.

63. Esson, J. and Johnston, W. (2010) Extending the impact of service-learning: teaching K-12 peer mentors. Proceedings of the 21st Biennial Conference on Chemical Education, University of North Texas, August 1–5, 2010, http://www.bcce2010.org/program_schedule/TuesdayAfternoonProgram.pdf (accessed 18 June 2013).

64. Essex, J. and Haxton, K. (2013) Behind the scenes at a Victorian pharmacy. *Educ. Chem.*, May, **50** (3), 26–29.

65. Sinclair, B. (2008) The undergraduate ambassador scheme implemented at St Andrews. *New Dir.*, **4**, 5–7.

66. Moss, K., Crowley, M., and Neale, N. (2009) Communicating science, engaging students, in *Proceedings of the 3rd Eurovariety in Chemistry Education Conference, University of Manchester, England, September 2–4, 2009*, Royal Society of Chemistry, London, http://www.heacademy.ac.uk/assets/documents/subjects/ps/variety_proceedings_2009.pdf (accessed 11 July 2013).

67. Bringle, R.G. (2013) in *Emerging Issues in Higher Education III: From Capacity Building to Sustainability* (eds C. O'Farrell and A. Farrell), EDIN, Athlone, pp. 225–226.

68. Bandy, J. (2013) Community Engaged Teaching Step by Step, http://cft.vanderbilt.edu/teaching-guides/teaching-through-community-engagement/community-engaged-teaching-step-by-step/ (accessed 12 October 2013).

69. Welch, M. (2010) O.P.E.R.A.: a first letter mnemonic and rubric for conceptualizing and implementing service-learning courses. *Aust. J. Educ. Res.*, **20** (1), 76–82.

70. UVU (2013) Utah Valley University Service-Learning Definition and Guiding Principles, https://www.uvu.edu/volunteer/faculty/resources.html (accessed 25 July 2013).

71. McEwen, L. and O'Connor, K.M. (2013) Building Staff/Faculty Capacity for University – Public/Community Engagement.

72. Kuh, G.D. (2008) *High-Impact Educational Practices: What Are They, Who has Access to Them, and Why They Matter*, Association of American Colleges and Universities, Washington DC.

73. Eyler, J.S., Giles, D.E. Jr., Stenson, C.M., and Gray, C.J. (2001) *At a Glance: What We Know About the Effects of*

Service-Learning on College Students, Faculty, Institutions and Communities, 1993-2000, Vanderbilt University, Nashville, TN.
74. Eyler, J. and Giles, D.E. Jr., (1999) *Where's the Learning in Service-Learning? Jossey-Bass Higher and Adult Education Series*, Jossey-Bass, Inc., San Francisco, CA.
75. Celio, C.I., Durlak, J., and Dymnicki, A. (2011) A meta-analysis of the impact of service-learning on students. *J. Experiential Educ.*, **34** (2), 164–181.
76. Thompson, M., Oakes, W., and Bodner, G. (2005) A qualitative investigation of a first-year engineering service-learning program. Proceedings of the 2005 American Society for Engineering Education Annual Conference and Exposition.
77. Corporation for National and Community Service (2007) The Impact of Service-Learning: A Review of Current Research, http://www.nationalservice.gov/sites/default/files/documents/issuebrief_servicelearning.pdf (accessed 7 November 2013).
78. Bringle, R.G., Hatcher, J.A., and Muthiah, R.N. (2010) The role of service-learning on the retention of first-year students to second year. *Mich. J. Community Serv. Learn.*, **16** (2), 38–49.
79. Caspersz, D., Kavanagh, M., and Whitto, D. (2012) Can service-learning be institutionalised? The case study of SIFE in Australia. *Australas. J. Univ.-Community Engagement*, 7 (10), 39–54.
80. Bowman, N., Brandenberger, J., Lapsley, D., Hill, P., and Quaranto, J. (2010) Serving in college, flourishing in adulthood: does community engagement during the college years predict adult well-being? *Appl. Psychol.: Health Well-Being*, **2** (1), 14–34.
81. Matusovich, H., Follman, D., and Oakes, W. (2006) Work in progress: a student perspective-why women choose service-learning. Proceedings of the 36th ASEE/IEEE Frontiers in Education Conference, San Diego, CA, October 28–31, 2006, pp. 7–8.
82. O'Connor, K.M., Lynch, K., and Owen, D. (2011) Student-community engagement and the development of graduate attributes. *Education + Training*, **53** (2/3), 100–115.
83. McCulloch, A. (2009) The student as co-producer: Learning from public administration about the student–university relationship. *Stud. Higher Educ.*, **34** (2), 171–183.
84. Dublin Institute of Technology (2012) Dublin Institute of Technology Students Learning With Communities Multidisciplinary Project Poster, http://www.dit.ie/media/ace/slwc/projectpostersinnovationdublin/Wells%20for%20Zoe.pdf (accessed 11 July 2013).
85. Sandy, M. and Holland, B.A. (2006) Different worlds and common ground: community partner perspectives on campus-community partnerships. *Mich. J. Community Serv. Learn.*, **13** (1), 30.
86. Boyle, M.E. (2007) Learning to neighbor? Service-learning in context. *J. Academic Ethics*, **5** (1), 85–104.
87. Holland, B.A. (2002) Every perspective counts: understanding the true meaning of reciprocity in partnerships. Keynote Address to the Western Regional Campus Compact Conference, http://www.coloradocollege.edu/dotAsset/c7cca881-4057-46a0-99d7-ba2f615dd6ec.pdf (accesssed 22 June 2013).
88. Caspersz, D., Olaru, D., and Smith, L. (2012) Striving for definitional clarity: what is service learning? Proceedings of the 21st Annual Teaching Learning Forum, Perth, Australia, February 2–3, 2012.
89. Fitch, A., Repman, A., and Schmidt, J. (1999) The Ethics of Community/Undergraduate Collaborative Research in Chemistry Monograph Chapter sponsored by the Association of Higher Education.
90. Rovner, S.L. (2004) Ethics 101 – Ethics education makes its way into the college chemistry curriculum, but is that too late to start? *Chem. Eng. News*, **82** (17), 33–35.
91. Norris-Tirrell, D., Lambert-Pennington, K., and Hyland, S. (2010) Embedding

service learning in engaged scholarship at research institutions to revitalize metropolitan neighborhoods. *J. Community Pract.*, **18**, 171–189.
92. Hill, S.D., Loney, R.K., and Reid, H. (2010) Notes from the field: brokering service learning between a rural community and large undergraduate class: insights from a case study. *J. Community Pract.*, **18**, 396–412.
93. Rosing, H. and Hofman, N.G. (2010) Notes from the field: service learning and the development of multidisciplinary community-based research initiatives. *J. Community Pract.*, **18**, 213–232.
94. Holland, B. (1997) Analyzing institutional commitment to service: a model of key organizational factors. *Mich. J. Community Serv. Learn.*, Fall, **4** (1), 30–41.
95. Saltmarsh, J., Giles, D.E. Jr., O'Meara, K., Sandmann, L., Ward, E., and Buglione, S.M. (2009) in *Creating Our Identities In Service-Learning And Community Engagement. Advances In Service-Learning Research* (eds B.E. Moely, S.H. Billig, and B.A. Holland), IAP Information Age Publishing, Charlotte, NC, pp. 3–29.
96. Sandmann, L.R. (2009) Community engagement: second-generation promotion and tenure issues and challenges, in *The Future of Service Learning* (eds J.R. Strait and M. Lima), Stylus Publishing, Virginia.
97. Kenworthy-U'Ren, A.L. (2008) A decade of service-learning: a review of the field ten years after JOBE's seminal special issue. *J. Bus. Ethics*, **81** (4), 811–822.
98. Steinke, P. and Fitch, P. (2007) Assessing service-learning. *Res. Pract. Assess.*, **1** (2), 1–8.
99. Rolfe, G., Freshwater, D., and Jasper, M. (2001) *Critical Reflection for Nursing and the Helping Professions: A User's Guide*, Palgrave, Basingstoke.
100. Winona State University (2013) Rubric for Assessing Service-Learning Reflections, http://course1.winona.edu/shatfield/air/reflectionrubric-1.pdf (accessed 26 July 2013).
101. Moon, J. (2005) *Learning Through Reflection*, Guide for Busy Academics, vol. **4**, The Higher Education Academy..
102. Ma, H.K. and Chan, C.M. (2013) A Hong Kong University first: establishing service-learning as an academic credit-bearing subject. *Gateways: Int. J. Community Res. Engagement*, **6**, 178–198.
103. Millican, J. and Bourner, T. (2011) Student-community engagement and the changing role and context of higher education. *Education + Training*, **53** (2/3), 89–99.
104. Crabtree, R.D. (2008) Theoretical foundations for international service-learning. *Mich. J. Community Serv. Learn.*, **15** (1), 18–36.
105. Crabtree, R.D. (2013) The intended and unintended consequences of international service-learning. *J. Higher Educ. Outreach Engagement*, **17** (2), 43–66.
106. Klak, T. and Mullaney, E.G. (2013) Levels and networks in community partnerships: a framework informed by our overseas partners. *Gateways: Int. J. Community Res. Engagement*, **6** (1), 1–21.
107. de Figueiredo, J.N., Keffer, A.M.J., Barrientos, M.A.M., and Gonzalez, S. (2013) A robust University-NGO partnership: analysing school efficiencies in Bolivia with community-based management techniques. *Gateways: Int. J. Community Res. Engagement*, **6** (1), 93–112.
108. Austin, D. (2010) Confronting environmental challenges on the US-Mexico border: long-term community-based research and community service learning in a binational partnership. *J. Community Pract.*, **18**, 361–395.
109. Ward, E., Buglione, S., Giles, D.W. Jr., and Saltmarsh, J. (2013) in *University Engagement With Socially Excluded Communities* (ed. P. Benneworth), Springer, pp. 285–308.
110. Lăzăroiu, G. (2012) The development of close linkages between universities and societal economic progress. *Econ., Manage. Financ. Mark.*, **4**, 251–257.
111. Sandmann, L., Saltmarsh, J., and O'Meara, K. (2008) An integrated model for advancing the scholarship of

engagement: creating academic homes for the engaged scholar. *J. Higher Educ. Outreach Engagement*, **12** (1), 47–64.

112. Waldner, L.S., Widener, M.C., and McGorry, S.Y. (2012) E-service learning: the evolution of service-learning to engage a growing online student population. *J. Higher Educ. Outreach Engagement*, **16** (2), 123–150.

15
The Role of Conceptual Integration in Understanding and Learning Chemistry
Keith S. Taber

15.1
Concepts, Coherence, and Conceptual Integration

This chapter explores the idea of conceptual integration in science, and in chemistry education, in order to suggest that conceptual integration should be a major focus of work in chemistry education (and science education more widely). There has been limited explicit research into this area, despite its clear significance for teaching chemistry and other subjects. It is considered that conceptual integration should be an explicit aim of science education and given a much higher profile in both research into student understanding of science and in science teaching itself. However, chemistry as a subject offers particular challenges in terms of demonstrating conceptual coherence (a prerequisite for effective integration), as it heavily draws upon both (i) concepts that have shifted over time and (ii) multiple models and representations that are used as complementary ways of understanding target concepts. As the theme of the present chapter is conceptual integration, it is useful to begin by explaining how that term is understood here.

15.1.1
The Nature of Concepts

There is extensive literature on the nature of concepts and related themes such as concept acquisition, formation, and development [1–3]. Gilbert and Watts [4] pointed out that the term "concept" is widely used to refer to both "an individual's psychological, personal, knowledge structure *and to* the organization of public knowledge systems" (pp. 64–65, present author's emphasis), and they suggested that the term "conception" should "be used to focus on the personalized theorizing and hypothesizing of individuals" (p. 69). This is an important consideration in science education where it is often important to distinguish learners' (often alternative) conceptions from the canonical concepts represented in the curriculum.

Yet, there is a potential problem with this distinction which becomes apparent if we ask what kind of "things" concepts are. Concepts are mental features – mental objects (categories) or actions (discriminations) – that are part of how humans

Chemistry Education: Best Practices, Opportunities and Trends, First Edition.
Edited by Javier García-Martínez and Elena Serrano-Torregrosa.
© 2015 Wiley-VCH Verlag GmbH & Co. KGaA. Published 2015 by Wiley-VCH Verlag GmbH & Co. KGaA.

make sense of the world. Consider as examples the concepts "acid" and "delocalization". Acids exist in the world, or at least there are entities in the world that chemists class as acids. However, arguably *the concept of* acid is not part of the physical world, but only exists in minds. Of course, many acids existed in the world before anyone came up with the idea of an acid, but without the concept no discriminations could be made to identify acids. The acid *concept* has shifted historically such that what precisely should be classed as an acid has changed over time [5]. The nature of particular substances that have been classified as acids has not changed, but whether they can canonically be considered acids certainly has.

So *there is a sense* in which acids only exist in human minds. That is not to adopt an idealist position that the material world is only a construction of the human mind. Rather, even if one adopts a realist position that the material world has existence independent of human minds (which is surely a prerequisite commitment for natural science), our thinking about it is structured through mental constructions such as concepts. There are regularities in the natural world (another basic commitment of natural science) and we might see the role of science as being to develop conceptual schemes that make the best sense of those regularities [3]. However, whilst the regularities are external to minds, their perception and conceptualization occur in the cognitive systems of individual human beings. Acids, as substances, exist regardless of what we think – acids *as acids* only exist when we make those discriminations within a conceptual framework.

Unlike "acid," "delocalization" does not refer to physical objects as such at all, but rather is a theoretical idea used as part of explanatory schemes to make sense of "observed" properties of certain substances: such as the relative stability of benzene, or the electrical conductivity of graphite or copper, for example. The term "observed" is placed in inverted commas because the descriptions used here (relative stability; electrical conductivity) do *not* refer to what can be observed in the normal meaning of the term. Rather, these are inferences from observed phenomena [6] – for example, the color being retained when benzene is mixed with dissolved bromine or the movement of an ammeter needle when a switch is closed in a test circuit. We infer that benzene does not readily undergo addition reactions, or that graphite has relatively low resistivity – but this is the theoretical language of chemistry, not the language of observed phenomena. This example reflects a good deal of the content of high school and college chemistry in that actual observed phenomena (which we try to make sense of as regularities in the natural world) are often conceptualized at two distinct levels – a theoretical redescription of phenomena at the "bench" level using chemical concepts, and at a further level of explanation based upon theoretical ideas relating to the structure of matter at submicroscopic scales [7–9]. This is represented in Figure 15.1. We explain the observations in terms of the formation of delocalized molecular orbitals from the overlap of unhybridized p orbitals – conjectured processes involving nonobservable theoretical entities.

Human conceptual development is interpretive and iterative (we make sense with and build upon our existing understanding), such that we do learn to see in terms of our concepts: over time, chemistry teachers are in effect "seeing" the

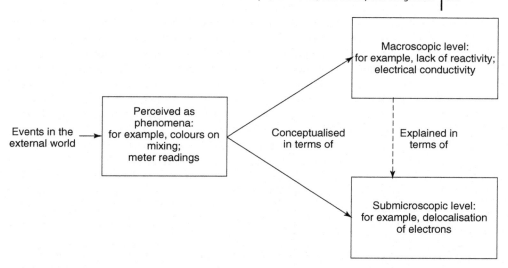

Figure 15.1 Chemists commonly conceptualize their observations at two distinct levels. (Adapted from Ref. [8], Figure 2, p.159.)

stability of benzene when a sample is shaken with bromine solution, or "seeing" the conductivity of graphite when a meter needle moves in a test circuit. That is, through developing expertise (discussed further below) over extended periods of engagement, these inferences have become automated in a somewhat similar manner to how many initially challenging motor skills can become automated, as for example, in riding a bicycle. Our students, especially when new to the subject, do not yet have the conceptual structures in place to be able to perceive phenomena as experts do. This kind of chunking in perception and learning to "see" inferences is not particular to chemistry, of course. Many readers of this chapter will have become experts in using the visual interface of their computers, and will now automatically "see" the representations on their screen *as* drives, folders, and documents, and will effortlessly copy files between computer drives by dragging (by moving a mouse or tracing on a track pad) icons across the screen.

15.1.2
Concepts and Systems of Public Knowledge

Concepts are best understood as mental entities, but as Gilbert and Watts [4] pointed out, the term "concept" is also widely used in referring to "the organization of public knowledge systems." That is, we commonly refer to canonical knowledge: the meaning of "acid" as understood in chemistry, for example. Indeed, a good deal of attention is paid in both chemistry teaching and chemical education research to aspects of student knowledge and understanding *in relation to* canonical target knowledge.

The notion of public systems of knowledge (such as canonical chemical knowledge) is therefore a very useful one: yet it is also problematic. Although we regularly refer to canonical knowledge – it is the key referent in teaching chemistry – it is rather difficult to locate. The primary literature is hardly a unified account of the subject, and in any case a constructivist perspective suggests that knowledge cannot be found in books and papers: rather they contain the (imperfect) representations of their authors' own knowledge, which have to be interpreted in the minds of readers.

We might suggest that scientific knowledge is not located in the literature but rather in the scientific community, instead. This seems a more viable suggestion if we consider that knowledge can only exist in minds [3], but the scientific community is not of one mind (literally, and on some particular matters figuratively as well); most scientists are only experts in some part of their discipline or field; and they can only at best represent their knowledge but not transfer it wholesale to anyone else.

A third option is that knowledge can be considered to exist in some "third world" of ideas and abstractions in and of themselves. This is World 3 in relation to World 1 being the material world, and World 2 that of subjective experience [10]. This Platonic notion finds a "place" to locate scientific knowledge, but – as we are not considered to have direct access to World 3 – leaves open the question of how anyone (scientist, teacher, or student) can ever really be said to know what canonical knowledge actually is. This is not the place to explore these issues in detail (see [3] for an extended discussion), but given the key role paid by canonical knowledge in education, it should be acknowledged that it ultimately proves an elusive referent. Arguably, canonical knowledge, such as scientific knowledge, is a fiction, albeit a very useful fiction – a kind of educational philosopher's stone that does useful work for us even if we can never actually firmly grasp it.

15.1.3
Conceptual Integration

We might consider concepts to be well integrated when they are organized into coherent and strongly linked structures. The idea of concepts being structured is well established, so Vygotsky wrote:

> Concepts do not lie in the child's mind like peas in a bag, without any bonds between them. If that were the case, no intellectual operation requiring coordination of thoughts would be possible, nor would any general conception of the world. Not even separate concepts as such could exist; their very nature presupposes a system [11, p. 197].

Yet this raises the question of precisely what is meant by a structure of concepts if concepts are themselves mental categories. One approach to this was taken by George Kelly, who developed the Personal Construct Theory [12] or PCT. Kelly referred to constructs but acknowledged that this was in effect what others often referred to as *concepts*. For Kelly, constructs are the basis of discrimination, and are organized into hierarchical structures. In PCT, the basic unit of perception/cognition is the bipolar construct that supports discriminations: good–bad, black–white, acid–base, and so on, and it is considered that each person develops his or her own personal system of such constructs.

Kelly's way of thinking about concepts is reflected in the methodology he developed for exploring an individual's systems of constructs: the construct repertory grid [13]. This involved a two-stage process of (i) eliciting constructs by asking respondents to discriminate a triad of three "elements" into two that were more similar in some way and one that was the "odd one out," and then repeating this for other triads, then (ii) asking respondents to suggest where each "element" was positioned on each of the elicited constructs. The outcome of the analysis is a tree-like representation (a dendrogram) of the person's constructs. In Kelly's work as a psychological counselor, the elements might have been significant others in the client's life, but the basic approach can be applied widely – so, for example, students can be asked to sort figures representing molecules and other submicroscopic chemical structures [14].

The idea that concepts (or constructs) can be seen as hierarchical and part of branching conceptual trees has informed work in conceptual analysis to inform curriculum development and planning teaching [15]. This way of thinking about concepts also has much in common with the way Chi and her colleagues have explored students' ontologies of the world [16, 17]. Chi has argued that one source of tenacious alternative conceptions among students is the development of concepts on the wrong ontological tree. In particular, when students are taught about processes, they may commonly think about these inappropriately in terms of substances: for example, heat as a substance, rather than heating as a process.

Kelly's methodology has proved a useful tool for eliciting aspects of people's implicit knowledge, but it is questionable whether hierarchical tree-like structures can capture the full complexity of our concepts. Rather, concepts have a strong associative aspect, in the sense that concepts are often better seen as embedded within propositional networks [4], akin to the kinds of concept maps that are sometimes used to represent student knowledge. That is, the "meaning" a person holds for any particular concept depends upon how that concept is understood in relation to their other concepts. As a simplistic example, if a person's concept of "element" is linked to their concept of "substance" because they understand an element to be a type of substance, then their understanding of element is clearly tied to *how they understand the notion of substance*. This is illustrated in Figure 15.2, which shows how the concept of "concept" might be understood in relation to a number of other concepts – this does include hierarchical relationships (e.g., implicit knowledge is a form of knowledge), but is not limited to these sorts of links.

The student who recognizes that antimony is an element, and whose conceptual structure includes the link that each element is a type of chemical substance, has a different conceptual understanding *of antinomy* than another student who has learnt that antimony is an element, but has not formed any further association *for element*. The same is true of whatever other links may or may not be formed in

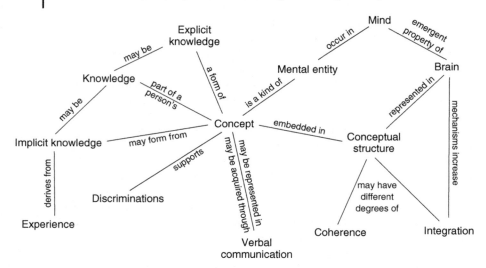

Figure 15.2 Concepts take their meaning within a network of other concepts.

individual learners' minds. Thus, in any particular class of students, there is going to be a sense in which every student has a somewhat unique concept of acid (or element, or delocalization, or metal, etc.). Consider, for example, two students in a class who when presented with the phrase "ethanoic acid + potassium hydroxide" are both able to identify this as an example of a neutralization reaction. If one of the students (but not the other) holds the alternative conception that a neutralization reaction always leads to a neutral product [18], then the two students have conceptualized the reaction differently even if their public statements (e.g., "that is a neutralization reaction") are the same. There is a kind of "conceptual inductive effect" at work here (see Figure 15.3), where specific propositions cannot be fully appreciated without knowing about adjacent linkages in conceptual structure.

The message for teachers is that it is important to both explore the links students hold for concepts under discussion, and to take time to explicitly map out the most important links whenever introducing, reviewing, or developing key chemical concepts. Given the likelihood of conceptual inductive effects, this should not be limited to primary links but also the most important indirect links. Time invested in regularly highlighting conceptual linkages in this way can help identify learners' alternative conceptions, reinforce the conceptual structure of chemistry, and model for students the value of thinking of chemical concepts within an overarching structure rather than as isolated ideas.

The complexity and subtlety of people's conceptual knowledge is such that any elicited mapping (cf. Figure 15.2) is only ever likely to be a very partial representation of some segment of a person's actual conceptual structure. The notion of conceptual integration can, however, be understood (and modeled) in such terms. Concepts are integrated to the extent to which they are strongly linked with other concepts.

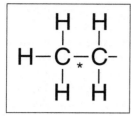
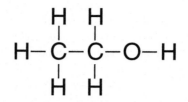

* This would seem to be a bond with symmetrical electron distribution along the bond axis...

... if we are not aware that one of the carbon centres is connected to an electronegative centre.

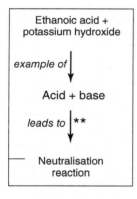
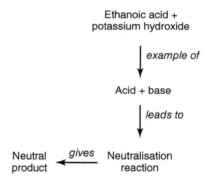

** If we elicited this conceptual link from a student we might think that they had understood the nature of the reaction. ....

... however, their other associations for the concept also influence their understanding.

Figure 15.3 The conceptual inductive effect. In a similar way to how effects are transmitted through the bonding framework of a molecule, a student's particular conceptualization depends on a network of propositional linkages.

15.2
Conceptual Integration and Coherence in Science

Science values conceptual integration and coherence. Science is a diverse activity, encompassing a range of disciplines and a vast domain divided into fields and subfields. Each specialist area has its own "disciplinary matrix" [19] with its particular key concepts and preferred theories and models. The number of "scientific concepts" that have currency in contemporary science must be vast.

Despite this diversity, coherence is considered very important in science. Science is built upon certain assumptions. Historically, early scientists may have espoused a realism and positivism that seems naive today, yet a more nuanced form of realism is still inherent is most scientists' work even if epistemologically a post-positivist understanding of the nature of science (NOS) is widely held [20]. Scientific work requires a worldview that includes some key metaphysical commitments [3, 21]. An obvious – ontological – one is that there is an external objective reality: different scientists interact with the same external physical world even if their own subjective experiences of it are individual [22]. Another ontological commitment is that at some level that external world exhibits a degree of stability which makes it feasible to undertake programmatic enquiry into its nature. We may observe change, but we assume that there is some more permanent level of stability to be found underpinning that change. It is also usually assumed that the most basic aspects of the physical world are invariant from place to place, as well as from time to time: thus the notion of the "laws" of nature. Newton proposed universal gravitation, and school children are taught that his law applies everywhere in the universe: however, that is of course a conjecture rather than a known fact. These ontological commitments are a kind of professional version of Pascal's wager. Where Pascal argued that it was rational to believe in a God who could offer you eternal life (as there was so much to lose by not believing, if this indeed transpired to be the case), scientists wager that the universe is stable and consistent enough to make enquiry into it worthwhile. Considering the universe to not have any underlying stability and order is a perfectly admissible perspective, but not one that can motivate scientific enquiry.

The accompanying epistemological commitment is that human beings are actually able to know the world, at least well enough to make the whole business of science viable. It is now widely accepted that scientific methodology necessarily underdetermines unique conclusions, and that human cognition and psychology do not allow direct knowledge of how the world is. Rather, we seek to build concepts, models, theories, and the like which – though necessarily human constructions – are constrained enough by empirical evidence (collected in as objective a way as possible) that we feel justified to consider it scientific knowledge: that is, not absolute knowledge but sufficiently well-supported knowledge to provisionally be reported as our best understanding to date. The logic and practical limitations of scientific investigation will always underdetermine certain scientific knowledge – but where there is clear inconsistency between different scientific theories or principles, this suggests a definite problem with our current best understanding. Scientists, then, test for coherence between different concepts and theories and feel more confident in those ideas that fit well with other well-established ones. The converse is also the case: where there are known problems of inconsistency between different scientific ideas, this provides a kind of collective "cognitive dissonance" that can motivate attempts to seek greater coherence.

Moreover, there is a common view in science that the most useful ideas are those that offer widely applicable frameworks for understanding and explaining broad ranges of phenomena. The rejection of evolutionary theory by students from some cultural backgrounds, including in some parts of the United States [23], is considered particularly problematic because evolution is the key unifying idea of modern biology [24]. Maxwell's work was lauded because it was seen to unify electricity, magnetism, and (what is now known as) electromagnetic radiation. Physicists today look to explore how quantum theory can be made consistent with other key physical theories, and seek a grand unifying theory which can subsume different areas of physics. The prominent display of periodic tables in so many schools and college chemistry laboratories reflects the value of the periodic system as an organizing principle in the subject.

The relationship between different disciplines and fields is a more complex one. A notion that once had currency was that one day biology would be reduced to chemistry, and chemistry to physics. This now seems rather naive. *In principle*, all chemical (and ultimately biological) concepts *could* be redescribed purely in terms of physics. That might be an immense intellectual achievement – akin to Alfred North Whitehead and Bertrand Russell's (ultimately incomplete) project to set out the principles on which all mathematics is ultimately based – but it would not negate the need for chemical and biological concepts in their own right. Whilst presumably any example of oxidation or of nucleophilic substitution or of a steroid (just to select a few of many possible examples) could in principle be described purely in terms of physics, that would not make such a level of description the most useful for many purposes. We might consider oxidation to be an emergent property of a chemical system, and the concept of "oxidation" therefore does useful mental work in thinking about chemistry. However, whilst it would usually be unhelpful to re-describe oxidation in purely physical terms, it would not be acceptable for such a concept to be *inconsistent with* physics. For example, electrical charge is expected to be conserved during oxidation processes: it is not exempt from the physical conservation principle because we class oxidation as a chemical rather than a physical process.

In the same way, bond-breaking is understood as an endothermic process in chemistry. So chemistry teachers become frustrated when their students read about "energy-rich" phosphate bonds in some biology texts, as bond breaking in adenosine triphosphate (ATP) cannot (of itself) release energy regardless of whether the context is the chemistry or biology class. It has been argued that, when understood *within a biological context*, it is defensible to describe how in photosynthesis "light energy becomes chemical bond energy" [25], but from a chemical perspective it is inappropriate to consider energy to be stored in bonds.

15.2.1
Multiple Models in Chemistry

Although we expect scientific theories to be consistent within and across scientific disciplines, our application of models offers more latitude for inconsistency. We acknowledge that our conceptual models are just that, models, and so may only relate to some aspects of the phenomena we study, and may have limited ranges of application. There are at least two different kinds of multiple models that will be met by students of the chemistry.

One area where students will come across apparently inconsistent models links to concept definitions that have shifted over time. There are different ways of defining oxidation or acids, for example. This is less a matter of uncertainty about the nature of the phenomena studied in chemistry than choices about how to usefully demarcate our concepts. It does not make good sense to think that, for example, defining oxidation in terms of addition of oxygen is "wrong" (because it misses some cases that are "really" oxidation) and that a definition in terms of loss of electrons is more accurate. Rather, as the theoretical apparatus of chemistry developed, it became useful to have a more inclusive category and the choice of a most useful definition shifted. Similarly, the Lewis model of acid is more inclusive than the Brønsted–Lowry model, but that does not equate to the Brønsted–Lowry notion of an acid being "wrong" because it misses some acids or the Lewis model being "wrong" because it includes non-acids. If our concept was meant to map onto a clear distinction in nature, then having two inconsistent sets of definitions would indicate error – but students need to understand that the acid concept is a human construction and the chemical community can choose how to restrict or extend it in relation to objective differences in the properties of different substances, and our theoretical models of why they react as they do.

A rather different situation is met when modeling chemical structures at the submicroscopic level. Students will meet various ways of modeling and representing atomic, molecular, and crystal structures. Whilst some models a student may meet are best now considered anachronistic and judged historical scientific models [26], much of the diversity arises from the difficulty posed by entities that cannot be fully represented in terms of what is familiar at the macroscopic scale. So, models of molecules showing atomic centers linked along major bond axes reflect some aspects of interest, whilst space-filling models reflect other aspects. Neither type of model effectively represents all the features of molecules we wish to highlight at various times. Rather, we maintain alternative models as part of our "conceptual tool kit", to be selected for particular cognitive work [27]. Some aspects of crystal structures are best shown by close-packing – other aspects make little sense when represented that way. Showing pairs of electrons around atomic cores is sometimes useful and sometimes unhelpful: ditto electron density maps; ditto electronic orbitals.

Where the teacher may be well aware of the limitations of the models presented in class, their ontological status as partial representations, and their epistemological status as thinking tools and sometimes elements of theories under test and development, none of this is necessarily obvious to students unless the teacher makes it explicit. From a NOS perspective, the teaching of any chemical model is necessarily inadequate unless the teaching makes it clear that what is being taught *is* a model and explores its limitations and range of application.

Where different models of the same target concept (e.g., of the atom, of the covalent bond, etc.) are met at different stages of education, teachers should take the opportunity to compare the strengths, limitations, and range of application of the new model being presented with the alternative model(s) students have been taught previously. This is important to avoid students being frustrated because they feel they are being told that what they learnt in earlier grades was wrong. However, this is also important from a NOS perspective both because (i) historical sequences of models can reflect how progress in chemistry involves a dialectic between theory and experiment, with proposed models suggesting empirical work which reveals their limitations and informs and motivates more sophisticated models and (ii) contemporary chemists often retain multiple alternative models in their "conceptual toolbox" from which to select according to the needs of particular tasks [27].

15.3
Conceptual Integration in Learning

Conceptual integration is an important factor in learning – both in relation to how students make sense of teaching and in terms of how they are subsequently able to apply learning. Ausubel [28] wrote about meaningful learning in terms of what was to be learnt being both relatable to some aspect of a learner's existing cognitive structure and being perceived as relevant. When teaching is not meaningful because it is not understood in terms of existing knowledge, it can only be learnt by rote. However, when it is meaningful, it can be associated with existing knowledge, which makes for more accessible and flexible learning.

Meaningful learning is more accessible because it is more likely to be activated when it is connected to other mental representations: in effect, there are multiple routes to reaching that specific knowledge representation. Meaningful learning is more flexible because it is understood more deeply (and so can more readily be applied in different contexts) by acquiring its meaning from its associations with other related concepts. To give a very simplistic example, knowing that ammonia is a compound, and that compounds are chemical substances, enables someone to use the information "ammonia is a compound" in more ways (in the most basic terms, as a response to the question "name a chemical substance," as well as a response to the question "name a chemical compound").

A common problem in much school and college learning is that, even when learning is meaningful in Ausubel's terms, it later proves rather fragile. This can be understood in terms of how the formation of memories commonly occurs in two stages [29]. When a learner associates new learning with existing representations in their conceptual structure, the association is itself represented within the brain by synaptic changes, but initially mediated by a particular brain structure that is involved in memory formation by acting as an intermediary connected to both the established and new representations. These linkages decay over time as a matter of course, such that the initial association may no longer be strongly represented. However, regular activation of the temporary association leads to new and more direct linkages being formed that provide a more permanent association once the temporary indirect link has decayed. Sufficient reinforcement of learning can lead to strong permanent links between representations such that there will be a low threshold for activating the link (so that, when thinking of one knowledge component represented in memory, we are likely to bring to mind the other).

From a constructivist perspective [30, 31], relevance must be understood within the context of an individual's conceptual ecology – as it is the learner who must perceive teaching as meaningful in terms of existing knowledge and understanding, and so form associations. However, part of the work of teaching is seeking to facilitate such linkages to "make the unfamiliar familiar" [32] by making explicit how new learning fits with previously taught material, and by seeking examples, analogies, and metaphors for new ideas that will help learners make sense of teaching.

15.3.1
The Drive for Coherence

The high value put on coherence between different theoretical ideas in science is reflected (or perhaps better, reflects) what seems to be a basic drive toward coherence within human cognition. This works at two levels. In perception, we are biased to see the world in terms of existing perspectives and frameworks. We seek to make sense, and the way to make sense is to interpret new information such that it fits with those categories, principles, and expectations about the world we have already formed. Much of this occurs at a preconscious level, so we are not even aware of how sensory data is filtered, selected, and arguably sometimes distorted by these processes. This means that students who seem to be making sense of teaching may not be understanding it as intended. For example, one response to being taught about the concept of electronic orbitals is to subsume the teaching under the existing concept of electron "orbits," so conflating two quite distinct models [33]. As suggested above, the teacher can preempt such problems to some extent by explicitly presenting the orbital model of the atom as a model and explaining its adoption in part by contrasting its features and characteristics with those of the alternative atomic model already familiar to the students.

There are also processes at work after initial learning of material that can lead to modifications in what has been learnt. It appears that the cognitive system has an in-built process for bringing material represented in the brain into greater coherence, even if this means modifying what we might consider memories. There seems limited explicit research on this effect in science education. Gauld [34] reported a study in physics education where learners who appeared to have been persuaded to shift from alternative conceptions about electric circuits by a teacher's demonstration later reverted to less scientific thinking *and* then recalled the demonstration in a distorted way inconsistent with the scientific model, but consistent with their alternative conceptions. Like Whig historians or Stalinist officials, we all (inadvertently) rewrite our personal histories to better suit our currently preferred narratives. Human memory does not seem to have evolved as a means of providing us with high-fidelity records of past experience: rather it seems to be an aspect of cognitive apparatus that has evolved to best help us form a coherent model of experience to support decisive action now (a tendency that may have been important in human evolution, as prevarication and waiting for the luxury of a strong evidence base are unlikely to have been helpful traits in the conditions under which most of our ancestors lived). It seems both that the way we perceive experience now is heavily influenced by the interpretive frameworks we have constructed from prior experience and that our representations of prior experience (memories) become modified over time to better maintain a coherent model of the world.

This is an important theme that does not seem to have been explicitly explored in much research in chemistry education. In one study, I found intriguing suggestions that some time (almost 4 years) after a student had studied a chemistry course he demonstrated a linkage between two ideas he had considered complementary but not related at the time of his studies [35]. This was despite my study participant reporting that he had not had reason to think about the material (i.e., consciously) since completing his course. However, this is one example. Given the importance of this theme to teaching and learning, it deserves more attention. Teachers are well aware of the importance of reinforcing new learning to support consolidation, but this aspect of pedagogic knowledge seems to largely be based on work on memory function undertaken in psychology, often using rather artificial target material, rather than studies in the context of authentic classroom learning.

15.3.2
Compartmentalization of Learning

This is particularly important because, despite the in-built drive to develop coherent models of the world, it is also a common classroom phenomenon that learners will fail to make the links teachers might hope, and sometimes even those that seem obvious to the teacher. One (about 17-year-old) student I interviewed was able to explain the origin of van der Waals' forces in terms of transient induced dipoles, but could not suggest a mechanism by which a balloon that had been charged by friction might remain attached to an uncharged wall [36]. Perhaps more mystifying to a teacher would be why a (about 11-year-old) student who demonstrated that he had learnt that everything was made of particles in one science topic would be unsure whether a new substance he came across in another topic, chlorophyll, might be made of particles; or why it was not obvious to a (about 14-year-old) student that the nucleus of a cell *must* be very much larger than the nucleus of an atom [37].

It is well recognized that the context of learning – where and when we learn something – can be significant for how readily we access and apply that learning. The issue of transfer of learning is a core issue in education research [38]. However, as teachers we may lose sight of this. I interviewed a (about 16-year-old) student who told me about how she had learnt about covalent bonds and ionic bonds. When shown an unfamiliar image of resonance forms of BF_3, Annie identified "single bonds." I asked her whether these were the same as covalent or ionic bonds, or something else. Her initial response was that "single bonds are different" [39]. She later acknowledged that actually this was probably just a different label, but still suggested that "they can probably occur in different, things like in organic you talk about single bonds more than you talk about covalent, and then like in inorganic you talk about covalent bond, more than you talk about single bonding or double bonding." That is, Annie went to classes with different lecturers, and in her classes on organic chemistry bonds were generally described differently from how they were discussed in her inorganic chemistry classes (because different teaching points were being emphasized). It would seem that, prior to my interviewing her, Annie had not considered relating the two ways of talking about bonds she met in her classes.[1]

Again, teachers can plan to teach in ways that minimize such problems by always thinking in terms of the potential links they want learners to make, and considering where terminology or styles of presentation in different branches of the subject can potentially obscure links. Chemistry offers many potential areas where such breakdown of communication can occur – outer shells sometimes but not always labeled valence shells; noble gases still sometimes referred to as inert gases; common use of traditional as well as systematic names for some substances (acetic acid, toluene, etc.); treatment of acids and bases in biochemistry/pharmacy; treatment of reaction mechanisms in inorganic and organic chemistry; and so on.

15.3.3
When Conceptual Integration Impedes Learning

The gist of this chapter is that conceptual integration is generally a positive thing – and that teachers should support students in developing highly integrated conceptual systems as these are more robust and offer greater facility to learners. However, major conceptual change may require us to entertain alternative inconsistent ways of thinking about a topic as an intermediate stage if we are to shift away from strongly held inadequate conceptions [2, 40].

A learner may have well-integrated conceptual knowledge of chemistry (or any other subject) without their conception matching canonical knowledge. Sometimes, learners acquire relatively isolated alternative conceptions – such as notions that all acids are dangerous or all metals are hard. However, learners can also develop well-integrated conceptual frameworks around alternative conceptions.

An example from chemistry learning is the "octet" conceptual framework [41], based around the explanatory principle that atoms "need" or "want" octets of electrons or full outer shells and that chemical processes happen so that atoms can meet this need. The framework was first identified among students in England, but

---

1) This example is presented in more detail on the ECLISE project Web site: see *http://www.educ.cam.ac.uk/research/projects/eclipse/* (accessed 8th January 2014).

research in other contexts suggests that it is widespread [42]. Students can develop quite extensive networks of related conceptions based around this explanatory principle – in terms of why reactions occur, the nature of bonds, what counts as a chemical bond, patterns of ionization, and which chemical species should be considered stable. These networks commonly consist of a mixture of alternative and more acceptable conceptions, but the ability to develop an extensive and coherent linked conceptual structure around the core principle is reflected in just how tenacious an alternative conception this can be [35].

A key feature of the "octet" framework is how it is highly anthropomorphic (in that it is described in terms of atoms as active agents that have desires), and this seems a strong feature of many students' explanations of chemistry at the submicroscopic scale, even at high school level [43] – and indeed beyond [44]. It was reported in one study [45] that substantial proportions of senior high school level students actually consider atoms to be alive. This reflects the difficulty for learners of acquiring the abstract theoretical models used at the submicroscopic scale to explain phenomena in chemistry (cf. Figure 15.1) – something well recognized as a core issue in chemistry education [46–50].

Teachers use metaphors and analogies to teach about abstract ideas, and this is often a sensible strategy to "make the unfamiliar familiar" to students. However, teachers' classroom talk that anthropomorphizes atoms and molecules offers a familiar way of thinking about the submicroscopic realm that many students not only readily adopt but also find difficult to move beyond. Teachers are advised that, if they do use such anthropomorphic language, they need to do so with care, always making sure that students appreciate that terms such as "the atom needs ... " do not suffice in scientific explanations, and to seek to shift classroom discourse from such "social" descriptions to more physical accounts (e.g., in terms of ideas such as charge, force, energy) before students begin habitually adopting anthropomorphic language.

15.3.4
Conceptual Integration and Expertise

It is considered that one of the most significant differences between a novice and an expert is the degree of conceptual integration of an expert's knowledge. An expert has a highly organized knowledge base to support his or her thinking [51]. Expertise in a field is considered to result from engagement with the field over an extended period – perhaps typically something like a decade of working in that area [52]. In that sense, teachers need to be aware that it is unreasonable to expect most of our students to demonstrate knowledge similar to that of experts, and this is especially so at school level where students follow a wide spread of curriculum subjects and have limited engagement with each. This does, however, raise an issue about the educational logic of standards and curriculum specifications which set out objectives across a diverse range of teaching topics. Regular shifting from topic to topic, especially if this is perceived by students as a complete disconnect and "moving on" to "something else" – such as when a (about 11-year-old) student was unclear why her class did a lesson on magnets when they were supposed to be studying electricity [37] – would not seem to support extensive engagement. Chemistry is underpinned by a range of core concepts and principles that are not only fundamental to the discipline but also to learning in the discipline: the student experience may not always reflect this unless teachers work hard to stress what is fundamental and how other material fits around that conceptual core.

15.4
Conclusions and Implications

The intention behind this chapter was to highlight the importance of the notion of conceptual integration in teaching and learning chemistry, given that it has not been subject to the attention in chemistry education that it might seem to warrant. What is known, or at least widely believed, about how learning occurs and how conceptual structures evolve can inform instruction; but further research is indicated to learn more about the details of how students integrate their chemistry knowledge, and what specific strategies teachers can employ to support them.

15.4.1
Implications for Teaching

At the most basic level, thinking about the importance of conceptual integration can inform teaching in ways that are recognized in constructivist literature [53, 54]. Teachers should always seek to ensure learners recognize which aspects of prior learning are intended to be related to new ideas presented in class. Teachers should revisit new learning regularly in a range of contexts to reinforce how new concepts "fit" into the scheme of the subject.

The present analysis also suggests that teachers need to emphasize the idea of the importance of integrating knowledge, and to model this by being explicit about when ideas being discussed today link with or exemplify other topics. This is important both for supporting students' metacognitive development (so they acquire "metaknowledge" related to the importance to learning of the degree of integration of their own knowledge structures, and metacognitive skills to monitor this aspect of their thinking) and for helping learners appreciate an important aspect of NOS – the epistemological commitments within science to finding coherence between ideas and to seek to subsume scientific concepts under a small number of fundamental principles.

In parallel with this, teachers need to emphasize the central role of models (in which I include typologies) within chemistry and help learners appreciate the epistemological status of models. Alternative models may not be consistent because our knowledge is uncertain, or (particularly in the case of the submicroscopic nature of matter) because we need to use complementary models, and different representations of those models, to highlight different features. This is important because research suggests that learners tend to commonly treat scientific models as realistic replicas, and they often fail to appreciate the nature and role of models in science subjects [55]. It is also important because chemistry students need to appreciate that, although we seek coherent and well-integrated conceptual schemes in chemistry and consider inconsistency between concepts as a sign of something being wrong, we also regularly present apparently inconsistent models for students to consider and apply. This is potentially very confusing for students unless we are explicit about the limitations of models [56], and why ambiguity and inconsistency often need to be tolerated in a subject which ultimately aspires to offer a coherent conceptual framework for understanding the material world.

Given the breadth of many school and college chemistry courses, where students meet a wide range of concepts – and may even experience different branches of chemistry as different subjects (taught in different styles by different subject experts, and sometimes apparently using a different specialist language) – it is important to identify core concepts which can act as organizing principles for learning and which students can cling to as anchors (or perhaps buoys!) when meeting new concepts. There is now a programme of work being undertaken to explore "learning progressions" in science that identify such key concepts and explore how they can be used as foci for planning teaching and assessing learning [57, 58], and this is one promising development.

Where possible, teaching staff should liaise over how they teach related topics. However, any chemistry teacher needs to analyze the content they are teaching in terms of desirable links with other topics, and potential points of confusion if other teachers (including those who taught our current students previously) may have adopted alternative terminology, metaphors, or approaches to presenting topics. Whilst many potential learning impediments may be identified from the research literature into student thinking, each student is unique, with his or her own somewhat idiosyncratic resources for interpreting new information, so effective teaching is always likely to be an iterative process where the teacher is regularly eliciting both the student's prior learning and how new teaching is being interpreted – checking both that intended links are being made, and that unhelpful misleading links are diagnosed and challenged before they are adopted and committed to by learners [37].

15.4.2
Directions for the Research Programme

The analysis presented here might be considered to be grounded upon a firm but patchy research base. Within science education, we have many studies that report aspects of how much different groups of students know about key chemistry topics and where they commonly have learning difficulties or alternative conceptions [31, 59, 60]. We also have a good understanding of notions such as meaningful learning, the importance of reinforcement of learning, consolidation and forgetting effects, and the like, from studies undertaken in psychology and the learning sciences. We have fewer studies about the nature of conceptual integration in chemistry students' knowledge structures at different levels, let alone how this may shift over time, or how it might relate to such things as student metacognition and study habits, curriculum structure, and teaching styles. Guidance to teachers is therefore largely in terms of general principles. There is much scope here for research that is able to look in detail at this aspect of chemistry learning across topics, over time, and in relation to diverse teaching and learning contexts.

References

1. Duit, R. and Treagust, D.F. (2012) in *Second International Handbook of Science Education* (eds B.J. Fraser, K.G. Tobin, and C.J. MacRobbie), Springer, Dordrecht, pp. 108–118.
2. Vosniadou, S. (ed.) (2008) *International Handbook of Research on Conceptual Change*, Educational Psychology Handbook Series, Routledge, London.
3. Taber, K.S. (2013) *Modelling Learners and Learning in Science Education: Developing Representations of Concepts, Conceptual Structure and Conceptual Change to Inform Teaching and Research*, Springer, Dordrecht.
4. Gilbert, J.K. and Watts, D.M. (1983) Concepts, misconceptions and alternative conceptions: changing perspectives in science education. *Stud. Sci. Educ.*, **10** (1), 61–98.
5. Oversby, J. (2012) in *Teaching Secondary Chemistry* (ed. K.S. Taber), Hodder Education, London, pp. 183–198.
6. de Jong, O. and Taber, K.S. (2014) The Many Faces of High School Chemistry, N. Lederman and S.K. Abell (Eds.), Handbook of Research in Science Education, Vol. 2, Routledge, New York, pp. 457–480.
7. Johnstone, A.H. (1982) Macro- and microchemsitry. *Sch. Sci. Rev.*, **64** (227), 377–379.
8. Taber, K.S. (2013) Revisiting the chemistry triplet: drawing upon the nature of chemical knowledge and the psychology of learning to inform chemistry education. *Chem. Educ. Res. Pract.*, **14** (2), 156–168.
9. Jensen, W.B. (1995) Logic, history and the teaching of chemistry, text of the Keynote Lectures given at the 57th Annual Summer Conference of the New England Association of Chemistry Teachers, Sacred Heart University, Fairfield, CT.
10. Popper, K.R. (1979) *Objective Knowledge: An Evolutionary Approach*, Revised edn, Oxford University Press, Oxford.
11. Vygotsky, L.S. (1934/1986) *Thought and Language* (ed. A. Kozulin), MIT Press, London.
12. Kelly, G. (1963) *A Theory of Personality: The Psychology of Personal Constructs*, W W Norton & Company, New York.
13. Fransella, F. and Bannister, D. (1977) *A Manual for Repertory Grid Technique*, Academic Press, London.
14. Taber, K.S. (1994) Can Kelly's triads be used to elicit aspects of chemistry students' conceptual frameworks? British Educational Research Association Annual Conference, Oxford, UK.
15. Herron, J.D. et al. (1977) Problems Associated with Concept Analysis. *Sci. Educ.*, **61** (2), 185–199.
16. Chi, M.T.H. (1992) in *Cognitive Models in Science* (ed. R.N. Giere), University of Minnesota Press, Minneapolis, MN, pp. 129–186.
17. Chi, M.T.H., Slotta, J.D., and de Leeuw, N. (1994) From things to processes; a theory of conceptual change for learning science concepts. *Learn. Instr.*, **4**, 27–43.
18. Schmidt, H.-J. (1991) A label as a hidden persuader: chemists' neutralization concept. *Int. J. Sci. Educ.*, **13** (4), 459–471.
19. Kuhn, T.S. (1996) *The Structure of Scientific Revolutions*, 3rd edn, University of Chicago, Chicago, IL.
20. Losee, J. (1993) *A Historical Introduction to the Philosophy of Science*, 3rd edn, Oxford University Press, Oxford.
21. Taber, K.S. (2013) in *Science Education for Diversity: Theory and Practice* (eds N. Mansour and R. Wegerif), Springer, Dordrecht, pp. 151–177.

22. Glasersfeld, E.v. (1989) Cognition, construction of knowledge, and teaching. *Synthese*, **80** (1), 121–140.
23. Long, D.E. (2011) *Evolution and Religion in American Education: An Ethnography*, Springer, Dordrecht.
24. Dobzhansky, T. (1973) Nothing in biology makes sense except in the light of evolution. *Am. Biol. Teach.*, **35** (3), 125–129.
25. Jin, H. and Anderson, C.W. (2012) in *Learning Progression in Science: Current Challenges and Furture Directions* (eds A.C. Alonzo and A.W. Gotwals), Sense, Rotterdam, pp. 151–181.
26. Justi, R. and Gilbert, J.K. (2000) History and philosophy of science through models: some challenges in the case of 'the atom'. *Int. J. Sci. Educ.*, **22** (9), 993–1009.
27. Taber, K.S. (1995) An analogy for discussing progression in learning chemistry. *Sch. Sci. Rev.*, **76** (276), 91–95.
28. Ausubel, D.P. (2000) *The Acquisition and Retention of Knowledge: A Cognitive View*, Kluwer Academic Publishers, Dordrecht.
29. Alvarez, P. and Squire, L.R. (1994) Memory consolidation and the medial temporal lobe: a simple network model. *Proc. Natl. Acad. Sci. U.S.A.*, **91**, 7041–7045.
30. Bodner, G.M. (1986) Constructivism: a theory of knowledge. *J. Chem. Educ.*, **63** (10), 873–878.
31. Taber, K.S. (2009) *Progressing Science Education: Constructing the Scientific Research Programme into the Contingent Nature of Learning Science*, Springer, Dordrecht.
32. Taber, K.S. (2002) *Chemical Misconceptions – Prevention, Diagnosis and Cure: Theoretical Background*, vol. 1, Royal Society of Chemistry, London.
33. Taber, K.S. (2005) Learning quanta: barriers to stimulating transitions in student understanding of orbital ideas. *Sci. Educ.*, **89** (1), 94–116.
34. Gauld, C. (1986) Models, meters and memory. *Res. Sci. Educ.*, **16** (1), 49–54.
35. Taber, K.S. (2003) Lost without trace or not brought to mind? – a case study of remembering and forgetting of college science. *Chem. Educ.: Res. Pract.*, **4** (3), 249–277.
36. Taber, K.S. (2008) Exploring conceptual integration in student thinking: evidence from a case study. *Int. J. Sci. Educ.*, **30** (14), 1915–1943.
37. Taber, K.S. (2014) *Student Thinking and Learning in Science: Perspectives on the Nature and Development of Learners' Ideas*, Routledge, New York.
38. Goldstone, R.L. and Day, S.B. (2012) Introduction to "New conceptualizations of transfer of learning". *Educ. Psychol.*, **47** (3), 149–152.
39. Taber, K.S. (1993) Stability and lability in student conceptions: some evidence from a case study. British Educational Research Association Annual Conference, University of Liverpool.
40. Thagard, P. (1992) *Conceptual Revolutions*, Princeton University Press, Oxford.
41. Taber, K.S. (1998) An alternative conceptual framework from chemistry education. *Int. J. Sci. Educ.*, **20** (5), 597–608.
42. Taber, K.S. (2013) in *Concepts of Matter in Science Education* (eds G. Tsaparlis and H. Sevian), Springer, Dordrecht, pp. 391–418.
43. Taber, K.S. and Adbo, K. (2013) in *Concepts of Matter in Science Education* (eds G. Tsaparlis and H. Sevian), Springer, Dordrecht, pp. 347–370.
44. Nicoll, G. (2001) A report of undergraduates' bonding misconceptions. *Int. J. Sci. Educ.*, **23** (7), 707–730.
45. Griffiths, A.K. and Preston, K.R. (1992) Grade-12 students' misconceptions relating to fundamental characteristics of atoms and molecules. *J. Res. Sci. Teach.*, **29** (6), 611–628.
46. Gilbert, J.K. and Treagust, D.F. (eds) (2009) *Multiple Representations in Chemical Education*, Springer, Dordrecht.
47. Tsaparlis, G. and Sevian, H. (eds) (2013) *Concepts of Matter in Science Education*, Springer, Dordrecht.
48. Harrison, A.G. and Treagust, D.F. (2002) in *Chemical Education: Towards Research-Based Practice* (eds J.K. Gilbert *et al.*), Kluwer Academic Publishers, Dordrecht, pp. 189–212.

49. Lijnse, P.L. et al., and Centre for Science and Mathematics Education (eds) (1990) *Relating Macroscopic Phenomena to Microscopic Particles: A Central Problem in Secondary Science Education*, University of Utrecht/CD-ß Press, Utrecht.
50. Taber, K.S. (2001) Building the structural concepts of chemistry: some considerations from educational research. *Chem. Educ.: Res. Pract. Eur.*, **2** (2), 123–158.
51. Dolfing, R., Bulte, A., Pilot, A., and Vermunt, J. (2012) Domain-specific expertise of chemistry teachers on context-based education about macro-micro thinking in structure-property relations. *Res. Sci. Educ.*, **42** (3), 567–588. doi: 10.1007/s11165-011-9211-z.
52. Gardner, H. (1998) *Extraordinary Minds*, Phoenix, London.
53. Taber, K.S. (2011) in *Educational Theory* (ed. J. Hassaskhah), Nova, New York, pp. 39–61.
54. Driver, R. and Oldham, V. (1986) A constructivist approach to curriculum development in science. *Stud. Sci. Educ.*, **13**, 105–122.
55. Treagust, D.F., Chittleborough, G., and Mamiala, T.L. (2002) Students' understanding of the role of scientific models in learning science. *Int. J. Sci. Educ.*, **24** (4), 357–368.
56. Taber, K.S. (2010) Straw men and false dichotomies: overcoming philosophical confusion in chemical education. *J. Chem. Educ.*, **87** (5), 552–558.
57. Alonzo, A.C. and Gotwals, A.W. (eds) (2012) *Learning Progressions in Science: Current Challenges and Future Directions*, Sense Publishers, Rotterdam.
58. Sevian, H. and Talanquer, V. (2014) Rethinking chemistry: a learning progression on chemical thinking. *Chem. Educ. Res. Pract.*, **15** (1), 10–23.
59. Duit, R. (2009) Bibliography – Students' and Teachers' Conceptions and Science Education, Kiel, http://www.ipn.uni-kiel.de/aktuell/stcse/stcse.html (accessed 8 January 2014).
60. Kind, V. (2004) *Beyond Appearances: Students' Misconceptions About Basic Chemical Ideas*, 2nd edn, Royal Society of Chemistry, London.

16
Learners Ideas, Misconceptions, and Challenge
Hans-Dieter Barke

"Look to the East – the sun is rising." One can hear that statement around the world – even if adults and educated persons know about the rotating earth and the fixed sun, no one is telling: "look how nice the earth is rotating to the East – just showing our sun"!

Even our scientists until the sixteenth century observed the same phenomenon and were seriously thinking that the sun was rotating around the earth. Copernicus was the first who interpreted the movement of our planets and published his heliocentric mental model in 1543: "the earth is rotating around his axis in one day, and rotates around the sun in one year." It took decades and decades for more and more scientists and citizens to accept that thinking – contrary to their everyday observations according to the "rotating sun."

So we cannot blame our children when they observe very carefully and derive the thinking of the geocentric model of earth. The physics teacher has to discuss intensively those observations. With a good spatial model of the sun in the middle, of rotating earth and moon, and of all the other planets, he or she can start to teach the heliocentric idea. The young students may realize a conceptual change and develop that idea – but at home with their family and friends they will not stay with that idea and will go on to describe their observations with the "rotating sun." They still keep both mental models in mind: for everyday life they talk about the "rising sun"; for the physics teacher or for the written test in science they will shift to the "rotating earth."

16.1
Preconcepts and School-Made Misconceptions

In chemistry, we have the same experiences according to the transformation of substances, to the explanation of combustion, and to the nature of gases. The students observe very well but cannot develop the scientific interpretation – they stay with prescientific ideas, with alternative ideas, or with preconcepts. Those ideas that derive from everyday life and that students bring into science lectures should be called *preconcepts* [1].

Chemistry Education: Best Practices, Opportunities and Trends, First Edition.
Edited by Javier García-Martínez and Elena Serrano-Torregrosa.
© 2015 Wiley-VCH Verlag GmbH & Co. KGaA. Published 2015 by Wiley-VCH Verlag GmbH & Co. KGaA.

If the science lectures are going on to the second and third year, one can experience that students – not knowing topics like equilibrium or donor–acceptor reactions from everyday life – do not develop scientific interpretations offered by the teacher, but often stay with mistakes, and with alternative ideas. Because those mistakes are mostly "school-made," we will call those ideas *school-made misconceptions* [1]. They can be explained as due to the difficulties of the topic or to insufficient teaching – but there is a chance to change the teaching process and to successfully prevent misconceptions. The preconcepts are developed by everyday life; one cannot prevent young students from getting those ideas. You have to accept them, and discuss and try to correct them to realize a conceptual change. But school-made misconceptions should not appear automatically – there is a chance to prevent students from acquiring misconceptions by good teaching.

16.2
Preconcepts of Children and Challenge

Many preconcepts have been listed and discussed with young students concerning the following:

- concrete-pictorial and magical-animistic ways of speaking [1] (pieces of wood *don't want* to burn, acids *attack* other substances, rust *eats up* iron, etc.)
- substance as a carrier of properties [1] (heated iron wool *turns* black, red-brown copper *changes* to green copper after time, etc.)
- mixing and unmixing elements in compounds [1] (silver sulfide *contains* silver and sulfur, water *consists* of hydrogen and oxygen, etc.)
- destruction theory versus conservation of mass [1] (water from puddles *is gone*, removing stains from clothes – *the fat is away*, etc.)
- combustion and destruction theory [1–4] (after burning on a grill, charcoal *is away*, wood and paper *are gone* after combustion, etc.)
- air and other gases [5, 6] (gases weigh *nothing*, hot air *raises even up*, water evaporates *to form air*, etc.).

In many publications [1–10], those preconcepts are reflected and the challenges discussed. Those concepts are unavoidable – they keep appearing with every new generation of kids. The school-made misconceptions are avoidable – therefore they are more important to discuss: in the following sections, they are reflected and the challenges are proposed.

16.3
School-Made Misconceptions and Challenge

"Without explicitly abolishing misconceptions of students it is not possible to integrate sustainable scientific concepts" [3]. For advanced topics in science, for example, acid–base reactions and proton transfer, students have hardly any preconcepts or misconceptions. The students know phenomena like the sour taste of juices or acidic chemicals in the bathroom, but there is no knowledge of protons being transferred from molecules or ions to others. It is remarkable that teachers may teach the proton transfer even with some key experiments – but empirical research shows that students mostly don't grab the idea sufficiently. Reasons and challenges are shown.

School-made misconceptions can be found in the following topics:

- ions as smallest particles of salt crystals and solutions
- chemical equilibrium
- acid–base reactions and proton transfer
- redox reactions and electron transfer.

16.3.1
Ions as Smallest Particles in Salt Crystals and Solutions

With Dalton's atomic model, mostly atoms and molecules are introduced, and teachers like to work with molecular symbols such as H_2O, NH_3, or CH_4 – the whole organic chemistry can be described by those or structural molecular symbols. If later ions are presented, it seems hard to handle ionic symbols – for the composition of salts and salt solutions, students tend to write molecular symbols as they are already used to: Na–Cl, Cl–Mg–Cl, Mg=O, and so on. The following empirical research will show it.

Symbols representing ions in a salt solution (see Figure 16.1, "before evaporation") were given to senior class students. Afterward, students were asked to describe what happens to the ions when the water evaporates. Apart from several correct answers regarding ions by crystallization of sodium chloride, a large number of answers were given based on the existence of NaCl molecules in crystals. These students started with ions in the solution, but when developing mental models for the evaporation of water, they argued with the "neutralization" of ions [4] and the continuous fusion of ions into molecules, and finally they imagined "NaCl molecules" as particles of solid sodium chloride crystals (see Figure 16.1).

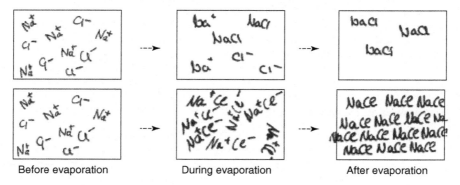

Before evaporation   During evaporation   After evaporation

Figure 16.1 Two examples for misconceptions concerning crystallization of sodium chloride [4].

In a questionnaire regarding the label on a bottle of "BONAQA" mineral water, students in upper grades were shown the names of salts contained in that water: calcium chloride, magnesium chloride, sodium chloride, and sodium bicarbonate [4]. The point of the questionnaire was to test their knowledge of the ions existing in mineral water. In order to note the correct ion symbols like Na^+, Cl^-, Ca^{2+}, Mg^{2+}, or HCO_3^-, most of the students suggested "salt molecules" (see Figure 16.2): by drawing their mental models, many students wrongly preferred "NaCl or $MgCl_2$ molecules" – even "$NaCl_2$ ions or molecules" (see Figure 16.2).

Despite the fact that all students had dealt with the term "ion" in the class, only 25% of them recognized "ions of various salts" as the correct alternative answer; about the same number of students chose "salt molecules." If one looks at the model drawings, a mere 4% of students actually included ion symbols in their drawings. Many of the test persons who, although they crossed off the ions as the correct answer, chose symbols for molecules (see Figure 16.2).

Most curricula introduce ions with ionic bonding and ion formation from elements. In the famous experiment according the sodium–chlorine reaction to form sodium chloride, teachers point out that sodium and chloride ions are formed by electron transfer, filling the outer electron shells like noble gas atoms and bonding in an ionic lattice by ionic bonds. All these new ideas are not easy to understand: different misconceptions arise if one introduces ions by ion formation and asks: "what holds the ions together" (see Figure 16.3).

16.3.1.1 Challenge of Misconceptions

Because of all new ideas about nucleus and shell, about electrons at different energy levels, about outer electrons, and about stable shells of noble gas atoms, many students are confused, and it seems better to introduce the idea about ions with the atomic model of Dalton. As soon as atoms and molecules are well known

Questionnaire:
BONAQA mineral water contains water, carbon dioxide, sodium bicarbonate, calcium chloride, magnesium chloride and sodium chloride.

Mark your answer:
(x) the water contains CO_2 molecules
( ) the water contains small salt crystals
( ) the water contains molecules of various salts
(x) the water contains ions of various salts
( ) Your answer: ....

(marked answers are the expected ones)

Draw your mental model:

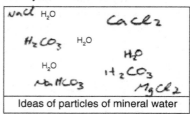

Ideas of particles of mineral water

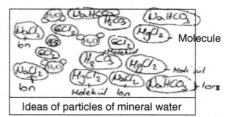

Ideas of particles of mineral water

Figure 16.2 Examples for misconceptions regarding particles in mineral water [4].

Figure 16.3 Empirical findings concerning students' misconceptions of ionic bonding [4].

and visualized by their symbols, also the third group of smallest particles should be introduced: the ions. One way is introducing the atoms of the periodic table with their atomic symbols and little spheres to visualize that every atom has a specific diameter. So it looks easy to symbolize also the corresponding ions with symbols and their specific diameter (see Figure 16.4): the charge number is given without comparing any protons in the nucleus and electrons in the shells – the ions are introduced without the differentiated atomic model! Remember: ions were discovered by Arrhenius in 1884 without knowing about electrons, and the salts existed millions of years before sodium or potassium was known!

Analogous to pointing out the composition of a water molecule by the H_2O symbol, one may state that sodium chloride is composed of Na^+ ions and Cl^- ions in an ionic giant structure, and that the ionic symbol for sodium chloride can be shown as $(Na^+)_1(Cl^-)_1$ or for magnesium chloride as $(Mg^{2+})_1(Cl^-)_2$. To shorten those formulae, it is possible to write NaCl and $MgCl_2$ – but the involved ions should be the mental model of students!

The composition of important salt crystals can be visualized by two-dimensional drawings of layers of the ionic lattice (see Figure 16.5) or by ionic symbols (see Figure 16.6): formulae of salts are easy to find by calculating equal numbers of + and − charges. If salt solutions are introduced at the same moment, the (aq) symbol should be added: Na^+(aq) ions and Cl^-(aq) ions for sodium chloride solution, and Mg^{2+}(aq) and Cl^-(aq) ions in the ratio 1:2 for magnesium chloride solution (see Figure 16.6). The (aq) symbol seems important because the charge of ions is nearly compensated by H_2O molecules: hydrated ions move freely without attraction in the solution. Ions in melted salts attract

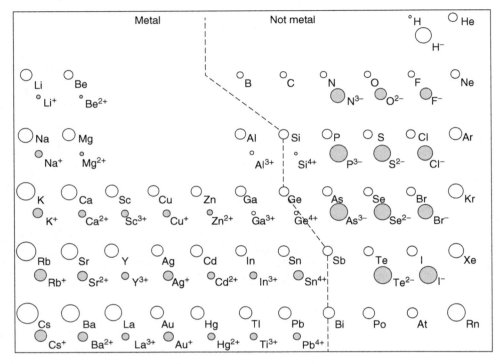

Figure 16.4 Periodic table of elements – depiction of a selection of atoms and ions and their spherical models [4].

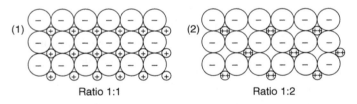

Figure 16.5 Two-dimensional models of ionic lattices in the ion ratio 1:1 (Na^+Cl^-) and the ion ratio 1:2 ($Mg^{2+}(Cl^-)_2$).

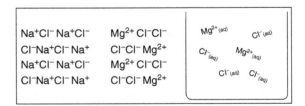

Figure 16.6 Two-dimensional symbolic models of solid salt crystals and magnesium chloride solution.

each other: beyond specific temperatures, they get together to form the ionic lattice in solid salt crystals.

If there is time to show the spatial arrangement of ions, the sphere-packing model of the NaCl structure can be built (see Figure 16.7a): with a triangle of 30-mm spheres the base packing is to be built, layers of 30-mm spheres are packed on top, finally 12-mm spheres are used to fill all big holes. The well-known elementary cube (see Figure 16.7b and c) is part of the giant structure: it can be shown as a part of the built sphere packing (Figure 16.7a) and can be taken out of the packing. It is also possible to build or to draw a crystal lattice with balls and sticks (Figure 16.7d) – it shows only the arrangement of ions but not the sizes of ions. Other structures are better to study by crystal lattices (Figure 16.7e,f): structure of lithium oxide $(Li^+)_2(O^{2-})_1$ or of zinc sulfide $(Zn^{2+})_1(S^{2-})_1$.

16.3.2
Chemical Equilibrium

In order to understand most of the basic concepts in chemistry, chemical equilibrium is enormously important. In this sense, Bergquist and Heikkinen [11] stated: "Yet equilibrium is fundamental to student understanding of other chemical topics such as acid and base behavior, oxidation–reduction reactions, and solubility. Mastery of equilibrium facilitates the mastery of these other chemical concepts."

Unfortunately, it seems to be difficult to teach this topic. Finley et al. [12] studied the level of difficulty of various themes in chemistry and reported the results of 100 randomly chosen teachers of chemistry from Wisconsin who chose chemical equilibrium as clearly the most difficult theme overall. Bergquist [11] noted: "Equilibrium, considered one of the more difficult chemical concepts to teach,

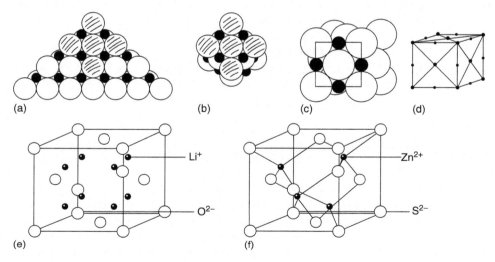

Figure 16.7 (a)–(f) Three-dimensional-structural models of sodium chloride, lithium oxide, and zinc sulfide.

involves a high level of students' misunderstanding." One can, therefore, expect a large variety of misconceptions due to the difficulties in teaching this subject as well as for understanding it.

16.3.2.1 Most Common Misconceptions

Tyson *et al.* [13], Banerjee and Power [14], and Hackling and Garnett [15] studied students' comprehension of chemical equilibrium. The following misconceptions were discovered in these studies: "You cannot alter the amount of a solid in an equilibrium mixture; the concentrations of all species in the reaction mixture are equal at equilibrium" [13]. "Large values of equilibrium constant imply a very fast reaction; increasing the temperature of an exothermic reaction would decrease the rate of the forward reaction; the Le Chatelier's principle could be used to predict the equilibrium constant" [14]. "The rate of the forward reaction increases with the time from the mixing of the reactants until equilibrium is established; a simple arithmetic relationship exists between the concentrations of reactants and products at equilibrium (e.g., concentrations of reactants equals concentrations of products); when a system is at equilibrium and a change is made in the conditions, the rate of the forward reaction increases but the rate of the reverse reaction decreases (...) the rate of forward and reverse reactions could be affected differently by addition of a catalyst" [15].

Kienast [16] carried out tests on chemical equilibrium with over 12 000 students in four test cycles. The following misconceptions were observed: "In equilibrium the sum of the amount of matter (concentrations) of reactants is equal to the sum of the amount of matter (concentrations) of the products; in equilibrium the amounts (concentrations) of all substances which are involved in equilibrium are the same; the sum of the amounts of matter (concentrations) remain the same during a reaction" [16]. Another questionnaire of Osthues [17] is shown for diagnosis and interpretation of the understanding of chemical equilibrium [4].

16.3.2.2 Challenge of Misconceptions

A first way to teach the concept of equilibrium may be the melting of ice with the thermometer, which shows 0 °C as long as a mixture of ice and water is present:

$$\text{Ice}(s, 0°C) \rightleftharpoons \text{water}(l, 0°C)$$

It doesn't matter if there is much ice or more water: if both substances are there, equilibrium between solid and liquid water exists. During heating, the energy is used to separate the water molecules from ice crystals – the temperature stays at 0 °C.

Another example shows the solubility of sodium chloride in water. If one observes a saturated sodium chloride solution together with solid sodium chloride at the bottom of the flask, and adds an additional portion of solid sodium chloride to it, this portion sinks down without dissolving. If one measures the density of the saturated solution before and after the addition of salt portions, one gets the same values. The concentration of the saturated solution does not depend on how much solid residue is present; equilibrium sets in between the

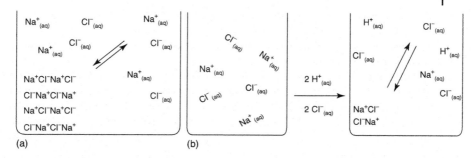

Figure 16.8 (a) model and (b) Beaker models for the solubility equilibrium of saturated sodium chloride solution.

saturated solution and arbitrary amounts of solid residue (see Figure 16.8a):

$$Na^+Cl^-(s, white) \rightleftharpoons Na^+(aq) + Cl^-(aq)$$

Even if concentrated hydrochloric acid is added to the saturated solution, the equilibrium stays: because of the high concentration of chloride ions, white solid sodium chloride crystals precipitate and decrease the concentration of sodium ions (see Figure 16.8b): an acidic sodium chloride solution remains.

One cannot see a *dynamic equilibrium,* that is, reactions from a saturated salt solution to solid salt and back. In order to have a better idea, it is possible to revert to a model experiment. Two similar measuring cylinders are prepared, 50 ml of water is placed in one of the cylinders, and the other one remains empty (see Figure 16.9). Using two glass tubes of equal diameter to transport water back and forth, water is continuously transported between the two cylinders. After several transports, 25 ml of water remains in each of the cylinders, and the water level does not change despite carrying constant volumes of water back and forth (not shown in Figure 16.9).

If two glass tubes with different diameters are used, then one cylinder would perhaps have the volume of 20 ml and the other would have 30 ml "in equilibrium":

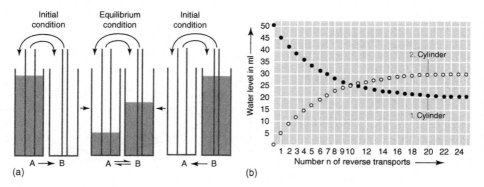

Figure 16.9 (a) and (b) Model experiment for the dynamic aspect of a chemical equilibrium [18].

the water level does not change because the same amount of water is continuously carried back and forth in the two different glass tubes (see Figure 16.9a). If one records the number of transports and the measured volumes in both cylinders a special graph results (see Figure 16.9b).

If calcium sulfate powder (gypsum) is mixed well with water and the suspension is left to stand, a white solid sinks down to the bottom. The question arising from the amount of solid substance is whether a part of the calcium sulfate dissolves or the substance is insoluble in water. Testing the electrical conductivity, however, shows a much higher value than with distilled water: calcium sulfate dissolves in very minute amounts; a dynamic equilibrium is formed between the solid residue and the saturated solution:

$$Ca^{2+}SO_4^{2-}(s, white) \rightleftharpoons Ca^{2+}(aq) + SO_4^{2-}(aq)$$

Magnesium sulfate and calcium sulfate solutions of equal concentrations show approximately the same electrical conductivity. If one compares the electrical conductivity of the saturated calcium sulfate solution with the conductivity of various standard solutions of soluble magnesium sulfate, one can find the unknown concentration of the saturated calcium sulfate solution at 30 °C:

$$c(\text{calcium sulfate}) = 10^{-2} \text{ mol l}^{-1}$$

Accordingly, for saturated calcium sulfate solution we know the ion concentrations (see Figure 16.10):

$$c(Ca^{2+}) = 10^{-2} \text{ mol l}^{-1} \text{ and } c(SO_4^{2-}) = 10^{-2} \text{ mol l}^{-1}$$

Now the solubility product can be defined in the following way (see Figure 16.10):

$$K_{sp}(CaSO_4) = c(Ca^{2+}) \times c(SO_4^{2-}) = 10^{-4}$$

If one is dealing with a diluted calcium sulfate solution, saturation can be attained in three different ways (see point A in Figure 16.10b): one continues to add solid calcium sulfate and reaches saturation (Point B). It is however also possible to add dropwise concentrated calcium chloride solution, thereby increasing the

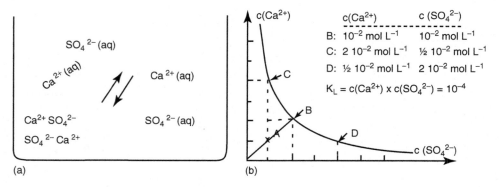

Figure 16.10 (a) and (b) Model drawing and mental model of solubility equilibrium of calcium sulfate.

concentration of Ca^{2+}(aq) ions until the first calcium sulfate crystals precipitate (Point C). It is also possible to add concentrated sodium sulfate solution, thereby increasing the concentration of SO_4^{2-}(aq) ions until the first solid calcium sulfate precipitates (Point D). In each case, we have a pair of values for the saturation equilibrium on the hyperbolic curve (see table in Figure 16.10); these pairs follow the solubility product. If one varies the concentrations of ions involved in equilibrium by adding the same kind of ions, then it is obvious that the product of ion concentrations is always constant, that this product has always, at constant temperature, the value $K_{sp} = 10^{-4}$. Tables and hyperbolic figures may demonstrate the concentration dependence of the related ions (see Figure 16.10).

The solubility equilibrium of calcium sulfate can also be demonstrated by supplementing portions of sodium sulfate and calcium chloride solutions: using highly concentrated solutions, solid calcium sulfate precipitates. In addition to Ca^{2+}(aq) ions and SO_4^{2-}(aq) ions, the solution also contains Na^+(aq) ions and Cl^-(aq) ions, and the equilibrium can be approached from the side of the dissolved ions:

$$c(Ca^{2+}) + c(SO_4^{2-}) \rightleftharpoons Ca^{2+}SO_4^{2-}(s, white)$$

16.3.3
Acid–Base Reactions and Proton Transfer

Examples of misconceptions are described by many authors around the world. In our institute, Musli [19] developed a questionnaire and gave it to about 100 students at senior classes of German high schools. Unusual and interesting statements from students have been quoted relating to acids, specifically on the differences between pure acids and acidic solutions, on neutralization, and on differences between strong and weak acids.

Acid Concepts Astonishingly, only acids are accredited with an "aggressive effect," although bases also have this attribute: "acids eat away, acids destroy, and acetic acid is a destructive and dangerous substance in chemistry, not used in normal everyday life" [19]. "An acid is something which eats material away or which can burn you; testing for acids can only be done by trying to eat something away, the difference between a strong and a weak acid is that strong acids eat material away faster than weak acids" [20]. Barker (Kind) [20] comments on these students' statements as follows: "no particle ideas are used here; the students give descriptive statements emphasizing a continuous, non-particle model for acids and bases, some including active, anthropomorphic ideas such as 'eating away'."

Regarding the question, "what do you understand by the term acid or base?," many students respond with a pH value ("acids have a small pH value"). Other statements describe acid concepts, which have been mainly learned and remembered: approximately 15% of the answers show the Arrhenius concept (acids contain H^+ ions); approximately 30% show the Broensted concept (acids release protons), whereby it is not certain if students correctly understand the notion of acids as acid particles. In the additional exercise, "give examples for atoms/ions/molecules that are acids or bases," mostly formulas for hydrochloric acid, sulfuric acid, and acetic acid are noted. Regarding the Broensted concept, the correct answers for base particles, that is, the hydroxide ions, were listed only in about 15% of the cases, at the same level as hydronium ions in dilute solutions of strong acids.

Sumfleth [21] shows that students accept the Broensted definition, but interpret bases mostly on Arrhenius' idea. Therefore, the knowledge about Broensteds' concept cannot be transferred to new contexts: "most students cannot really apply acid–base theories; this is also evident for students who have chosen chemistry as their major." Students also have a lot of difficulties with the idea of an acid. They tend to think in three directions:

1) Acids as pure substances like the gas hydrogen chloride, HCl;
2) Acids as solutions like hydrochloric acid, containing $H^+(aq)$ ions and Cl^- (aq) ions;
3) Acids as particles like hydronium ions, $H_3O^+(aq)$.

Mostly, students mix up all three ideas. They speak of substances: "hydrochloric acid gives one proton." They think protons come out of the nucleus of atoms or ions: "the other particle should be radioactive," and so on. Students have problems switching from the level of substances to the level of particles and they like – even in advanced classes – to stay on the level of substances: "hydrogen chloride plus acid gives hydrochloric acid." When discussing corresponding acid–base pairs, students do not deal appropriately with the level of particles, they prefer to state: "hydrogen chloride and water form the corresponding acid–base pair."

Pure Acids and Acidic Solutions In another exercise [19], the students are supposed to state the similarities and differences between pure sulfuric acid and the 0.1 M solution, and to schematically draw the smallest particles in two model beakers (see Figure 16.11). Correct answers regarding hydronium ions and sulfate ions in dilute solutions could be found in 10% of the answers or model drawings. Approximately 45% of the answers approached it from the dilution effect: symbols for

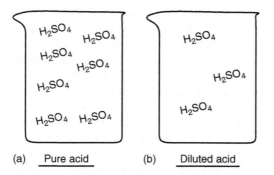

Figure 16.11 (a) and (b) Beaker models of pure and diluted sulfuric acid [4].

sulfuric acid molecules for the diluted solution were written with larger distances (see Figure 16.11).

Many other answers offer different claims: "pH value of pure acid is less; pH values are different for acids and acidic solutions (without mentioning pH value or differences); the densities vary; pure acids are much more corrosive, are more amenable to reactions than the solution." Only about 10% of the students gave the correct verbal answers and included appropriate model drawings with the expected ion symbols for the diluted solution. A surprising fact is that two students who gave a correct verbal answer regarding the "dissociation in diluted sulfuric acid solution" did not note any ion symbols.

Neutralization In this exercise [19], it was stated that "hydrochloric acid reacts with sodium hydroxide solution." The students were asked first to show chemical equations using the types of involved particles. Approximately 80% of the students were able to write the common equation: $HCl + NaOH \rightarrow NaCl + H_2O$. Half of the students noted the reaction equation with ion symbols and expressed that the $H^+(aq)$ ions and the $OH^-(aq)$ ions react to produce H_2O molecules. Most of the students stated that "NaCl" is formed without showing sodium ions and chloride ions; some even offered "NaCl molecules," "solid NaCl" or "NaCl crystals" as reaction products. Sumfleth [21] found that students think along the lines of acid–base equilibrium: "after neutralization, sodium chloride solution contains the same amount of hydrochloric acid and sodium hydroxide solution; with neutralization there exists equilibrium of acid and base."

Strong and Weak Acids Sumfleth [21] describes the common misconception that, for most students, acid strength is solely based on the pH value of solutions. Thus, it is possible for them to determine the acid strength in an experiment by using acid–base indicators. Students overlook the fact that by taking a 1 M hydrochloric acid solution with a pH value of 0, one can dilute to every larger pH value up to almost 7. The acid strength as equilibrium and as different concentrations of molecules or ions and mixing those ideas causes confusion.

In our questionnaire [19], students were asked to compare and contrast 0.1 M solutions of hydrochloric acid (HCl) and acetic acid (HAc); and in addition students were requested to draw schematic beaker models of the involved atoms, ions, or molecules. Approximately half of the students gave no answers concerning similarities and differences, 20% mentioned the acid strength, and 10% noted the pH value as differences. Acetic acid was regarded as "the stronger acid because a larger I-effect of the methyl group can be registered at CH_3COOH molecules and therefore the proton can more easily split off." This quotation shows that the treatment, which coincidentally took place in the half year of the studies in organic chemistry, led the students to associations on arbitrary contents, which they did not properly understand. Only 15% of the students showed appropriate acetic acid molecular models and the related ions in their model drawings (see Figure 16.12a).

To the same degree, students have drawn correct ion symbols but no molecule symbols, or they merely imagine only molecules and no ions (see Figure 16.12).

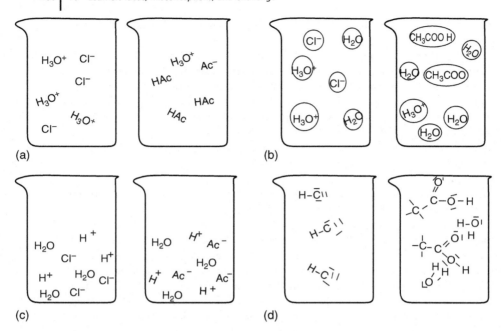

Figure 16.12 (a)–(d) Examples for appropriate and inappropriate mental models on weak acids [4].

From this data one can easily conclude that these students have not understood the differences between strong and weak acids; they know about equilibria but do not apply the knowledge on the equilibrium of molecules and ions in weak acids.

Challenge of Misconceptions Because acids are known as *solutions* which are "destroying other material," those statements support the destruction concept of students. To challenge this misconception, one can show that acidic household cleaners remove lime deposits but produce salt solutions and carbon dioxide: all changes of material by acids or bases are chemical reactions producing other special products. But the most important challenges are misconceptions according the Broensted concept, neutralization, and weak acids.

Broensted Concept After knowing some phenomena and the facts that acidic solutions contain H^+(aq) ions and basic solutions OH^-(aq) ions, it is important to convince learners that the proton transfer idea is the broader concept for acids and bases. Because one proton can only go from one particle to another one, this Broensted idea is based on acidic particles which give protons, such as HCl molecules, H_2SO_4 molecules, H_3O^+(aq) ions, or HSO_4^- (aq) ion.

One example of a typical proton transfer reaction is the formation of hydrogen chloride gas by sodium chloride and pure sulfuric acid. Both are put into a gas developer, the acid is dropped to the salt: gaseous hydrogen chloride can be filled into a gas syringe or a cylinder. By this reaction, H_2SO_4 molecules donate protons

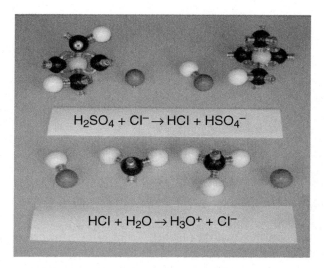

Figure 16.13 Visualization of two acid–base reactions in the sense of Broensted's theory
Photo by Ulrike Henkel, University of Muenster.

(H^+ ions) to Cl^- ions of sodium chloride to form HCl molecules and HSO_4^- ions, sodium hydrogensulfate remains (see Figure 16.13):

$$H_2SO_4 \text{ molecule} + Cl^- \text{ ion} \rightarrow HCl \text{ molecule} + HSO_4^- \text{ ion}$$

The produced hydrogen chloride gas can be mixed with water: the indicator changes colors, and electric conductivity raises. This well-known reaction forms hydrochloric acid solution; HCl molecules give protons to H_2O molecules, and the following ions are obtained (see Figure 16.13):

$$HCl \text{ molecule} + H_2O \text{ molecule} \rightarrow H_3O^+(aq) \text{ ion} + Cl^-(aq) \text{ ion}$$

In both cases, the molecules are acids or acidic particles which donate protons; Cl^- ions and H_2O molecules are bases or basic particles which accept protons. In hydrochloric acid, the $H_3O^+(aq)$ ion reacts as a proton donor, but also in diluted sulfuric acid the $H_3O^+(aq)$ ion is the acidic particle – not the H_2SO_4 molecule. For all acid–base reactions, one has to look at those particles which give protons and at those which take protons.

Neutralization Taking solutions of strong acids and bases, the $H_3O^+(aq)$ ions are the acidic particles and $OH^-(aq)$ ions the basic particles, and both react to form water molecules:

$$H_3O^+(aq) \text{ ion} + OH^-(aq) \text{ ion} \rightarrow 2 \; H_2O \text{ molecules}$$

After their reaction, the other ions remain: in case of the reaction of hydrochloric acid and sodium hydroxide solution, $Na^+(aq)$ ions and $Cl^-(aq)$ ions remain as "spectator ions," they are no reacting partners. No "solid salt" or "NaCl molecules"

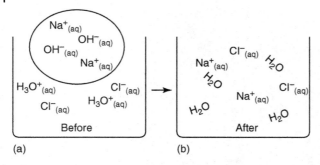

Figure 16.14 Beaker model of the neutralization of hydrochloric acid by sodium hydroxide.

are produced, but sodium chloride solution remains – it is good for understanding to visualize this by ion symbols (see Figure 16.14).

It is also advantageous to visualize that the number of ions is the same before and after neutralization: four ions in this model (see Figure 16.14) are there before neutralization, four ions are there afterward. So, the H_3O^+(aq) ions are replaced by Na^+(aq) ions, and the electric conductivity goes down during neutralization because H_3O^+(aq) ions have a higher specific conductivity compared to the Na^+(aq) ions after neutralization.

Weak Acids The term *"weak"* suggests itself the following most common misconception: weak acids are "weakly concentrated." It may well be that during students' lessons, protolysis equilibrium of acetic acid was used as an example, maybe even equilibrium constants came into play, and pH values of specific acetic acid solutions were measured or calculated – however, only a few students are able to comprehend and connect all these facts to develop the scientific idea about weak acids. In order to look at the degree of protolysis, it is advisable to use convincing experiments. If the pH values of 1.0 and 0.1 M solutions of two acids, say, hydrochloric acid and acetic acid, are measured with a calibrated pH meter, one gets the expected pH values of 0 and 1 for hydrochloric acid solutions – but not for the acetic acid solutions: approximate pH values of 2.4 and 2.9 can be measured.

When this happens, a classic cognitive conflict arises: "what is so different about acetic acid"? If the 0.1 M acetic acid solution shows a pH of nearly 3, the concentration of the H^+(aq) ions should be 10^{-3} mol l^{-1}. Because the concentration of HAc molecules starts with c(HAc) = 10^{-1} mol l^{-1}, only 1% of the HAc molecules protolyze into ions. In a beaker model, one should draw 99 models of HAc molecules compared to only 1 H_3O^+(aq) ion and 1 Ac$^-$ (aq) ion – in every case the number of molecule models must be higher than the number of ions (see Figure 16.15b). If the aspect of a dynamic equilibrium is connected and k_S constants are discussed carefully, the understanding will rise.

Additionally, electrical conductivity measurements help in the understanding of protolysis equilibrium for weak acids. The comparison of equally concentrated strong and weak acids supplies the much lesser conductivity for weak acid solutions. If one carries out a conductivity titration, one gets very different

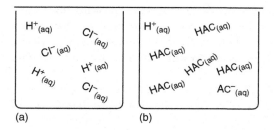

Figure 16.15 (a) and (b) Beaker models of a strong and a weak acid.

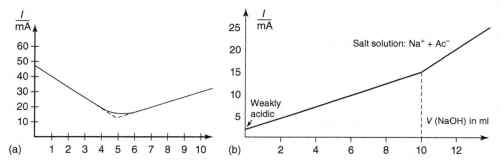

Figure 16.16 Diagrams of conductivity titrations of hydrochloric acid (a) and acetic acid (b) [18].

forms of conductivity curves in comparison to the titration of strong acids (see Figure 16.16). On titrating with sodium hydroxide solution, the measured values do not decrease but they rather increase. In this titration, a very low concentration of hydronium ions reacts with hydroxide ions, but mostly the large number of HAc molecules are transferring protons to OH$^-$(aq) ions: HAc molecules are replaced by Ac$^-$(aq) ions and, therefore, the increase in conductivity is explained. Later, after the equivalent point is reached and an excess of hydroxide ions appear, the curve rises more steeply. For the description of this neutralization, there are two kinds of acid–base reactions (see also Figure 16.15b):

$$HAc(aq) + OH^-(aq) \rightarrow H_2O(aq) + Ac^-(aq)$$
$$H_3O^+(aq) + OH^-(aq) \rightarrow 2\,H_2O(aq)$$

If all acid–base reactions are interpreted consequently with atoms, molecules, or ions as acidic and basic particles, students may get a scientific understanding of the Broensted concept and will not develop misconceptions as presented.

16.3.4
Redox Reactions and Electron Transfer

As in the historical development, the Lavoisier definition of oxygen transfer is often used in beginners' lessons ("metals take oxygen, metal oxides are formed"). Later, as soon as the differentiated atomic model is introduced, the redox reaction regarding electron transfer is applied in advanced lessons. Knowing the oxygen transfer and the idea of the redox reaction, there is often the belief that oxygen has to be involved in *every* redox reaction. The reason for this may be the syllable −ox, which is semantically strongly associated with the name oxygen (oxidation, metal oxide, or nonmetal oxide).

Schmidt [22] described studies with almost 5000 students who were asked to decide which of his listed reactions belonged to redox reactions: the reaction of diluted hydrochloric acid with (i) magnesium (Mg), (ii) magnesium oxide (MgO), and (iii) magnesium hydroxide (Mg(OH)$_2$). We know of course that (i) is to be identified as a redox reaction and that (ii) and (iii) are acid–base reactions: in (ii), H_3O^+(aq) ions react with O^{2-} ions of magnesium oxide, and in (iii) H_3O^+(aq) ions react with OH^- ions of magnesium hydroxide.

Approximately half of the students in advanced courses chose the correct answer. The remaining students marked one or both oxygen-related reactions and gave explanations like: "(ii) and (iii) contain oxygen, which is absolutely necessary for redox reactions; oxygen is necessary for every redox reaction, so (i) cannot be a redox reaction; (ii) and (iii) are redox reactions because in both cases oxygen and electron transfer takes place; oxidation means a reaction in which oxygen is involved. The ending "oxide" shows that (ii) and (iii) are redox reactions" [22].

According to the oxygen concept, Schmidt [22] cited the following study about a typical acid–base reaction: "Garnett and Treagust, in 1992, asked senior high school students whether or not the equation $CO_3^{2-} + 2 H^+ \rightarrow H_2O + CO_2$ represents a redox reaction. All students with correct answers used the oxidation number method. Those who answered incorrectly had two reasons. One was to assume that the carbonate ion donates one oxygen atom to form carbon dioxide and was, therefore, reduced. The other was to assign the oxidation number to polyatomic species by using their charge number. CO_3^{2-} was given the oxidation number −2, and CO_2 the oxidation number 0. Consequently, the reaction $CO_3^{2-} \rightarrow CO_2$ was identified as an oxidation. In a similar manner, the reaction $H_3O^+ \rightarrow H_2O$ can be identified as a reduction: the hydronium ions must have gained electrons and so should have been reduced" [22].

Sumfleth [21] asked students in grades 6–12 in Germany to provide an explanation regarding the popular reaction of an iron nail in copper sulfate solution. She found incorrect answers, which could be traced back to preconcepts and school-made misconceptions.

Especially, students in grades 6–8 described the formation of a copper-colored coating with "sedimentation, clinging to, sticking to, or color fading of a material on an iron nail" or "the copper sulfate colors the iron nail, the copper sulfate sticks on to it, like when a piece of wood is placed in a dye and is then dried." Half of the 7th grade students guessed "an attraction of the substances" as the reason; the other students mentioned a pre-existing magnetism – probably because of the iron nail. These students, however, only described their observations with words, and one cannot admonish them for their preliminary ideas. Even in senior high school classes, these discussions remain: "copper sulfate is reduced; copper atoms attract electrons; iron nails can absorb ions from the solution" [21].

Heints [23] carried out new studies in grades 10–12 at German high schools where redox reactions have been introduced as electron transfer; the found school-made misconceptions are similar to those that were mentioned already. Many other references show misconceptions in the area of redox reactions, especially with the interpretation of voltage and electric current in electrolysis or Galvanic cells. Marohn [24] looked for the mental models that students develop by discussing Galvanic cells. In addition, Garnett and Treagust discovered conceptual difficulties in the area of electric circuits [25] and electrolytic cells [26]. The same was the case with Ogade and Bradley working on electrode processes [27], and Sanger and Greenbowe investigating common misconceptions in electrochemistry [28] or current flow in electrolyte solutions and the salt bridge [29].

Challenge of Misconceptions Nevertheless, these topics are so difficult to understand that misconceptions can hardly be avoided – especially concerning the nature of electrons as waves and/or particles, and concerning the electromagnetic fields and their forces. Therefore, the only challenge is to look to the basic definitions of the redox reaction and to discuss common experiments to gain scientifically accepted mental models of redox reactions.

Oxygen Transfer If the students in beginner classes of chemistry should know about the production of iron, copper, or other metals from ores and metal oxides, one can demonstrate the reaction of copper oxide with carbon or with magnesium. One should stay on the macro level of substances and their reactions, and describe the observations only by words:

Copper oxide(s, black) + carbon(s, black)

→ copper(s, red) + carbon dioxide(g)

Copper oxide(s, black) + magnesium(s, metallic)

→ copper(s, red) + magnesium oxide(s, white)

It can be stated that copper oxide is reduced to copper, and that carbon is oxidized to the compound carbon dioxide – but perhaps one can avoid calling this reaction a redox reaction. Because of all misconceptions mixing the oxygen and electron definition, one can wait and name in higher classes of chemistry only the electron transfers with the idea of redox reaction.

Electron Transfer For the same reason, one starts that topic with reactions where no oxygen is involved, for example, with the precipitation of copper from a copper sulfate or, better, a copper chloride solution. Because some students argue with "iron takes oxygen from sulfate ions" [23], it seems more acceptable to use copper *chloride* solution. A prerequisite for the interpretation of metal precipitations is the term "*ion*" and the atomic structure by nucleus and differentiated electron shells. So the blue color of a diluted copper chloride solution can be explained by the presence of Cu^{2+}(aq) ions. Armed with this information, there are good ways for the problem-oriented interpretation of the following experiments.

An iron nail is dipped into copper chloride solution and taken out after 20 s: a copper-colored coating appears on the iron nail. If iron wool is placed in copper chloride solution, the wool turns red, the solution warms up, and the blue color of the solution disappears. The discoloration of the solution almost forces an interpretation that Cu^{2+}(aq) ions from the solution "disappear," or have reacted. This question leads to the supposition that they have deposited as Cu atoms on the iron and have formed copper crystals.

If a helix-shaped copper wire is placed in a diluted silver nitrate solution and one waits for a few minutes, then the development of silver crystal needles can be observed and also the change in the color of the initially colorless solution to blue. With this reaction, one observes that Cu^{2+}(aq) ions have appeared and that copper metal has partially dissolved. From this reaction, one concludes that, with experiences gathered from the first experiment, metal atoms dissolve as ions, accompanied by the release of electrons. Along with this, metal cations of the salt solution take electrons, forming metal atoms and crystallizing to needles of pure silver:

Cu atom $\rightarrow Cu^{2+}$(aq) ion + 2 e^-

2 Ag^+(aq) ions + 2 $e^- \rightarrow$ 2 Ag atoms

Describing the half reactions, it should be made apparent to the students that the term "+ 2e^-" should be placed on the correct side of the equation: one Cu atom can become one Cu^{2+} ion only if it simultaneously releases two electrons. It is advisable to suggest to students that the number of atoms and the number of charges should be the same "left and right of the arrow." In the given examples, the number of the charges on both sides is zero in each case.

It should be concluded that the ions from the more noble metals are changed into atoms and crystallized from the solution. Simultaneously, as a result of electron transfers, the atoms of active metals dissolve through the formation of ions. This hypothesis can systematically be tested with other metal pairs; the observations are noted by the precipitation sequence of metals.

Of course, those reactions should be visualized, for example, by a beaker model (see Figure 16.17): each Cu^{2+} ion from the solution is taking two electrons, and an iron atom of the nail is delivering them, dissolving as an Fe^{2+} ion. The chloride or sulfate ions are not reacting, so they can be called "*spectator ions.*"

The conversion of metal compounds to pure metals is historically known as reduction; so the reduction of metal ions with the gaining of electrons is thereby explained:

2 Ag^+(aq) ions + 2 $e^- \rightarrow$ 2 Ag atoms : gain of electrons; reduction

The gained electrons come from the reacting metal atoms, which form ions by losing electrons:

Cu atom $\rightarrow Cu^{2+}$(aq) ions + 2 e^- : loss of electrons; oxidation

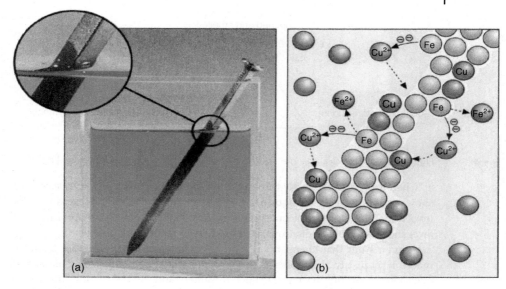

Figure 16.17 (a) and (b) Reaction of an iron nail with copper chloride solution and mental model [18].

Altogether, an electron transfer takes place from Cu atoms of copper to Ag^+ ions of the solution:

$$Cu + 2\,Ag^+(aq) \rightarrow Cu^{2+}(aq) + 2\,Ag \; : \; \text{electron transfer; redox reaction}$$

The term *oxidation* can now be associated with well-known metal–oxygen reactions; also in these reactions, the metal atoms are oxidized into their corresponding metal ions, and oxygen atoms are taking electrons and are reduced to oxide ions. Oxygen reactions can be called *special types of redox reactions* – but most other redox reactions deal without oxygen as a reaction partner!

If one argues consequently by all further redox reactions with the atoms, ions, and molecules, this topic can be understood and the definition by "oxygen transfer" should not interfere with the idea of "electron transfer." Later, redox reactions can be explained by oxidation numbers too – but pay attention: by this mental model the oxidation number of *atoms* or of *atoms in molecules* is involved, not any oxidation number of *substances*.

16.4
Best Practice to Challenge Misconceptions

Acid–base reactions and proton transfer can only be explained if consequently the atoms, ions, or molecules are pointed out that give or take a proton. With redox reactions and electron transfer, it is the same: atoms, ions, or molecules are giving or accepting one or two electrons – not substances! Johnstone [30] created

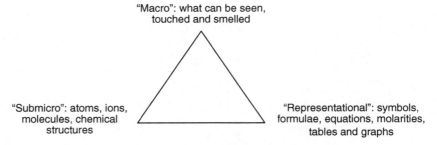

Figure 16.18 "Chemical triangle" for teaching chemistry according to Johnstone [30].

a "Chemical Triangle" with three corners (see Figure 16.18): The macro level shows phenomena such as substances and reactions; the submicro level shows the involved atoms, ions, molecules, and chemical structures; the representational level shows formulae, equations, stoichiometric calculations, and so on.

He points out that chemistry is hard to understand if one switches from the macro level just to the representational level: students memorize formulae and equations, and don't have the chance to understand. By instructing first – after showing some phenomena – the submicro level and the involved atoms, ions, molecules, and chemical structures of involved substances, the learners will understand in a better way. One example: conducting titrations in the neutralization topic, one shows and discusses the beaker model first (see Figure 16.14), and after that one can develop equations to show the reaction of hydronium ions with hydroxide ions to water molecules. Going this way, students will accept that formulae and reaction equations are shortening models of all theoretical explanation, also of the shown beaker models.

16.4.1
Misconceptions

Many misconceptions appear when the "submicro level" [30] is introduced. Students transfer properties of substances to properties of particles [4]:

- S atoms are yellow, Cu atoms are red
- P atoms are poisonous, they ignite themselves
- one Cu atom is the smallest portion of copper
- sugar molecules are sweet
- sugar molecules disappear by dissolving sugar in water, but the water tastes sweet
- particles can disappear by dissolving crystals, but they appear again by crystallization
- water has an angle of 109°
- water molecules are liquid
- O atoms have two arms, H atoms only one arm
- C atoms are destroyed by combustion of charcoal

- magnesium contains of two kinds of particles: one kind evaporates by combustion, the others remain as ashes
- between molecules of gases there must exist some invisible material, there cannot be a vacuum (horror vacui)
- gas molecules have no mass.

Teachers and students can avoid the mixing up of those misconceptions if they differentiate three levels of terminology concerning the three levels of Johnstone's triangle (see Figure 16.18):

- *macro level (reality):* substances and their properties such as density, melting temperature, boiling temperature, electric conductivity, pH values, and so on; chemical reactions, substances before and after reactions, energy changes, and so on.
- *submicro level (mental and concrete models):* experts investigate substances and get mental models about chemical structures by their scientific theories and measurements; learners cannot go this way – they need sphere-packing and lattice models for giant structures or molecular models for the structure of the involved molecules as concrete models concerning the arrangement of atoms, ions, or molecules; by those models they can develop suitable mental models.
- *representational level (symbolic level):* formulae, chemical equations, mole idea, stoichiometric calculations, equilibrium constants and their use, calculations of pH values or redox potentials, thermodynamics and calculations of energy changes, and so on.

Besides all the "preconcepts" brought from every-day life [4] and those misconceptions concerning the chemical terminology, students also develop "school-made misconceptions" by insufficient teaching in the area of difficult topics [4]:

- chemical equilibrium and the use of equilibrium constants
- acid–base reactions and proton transfer from one particle to another
- redox reactions and electron transfer from one particle to another
- complex reactions and ligand transfer from one particle to another
- energy transfer, specially concerning chemical energy.

If teachers know those misconceptions, they can plan all instruction based on this knowledge and can prevent students from school-made misconceptions or can even integrate misconceptions into instruction for a better understanding.

16.4.2
Integrating Misconceptions into Instruction

In older times, teachers perceived the students like "blank pages" and thought that teachers only have to fill the "blank pages" with contents of science. Today we know that at a very early stage, students develop their own preconcepts about properties of substances and their changes, and about combustion processes and the role of gases. Today, empirical studies show that we have more success in teaching and learning when we integrate those alternative models into instruction: the *conceptual change* seems more realistic if students discuss their conception, feel uncomfortable with it, feel that the new scientific concept can explain better, and can do a conceptual change more successful [31]. Also school-made misconceptions should be reflected and compared with the scientific explanation.

One way for the comparison of own concepts and scientific ones is *concept cartoons* [10]: the right answer is shown by a statement of a boy or girl – and a lot of alternatives are shown too. In an example (see Figure 16.19), students are asked: "what species are present in hydrochloric acid"? By this way the teacher may diagnose misconceptions about the composition of the acidic solution and will find how students are thinking.

With the preparation of this topic, the teacher can challenge those misconceptions by convicting experiments, suitable models, and problem-solving teaching. After finishing the topic, the teacher may show the same cartoon another time: students will discover the right answer and will explain what is wrong with the other alternatives. With this knowledge, students will write a better test or will give the correct answer more easily.

The American scientist Ausubel [32] has written a big book about educational psychology. In an interview, he was asked to mention only one sentence which seems the most important for education. Ausubel stated: "Ask your students what

Figure 16.19 Concept cartoon concerning the composition of hydrochloric acid [10].

they know about a topic. Take those answers and plan your instruction on the basis of that knowledge" [32]. Also the reflected misconceptions are part of the knowledge that students bring to class: teachers should know this knowledge and should integrate it in his lectures!

16.5
Conclusion

Chemistry, physics, and biology seem to be subjects in school that are notorius for the development of misconceptions by students. They bring a lot of own experiences and observations of their every-day life into the class: alternative explanations or "preconcepts." Teachers cannot avoid those preconcepts according structure of matter, combustion, or the nature of gases - they have to convince students by good experiments and mental models to open the way for understanding scientific explanations; students should realize a conceptual change.

Another kind of misconceptions is also discussed: "school-made misconceptions." They are developed by insufficient teaching in the area of difficult topics such as equilibrium, acid-base, redox, and complex reactions, or energy transfer. This kind of misconceptions can be prevented by teaching based on the knowledge of those misconceptions or even by including the known misconceptions into discussions with young learners: they may grab the scientific idea better if they reflect not only the scientific concept but also compare it with unsuitable explanations. Some examples are described in the area of ionic compounds, equilibrium, acid-base reactions, and redox reactions. Chemistry.

References

1. Barke, H.-D., Harsch, G., and Schmid, S. (2012) *Essentials of Chemical Education*, Springer, Berlin, Heidelberg.
2. Driver, R. (1985) *Children's Ideas in Science*, University Press, Philadelphia, PA.
3. Pfundt, H. (1975) Urspruengliche Vorstellungen der Schueler fuer chemische Vorgaenge. *MNU*, **28**, 157.
4. Barke, H.-D., Hazari, A., and Yitbarek, S. (2009) *Misconceptions in Chemistry*, Springer, Berlin, Heidelberg.
5. Muench, R. (1982) Luft und Gewicht. *NiU-P/C*, **30**, 429.
6. Weerda, J. (1981) Zur Entwicklung des Gasbegriffs beim Kinde. *NiU-P/C*, **29**, 90.
7. Voss, D. (1998) *Der Gasbegriff in den Vorstellungen der Schueler und Schuelerinnen*, University of Muenster.
8. Piaget, J. and Inhelder, B. (1971) *Die Entwicklung des raeumlichen Denkens beim Kinde*, Klett, Stuttgart.
9. Duit, R. (1996) Lernen als Konzeptwechsel im naturwissenschaftlichen Unterricht, in *Lernen in den Naturwissenschaften*, IPN, Kiel.
10. Temechegn, E. and Sileshi, Y. (2004) *Concept Cartoons as a Strategy in Learning, Teaching and Assessment in Chemistry*, Addis Ababa, Ethiopia.
11. Bergquist, W. and Heikkinen, H. (1990) Student ideas regarding chemical equilibrium: what written test answers do not reveal. *J. Chem. Educ.*, **67**, 1000.
12. Finley, F.N., Stewart, J., and Yarroch, W.L. (1982) Teachers' perceptions of important and difficult science content. *Sci. Educ.*, **4**, 531–538.

13. Tyson, L., Treagust, D.F., and Bucat, R.B. (1999) The complexity of teaching and learning chemical equilibrium. *J. Chem. Educ.*, **76**, 554.
14. Banerjee, A.C. and Power, C.N. (1991) The development of modules for the teaching of chemical equilibrium. *Int. J. Sci. Educ.*, **13**, 358.
15. Hackling, M.W. and Garnett, P.J. (1985) Misconceptions of chemical equilibrium. *Eur. J. Sci. Educ.*, **7**, 205.
16. Kienast, S. (1999) *Schwierigkeiten von Schuelern bei der Anwendung der Gleichgewichtsvorstellung in der Chemie: Eine empirische Untersuchung ueber Schuelervorstellungen*, Shaker, Aachen.
17. Osthues, T. (2005) Chemisches Gleichgewicht. Empirische Erhebung von Fehlvorstellungen im Chemieunterricht. Master thesis. University of Muenster.
18. Asselborn, W., Jaeckel, M., and Risch, K.T. (1998) *Chemie heute Sekundarstufe II*, Schroedel, Hannover.
19. Musli, S. (2004) *Saeure-Base-Reaktionen*: Empirische Erhebung zu Schuelervorstellungen und Vorschlaege zu deren Korrektur. Master Thesis, University of Muenster.
20. Barker, V. (Kind V.) (2000) *Beyond Appearances – Students' Misconceptions about Basic Chemical Ideas*, Royal Society of Chemistry, London.
21. Sumfleth, E. (1992) Schuelervorstellungen im Chemieunterricht. *MNU*, **45**, 410.
22. Schmidt, H.J. (2003) Shift of meaning and students' alternative concepts. *Int. J. Sci. Educ.*, **25**, 1409.
23. Heints, V. (2005) Redoxreaktionen: Empirische Erhebung zu Schuelervorstellungen und Vorschlaege zu deren Korrektur. Master thesis. University of Muenster.
24. Marohn, A. (1999) Falschvorstellungen von Schuelern in der Elektrochemie – eine empirische Untersuchung. Dissertation. University of Dortmund.
25. Garnett, P.J. and Treagust, D.F. (1992) Conceptual difficulties experienced by senior high school students of electrochemistry: electric circuits and oxidation-reduction equations. *J. Res. Sci. Teach.*, **29**, 121.
26. Garnett, P.J. and Treagust, D.F. (1992) Conceptual difficulties experienced by senior high school students of electrochemistry: electrochemical and electrolytical cells. *J. Res. Sci. Teach.*, **29**, 1079.
27. Ogade, A.N. and Bradley, K.H. (1996) Electrode processes and aspects relating to cell EMF, current, and cell components in operating electrochemical cells. *J. Chem. Educ.*, **73**, 1145.
28. Sanger, M.J. and Greenbowe, T.J. (1997) Common students' misconceptions in electrochemistry: galvanic, electrolytic and concentration cells. *J. Res. Sci. Teach.*, **34**, 377.
29. Sanger, M.J. and Greenbowe, T.J. (1997) Common students' misconceptions in electrochemistry: current flow in electrolyte solutions and salt bridge. *J. Chem. Educ.*, **74**, 819.
30. Johnstone, A.H. (1997) Chemistry teaching – science or alchemy? *J. Chem. Educ.*, **74**, 268.
31. Posner, G.J., Hewson, P.W., and Gertzog, W.A. (1982) Accommodation of a scientific conception. Towards a theory of conceptual change. *Sci. Educ.*, **66** (2), 211.
32. Ausubel, D.P. (1974) *Educational Psychology. A Cognitive View*, Rinehard and Winston, New York, Holt.

17
The Role of Language in the Teaching and Learning of Chemistry
Peter E. Childs, Silvija Markic, and Marie C. Ryan

17.1
Introduction

> Do you wish to learn science easily? Then begin by learning your own language.
>
> *Étienne Bonnot (1714–1780)*
>
> Essai sur l'origine des connasissances humaines.

There can be no teaching, learning, thinking, or understanding in any subject without a basic proficiency in language. Mammino has emphasized this strongly in her work, identifying the importance of language mastery in science learning: *Students need to be guided to realise that science learning and language mastering cannot be separated, that understanding depends largely on language mastering and that incorrect language unavoidably results in incorrect science* [1, p. 142]. Poor language proficiency affects both understanding (reading and listening) and expression (speaking and writing). The level of achievement of a student in science is probably limited by their linguistic and/or mathematical ability. These hurdles faced by a novice learning science are one of the main reasons why many students (and adults) think science, and particularly chemistry and physics, is hard.

Science as a discipline, including chemistry, depends on two supporting pillars – language and mathematics. Both are essential for understanding, using, and communicating scientific ideas, and an initial proficiency in both language and mathematics is a prerequisite for studying science.

> There are two essential pre-requisites for science education: literacy and numeracy. You cannot study any science without being able to read and write and also use numbers. These are the foundations of science education and weaknesses in either area will seriously disadvantage a student of science [2].

There has been a growing recognition in the last 20–30 years of the importance of language in science education; however, it is still a relatively neglected area of research. Since the influential book by Wellington and Osborne *Language and Literacy and Science Education* [3], there have been chapters on language in science education reference works such as in the *Second International Handbook of Science Education* [4] and the *Encyclopedia of Science Education* [5]. Language also features now in handbooks for science teachers such as *Science Learning, Science Teaching* [6], and *Good Practice in Science Teaching: what research has to say* [7]. Recently, Markic *et al.* [8] have included a chapter on language in *Sourcebook for Science Teaching* aimed at trainee science teachers.

The role of language in teaching and learning chemistry (or any science) is a possibly a bigger problem today than it was even 20 or 30 years ago. Since then, the nature and role of education has changed in many countries, in addition to profound changes in society and technology, which have impinged on the place of language in schools. We suggest some factors here that have reduced overall literacy in relation to science:

1) There has been a move from science as an elitist and specialist subject to "science for all," both in secondary school and at university. This means that science is now being taught almost always in a mixed ability context, to students with a range of linguistic skills and greater heterogeneity of ability.
2) In many countries in Europe (and worldwide), there are more non-nationals and second-language learners (SLLs) in classrooms, so there is more linguistic heterogeneity. Many students are trying to learn science in a second language, often without home support.
3) There is some evidence of an overall decline in literacy levels due to changes in the education system, with less emphasis on formal methods of language instruction, changes in assessment, greater diversity, and social changes, for example, the growth of text messaging.
4) Since the 1950s, there has been a decline in the study of classical languages at school (Latin and Greek), which had been part of the core curriculum for centuries. Thus modern students are no longer familiar with the roots of much scientific language, for example, the use of suffices and prefixes drawn from Latin and Greek. Hogben [9] commented on this in relation to intending scientists and doctors:

   > From the start, students of natural science thus held all the clues to decoding it. Most of them now have no knowledge of Latin, still less of Greek. To them, a vocabulary adequate to the needs of many branches of scientific enquiry is forbidding and mysterious.

5) There has been a change in science assessment procedures to more objective-style and structured questions, requiring little writing and explanation, and less emphasis on extended essays and comprehension.

Taken together, these changes have made students less prepared to study and succeed in science. As Mammino [1, p. 147] said: *The development of abilities that are fundamental to science learning, like logical and reasoning abilities, visualization ability, or the ability to recognise and utilise the descriptive roles of mathematics, closely depend on the degree of language-mastering.* If you don't have it, you can't use it!

The language of chemistry is rich, diverse, and complex, with roots in alchemy, everyday language, and phenomena. Teaching and learning chemistry is not just about learning the language, though it plays a key role, and learning the language, facts, and concepts of chemistry must go hand in hand. Just as in learning one's own language, there should be a natural acquisition of chemical language on a need-to-know basis. The use of games, activities, and other teaching strategies can help to make the process more enjoyable for the student and the teacher. The main thing for the teacher is to recognize that the problem of language is bigger than just the technical terms and symbols, and to be aware of the areas where students find difficulty. Language in teaching and learning chemistry, as in any subject, is crucial in thinking, visualization, and understanding.

17.2
The History and Development of Chemical Language

> To write a full description of the origin, growth and misadventures of the language of chemistry is to write a history of the subject [10].

The language of chemistry, like all the sciences, has developed over the centuries. Prior to the late eighteenth century, there was no systematic chemical nomenclature. The names of the elements and chemical compounds as well as the symbols used varied from person to person and country to country. The early (al)chemists had their own individual languages, designed to conceal as much as to communicate. The same substance would have different names even in the same language and there was no order or system. However, there was a common linguistic foundation in the classical languages, as Hogben [9, p. 3] points out:

> Western Christendom has equipped what is now world-wide science with a world-wide and constructed vocabulary ... based on two dead languages (Latin and Greek) which were a routine part of western education until the 1950s/60s.

17.2.1
Chemical Symbols: From Alchemy to Chemistry, from Dalton to Berzelius

The old alchemical symbols for elements and compounds were obscure and idiosyncratic, designed as much to conceal as to communicate. There were some common symbols but no universal system and each alchemist was a law unto himself. Figure 17.1 shows the complexity of these symbols [11].

Figure 17.1 Symbols of the alchemists [11].

⊙	Hydrogen	⊕	Soda	⊙⊕	Ammonia
⊕	Nitrogen	⊕	Pot Ash	⊙●	Olefiant
●	Carbon	○	Oxygen	○●	Carbonic oxide
⊕	Sulphur	©	Copper	○●○	Carbonic acid
⊗	Phosphorus	Ⓛ	Lead		Sulphuric acid
⊙	Alumina	⊙○	Water		

Figure 17.2 Dalton's 1808 symbols and formulae for elements and compounds.

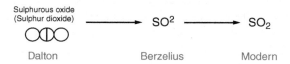

Figure 17.3 Various notations for sulfur dioxide.

When John Dalton devised his symbols for elements and compounds as part of his new system of chemistry, he was closer to the alchemists than to modern notation. Figure 17.2 shows some of Dalton's symbols.

The problems with Dalton's imaginative symbols were that they were cumbersome to write and difficult to remember and presented problems for the printer. Berzelius' elegant solution (1813) was to use ordinary letters, linked to the Latin name of the elements, making them easy to remember and write, and presenting no problems to the printer. Dalton's symbols never really caught on, and even the simpler system of Berzelius was slow to be adopted. "As late as 1837, Dalton complained that 'Berzelius; symbols are horrifying; a young student of chemistry might as well learn Hebrew as make himself acquainted with them. They appear like a chaos of atoms ... to equally perplex the adept of Science, to discourage the learner as well as to cloud the beauty of the Atomic Theory [12, p. 115].'" Figure 17.3 compares the formula for sulfur dioxide as written by Dalton, Berzelius, and in modern usage.

17.2.2
A Systematic Nomenclature

It can be argued that the development of chemistry as a science could not happen until the language was systematized. There were several stages in this process.

1) *The new nomenclature (1789)*: at the end of the eighteenth century, French chemists proposed a new system of chemical nomenclature in *Méthode de Nomenclature Chimique*, based on the structure of the French language

> As ideas are preserved and communicated by means of words, it necessarily follows that we cannot improve the language of any science,

without at the same time improving the science itself; neither can we, on the other hand, improve a science without improving the language or nomenclature which belongs to it. [13, Preface, xiv–v]

This new nomenclature was not accepted overnight, but the similarity of English and French meant that it was readily adopted by English chemists. The international reputation of Lavoisier also helped to give the proposals credibility. Hogben [9, p. 36] commented:

> Prompt acceptance of the new nomenclature by Britain, at a time when Britain and France were in the vanguard of chemical discovery and chemical industry, confronted the international scene with a fait accompli. It did so by the unforeseen accident that two national languages had the same battery of suffixes.

2) *A symbolic language*: the next important development was the development of a universal symbolic language to represent elements and compounds. The use of symbols went back to the alchemists, but each person invented their own symbols and there was no universal agreement. Each national language has its own names for the elements and compounds, so that a chemist cannot read books or papers in another language. John Dalton invented his own symbols to represent elements and compounds, but this was not based on the alphabet. However, in 1813 the Swedish chemist Berzelius proposed new symbols and chemical notation based on the standard alphabet, which would eventually develop into a universal language for chemistry. The international stature of Berzelius helped to get these proposals accepted and they still form the foundation of the symbolic language of chemistry.

3) *International agreements*: the third major stage was the initiation of international meetings to discuss, formalize, and agree on the language of chemistry. The first of these was held in 1860 in Karlsruhe. This congress dealt with the systematic structural representation of chemical formulae, a language in its own right. At this time, there was no agreement on the representation of chemical formulae as Figure 17.4 shows [14], and with such diversity, progress in chemistry or the teaching of chemistry was almost impossible. A famous chemist who attended the Congress commented afterwards, "*Symbolic formulae ... would deserve to rank among the chemist' most powerful instruments of research* [15, p.87]." These congresses, convened to gain international agreement in matters of nomenclature and units, developed into the International Union of Pure and Applied Chemistry (IUPAC), which was founded in 1919.

4) *Computer-readable formulae*: the fourth stage in the development of chemical language has been the use of computer-readable conventions for uniquely identifying chemical substances, allowing for searchable databases. This was pioneered in the United States by the American Chemical Society from the early 1960s.

$C_4H_4O_4$	. . . .	Empirische Formel.
$C_4H_3O_3 + HO$	. . . . . .	Dualistische Formel.
$C_4H_3O_4 \cdot H$	. . . .	Wasserstoffsäure-Theori.
$C_4H_4 + O_4$	. . . .	Kerntheorie.
$C_4H_3O_2 + HO_2$	. . . . .	Longehamp's Ansich.
$C_4H + H_3O_4$	. . . . . .	Graham's Ansicht.
$C_4H_3O_2 \cdot O + HO$	. . . . .	Radicaltheorie.
$C_4H_3 \cdot O_3 + HO$	. . . . .	Radicaltheorie.
${C_4H_3O_2 \atop H}\}O_2$	. . . .	Gerhardt. Typentheo
${C_4H_3 \atop H}\}O_4$	. . . .	Typentheorie (Schisch)
$C_2O_3 + C_2H_3 + H\,O$	. . . .	Berzelius' Paarlingsth
$H\,O \cdot (C_2H_3)C_2, O_3$	. . . . .	Kolbe's Ansicht.
$H\,O \cdot (C_2H_3)C_2, O \cdot O_2$	. . . .	ditto
${C_2(C_2H_3)O_2 \atop H}\}O_2$	. . . . . .	Wurtz.
${C_2H_3(C_2O_2) \atop H}\}O_2$	. . . . . .	Mendius.
${C_2H_2 \cdot HO \atop HO}\}C_2O_2$	. . . . .	Geuther.
$C_2\!\left\{{C_2H_3 \atop O \atop O}\right\}O + HO$	. . . .	Rochleder.
$\left(C_2\dfrac{H_3}{CO} + CO_2\right) + HO$	.	Persoz.
$C_2\!\left\{{C_2\{{O_2 \atop H}\} \atop H \atop {H \over H}}\right\}O_2$	. . . . .	Buff.

Figure 17.4 Representations of acetic acid in 1861 [14].

5) *The dominance of English*: the fifth stage in this process has been the growth of English as the universal scientific language. The symbolic language of chemistry is universal, just as is the symbolic language of mathematics, but language has its own names for elements and compounds. Scientific journals are still published in national languages but there is a growing trend to publish in English allowing for the universal communication of scientific ideas.

The history of chemical nomenclature has been covered by Crosland [16] in *Historical Studies in the Language of Chemistry*. Aspects of chemical nomenclature were covered in Thurlow [17], including the naming of the elements [18].

The language of chemistry has gone from individualistic alchemical names to systematic names, which in turn have developed to become a universal naming system for any chemical substance. However, old names die hard, and trivial names are still in widespread use in industry and in society. For example, vinegar became acetic acid and then ethanoic acid, but it is usually only in the school, university, or research laboratory that CH_3COOH is called *ethanoic acid*; industry still uses acetic acid and everyone pours vinegar on their food. Older names often remain as linguistic fossils long after systematic names have been agreed, and this adds to the confusion of students. Trivial names are easier to remember but harder to translate into formulae without rote learning; the advantage of systematic nomenclature is that the rules derive the name from the formula, and vice versa, for any compound. One pedagogical issue is that teachers were often taught and learned an older nomenclature and now have to teach a newer, unfamiliar one.

17.3
The Role of Language in Science Education

> Science without literacy is like a ship without a sail. So just as it is impossible to construct a house without a roof, it is impossible to build understanding of science without exploring how the multiple languages of science are used to construct meaning [19].

To understand the importance of language in chemical/science education, one must first understand the meaning and the importance of language in human culture. The *Report of the Board of Studies for Languages* in Ireland [20] defines the curriculum category "language" as follows:

Language is

- the chief means by which we think – all language activities, in whatever language, are exercises in thinking;
- the vehicle through which knowledge is acquired and organized;
- the chief means of interpersonal communication;
- a central factor in the growth of the learner's personality;
- one of the chief means by which societies and cultures define and organize themselves and by which culture is transmitted within and across societies and cultures.

Language is undoubtedly central to the learning process. Almost all teaching and learning takes place using the medium of language, written and spoken. However, its importance and impact on student learning of science is often somewhat neglected. According to Ford and Peat [21], the traditional view of language in science is that has a passive role in the learning process, and is simply the transmission medium for meaning and information. However, this view is disputed, and others have argued that language plays a more active role in the development of scientific ideas and thought [21, 22, p. 679]. Language is an integral part of science and science literacy. According to Yore *et al.* [23], *language is a means of doing science and of constructing science understandings; language is also an end, a fundamental goal of science literacy, in that it is used to communicate about inquiries, procedures, and science understandings to other people so that they can make informed decisions and take informed actions.* Thus, science is a process of inquiry conducted through the use of language, which facilities debates, discourse, and the creation of understanding. As Henderson and Wellington put it, ... *the quality of language is bound up with the quality of learning* [24, p. 36] and Wellington and Osborne [3, p. 6] further stated that *language development and conceptual development are inextricably linked. Thought requires language, language requires thought.* Language is thus one of the vital and irreplaceable pillars on which science rests, along with mathematics.

Postman and Weingartner [25] expressed a similar view regarding the role of language in science:

> Almost all of what we customarily call "knowledge" is language, which means that the key to understanding a subject is to understand its language. A discipline is a way of knowing, and whatever is known is inseparable from the symbols (mostly words) in which the knowing is coded. What is biology (for example) other than words? If all the words that biologists use were subtracted from the language, there would be no biology. This means, of course, that every teacher is a language teacher: teachers, quite literally, have little else to teach, but a way of talking and therefore seeing the world.

According to Osborne [19], this conception of the role of language in science education is not held by many science teachers. In contrast, the general perception amongst science teachers is that discourse in science is effectively transparent and that language unambiguously represents the physical world [26]. In this view, the main obstacle to the acquisition of key scientific concepts is the language used to convey meaning [27].

Language in the classroom permeates every activity and encompasses both written and spoken language (see Figure 17.5). Classroom language is often dominated by the teacher, and the student may be fairly passive. Our aim should be to move

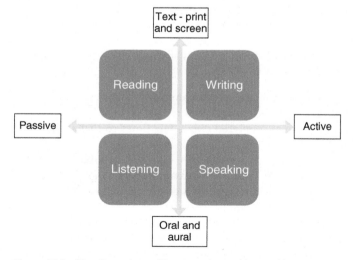

Figure 17.5 The dimensions of language in teaching and learning.

the student's language use toward the active area while still using the whole range of language skills.

17.4
Problems with Language in the Teaching and Learning of Chemistry

> Science does not speak of the world in the language of words alone, and in many cases it simply cannot do so. The natural language of science is a synergistic integration of words, diagrams, pictures, graphs, maps, equations, tables, charts, and other forms of visual mathematical expression [28].

Science has its own language, and learning this language acts as an impediment to many students' acquisition of scientific knowledge and understanding. Wellington and Osborne [3, p. 10] highlighted the fact that *one of the important features of science is the richness of the words and terms it uses* and Lemke [28] has pointed out the diversity of breadth of language in science. According to Rutherford [29], the language of science transcends other language differences.

Chemistry has been described as a language in its own right, with its own alphabet (the symbols for the chemical elements), a massive vocabulary (the formulae of chemical substances), and sentences and syntax (chemical equations and the rules of chemical combination). Indeed, a recent article by Laszlo [30] has suggested that we should teach chemistry as a language.

> We want also students to gain familiarity with the nanoworld and we ask them to interpret data in terms of entities, such as molecules, existing in that microcosm. This entails mastery for them of a new language. Chemistry teachers are linguistic guides, they are interpreters, they teach their students how to craft well-formed chemical sentences [30, p. 1682].

The language of science, and chemistry in particular, is multifaceted and is more complex than ordinary, everyday language:

- it has a large, specialized, precise, and unfamiliar vocabulary, for example, *hydrophilic, autotroph, cytokinesis, amphoteric*;
- many technical words are encountered only a few times so that they do not become familiar;
- the roots of many scientific words and terms are based on Greek and Latin, which are unfamiliar to today's students, for example, *poly-, -mer, hyper-*;
- it uses terms with different meanings in a science context to their everyday use, for example, *solution, force, current*;
- it has challenging written and oral demands, with polysyllabic words and complex sentences, for example, *oligosaccharide, hypothesis, photosynthesis*;
- it uses symbolic language, which contributes to students' difficulties with science, for example, CH_4, NH_4^+;

17.4 Problems with Language in the Teaching and Learning of Chemistry | 431

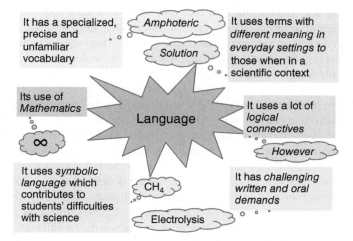

Figure 17.6 The multifaceted nature of the language of science.

- it uses diagrams to represent structures, which are a vital part of chemical discourse;
- it uses many logical connectives and command words, for example, *therefore, because, explain, describe*;
- mathematics and its symbolic language are a key part of science, for example, $PV = nRT$;
- there is a lack of practice in using the language of science in writing/reading or speaking/listening except in school;
- there is a greater requirement for accuracy and rigor in scientific discourse, particularly spelling, for example, *alkane versus alkene*.

Although some of these aspects are found in everyday language and in other school subjects, science, and chemistry in particular, is unique in the richness, density, and frequency of such features in everyday classroom language. Some of these features are summarized in Figure 17.6. For example, it has been suggested that students encounter more new words in learning science in lower secondary school than they do in learning a foreign language!

According to Osborne [19], all aspects of the multifaceted language of science are used in the construction of meaning and understanding. In light of this complexity, it is not surprising that many teachers have argued *that it is the language which is the main stumbling block for learning science rather than the science content itself* [31]. The language of science is a problem that students face at both the second and third level. Wellington and Osborne [3, p. 2] claimed that *Language is a major barrier (if not the major barrier) to most students in learning science.* The main aim of science education is to assist students in utilizing the multiple languages of science to construct and interpret meaning. From a linguistic and semiotic perspective, in science education the concern is not solely in ensuring that students understand the concept but also that they can move back and forth between the different representations – verbal, symbolic, diagrammatic – and that they can begin to recognize, use, and interpret the equivalencies between these forms [19]. Lemke [28] has stated that complete understanding of a concept occurs only when each of the facets of language, which represent that concept, are utilized in an integrated and overlapping manner. In light of such linguistic complexity, it is hardly a surprise that many students find it difficult to fully understand scientific concepts.

In order to assist students in their acquisition of scientific knowledge and understanding, it is imperative that teachers are aware of the complexities that students confront when learning science. Teachers must realize that within every subject or knowledge domain there is a specific "register," which is defined as "a set of meanings that is appropriate to a particular function of language, together with the words and structures which express these meanings" [32, p. 65]. The science register includes both technical and nontechnical terms and words. To differentiate between the two, the technical vocabulary is specific to the science context and serves a more specialized purpose, whereas the nontechnical vocabulary is seen as "everyday" language used in normal communication. A problem arises when students are familiar with the "everyday" meaning of these nontechnical words, which in some cases assume a more specialized meaning when used in the science context. It is this "double meaning" of the words that causes difficulty, because students are not aware of both meanings. This affects students' learning and consequently leads to misconceptions.

The technical language of science poses a problem of familiarity, but students seem to be able to cope reasonably well with this. A more acute problem lies in the use in science of normal, familiar language in a highly specific, different, and unfamiliar way [33–38]. A simple word like "mass" (a large amount or even a religious service) becomes something different in physics. "Concept or conception" in some contexts is an idea, but in others it is connected to birth. Even a word like "audible," which one would expect to remain stable in meaning, can be linked by students with a car brand (Audi) or with the sound-alike word "edible." When teachers are asked to identify difficult words in a science text, they only notice the technical words and not other problematic words [29].

17.4.1
Technical Words and Terms

This component of scientific language deals with the technical words and terms, specific to a science subject. According to Gardner [34, p. 7], technical words include such things as physical concepts (mass, force), names of chemical elements, minerals, plants, organs, processes, apparatus, process words (filtrate, distil, agitate), and so on. Science students must become familiar with a wide range of specialist vocabulary. These words are more than likely new to the student and often are complex and difficult to spell. Table 17.1 lists some technical words that are found in science textbooks. Many of these are introduced for the first time in

Table 17.1 Examples of technical words in science.

Photosynthesis	Refraction	Tropism
Cytoplasm	Catalyst	Joule
Electron	Enzyme	Respiration
Nucleus	Isomer	Chloroplast

an introductory secondary science course and are more complex than everyday language.

All teachers recognize the new words that are introduced in even lower level science courses. Chemistry is particularly rich in new words, and the number increases with the level. Identifying new words to be introduced in a chapter or lesson at the start is helpful, through word lists or using a word bank or glossary. However, this can be a passive activity for the student, and frequent practice is needed in using new words in written and oral contexts. Spelling matters in chemistry, where one letter can change the meaning of a word, for example, alkene, alkane, alkyne. A variety of activities should be used to allow students to use new words and become familiar with their meaning in context. Although there is value in a preprepared word bank or glossary, it is also helpful for students to create their own list word by word, on a need-to-know basis, as they progress.

17.4.2
Nontechnical Words

This component of the science register encompasses the words that are used during classroom discourse but are not classified as technical words, or that have double meanings. For example, let us consider the following extract from Gardner [33, p. 7] to illustrate what is implied by nontechnical words:

> Gas molecules display **random** motion; we may **predict** their **behavior** from **theoretical** considerations: the **actual** volume of the molecules may be **neglected**.

In the above extract, the six words: *random, predict, behavior, theoretical, actual,* and *neglected*, are referred to as *nontechnical terms*, although they play an intrinsic role in understanding the scientific concept. The words *diversity, reaction, disintegrate, alloy, solution* are all words that are used in an everyday context but they become part of technical language when used in a science context, often with a different meaning [39].

It is acknowledged that with every science class, students will encounter new words, some technical and others nontechnical. However, research shows that students often see or hear a word and assume its meaning, without assessing the context in which it is used. This creates a barrier to achieving a shared meaning between the teacher and student of the words in a scientific context [33–35, 37, 40–43]. We could say they are divided by a common language.

Table 17.2 Examples of technical and nontechnical words in science with dual meaning.

Technical words with dual meaning	Nontechnical words with dual meaning
Alloy	Volume
Work	Effect
Efficient	Complex
Solution	Transfer
Conduct	Initial
Reflection	Substitute
Law	Dependent
Contract	Tendency
Field	Agent
Circuit	Rate
Charge	
Cycle	

Table 17.2 illustrates both the technical and nontechnical words used in science that have dual meanings and may be causes of confusion in the science classroom.

These present a greater problem as teachers are often not aware that they present a problem, and they assume a common understanding between the teacher and students.

> The problem lay, not so much in the technical language of science, but in the vocabulary and usage of normal English in a science context. Students and teachers saw familiar words and phrases which both "understood," but the assumption that both understandings were identical was just not tenable [37].

The first step is for science teachers to become aware of the most problematic words used in their national language (as this problem is language-specific). When they first occur in a lesson or textbook, the teacher must deliberately address the issue and try to ensure that the teacher and students share a common understanding. This will need to be continually reinforced, so that students become familiar with the idea that the meaning depends on the context.

17.4.3
Logical Connectives

> Logical connectives are words or phrases which serve as links between sentences, or between propositions within a sentence, or between a proposition and a concept [44].

Unless you understand these connective words used in science texts, you cannot make sense of the sentences. Experts take their meaning and use for granted, but they may present great difficulty to novices. Gardner *et al.* [44] identified 75

Table 17.3 Logical connectives used in science education.

In spite of	Conversely
Although	Nevertheless
Because	In addition
However	Moreover

logical connectives that present problems to students starting to study science, and Table 17.3 shows some of these logical connectives.

It can be argued that these are an intrinsic part of science, as they relate to causality and sequence and relationships. Eliminating them from science education by using simpler language would thus remove an important part of science, used, for example, in argumentation. It is better for the teacher to be aware that they present problems, and then to identify them clearly for students, and explain their meaning and use them consistently and often.

17.4.4
Command Words

One of the problems with assessment, whether formative or summative, is the language used to formulate questions. Often, the language used can be a barrier to understanding what the question requires. Despite many improvements in teaching and learning strategies, many students still do not do well in examinations. Literacy is a basic requirement for any form of written assessment, and poor performance may be due to literacy problems rather than the content being examined. All written assessments also test the students' literacy as well as their knowledge and understanding of science. One specific aspect of language in assessment is the use of command words to indicate what students have to do in answering a question, for example, describe, explain, calculate, compare, and so on, (see Table 17.4 for more examples) [67]. To the teacher and the expert, these present no problem, as we understand what the words mean and what is required. Students often do not understand fully these command words and may answer the question in the wrong way, for example, a question says "Write brief notes on something," requiring a short summary of the topic, but the student writes all they know on the topic, or when asked to "compare" they in fact "describe."

Examiners have a portfolio of such command words which they use in examinations. However, when teachers examine their own students in tests and end-of-term examinations, they may use command words different from those used in state examinations and they may not explain what the words mean. Because of this lack of consistency between different forms of assessment, students often go into their examinations underprepared.

Students should be provided with a list of the command words that will be used in school tests and examinations, and also in the state examinations, along with a definition of what each command word requires. However, just giving out a list is not enough. The teacher needs to explain and give examples of what they

Table 17.4 Some examples of command words used in assessment and their link to cognitive demand.

Knowledge and understanding	Application	Analysis	Evaluation
Define	Apply	Analyze	Advise
Describe	Calculate	Explain	Evaluate
List	Examine	Discuss	Assess
Identify	Give an example	Organize	Consider

mean, should use the command words in teaching, give exercises that practice using them, and test students' understanding of the terms. The command words need to be used consistently in all school assessments by all teachers. The command words need to be continuously used and reinforced so that students become totally familiar with them and confident in their use. Making sure your students understand and can apply these command words is as important in their examination preparation as learning the content. Helping students to master and decode the command language used in examinations enables them to demonstrate their ability, knowing what the questions require of them. This allows them to present the information they do possess [45].

17.4.5
Argumentation and Discourse

In recent years, many researchers have looked at the role of argumentation and discourse in science teaching [46–50]. Argumentation is what scientists do to explore, refine, and test ideas and is an important aspect of language use in science, but one which is largely missing from school science lessons. In addition, argumentation gives students practice in using scientific language and ideas, and thus develops communication skills. Driver *et al.* [46] concluded that *in the light of our emerging understanding of science as social practice, with rhetoric and argument as a central feature, to continue with current approaches to the teaching of science would be to misrepresent science and its nature. If this pattern is to change, then it seems crucial that any intervention should pay attention not only to ways of enhancing the argument skills of young people, but also to improving teachers' knowledge, awareness, and competence in managing student participation in discussion and argument.* The more recent study by Osborne *et al.* [50] has shown how difficult it is to change teachers' practices in this area in a short intervention.

17.4.6
Readability of Texts

There are various measures of the readability of texts, often expressed as a grade level or reading age, for example, the Gunning "FOG" Readability Test, the Flesch–Kincaid Formula, or the McLaughlin "SMOG" (Simple Measure of

Gobbledygook) Formula. (The Flesch–Kincaid Formula can be applied using Microsoft Word.) However, the various methods depend on counting the sentence length and the number of syllables in words. This means that science texts invariably give high reading ages because of the polysyllabic nature of science vocabulary and, often, the use of long and complex sentences [51]. Readability is more complex than just this factor and depends on:

1) Interest and motivation
2) Clarity of diagrams and pictures
3) Readability of the text.

The readability of text (whether in textbooks, examinations, worksheets, etc.) is likely to be a greater problem for students of low language proficiency and SLLs, and should be kept in mind when choosing textbooks and producing worksheets and tests.

17.5
Language Issues in Dealing with Diversity

Diversity in the classroom has major implications for the use of language in teaching and learning chemistry. We can identify three aspects of diversity:

1) Mixed ability: there is a range of abilities within the comprehensive classroom, which also includes a range of language proficiency;
2) Second language learners (SLLs): most countries now have increasing numbers of SLLs in their schools and this presents major problems in using language;
3) Students with special needs and disabled students: the move to inclusive education in many countries means that classrooms contain a significant proportion of students with special needs (physical, mental, behavioral, and emotional), and many of these will have issues relating to language.

The main focus of this section will be the issue of SLLs, as this is a relatively new and increasing problem in many countries.

17.5.1
Second Language Learners

Through the movement of people from one country to another, the population in schools has started to become more and more heterogeneous. This factor has become more obvious since the initial PISA (programme for international student Assessment) and TIMMS (Trends in international mathematics science study) studies [52]. The multilingualism of the students (which is mainly bilingualism or semi-bilingualism) is mostly a disadvantage in the school context [53]. In many cases, non-native students reach lower levels of linguistic competence compared to the national norm and thus lower levels of education than native-speaking students [54]. There are different factors that influence the second language development of students. Collier [55] listed seven factors: (i) students' age of arrival, (ii) length of residence in the country, (iii) grade of entry into the new school, (iv) first language reading and writing literacy skills, (v) formal educational background, (vi) family's educational and socioeconomic background, and (vii) students' former exposure to the new country's lifestyle. Additionally, Reich and Roth [56] found that students who possess high linguistic skills in their first language, when entering into the new school system, also achieve good proficiency when learning their new "official" language, and the development of the mother tongue becomes secondary.

Most of the research about language focusing on student literacy among non-native speakers in science classes has been done in elementary schools and with English language learners (ELLs) [57]. Lee *et al.* [58] show that elementary teachers rarely discuss student diversity in their own teaching with other colleagues in their schools. They also show that teachers do pay attention to linguistic issues among their students, but they tend to do so quite randomly. The study by Verplaetse [59] shows that teachers tend to talk differently to ELLs, compared to native speakers. They usually speak slowly to ELLs, use simpler words, and tell students exactly what to do, rather than asking questions to elicit information. When they do employ questions, they use simple yes/no questions, instead of asking questions that demand high-level thinking and linguistic skills. Thus, students in such a situation do not have an opportunity to improve their language and to engage themselves fully in chemistry lessons [60]. However, all students with less developed linguistic skills in a country's official school language tend to have difficulties in learning subject-matter content. This is because they often have an insufficient grasp of the rules governing the spoken and written language, which they are expected to master and use [53, 61]. In doing so, they lose the motivation to study chemistry and – even worse – the difference between their linguistic skills in the new language and those of native students get wider.

One explanation for science teachers behaving like this is that they do not feel responsible for teaching language in their chemistry classes and they simply assume (since the students are regular in classes) that they are all capable and efficient users of the official language of the country [62]. Chemistry teachers do not see *teaching for diversity* as their responsibility or as an aim of chemistry lessons. Bryan and Atwater [62] concluded this for secondary school science teachers and showed that the linguistic heterogeneity in the classroom often slips in under the radar, since many teachers just accept it as a given component in their classroom. Chemistry teachers are mainly not aware of the influence of deficient linguistic skills in learning chemistry. A not-infrequent opinion, which researchers constantly hear from teachers, is that the severe time shortages for covering the chemistry curriculum are a good reason not to deal with linguistic heterogeneity and its attendant difficulties [63]; their quantitative research study of 33 teachers revealed that the language barriers and ELLs' lack of foundational science knowledge represented the largest challenges to science teachers.

Not only is the teacher's sensitivity toward this issue important but so also are the strategies for dealing with it. The teachers in Cho and McDonnough's [63] study stated that the most common accommodation made by the teachers was giving ELLs additional time to complete assigned tasks. The second-most popular strategy was slowing down the teachers' rate of speaking to aid in understanding, followed closely by the strategy of grouping ELLs together so that they could help one another. It is interesting to note that alternatives such as providing different tasks and assignments, substituting differentiated instructional materials, or using other grading/assessing methods were the least adopted methods on the list, occurring only rarely or never. Discussing the results of their study, Cho and McDonnough [63] assume that the lack of supplementary teaching materials in ELL classrooms and the lack of school resources actually available to teachers could be the main reasons for this behavior. Another reason for not using corrective strategies may be that teachers do not know how to properly adjust their instruction, instructional materials, assignments, and tasks for the needs of ELLs. The participants also reported that they rarely graded ELLs any differently from other students. Nor did they consult directly with ELL teachers to address linguistic heterogeneity. This may seem surprising since there is often an ELL teacher as an in-house expert in many schools, but they tend to be quite generally ignored by their colleagues. The very limited use of accommodation strategies and tools found in this study must lead us to the conclusion that there is a pressing need for targeted professional development on this issue.

Thus, there is a great need to make chemistry teachers more sensitive to the linguistic issues, and to heterogeneity in general, in their classes. Starting from the above-mentioned results, we need to pay more attention to changing teachers' attitudes and beliefs concerning this topic. One possible method to make teachers more aware of this issue is to let them write a laboratory report in a second language, which they themselves studied in school. Usually, the knowledge in the second language of (student) teachers is not good enough to enable them to do so, even if it is a report on a simple experiment such as, for example, simple filtration. Furthermore, they could answer a "test" using a text that is drawn from another subject domain such as law or psychology, where the language is unfamiliar. By doing so, it is possible to put the (student) teachers in the same situation as some of their students. Additionally, examples from everyday school lessons may make (student) teachers aware of the problems that are caused by the lower language skills of the students in their classes. Finally, different methods for dealing with and teaching in linguistically heterogeneous classes must be given. Here, we will give only a short overview of different methods. A list and more detailed description of the methods are given in Markic *et al.* [8]. Also Markic [64, 65] concluded that the cooperation between chemistry/science teachers and German second language teachers (in a German school context) offers a good opportunity to develop new teaching materials for dealing with the linguistic heterogeneity of chemistry/science lessons. Furthermore, this project shows that the collaborative development of chemistry lesson plans by chemistry/science teachers and German second language teachers seems to be a promising way to create motivating and attractive learning environments, which allow teachers to help students not only to learn chemistry but also to improve their knowledge and competencies in the German language. These methods should be transferrable to other languages.

17.5.2
Some Strategies for Improving Language Skills of SLLs

In the literature, there are some special pedagogies and tools available to support students with linguistic problems in their learning of chemistry (or science in general). General strategies that can be used are the following:

1) The use of pictorial explanation (e.g., laboratory tasks by a sequence of pictures instead of text);
2) The use of different methods as an aid for formulating sentences;
3) Support for easier writing and understanding of texts;
4) The use of different modern and innovative teaching methods such as cooperative learning.

Additionally, the following tools can be used in chemistry classes to make the teaching more language-sensitive (see also Henderson and Wellington [24]):

- *Use activities, visual tools, and vocabulary/semantic tools for learning words and terms*: Games, quizzes, or written tasks can help students to learn and memorize new words and terms, pictorial explanations, and so on.
- *Use new terms at the right dose and train in the use of them*: Do you really need every scientific word that is written in a book? It is important not to introduce too many technical terms to the students at once. After implementing the new terms for students, it is important to use the word in different contexts as well.
- *Help in safeguarding new terms and structures*: It is helpful for students if the teacher lists all new words covered in a lesson on the board (or on a poster), making sure students can spell them correctly, know their meaning, and, where possible, know something about the structure of the words.
- *Noticing and dealing with linguistic mistakes*: In chemistry classes, it is important that the teacher notices mistakes made by the students and addresses them. One method is for the students to say or read the sentence again, and maybe then they will notice the mistake themselves. Also, the teacher can repeat the sentence and follow it with the corrected one. The students can analyze their mistake and reflect on it.
- *Presentation of texts:* Less is often more. Keep sentences short, and reduce the amount of text to the necessary minimum. Give the text a clear structure: organize the text clearly through the use of paragraphs and subheadings. Leave space for students' comments.

Many of these strategies will prove useful with native speakers to reduce the language barrier and are not confined to use with SLLs.

17.5.3
Special-Needs Students

The main focus in this section has been on SLLs, as these are becoming an increasingly prominent feature of many classrooms. All the problems of language in science will be increased for SLLs. However, the move toward the inclusion of students with special needs and disabled students in ordinary classrooms [66] also has implications for language in teaching and learning chemistry. Each specific disability will have its own problems and needs, and we are not able to address these here. Hearing-impaired or sight-impaired students may have good language skills, but each will need special provision in the classroom. Writing may be difficult for students who are physically impaired or dyspraxic, but they may be assisted by special technologies, for example, adapted computers. Increasing numbers of students are being diagnosed as dyslexic and need specific help in reading and writing. This problem is now better recognized in education, and resources and strategies for dealing with it have been developed. The presence of students with special needs in science classrooms presents the teacher with a specific set of problems, of which language is part, and makes their job more challenging.

17.6
Summary and Conclusions

> The impossibility of separating the nomenclature of a science from the science itself, is owing to this, that every branch of physical science must consist of three things; the series of facts which are the objects of the science, the ideas which represent these facts, and the words by which these ideas are expressed. Like three impressions of the same seal, the word ought to produce the idea, and the idea to be a picture of the fact [13, Preface, xiv].

It is clear, we hope, from this brief review on the role of language in the teaching and learning of chemistry that attention to the language aspects of chemistry is vital to chemical education at all levels. This is particularly true when chemistry is first introduced, usually in lower secondary school. Chemistry teachers traditionally focus on the technical and symbolic language of chemistry, which in itself is like learning a strange, new, and difficult language. Teachers often neglect or are unaware of other issues, such as nontechnical words, logical connectives, and command words in assessment, which also act as barriers to the understanding and mastery of chemical ideas. The barrier to learning presented by language is worse because it is largely unrecognized. Part of the solution is for the teacher to become aware of the various dimensions and chemical language and the problems they present to novices. Teachers need to adopt a range of teaching and learning strategies to develop language skills alongside, and integrated with, the content of chemistry. Facility and accuracy in the use of scientific language is an essential element of scientific literacy, which is more than knowing about science or the facts of science, but also requires students to be able to argue, evaluate, and apply scientific ideas in both scientific and everyday contexts. This is impossible without mastering the language of science in all its richness, diversity, and complexity.

References

1. Mammino, L. (2010) The essential role of language mastering in science and technology education. *Int. J. Educ. Inf. Technol.*, **4** (3), 139–148.
2. Childs, P.E. (2006) The problems with science education: "The more things change, the more they are the same", in *Proceedings, SMEC 2006*, DCU, Dublin, http://main.spd.dcu.ie/main/academic/mathematics/documents/smec06keynote1.pdf (accessed 8 December 2013).
3. Wellington, J. and Osborne, J. (2001) *Language and Literacy and Science Education*, Open University Press, Buckingham.
4. Fraser, B.J., Tobin, K., and McRobbie, C.J. (eds) (2012) *Second International Handbook of Science Education*, Springer, London, New York.
5. Gunstone, R. (ed.) (2014) *Encyclopedia of Science Education*, London, New York, Springer.
6. Wellington, J. and Ireson, G. (2012) Language in science teaching and learning, in *Science Learning, Science Teaching*, Chapter 11, 3rd edn, Routledge, Abingdon.
7. Osborne, J. and Dillon, J. (2010) in *Good Practice in Science Teaching: What Research Has to Say*, Chapter 7, 2nd edn (eds M. Evagorou and J. Osborne), Oxford University Press, pp. 135–157.
8. Markic, S., Broggy, J., and Childs, P. (2013) in *Teaching Chemistry – A Studybook* (eds I. Eilks and A. Hofstein), Sense, Rotterdam, pp. 127–152.
9. Hogben, L. (1969) *The Vocabulary of Science*, Heinemann, London.
10. Pattison Muir, M.M. (1907) *A History of Chemical Theories and Laws*, John Wiley & Sons, Inc., NewYork.
11. De Laurence, L.W. (ed.) (1913) *The Philosophy of Natural Magic*, by, Henry Cornelius Agrippa, http://www.sacred-texts.com/eso/pnm/img/28400.jpg (accessed 28 July 2014).
12. Ihde, A.J. (1964) *The Development of Modern Chemistry*, Harper & Row, New York, Evanston, London.
13. Lavoisier, A.-L. (1790) *Elements of Chemistry* (trans. R. Kerr).
14. Kekulé, A. (1861) *Lehrbuch der Organischen Chemie*, Enke, Erlangen, p. 58 Wikipdedia Commons, http://en.wikipedia.org/wiki/File:Kekule_acetic_acid_formulae.jpg) (accessed 28 July 2014).
15. Hofmann, A.W. (1865) *Introduction to Modern Chemistry Experimental and Theoretic*, Walton and Maberley, London.
16. Crosland, M.P. (1962) *Historical Studies in the Language of Chemistry*, Heinemann, London.
17. Thurlow, K.J. (ed.) (1988) *Chemical Nomenclature*, Springer, London, New York.
18. Childs, P.E. (1988) in *Chemical Nomenclature* (ed. K.J. Thurlow), Springer, London, New York, pp. 27–66.
19. Osborne, J. (2002) Science without literacy: a ship without a sail? *Cambridge J. Educ.*, **32** (2), 203–218.
20. CEB (1987) *Report of the Board of Studies for Languages*, Curriculum and Examinations Board, Dublin.
21. Ford, A. and Peat, D. (1988) The role of language in science. *Found. Phys.*, **18**, 1233–1242.
22. Hogan, K. and Maglienti, M. (2001) Comparing the epistemological underpinnings of students' and scientists' reasoning about conclusions. *J. Res. Sci. Teach.*, **38**, 663–687.
23. Yore, L., Hand, B., Goldman, S., Hildebrand, G., Osborne, J., Treagust, D., and Wallace, C. (2004) New directions in language and science education research. *Reading Res. Q.*, **39** (3), 347–352.
24. Henderson, J. and Wellington, J. (1998) Lowering the language barrier in learning and teaching science. *Sch. Sci. Rev.*, **79** (288), 35–46.

25. Postman, N. and Weingartner, C. (1971) *Teaching as a Subversive Activity*, Penguin, Harmondsworth.
26. Lemke, J.L. (1990) *Talking science: Language, Learning and Values*, Ablex, Norwood, NJ.
27. Shayer, M. and Adey, P. (1981) *A Science of Science Teaching*, Heinemann, London.
28. Lemke, J. L. (1998) Teaching All the Languages of Science: Words, Symbols, Images and Actions, *http://academic.brooklyn.cuny.edu/education/jlemke/papers/barcelon.htm* (accessed 8 December 2013).
29. Rutherford, M. (1993) Making scientific language accessible to science learners. Paper presented at the First Annual Meeting of the Southern African Association for Research in Mathematics and Science Education, Grahamstown, South Africa.
30. Laszlo, P. (2013) Towards teaching chemistry as a language. *Sci. Educ.*, **22**, 1669–1706.
31. Chui Seng Yong, B. (2010) Can pupils read secondary science textbooks comfortably? *Brunei Inst. J. Sci. Math. Educ.*, **2** (1), 59–67.
32. Halliday, M.A.K. (1975) *Learning How to Mean: Explorations in the Development of Language*, Edward Arnold, London.
33. Gardner, P.L. (1972) Difficulties with non-technical vocabulary amongst junior secondary school students: the words in science project. *Res. Sci. Educ.*, **2** (1), 58–81.
34. Gardner, P.L. (1972) *Words in Science*, Australian Science Education Project, Melbourne.
35. Cassels, J.R.T. and Johnstone, A.H. (1980) *Understanding of Non-Technical Words in Science*, Royal Society of Chemistry, London.
36. Cassels, J.R.T. and Johnstone, A.H. (1983) Meaning of the words and the teaching in chemistry. *Educ. Chem.*, **20**, 10–11.
37. Cassels, J.R.T. and Johnstone, A.H. (1985) *Words that Matter in Science*, Royal Society of Chemistry, London.
38. Ali, M. and Ismail, Z. (2006) Comprehension level of non-technical terms in science: are we ready for science in english. *J. Pend. dan Pend., Jil. since 2010.*, **21**, 73–83 *http://apjee.usm.my/APJEE_21_2006/5%20Maznah%20(73-83).pdf* Accessed 22/10/14.
39. Barnes, D., Britton, J., and Torbe, M. (1986) *Language, the Learner and the School*, 3rd (New) edn, Penguin Books Ltd, Harmondsworth.
40. Marshall, S., Gilmour, M., and Lewis, D. (1991) Words that matter in science and technology: a study of Papua New Guinean students' comprehension of non-technical words used in science and technology. *Res. Sci. Technol. Educ.*, **9** (1), 5–16.
41. Farell, M.P. and Ventura, F. (1998) Words and understanding in physics. *Lang. Educ.*, **12** (4), 243–254.
42. Prophet, B. and Towse, P. (1999) Pupils' understanding of some non-technical words in science. *Sch. Sci. Rev.*, **81** (295), 79–86.
43. Oyoo, S.O. (2000) Understanding of some non-technical words in science and suggestions for the effective use of language in science classrooms. Unpublished M.Ed. (Science Education) dissertation. School of Education, University of Leeds, Leeds.
44. Gardner, P.L., Schafe, U., Myint Thein, U., and Watterson, R. (1976) Logical connectives in science: some preliminary findings. *Res. Sci. Educ.*, **6** (1), 97–108.
45. Staver, J. (2007) *Teaching Science*, Educational Practices Series, and the International Bureau of Education (UNESCO), Geneva *http://www.ibe.unesco.org/publications/EducationalPracticesSeriesPdf/Practice_17.pdf* Accessed 22/10/14.
46. Driver, R., Newton, P., and Osborne, J. (2000) Establishing the norms of scientific argumentation in classrooms. *Sci. Educ.*, **84**, 287–312.
47. Erduran, S. and Jiménez-Aleixandre, M.P. (eds) (2007) *Argumentation in Science Education: Perspectives from Classroom-Based Research*, Springer, London, New York.
48. Cavagnetto, A.R. (2010) Argument to foster scientific literacy: a review of

argument interventions in K–12 science contexts. *Rev. Educ. Res.*, **80** (3), 336–371.
49. Khine, M.S. (ed.) (2012) *Perspectives on Scientific Argumentation, Theory, Practice and Design*, Springer, London, New York, Dordrecht, Heidelberg.
50. Osborne, J., Simon, S., Christodoulou, A., Howell-Richardson, C., and Richardson, K. (2013) Learning to argue: a study of four schools and their attempt to develop the use of argumentation as a common instructional practice and its impact on students. *J. Res. Sci. Teach.*, **50** (3), 315–347.
51. Johnson, C. and Johnson, K. (2013) Readability and Reading Ages of School Science Textbooks, http://www.timetabler.com/reading.html (accessed 8 December 2013).
52. Lynch, S. (2001) "Science for all" is not equal to "one size fits all": linguistic and cultural diversity and science education reform. *J. Res. Sci. Teach.*, **38**, 622–627.
53. Johnstone, A. and Selepeng, D. (2001) A language problem revisited. *Chem. Educ. Pract. Eur.*, **2** (1), 19–29.
54. Lee, O. (2001) Culture and language in science education: what do we know and what do we need to know? *J. Res. Sci. Teach.*, **38**, 499–501.
55. Collier, V. (1987) Age and rate of acquisition of second language for academic purposes. *TESOL Q.*, **21**, 617–641.
56. Reich, H.H. and Roth, H.-J. (2002) *Spracherwerb zweisprachig aufwachsender Kinder und Jugendlicher*, Behörde für Bildung und Sport, Hamburg.
57. Lee, O. (2005) Science education with English language learners: synthesis and research agenda. *Rev. Educ. Res.*, **75**, 491–530.
58. Lee, O., Maaerten-Rivera, J., Buxton, C., Penfield, R., and Secada, W.G. (2009) Urban elementary teachers' perspectives on teaching science to English language learners. *J. Sci. Teach. Educ.*, **20**, 263–286.
59. Verplaetse, L.S. (1998) How content teachers interact with English language learners. *TESOL Q.*, **7**, 24–28.
60. Seedhouse, P. (2004) *The International Architecture of the Language Classroom: A Conversation Analysis Perspective*, Blackwell, Malden.
61. Howe, M. (1970) *Introduction to Human Memory: A Psychological Approach*, Harper and Rowe, New York.
62. Bryan, L.A. and Atwater, M.M. (2002) Teacher beliefs and cultural models: a challenge for science teacher preparation programs. *Sci. Educ.*, **86**, 821–839.
63. Cho, S. and McDonnough, J.T. (2009) Meeting the needs of high school science teachers in English language learner instruction. *J. Sci. Teach. Educ.*, **20**, 385–402.
64. Markic, S. (2011) Lesson plans for student language heterogeneity while learning about "matter and its properties". Paper presented at the 9th ESERA Conference, Lyon, France, http://www.esera.org/media/ebook/strand3/ebook-esera2011_MARKIC-03.pdf (Accessed 22/10/14 December 2013.).
65. Markic, S. (2012) Lesson plans for students language heterogeneity while learning science, in *Heterogeneity and Cultural Diversity in Science Education and Science Education Research* (eds S. Markic, D. di Fuccia, I. Eilks, and B. Ralle), Aachen, Shaker.
66. European Commission (2013) Support for Children with Special Education Needs (SEN), http://europa.eu/epic/studies-reports/docs/eaf_policy_brief_-_support_for_sen_children_final_version.pdf (accessed 2 December 2013).
67. Childs, P.E. and Ryan, M.C. (2013) Using command words in assessment. *Resour. Res. Guides*, **4** (5), NCE-MSTL, Limerick, www.nce-mstl.ie (accessed 8 December 2013).

Further Reading

General

Buxton, C.A. and Lee, O. (2014) in *Handbook of Research on Science Education*, Chapter 11, vol. 2 (eds N. Lederman and S.K. Abel), Routledge, Abingdon.

Herr, J. (2008) *The Sourcebook for Teaching Science, Grades 6-12: Strategies, Activities, and Instructional Resources*, Jossey-Bass (Section 1: Developing Scientific Literacy, pp. 1-57).

Mansour, N. and Wegerif, R. (2013) *Science Education for Diversity. Theory and Practice*, Springer, Dordrecht.

Miller, J., Kostogriz, A., and Gearon, M. (2009) *Culturally and Linguistic Diverse Classrooms – New Dilemmas for Teachers*, Multilingual Matters, Clevedon.

18
Using the Cognitive Conflict Strategy with Classroom Chemistry Demonstrations
Robert (Bob) Bucat

18.1
Introduction

The target audience of this chapter is taken to be the practitioners of the chemistry education enterprise – secondary school teachers and university lecturers. Consequently, the author's intent has not been to produce an academic treatise on the research scholarship associated with the cognitive conflict strategy and the notion of conceptual change. Rather, the author has taken a more practical line, calling upon those findings of research, and particularly his own experiences and insights, that might enhance teachers' awareness of the issues associated with the design of teaching elements based on challenging students' prior conceptions through inducement of cognitive conflict.

It is reasonable to claim that a significant factor that contributes to the qualities of excellent teachers – whether in chemistry or in any other discipline – is their ability to make sound decisions both before the class and "on the run" during the class. The author knows of no research that has explored this quantitatively, but the very good teacher certainly makes many decisions associated with the presentation of each class. These decisions are his/her answers to a myriad of self-analytical questions, such as the following: To what level in this topic do I want to take the students in this class? What are the few absolutely key ideas that I hope the students will come to understand? What will they need to know in advance of the class? What do they know? How do I know that? What will be my balance of informing and student-generated learning? How will I attempt to motivate the students to seek meaningful understanding? What are the particular challenges of teaching these concepts that may be unique to these concepts? What particular strategies will I use to address those specific challenges? What do I know about common student misconceptions in this area? How can I explicitly address the possibility that my students will go down paths to recognized undesirable learning? What questions should I ask? What tasks should I set? How will I respond to particular expected (from experience) student questions? How can I "use" the students to assist each other's learning? How can

I assess the progress of their learning during the class? How can I consolidate learning?

By and large, these decisions are judgmental: there can be no absolutely correct answers. And of course, the good teacher will be able to justify each of the decisions made by sensible reasoning and sound logic. The better informed the teacher is about what classroom research findings have to say (as distinct from theoretical discussion) and about the experiences of other teachers in similar situations, the more soundly based will be the decisions. For those teachers who are already trying to provoke the students to challenge their own previous conceptions, as well as for those who are contemplating doing so, the following discussions may be of practical value.

18.2
What Is the Cognitive Conflict Teaching Strategy?

In the context of classroom chemistry demonstrations, the cognitive conflict teaching strategy involves demonstration of a phenomenon during which the observations made by the students are at odds with what they expect. Perhaps the simplest description is that of Appleton [1]: "a puzzling situation which is counter-intuitive."

The terms *cognitive conflict, cognitive dissonance, conceptual conflict,* and *discrepant event* are used in the literature more or less interchangeably, although the latter obviously refers to a phenomenon observed, while the first three apply to a state of mind that might arise from experiencing a discrepant event. Kang *et al.* [2] make the distinction clearly: "A cognitive conflict strategy emphasises destabilising students' confidence in their existing conceptions through contradictory experiences such as discrepant events and then enabling students to replace their inaccurate preconceptions with scientifically accepted conceptions."

Wright [3], not necessarily referring to classroom situations, defines a discrepant event as "a phenomenon which occurs that seems to run contrary to our first line of reasoning," and claims that such events are "a good device to stimulate interest in learning science concepts and principles."

We have all experienced a furrowing of the brow signifying intellectual disharmony when confronted with a discrepancy between what we expected and what we observed. One common reaction to this situation is to wonder whether the phenomenon observed was a trick, or whether it somehow "went wrong." In a teaching situation, a discrepant event might be used to create a degree of cognitive conflict, which motivates the students to resolve the disharmony in such a way as to lead to the learning intended by the teacher.

Longfield [4] reviews the use of discrepant teaching events, although not specifically in the chemistry discipline.

18.3
Some Examples of Situations with Potential to Induce Cognitive Conflict

To provide some examples of the subject matter of this chapter, the following is a list some of teaching situations in chemistry which might be used to induce cognitive conflict as a means of arriving at deeper student understandings through resolution of the conflict. The examples might help readers to design, or to identify, others. Firstly, some classroom demonstrations are listed. The appropriateness of each of these depends on the prior knowledge of the students: what is an unexpected outcome for one person is not necessarily so for another, and teachers need to make judgments about what is appropriate, for whom, when, and how.

- (For students who are not aware that the boiling point of water depends on the external pressure). "Boil an egg" (that is, place an egg in boiling water) at reduced pressure for an extended period. After, say, 30 min, students expect that it will be a hard-boiled egg. On cracking the egg open, it is seen to be uncooked.
- (For students who have been introduced to the concept of acids in aqueous solution, acidity in terms of $H^+(aq)$ ion concentration, and pH, but not the distinction between strong and weak acids). Measure the similar pH values (2.0) of $0.01\,mol\,l^{-1}$ solutions of hydrochloric acid and nitric acid. Students predict the same value for an acetic acid solution of the same concentration, and are surprised, and perhaps puzzled, that the pH is about 4.
- (In the context of learning about the amphiprotic nature of the singly deprotonated salts of diprotic acids, and the relative extents of their acidic and basic reactions with water). Solutions of sodium hydrogensulfate are shown to be acidic, but solutions of sodium hydrogencarbonate are basic.
- (For students who have no previous knowledge about buffer solutions). The pH change brought about by addition of, say, 5 ml of a base solution to 100 ml of pure water is compared with the pH change on addition of the same amount of base solution to 100 ml of buffer solution with the same pH as the water.
- (To heighten students' awareness of the species needed in aqueous solution to form a buffer solution, given that they may have been told that a buffer solution should contain a weak acid species and its conjugate base species). Explore the students' reactions to the proposition that a solution into which we put equal amounts (moles) of carbonic acid and carbonate ions is a buffer solution. Experiment demonstrates that the pH is extremely sensitive to additions of small amounts of strong acid or base. This is not a buffer solution because carbonate ions are not the conjugate base of carbonic acid. In fact, these species will react and the solution is identical to a solution made by adding a hydrogencarbonate salt to water. The steep titration curve near the first equivalence point in titrations of carbonic acid with base is illustrative of the lack of buffering ability of this mixture.

Xie [5] has opened up a new world of observation through infrared imaging to show the temperature distribution within systems. Using this thermographic technique, Schönborn et al. [6] report a simple experiment that can induce cognitive conflict between visual and tactile inputs in relation to the common misconception that metals are colder than wood.

Cognitive conflict situations that can be used as teaching strategies outside of the laboratory include those listed below.

- A $0.1\,\text{mol}\,\text{l}^{-1}$ hydrochloric acid solution has pH 1. Dilute by a factor of 10. The resulting $0.01\,\text{mol}\,\text{l}^{-1}$ solution has pH 2. A $0.001\,\text{mol}\,\text{l}^{-1}$ solution has pH 3, and a $0.0001\,\text{mol}\,\text{l}^{-1}$ solution has pH 4. Suggest that if we continue 1:10 dilutions, we will arrive at a $1\times 10^{-9}\,\text{mol}\,\text{l}^{-1}$ solution with pH 9. Of course, dilution of a solution of acid cannot give rise to a basic solution. The pH will not change beyond that of water (or a very, very dilute solution of the acid) – pH 7.
- (For students who have not encountered the idea that the self-ionization constant of water K_a (and all equilibrium constants) change with temperature). Ask "What is the pH of pure water at 80 °C?" Usually some students are puzzled that the pH of water (or any neutral solution) is not defined by pH 7. This is the value appropriate only to 21 °C. Neutral solutions are defined as those in which $[\text{H}_3\text{O}^+] = [\text{OH}^-]$, and the equal concentrations of these two species varies as the temperature of a neutral solution is changed. Approximately, $[\text{H}_3\text{O}^+]^2 = K_a$, and K_a increases as temperature is raised.
- Ask students to construct a Venn (relational) diagram to illustrate the distinctions among the meanings of the terms *concentrated*, *dilute*, *strong*, and *weak*. Often students confuse the term weak (a property of electrolyte solutes) with dilute (a property of solutions related to the amount of solute per unit volume).
- (For reasonably advanced students ready to go beyond unthinking substitutions when calculating solubilities from solubility products). Ask students to estimate the solubility of iron(III) hydroxide in water at 25 °C, given that $K_{sp} = 3\times 10^{-39}$. By reference to a template commonly used to solve such questions, some students might make the assumption that in a saturated Fe(OH)_3 solution, $[\text{Fe}^{3+}] = x\,\text{mol}\,\text{l}^{-1}$ and so $[\text{OH}^-] = 3x\,\text{mol}\,\text{l}^{-1}$. Substitution into the standard expression for the solubility product leads to the "solution" that $[\text{Fe}^{3+}] = 1\times 10^{-10}\,\text{mol}\,\text{l}^{-1}$. An alert student might realize that in that case, $[\text{OH}^-] = 3\times 10^{-10}\,\text{mol}\,\text{l}^{-1}$, which is lower than in pure water, as a result of dissolving a hydroxide! The flaw in this solution is that Fe(OH)_3 is so slightly soluble that $[\text{OH}^-]$ remains negligibly different from $1\times 10^{-7}\,\text{mol}\,\text{l}^{-1}$. If we use this value in the solubility product equality, we derive that $[\text{Fe}^{3+}] = 3\times 10^{-18}\,\text{mol}\,\text{l}^{-1}$ at equilibrium.
- Ask students whether air is a homogeneous mixture or a heterogeneous one. Most students will immediately declare that it is homogeneous. Reference to the changing concentration of the atmosphere with altitude, along with the usual definition of "homogeneous," will probably cause cognitive conflict. The desired outcome here is not whether the correct answer is "homogeneous" or "heterogeneous," but that seldom are human-designed categorizations absolute: there are "gray areas."
- Some students have a limited view of the law of equilibrium (or the law of mass action): they consider only the equality between the reaction quotient Q and

the equilibrium constant K in a single vessel. For example, in a vessel in which $N_2O_4(g)$ and $NO_2(g)$ are in equilibrium, students asked about the value of the quotient $[NO_2]^2/[N_2O_4]$ invariably correctly claim it to be constant, to be consistent with the law of equilibrium. Given a discussion that confirms the students' knowledge that both $[N_2O_4]$ and $[NO_2]$ are constant, the teacher can ask about the constancy (over time) of the different functions $[NO_2]^7/[N_2O_4]^{0.5}$ or $[NO_2]^1/[N_2O_4]^3$, or any other function involving only $[N_2O_4]$ and $[NO_2]$ vessel of interest. Now we have potential for cognitive dissonance. What is special about the function $[NO_2]^2/[N_2O_4]$? Why does the law of equilibrium only make reference to this particular function? Consideration of the values of various functions of species concentrations in many vessels has the potential to develop a much deeper meaning of the law of equilibrium.
- Tell students that the answer to an acid–base question is pH = 5.2, and ask them to suggest what the question might have been. This can initially be very challenging to students. There are very many ways that this pH could be achieved.

Wright and Govindarajan [8] have published a list of common conceptual discrepancies in biology, some overlapping with chemistry. Most do not lend themselves to classroom demonstrations.

18.4
Origins of the Cognitive Conflict Teaching Strategy

Recommendations concerning provocation of cognitive conflict as a means of promoting deep learning have a long history. More than 50 years ago, Suchman [9] urged the use of situations that are "sufficiently surprising or contrary in outcome to capture students' attention."

A source of very considerable inspiration to science educators and curriculum developers in the 1970s and 1980s, and a catalyst for the inquiry learning movement, was Jean Piaget. In discussions of his studies of children's learning patterns, Piaget [10] claimed that "every new problem provokes a disequilibrium … the solution of which consists in a re-equilibration … " If a learner can *assimilate* an observation of new idea into his or her current cognitive structure, no disequilibrium is experienced. However, if unsatisfying attempts to assimilate lead to disequilibrium, the learner is forced to *accommodate* his or her cognitive structure (by modification or replacement) so that the learner achieves a new condition of equilibrium.

During the 1970s and 1980s began an intense wave of science education research that focused on identifying student's understandings of the natural world and scientists' conceptual explanations of it. This community was astonished by the number of surprising misconceptions (or *alternative conceptions*) of fundamental and basic scientific ideas that were identified. These seemed to know no boundaries of age or geography. Even many students judged to be very able by usual testing procedures have shown inadequate or unstable concepts when researchers used probes that require students to apply their conceptual understandings to situations or events – real or imagined. Just a few of the myriad of examples of such research include investigations of student understandings of chemical equilibrium (Hackling and Garnett [11], Tyson et al. [12]), stoichiometry (Mitchell and Gunstone [13]), changes of state (Osborne and Cosgrove [7]), chemical change (Hesse and Anderson [14]), and the nature of matter (Renstrom et al. [15]). Ozmen [16] has published a detailed literature review of research into students' conceptions of chemical bonding, and Kind [17] has reviewed research findings across many topics. There are more detailed discussions in other chapters in this book.

This exposure of students' inadequate understandings of so many concepts caused many in the chemical education community to jump with alacrity to criticize the transmissive teaching style based on the assumption that knowledge can be transferred intact from the teacher's mind to the students' minds. This gave rise to a very strong movement advocating "constructivism" as a model on which teaching and learning should be based. There are a variety of versions of constructivism, and much debate about its efficacy, but there can be no doubt that this movement has initiated change in the way that teaching and learning are perceived. From the perspective of teaching chemistry, the constructivist model of learning is nicely encapsulated by Bodner [18]. To Bodner, the kernel of this model is that "knowledge is constructed in the mind of the learner," rather than transferred to the mind of the learner. Construction of knowledge from experiences (such as what we see and what we are told) presumes an interaction between our prior knowledge and the new sensory inputs, which must create some disequilibrium for new knowledge to be constructed. It is obvious then how disciples of the constructivist view of learning urged the design of teaching strategies that caused the students to resolve cognitive dissonance to construct understanding.

In this same environment, scholars were theorizing about, and researching into, strategies most likely to bring about conceptual change in students – partly by consideration of the circumstances under which scientists have arrived at changes of worldviews. In a landmark paper, Posner et al. [19] proposed that four conditions are necessary for accommodation of one's "conceptual ecology": (i) there must be dissatisfaction with existing conceptions; (ii) a new conception must be intelligible; (iii) a new conception must appear initially plausible; and (iv) a new concept should seem fruitful. In common language, new experiences must create some puzzlement, and, as we search for sense-making, an alternative explanation should be understandable, should seem sensible, and must lead to a cognitive "payoff." Notwithstanding modifications of this proposal, and debates about what is meant by conceptual change, and whether this scheme applies to major or relatively minor cognitive accommodation, there is an obvious parallel logic that has given a sense of validity to the use of the cognitive conflict teaching strategy: we need to create conditions in which our students experience dissatisfaction with their current knowledge. It is easy to imagine that a student engaged in making sense of a discrepant event might experience stages that correspond with the four conditions of Posner *et al.*, leading to an "Aha!" experience.

Hewson and Hewson [20] provide an excellent analysis of how a conceptual change model can justify the use of conceptual conflict in the design of instruction.

One of the most significant outcomes of the "misconceptions identification" research is the awareness that conceptions – whether "good" or not – are resistant to change. Bodner [18] wrote "Each of us constructs knowledge that 'fits' our experiences." Once we have constructed this knowledge, simply being told that we are wrong is not enough to make us change our (mis) concepts. Let me give just one example (of thousands) of the robustness of what the reader may think to be a rather extraordinary student misconception. When probing first-year university students' conceptions about atomic structure, Keogh [21] asked them what they would purchase if they could go to a shop to obtain the components to build a boron atom. One student replied "a boron nucleus, five electrons and an electron cloud." When asked about the electron cloud, the student said that it was "the place where the electrons go." Feeling a moral duty to her research "subjects," Keogh sought out this student and gave a short tutorial, emphasizing that in the quantum mechanical model, the electron cloud is the electrons. Four weeks later, in a second round of interviews, the student again informed Keogh that he would buy an electron cloud in which to put the five electrons! Well, if conceptions cling so strongly, doesn't it make sense that we might have to vigorously disturb the cognitive equilibrium (that is, to induce cognitive dissonance through a discrepant event) to loosen the ties?

18.5
Some Issues Arising from *A Priori* Consideration

While, according to Piaget [10], disequilibrium is an inherent part of learning, it cannot guarantee learning. Critical to the cognitive conflict teaching strategy is the requirement that the teacher designs a discrepant event to create a level of dissonance that the students are able to resolve in the light of their previous knowledge – or, more probably, being aware of the level of prior knowledge of the students, to choose a discrepant event that suits the teaching/learning purpose.

Just on logical grounds, we can recognize that a significant feature of the design of a discrepant event is the adjudged match of the demands of the phenomenon to the prior knowledge level of the students. The same phenomenon shown to different people can be expected to induce a range of levels of cognitive dissonance. A phenomenon demonstrated in a certain way that brings about a level of dissonance that some students can resolve might cognitively "drown" some other students whose lesser prior knowledge means that there cannot be a discrepancy between expectation and observation because they have no basis for any expectation at all. An extreme illustration of this might be to ask a 6-year-old student by how much the pH of a buffer solution would change in response to addition of a specified amount of base. At the other end of the scale of observers, what is a discrepant event for a first-year university student might give entirely expected outcomes, with no cognitive dissonance, for a third-year student, or to a chemistry professor. In technical terms, according to Driver [22], "If dissonance is too great, assimilation will not take place at all. If there is not sense of dissonance, information is assimilated without change."

Accordingly, it is not sensible to consider the effectiveness of a classroom demonstration in some absolute way. For example, the question "Is 'boiling an egg' under reduced pressure effective in helping students to realize that the boiling points of liquids depend on the external pressure?" is invalid. The answer surely depends on so many factors, such as the prior knowledge "readiness" of the students, how the demonstration is prefaced by teacher talk or student interaction, how the students make a commitment to a prediction, whether there is a discrepancy between student prediction and observation, whether the students recognize the discrepancy and experience cognitive conflict, how the teacher helps the students to come to the intended resolution, and whether the students experience an "Aha!" moment from satisfaction with the resolution.

A consequence of the preceding issues is that it is not sensible to discuss a particular demonstration as a discrepant event, whether in a teaching or a research environment, without detailed specification of the context and manner of its presentation. Only then can other readers contemplate the appropriateness of the discussion to their particular contexts, and the manner of their presentation of the demonstration – which will be at least partly influenced by their personality.

In simple terms, we might imagine that cognitive conflict could arise for either of two reasons: (i) the students have no prior knowledge of the concept to be introduced through a demonstrated phenomenon, or (ii) the students have a conception, derived either from previous education or from out-of-school everyday experience which leads them to arrive at a prediction different from that observed. In the first category, any observation at all might lead to some desire to construct meaning to explain the observation. An example might be if students with no experience at all of buffer solutions are asked to predict what the pH change will be on addition of some acid solution to a buffer solution. The second category is nicely exemplified by the imagined response of the leading theoretical scientists nearly 100 years ago when presented with the evidence that electrons (previously "known" to be particles) undergo wave-like diffraction when passed through a thin crystal section. Or, to go to an earlier period in history, outside of the domain of chemistry, when the conception that the earth was flat was challenged by sailors who traveled in one direction and came back to where they began. In such cases, the previously held conception must at least be modified, if not rejected entirely, for cognitive conflict to be assuaged.

Over the course of formal education (primary school to perhaps university graduation), there are many instances where students can encounter puzzlement of the first category mentioned above, when they have no prior knowledge, and have no reason to expect any outcome of a demonstration. Students are frequently introduced to concepts about which they may have no previous knowledge, either from school or out-of-school experience: such concepts include chemical equilibrium, entropy, free energy, atomic orbital, electronegativity, catalyst, amount of substance, nucleophile, critical point, bond order, and delocalized electrons.

And every science teacher is aware that we sometimes require our students to reject a conception that "we" previously introduced in favor of another: we introduced the first conception because it was judged to be learnable by the students, and sufficient for their immediate needs. We do this in full knowledge that we will require them to change conceptions when the first proves to be too limiting at a later stage of chemistry. For example, we deliberately introduce the concept of base as a species that ionizes to form $OH^-(aq)$ ions in aqueous solution, but later expect students to accept a model of base as a species that competes for and gains H^+ ions. It is very common to develop students' first ideas of the nature of atoms through the Bohr model with its assumption of electrons as particles moving in orbits, but later on to expect a transition to a quantum mechanical model in which electrons in atoms are conceived as waveforms. And we all know of the students' abhorrence of why we don't "teach them what is right" the first time. The science education research literature has much to say about the difficulty of changing one's conceptions, as discussed earlier.

18.6
A Particular Research Study

In an action research project in which Baddock explored the effectiveness of her own teaching using a cognitive conflict strategy, the details of her findings (Baddock and Bucat [23]) expose some potential pitfalls of this teaching methodology. Her year-11 students, aged 15 and 16, in an Australian high school had been introduced to the concept of acids in aqueous solution and their characteristic property of forming $H^+(aq)$ ions by ionization in aqueous solution. By reference to the equations for ionization of HNO_3 and HCl in solution, the teacher's exemplars were strong acids and there was an implication that all acid molecules ionized in solution. There had been no previous reference to acids for which this is not the case, and so, of course, the students had no awareness of the distinction between strong acids and weak acids. In fact, the purpose of the classroom demonstration was to lead the students to construct a meaningful differentiation between the concepts *strong acid* and *weak acid*. The concept of pH had been presented to the students in the class previous to that in which the demonstration was shown.

Baddock's findings were so significant that it worth describing here some aspects of the previously published research study. This will also provide the context for other discussions.

HCl solutions of concentrations 1, 0.1, 0.01, and 0.001 $mol\,l^{-1}$ were made by successive dilutions in front of the class, and some of each put into different Petri dishes on an overhead projector. Methyl violet added to each was observed to be yellow, green, blue, and purple, respectively. Students were then challenged to predict in writing what the color would be if methyl violet was added to another Petri

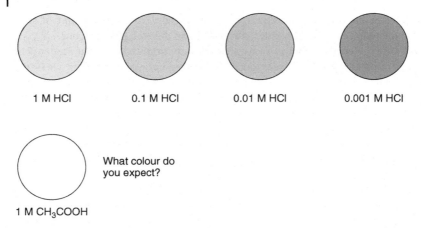

Figure 18.1 The task, designed as a discrepant event presented to students by Baddock.

dish containing some 1 mol l^{-1} acetic acid solution. For clarity, the task presented to the students is shown in Figure 18.1.

Then, with an attempt to induce some suspense, methyl violet was added to the 1 mol l^{-1} acetic acid solution. It turned blue – similar to its color in the 0.01 mol l^{-1} HCl solution.

In a general sense, Baddock used the predict–observe–explain (POE) technique described by White and Gunstone [24] and which has been used to probe students' understandings of situations and events. Before adding methyl violet to the acetic acid solution, students were asked to predict in writing what its color would be.

Before and during the demonstration, this highly regarded teacher set the scene to try to maximize the likelihood of surprise when the indicator was added to the CH_3COOH solution. After the observations, she engaged the students in class discussion designed to arrive at explanations that resolved any discrepancy between expected and observed outcomes. During discussion, she referred to the HCl solutions by their pH values (0–3), rather than their HCl concentrations. The five solutions were not labeled. The formulas of the acids (HCl and CH_3COOH) were assumed to be known to the students. The teacher expressed a very broad purpose of the exercise: "I want to look at some different acids in more detail."

In the first year of this action research study, 18 of the 22 students predicted that the indicator would turn yellow in the 1 mol l^{-1} CH_3COOH solution – the same as in the 1 mol l^{-1} HCl solution, as we might expect. The other four predicted different colors on the basis of suspicion that they were being "tricked."

During the next chemistry class, the students were asked to respond to specific questions about the demonstration in writing. The responses were extremely disappointing to the researcher as a teacher, and caused her to feel quite inadequate.

To check the students' knowledge of the procedural aspects of the demonstration, they were asked to describe the demonstration. Five students made no reference to the acetic acid solution. Of the other 17, 6 did not refer to the concentrations of the CH_3COOH solution, and none specified the concentration of H^+ ions in any solution.

To the question "What was the aim of the demonstration?" only one student gave an answer reasonably aligned with the teacher's purpose. Almost all said that it was an investigation either of the HCl solutions or of the indicator. The intended dissonance was not achieved. One brutal response was: "The aim was to completely confuse us. It worked."

Given the previous answers, it is not surprising that responses to "Explain the observations" were less than hoped for. None of the students related the indicator color to the hydrogen ion concentration. References to the pH of the solutions exposed poor understanding of the concept of pH.

The teacher-researcher's self-analysis of these unsatisfactory outcomes included the following issues, all of which probably seem obvious with hindsight. But recall that this is an experienced, highly regarded teacher whose in-class judgments and decisions should be respected.

- During the demonstration, the teacher only specified the pH of the HCl solutions, rather than the $H^+(aq)$ ion concentrations. So, in order to relate the $H^+(aq)$ ion concentration of the CH_3COOH solution to those of the HCl solutions, students had to perform a logic step that could have been avoided. This is at odds with the advice of Johnstone [25] and many others to minimize the possibility of cognitive "overload", because of the limited capacity of short-term working memory.
- In keeping with the demanding school syllabus, the concept of pH had only been introduced during the lesson prior to that involving the demonstration. It seems that the students had not yet had sufficient time nor practice to understand the meaning of pH well, nor to reach a level of facile conversion between solution pH and $H^+(aq)$ ion concentration.
- On the projector, there were no labels to indicate the identities of the solutions, the solution concentrations, the $H^+(aq)$ ion concentrations, or the pH of HCl solutions. This also probably presented a challenging demand on the students' mental working space capacity proposed by Johnstone [25].
- The teacher had presumed that the intended discrepancy would "hit the students in the face." Obviously, this was not the case. Cognitive overload can make it difficult for students to discriminate from all the knowledge and sensory inputs those that are important and those that are not. Just as on the way to work we do not regard as significant every face we see, nor every car number plate, nor every advertising hoarding – unless there is a link between any of these and something in our long-term memory – we should not expect students to regard as significant the key features of a demonstration. The information-processing model of learning proposed by Johnstone [26] includes a selection filter which we use to choose from all of the sensory inputs. White [27] has suggested that

the first step in making sense of an event is to select for attention the essential features from among all of the incoming sensory signals. The teacher concluded that she needed to be more explicit, more repetitive, and perhaps more dramatic in directing the students to awareness of the intended discrepancy.

- Gunstone [28] has referred to the importance of students' awareness of the aim of practical activities. In the light of this, the almost universal inability of the students to reasonably express the teacher's aim of the demonstration is worrying. Here we have a major issue in the manner of presentation of discrepant events. If the teacher provides to the students an explicit statement of her aims (I am going to demonstrate that not all acetic acid molecules ionize in aqueous solution, and this is a property of some acids called *weak acids*), then the potential for cognitive dissonance is lessened. In the case reported here, the teacher gave a very bland expression of intent. In any use of the cognitive conflict teaching strategy, perhaps design of how the aim is expressed is as important as any other factor. In this case, perhaps a satisfactory statement of intent might be via a question such as "Do all of the molecules of all acids ionize in aqueous solution?"

In each of the 2 years following the reported case above, this teacher-researcher repeated her study with the new cohort. Consistent with the action research process, she modified her presentation, in each case taking into account her analysis of previous findings. By and large, although the students showed evidence of much keener awareness of the intended discrepancy, resolution of the resulting dissonance did not necessarily follow.

A rather interesting student reaction issued from the teacher's attempts to help the students to arrive at cognitive conflict resolution through peer discussions. When the students did not converge on an explanation quickly, and the teacher did not "round up" for them, some students became frustrated and even engaged in aggressive behavior, with comments such as "This is stupid!" This is consistent with the previously reported research of Mitchell [29] in situations where the expectations of the students are that the teacher is a provider of information, rather than a facilitator of learning. Mitchell found that for some time students felt out of their comfort zone when required to propose and defend ideas. On the other hand, Mitchell reported that, as time passed, students developed more trust in the teacher and in peer group discussions as a way of learning, and eventually enthusiastically adopted this as the norm.

We haven't yet dealt with all of the potholes in the road of teaching by arousing cognitive conflict. In each year that Baddock presented the demonstration of acetic acid as a weak acid, she interviewed a few students, in each case showing them a videoclip of the class demonstration to stimulate their recall of events. This was a research strategy taken from O'Brien [30]. She found two fascinating examples of unintended learning that had occurred – both undesirable:

- Two students arrived at a conclusion that the demonstration was intended to show them that the lower the hydrogen ion concentration in aqueous solution, the darker is the indicator color. While this conclusion was consistent with the

observations in this particular demonstration, these students took it to be a generalization across all indicators.
- A few students had "Aha" moments when they thought that they had resolved the discrepancy: they suggested that of course the concentrations of H^+(aq) ions in HCl and CH_3COOH solutions would be different because each CH_3COOH molecule has four H atoms, while each HCl molecule has only one. On one hand, consideration of the number of H atoms in each acid molecule is absolutely a reasonable course of action for these students: how are they to know which of the H atoms in each molecule are ionizable upon dissolution in water? On the other hand, they jumped with alacrity to an explanation that is at odds with the observations: the concentration of H^+(aq) ions is higher in HCl solutions than in CH_3COOH solutions of the same concentration.

All in all, Baddock's research has identified numerous pitfalls in her presentation of this demonstration to her students in the context of their prior knowledge base – sufficient that we might wonder if a more effective method for this purpose might be to present the information about weak acids in a rather transmission-of-knowledge approach, and then confirm the presented information by the demonstration carried out in a way to illustrate its veracity. Then again, we can't know the answer to that supposition without more research: evidence-based classroom research has a habit of surprising both teachers and logic.

In order to bring into sharp focus that there is perhaps more than meets the eye in presenting a demonstration such as the one discussed, it should be borne in mind that the above discussion rather focused on the negative research findings, and largely ignored positive outcomes of the demonstration. Also, this story has been about but one particular demonstration used for one particular purpose. Readers might be able to engage in action research, no matter how formal, to explore the conditions under which other discrepant events work most efficiently, for their students.

Baddock's work might also cause us to wonder more generally if our students learn what we hope for from our classroom demonstrations – whether we use the cognitive conflict strategy or not – and whether unintended learning takes place.

18.7
The Logic Processes of Cognitive Conflict Recognition and Resolution

Let's presume that situational factors such as students' prior knowledge "readiness," the labeling of solutions, and the manner of the teacher's presentation are not hindrances to the effectiveness of Baddock's demonstration, described above. Let's also presume that the potentially confounding concept of pH is not used: all solutions are defined by their H^+(aq) ion concentrations. Then it is instructive to contemplate the minimum amount of intellectual logic that a student must engage in to arrive at conflict recognition, and then at conflict resolution. We could imagine all of the stages of inference, deduction, and resolution arising from interactions between prior knowledge and the observations shown in Table 18.1.

By no means is the suggestion made that all students would need to engage in precisely these steps. For example, some students might deal with three of these suggested stages as one "chunk." Neither is any suggestion intended about a

Table 18.1 Imagined logic steps in which students might need to engage for conflict recognition and conflict resolution in the Baddock demonstration.

Prior knowledge	HCl is an acid which in aqueous solution ionizes to form H^+(aq) ions
Assumption (probably implicit)	All of the HCl molecules ionize to form H^+(aq) ions in solution
Prior knowledge	The colors of indicators in aqueous solution indicate the concentration of H^+(aq) ions in the solution
Observations	In 1 mol l^{-1} HCl solutions, methyl violet indicator is yellow
	In 0.1 mol l^{-1} HCl solution, methyl violet indicator is blue-green
	In 0.01 mol l^{-1} HCl solution, methyl violet indicator is dark blue
	In 0.001 mol l^{-1} HCl solution, methyl violet indicator is purple
Deduction	In solutions with [H^+] = 1 mol l^{-1}, the indicator is yellow
	In solutions with [H^+] = 0.1 mol l^{-1}, the indicator is blue-green
	In solutions with [H^+] = 0.011 mol l^{-1}, the indicator is dark blue
	In solutions with [H^+] = 0.0011 mol l^{-1}, the indicator is mauve-purple
Observation	In 1 mol l^{-1} CH_3COOH solution, the indicator is purple
Conflict	The color of the indicator in 1 mol l^{-1} CH_3COOH solution is different from that in 1 mol l^{-1} HCl solution
Deduction at level 1	From the colors of the indicator in 1 mol l^{-1} HCl solution (yellow) and in 1 mol l^{-1} CH_3COOH solution (purple), [H^+] is less in the CH_3COOH solution than in the HCl solution
Deduction at level 2	Because the purple color of the indicator in the 1 mol l^{-1} CH_3COOH solution is the same as that in 0.001 mol l^{-1} HCl solution, [H^+] is similar in these two solutions
Deduction at level 3	Since we assume that in 0.001 mol l^{-1} HCl solution [H^+] = 0.001 mol l^{-1}, we can deduce that in the 1 mol l^{-1} CH_3COOH solution also [H^+] = 0.001 mol l^{-1}
Conflict	How can it be that [H^+] = only 0.001 mol l^{-1} in 1 mol l^{-1} CH_3COOH solution?
Resolution at level 1	Only some of the CH_3COOH molecules ionize in aqueous solution
Resolution at level 2	In the 1 mol l^{-1} acetic acid solution, approximately 1 of every 1000 CH_3COOH molecules ionizes
Generalization	There are other acids, such as acetic acid, of which only some of the molecules are ionized in aqueous solution, and these are called *weak acids*

necessary sequence of stages. Nonetheless, this analytical process is informative in exposing the high level of intellectual demand that this demonstration places on students. And this demand may not be unique to this particular demonstration, regardless of how simple it seems to the teacher. Furthermore, we shouldn't presume without further consideration that this demand is greater than that in classroom demonstrations that do not aim for resolution of an induced dissonance. Of course, and once again, we should recognize that the demand depends upon the level of outcomes intended, the prior knowledge of the students, their abilities, and the degree of acceptance by them of participation in a teaching situation in which they are largely responsible for their own learning.

18.8
Selected Messages from the Research Literature

From this point on, we will be concerned with what teachers and researchers have found in relation to the use of discrepant events. The reader should be aware that publications tend to focus on the negative, rather than the positive: authors often point out the hazards and pitfalls of what they have done by referring to those (usually few) students who have demonstrated that the strategy was unsuccessful for them. This is with the good intention of alerting other practitioners to the care that needs to be taken. Sometimes we are not even told about the students for whom the strategy was effective. This bias may show up in the summaries and discussions that follow.

De Vos and Verdonk [31] describe a discrepant event designed to help students achieve the understanding that new substances are formed during a chemical reaction. Some solid crystals of lead nitrate and sodium iodide, both white, are added to a mortar. Even the slightest grinding leads to a brilliant yellow, which astonishes the students. The description of how the authors had thought through this presentation is quite brilliant. So too is the discussion of the students' reactions and how they students were guided toward the desired conclusion, even if the evidence is anecdotal. It is evident that here are experts at work, and their article should be compulsory reading for all chemistry teachers.

De Vos and Verdonk are keenly aware that resolution of the conflict between expectation and observation, which they describe as "both an intellectual and an emotional experience," is not a simple matter. They report that "when the observed facts do not fit in with the existing theories (of substance conservation), so much the worse for facts. Some students actually attempt to deny the existence of the yellow color or to explain it away." Also, a group of four "together mention the yellow color 22 times before they dare to write their observation down." It is very instructional to read how the authors take precautions in the design of this seemingly simple demonstration to leave the students with no "escape route": only one way of resolving the conflict is inevitable, if not simple, for them.

Wright and Govindarajan [8] describe how in a biology class spontaneous statements that exhibit students' beliefs can be used to motivate students to learn through dissonance resolution.

Gunstone [28] is the author of an insightful article on applying constructivist principles in practical work in science. In the context of students' prior beliefs and the science that we hope they will come to understand, he says: "At first glance it seems most tempting to see the changing of contrary student conceptions, the restructuring of ideas and beliefs so that these become ideas and beliefs of science, as being achieved simply by using practical work to show directly the inconsistency between the contrary conceptions and science. This view suggests that using the laboratory to have students restructure their personal theories is straightforward. However, such a view is a naive one. The real picture is more difficult, and the messages for practical work to be derived from constructivism rather more complex."

Gunstone illustrates his claim by reference to practical classes in physics for students in education who had already graduated in chemistry or biology. Tasks were conducted using a POE strategy (White and Gunstone [24]): the students are expected to write down their predictions, their observations, and their reconciliation explanations. In this way, commitments are required from the students. Even though these were not chemistry-based tasks, his evidence about the nature of student observations is valuable:

- Observations are theory-dependent. In a task involving DC electricity, students were asked to predict the relative current at two ammeters in a circuit. The reading at meter 1 was only slightly higher (3%) than at meter 2. During the task, some saw, and recorded, equal meter reading, and some recorded that the reading on meter 1 was higher than that on meter 2. There was a high correlation between what was observed and what had been predicted. In another task, a blackboard duster was dropped from a height of 2 m. Students were asked to compare, by naked eye, its falling speed at 1 m, and just before hitting the floor. Gunstone could not see a difference, but all of the students claimed to have done so. Some gave semiquantitative estimates of how much faster the duster was falling at the lower level than at 1 m, and one even recorded as an observation that the speed at the lower point was 1.414 times faster than at the higher point.
- Students' personal theories can lead them to reject observations via a denial of the legitimacy of the observations. Students plotted voltage versus current in a circuit containing an incandescent globe. Even though the points appeared to fall on a distinct curve, many students, in a desire to confirm Ohm's law, drew a straight line plot of the relationship. When a piece of rotting liver was left in a well-sealed jar for 2 weeks, and its weight found to be constant, some students claimed that the seal was not airtight: air must have got in. A block of wood and a bucket of sand were connected by a rope over a pulley wheel – at rest at the same height. When the block of wood was held lower, about half of the 470 students predicted that it would return to its original position when released. It

did not. Of those faced with a discrepancy between prediction and observation, most denied the observation, usually by arguing that the wood had been held down too long and had "got used to its new equilibrium position." So, even when the same observation is made by the students, they do not all accept the validity of that observation.
- The inferences drawn from observations are influenced by personal theories. In a group of four using the same set of current versus voltage data, each produced a different graph to represent the relationship. So, even when all students make the same observations, and the validity of these is accepted, interpretation will not necessarily proceed in the direction intended.

These findings by Gunstone have some commonality with the later work of Chinn and Brewer [32] who proposed eight categories of student responses to contradictory information: ignoring the data, rejecting the data, professing uncertainty about the validity of the data, excluding the data from the domain of the current theory, holding the data in abeyance, re-interpreting the data, accepting the data and making peripheral changes to the current theory, and accepting the data and changing theories.

That the optimal level of cognitive conflict induced by a task varies from student to student is well illustrated by Shapiro [33], again in a physics context – learning about light. Four primary school students were seated around a table on which was a saucer containing a coin. The students crouched down until the coin was just no longer visible below the saucer edge. As water was slowly added to the saucer, suddenly the coin became visible again. One student (Mark) exalted in an "Aha!" moment, describing to the others how this could be explained by refraction of light, while they looked at him perplexed. Another student copied Mark's written explanation word for word.

Limon [34] has reviewed the published results obtained in the application of the cognitive conflict strategy in the classroom and concludes that these are ambiguous and inconclusive. In her reflection on reasons for this, she comments that " … many of the difficulties found in the application of the cognitive conflict strategy in the classroom are closely related to the complexity of factors intervening in the context of school learning" (p. 364) and " … the cognitive conflict paradigm as an instructional strategy centred on students' cognitive aspects neglects many other variables that influence students' learning in the school setting." (p. 365). In Table 18.2 is a list of factors that she considers may be vital to the induction and resolution of meaningful cognitive conflict.

Limon provides an in-depth analysis of how several of these potentially significant factors might influence learning from discrepant events.

Kang et al. [2] also have recognized the importance of factors other than cognitive issues – especially situational interest, student effort, and attention – in the effectiveness of discrepant events to bring about conceptual change. These workers conducted a quantitative study designed to elucidate the relative significance of components in a proposed path model of learning by cognitive conflict resolution, shown here in Figure 18.2.

Table 18.2 Variables other than cognitive aspects considered by Limon that might intervene in the successful application of learning through a cognitive conflict strategy.

Variables that might contribute to inducing a meaningful cognitive conflict

Variables related to the learner	Prior knowledge
	Motivation and interests
	Epistemological beliefs (about learning and teaching and about the subject-matter to be learned)
	Values and attitudes toward learning
	Learning strategies and cognitive engagement in the learning tasks
	Reasoning abilities
Variables related to the social context in which learning takes place	Role of peers
	Teacher–learner relationships
	Teacher–learners relationships
Variables related to the teacher	Domain-specific subject-matter knowledge
	Motivation and interests
	Epistemological beliefs about learning and teaching and about the subject-matter taught
	Values and attitudes toward learning and teaching
	Teaching strategies
	Level of training to be a teacher

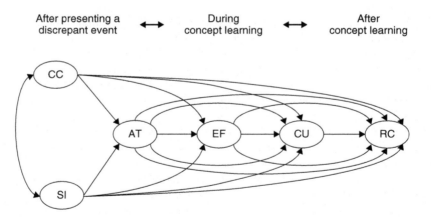

Figure 18.2 Proposed path model predicting the influences of cognitive conflict, situational interest, attention, and effort on conceptual understanding and the retention of the conception. Abbreviations: CC, cognitive conflict; SI, situational interest; AT, attention; EF, effort; CU, conceptual understanding; RC, retention of the concepion. Taken from Kang et al. [2].

In a computer-assisted instruction unit designed to learn the concept of density of materials, and, in particular, to differentiate density from mass, Kang *et al.* report that the effect of situational interest on conceptual learning was considerably greater than that due to the cognitive conflict. The authors do, however, acknowledge that the presence of a teacher might give rise to different results than those found in their teacherless instruction situation. Nonetheless, we have confirmatory evidence that the use of a cognitive conflict strategy is complex, and the presentation of a discrepant event without attention to many other classroom factors is unlikely to be successful.

18.9
A Personal Anecdote

The author can reflect on his own experience in a first-year tertiary-level class during lectures on the topic of phase changes – in particular, the dependence of the boiling temperature of water on the pressure acting on the surface of the water. For many years, at the beginning of a lecture I would put some water into a wide-mouthed flask, and lower an egg into it. A tube passing through the stopper was connected by plastic tubing to a vacuum pump, and the flask set onto a gauze mat above a Bunsen burner with very low flame. The water was boiled for about 30 min – unbeknown to the students, at about 35 °C. Near the end of the lecture period, I removed the egg, making reference to the "30-min hard-boiled egg." This was accompanied by an overt display of pretended "burning" fingers as I gave the impression that the flask and the water were very hot – in order, of course, to enhance the level of cognitive conflict. When the eggshell was broken open, out flowed an uncooked egg. The obviously puzzled students were sent on their way with the question "How can that be?" and no further discussion. Of course, I presumed that the animated discussion among the students as they exited to go to their next lecture was all about resolution of the cognitive dissonance. A few – very few – students over the years came to the front after the lecture and asked to feel how hot the flask was.

In the light of Baddock's findings, I had begun to think about the efficacy of my own demonstrations, on which my lectures were heavily dependent. So, in the course of supervising laboratory classes, I asked students about their interpretation of this demonstration (and several others). Universally, the students were amazed, and quite puzzled, at the observation of the raw egg. Almost universally, the students had given up rather quickly on trying to resolve the discrepancy between expectation and observation. Further probing exposed the consistent view that since the water was very hot, the egg would certainly have cooked, and so the lecturer must have performed some "magic trick." I presumed that the students were wrongly associating the phenomenon of boiling water with hotness, and why shouldn't they think that, since that was their lifetime experience at home? I assured them that in fact the water was not hot. This they refused to believe. "The water was obviously hot. You could hardly touch the flask without flinching!" When eventually they were convinced that the water was indeed not hot, they were somewhat aggravated at what they called my "deception" – probably warranted.

What a lesson for me! Misplaced dramatic effect! How could I make it up to 30 or more cohorts of very bright students whom I had confused, or at least whose time I wasted? I had misled them by dramatically giving the impression that the water was hot. The conflict that I had been inducing in the students was that the water was hot, when indeed it was not. The intended discrepancy to be resolved was that the boiling water was only lukewarm, although in the students' experience boiling water is hot.

In consideration of these findings, as well as what we now understand about the influence of cognitive "overload" from Johnstone [25], the demonstration of this phenomenon has now been drastically redesigned. In its original form, the presentation required the students to make an inference about the temperature of the water boiling under reduced pressure in an indirect way: from the observation that the egg was uncooked, students had to infer that the boiling water was not hot. And then they could focus on why, and how, water would boil when it was not hot, by making some link to the fact that the pressure was reduced. To invoke the egg is wonderfully dramatic, and perhaps a nice way to consolidate meaning for those who have had some initial exposure of the boiling point versus pressure relationship. But can we not eliminate a step from the resolution of conflict process by a more direct indication of the temperature of the boiling water? Nowadays, depending on the context of the presentation, I either invite the students to touch the flask, or use a thermistor to measure the water's temperature which is displayed on a projector. Now the students can focus on the real discrepancy unfettered by distracting "baggage." And so of course they understand .... Or do they?

18.10
Conclusion

All of the evidence, and all of our reflection upon the use of the cognitive conflict teaching strategy, tells us that we need to design our presentation of the discrepant carefully, so that the demonstrated phenomenon is specifically suited to the teacher, and to the students, to achieve the hoped-for outcome. We need to make considered and well-based judgments at every step of our planning and execution.

But isn't this true of any strategy that we propose to use? Before you decide that inducing cognitive dissonance has too many pitfalls, consider that of course we don't know what understanding and misunderstandings might have arisen in a "standard" presentation, without attempting to induce dissonance, designed to introduce students to the distinction between strong acids and weak acids – or for any other purpose. In fact, can there be learning without disequilibrium? After all, the raft of student "misconceptions that have been identified mostly with students taught by the transmission mode of teaching."

While teaching by use of the cognitive conflict strategy undoubtedly has its complexities, and the research reports are not conclusive about its efficacy, Zohar and Aharon-Kravetsky [35] make the sensible suggestion that "A challenge of current research is to study the conditions under which cognitive conflict is effective." We could try to find out in our own classrooms: attempts to research these questions could be quite specific to particular demonstrations with particular students at a particular time, for a particular purpose, and they need not be too ambitious.

In the meantime, there are some general conclusions and recommendations that can be made from the above discussions:

- A cognitive conflict strategy should be chosen only if particular benefits can be expected that are otherwise difficult to achieve.
- Resolution of the cognitive conflict arising from a discrepant event should be as direct as possible; that is, the more steps of interpretation and resolution that are required, the less likely is resolution.
- The teacher cannot assume that the intended conflict arising from a discrepant event is perceived by the students. Effort may need to be invested by the teacher to ensure that the conflict is recognized as such – even to the point of explicit indication of the conflict.
- Teacher planning and effort may need to be invested in inducing in students a motivation to resolve the conflict.
- Monitoring the process of conflict resolution and evaluating whether the target conception has been adopted may be important components of this strategy.
- Student–student discussion, as well as teacher–student discussion, is likely to enhance effectiveness of attaining the teaching–learning goal.

References

1. Appleton, K. (1993) What makes lessons different? A comparison of students' behaviour in two science lessons. *Res. Sci. Educ.*, **23**, 1–9.
2. Kang, L.C., Scharmann, H., Kang, S., and Noh, T. (2010) Cognitive conflict and situational interest as factors influencing conceptual change. *Int. J. Environ. Sci. Educ.*, **5** (4), 383–405.
3. Wright, E.L. (1981) Fifteen simple discrepant events that teach science principles and concepts. *Sch. Sci. Math.*, **81**, 575–580.
4. Longfield, J. (2009) Discrepant teaching events: using an inquiry stance to address students' misconceptions. *Int. J. Teach. Learn. High. Educ.*, **21**, 266–271.
5. Xie, C. (2011) Visualizing chemistry using infrared imaging. *J. Chem. Educ.*, **88**, 881–885.
6. Schönborn, K., Haglund, J., and Xie, C. (2014) Pupils' early explorations of thermoimaging to interpret heat and temperature. *J. Balt. Sci. Educ.*, **13**, 118–132.
7. Osborne, R. and Cosgrove, M.M. (1983) Children's conceptions of the changes of state of water. *J. Res. Sci. Teach.*, **20**, 825–838.
8. Wright, E.L. and Govindarajan, G. (1992) Stirring the biology teaching pot with discrepant events. *Am. Biol. Teach.*, **54**, 205–210.

9. Suchman, J.R. (1960) Inquiry training in the elementary school. *Sci. Teach.*, **27**, 42–45.
10. Piaget, J. (1961) The genetic approach to the psychology of thought. *J. Educ. Psychol.*, **52**, 151–161.
11. Hackling, M.W. and Garnett, P.J. (1985) Misconceptions of chemical equilibrium. *Eur. J. Sci. Educ.*, **7**, 205–214.
12. Tyson, L., Treagust, D.F., and Bucat, R.B. (1999) The complexity of teaching and learning chemical equilibrium. *J. Chem. Educ.*, **76**, 554–558.
13. Mitchell, I. and Gunstone, R.F. (1984) Some student conceptions brought to the study of stoichiometry. *Res. Sci. Educ.*, **14**, 78–88.
14. Hesse, J.J. and Anderson, C.A. (1992) Students' conceptions of chemical change. *J. Res. Sci. Teach.*, **29**, 277–299.
15. Renstrom, L., Andersson, B., and Marton, F. (1990) Students' conceptions of matter. *J. Educ. Psychol.*, **82**, 555–569.
16. Ozmen, H. (2004) Some student misconceptions in chemistry: a literature review of chemical bonding. *J. Sci. Educ. Technol.*, **13**, 147–159.
17. Kind, V. (2004) *Beyond Appearances: Students' Misconceptions About Basic Chemical Ideas*, 2nd. A publication of the edn, Royal Society of Chemistry, http://www.rsc.org/images/Misconceptions_update_tcm18-188603.pdf (accessed 30 July 2014).
18. Bodner, G. (1986) Constructivism: a theory of knowledge. *J. Chem. Educ.*, **63**, 873–878.
19. Posner, G.J., Strike, K.A., Hewson, P.W., and Gertzog, W.A. (1982) Accommodation of a scientific conception: toward a theory of conceptual change. *Sci. Educ.*, **66**, 211–227.
20. Hewson, P.W. and Hewson, M.G.A.'B. (1984) The role of conceptual conflict in conceptual change and the design of science instruction. *Instr. Sci.*, **13**, 1–13.
21. Keogh, L. (1989) Students understandings of atomic structure. Unpublished Honours-level dissertation. The University of Western Australia.
22. Driver, R. (1983) *The Pupil as Scientist*, Open University Press, p. 53.
23. Baddock, M. and Bucat, R. (2008) Effectiveness of a classroom chemistry demonstration using the cognitive conflict strategy. *Inter. J. Sci. Educ.*, **30**, 1115–1128.
24. White, R. and Gunstone, R. (1992) in *Probing Understanding* (eds R. White and R. Gunstone), Falmer Press, pp. 44–64.
25. Johnstone, A.H. (1986) Capacities, demands and processes – A predictive model for science education. *Educ. Chem.*, **23**, 80–84.
26. Johnstone, A.H. (1997) Chemistry teaching – Science or alchemy? *J. Chem. Educ.*, **74**, 262–268.
27. White, R.T. (1991) in *Practical Science* (ed. B.E. Woolnough), Falmer Press, pp. 78–86.
28. Gunstone, R.F. (1991) in *Practical Science* (ed. B.E. Woolnough), Falmer Press, pp. 67–77.
29. Mitchell, I. (1986) in *Improving the Quality of Teaching and Learning: An Australian Case Study – The PEEL Project* (eds J.R. Baird and I.J. Mitchell), The PEEL Group, Faculty of Education, Monash University, Melbourne, pp. 45–85.
30. O'Brien, J. (1993) Action research through stimulated recall. *Res. Sci. Educ.*, **23**, 214–221.
31. De Vos, W. and Verdonk, A.H. (1985) A new road to reactions, part 1. *J. Chem. Educ.*, **62**, 238–240.
32. Chinn, C.A. and Brewer, W.F. (1998) An empirical text of a taxonomy of responses to anomalous data in science. *J. Res. Sci. Teach.*, **35**, 623–654.
33. Shapiro, B. (1988) in *Development and Dilemmas in Science Education* (ed. P. Fensham), Falmer Press, pp. 96–120.
34. Limon, M. (2001) On the cognitive conflict as an instructional strategy for conceptual change: a critical appraisal. *Learn. Instr.*, **11**, 357–380.
35. Zohar, A. and Aharon-Kravetsky, A. (2005) Exploring the effects of cognitive conflict and direct teaching for students at different academic levels. *J. Res. Sci. Teach.*, **42**, 829–855.

19
Chemistry Education for Gifted Learners
Manabu Sumida and Atsushi Ohashi

19.1
The Gap between Students' Images of Chemistry and Research Trends in Chemistry

Whenever we ask students to draw a picture of scientist at work, they usually draw an elderly male scientist wearing a white coat in a laboratory, shaking a test tube, and causing a chemical reaction [1–3]. The students' stereotypical images of science and scientists reflect research in chemistry rather than physics or biology. The students' images of science are not formed through personal experiences in science, such as reading science journal articles and/or operating advanced research equipment in an institution. There are only a few students who have a chance to talk with scientists in everyday life or experience reading some of their published works. Although we use many chemical products, there are very few who observed production in the factories. In other words, students construct pseudo images of chemistry without studying what chemistry is all about.

Scerri and McIntyre deplore that the philosophy of chemistry has been sadly neglected by most contemporary literature in the philosophy of science, and they proposed four important themes to discuss the philosophy of chemistry: Reduction, Laws, Explanations, and Supervenience [4]. Paul and Elder summarize the goal of chemistry; that is to study the most basic elements out of which all substances are composed and the conditions under which, and the mechanisms by which, substances are transformed into new substances [5, p. 30]. They note a specific key assumption of chemistry; that is all (or most) of the changes in identity of substances, as they react with other substances, can be accounted for by the theories and laws of modern chemistry [5, p. 30].

The turn of the twentieth century was a turning point in the history of chemistry. Chemistry has a position in the center of the sciences, bordering onto physics, which provides its theoretical foundation, on one side, and onto biology on the other, living organisms being the most complex of all chemical systems [6, p. 73]. Malmstrom and Anderson listed 13 fields that reveal the development of chemistry including breakthroughs in all of its branches during the twentieth century [7]. These are "General and Physical Chemistry," "Chemical Thermodynamics," "Chemical Change," "Theoretical Chemistry and Chemical Bonding," "Chemical

Chemistry Education: Best Practices, Opportunities and Trends, First Edition.
Edited by Javier García-Martínez and Elena Serrano-Torregrosa.
© 2015 Wiley-VCH Verlag GmbH & Co. KGaA. Published 2015 by Wiley-VCH Verlag GmbH & Co. KGaA.

Structure," " Inorganic and Nuclear Chemistry," "General Organic Chemistry," "Preparative Organic Chemistry," "Chemistry of Natural Product," " Analytical Chemistry and Separation Science," "Polymers and Colloids," "Biochemistry," and "Applied Chemistry."

Learning chemistry includes familiar and concrete phenomena and events but it requires higher order and complex thinking (e.g., [8]). Holyoak and Thagard listed 16 scientific analogies to analyze higher order thinking including "Benzene/Snake" by Friedrich Kekule and "Respiration/Combustion" by Antoine Lavoisier [9]. In fact, Lavoisier's achievement in the eighteenth century appear, not as the goal of a teleological progress toward modern science but rather, as a brilliantly creative response to the need to fulfill chemists' task of organizing information about substances, their properties and behavior, in the face of unsettling new phenomena [10, p. 377]. His instrumentation and techniques of precision measurement were striking innovations [11, p. 396].

In this chapter, the world trends of Nobel Laureates in chemistry from 1901 to 2012 are summarized to illustrate how giftedness in chemistry was required in the new century. Information about identification, curriculum development, and the implementation of gifted education in chemistry from diverse contexts is also considered.

19.2
The Nobel Prize in Chemistry from 1901 to 2012: The Distribution and Movement of Intelligence

In this chapter, the Nobel Prize in Chemistry is used an exemplar to consider international trends in chemistry research after the twentieth century. The Nobel Prize in chemistry is awarded to "the person who shall have made the most important chemical discovery or improvement." The Prize is awarded by the Royal Swedish Academy of Sciences and has been awarded to 163 Laureates from 1901 to 2012. One of the reasons why the Nobel Prize is one of the most prestigious awards is that this was the first international science award [12].

First, we divide the history of 1901–2012 into three timeframes; (i) the first half of the twentieth century (from 1901 to 1950), (ii) the second half of the twentieth century (from 1951 to 2000), and (iii) the twenty-first century (from 2001 to 2012). Figure 19.1 illustrates the number of Chemistry Laureates and the countries of their affiliation at the time of the award in the three different timeframes.

Even though laws and concepts of chemistry are common across languages and cultures, the center of chemistry research has been changing through time. It is clear in Figure 19.1 that the Nobel scene was Eurocentric in the first half of the twentieth century. In the second half of the twentieth century, 44 scientists in the United States have received the Nobel Prize for chemistry. Awards in East Asia and Middle Eastern countries seem to be on the rise in the twenty-first century; Japan has five Laureates in chemistry and Israel has four.

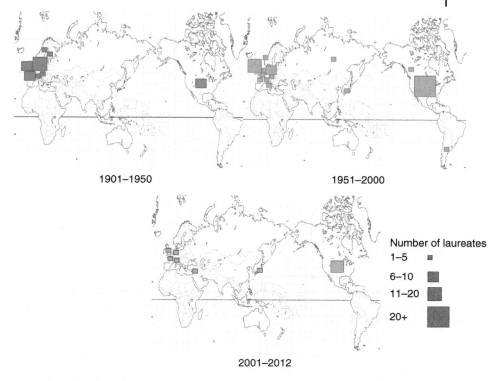

Figure 19.1 The number of Nobel Laureates in chemistry and the country of their affiliation at the time of the award.

Table 19.1 The distribution of fields of Chemistry Laureates.

1901–1950 ($n = 73$)	1951–2000 ($n = 65$)	2001–2012 ($n = 18$)
Physical chemistry (20.0)	Biochemistry (21.9)	Biochemistry (33.3)
Organic chemistry (20.0)	Physical chemistry (13.7)	Structural chemistry (22.2)
Nuclear chemistry (13.8)	Organic chemistry (12.3)	Organic chemistry (16.7)
Natural products chemistry (12.3)		

( ):% Source: *http://www.nobelprize.org/nobel_prizes/chemistry/fields.html*.

Next, in the three timeframes (1901–1950, 1950–2000, and 2001–2012), the fields of the Chemistry Laureates were sorted and listed over 10% in each timeframe in Table 19.1. The table shows that the trend fields have been changing with time as the center of chemistry researches does. The emergence of biochemistry in the second half of the twentieth century and structural chemistry in the twenty-first century was notable.

Table 19.2 The distribution of shared and unshared Nobel Prizes in chemistry (%).

		1901–1950 (n = 42)	1951–2000 (n = 50)	2001–2012 (n = 12)
One Laureate only		81.0	52.0	25.0
More than one Laureate	From the same country	11.9	14.0	25.0
	From different countries	7.1	34.0	50.0

Table 19.2 summarizes the distribution of shared and unshared Nobel Prizes in chemistry in the three timeframes from 1901 to 2012. For cases in which more than one Laureate were awarded in a year, they were classified as to whether they belonged to a research institution in the same country or not. Table 19.2 shows that the percentage of individual; Laureate has been decreasing and the percentage of shared awards has been increasing. Especially, about half of the Nobel Prizes in chemistry were awarded to more than one scientist from different countries in the twenty-first century.

In summary, the Eurocentric chemistry research was dominant in the first half of the twentieth century; the United States led the second half and Asian countries have been closely following. Interdisciplinary fields such as biochemistry and structural chemistry are rising and modern chemistry research is international and collaborative. There were only four female Laureates in chemistry among 163 Laureates from 1901 to 2012. Two of these four Laureates, Marie Curie and Dorothy Crowfoot Hodgkin, were awarded with unshared Prizes. Gender issue in chemistry education for the gifted is a crucial perspective.

19.3
Identification of Gifted Students in Chemistry

19.3.1
Domain-Specificity of Giftedness

IQ Intelligence Quotiens tests have been the classical method for identifying a gifted child in many countries. The IQ test score (e.g., a score of 130 or higher) is still very commonly used at least as a partial basis for identification. In some cases, some upper percentage of a range of scores (e.g., top 10%) is still used as the standard, based on the notion that a gifted child is someone who performs better than other children of the same age. Researchers and practitioners have come to agree that a more diverse system that incorporates measures other than IQ must be used to identify gifted children.

To identify gifted children, it may be helpful to use general behavioral characteristics of the individual, such as having a large vocabulary, the ability to express themselves well, mental agility, a sense of humor, and the ability to concentrate on one thing for a long period of time. A child may systematically memorize or tell the names and characteristics of hundreds of animated cartoon characters, however, this does not mean he or she will easily be able to learn and retain the names and properties of the 118 chemical elements or understand the power of the Periodic Table for organizing these elements. Similarly, even children who can focus for hours at a time on an activity such as catching insects may not demonstrate any interest in the intricacies of research or the creative arts such as painting or music. It is normal for people to be stronger in some areas and weaker in others. The domain-specific, dynamic nature of science, with its encompassing wealth of fields of study, can accommodate children's varied areas of interest, and be an ideal subject area for children to show their giftedness and in which educators can identify the giftedness. A great deal of attention has recently been paid to science education in research and practice related to specific domain of giftedness.

As found in reviews (e.g., [13, 14]), a large number of research studies citing specific data have been reported to justify assumptions of naïve physics and naïve biology as innate constraints or domains in which infants acquire scientific knowledge and change their understanding in science. Children are able to do abstract scientific thinking in specific domains from a very early phase of their development. Gopnik *et al.* noted that young children and scientists are the best learners in the world; they are highly similar in the ways in which they adjust and change existing knowledge as they interact with the outside world, and they can perform computations that cannot be duplicated even by the most advanced computers [15]. Gopnik also shows that while careful observation of young children will reveal differences in their level of precision or competence, this will also reveal that young children and researchers share a sense of curiosity that makes them want to know more about the world through experimentation and observation, and that they seem to be born with this characteristic [16].

Taber discussed gifted education practically from different perspectives in the context of formal secondary school level science education in the UK. He proposed four clusters of characteristics of what he termed, "able science learners:" "scientific curiosity," "cognitive abilities," "metacognitive abilities," and "leadership" [17, p. 9]. Sumida has developed an original behavior checklist that can be used for Japanese primary school children in science classrooms in a non-Western context, including 60 items such as "reports clearly the result of an observation and experiment" and "tries to do things in his/her own way, not according to the instructions given" [18, p. 2103]. As a result of his analysis, three gifted styles in science were identified: "spontaneous style," "expert style," and "solid style."

There are many science-specific programs for gifted children in and out of school from early childhood to high school worldwide. In the Philippines, the Special Science Elementary School Project (SSES Project) was implemented in 2007 in response to the growing importance of science and gifted children and the number of public schools involved in the project increased to 100 in 2010 [19].

In the United States, the Centre for Gifted Education at the College of William and Mary analyzed numerous commercial science programs for gifted children (e.g., "Full Option Science System (FOSS)" and "Great Explorations in Math and Science (GEMS)") and found that very few were appropriately challenging for gifted learners [20]. The Centre developed original problem-based learning (PBL) science units for high-ability learners from Grades 1 to 8 including chemistry (e.g., [21–23]). However, there has been little study about giftedness specific to "chemistry." How many domains should be included in chemistry education? Is it important to consider creative interdisciplinary fields such as physical chemistry and biochemistry as independent domains? New perspectives are needed to discuss its educational implication and practical challenges in gifted education specific to chemistry.

19.3.2
Natural Selection Model of Gifted Students in Science

If a country or district has a formal system of gifted education, it might be possible to screen students formally and systematically, and to differentiate chemistry curriculum for meeting the needs of these scientifically gifted students. It might be also possible to introduce gifted students to informal chemistry education such as chemistry camp or a chemistry Olympiad. We may be able to name this kind of gifted education system, which is institutionally and conceptually organized as the Mode 1 of gifted education.

On the other hand, there may be a different mode of gifted education that is an implicit system, which may be named the Mode 2 of gifted education. For example, as in Japan, students' science achievement and science literacy are very high on international standardized tests, although Japan does not have any formal education system for the gifted [24]. Table 19.3 is a gifted behavioral checklist in science for the Japanese junior high school students revised from the Sumida's checklist for the Japanese elementary school students [18]. Figures 19.2 and 19.3 compare the gifted behavioral scores in different students groups in public junior high schools in a city in Japan using the checklist.

The junior high school students who participated in a special science program at the university were not formally identified as gifted because there is no formal education system for the gifted in Japan. The coordinators accepted their applications on the basis of their willingness to be part of the program and the recommendations from their science teachers. Nevertheless, it is obvious that the students who participated in the program showed higher gifted behavioral scores in all 30 items than those students who participated only in local science events and regular students who never participated in any science events. Between the students who only participated in science events and the students who never participated in any science events, the score for the item 26 of the former is higher than that of the latter. The average score of these 30 items for the students who participated in a special science program at university, the students who participated only in local science events, and the students who never participated in any science events are 3.18, 2.53, and 2.36 respectively.

Table 19.3 Gifted behavioral checklist in science for junior high school students.

Number	Items	Scale
1	Is knowledgeable about science	1-2-3-4
2	Takes care of animals or grows plants according to their ecology	1-2-3-4
3	Collects many natural things and classifies them by their characteristics	1-2-3-4
4	Observes natural events and phenomena in detail	1-2-3-4
5	Becomes persistent in exploring specific topics and tasks	1-2-3-4
6	Makes his/her own modifications and improvements to equipment used in making things	1-2-3-4
7	Comes up with many ideas and answers about a question	1-2-3-4
8	Is dissatisfied with an explanation acceptable easily to others	1-2-3-4
9	Finds patterns in natural events and phenomena	1-2-3-4
10	Shows interest in analysis and explanation using numbers	1-2-3-4
11	Makes persuasive reasoning	1-2-3-4
12	Handles equipment used in an observation and experiment correctly and adeptly	1-2-3-4
13	Express his/her own ideas and findings clearly to others	1-2-3-4
14	Expresses what she/he learned in his/her own words	1-2-3-4
15	Applies patterns and trends discovered in science in other situations	1-2-3-4
16	Understands quickly what she/he has learned in science	1-2-3-4
17	Confident about his/her knowledge and understanding of science	1-2-3-4
18	Links up and inter-relates the various topic in science	1-2-3-4
19	Understands the causal relations of a natural phenomenon	1-2-3-4
20	Retains things studied in science classes in detail for a long time	1-2-3-4
21	Shows different ways of doing and thinking from others without caring	1-2-3-4
22	Tries to do things in his/her own way, not by the instruction	1-2-3-4
23	Dares to make an observation and experiment without fear of failure	1-2-3-4
24	Asks strange questions which puzzle the teacher and parents	1-2-3-4
25	Has interest in things different from peers	1-2-3-4
26	Becomes too absorbed in an observation and experiment to finish the task in time	1-2-3-4
27	Solves science problems in cooperation with others	1-2-3-4
28	Offers a challenge to something new autonomously	1-2-3-4
29	Prefers to challenge uneasy tasks in science	1-2-3-4
30	Recognizes to appreciate others' unique ideas	1-2-3-4

1: Strongly disagree; 2: disagree; 3: agree; and 4: strongly agree Adapted table from Ref.[18]
Reproduce with the permission of IJSE.

Figure 19.3 shows that the students who received any science awards had higher gifted behavior scores for all 30 items than those did not. Especially, the former had higher scores for the items 1, 3, 4, 7, 17, 19, 24, and 25 than the latter did. The average score for these 30 items of the students who received any science award and who did not any were 2.77 and 2.45, respectively.

The data in Figures 19.2 and 19.3 are from the Japanese students in urban public schools. Even in the context of Japan where gifted education is not formally institutionalized, it seems possible to identify gifted students through (i) their motivation in informal special science programs, (ii) the experiences of science contests, and (iii) their behavioral characteristics in science. We would like name this kind of

476 | *19 Chemistry Education for Gifted Learners*

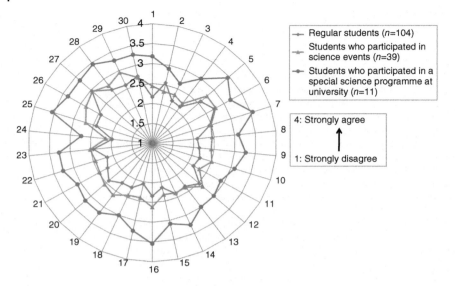

Figure 19.2 A comparison of gifted behavioral scores between different participation groups.

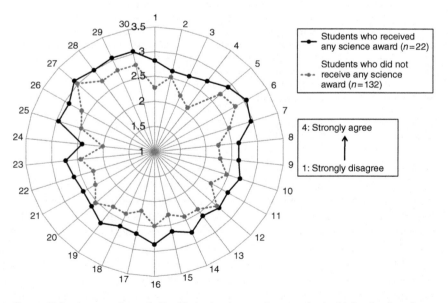

Figure 19.3 A comparison of gifted behavioral scores between junior high school students who received any science award and who did not.

identification of gifted students as a natural selection model of gifted students in science, which is a bottom-up approach to gifted education.

19.4 Curriculum Development and Implementation of Chemistry Education for the Gifted

19.4.1 Acceleration and Enrichment

There are two main forms of gifted education programs: "acceleration" and "enrichment." Through acceleration children are given opportunities to study contents in the upper grade curriculum and gain (extra) credits. In general, acceleration programs match the chemistry curriculum and instruction with the readiness and motivation of students. Skipping grades, special grouping in a specific subject, forms of Advanced Placement (AP) courses provided in some countries, and dual enrollment are examples of acceleration.

Enrichment provides children with opportunities to study interdisciplinary and/or extended content. Forms include personal learning, project learning, center approaches, weekend/vacation programs, and contests. Table 19.4 summarizes the key characteristics of acceleration and enrichment.

Wai *et al.* studied participation among 1467 individuals who had been identified as gifted in mathematics at age 13 in various educational opportunities such as academic competitions, research apprenticeships, academic clubs, summer programs, and accelerated classes [25]. They found that those who had been involved in more of these educational opportunities (a higher "STEM dose" Science, Technology, Engineering, and Mathematics (STEM)) had, at age 33, higher rates of notable accomplishments in STEM, such as earning a Ph.D., writing publications, obtaining patents, or securing an academic career.

In the Ehime University Science Innovation Program, we have developed and implemented the content of Gibbs Free Energy and chemical kinetics as accelerated content for first and second grade gifted junior high school students. In terms of serving as enrichment, the activities overlapped with advanced biochemistry, and students were encouraged to participate in science tournaments and observe a research institution of a company.

Table 19.4 Two forms of gifted education program in science.

Forms	Possible examples of gifted education program
Acceleration	Skipping grade, early entrance, special class, advanced learning in a specific subject, dual enrolment, AP/college credits, differentiation dual enrolment, home tutoring, science club
Enrichment	Personal learning, project learning, learning center approach, Saturday/Summer/Winter science programme/camps, science fair/contest, science olympic, special program in companies, or museum/zoo

19.4.2
Higher Order Thinking and the Worldview of Chemistry

Many children display interest in playing with water, mixing liquids, dissolving solids, and blowing into balloons from an early age. However, it is important in science to bridge the phenomenal world to the world of ideas [26]. In recent years, many new views have been aired on cognitive development, and many of them challenge Piaget's theory on children's acquisition of scientific knowledge and conceptual change in science [27].

The cognitive sequence in a traditional curriculum, which is from sensorimotor, concrete to abstract, may not ignite the potential of gifted children. Gifted children are more able to do abstract scientific thinking in specific domains from an early phase of their development. The traditional science curriculum should be reconstructed from its presupposition for meeting the needs of gifted children. Figure 19.4 compares two models of cognitive sequence in curriculum. We are proposing the Hybrid model to consider the optimum balance of different cognition for each level, be it in school, grade or personal, respectively. Especially for gifted children who show higher levels of thinking than regular children, developing a science curriculum requires combining different levels of cognition, including abstract thinking, to establish a significant intellectual network. Meijer, Bulte, and Pilot propose a unique thinking model to connect between the scale, structures, and properties [28].

To consider different levels of cognition in chemistry, Figure 19.5 shows an example of four different levels of cognition and its cycle of learning. The four stages of cognition are "Phenomena," "Descriptive Models," "Extensive Models," and "Mathematical Models." They are not hierarchical but heterarchical and interrelated. The cycle can be spiral to elaborate learning. And the agent and presupposition of the cycle and cognition is the worldview of chemistry that is a kind of epistemological thinking. Sharing the worldview, "Natural phenomena can be described by the property, composition, and conditions of the most basic elements," children's thinking could be elaborated.

The appropriate equipment, including Information and Communications Technology (ICT), must be effective in enhancing gifted children's thinking. There are many technology options such as computer programming language, simulations and WebQuests, Ask-an-expert, Telementoring, Robotics, Web Page creation on

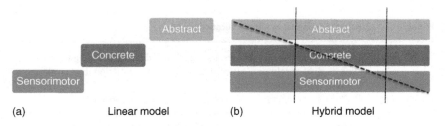

Figure 19.4 (a,b) Two models of cognitive sequence.

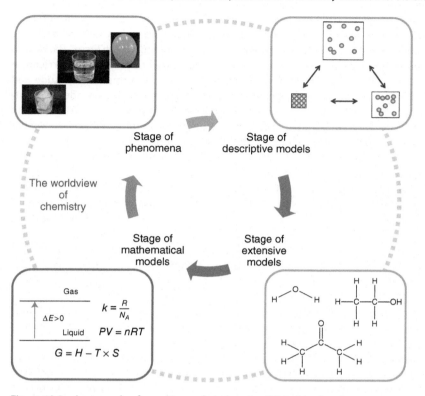

Figure 19.5 An example of cognitive cycle in learning "Water" in chemistry.

topic of interest, Service learning option, and many general options such as Computer assisted instruction, Distance learning, Discussion boards and chats, Virtual field trips [29]. Even a simple molecular model kit is now very useful and reasonable even for young children to study the structure of a simple matter. Although spatial cognition is difficult for adults, playing a molecular model kit has no difference for children who play Lego. Once the children share the worldview of chemistry such as "the structure of matter influences its property and reaction surrounding us," children will further be absorbed in chemistry thinking and be able to realize its connection with their everyday life. In fact, a gifted Japanese primary child developed a card game on the property of elements and its reactions, and established an enterprise (He is the president of the company).

19.4.3
Promoting Creativity and Innovation

Li and Kaufman introduced the two primary elements common to most explicit definitions of creativity as "originality (also known as *novelty*)" and "task-appropriateness (sometimes called *usefulness*)" [30]. Chemistry education must offer gifted students opportunities to develop their potential into creative performance and achievement. Likewise, schools can provide and facilitate various activities to enhance creativity of their students. VanTassel Baska and Stambaugh propose nine options for the application of creativity to curriculum and instruction especially for gifted students; creative expression, aptitude and interest matches, links to the professions, reading of biographies, academic counseling, multiple options and outlets for creativity, emphasis on metacognition, open-ended activities and approaches, and emphasis on targeted extracurricular options [31]. Reghetto highlights the needs for more content specific insights that conceptualize the teaching of creativity and academic subject matter with respect to classroom teaching [32].

Chemistry curriculum and instruction for gifted students should include activities, which let students express their own ideas, ways of doing, performance, and achievement to share with others. Paik recounts the characteristics of Pablo Picasso, Thomas Edison, and Johann Sebastian Bach, and proposes the idea of creativity and productive giftedness [33]. Making things in science include creative thinking and innovation, and integrated everything to everyday life as well.

As Paik suggests [34], it is effective to develop and instill the importance of productive giftedness in creative and talented individuals, especially in the early years. Figure 19.6 shows the science notebook of a gifted kindergarten child at Ehime University Kids Academy. The Academy has implemented a project-based science program for 5–8 year old gifted children at Ehime University.

During the project "Water," the children produced a variety of original opinions in an activity when they were asked to identify various forms of water they encountered in their daily lives. In the worksheet, a first-grader listed more than 20 different forms of water, including rivers, toilets, tears, baths, lakes, pools, running noses, and sweat. The child divided these different forms of water into clean and dirty water and was able to explain and share how she determined which was which. At the end of the project, the children created a filtering device. They came up with the materials, amounts, and procedures on their own and tested the results of their arrangements. They used pictures as evidence and explained how they arranged their devices and what the results were. A K-grade child disassembled the water filter in her yard at home after the program and systematically performed additional experiment. She tried to improve its performance by changing how it was put together and adding different materials, and recorded the process and results in detail (Figure 19.6).

19.4.4
Studying Beyond the Classrooms

In the context of heterogeneous groups in science classrooms, significant changes in curriculum, and instruction such as differentiation strategies(e.g., [35–37]) and parallel curriculum (e.g., [38, 39]) have been attempted to meet diverse needs of students. Matthews summarizes issues related to science education in school and ways to resolve them, and introduces informal science education

(a)
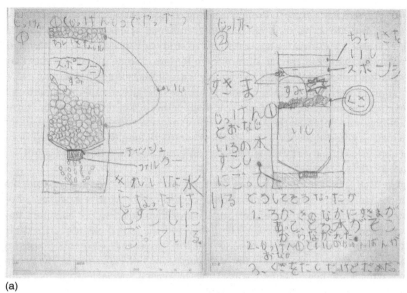

(b)
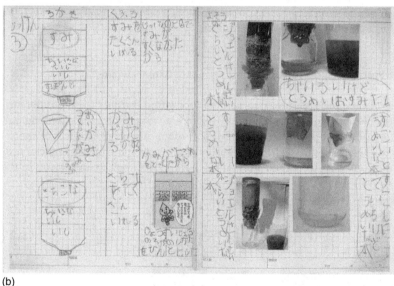

Figure 19.6 (a,b) Science notebook of a gifted kindergarten child at Kids Academy.

for gifted students such as dual enrollment, science club, summer program, and science fair/contest [40]. Cambell and Walberg found out that 52% of adult Science Olympians in the US earned doctorates and 345 Olympians produced a total of 8629 publications including 132 publications in chemistry [41]. The 46th International Chemistry Olympiad will be held on 20–29 July 2014 in Vietnam (*http://icho2014.hus.edu.vn/*). Project learning to solve a real life issue in the local context is also attractive to gifted students. Informal science education plays a significant role as it provides opportunities to study advanced science with other students who exhibit characteristics of the giftedness. Some gifted students may show his/her full potentials only in working with similar gifted students.

As another example, science contests can provide gifted students a sense of achievement, confidence, self-efficacy, and opportunities to reflect on their learning, abilities, and potential through preparation process and discussion with other participating gifted students. It will also be meaningful for gifted students to establish a network with other gifted students in different classes, school, and countries (Figure 19.7).

Figure 19.8 shows a special lecture for the gifted junior highs school students of Ehime University Science Innovation Program by Dr. Morita Kosuke, Laboratory Head of Research Group for Superheavy Element, RIKEN Nishina Centre for Accelerator-Based Science. He discovered the 113rd element. His study on the physical and chemical properties of the superheavy elements possesses the advanced science innovation and the persistence of experiment. His group observed the super heavy element three times in 9 years. Special lectures by a leading scientist could contribute to establishing a sound image of science/scientist; leading scientists can also play as good role models who can inspire students to be actively involved in a scientific community.

19.4.5
Can the Special Science Program Meet the Needs of Gifted Students?

In general, an education program designed for gifted learners tries to differentiate curriculum for the gifted students in different ways, for example, pulling out, differentiation, and parallel curriculum. This often obscures the effect of special program specific to gifted students, and how different the effects of special program are from that of regular program. Figure 19.9 shows a comparison of the

Figure 19.7 The national science contest.

Figure 19.8 A special lecture by Dr. Morita.

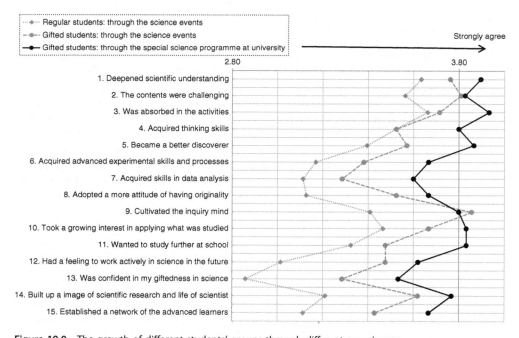

Figure 19.9 The growth of different students' groups through different experiences.

average self-evaluation score of the participating students through two different experiences at local science events and a special science program at university.

As shown in Figure 19.2, the gifted junior high school students participated both in the Ehime University Science Innovation Program and the local science events. The local science events were planned and implemented for junior high students from all schools by science teachers in public junior high schools in the city. The main purpose of the events is to provide junior high school students experiences in science observation/experiment, and making things and encountering advanced science and technology, to enhance their interest in science and technology, and develop basic knowledge/understanding/skills in science and technology. The events also aim to enhance friendship among students in different schools.

Moreover, the Ehime University Science Innovation Program targets gifted students who participated in science events planned and implemented by university professors in the weekend at a university. The program has four purposes; (i) raising a challenging question, (ii) working persistently from multi-perspectives, (iii) deepening understanding collaboratively, and (iv) communicating results of the study and its significance. The purpose of the program was also to nurture the next generation of leading scientist who can challenge scientific innovation.

Comparing the effect of science events for all students and the effect of special science program for the gifted will help us to consider the characteristics of gifted students and the educational implications of different science education. We conducted a survey about the growth of students in different abilities both through local science events for all and special science program for the gifted. In the comparison, the regular students participated in the local science events, and the gifted students participated in both the local science events and the special science program (Ehime University Innovation Program). Both student group members self-evaluated their growth in abilities through their own experience. All the gifted students who participated in the special science program participated in local science events with the regular students too. The local science events were mainly conducted by science teachers in the city; the number and contents of local science events that the regular students and the gifted students experienced were all the same. The result of the comparison is summarized in Figure 19.9.

Figure 19.9 clearly shows that the gifted students experienced more personal/scientific more growth than the regular students even in the local science events for all junior high students. The gifted students felt more significant growth through the special science program for the gifted than through the local science events for all. Especially, the science innovation program seemed attractive and meaningful to the gifted students for acquiring thinking skills and being better in discovery. The special science program such as the Ehime University Science Innovation Program has a more significant effect on nurturing different aspects of knowledge, attitudes, and skills in science than popular science events for all students. Developing a gifted science program must be sensitive to characteristics of gifted students, and the purpose of the program should be clearly considered and identified.

19.5
Conclusions

Chemistry has been changing dramatically since the twentieth century. The enterprise has been diverse, complex, and involves international collaboration.

The development of chemistry carries with its important implications for the quality of life. Chemical knowledge has had significant implications in medicine, agriculture, engineering, physics, and biology. Many new substances and materials have been produced through chemistry. However, these characteristics of modern chemistry do not seem to be reflected in education, although many children show a strong interest in natural phenomena from an early age and demonstrate an outstanding ability to think creatively and in abstract terms. Young children also love to operate science equipment and acquire skills of experiment and a modeling kit and ICT could enhance their chemistry learning.

The domain-specific and dynamic nature of science, with its encompassing wealth of different fields of study, can accommodate children's varied areas of interest and for this reason makes science learning ideal for the development of a checklist for giftedness. Hands-on activities in chemistry learning can promote creative ideas and develop a persistence in children that often surpasses their teachers' expectations. The physical and creative intellectual activity that chemistry offers – the opportunity to observe, experiment, and make things – also provides an appropriate context for teachers to find out children's potentials, and for children to show their giftedness. The skills of recording observations, interpreting experimental data, and making models in chemistry can potentially deepen and extend the learning process into other areas.

Among those scientists who have made their mark on history, quite a few are known to have lived with not only outstanding talent and brilliance but also some kind of learning difficulty [42]. Even when students are identified as gifted, it is inaccurate to view each of them as a perfect student who will demonstrate excellence in every field. Exceptional students will also need support for their unique socio-emotional development. In reality, major issues being faced by gifted educators are problems like students' loss of self-confidence, the pressures related to the perfectionism typical of gifted students, and the underachievement of gifted students in areas outside their interests. An appropriate balance is needed between a focus on acquiring knowledge efficiently and developing skills appropriately within the expected context and on demonstrating independence and collaborative creativity. Belardo and Sumida [43] has developed a program for gifted female high school students for balancing the socio-emotional and intellectual needs in a Philippine setting. This program also emphasizes that the cultural context in which the children are identified as gifted should not be ignored.

It is imperative to provide opportunities where all children can demonstrate their giftedness. Support for highly gifted social minorities is an important issue in the world. Opportunities need to be created where all children can receive high quality education from early years and develop their giftedness throughout their lifetime. Chemistry encompasses collaborative learning activities in the laboratory and in the field. Chemistry endeavor is an international team effort, with scientists working within self-established norms and sharing basic attitudes and ways of thinking. The development of chemistry curriculum and teaching materials that accommodate special needs of gifted children and the implementation of related teaching methods and assessment are relevant to all teaching subjects, school types, and education in general, and can be used in the educational activities of communities and global societies as well.

References

1. Mead, M. and Metraux, R. (1957) *Science*, **126**, 384–390.
2. Schibeci, R.A. (1986) *Sci. Educ.*, **70**, 139–149.
3. Sjoberg, S. (2000) *Science and Scientist: The SAS-Study*, University of Oslo.
4. Scerri, E.R. and McIntyre, L. (1997) *Synthesis*, **111**, 213–232.
5. Paul, R. and Elder, L. (2006) *A Miniature Guide for Students and Faculty to Scientific Thinking*, Foundation for Critical Thinking, Dillon Beach, CA.
6. Malmstrom, B.G. and Andersson, B. (2001) in *The Nobel Prize: The First 100 Years* (eds A.W. Lavinovitz and N. Ringertz), Imperial College Press, London, pp. 73–110.
7. ibid.
8. Gilbert, J.K. and Treagust, D. (eds) (2009) *Multiple Representations in Chemical Education*, Springer, Milton Keynes.
9. Holyoak, K.J. and Thagard, P. (1996) *Mental Leaps: Analogy in Creative Thought*, MIP Press, Cambridge, MA.
10. Golinski, J. (2003) in *The Cambridge History of Science*, Eighteenth-century Science, vol. **4** (ed. R. Porter), Cambridge University Press, Cambridge, pp. 375–396.
11. ibid.
12. Larsson, U. (2001) *Culture of Creativity: The Centennial Exhibition of the Nobel Prize*, Science History Publications.
13. Wellman, H. and Gelman, S. (1992) *Annu. Rev. Psychol.*, **43**, 337–375.
14. Hirschfeld, L.A. and Gelman, S.A. (1994) *Mapping the Mind: Domain Specificity in Cognition and Culture*, Cambridge University Press, New York.
15. Gopnik, A., Meltzoff, A.N., and Kuhl, P.K. (1999) *The Scientist in the Crib: What Early Learning Tells Us about The Mind*, HarperCollins Publishers Inc., New York.
16. Gopnick, A. (2009) *The Philosophical Baby*, Picador, New York.
17. Taber, K.S. (2007) *Science Education for Gifted Learners*, Routledge, Oxon.
18. Sumida, M. (2010) *Int. J. Sci. Educ.*, **32**, 2097–2111.
19. Faustino, J. and Hiwatig, A.D.F. (2012) *J. Sci. Educ. Jpn.*, **36** (2), 131–141.
20. Johnson, D.T., Boyce, L.N., and VanTassel-Baska, J. (1995) *Gifted Child Q.*, **39** (1), 36–44.
21. The College of William and Mary School of Education Center for Gifted Education (2008) *Water Works*, Prufrock Press, Waco, TX.
22. The College of William and Mary School of Education Center for Gifted Education (2008) *What's the Matter?*, Prufrock Press, Waco, TX.
23. The College of William and Mary School of Education Center for Gifted Education (2007) *Acid, Acid Everywhere: Exploring Chemical, Ecological, and Transportation Systems*, 2nd edn, Kendall Hunt Publishing Company, Dubuque, IA.
24. Sumida, M. (2013) *J. Educ. Gifted*, **36**, 277–289.
25. Wai, J., Lubinski, D., Benbow, C.P., and Steiger, J.H. (2010) *J. Educ. Psychol.*, **104**, 860–871.
26. Sumida, M. (2012) in *Debates on Early Childhood Policies and Practices: Global Snapshots of Pedagogical Thinking and Encounters* (ed. T. Papatheodorou), Routledge, Oxon, pp. 123–135.
27. Metz, K. (1997) *Rev. Educ. Res.*, **67**, 151–163.
28. Meijer, M.R., Bulte, A.M.W., and Pilot, A. (2009) in *Multiple Representations in Chemical Education* (eds J.K. Gilbert and D. Treagust), Springer, Keynes, pp. 195–213.
29. VanTassel-Baska, J. and Stembaugh, T. (2006) *Comprehensive Curriculum*

for Gifted Learners, 3rd edn, Pearson Education Inc, Boston, MA.
30. Li, Q. and Kaufman, J.C. (2014) in *Critical Issues and Practices in Gifted Education*, 2nd edn (eds J.A. Plucker and C.M. Callahan), Prufrock Press, Waco, TX, pp. 173–182.
31. VanTassel-Baska, J. and Stembaugh, T. (2006) *Comprehensive Curriculum for Gifted Learners*, 3rd edn, Pearson Education Inc., Boston, MA.
32. Beghetto, R.A. (2014) Creativity-development and enhancement, in *Critical Issues and Practices in Gifted Education*, 2nd edn (eds J.A. Plucker and C.M. Callahan), Prufrock Press, Waco, TX, pp. 183–196.
33. Paik, S.J. (2013) in *Creatively Gifted Students are Not Like Other Gifted Students* (eds K.H. Kim, J.C. Kaufman, J. Baer, and B. Sriraman), Sense Publishers, Rotterdam, pp. 101–119.
34. ibid.
35. Heacox, D. (2002) *Differentiating Instruction in the Regular Classroom*, Free Spirit Publishing Inc., Minneapolis, MN.
36. Westphal, L.E. (2007) *Differentiating Instruction with Menus: Science*, Prufrock Press, Waco, TX.
37. Faustino, J. (2013) *Harris J. Educ.*, **1** (1), 32–46.
38. Tomlinsonm, C.A., Kaplan, S.N., Renzulli, J.S., Purcell, J., Leppien, J., and Burns, D. (2002) *The Parallel Curriculum: A Design to Develop High Potential and Challenge High-Ability Learners*, Corwin Press, Thousand Oaks, CA.
39. Faustino, J., Sumida, M., Fajardo, A., and Pawilen, G. (2010) *Bull. Center Educ. Educ. Res. Faculty Educ. Ehime Univ.*, **28**, 51–65.
40. Matthews, M. (2006) *Encouraging Young Child's Science Talent: The Involved Parents' Guide*, Prufrock Press, Waco, TX.
41. Campbell, J.R. and Walberg, H.J. (2011) *Roeper Rev.*, **33**, 8–17.
42. West, T.G. (1991) *In the Mind's Eye: Visual Thinkers, Gifted People with Learning Difficulties, Computer Images, and the Ironies of Creativity*, Prometheus Books, New York.
43. Belardo, F.C. and Sumida, M. (2010) *Bull. Center Educ. Educ. Res. Faculty Educ. Ehime Univ.*, **57**, 139–146.

20
Experimental Experience Through Project-Based Learning
Jens Josephsen and Søren Hvidt

In "Chemistry Education: Best practice, Innovative Strategies and New Technologies"
Javier Garcia-Martinez and Elena Serrano-Torregrosa, Eds.

20.1
Teaching Experimental Experience

20.1.1
Practical Work in Chemistry Education

Chemistry as a scientific discipline developed from practical and experimental endeavors, and practical work [1] at the laboratory bench is at the very heart of chemistry. Laboratory training is therefore a compulsory part of any contemporary bachelor or master program in chemistry. This is recognized by international and national bodies and laid down in legislation, rules, and recommendations. In 2010 The European Union adopted *A Framework for Qualifications of the European Higher Education Area* [2]. Outcomes have been formulated for the first cycle, the bachelor level, in five qualifications [2], one of which states that (any) bachelor should have "the ability to gather and interpret relevant data (usually within their field of study) ... " Of course, the handling of data is highly relevant in chemistry and although the amount of data available in the literature is immense, many problems at this level typically require new data generated in the laboratory by a professional chemist.

Also the Chemistry "Eurobachelor" [3] "has built up practical skills in chemistry during laboratory courses...." and "has the ability to gather and interpret relevant data...." cf. the formulation above. In addition it is stated [3] that "practical courses must continue to play an important role in university chemistry education...."

Chemistry Education: Best Practices, Opportunities and Trends, First Edition.
Edited by Javier García-Martínez and Elena Serrano-Torregrosa.
© 2015 Wiley-VCH Verlag GmbH & Co. KGaA. Published 2015 by Wiley-VCH Verlag GmbH & Co. KGaA.

In Britain, the Royal Society of Chemistry (RSC) further points out that "Practical skills" should come from at least 300 time-tabled hours in the lab [4]. The practical skills to be obtained through this effort are specified in the *Subject benchmark statement* for chemistry [5], and include the handling of safety-hazards aspects, conduct of documented laboratory procedures, monitoring, and recording, operation of standard instruments, and dealing with accuracy of own experimental data.

In papers from the American Chemical Society (ACS) [6], the certified curriculum also quantifies the laboratory experience to a comparable extent (400 laboratory hours), and it is stated that with regard to problem-solving skills: "students should use appropriate laboratory skills and instrumentation to solve problems, while understanding the fundamental uncertainties in experimental measurements." In addition necessary laboratory safety skills are described in some detail. Also the distinction between hands-on activities and simulations is emphasized by ACS [7].

20.1.2
Why Practical Work in Chemistry Education?

This concurrent demand for laboratory work in chemistry education is in accord with the tenets of the worldwide chemistry community, but it is not necessarily clear why it is worthwhile to take the effort, time, and cost to include laboratory work in the curriculum. Several papers point at the lack of clear evidence for what can actually be achieved from laboratory work [8–12]. It therefore seems relevant to deal with the following question: "Currently we have laboratory work in the curricula per tradition. Are there any evidence-based arguments for keeping it there at all?" Such arguments should make it clear that laboratory work serves important purposes, which would be missing if it were absent. In addition, if laboratory work is proved to be an important part of the curriculum, how much laboratory work is needed?

The first part of the question is: Can the presence of laboratory work in a chemistry program be justified?

While the arguments depend on the level of teaching, there is an increasing understanding that different types of practical work may play different roles in supporting educational goals and that not all goals can be achieved by all kinds of practical work [13–16]. We have to be more specific about which skills laboratory work should develop and devise practical work that gives the opportunity to develop these skills.

Dealing with school science in general, Jenkins [17] presents two arguments for practical work, which also seem valid for the tertiary level. The first argument is, "that only work at the bench ... can give ... a feel for phenomena, a building up of experience about the natural world ... " and the second one is, that "some practical activities ... help students to understand how difficult it is to obtain secure knowledge about the natural world." The "feel for phenomena" reason which has previously been formulated as experiences [13] seems unquestioned and has for some time been seen as a valid justification [9, 13, 14, 16] for practical work in science education. The second argument about the nature of "gathering data" is perhaps a little vague (or complex) when it comes to tertiary education, although it is in line with the above ACS 2008-formulation of the problem-solving skills [6]. It has recently been announced, that ACS expects to increase the emphasis on student skill development including the problem-solving skills as part of the coming revision of the certification guidelines [18]. A sharper reformulation separates the argumentation into two parts. The first part refers to the exercise-type [13] of practical work, where the goal is to develop practical skills (if not to become a professional) such as how a given experimental procedure functions in the learner's hands or how a piece of equipment or an instrument is used. The second part of the argument refers to the investigation-type [13] of practical work. The goal is to learn academic problem solving [14] or the processes of scientific inquiry with its several stages and a series of skills including some skills in the laboratory [19] where data are obtained (as an example, think of experimental design). Still the "understand[ing of] how difficult it is [and how] to obtain secure knowledge...." [17] is generally recognized as valid justification of practical work in science (chemistry) education [9, 13, 14, 16] and the original terminology *exercises, investigations,* and *experiences* [13] for different types of practical work seems both valid and useful.

When it comes to the volume of practical work in the laboratory, ACS, and RSC have dared to set a lower limit of laboratory time to 400 or 300 scheduled hours. It has been questioned if a minimum number of laboratory hours *per se* ensures relevant learning [20]. In spite of this, it seems to be a rather modest requirement when considering that it only represents a small fraction of the total 5000 study hours during a three-year (180 ECTS) program. Other institutions dealing with accreditation will, of course, also consider and evaluate whether or not the extent of laboratory work in the entire program is a sufficiently important part of the curriculum.

20.1.3
Practical Work in the Laboratory

The questioning of practical work in chemistry teaching also comes from disappointing experiences. We might have known dissatisfaction when laboratory time seemed to be a waste of time for certain students. And we might sometimes have had the feeling that the efforts in terms of staff and of equipment and chemicals expenses did not appear to match the students' outcome. Such experience has been inspiring and led to a more close analysis of discrepancies between what we think and hope students learn from practical work and what they actually learn [21] as outlined above.

The information technology age also offers opportunities to optimize the teaching and has led to the development of IT-based teaching material dealing with experimental chemistry. The demonstration of certain experiments and the use of equipment on film were early examples. At present they are replaced by a great variety of instruction videos, presentation, and animation tools, isolated or on-line simulation packages and textbook companion web sites containing interactive data bases and graphs, multiple-choice puzzles, animated molecular structures, and so on. This type of material might be very helpful for different learning purposes, but are not suited to give experiences from practical work *in the laboratory* as outlined above. Thus ACS stresses the "Importance of Hands-on Laboratory Activities" [7]. Simulations of experiments in the laboratory may very well be useful in reinforcing outcomes from the laboratory as pre-labs or post-labs (elaborated in Section 20.3.2). However, such representations of practical activities cannot replace the real thing: hands-on. While the authors acknowledge the benefits and potentials of simulations and other IT-based learning materials and support its use as a real opportunity to make the time in the laboratory more effective [22, 23], this presentation will focus on hands-on activities.

Most of the above arguments are qualitative in nature, and they have to be elaborated and specified further in order to justify the effort it takes to have practical work in the laboratory and, most important, in order to be informative to curriculum reform and to teaching. Different types of laboratory work definitely serve different purposes.

20.2
Instruction Styles

20.2.1
Different Goals and Instruction Styles for Practical Work

The three terms: exercises, investigations, and experiences [13], point at quite different goals for practical work as has been discussed [9–11, 13–17], and they depend on different instruction styles, and, in turn, a given laboratory instruction style paves the road to a specific goal (or specific goals) while it blocks the route to others. Domin [24] has identified four instruction styles characterized by three descriptors, each of which has two positions. These descriptors are the intended outcome (which may be predetermined or undetermined), the approach (which may be deductive or inductive), and the procedure (which may be given or student-generated). The most common style (where the instructor exerts a high degree of control) is expository with given procedures and outcomes and use of a deductive approach. This style may be very efficient for obtaining certain goals, especially when the students' actual learning turns out to be in accord with the intended goals. This is, however, not necessarily the case, which indicates the need for a careful analysis and declaration of the intended goals and a clever design of the laboratory exercises to pursue the goals [21]. The expository style is somewhat cookbook-like and makes the students follow a recipe. This situation is suited for supporting manipulative skills, for example, demonstrating how to perform a standard procedure and having the students practice it. One feature, which is definitely missing here, is experimental design [11, 15, 16, 19, 20, 24]. This possible goal is in focus in the problem-based [24] style. While the problem to be solved is given, the procedures with which a solution can be found are the responsibility of the students. Here the instructor acts as a facilitator and may help students mostly on request and eventually confirm their design to be adequate for the purpose. Still the problem-solving laboratory presupposes that the students apply their understanding. In that situation they practice Higher Order Cognitive Skills, HOCSs [25], most often based upon their prior mastering of some Lower Order Cognitive Skills, LOCSs. In a study of the effects of the introduction of a whole problem-based learning, PBL, laboratory module [26] instead of the traditional laboratory course, the students generally turned out to be very positive about this set up, but those having difficulties doing calculations had a hard time. Also, those students with only a minor background in chemistry had to struggle harder because of their weak LOCS.

The problem-based instruction in the laboratory is a special case of the radical PBL programs which are found especially in teaching medical students. In a meta-analysis of the efficiency of this approach [27], 40 studies of problem-based, mainly medical programs, were reviewed. The analysis measures a possible advantage of PBL-programs over conventional programs in three measurable areas of understanding. The first area, the understanding of concepts, did not show any difference between the two types of programs, but the second area, the understanding of principles that link concepts, was significantly better for the PBL-students. A slight advantage in favor of the PBL approach was suggested with respect to mastering the linking of concept and principles to conditions and procedures, the third area.

20.2.2
Emphasis on Inquiry

The two inductive instruction styles [24] are inquiry-types. The discovery (guided inquiry) instructor knows the intended outcome and devises the method to obtain data. Here the idea is that the student discovers a principle based on the interpretation of own experimental data. In the more advanced form of (open-) inquiry, an area of (scientific) concern is used as the starting point of a (longer) process, where neither the outcome nor the methods are given. It appears that the processes of scientific inquiry or experience with experimentation (including thinking taking place outside the laboratory) are considered very important qualifications to aim for [11, 14–16, 20, 24]. Accordingly, elements of inquiry are gradually introduced and welcomed, for example, during or at the end of an expository-style laboratory course [15, 20]. The expression "Problem-based learning" in its most literal meaning suggests that we start with a problem, and, while we solve the problem, learning takes place. Another popular formulation is "learning by doing." Following the above taxonomy [24] the instructor gives the problem and knows the answer, while students select an appropriate method to obtain the relevant data in the process. However, the term "problem" is biased and is used for many things in education. In a textbook solved problems and end-of-chapter problems serve two rather different purposes, while "problem solving skills" often refers to more complex qualifications and "problem-oriented studies" is relevant on the curriculum level. Recently the term "inquiry" (e.g., in Inquiry Based Learning, IBL, or Inquiry Based Science Education, IBSE) re-emerges in the science education literature [28–31]. In relation to practical work, an alternative way to characterize student work in the laboratory in terms of "levels of inquiry" has been suggested [32]. In short, three elements "Problem/Question," "Procedure/Method," and "Solution" are either "Provided to student" or to be "Constructed by student." For level 0 experiments all three elements are provided to student, while the "Solution" is only intended to be constructed by the student at level 1. Level 2 only provides the "Problem/Question," while only a "raw" phenomena description is provided to the student at level 3. Using this model, 27 literature or commercially available undergraduate laboratory experiment manuals in general and organic chemistry were analyzed. Although two thirds of the experiments [32] belonged to level 0 or 1, as much as one out of eight manuals were at level 3. This study did not include an analysis of the different teacher roles which best support the development of student learning at each of the four levels, but they are obviously different [24].

20.3
Developments in Teaching

20.3.1
Developments at the Upper Secondary Level

The importance of an active, teaching teacher is strongly advocated [33] instead of less guided instruction styles. The discussions and studies of upper secondary school science teaching are extensive, which might be a consequence of the amount of upper secondary science teaching compared to that at the tertiary level. The harsh rejection [33] of "minimal guided instruction" was opposed [34] by stressing the important role of the teacher in the scaffolding process in problem-based and inquiry learning. The main criticism [33] is that there is a lack of convincing evidence for the benefits of alternative instruction methods compared to the conventional expository instruction [24] and that there are a lot of other arguments for the need for an active teacher involvement. Many studies have tried to show effects of specified changes in teaching toward IBSE. Three meta-analyses of research results of the possible benefits of IBSE teaching (at the secondary level) have appeared lately [35–37]. In short, the first study [35] "indicates that, having students actively think about and participate in the investigation process increases their conceptual learning. Additionally, hands-on experiences with scientific or natural phenomena were also found to be associated with increased conceptual learning." However, no evidence was found [35], that an increased amount of inquiry in instruction gave more learning outcomes for the students. As is also stated, "these findings are consistent with what constructivist learning theory would predict – active construction of knowledge is necessary for understanding..." This constructivist approach was embedded in a recent study reporting the experience with practical activities to be completed during a 2-h period in teams of three collaborating A-level students [38]. The design had deliberately two aims. One was to prepare the students for practical activities at university. The second aim was to challenge their thinking to a point where they got "stuck," a situation "where they are about to learn something new." The second meta-analysis [36] had two foci. First, the way the degree of teacher involvement in IBSE instruction (compared to traditional instruction) influences the positive learning effect has direct relevance to the apparent controversy about this [33, 34]. The second focus is what type of cognitive activity is included in the teaching. Twenty-two reports of studies of experimental teaching were listed according to the involvement of one or more of four domains of inquiry in the experiment. The four domains were: (i) procedural (those normally included when describing the processes of scientific inquiry, e.g., experimental design); (ii) epistemic (e.g., drawing conclusion based on evidence, reflection on the nature of science); (iii) conceptual (e.g., drawing on prior knowledge); and (iv) social (e.g., working collaboratively). The largest positive effect in learning compared with conventional teaching was measured when the instructional strategy included procedural, epistemic, and social activities. Although IBSE strategies are beneficial, strategies other than IBSE seem more effective in school science teaching. The review of experimental teaching strategies in US identified eight strategies, including IBSE as one of them [37]. Context-introducing and cooperative learning strategies were, however, found to have the highest impact on student performance, while experience with hands-on and IBSE strategies also had a positive effect. Overall from the practical work perspective, the results indicate that teacher supported hands-on activities should be combined with a number of cognitive and social activities.

As noted earlier [16] thinking with regard to school science teaching and learning is rather extensive and knowledge accumulates about what works and what does not. At the tertiary level there are for sure a number of lessons to be learned from the activities at the upper secondary level.

20.3.2
Trials and Changes at the Tertiary Level

During the last 50 years we have witnessed a transition from elite to mass universities, and from time to time, when teaching has proven ineffective, developments in school science teaching have been modified and implemented at universities. The expository style of instruction through lectures (discourse before audience or class on given subject [39]) and laboratory exercises have been challenged and new ways are gradually being introduced. Thus in relation to practical work (in the curriculum), inquiry elements are gradually being appreciated by students and recommended by course managers and researchers in chemical education. Different strategies have been used, for example, introduction of open-ended tasks during or at the end of an expository-style laboratory course [15, 20, 40, 41], sometimes called mini-projects [42], or a "less organized environment" [43]. In such cases the response from the students is almost always positive, and as such a motivating factor, the importance of which should not be neglected. In the mini-project case [42] the instructors met with these first year students later in their studies and had the impression that they approached their third year project more easily. In the "less organized environment" laboratory course the students developed self-confidence by collaborating in groups and they performed satisfactorily at the end of the course. Process-Oriented, Guided-Inquiry Learning (POGIL) is another approach based on "A Philosophical and Pedagogical Basis" [29] and introduced in a college general chemistry introductory course [30]. The rather radical change away from traditional lecturing and "cookbook" laboratories to the use of well-organized group work, critical thinking questions, question of the day (chemical properties question connected to the laboratory work) have been appreciated by students and may have a different effect on retention, grades, and so on, depending on the extent to which POGIL-elements are introduced [44]. Changes from established routines always take time, but it is of greater concern whether changes in teaching strategies implies more work for the teaching staff on the daily basis or not. This aspect was addressed for a first year chemistry course for students who were not science majors [45]. Rather than thinking of new laboratory tasks the written instruction was changed from "cookbook" to a lab manual, dealing with the same laboratory activities as earlier, but only giving general procedural guidelines. Pre-laboratory assignments [46] (a pre-lab [15, 29, 42, 46]) were a means to make students prepare for the laboratory. A day before the pre-lab session students had to hand in a report on the pre-lab assignment, including the development of the laboratory procedures in detail (to a "personal recipe"). Details were adjusted during the pre-lab discussion in class. This set up did not take more staff time, but forced students to think and enabled them to practice experimental design. An instruction has also been reported [47], where the preparation for the laboratory was switched during the course to become a post laboratory activity (a kind of post-lab [48]), where the laboratory "experiments" were explained in full.

An additional aspect influencing the students' cognitive development is to which degree the intended learning outcomes in terms of other skills (such as communication and team working skills, and skills relating to the processes of scientific inquiry [11, 19, 20, 28]) are valued and assessed. In advocating the development of open-ended problem solving skills this aspect was addressed precisely: "If academics value these skills, they must be properly addressed within the curriculum and rewarded within the assessment system" [49]. Fair approaches to this aspect have been given [50] in which laboratory diaries were marked, in addition to the laboratory reports. Also the presentation of the obtained results of the laboratory work was assessed and contributed to the final grade.

20.3.3
Lessons Learned

Increasing attention has been given to the quality of instruction at the bachelor level. We have passed the position expressed by Edward Gibbon [51] commenting on the lack of success that the Roman emperor Marcus Aurelius had with his son Commodus " … the power of instruction is seldom of much efficacy, except in those happy dispositions where it is almost superfluous." We are aware that during the freshman year we cannot rely on the assumption that the instruction *per se* leads to a metamorphosis of the freshman student from a school student (a pupil to teach) to a university student (a self-paced, determined person, who knows how to study independently). Instruction matters. In summary, evidence emerges supporting the belief that we could help students in their cognitive development and to maximize the gain from practical work (practicals, experimental work, laboratory work, experiments, mini-projects, etc.) by taking a number of factors into account:

- By formulating the practical work within a *context*, students may realize that the learning activity has some *relevance* in real life outside the chemistry laboratory or the lecture theater. If students feel *ownership* to the assignment *motivation* and engagement may very well increase [11, 26, 52–54].
- The declaration of *clear aims* of practical work [11, 55] helps students to focus on the possible learning outcomes.
- Pre-laboratory activities (*pre-labs*) outside the laboratory help students to prepare for the laboratory and cut the cognitive load in the laboratory [11, 15, 20, 26, 42, 46]. These activities are part of the inquiry processes to be developed in and outside the laboratory.
- By having students work on open-ended investigations a number of *inquiry processes*, including experimental design [15, 16, 19, 20, 26] are needed and thus developed. Having students *think* [35, 38, 46] is the short formulation. In some cases students get "stuck" [38, 56], which in the less formal laboratory environment is a favorable basis for thinking and "learning something new" [38].
- Post-laboratory activities (*post-labs*) [11, 15, 20, 22, 48, 57] are very important and may take the form of writing a *report* on the activity according to a template. Such a report may include important *observations* experienced in the laboratory, a discussion of the *validity of data* obtained, the formulation of a *conclusion from experimental results*, and answers to other questions relating to procedures and methods applied in the laboratory.
- *Fair and declared assessment* of performance in the laboratory [26, 49, 50] may help students to give the work in the laboratory enough attention and time to learn something from it rather than just having completed the task and having their report accepted.
- Having students work in pairs [22, 36, 37, 43] or in *small teams* help students to discuss and develop their understanding and to develop transferable skills such as team work and communication.

20.4
New Insight and Implementation

20.4.1
Curriculum Reform and Experimental Experience

While each of the above measures can be, and have been, implemented more or less independently in given courses, most, or all of them, may be elements of a sweeping reform. This is most often within reach of the course coordinator in charge, who should have a good case in believing that improvements will be seen when measuring student satisfaction, retention, pass ratios, and marks (achieved learning goals).

The next question is at the curriculum level (See also Chapter 9 of this monograph) and out of the hands of the course coordinator: How is the interplay and synergism between laboratory courses, lecture courses, tutorials, and other types of instruction, and how should these elements be balanced in an entire program? Because, at the end of the day, the important thing is what the student is able to do after graduation. Here the European Qualifications Framework [2] or other national and international standards [3–6] are helpful to the faculty/university in developing programs to meet evaluations and accreditations. Such developments will probably seldom lead to a radical curriculum reform in the sense that everything starts from scratch. Tradition is a safeguard against both visionary thoughts and mere fantasies. It might be easier when a new university or a brand new program is to be designed from the ground up.

Around 40 years ago, there appeared to be such an opportunity at Roskilde University [58]. At that time the understanding of teaching strategies discussed above was not clear to those in charge of the curricula and programs, but the principle of broad introductions to a number of allied subjects (much like in some American colleges) resulted in a natural sciences faculty with chemistry as a part of the program in all three levels in the higher education area. Other general principles in the programs were problem orientation, project organization, and student involvement instead of participant direction [58]. The programs have gone through changes over the years and a picture of the current development is given below.

20.4.1.1 Problem-Based Group-Organized Project Work
Students in the bachelor program in natural sciences at Roskilde University spend the first three semesters of the six-semester program in a special building, called a "house," with many group rooms and a large meeting room but no laboratories. A house typically has 60–80 students and about 10–12 supervisors (on part-time) from all the different areas of science. Half of the workload in each semester (15 ECTS) is reserved for project work and the remaining 15 ECTS for a broad range of courses in different areas of science. Students choose three 5 ECTS courses determined by their interest and level of prior knowledge. During the second year students choose their two bachelor degree subjects and the respective course

packages. In order to ensure all students a broad exposure to different aspects of science the three semester projects have to satisfy semester themes, which are different perspectives on science rather than defined by a science subject:

- *First semester*: Application of science in technology and society
- *Second semester*: Interaction between model, theory, experiment, and simulation in natural sciences
- *Third semester*: Natural sciences and theory of science.

Small groups of students (typically 4–6 students in each group) and an allocated supervisor work together the entire semester. Each group completes the project work with a project report followed by an exam. The process of group formation and group work will be illustrated here for a second semester class.

20.4.1.2 Second Semester Project Work

This semester has to introduce the students to how new insight is obtained in science and illustrate similarities and differences between different scientific disciplines, methods, and traditions. The allocated supervisors meet prior to the arrival of students and discuss the semester theme and make plans for the semester. The organization of the progress of the project work through the semester is illustrated in Figure 20.1

When the students arrive after winter break the group formation process starts. On the first day they are introduced to the specific second semester theme and

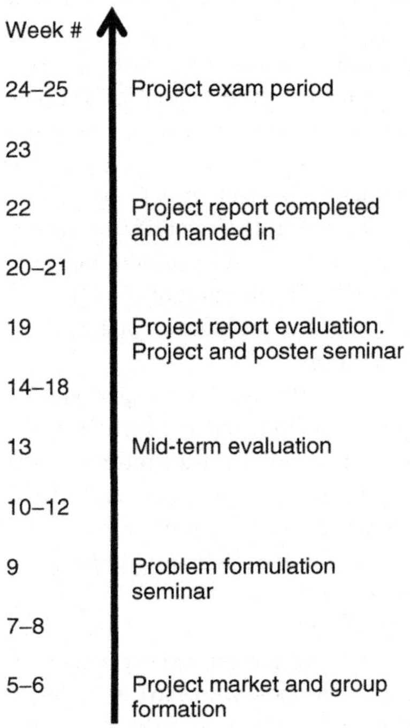

Week #	
24–25	Project exam period
23	
22	Project report completed and handed in
20–21	
19	Project report evaluation. Project and poster seminar
14–18	
13	Mid-term evaluation
10–12	
9	Problem formulation seminar
7–8	
5–6	Project market and group formation

Figure 20.1 Project–organization: important events during the semester.

all supervisors – especially those who did not work with this group of students in their first semester – present themselves briefly. Students from previous years tell about their second semester project work. The students are also introduced to the project market process and are encouraged to think about their own project ideas. Special bulletin boards are reserved for student proposals. The written proposals give a short introduction to the project ideas often inspired by empirical observations or news from various media. Another bulletin board is reserved for general project ideas from supervisors. These project ideas have to be broad and general, and not specific. Students are welcome to be inspired by supervisor ideas but need to formulate their own project ideas and post them on the student bulletin board in order to ensure their ownership of the project thereby, hopefully, enhancing their motivation. The initial student project proposals are often broad and unfocused, for example, "I would like to investigate food additives and allergies" or "why is it difficult to control sour doughs in a reproducible manner."

During the first 2 weeks of the semester students discuss project ideas and project markets are organized. Students present their proposals from the bulletin board for the other students and supervisors, and the proposals are discussed. Comments on how the project idea lives up to the semester theme are given, and students and supervisors suggest alternative directions and which supervisors may assist the students. Other students can join existing proposals, and discussion meetings with students and supervisors are planned. Four project markets are held before the final groups are formed. Project ideas become more specific and qualified during this process. Once the groups with typically four to six students are formed each group will be assigned a supervisor.

The project group and their supervisor typically meet at least once a week. Prior to these meetings students write an agenda for the group meeting and send written materials (short statements, contributions to the project procedure, questions regarding the understanding of difficult concepts, etc.) to be discussed at the meeting. Students gather information about the project idea from various sources (textbooks, other monographs, technical reports, the scientific literature, etc.) and together with the supervisor search for a more specific problem formulation the following weeks. The students work toward a declaration of clear aims of the project work and focus on a clear formulation of the problem as one or more questions to study, the so called "problem formulation."

All refined project ideas are presented for the other students and supervisors at a problem formulation seminar. At this seminar the groups tell about their intentions, which problem formulation they have reached, and how they intend to find a solution to their problem as part of their inquiry process. Students and supervisors then comment on and discuss the problem formulation, and how semester theme requirements can be met. Proposed experimental design and methods are discussed, and alternative possibilities are suggested.

About a month later mid-term evaluations take place, as shown in Figure 20.1. Each group writes a short progress report (about 10–15 pages) and presents their work for another group of students and their supervisor, who have the role of "opponents." The opponent group gives comments on the written material and an oral presentation. The problem formulation is discussed again, adjustments and alternative versions are proposed, and it is checked that the outline of the final report and the chosen methods are logically connected to the problem formulation. Experimental results and experiences obtained at this stage are also presented, and strategies and complications are discussed. The inquiry process with choice of experimental procedures and problems encountered are important elements in this evaluation, and, in addition, students practice oral presentations.

A preliminary version of the final written report is printed about a month later and evaluated a week later with the same group and supervisor present as at the mid-term evaluation (see Figure 20.1). The opponent group and their supervisor have examined the material and are prepared to point out weaknesses and suggest corrections and changes. The group has almost 2 weeks to work full-time on the final version of the report.

All final projects are presented at a seminar for all students and supervisors in the "house," as shown in Figure 20.1. All groups communicate their report and results at the seminar in the form of posters. Communicative skills and scientific contents are evaluated for each poster. Differences as reflected in the posters between scientific work, methods, and traditions in various areas of science are exemplified and discussed in this way. The individual level of each student is finally assessed at oral exams led by the supervisor and an external examiner from another university. It is tested that students are able to explain all aspects of the project report and are able to reflect on details in the report.

Students spend half of their time on course work and half on project work as mentioned above and the stipulated workload through the semester is illustrated in Figures 20.2 and 20.3. As seen from Figure 20.2 the project workload varies greatly during the semester. The first intensive period on the project work is during the first weeks when project groups are formed and project themes are found. The activity is also high in connection with the problem formulation seminar and the mid-term evaluation. The highest activity is toward the end of the project where all course activities are completed and the students have three to four weeks entirely

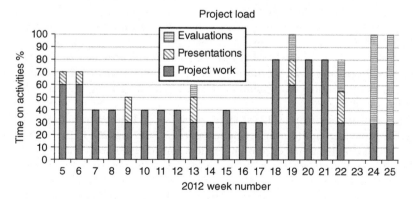

Figure 20.2 Characteristic times spent on project work, presentations, and project evaluations during spring semester 2012. Important events: see also Figure 20.1 above.

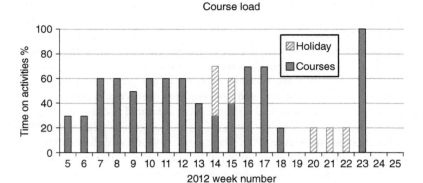

Figure 20.3 Characteristic times spent on courses during spring semester 2012. Holidays are marked as well.

devoted to project work (compare Figures 20.2 and 20.3). During this period both a preliminary and the final versions of the project reports are completed.

20.4.2
Analysis of Second Semester Project Reports

A written project report is intended to communicate to peers and examiners the results of the project group's efforts to solve the problem they formulated in the early stages of the project. It should also give some clues as to the level and area of knowledge that was in play in the group during the project period. Further, it should contain descriptions of what steps were taken in order to solve the problem by including which methods and techniques were used, and also how data were analyzed and interpreted. Also to what extent the students master scientific notation and argumentation can be seen in a report. Of course, the supervisor and examiners would have a better idea and more detailed knowledge (also at the individual level) of which skills and competencies students have (obtained) at the end of the freshman year. Still, a project report is a fair documentation of the work done and bears witness what the student has learned. In any case, the project reports lend themselves to an analysis of what students, on average, achieved through this type of teaching strategy and is a more reliable source of information than written intentions and goals (in a curriculum or syllabus) regarding what had happened and what was learned.

The present analysis includes 123 students of two second semester "houses" who formed 24 project groups (during spring semesters in 2012 and 2013). Each report was on average 50 (±20) pages plus appendices (mostly laboratory manuals and diaries with detailed observations and measured data). The reports follow more or less a common template starting with an abstract (in Danish and in English) and an introduction outlining the context and leading toward a problem formulated in a way that allows a scientific investigation. Next section contains an account of the context to which the problem belongs and a description of the key concepts with regard to the question. It is followed by the methods section, often with a discussion of why and how the method(s) work(s). The materials and methods are next described in detail, and actual and relevant observations and results obtained are given. Further, the result of data processing is presented, and finally, these results are discussed (including retrospective reflections – a kind of post-lab) and related to existing knowledge. A list of literature which has been referred to in the report is given according to one of the normal citation systems.

20.4.2.1 Analysis of Reports from a Chemistry Point of View

The 24 project reports have been evaluated following the categories described in a previous study [16]. The contents and levels of key chemistry elements (synthesis, separation, detection, quantification, identification, and characterization) in all reports were identified. Most of the reports are on problems and topics in other natural science disciplines than chemistry as indicated by titles and general subject areas discussed below.

Four reports have only used simulations and did not get experience from working in the "wet" laboratory. They have been excluded from the analysis and are not included in the evaluation of evidence in the reports of key chemistry elements.

In accordance with earlier findings [16] "synthesis" was virtually absent in the reports. Only one project reported a very simple synthesis-like experiment where a few components at a time of a complex cosmetic product were heated gently to tell whether or not an unpleasantly smelling compound had been formed. This was the case when all of the components of the cosmetic product were heated together.

Techniques to quantify known components in an untreated sample or in a biochemical assay were used in almost half of the projects. Also, "separation" and "identification" were used mainly as biotechnology tools in one out of three projects, but only described according to a chemistry tradition in a few projects. The application of such techniques is widespread in science, but they appeared to have been used as "standard techniques" as such, and the design aspect was almost never part of the reporting of the investigations. Although "characterization" of chemical substances or well-defined mixtures is a rather broad category this type of experiments was only present in a few reports.

In line with earlier findings [16] the experimental experience obtained by these students, when observed through classical "chemistry glasses" is not very impressive, but, as was also suggested earlier [16], other perspectives ought to be used in the assessment of the experience obtained through this instruction type for its relevance to a chemistry program.

20.4.2.2 Elements of Experimental Work

The reports were also analyzed for more general science aspects. The template of analyses used [16] has been designed to include a broad range of aspects of experimental work and seems also to be in accord with later lists formulated in terms of, for example, skills [11] or steps [59] perhaps with a different weighting of some points on the list.

The analysis of the project reports is summarized in Table 20.1. The numbers allocated to each project report are given on a semi quantitative scale, where 0 signifies that the given element is virtually absent in the report; 1 means that there are indications, that the given element was present (mostly in a tool-like way) in the processes during the project period; and 2 is entered when it is clearly demonstrated in the report, that the element had provided an intellectual challenge in the students' work. Both authors marked independently and upon disagreement, a process of closer convergence analysis followed.

The titles of the project reports are given in the first column. The title is normally significant and together with the abstract it always gives a fair idea of the problem studied.

> *Example 1*: Pine mouth syndrome. This group of students became interested in an observation that the intake of certain pine nuts resulted in dysgeusia (a distortion of the sense of taste). It had been noted that a bitter-metallic sense appeared with *Pinus armandii* nuts lasting a couple of days, whereas other pine nuts did not have this effect. Do these nuts contain a problematic compound not found in other nuts?
>
> *Example 2*: Solute release from poloxamer gels. The group was interested in delayed drug release. Gels offer such possibilities and poloxamer triblock copolymers have unique and promising thermal properties. The group wanted to see if constant release rates could be obtained from such gels with model compounds. Do different model compounds have varying release rates?
>
> *Example 3*: The effect of temperature on survival and growth of the brown algae *Saccharina latissima*. The students were concerned about possible effects of increasing temperature in oceans on living organisms. They undertook a study of how a test organism responded to different temperatures. Are temperatures becoming too high in Scandinavian coastal waters to allow optimal growth?

In the next column of Table 20.1 the area of study is characterized as a sub-discipline of classical subjects. It appears that the great majority of problems relates to some biological phenomenon. Most of these projects involve chemistry-based biochemical techniques where handling of chemical matrixes, reagents or mixtures of chemical substances is fundamental.

> *Example 1*: Pine mouth syndrome. This is a typical analytical chemistry project. The supervisor had routine (advanced) chromatographic techniques in his laboratory, and standard and more special extraction methods within reach. The inspiration, however, came from the "clinic."
>
> *Example 2*: Solute release from poloxamer gels. This is a medico chemistry project. The supervisor had extensive knowledge of the handling of poloxamer gels and had access to a variety of routine detection methods, for example, ordinary UV-VIS spectrophotometry. The pharmacy inspiration was, however, evident.

Table 20.1 Titles and area of study of 20 experimental project reports from second semester 15 ECTS projects.

Title of report	Area of study	O1	O2	D1	D2	E1	E2	R1	R2	I1	I2
A study of physiological tolerance in the native *Nereis diversicolor* and the invasive *Marenzelleria viridis*	Aquatic ecology	2	0	1	2	1	1	1	2	1	0
Pine mouth syndrome. Differences in the extraction profiles of *Pinus armandi* and *Pinus koraiensis*	Analytical chemistry	2	2	2	1	1	1	1	2	2	1
Green roofs	Urban ecology	2	1	2	1	0	1	1	2	1	2
Shear jamming of granular materials in three dimensions	Applied physics	2	1	2	2	1	1	2	2	2	1
Solute release rates form poloxamer gels	Medico chemistry	2	1	1	2	1	1	1	2	2	1
Gene expression of Metallothionine and Cytochrome C oxidase in *Lumbriculus variegatus*-Influence of Cu nanoparticles and ions	Molecular toxicology	1	0	1	1	0	1	1	0	1	0
The influence of duration and route of exposure and form of copper on bioaccululation of copper in *Lumbriculus variegatus*	Ecotoxicology	1	0	1	1	1	2	1	2	1	0

(continued overleaf)

Table 20.1 (Continued)

Title of report	Area of study	O1	O2	D1	D2	E1	E2	R1	R2	I1	I2
The possible involvement of microRNA-29a in the regulation of mitochondrial endcoded genes in β-cells	Molecular biology	2	1	1	1	0	0	1	2	1	1
NDRG2's interaction with Api5	Molecular biology	2	2	2	1	1	2	1	2	2	1
The antihistamine diphenhydramine inhibits the phototaxis of Daphnia magna	Aquatic ecology	2	2	2	1	2	2	1	2	2	1
The effect of temperature on survival and growth of the brown algae Saccharina latissima	Aquatic ecology	2	2	2	2	2	1	2	2	2	2
Influence of pH on hatching out of eggs from Acartia tonsa	Aquatic ecology	1	1	2	1	1	0	2	2	2	0
Cultivation/expression of the anti freeze protein (RmAFP#1) in Bacillus subtilis	Molecular biology	2	1	1	1	1	1	1	2	2	1
HOXA9 and CDX2 and leukemia	Molecular biology	1	1	2	1	1	1	1	1	1	0
Isoforms of CDX2	Molecular biology	1	2	2	2	1	1	1	2	2	0

20.4 New Insight and Implementation

Topic	Category										
Viscoelastic properties of Bitumen	Materials physics	1	1	2	2	1	1	1	2	2	1
Spectroscopy of the molecular rotor 1,4-diethynylbenzene	Spectroscopy	1	0	1	2	1	0	1	2	1	0
Is the growth rate of *Fucus vesicolosus* affected by exposure to herbicides at different salinities?	Aquatic ecology	2	2	1	1	2	1	1	1	2	1
Regulation of in vitro cell proliferation by the stabilization of H1F1α by $CoCl_2$	Molecular biology	2	0	1	1	0	0	1	0	1	2
Chasing sotolon in skin tonic	Analytical chemistry	2	1	2	1	1	0	0	1	1	1

Entries for the last 10 columns refer to elements of experimental experience (see text for an explanation of categories in Section 20.4.2.2).

Example 3: The effect of temperature on survival and growth of the brown algae *Saccharina latissima*. This is a biogeographic study. The supervisor was accustomed to handling macroalgae in the laboratory and had access to a variety of techniques for the measurement of growth, photosynthetic activity, stress, and so on.

The next 10 column entries refer to the above elements of science experimental experience (Objectives, Design, Experimental, Results, and Interpretation). It was investigated how the students argue for their objectives and give reasons for the chosen project formulation (O1). The student's reflections on possible outcomes of their work were also evaluated (O2). The arguments for choice of design of possible experimental study were also studied (D1) and whether they use standard techniques or novel techniques (D2). Discussions of experimental aspects including calibration and standardization (E1) and reproduction and adequate presentations of experimental results (E2) were analyzed. The student's reflections on accuracy and precision (R1) and result presentation (R2) were investigated. Finally the level of interpretation and discussion with own expectations (I1) and with existing results in the literature (I2) were evaluated.

The four simulation projects are not included in the evaluation of evidence in the reports of elements of experimental experience, although these students have indeed experienced the handling of data and several of the other elements taking place outside the laboratory.

The results of the analysis of the remaining 20 reports are summarized in Figure 20.4.

Furthermore some results are worth mentioning:

Objectives As is seen from the table almost all students have participated in a discussion of the project context. Why this particular question is worth studying

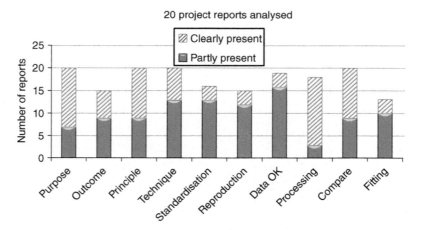

Figure 20.4 Presence of elements of experimental experience found in 20 s semester project reports.

(apart from its educational value) had apparently been addressed appropriately. Only one third of the projects, however, documented explicit intellectual efforts on possible results of the experiments, while only suggestions concerning that question were implicit in many of the other project reports.

> *Example 1*: Pine mouth syndrome. The objectives are clearly stated in the introduction of the report and it is argued why it is relevant to study this further (outside the clinic) using analytical chemical techniques. The working hypothesis was that nuts from *Pinus armandii* contain a problematic compound not found in other pine nuts.
> *Example 2*: Solute release from poloxamer gels. The introduction clearly explains the context. Hypotheses about release rates under different gel concentrations and the hydrophobicity of solutes were formulated. The possible outcomes were, however, not discussed in detail.
> *Example 3*: The effect of temperature on survival and growth of the brown algae *Saccharina latissima*. It is explained in the introduction that some macroalgae had been observed to withdraw from certain coastal areas in Scandinavia. The students anticipated that the growth temperature profile for algae growth has an optimum below the current temperatures.

Design All project reports witness some consciousness of which principle(s) of investigation has been used, although in half of the cases the argumentation is not very elaborate. Not surprisingly, examples of methods relevant to the problem in question have been found. Since much school practical work tries to illustrate principles and concepts students are accustomed to this linking between "theory and practice." Likewise, there is always an account of which techniques have been chosen, although it is only every third report that argues why the experimental procedure was chosen over others. It probably reflects that standard techniques currently running in the supervisor's laboratory are obvious choices.

> *Example 1*: Pine mouth syndrome. The group found a number of extraction procedures, for example, Soxhlet and supercritical CO_2 extraction, while considering properties like solubility and thermal stability of possible compounds like terpenes. They also identified a number of combined separation and detection methods available to the supervisor.
> *Example 2*: Solute release from poloxamer gels. The report has little consideration of different established methods but they developed a novel test method.
> *Example 3*: The effect of temperature on survival and growth of the brown algae *Saccharina latissima*. Different possible indicators of optimum growth of this alga were identified and methods for some of these were selected. Several techniques were adapted and optimized for the purpose.

Experimental For the same reason most reports only report very briefly on the reliability of the laboratory method used. While a few reports do not deal with this question, another minority of reports contains a sufficient demonstration of control over experimental details. In addition only few reports address reproducibility of experimental procedures and measurements properly, while quite a few seem to have run through procedures only once. Admittedly, some methods and procedures are advanced and lengthy and time limits have determined this lack of perfection in the experimental work. Greater emphasis on this aspect could improve the understanding of this important aspect of experimental investigations.

> *Example 1*: Pine mouth syndrome. The extraction procedures were described in detail and the use of GC-MS and RP-HPLC was documented and optimized. Tests of reproducibility and accuracy were, however, minimal.
> *Example 2*: Solute release from poloxamer gels: Different gel geometries were tested and calibration curves were obtained on different dyes.
> *Example 3*: The effect of temperature on survival and growth of the brown algae *Saccharina latissima*. Growth measurements, pulse amplitude modulation fluorescence (indicating stress level) experiments, photometric measurements of pigment levels were described in detail, whereas test of reproducibility was indicated only partly.

Results Very few reports have addressed data reliability up front, but most reports have this feature safely buried in the trust on the standard methods and procedures in the supervisor's laboratory routines. The reports working with organisms have included simple statistics and present variance or standard deviations. Most projects report relevant transformation of data to experimental results according to models and relations between variables in diagrams.

> *Example 1*: Pine mouth syndrome. The students found a unique compound in HPLC chromatograms after Soxhlet and supercritical CO_2 extraction. They were not able to identify the unique compound. The results were presented in tables and figures with self-explanatory figure legends.
> *Example 2*: Solute release from poloxamer gels. The students found that release rates were independent of the various model compounds but dependent on the gel strength. Results are presented in photographs, figures, and tables with some calibration tests but little reproducibility and accuracy reflections. Figures are well chosen and presented.
> *Example 3*: The effect of temperature on survival and growth of the brown algae *Saccharina latissima*. A temperature optimum between 5 and 15 °C was found for several of the measured response parameters. The reliability of the results was discussed and were presented in figures and graphs with simple statistical treatment included.

Interpretation All reports try to conclude from their findings, but only in half of the cases is a detailed and balanced interpretation of the obtained results presented. With freshman students this is perhaps not surprising since the art of making valid conclusions from scientific findings matures with growing individual knowledge and experience. Also, most students appear to forget to discuss and fit their results into existing knowledge – again not unexpected, since freshman students have less knowledge of a broader perspective to fit own results into.

Example 1: Pine mouth syndrome. In the report there is a clear discussion of experimental problems and results but the comparison with literature results is minimal.

Example 2: Solute release from poloxamer gels. The obtained results were compared with their original hypothesis but little comparison with literature results was found.

Example 3: The effect of temperature on survival and growth of the brown algae *Saccharina latissima*. The results were compared with the hypotheses and the relation to relevant literature findings was discussed.

20.5 The Chemistry Point of View Revisited

In summary, through the described organization of the project work all students have experienced the processes of scientific inquiry and have participated in several seminars with discussions of context and details of their work, and they have participated in the preparation of a written report and a poster. The performance of the individual student has been assessed through an oral examination, where any detail in the report may be defended and discussed. The exam results are in general rather satisfying. With the present sample of 123 students, 25 obtained "12," 62 got "10," 28 received "7," 7 obtained "4," and one got "02," using the ECTS marking scale.

Most, if not all, of the elements of instruction considered to be beneficial in Section 20.3.3 have been incorporated in the reported instructional organization of the program which later leads to a bachelor degree in natural science – in some cases with a major in chemistry. Especially the pre-lab and post-lab dimensions generally played major roles in the reports (and in learning as far as this is represented by the marks given) as integral parts of scientific work.

It is our experience that very few students drop out, and that the students seem motivated and work hard to reach a conclusion and prepare a good report. The present analysis suggests that the first year students in general have produced reports which represent a fair coverage of the elements of scientific inquiry generally at a high undergraduate level, including both practical skills and other skills needed by an experimental chemist (or scientist in general) [3, 5, 6, 11, 19], even in cases where projects do not deal with the study of chemical substances or problems. Learning experiences central to the education of an experimental chemist are practiced through this project-organized set up.

From the list of project titles in Table 20.1, it is obvious that most of the projects were inspired by a more or less clearly biological phenomenon. At the start of the semester when students are asked to suggest fields of interests and problems to study, they often think within a biology frame like curing diseases, healthy diets, pollution, and environment protection. Such areas are easily accessible through newspapers and their own personal experience and more obvious to think of than how, for example, powder (flour) or other small particles in large amounts (grain) behave when transferred from one container to a smaller one. However, it is also clear from the analysis of the reports that most of the practical work includes analytical chemistry or chemistry-based techniques in biochemistry, molecular biology, or toxicology. Here the (safe) handling of chemical matrixes, reagents, or mixtures of chemical substances, equipment and instruments, and the mastering of stoichiometric concepts and calculations in experimental work are fundamental skills. However, it does not necessarily mean, that the students at this stage have a deep knowledge of why or how all the things they encounter in the laboratory work. Hopefully most of them experience that practical and conceptual chemistry is essential for and interwoven with many other areas of science and technology and should be mastered in order to fully benefit from it.

20.6
Project-Based Learning

During the last half decennium the understanding of learning mechanisms has improved, new teaching strategies and methods have been introduced and the old ones refined. The main challenge has been the gap between the intended learning – the curriculum – and the actual learning – student qualifications. The developments have taken place on course level and also on degree level.[1] New teaching principles and methods and corrections to existing practices have been introduced under different headlines, some of which have been given above: constructivist approaches, PBL, IBSE, IBL, POGIL, flipped classroom [60], and so on.

Project-based learning is another term that signals a way of organizing teaching (and, hence, learning[2]), and many teaching activities have furthermore "project" as a label. One possible way to characterize different types of such projects is to consider the degree of student influence on the choice of subject matter and of the student's responsibility to manage the projects process.

Many project-based activities are well-planned "standard-projects" often with an experimental dimension. Such student-activities which may be repeated with the next group of students are often valuable supplements to textbook reading. These case studies may help students to comprehend (possibly at a higher cognitive level) the subject matter in cases where the project applies concepts and theories and illustrates facts in the syllabus. Still the teacher is pulling the strings

---

1) Chapter 9 of this monograph deals with "New perspectives on the undergraduate chemistry curriculum in universities"
2) Chapter 13 of this monograph deals with "Student centered learning"

keeping the students on track through the intended subject matter toward the correct answer.

In the pure form of *project-organization* students are organized in project groups with a supervisor around a task which offers an opportunity to learn. When *problem-oriented*, the task is a real problem, which the student group formulates within an area of interest under guidance as a question which a scientific investigation can throw light upon. In this case the supervisor does not know the full "answer" to the question, but knows relevant methods and procedures to approach and deal with the problem. This is similar to supervision of a student working on his or her thesis. In such cases the problem dictates that some areas of subject matter (method or model) have to be studied during the process in order to make it possible to solve the problem. In contrast to traditional teaching methods this teaching may also take place in early parts of a bachelor program. Thus an obvious challenge is that there is more necessary subject matter to learn during the processes (of scientific inquiry) relative to what a student already should master at the end of an entire program. In turn, it implies that the progress of the project is slower and the ambitions lower. Other general professional skills like cooperation abilities, independence and inter-disciplinarity are, however, required and developed during work in such group based projects.

A degree program in chemistry invariably contains certain subject areas of "core chemistry" which should be covered in any case. The EU Qualification Framework formulations [2] of more general outcomes contribute, however, to a more complete description of a contemporary curriculum and these aspects should simultaneously be reflected in actual teaching practices.

The systematic coverage of subject matter has a very long tradition in teaching, but it may in part be organized in the above mentioned "standard-projects" or miniprojects being elements of lecture-courses and laboratory courses or both and in many other ways. Developments in this area are important. The desired qualifications (in terms of knowledge, skills, and competences) should also be addressed for many reasons. Experimental work as part of problem-oriented learning activities also offers great opportunities to practice skills and competences. Introduction of project-organized group work is indeed a possibility to support such goals.

We have tried to balance these two "what to study"-principles (cf. Section 20.4.1.1. above) as shown in Figures 20.2 and 20.3: In our program half of the study load is organized in lecture- and lab-courses (what should you know to cover core areas in chemistry) and the other half in projects (what does it take to solve a problem using chemistry methods). Other approaches including project-based learning activities also in earlier parts of the curriculum are being developed at many universities and the trend is changing away from teaching strictly based on the proverb "Don't try to walk before you can crawl." Students should be invited to take ownership of their learning processes. This is the intention in flipping the classroom [60] and easily obtained for the vast majority of students in a project-organized environment.

References

1. In the following the expressions "practical work", "laboratory work", "experimental work", and "experiments" will be used essentially as synonyms in line with Hodson, D. (2001) "Research on practical work in school and universities: in pursuit of better questions and better methods." Paper presented at the 6th ECRICE, Aveiro, Portugal.
2. Ministry of Science and Innovation (2005) A Framework for Qualifications of The European Higher Education Area, *http://www.ond.vlaanderen.be/ hogeronderwijs/bologna/documents/ 050218_QF_EHEA.pdf* (accessed 23 September 2013).
3. The Chemistry Eurobachelor (2007) *http://ectn-assoc.cpe.fr/chemistry-eurolabels/doc/officials/Off_EBL070131_Eurobachelor_Framework_2007V1.pdf* (accessed 23 September 2013).
4. RSC Accreditation of Degree Programmes (2012) *http://www.rsc.org/ images/Accreditation%20Criteria %20January%202012_tcm18-151306.pdf* (accessed 23 September 2013).
5. QAA (2007) Subject Benchmark Statement: Chemistry, Quality Assurance Agency, *http:// www.qaa.ac.uk/Publications/ InformationAndGuidance/Pages/Subject-benchmark-statement-Chemistry.aspx* (accessed 23 September 2013).
6. ACS Guidelines and Evaluation Procedures for Bachelor's Degree Programs (2008) Undergraduate Professional Education in Chemistry, *http://www.acs.org/content/dam/acsorg/ about/governance/committees/training/ acsapproved/degreeprogram/2008-acs-guidelines-for-bachelors-degree-programs.pdf* (accessed 23 September 2013).
7. ACS Public Policy Statement (2011-2014) Importance of Hands-on Laboratory Activities, *http:// www.acs.org/content/dam/acsorg/policy/ publicpolicies/invest/computersimulations/ 2011-04-importance-of-hands-on-laboratory-activities.pdf* (accessed 23 September 2013).
8. Hofstein, A. and Lunetta, V.N. (2004) The laboratory in science education: foundations for the twenty-first century. *Sci. Educ.*, **88**, 28–54.
9. Hodson, D. (1993) Re-thinking old ways: towards a more critical approach to practical work in school science. *Stud. Sci. Educ.*, **22**, 85–142.
10. White, R.T. (1996) The link between the laboratory and learning. *Int. J. Sci. Educ.*, **18**, 761–774.
11. Reid, N. and Shah, I. (2007) The role of laboratory work in university chemistry. *Chem. Educ. Res. Pract.*, **8**, 172–185.
12. Hawkes, S.J. (2004) Chemistry is not a laboratory science. *J. Chem. Educ.*, **81**, 1257–1257.
13. Woolnough, B.E. (1983) Exercises, investigations and experiences. *Phys. Educ.*, **18**, 60–63.
14. Kirschner, P.A. (1992) Epistemology, practical work and academic skills in science education. *Sci. Educ.*, **1**, 273–299.
15. Johnstone, A.H. and Al-Shuaili, A. (2001) Learning in the laboratory; some thoughts from the literature. *Univ. Chem. Educ.*, **5**, 42–51.
16. Josephsen, J. (2003) Experimental training for chemistry students: does experimental experience from the general sciences contribute? *Chem. Educ. Res. Pract.*, **4**, 205–218.
17. Jenkins, E.W. (1999) in *Practical Work in Science Education – Recent Research Studies* (eds J. Leach and A.C. Paulsen), Klüwer Academic Publishers and Roskilde University Press, Frederiksberg, pp. 19–32; ISBN 987-7867-079-9.
18. McCoy, A.B. and Darbeau, R.W. (2013) Revision for the ACS guidelines for Bachelor's degree programs. *J. Chem. Educ.*, **90**, 398–400.

19. Garrat, J. (2002) Laboratory work provides only one of many skills needed by the experimental chemist. *Univ. Chem. Educ.*, **6**, 58–64.
20. Bennett, S.W. and O'Neale, K. (1996) Skills development and practical work in chemistry. *Univ. Chem. Educ.*, **2**, 58–62.
21. Millar, R., Le Maréchal, J.-F., and Tiberghien, A. (1999) in *Practical Work in Science Education – Recent Research Studies* (eds J. Leach and A.C. Paulsen), Klüwer Academic Publishers and Roskilde University Press, Frederiksberg, pp. 33–59; ISBN 987-7867-079-9
22. Josephsen, J. and Kristensen, A.K. (2006) Simulation of laboratory assignments to support students' learning of introductory inorganic chemistry. *Chem. Educ. Res. Pract.*, **7**, 266–279.
23. Heilesen, S. and Josephsen, J. (2008) E-learning: between augmentation and disruption? *Comput. Educ.*, **50**, 525–534.
24. Domin, D.S. (1999) A review of lsaboratory instruction styles. *J. Chem. Educ.*, **76**, 543–547.
25. Zoller, U. and Pushkin, D. (2007) Matching HOCS promotion goals with problem-based laboratory practice in a freshman organic chemistry course. *Chem. Educ. Res. Pract.*, **8**, 153–171.
26. Kelly, O. and Finlayson, O. (2009) A hurdle too high? Students' experience of a PBL laboratory module. *Chem. Educ. Res. Pract.*, **10**, 42–52.
27. Gijbels, D., Dochy, F., Van den Bossche, P., and Segers, M. (2005) Effects of problem-based learning: a meta-analysis from the angle of assessment. *Rev. Educ. Res.*, **75**, 27–61.
28. National Research Council (2000) *Inquiry and the National Science Education Standards*, National Academy Press, Washington, DC.
29. Spencer, J.N. (1999) New directions in teaching chemistry: a philosophical and pedagogical basis. *J. Chem. Educ.*, **76**, 566–569.
30. Farrell, J.J., Moog, R.S., and Spencer, J.N. (1999) A guided inquiry general chemistry course. *J. Chem. Educ.*, **76**, 570–574.
31. Rushton, G.T., Lotter, C., and Singer, J. (2011) Chemistry teacher's emerging expertise in inquiry teaching: the effect of a professional development model on beliefs and practice. *J. Sci. Teach. Educ.*, **22**, 23–52.
32. Fay, M., Grove, N.P., Towns, M.H., and Bretz, S.L. (2007) A rubric to characterize inquiry in the undergraduate chemistry laboratory. *Chem. Educ. Res. Pract.*, **8**, 212–219.
33. Kirschner, P.A., Sweller, J., and Clark, R.E. (2006) Why minimal guidance during instruction does not work: an analysis of the failure of constructivist, discovery, problem-based experiential, and inquiry-based teaching. *Educ. Psychol.*, **41**, 75–86.
34. Hmelo-Silver, C.E., Duncan, R.G., and Chinn, C.A. (2007) (2006) Scaffolding and achievement in problem-based and inquiry learning: a response to Kirschner, Sweller and Clark. *Educ. Psychol.*, **42**, 99–107.
35. Minner, D.D., Levy, A.J., and Century, J. (2010) Inquiry-based science instruction-what is it and does it matter? Results from a research synthesis years 1984-2002. *J. Res. Sci. Teach.*, **47**, 474–496.
36. Schroeder, C.M., Scott, T.P., Tolson, H., Huang, T.-Y., and Lee, U.-H. (2007) A meta-analysis of national research: effects of teaching strategies on student achievement in science in the United States. *J. Res. Sci. Teach.*, **44**, 1436–1460.
37. Furtak, E.M., Seidel, T., Iverson, H., and Briggs, D.C. (2012) Experimental and quasi-experimental studies of inquiry-based science teaching – a meta-analysis. *Rev. Educ. Res.*, **82**, 300–329.
38. Smith, C.J. (2012) Improving the school-to-university transition: using a problem-based approach to teach practical skills whilst simultaneously developing students' study skills. *Chem. Educ. Res. Pract.*, **13**, 490–499.
39. Fowler, H.W. and Fowler, F.G. (1977) *The Concise Oxford Dictionary of Current English*, Oxford University Press, Oxford.
40. Frerichs, V.A. (2013) ConfChem conference on case-based studies in chemical education: use of case-study for introductory chemistry laboratory environment. *J. Chem. Educ.*, **90**, 268–270.

41. Hunter, C., Wardell, S., and Wilkins, H. (2000) Introducing first-year students to some skills of investigatory laboratory work. *Univ. Chem. Educ.*, **4**, 14–17.
42. Mc Donnel, C., O'Connor, C., and Seery, M.K. (2007) Developing practical chemistry skills by means of student-driven problem based learning mini-projects. *Chem. Educ. Res. Pract.*, **8**, 130–139.
43. Lyall, R.J. (2010) Practical work in chemistry: chemistry students' perception of working in a less organised environment. *Chem. Educ. Res. Pract.*, **11**, 302–307.
44. Chase, A., Pakhira, D., and Stains, M.J. (2013) Implementing process-oriented, guided-inquiry learning for the first time: adaptations and short-term impacts on students' attitude and performance. *J. Chem. Educ.*, **90**, 409–416.
45. Laredo, T. (2013) Changing the first-year chemistry laboratory manual to implement a problem-based approach that improves student engagement. *J. Chem. Educ.*, **90**, 1151–1154.
46. Johnstone, A.H. and Letton, K.M. (1991) Practical measures for practical work. *Educ. Chem.*, **28**, 81–83.
47. Jalil, P.A. (2006) A procedural problem in laboratory teaching: experiment and explain, or vice-versa? *J. Chem. Educ.*, **83**, 159–163.
48. Johnstone, A.H., Watt, A., and Zaman, T.U. (1998) The students attitude and cognition change to a physics laboratory. *Phys. Educ.*, **33**, 22–29.
49. Overton, T.L. and Potter, N. (2011) Investigating students' success in solving and attitudes towards context-rich open-ended problems in chemistry. *Chem. Educ. Res. Pract.*, **12**, 294–302.
50. Hunter, C., McCosh, R., and Wilkins, H. (2003) Integrating learning and assessment in laboratory work. *Chem. Educ. Res. Pract.*, **4**, 67–75.
51. Gibbon, E. (1776) *The Decline and Fall of the Roman Empire*, Chapter 4, vol. 1, http://sacred_texts.com/cla/01/daf01013.htm (accessed November, 12. 2014).
52. Hopkins, T.A. and Samide, M. (2013) Using a thematic laboratory-centered curriculum to teach general chemistry. *J. Chem. Educ.*, **90**, 1162–1166.
53. Tomasik, J.H., Cottone, K.E., Heethuis, M.T., and Mueller, A. (2013) Development and preliminary impacts of the implementation of an authentic research-based experiment in general chemistry. *J. Chem. Educ.*, **90**, 1155–1161.
54. Duis, J.M., Schafer, L.L., Nussbaum, S., and Stewart, J.J. (2013) A process for developing introductory science laboratory learning goals to enhance student learning and instructional alignment. *J. Chem. Educ.*, **90**, 1148–1150.
55. Marsh, P.A. (2007) What is known about student learning outcomes and how does it relate to the scholarship of teaching and learning? *Int. J. Scholarship Teach. Learn.*, **1**, 1–12.
56. Raine, D. and Symons, S. (2005) *PossiBiLities: A Practice Guide to Problem-Based Learning in Physics and Astronomy*, The Higher Education Academy, Physical Sciences Centre, http://www.heacademy.ac.uk/assets/ps/documents/practice_guides/ps0080_possibilities_problem_based_learning_in_physics_and_astronomy_mar_2005.pdf (accessed 7 October 2013).
57. Byers, W. (2002) Promoting active learning through small group laboratory classes. *Univ. Chem. Educ.*, **6**, 28–34.
58. Olesen, H.S. and Jensen, J.H. (eds) (1999) *Project Studies: A Late Modern University Reform*, Roskilde University Press, Frederiksberg.
59. Bennett, S.W. (2008) Problem solving: can anybody do it? *Chem. Educ. Res. Pract.*, **9**, 60–64.
60. Smith, J.D. (2013) Student attitudes toward flipping the general chemistry classroom. *Chem. Educ. Res. Pract.*, **14**, 607–614.

21
The Development of High-Order Learning Skills in High School Chemistry Laboratory: "Skills for Life"
Avi Hofstein

21.1
Introduction: The Chemistry Laboratory in High School Setting

Since the ninetieth century when schools began to teach science systematically, the laboratory became a distinctive feature of science education [1]. After the First World War, with rapid increase of science knowledge, the laboratory was used mainly as a means for confirmation and illustration of the information gained previously from a lecture or textbooks. Regarding skills, the main goal was to teach students practical abilities such as manipulation of laboratory chemistry equipment and handling materials. Radical change regarding laboratory work occurred with the reform in science education in the 1960s and early 1970s in the United States (e.g., CHEMStudy) and United Kingdom (e.g., Nuffield O-level and A-level chemistry), and other countries followed this reform (e.g., Israel). The idea was to engage students with investigations, discoveries, inquiry, problem-solving activities, and as an instructional technique to demonstrate to the learner ideas related to the *nature and history of science*. This period is associated with the many curriculum projects that were developed to renew and improve science teaching and learning. The projects started in the late 1950s with focus on updating and reorganizing content knowledge in the science curricula, but soon reformists turned their attention toward *science process* as a main aim and the organizing principle for science education, as expressed by Klainin [2] in Thailand, who wrote:

> Many science educators and philosophers of science education (e.g., in the USA: Schwab, 1962; Rutherford and Gardner, 1970) regarded science education as a process of thought and action, as a means of acquiring new knowledge, and a means of understanding the natural world (p. 171).

Thus, the laboratory became the core of the science learning process and of science instruction [3–5].

Over the years, the science laboratory was extensively and comprehensively researched, and hundreds of research papers and doctoral dissertation were published all over the world [3, 5–7]. This embracement of practical work, however,

Chemistry Education: Best Practices, Opportunities and Trends, First Edition.
Edited by Javier García-Martínez and Elena Serrano-Torregrosa.
© 2015 Wiley-VCH Verlag GmbH & Co. KGaA. Published 2015 by Wiley-VCH Verlag GmbH & Co. KGaA.

has been contrasted with challenges and serious questions about its efficiency and benefits for learning and teaching in the laboratories [6–9]. For many chemistry teachers (and often chemistry curriculum developers), practical work still meant simple recipe-type activities that students are following and, using practical skills, without the necessary mental engagement and/or development of high-order learning skills: in other words, "hands-on" than "mind-on" laboratory activities. The aimed-for ideal of open-ended inquiry in which students have opportunities to plan an experiment, to ask questions, to hypothesize, and to plan an experiment again to verify or reject their hypothesis happens more rarely – and when it does, the learning outcome is much discussed. This chapter will review and discuss research on practical work in order to demonstrate its potentials for development of high-order learning skills in the context of teaching and learning in the chemistry laboratory. The emphasis on processes rather than the products of science was fueled by many initiatives and satisfied different interests. Some educators wanted a return to a more student-oriented pedagogy after the early reform projects which they thought paid too much attention on the subject knowledge. Others regarded science process as the solution to the rapid development of knowledge in science and technology: learning the science process was seen as more sustainable and therefore a way of making students prepared for the unknown challenges of the future. Maybe most important, however, was the stark development in cognitive psychology, which drew the attention toward reasoning processes and scientific thinking. Psychologists such as Bruner, Piaget, and Gagne helped to explain the thinking involved in the science process and inspired the idea that science teaching could help develop this type of thinking in young people.

Although this development originated in the United States, it was soon echoed in many other nations [3, 10]. Everywhere, the laboratory and practical work were put in focus. In 1963, Kerr [11] in the United Kingdom suggested that practical work should be integrated with theoretical work in the sciences and should be used for its contribution to provide facts by investigations, and as a result to arrive at principles that are related to these facts. This became a guiding principle in the many of *Nuffield Curriculum Projects* that developed in the late 1960s and early 1970s.

The interest in practical work in science education research in this period is clearly presented by Lazarowitz and Tamir [5] in their review on laboratory work which identified 37 reviews on issues of the laboratory in the context of science education [3, 6, 12, 13]. These reviews express a similar strong belief regarding the potential of practical work as the curriculum projects, but also admit strong difficulties in obtaining convincing data on the educational effectiveness of such teaching. Not surprisingly, the only area in which laboratory work showed real advantage (when compared to the non-practical learning modes) was in the development of laboratory manipulative skills. Conceptual understanding, critical thinking, and understanding the nature of science showed little or no differences. Lazarowitz and Tamir [5] and Hofstein and Lunetta [3, 6] suggested that one of the reasons for this relates to the use of inadequate assessment and research procedures.

It was also an issue that the practice in the laboratory did not change as easily toward an open-ended style of teaching as the curriculum projects suggested. Teachers rather preferred a safer "cook-book" approach. In addition, Johnstone and Wham [14] (mainly in the context of chemistry teaching and learning) claimed that educators underestimated the high cognitive demand of practical work on the learner. In practical work, the student has to handle a vast amount of information regarding names of equipment and materials, instructions regarding the process, data and observations, overloading the student's working memory. This makes laboratory learning complicated rather than a simple and safe way toward learning.

To sum up, based on the important publication related to science laboratories titled *America's Lab Report* published by the National research Council [15]

> The science learning goals of laboratory experiences include enhancing mastery of science subject matter, developing scientific reasoning abilities, increasing understanding of the complexity and ambiguity of empirical work, developing practical skills, increasing understanding of the nature of science, cultivating interest in science and science learning, and improving teamwork abilities. The research suggests that laboratory experiences will be more likely to achieve these goals if they (1) *are designed with clear learning outcomes in mind,* (2) *are thoughtfully sequenced into the flow of classroom science instruction,* (3) *integrate learning of science content and process, and* (4) *incorporate ongoing student reflection and discussion* (p. 13).

In this chapter, I have chosen to discuss in depth three themes (skills) that emerged in the last 10–15 years or so years, namely metacognition, argumentation, and asking questions, in the context of high school chemistry laboratories. It is suggested that these learning skills are aimed at educating future citizens through the active involvement of students in the chemistry laboratory.

21.2
The Development of High-Order Learning Skills in the Chemistry Laboratory

21.2.1
Introduction

The key point to be made regarding the development of high-order learning skills is that practical work is not a static issue but something that has evolved gradually over the years, and which is still developing. The development relates to changing aims and goals for science (in our case chemistry) education, to the developments in the understanding of science learning, to changing views and understanding of science inquiry, and to more recent developments in educational technologies. In reviewing the literature in the last 50 years, it is clear that high-order learning skills in the chemistry laboratory emerged only in the last 15 or so years influenced mainly by among others integrating new technologies (ICT) into the chemistry classroom in general and into the chemistry laboratory, in particular [16, 17], and also in our knowledge (based on research) regarding students' learning in laboratories and change of goals for learning chemistry in laboratories. The "new" goals are very often influenced by the movement *from learning science* (chemistry) for those who are going in the future to embark in a career in the sciences (chemists, engineers, and medical doctors) to the movement *to teach science* (in this case chemistry) "for all" [18] – in other words, chemistry program for those students who in the future will have to operate as literate citizens in the society in which they live and operate. One of the key goals of contemporary science education is to provide students with the opportunity to develop scientific literacy in general and chemical literacy in particular [19]. Examples of chemical literacy components include understanding the particulate nature of matter, knowledge of chemical reactions between substances to create new ones, the ability to use laws and theories to explain various phenomena, to understand the application of chemistry knowledge to students' personal lives and to the society in which they live, and finally to develop the ability to implement high-order learning (or thinking) skills in a regular manner. The uniqueness of chemistry is reflected in the strong relationship between the four chemistry understanding levels: macroscopic, microscopic, symbolic, and process [20]. In addition, in many countries around the world, achieving scientific literacy has become a central goal for education [19]. As said before in this chapter, the target population is those who will eventually be thoughtful citizens in a scientific/technological-oriented society as well as those who will eventually embark on a career in the sciences. They will be required to ask critical questions and seek answers upon which they will need to make valid decisions. Thus, the development of students' abilities to ask questions and pose critical thinking is seen as an important component of their chemistry literacy. It is recognized that citizens' needs include more than just an understanding scientific knowledge. In everyday life, science is often involved in public debate and used as evidence to support political views. Science also frequently presents findings and information that challenge existing norms and ethical standards in society. Mostly it is "cutting edge" science and not established theories that are at play. For this reason, it does not help to know "text book science," but rather it is necessary with knowledge *about* science. Citizens need to understand the principles in scientific inquiry and how science operates at a social level [21]. The natural question, of course, is to what degree and in what ways the science laboratory can help provide students with such understanding.

21.2.2
What Are High-Order Learning Skills?

Higher order thinking/learning skills and activities in the context of learning science are considered to be complex and non-algorithmic, and involve applications of multiple criteria instead of memorizing facts [22]. These activities include asking research questions, solving authentic problems, argumentation, metacognitive skills, drawing conclusions, making comparisons, dealing with controversies, and taking a stand [23]. Gunstone and Champagne [24] claimed that meaningful learning in the laboratory occurs when students are given ample opportunities for interaction and reflection in order to initiate discussion. It is suggested that some of these skills could be developed as part of inquiry laboratories (more details will be provided in the next few paragraph). Before elaborating on each of the learning skills that might be developed in the inquiry laboratory, let us elaborate on the meaning of inquiry.

The *National Science Education Standards* published by the NRC [25] as well as the 2061 project [26] reaffirm the claim that inquiry is central to the achievement of scientific literacy.

The *National Science Education Standards* use the term "inquiry" in two ways [4, 27]:

1) Inquiry as *content understanding*, in which students have opportunities to construct concepts and patterns, and to create meaning about an idea in order to explain what they experience; and
2) Inquiry in terms of *skills* and *abilities*. In this category, Bybee included identifying and posing scientifically oriented questions, forming hypotheses, designing and conducting scientific investigations, formulating and revising scientific explanations, and communicating and defending scientific arguments. It is suggested that many of these abilities and skills are in alignment with those that characterize inquiry-based chemistry laboratory work, an activity that places the student at the center of the learning process [9, 20, 28, 29]. Researchers claim that learning in the laboratory might provide a constructivist environment that fosters higher order thinking and metacognitive skills [9, 30], including scientific thinking and inquiry skills [6, 7]. An example of an inquiry approach is illustrated in Figure 21.1.

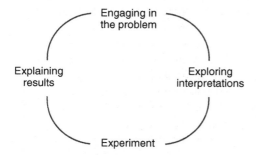

Figure 21.1 The learning cycle in an inquiry-type experiment.

21.3
From Theory to Practice: How Are Chemistry Laboratories Used?

The question to be asked before elaborating on each of the high-order learning skills is to what extent the use of practical work has changed at schools. I shall be looking at research describing how laboratories are used by teachers and students, as well as the nature of laboratory activities and facilities.

Hofstein and Lunetta [6] and Lunetta et al. [7] in their comprehensive reviews wrote that, although science teachers' articulated philosophies appeared to support a hands-on investigative approach with authentic learning experiences, the classroom practice of those teachers did not generally appear to be consistent with their stated philosophies. Several studies have reported that very often teachers involve students principally in relatively low-level, routine activities in laboratories and that teacher–student interactions focused principally on low-level procedural questions and answers. Marx et al. [31] reported that science teachers often have difficulty helping students ask thoughtful questions, design investigations, and draw conclusions from data. Similar findings were reported regarding chemistry laboratory settings by De Carlo and Rubba [32]. More recently, Abrahams and Millar [33] in the United Kingdom investigated the effectiveness of practical work by analyzing a sample of 25 "typical" science lessons involving practical work in English secondary schools. They conclude that the teachers' focus in these lessons was predominantly on making students manipulate physical objects and equipment. Hardly any teachers focused on the cognitive challenge of linking observations and experiences to conceptual ideas. Neither was there any focus on developing students' understanding of scientific inquiry procedures.

These are findings that echo the situation at any time in the history of school science. Basic elements of teachers' implementation of practical work seem not to have changed over the last century: students still carry out recipe-type activities which are supposed to reflect science procedures and teach science knowledge, but which in general fails on both. This is not to say everything is the same. Science education has moved forward in the last decades and improved teachers' professional knowledge and classroom practice, but this improvement has not sufficiently caught up with the challenges of using laboratory work in an efficient and appropriate way. Teachers still do not perceive what is required to make that laboratory activities serve as a principal means of enabling students to construct meaningful knowledge of science, and they do not engage students in laboratory activities in ways that are likely to promote the development of science concepts. In addition, many teachers do not perceive that helping students understand how scientific knowledge is developed and used in a scientific community is an especially important goal of laboratory activities for their students.

The reviews discussed earlier in this chapter reported a mismatch between the goals articulated for the school science laboratory and what students regularly do in those experiences. Ensuring that students' experiences in the laboratory are aligned with stated goals for learning demands that teachers explicitly link decisions regarding laboratory topics, activities, materials, and teaching strategies to desired outcomes for students' learning. The body of past research suggests that far more attention to the crucial roles of the teacher and other sources of guidance during laboratory activities is required, and researchers must also be diligent in examining the many variables that interact to influence the learning that occurs in the complex classroom laboratory. The science learning goals of laboratory experiences include enhancing mastery of science subject matter, developing scientific reasoning abilities, increasing understanding of the complexity and ambiguity of empirical work, developing practical skills, increasing understanding of the nature of science, cultivating interest in science and science learning, and improving teamwork abilities.

The research suggests that laboratory experiences will be more likely to achieve these goals if they (i) are designed with clear learning outcomes in mind, (ii) are thoughtfully sequenced into the flow of classroom science instruction, (iii) integrate learning of science content and process, and (iv) incorporate ongoing students' reflections and discussion.

21.4
Emerging High-Order Learning Skills in the Chemistry Laboratory

21.4.1
First Theme: Developing Metacognitive Skills

It is suggested that the inquiry laboratory, in addition to the cognitive and affective variables mentioned above, if designed properly can provide the students with opportunities to develop metacognitive learning skills. *Metacognition* refers to higher order thinking skills that involves active control over the thinking processes involved in learning. Activities such as planning how to approach a given learning task, monitoring comprehension, and evaluating progress toward the completion of a task are *metacognitive* in nature [34]. White and Mitchell [35] specified students' behaviors that are characterized as *good learning behaviors* for students who developed certain *metacognitive* skills. A large part of these behaviors (and skills) are actions that constitute an integral part of the inquiry laboratory activity, such as asking questions, checking work against instructions, correcting errors and omissions, justifying opinions, seeking reasons for aspects of current work, suggesting new activities and alternative procedures, and planning a general strategy before starting to work. Students participating in the inquiry laboratory activities are required to evaluate the experiment they had designed and monitor their thinking processes, thereby developing their metacognitive skills. Baird and White [36] as well as Kuhn *et al.* [37] argued that students who experience inquiry activity attain a desirable level of metacognition.

Metacognition refers to higher order thinking skills that involves active control (and awareness) over the thinking processes that emerge in learning. Activities such as planning how to approach a given learning task, monitoring comprehension, and evaluating progress toward the completion of a task are metacognitive in nature [34]. There is no single definition used for metacognition, and its diverse meanings are represented in the literature that deals with thinking skills. Schraw [38], for example, presents a model in which metacognition includes two main components: "knowledge of cognition" and "regulation of cognition." *Knowledge of cognition* refers to what individuals know about their own cognition or about cognition in general. It includes at least three different kinds of metacognitive knowledge: *declarative knowledge*, which includes knowledge about oneself as a learner and about factors that influence one's performance (knowing "about" things); *procedural knowledge*, which refers to knowledge about doing things in terms of having heuristics and strategies (knowing "how" to do things); and *conditional knowledge*, which refers to knowing when and why to use declarative and procedural knowledge (knowing the "why" and "when" aspects of cognition). *Regulation of cognition* refers to a set of activities that help students control their learning. Although a number of regulatory skills have been described in the literature, three essential skills are included in all accounts: *planning*, which involves the selection of appropriate strategies and the allocation of resources that affect performance; *monitoring*, which refers to one's online awareness of comprehension and task performance; and *evaluating*, which refers to appraising the products and efficiency of one's learning.

Other researchers have made different divisions and categorizations of metacognition (see, for example, Baird and White [36], Kuhn, 1999, [37]).

When applied to science learning, generally metacognition is related to *meaningful learning*, or learning with understanding [35–37, 39, 40] which includes being able to apply what has been learned in new contexts. Metacognition is also related to developing *independent learners* [25], who typically are aware of their knowledge and of the options to enlarge it. One key component is that of *control* of the problem-solving processes and the performance of other learning assignments. Researchers link this *control* to the student's *awareness* of his physical and cognitive actions during the performance of the tasks [36]. Another element is the student's *monitoring* of knowledge [39]. A learner who properly monitors his knowledge can distinguish between the concepts that he knows and the concepts that he does not know and can plan his learning effectively.

The link between metacognition and scientific inquiry for many seems to be an obvious one. Scientists depend on their ability to control reasoning when working out new ideas and weighing up the evidence confirming or contrasting these. Kuhn et al. [37] argue that students who experience inquiry activities in a similar way

> come to understand that they are able to acquire knowledge they desire, in virtually any content domain, in ways that they can initiate, manage, and execute on their own, and that such knowledge is empowering (p. 496).

Table 21.1 Inquiry-type chemistry laboratories.

Phase	Abilities and skills
Phase 1: Pre-inquiry • Describe in detail the apparatus in front of you • Add drops of water to the small test tube until the powder is wet. Seal the test tubes immediately	• Conducting an experiment
• Observe the test tube carefully, and record all your observations in your notebook	• Observing and recording observations
• *Phase 2: The inquiry phase of the experiment* *1. Hypothesizing* • Ask relevant questions. Choose one question for further investigation • Formulate a hypothesis that is aligned with your chosen question	• Asking questions and hypothesizing
2. Planning an experiment • Plan an experiment to investigate the question • Ask the teacher to provide you with the equipment and material needed to conduct the experiment	• Planning an experiment
• Conduct the experiment that you proposed	• Conducting the planned experiment
• Observe and note clearly your observations • Discuss with your group whether your hypothesis was accepted or you need to reject it	• Analyzing results, asking further questions, and presenting the results

Baird and White [36] claim that four conditions are necessary in order to induce the personal development entailed in directing purposeful inquiry: time, opportunity, guidance, and support. The science teacher should provide the students with experiences, opportunities, and the time to discuss their idea about the problems that they have to solve during the learning activity. The role of the teacher is to provide continuous guidance and support to ensure that the students develop control and awareness over their learning. This, it is suggested, can be accomplished by providing the students with more freedom to select the subject of their project and to manage their time and their actions in the problem-solving process.

An application of these perspectives is demonstrated in a chemistry laboratory program titled "Learning in the chemistry laboratory by the inquiry approach," [28] developed in at the department of Science Teaching at the Weizmann Institute of Science in Israel. For this program, about 100 inquiry-type experiments were developed and implemented in 11th and 12th grade chemistry classes in Israel (For the nature and dynamics of these experiments, see Table 21.1.).

A two-phased teaching process was used, including a guided pre-inquiry phase followed by a more open-ended inquiry phase. Based on their research Kipnis and

Table 21.2 Matching *the metacognitive* components that were observed with the various inquiry stages.

The inquiry stage	The students' *metacognitive* activity	The *metacognitive* component that is expressed
Asking questions and choosing an inquiry question	The students revealed their thoughts about the questions that were suggested by their partners and about their own questions	*Metacognitive declaration knowledge*
Choosing the inquiry question	The students choose the question that leads to conclusions	*Metacognitive procedural knowledge*
Performing their own experiment	The students plan changes and improvements to the experiment	*Planning* component of *regulation of cognition*
Drawing conclusions and writing the final report	At the final stage of the inquiry activity, the students use knowledge that was acquired in a different context to write their report and to draw conclusions	*Metacognitive conditional knowledge*
During the whole activity	The students examine the results of their observations in order to decide whether the results are logical	*Monitoring* and *evaluating* components of *regulation of cognition*

Hofstein [30] have linked metacognitive skills (based on Schraw's 1998 model) [38] to various stages of the inquiry-oriented experiments: (i) While asking questions and choosing an inquiry question, the students revealed their thoughts about the questions that were suggested by their partners and about their own questions. In this stage, the *metacognitive declarative knowledge* is expressed. (ii) While choosing the inquiry question, the students expressed their *metacognitive procedural knowledge* by choosing the question that leads to conclusions. (iii) While performing their own experiment and planning changes and improvements, the students demonstrate the *planning* component of *regulation of cognition*. (iv) At the final stage of the inquiry activity, when the students write their report and have to draw conclusions, they utilize *metacognitive conditional knowledge*. (v) During the whole activity, the students made use of the *monitoring* and *evaluating* components concerned with *regulation of cognition*. In this way, they examined the results of their observations in order to decide whether the results were logical (for a summary see Table 21.2).

Knowledge-centered learning environments encourage students to reflect on their own learning progress (metacognition). Learning is facilitated when individuals identify, monitor, and regulate their own thinking and learning. To be effective problem solvers and learners, students need to determine what they already know and what else they need to know in any given situation, including when things are not going as expected. For example, students with better developed metacognitive strategies will abandon an unproductive problem-solving strategy very quickly and substitute a more productive one, whereas students who possess less effective metacognitive skills will continue to use the same strategy long after it has failed to produce results [41].

21.4.2
Second Theme: Scientific (Chemical) Argumentation

21.4.2.1 The Nature of Argumentation in Science Education

When Driver *et al.* [42] presented their introduction to *argumentation* in the context of learning science, they pointed toward the relevance for practical work (see Figure 21.2). They saw argumentation as correcting the misinterpretation of scientific method which has dominated much science teaching in general and practical work in particular. Rather than focusing on the stepwise series of actions carried out by scientists in experiments, they claimed that focus should be directed toward the *epistemic practice* involved when developing and evaluating scientific knowledge. We sense two overlapping learning aims: first, that students should understand the scientific standards and their guiding epistemologies, and, next, that they should be able to apply these standards in their own argumentation.

We find many ways of approaching research of students' epistemological understanding and argumentation skills. One contribution comes from psychologists who identify scientific argumentation as the key element of scientific thinking [37] and work from the perspective that certain reasoning skills related to argumentation are domain-general. People who are good at scientific argumentation are (i) able to think *about* a scientific theory, rather than just think *with* it; (ii) able to encode and think about evidence in a similar way, and by this distance evidence from the theory; and (iii) able to put aside their personal opinions about what is "right" and rather weigh the theoretical claim against the evidence.

Figure 21.2 Working in chemistry laboratories.

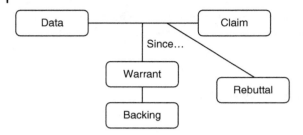

Figure 21.3 Toulmin's [53] model of argumentation.

Several research studies indicate that the development of students' argumentation skills and science epistemologies is rather complicated. Students, for example, may hold some beliefs about professional science and very different beliefs about their own practices with inquiry at school: that is, have one set of *formal* epistemologies and another set of *personal* epistemologies [43]. Many years of teaching "ideas and evidence" in the United Kingdom through practical investigations illustrates this complexity [44, 45]. The overall picture has been that students become good at doing specific types of routine experiments, and solve these from school-based strategies rather than a general understanding of formal scientific epistemologies [46].

A way forward to understand how practical work may contribute toward the development of students' epistemological understanding and argumentation skills may be to look closer at the "teaching ecology" of the laboratory [47]. It is strongly argued that bringing argumentation into science classrooms requires the enactment of contexts that transform them into knowledge-producing communities, which encourage dialogic discourse and various forms of cognitive, social, and cultural interactions among learners [48, 49]. The right ecology inviting this practice is created through the social and physical environment [50], the laboratory tasks [51], and the organization principles used by the teacher [52]. A reconsideration of all these factors is therefore needed for the science laboratory to contribute meaningfully and effectively toward the new learning goals.

Building an argument has significant social importance for students, in addition to their learning scientific concepts and high-order learning skills. While students are engaged in activities in which they are provided with opportunities to develop argumentative skills, they learn how to conduct a meaningful conversation with peers. Needless to say, these skills are useful for overcoming life's challenges and are not used solely in the context of science learning [47]. Toulmin's model [53] (see Figure 21.3) has been used to characterize argumentation in science lessons and is implicit in using a coding system of others [42, 43, 47, 54, 55].

21.4.2.2 Argumentation in the Chemistry Laboratory

In recent years, two studies have appeared in the literature regarding the nature of the experiments as a platform for evoking argumentation both quantitatively (number of arguments) and qualitatively (level of arguments). Kind *et al.* [56] in the United Kingdom investigated the quality of argumentation among 12–13-year-old students in the United Kingdom in the context of lower secondary school physical science program. Their study explored the development of argumentation in students who undertook three different designs of laboratory-based tasks. The tasks described in their paper involved the students in the following: collecting and making sense of data; collecting data for addressing conflicting hypothesis; and paper-based discussions in the pre-collected data phase about an experiment. Their finding showed that the paper-based tasks generated larger number of arguments in a unit of time compared to the two other (above-mentioned) tasks. In addition, they found that, in order to encourage the development of high-level and authentic argumentation, there is a need to change the practice that generally exists in the science laboratories in England. They suggested that more rigorous and longitudinal research is needed in order to explore the potential of the science (chemistry) laboratory as a platform for development of students' ability to argue effectively and in an articulated way.

The second study was conducted in Israel by Katchevich *et al*. [55] in the context of 12 years research and development of inquiry-type chemistry laboratories in upper secondary schools in grades 10–12 (for more details about the philosophy and rationale of the project, see Hofstein *et al*. [28]). The study focused on the process in which students constructed arguments in the chemistry laboratory while conducting different types of experiments. It was found that *inquiry-type* experiments have the potential to serve as an effective platform for formulating arguments because of the special features of this learning environment. The discourse was analyzed according to the following criteria – the components of the basic argument: claims, evidence, and scientific explanations. The analysis to identify the components of the argument was performed using Toulmin's model (Figure 21.3) [53]. Katchevich *et al*. [55] implemented Toulmin's framework to investigate on the epistemic and argumentative operations adopted by students. During the discourse, the students suggest different explanations for the various phenomena that they observed during the experimental procedure and then analyze the data and present arguments. The discourse conducted during *inquiry-type* experiments was found to be rich in arguments, whereas that during *confirmatory-type* experiments was found to be sparse in arguments. (For a comparison of inquiry- and confirmatory-type chemistry experiment, see Table 21.3.)

In addition, it was found that the arguments, which were developed during the discourse of an inquiry-type experiment, were generated during the following stages of the inquiry process: hypothesis-building analysis of the results and drawing appropriate conclusions. On the other hand, confirmatory-type experiment revealed fewer arguments. In addition, the arguments that were posed in the confirmatory-type experiments were of law level in their characteristic (for the various levels of the arguments see Table 21.4).

Table 21.4 presents the five levels of argumentation. The lowest level is "claim," when the students just provide a simple claim regarding experimental phenomena. The second level in addition to claim includes also claim based on experimental

Table 21.3 Skills developed in two types of experiments.

Learning skills that are involved during the experiment	Confirmatory-type experiment	Open-ended-type inquiry experiment
Conducting an experiment according to the teacher's instructions	✓	✓
Asking questions		✓
Formulating research questions		✓
Constructing a rational hypothesis		✓
Designing an appropriate inquiry experiment		✓
Conducting the experiment that was planned by the students		✓
Organizing the results	✓	✓
Analyzing the results	✓	✓
Drawing conclusions	✓	✓
Summarizing the experiment's procedure	✓	✓

data based on scientific principle (a warrant). The highest level (level 5) includes all the argument's components, namely a claim that includes rebuttal (untrue) based on data and warrant.

Based on a detailed analysis of the discourse that was conducted in the chemistry laboratory, the researchers concluded that the open-ended inquiry experiments stimulate and encourage the construction of arguments, especially the stages of hypotheses definition, analysis of the results, and drawing of conclusions. Some arguments are raised by individuals and some by the group. Both types of arguments consist of explanations and scientific evidence, which link the claims to the evidence. Therefore, it is suggested that the mentioned learning

Table 21.4 Levels of arguments in the inquiry chemistry laboratory.

The components of the argument	Symbol	The levels of the argument
Claim	C	1
Claim + Data or Claim + scientific basis-Warrant	CD/CW	2
Claim + Data + Warrant or Claim + Data + Rebuttal or Claim + Warrant + Rebuttal	CDW/CDR/CWR	3
Claim + Data + Warrant + Backing	CDWB	4
Rebuttal that includes Claim + Data + Warrant	CDWR	5

Based on: Erduran et al. [54]).

Table 21.5 The Criteria for sorting complex/simple experiments.

The type of experiment	Alignment with the concept or topic	Including concepts beyond the curriculum
Complex open-ended inquiry	No	Yes
Simple open-ended inquiry	Yes	No

environments of open-ended inquiry experiments is an effective platform for raising chemistry-based arguments.

More recently, Katchevich et al. [55] investigated the issue of the complexity of the chemistry inquiry experiments and posing arguments. The complexity of the inquiry experiments is presented in Table 21.5.

An analysis of the discourse clearly showed that the more complex the experiment, the more are the arguments raised by the groups of students. In addition, the complexity of the experiments derived higher level of arguments.

21.4.3
Asking Questions in the Chemistry Laboratory

In attempt to develop scientific literacy among students, teachers must create effective learning environments in which students are given opportunities to ask relevant and scientifically sound questions [57]. Usually, questions asked during a lesson are those initiated by the teacher and only rarely by the students; also, questions do not emerge spontaneously from students, but rather they have to be encouraged. In addition, it was reported that in cases where students do ask questions during the lessons, they are usually informative ones. The content of a question can indicate the level of thinking of the person who raised it. It should be noted that, in general, the cognitive level of a certain question is determined by the type of answer that it requires [58]. Several studies have noted the importance (and value) of questioning skills. For example, Zoller [59] in the context of learning high school chemistry, suggested that questioning is an important component in a real world, involving problem-solving and decision-making processes. Asking critical-type questions regarding a specific phenomenon posed to the students through a certain experiment or an article can avoid the general phenomenon of students asking factual-type questions [60]. It has been found that in science education, providing students with the opportunities to ask questions has the potential to enhance their creativity as well as their high-order thinking skills. More recently, Cuccio-Schirripa and Steiner [61] suggested that

> Questioning is one of the thinking processing skills which is structurally embedded in the thinking operation of critical thinking, creative thinking, and problem solving (p. 210).

This quote is in alignment with the results of a study conducted in chemistry by Dori and Herscovitz [62], who found that fostering 10th grade students' capabilities to pose questions improved their problem-solving ability. Hofstein *et al.* [29] conducted a research study that focused on the ability of high school (11th and 12th grade) chemistry students who learn chemistry through the inquiry approach (see Table 21.1 in this chapter and in Hofstein *et al.* [28]) to ask meaningful and scientifically sound questions. Two aspects were investigated in this study: (i) the ability of students to ask questions related to their observations and findings in an inquiry-type experiment (a practical test), and (ii) the ability of students to ask questions after critically reading a scientific article. The student population consisted of two groups: an inquiry-laboratory group (experimental group) and a traditional laboratory-type group (control group). Three common features were researched: (i) the number of questions that were asked by each of the students, (ii) the cognitive level of the questions, and (iii) the nature of the questions that were chosen by the students, for the purpose of further investigation. Importantly, it was found that students in the inquiry group who had experience in asking questions in the chemistry laboratory outperformed the control group in their ability to ask more and better questions. As seen from Table 21.2, the activity of asking inquiry-type questions (which are, by definition, high-level questions) is one of the operations that the students are required to do during every full inquiry experiment. In contrast, the students of the control group, who had learned the traditional-type program that does not contain the inquiry experiments, did not have any opportunity to practice the activity of asking questions and specifically asking inquiry questions, which are higher level questions, and therefore their skills in asking questions, as was indicated by the test, were lower.

21.5
Summary, Conclusions, and Recommendations

In this chapter, several research studies conducted in the chemistry laboratory were presented. The genuine attempt made to provide chemistry students with opportunities to learn and assume responsibility for their own learning as a result of conducting inquiry-type chemistry experiments was described. Evidence was presented that clearly shows that the students improved their ability to develop metacognitive skills, to sharpen their argumentative abilities, and to ask better (i.e., high-level) and more relevant questions (i.e., related to the chemistry concept learned). This happened because the students got the opportunity to develop such (and similar) skills. The three themes were chosen because I sincerely believe that these skills (metacognition, argumentation, and asking "good" questions) not only enhance effective chemistry learning but also will educate future literate citizens in their daily lives. In addition, the students who were involved in the inquiry experiences were more motivated to pose questions regarding scientific phenomena that were presented to them via an article. These findings are not surprising because during the inquiry activity, which was an integral part of their chemistry laboratory activities, these students had practiced asking questions and formulating inquiry questions. As can be seen from Table 21.1, the activity of asking inquiry-type questions (which are, by definition, high-level questions) is one of the operations that the students are required to do during almost in every inquiry experiment. In contrast, the students of the control group who had learned the traditional-type program, which does not contain the inquiry experiments, did not have any opportunity to practice the activity of asking questions and specifically asking inquiry questions that are of higher level, and therefore their skills in asking questions, as was indicated by the test, were lower. The activities in which the students were involved in this project are very much in alignment with the claim made by Tobin [63] who wrote:

> A crucial ingredient for meaningful learning in laboratory activities is to provide for each student opportunities to reflect on findings, clarify understanding and misunderstanding with peers, and consult a range of resources, which include other students, the teacher, and books and materials (p. 415).

In many countries around the world, achieving chemistry literacy for all students has become a central goal of education. Although admirable, this goal represents a challenge for both science curriculum developers and teachers who cooperatively work to attain this goal. The target population is not only those who will eventually embark on a career in the sciences but also those who are often called "future citizens." As such, they will often find themselves in situations in which they will need to ask critical questions and seek answers based on which they will need to make valid decision. Thus, the development of students' ability to ask questions should be seen as an important component of scientific literacy and should not be overlooked.

In recent years, there has been substantive growth in understanding associated with teaching, learning, and assessment in school chemistry laboratory work. At the beginning of the twenty-first century, when many are again seeking reforms in science education, the knowledge that has been developed about learning based upon careful scholarship should be incorporated in that reform. The "*The less is more slogan*" in "*Benchmark for Science Literacy*" (AAAS, 1990, p. 320) [26] has been articulated to guide curriculum development and teaching consistent with the contemporary reform. The intended message is that formal teaching results in greater understanding when students study a limited number of chemistry topics, in depth and with care, rather than a large number of topics much more superficially, as is the practice in many upper secondary school chemistry classrooms. In order to make room for the inquiry laboratories in general and development of high order learning skills in particular, the syllabus (content) should be reduced profoundly. Well-designed, inquiry-type laboratory activities can provide learning opportunities and help students develop high-level learning skills. They also provide important opportunities to help students to learn to investigate (e.g., ask questions), to construct scientific assertions, and to justify those assertions in classroom community of peer investigators in contact with more expert scientific community. There is no doubt that such activities are time consuming, and therefore the education system must provide time and opportunities for teachers to interact with their students and also time for students to perform and reflect on such and similar complex inquiry and investigative tasks. Such experiences should be integrated with other science classroom learning experiences in order to enable the students to make connections between what is learned in the chemistry classroom and what is learned and investigated in the laboratory. This is highly based on the growing sense that learning is contextualized and that learners construct knowledge by solving genuine and meaningful problems [64]. One of the most crucial problems regarding the implementation of inquiry-type chemistry laboratory experiment is the issue of assessing students' achievement and progress in such a unique learning environment. In general, large numbers of science teachers are not using authentic and practical assessment on a regular basis.

In addition to the organizational factors regarding the volume of the content taught and the context in which laboratory experiences are conducted, we have to pay a lot more attention to the science teacher. Clearly, serious discrepancies exist between what is actually occurring in the laboratory classroom and what is recommended for high-level science teaching.

Unfortunately, many science teachers do not utilize or manage the laboratory effectively.

In their recent review of the literature, Hofstein and Lunetta [6, p. 45] noted that

> Conditions are especially demanding in science laboratories in which the teacher is to act as facilitator who guides inquiry that enables students to construct more scientific concepts ... Teachers are often not well informed about new modes of learning and their implications for teaching and curriculum. While excellent examples of teaching can be observed, the classroom behaviors of many teachers continue to suggest conventional beliefs that knowledge is directly transmitted to the students and that it is to be remembered as conveyed.

This is, in fact, a call for changing the strategies that are employed in preservice and in-service professional development courses provided to the science teachers. It is suggested that, in order to implement similar learning strategies described in this chapter, teachers need to undergo continuous and long-term professional development aimed at enhancing both their content knowledge as well as their pedagogical content knowledge. Such professional development experiences have the potential to help teachers develop skills and confidence to construct effective learning environment to include substantive and meaningful science laboratory experiences. More research is needed to investigate the effectiveness of different models for science teachers' professional development that are used to provide teachers with the skill to implement student-centered instructional techniques in general and inquiry-type experiences in the chemistry laboratory in particular.

We operate in an era in which we can provide the students with ICT tools (including the Internet and computer simulations). Inquiry-empowering technologies have been developed and implemented to assist students in many of the skills in which they are involved before, during, and after the experimentation phase. These includes, among others, gathering information, planning an experiment, as well as interpreting, analyzing, and reporting data. The ICT tools used and integrated in the chemistry laboratory, if developed effectively, can help in reducing the time during which students might be involved in law-level skills and thus provide time for the development of high-order learning skills, which were discussed in detail in this chapter.

In conclusion, we are entering an era in which high-order learning (and thinking skills) in the context of teaching and learning chemistry is seen to be as important as content and conceptual skills. Thus, research that will eventually enhance our understanding of how these skills are developed and in which pedagogical context should not be overlooked. Over the period of almost 60 years, the goals for practical work in the context of chemistry changed periodically. Aligned with these goals, the community in chemistry education also changed the skills that the learners had to develop. In the 1960s and 1970s, in many countries around the world, students learned chemistry for future science-based careers. Thus, it was assumed that the manipulative skills were an important ingredient of practical work. In the last 20 years, the idea of teaching chemistry for all students has become one of the key challenges. As a result, the nature of the experiments as well as the role of the students and their chemistry teachers has changed. Manipulating equipment and complex calculations are seen as a tool to support the experimentation.

In the turn of the century, we can claim that chemistry education is in a better position than it was earlier for developing meaningful and appropriate practices for chemistry-based laboratory work. The research community in chemistry education has accumulated a lot of information (based on research) regarding learning and teaching skills. This information should be used for the development of new laboratory strategies and professional skills for teachers. Thus, this chapter is in fact a call for chemistry education researchers to engage in practical work in order to further develop and enhance this area.

References

1. Rosen, S.A. (1954) *Am. J. Phys.*, **22**, 194–204.
2. Klainin, S. (1988) in *Developments and Dilemmas in Science Education* (ed. P. Fensham), The Falmer Press, London, pp. 169–188.
3. Hofstein, A. and Lunetta, V.N. (1982) *Rev. Educ. Res.*, **52**, 201–217.
4. Lunetta, V.N. (1998) in *International Handbook of Science Education* (eds B. Fraser and K. Tobin), Kluwer Academic Publishers, Dordrecht, pp. 249–262.

5. Lazarowitz, R. and Tamir, P. (1994) in *Handbook of Research on Science Teaching* (ed. D.L. Gabel), Macmillan, New York, pp. 94–127.
6. Hofstein, A. and Lunetta, V.N. (2004) *Sci. Educ.*, **88**, 28–54.
7. Lunetta, V.N., Hofstein, A., and Clough, M.P. (2007) in *Handbook of Research on Science Education* (eds S.K. Abell and N.G. Lederman), Lawrence Erlbaum, Mahwah, NJ, pp. 393–441.
8. Hodson, D. (1993) *Stud. Sci. Educ.*, **22**, 85–142.
9. Hofstein, A. and Kind, P.M. (2012) in *Second International Handbook of Science Education* (eds B. Fraser, K. Tobin, and C. McRobbie), Springer, Dordrecht, pp. 189–209.
10. Bates, G.R. (1978) in *What Research Says to the Science Teacher*, vol. 1 (ed. M.B. Rowe), National Science Teachers Association (NSTA), Washington, DC, pp. 58–82.
11. Kerr, J.F. (1963) *Practical Work in School Science*, Leicester University Press, Leicester.
12. Shulman, L.S. and Tamir, P. (1973) Research on teaching in the natural sciences, in *Second Handbook of Research on Teaching* (ed. R.M.W. Travers), Rand McNally.
13. Bryce, T.G.K. and Robertson, I.J. (1985) *Stud. Sci. Educ.*, **12**, 1–24.
14. Johnstone, A.H. and Wham, A.J.B. (1982) *Educ. Chem.*, **19** (3), 71–73.
15. National Research Council (1996) National Science Education Standards, http://www.nap.edu/openbook.php?record_id=4962
16. Hofstein, A., Kipnis, M., and Abrahams, I. (2013) in *Teaching Chemistry a Study Book* (eds I. Eilks and A. Hofstein), Sense Publishers, Rotterdam, pp. 153–183.
17. Dori, Y.J., Rodrigues, S., and Schanze, S. (2013) in *Teaching Chemistry a Study Book* (eds I. Eilks and A. Hofstein), Sense Publishers, Roterdam, p. 213.
18. Fensham, P.J. (1983) *Sci. Educ.*, **67**, 3–12.
19. Schwartz, Y., Ben-Zvi, R., and Hofstein, A. (2006) *Chem. Educ. Res. Pract.*, **7**, 203–225.
20. Dori, Y.J. and Sasson, I. (2008) *J. Res. Sci. Teach.*, **45** (2), 219–250.
21. Millar, R. and Osborne, J. (1998) *Beyond 2000: Science Education for the Future*, King's College, London.
22. Resnick, L.B. (1987) *Education and Learning to Think*, National Academy Press, Washington, DC.
23. Zohar, A. and Dori, J.Y. (2003) *J. Learn Sci.*, **12**, 145–182.
24. Gunstone, R.F. and Champagne, A.B. (1990) in *The Student Laboratory and the Science Curriculum* (ed. E. Hegarty-Hazel), Routledge, London, pp. 159–182.
25. National Research Council NRC (1996) *National Science Education Standards*, National Academy Press, Washington, DC.
26. American Association for the Advancement of Science (AAAS) (1990) *Science for all Americans*, American Association for the Advancement of Science, Washington DC.
27. Bybee, R. (2000) in *Inquiring into Inquiry Learning and Teaching* (eds J. Minstrel and E.H. Van Zee), American Association for the Advancement of Science, Washington, DC, pp. 20–46.
28. Hofstein, A., Shore, R., and Kipnis, M. (2004) *Int. J. Sci. Educ.*, **26**, 47–62.
29. Hofstein, A., Navon, O., Kipnis, M., and Mamlok-Naaman, R. (2005) *J. Res. Sci. Teach.*, **42**, 791–806.
30. Kipnis, M. and Hofstein, A. (2008) *Int. J. Sci. Math. Educ.*, **6**, 601–627.
31. Marx, R.W., Freeman, J.G., Krajcik, J.S., and Blumenfeld, P.C. (1998) in *International Handbook of Science Education* (eds B. Fraser and K. Tobin), Kluwer Academic Publishers, Dordrecht, pp. 667–680.
32. De Carlo, C.L. and Rubba, P. (1994) *J. Chem. Educ.*, **76**, 109–111.
33. Abrahams, I.Z. and Millar, R. (2008) *Int. J. Sci. Educ.*, **30**, 1945–1969.
34. Livingston, J.A. (1997) Metacognition: An Overview. State University of New York at Buffalo, http://www.gse.buffalo.edu/fas/shuell/cep564/Metacog.htm (accessed 10 April 2004).
35. White, R.T. and Mitchell, I.J. (1994) *Stud. Sci. Educ.*, **23**, 21–37.

36. Baird, J.R. and White, R.T. (1996) in *Improving Teaching and Learning in Science and Mathematics* (eds D.F. Treagust, R. Duit, and B.J. Fraser), Teachers College, Columbia University Press, New York, pp. 190–200.
37. Kuhn, D., Black, J., Keselman, A., and Kaplan, D. (2000) *Cognit. Instr.*, **18**, 495–523.
38. Schraw, G. (1998) *Instr. Sci.*, **26**, 113–125.
39. Rickey, D. and Stacy, A.M. (2000) *J. Chem. Educ.*, **77**, 915–920.
40. Thomas, G.P. and McRobbie, C.J. (2001) *J. Res. Sci. Teach.*, **38**, 222–259.
41. Gobert, J.D. and Clement, J.C. (1999) *J. Res. Sci. Teach.*, **36**, 39–53.
42. Driver, R., Newton, P., and Osborne, J. (2000) *Sci. Educ.*, **84**, 287–312.
43. Sandoval, W.A. (2005) *Sci. Educ.*, **89**, 634–656.
44. Driver, R., Leach, J., Millar, R., and Scott, P. (1996) *Young Peoples' Images of Science*, Open University Press, Buckingham.
45. Solomon, J., Duveen, J., and Scott, L. (1994) *Int. J. Sci. Educ.*, **16**, 361–373.
46. Kind, P.M. (2003) *Sch. Sci. Rev.*, **85** (311), 83–90.
47. Jimenez-Aleixandre, M.P., Rodriguez, A.B., and Duschl, R.A. (2000) *Sci. Educ.*, **84**, 757–792.
48. Duschl, R.A. and Osborne, J. (2002) *Stud. Sci. Educ.*, **38**, 39–7.
49. Newton, P., Driver, R., and Osborne, J. (1999) *Int. J. Sci. Educ.*, **21**, 553–576.
50. Roth, W.M., Bowen, M.K., and McGinn, W.M. (1999) *J. Learn. Sci.*, **8**, 293–347.
51. Chinn, C.A. and Malhotra, B.A. (2002) *Sci. Educ.*, **86**, 175–218.
52. Scott, P. (1998) *Stud. Sci. Educ.*, **32**, 45–80.
53. Toulmin, S. (1958) *The Uses of Argument*, Cambridge University Press, Cambridge.
54. Erduran, S., Simon, S., and Osborne, J. (2004) *Sci. Educ.*, **88**, 915–933.
55. Katchevich, D., Hofstein, A., and Mamlok-Naaman, R. (2013) *Res. Sci. Educ.*, **43**, 317–345.
56. Kind, P., Wilson, J., Hofstein, A., and Kind, V. (2011) *Int. J. Sci. Educ.*, **33**, 2527–2558.
57. Penick, J.E., Crow, L.W., and Bonnsteter, R.J. (1996) *Sci. Teach.*, **63**, 26–29.
58. Yarden, A., Brill, G., and Falk, H. (2001) *J. Biol. Educ.*, **35**, 190–195.
59. Zoller, U. (1987) *J. Chem. Educ.*, **64**, 510–511.
60. Shodell, M. (1995) *Am. Biol. Teach.*, **57**, 278–281.
61. Cuccio-Schirripa, S. and Steiner, H.E. (2000) *J. Res. Sci. Teach.*, **37**, 210–224.
62. Dori, Y.D. and Herscovitz, O. (1999) *J. Res. Sci. Teach.*, **36**, 411–430.
63. Tobin, K. (1990) *Sch. Sci. Math.*, **90**, 403–418.
64. Brown, J.S., Collins, A., and Duguid, P. (1989) *Educ. Res.*, **18** (1), 32–41.

22
Chemistry Education Through Microscale Experiments*
Beverly Bell, John D. Bradley, and Erica Steenberg

22.1
Experimentation at the Heart of Chemistry and Chemistry Education

In the popular, uneducated mind, chemistry is symbolized by the white-coated man in a laboratory crowded with large glass vessels containing liquids of assorted colors, and despite indicators of modernity, the scene is often evocative of older times with its hint of madness, magic, and the exotic. Chemists react against this stereotype, yet like all stereotypes it contains fragments of the truth. And chemistry education is concerned with learning about this scene and its background and/or about learning to be like and to work like that man.

The essential truth in the image is of course that experimentation is at the heart of chemistry – a view expressed succinctly in an IUPAC report [1]:

> ...chemistry is fundamentally an experimental subject... education in chemistry must have an ineluctable experimental component.

This truth still holds, despite the virtual, computer-based activities that many chemists engage in today. The other fragments of the image – the gender, the large glass vessels, and so on – may no longer be truthful to the reality. Similarly, the popular image of the hazards of chemistry experimentation, with its explosions and evil smells polluting the air, is now far from the truth and worth no more than a condescending laugh by the professional chemist.

Education aims to prepare young people for adult life. Thus it is unsurprising that experimentation is generally seen as an integral part of chemistry education. This is commonly reflected in curriculum documents published by Ministries of Education, and in the views expressed by teachers. This is logical, whether the overall curricular aim is to learn about chemistry or to learn to be a chemist.

* This chapter is dedicated to Dr Erica Steenberg (1953–2013), whose valuable contributions to Chemistry Education were many, especially through Microscale Experimentation.

Chemistry Education: Best Practices, Opportunities and Trends, First Edition.
Edited by Javier García-Martínez and Elena Serrano-Torregrosa.
© 2015 Wiley-VCH Verlag GmbH & Co. KGaA. Published 2015 by Wiley-VCH Verlag GmbH & Co. KGaA.

22.2
Aims of Practical Work

Accepting that practical work should be part of the chemistry curriculum, it remains to clarify more specifically what purposes it might serve. Many authors have engaged in debate about this, either from a philosophical/educational view or because they have to justify the costs inherent in providing for practical work. As the push to provide education for all has strengthened, so the cost burden of this provision has become increasingly apparent. Whilst science must be in the curriculum, is it really essential to include practical activities? [2]

Among the many published lists of aims to be found in the literature, the relatively brief one of Woolnough and Allsop [3] has much general support:

- Motivation
- Developing Practical Skills
- Learning the Scientific Approach
- Gaining a Better Understanding of Theoretical Aspects of the Subject.

These may be seen to reflect the alternative overall curricular aims mentioned above – learning about chemistry or learning to be a chemist.

The first aim is distinctive, implying that practical work has potential benefits for the entire subject curriculum. Anybody concerned with education would prize this attribute if the aim can be achieved in the local circumstances. Teachers must surely prize motivation and should thus be motivated to include practical work. They would also value the better understanding of theoretical aspects that may be achieved through practical work, because this understanding is usually the main focus of national assessment.

22.3
Achieving the Aims

The realization of these aims is the challenge. Starting with the teachers who lack the skills and methods required, the lack of physical resources (equipment and chemicals), the lack of technical assistance, the threats and constraints of health and safety legislation, and ending with the crowded classrooms, the challenge has proved, more often than not, too great.

Furthermore, in some countries practical work is not assessed, while in others it is rigidly defined within an examination framework that is primarily skills-focused and disadvantages learners from poorer schools. For some of these reasons, written practical examinations have been introduced, which may be attempted without ever having enjoyed practical work at all.

22.4
Microscale Chemistry Practical Work – "The Trend from Macro Is Now Established"

Given the undeniable reason for chemistry experimentation to be part of a chemistry curriculum and the impressive aims that may be achieved by undertaking it, the widespread failure has provoked many into searching for solutions. The adoption of microscale techniques has been one proposed solution. Several advantages are claimed for these techniques: lower cost of equipment and consumables (chemicals), easier and more user-friendly equipment, greater safety and lesser environmental impact, and so on. These advantages are attractive and they are behind the trend [4].

Microscale chemistry has deep roots. A principal root was the microscale quantitative analysis that developed into a highly specialized professional service. Within the research and teaching laboratories, however, it followed only slowly. Spot tests and semi-micro qualitative analysis had their minority followings but they never overwhelmed the traditional scale work. It was when microscale glassware for organic chemistry (first for research and then for teaching) came in that one could say a trend developed. From the 1980s onwards, this has spread across the majority of educational institutions where the curriculum requires organic chemistry practical work. Inorganic chemistry has followed suit.

Pike [5] has summarized the reasons why microscaling suddenly caught on. Essentially it was driven by escalating costs, which in turn originated from the chemicals used, glassware breakages, and safety and environmental requirements. Acceptance by traditionalists was probably eased by the recognition that much of the equipment looked like miniaturized versions of what they were familiar with.

At somewhat lower educational levels, traditional scale glassware remained for some time, but gradually here too microscale equipment (mostly plastics) has been penetrating. The driver is again escalating costs. In addition, felt more strongly at these levels, is the huge expansion in access to science education. This wonderful development remains in many countries unsatisfied in the sense that after access there are inadequate resources (human and physical) to serve all in the traditional ways. Slowly this unsatisfactory provision is being overcome by recognizing the potential of microscale chemistry experimentation. The UNESCO Global Microscience project has done much to create awareness, especially in developing countries, and hence to unlock this potential [6, 7].

Simple logic suggests there is no reason why any of the four aims would in some way be thwarted simply by miniaturizing the equipment. Only those who interpret the aim of developing practical skills in terms of using the specific traditional equipment (especially the glassware), beloved of the stereotypical chemist, will argue otherwise.

The aims of practical work in chemistry are surely general and do not specify particular equipment, nor should they in general education. On-the-job training is another matter, and is very likely to include training on equipment that is well beyond the reach of any school system. An unfortunate aspect of some examining boards and curriculum advisers is that they have not clarified this distinction!

22.5
Case Study I: Does Scale Matter? Study of a First-Year University Laboratory Class

University first-year chemistry laboratory classes are generally viewed by academic staff as a kind of necessary evil. The classes are often large and populated by students with no intention of majoring in the subject, and are a substantial cost burden on the department. In addition, especially in developing countries like South Africa, the majority of students have had no previous hands-on practical experiences at school. In this context, Sebuyira [8] compared attitudes, knowledge gains, and views about traditional macroscale and microscale versions of the same practical experiments. Sixty-five students completed four experiments, two on macroscale and two on microscale, and comparisons were based upon these experiences:

1) Properties of oxides
2) Metals: aluminium, copper, lead, and zinc
3) Acids and bases
4) Physical and chemical properties of organic compounds.

The experiments involved qualitative or semi-quantitative observations. Demonstrators, laboratory managers (academic), and laboratory technical staff were also interviewed. None of the participants had previous experience with the microscale equipment but, being in the second half of the course, all students had prior experience of the macroscale equipment.

The general attitude toward microscale experiments was positive across all types of individuals involved. Students found that using the microscale equipment was more demanding of their attention, and was safer and quicker. Technical staff commented on the lesser mess created by students using microscale.

As regards knowledge gains, pre and post testing was conducted for two of the experiments. In both cases, the knowledge gains by the microscale students were superior to the others. This surprising result was linked in one experiment with the microscale students avoiding contamination of test tubes, which the macroscale students suffered from. In the other experiment, the difference could be related to the superior clarity of the microscale conductivity indicator (Bar Light Emitting Diode or LED) used, as compared to the estimation of the light intensity of a bulb.

The superiority of the knowledge gains of microscale students may be due to the ancillary influences mentioned, but they are not to be despised because of that. More importantly, they counter the possible interpretation of "more demanding of their attention" as meaning observation is impaired. Student responses included statements like "more difficult to see," but this should not be interpreted as "unable to observe."

Cost comparisons were very favorable to the microscale experiments, with the required equipment being some six-fold cheaper than the macroscale and the chemicals about one-third the cost of macroscale.

Table 22.1 Does scale matter?

Does scale matter?	Macro vs micro
Results/observations	Approximately same, but micro better
Learning	Micro somewhat better
Safety/environment	Micro better
Speed	Micro faster
Cost	Micro much cheaper

Considering all the circumstances, the answer to the question – Does Scale Matter? – depends upon what you mean. Table 22.1 refers to the possible different aspects being queried.

So scale does matter, but with a clear indication that micro has the advantage. This was in a context already described, and where the instructional style was the much-maligned expository style [9]. It is a matter for speculation what conclusions might be reached when other styles of laboratory instruction (inquiry, discovery, problem-based) are employed. However, it must surely be that the favoring of microscale in regard to safety, speed, and cost should apply regardless of style, and should foster the conduct of more experimentation, which lies at the heart of chemistry education.

In conclusion, this case study showed us that three of the aims (motivation, practical skills, and better understanding) were achievable at least as well on microscale as on the macroscale.

22.6
Case Study II: Can Microscale Experimentation Be Used Successfully by All?

Case Study I involved first-year university students at an urban university in South Africa. Case Study II focuses on the evidence from the school system as derived from an earlier study (1995) by Vermaak [10] in South Africa. In this country, there is a wide variety of schools, learners, and teachers, so a number of comparisons can be made that can answer the question posed. The total sample used consisted of 30 schools and involved about 700 learners in Grade 11, carrying out five different experiments. Significant improvements in subject knowledge and understanding on certain topics could be attributed to the microscale chemistry experimentation (pre/post testing). Furthermore, no significant differences were found between boys and girls or between different cultural groups. Irrespective of pupils' background or prior knowledge, they equally improved their subject understanding. Differences were, however, found between pupils from different geographical areas – perhaps reflecting teacher differences.

Attitudes were also studied, and this revealed a very strong positive attitude to practical work in general at the outset. This was enhanced after the microscale experience. There was no difference between boys and girls in this regard, but there were significant differences between cultural groups and geographical areas.

The above findings provide evidence, therefore, that, in secondary schools also, most learners can achieve the aims of practical work with microscale equipment. It may be acknowledged that the quality of the teacher is a major determinant of outcomes, and this is a significant limitation in many countries to getting value from practical work, regardless of the scale.

The influence of teacher quality is likely to be particularly keenly felt in primary schools, where many teachers may lack subject knowledge and experience of experimentation. Nevertheless, where the quality is good, the outcomes from microscale chemistry experimentation can again be good. The Kenyan study by Michieka and Wasika [11] used 92 learners from four schools and found skills development and knowledge gains attributable to this.

In conclusion, there is evidence that microscale chemistry (or science) experimentation can be used successfully with a variety of school learners to achieve motivation, develop practical skills, and gain a better understanding of theoretical aspects of the subject.

22.7
Case Study III: Can Quantitative Practical Skills Be Learned with Microscale Equipment?

Lord Kelvin (1824–1907) said: "when you can measure what you are speaking about and express it in numbers, you know something about it." This implies that the aim of developing practical skills must include quantitative ones.

But can microscale chemistry be used quantitatively? This question is frequently asked, probably because there has been so much attention given to the wonderfully quick and simple qualitative microscale experiments previously mentioned. Traditionalists are inclined to say, "microscale is all very well, BUT when it comes to 'real chemistry' (otherwise known as *volumetric analysis*) you must use traditional glassware." In this case study, we show that in the context of chemistry education this is quite wrong. Besides, quantitative chemistry is not only volumetric analysis.

22.7.1
Volumetric Analysis – Microtitration

Serious studies have revealed what can be achieved on microscale in the area of volumetric analysis. We present here the microtitration and demonstrate the precision and accuracy of this technique. This may use plastics rather than glass and typically involves volumes that are 10% of traditional ones.

For our purposes, a microtitration can be described as a small-scale titration that is carried out in the large wells of a comboplate (a microwell plate with a combination of 48 small wells and 12 large wells) using a microburette. The microburette consists of a plastic 2-ml (2 ml × 0.01) pipette with a tip at the end. This in

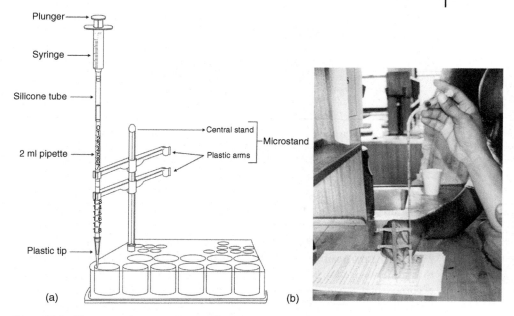

Figure 22.1 Diagrammatic representation of a microtitration setup (a) and educators performing a microtitration in the comboplate (b).

turn is attached to a plastic syringe, which is used as the means by which a solution is drawn into the burette as well as added to a microwell. The microburette is supported by a special plastic stand referred to as a *microstand* (Figure 22.1). The microburette clips into the microstand, which fits snugly into the small wells of the comboplate. A second identical pipette with tip and syringe mimics a conventional pipette used to accurately add aliquots of analyte solution to the large microwells.

These microtitrations use very small volumes of titrant and analyte solutions (maximum volume of titration mixture is about 2 ml). Since there are several microwells in which microtitrations can be performed, students can repeat their titrations quickly and easily, simply by moving the position of the microburette in the comboplate.

Comparisons have been made with conventional methods of titration by a number of groups. Table 22.2 shows the molar concentration calculated for a sodium hydroxide solution after performing macroscale and microscale standardizations of the solution using 0.10 M hydrochloric acid. A normal 50-ml burette was used for the conventional titrations and the microburette (2 ml × 0.01) described above for the microtitrations. The relative percentage difference in the microtitration technique is 0.17%. Teachers and students can therefore be confident that microtitration can yield sufficiently accurate results.

The results in Table 22.2 are based on our own study (B. Bell, 2001, unpublished results) carried out by an individual who was familiar with the microscale

Table 22.2 Comparison of macroscale versus microscale titration for the standardization of NaOH(aq) with standard HCl(aq).

Method of titration	Volume of NaOH(aq) (ml)	Number of determinations	Average volume of standard HCl(aq) titrated after blank (ml)	Concentration of NaOH(aq) calculated (mol dm^{-3})	Relative % difference
Macro	20.00	4	19.24	0.1149	—
Micro	1.000	7	1.006	0.1151	0.17

$$\text{Relative \% difference} = \frac{[\text{concentration in microtitration} - \text{concentration in macrotitration}]}{\text{concentration in macrotitration}} \times 100\%$$

procedure at the time. Other research has focused on microtitrations where the participants have not previously been exposed to the small-scale technique, such as that carried out in 2006 with Malaysian secondary school teachers [12] and in 2003 with American university-level students [13]. Both of these groups found that students need to be familiar with handling the microburette, but once this is accomplished, the microburette can give precise measurements: in the 2006 study, the relative standard deviation of a microscale acid–base titration was found to be 2.0% compared with 1.3% using the traditional technique. In terms of accuracy, the relative error of the microscale technique was shown to be 0.86% [12]. In the 2003 university-level project [13], the results also showed that the microtitration procedure was comparable in accuracy and precision to the traditional one.

Recently (2013), Abdullah *et al.* introduced microscale volumetric analysis experiments into the Malaysian preservice teachers (PPISMP) program [14] and again found that the precision achieved was acceptable for quantitative analysis. Moreover, the results revealed that, by using microscale titrations, the apparatus cost can be reduced by 82%, waste can be reduced by 99%, and up to 60% time can be saved on volumetric analyses. This substantiates the findings of Case Study I.

Can quantitative practical skills be learned using microscale titration equipment? Yes!

22.7.2
Gravimetric Measurements

It has been stated that modern chemistry began with the use of a precise scale and that Lavoisier could demonstrate in 1775 already that the mass of solid increases when a metal reacts with oxygen [15]. So also, learners should be able to make mass measurements and interpret these in different contexts, for example, conservation of mass, stoichiometry, density determinations, percentage yield of product, analytical chemistry, and so on. There are various options available for the small-scale chemist.

The use of an Egyptian 25-g two-pan scale for microscale chemistry practical work involving beginners of junior high school has been documented [15], and a similar scale has been used in a special low-cost experimental kit developed for junior high school students in Germany, participating in a project called "Chemistry in Context" [16].

Others have used a portable mass balance or a Digital Pocket Scale (DPS). These pocket scales are much lower in cost than their traditional laboratory counterparts (20 times the value of a typical DPS) [17]. They use batteries (ordinary and/or rechargeable) and can be used to teach learners weighing by difference, although they also have a tare function.

There are various types and models of DPS. Those with a capacity of $150-200 \pm 0.05$ g and reading to two decimal places are suitable for small-scale gravimetric experiments. Those with a capacity of 500 g can be used by the teacher to prepare stock solutions of chemicals required for microscale practical work. The DPS has been used for several experiments at school level, which include measurements of mass – especially conservation of mass activities described in Case Study IV. In their book *Microscale Chemistry Experiments Using Water and Disposable Materials*, El-Marsafy, Schwarz, and Najdoski use a similar DPS for some experiments, such as "Heating a Cola beverage to dryness" in order to weigh the solids separated from small volumes of Cola and Cola Light [18].

Bell *et al.* tested the suitability of the DPS for gravimetric measurements related to the electrowinning of copper [19]. The copper mining industry is important in South Africa and is therefore included in the school science curriculum. One of the methods for purifying copper is electrowinning, where the impure copper is used as an anode in an electrolytic cell containing acidified copper sulfate solution as electrolyte. During electrolysis, copper ions are formed in solution as a result of oxidation at the anode, and pure copper forms at the cathode (reduction). The cathode is a pure copper electrode.

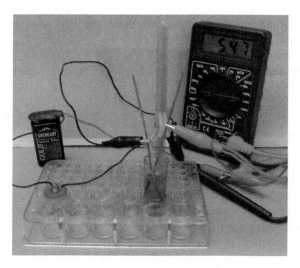

Figure 22.2 Microscale setup for the electrowinning of copper.

Table 22.3 Gravimetric results for small-scale electrowinning of copper.

Experimental conditions	Plating time (h)	Mass loss (g)	Mass gain (g)
1 × 1.5 V zinc/carbon cell 1 M copper sulfate electrolyte in large well of comboplate 1 mm dia. copper rod immersed ±10 mm Distance between electrodes ±10 mm No electrode surface preparation	3	−0.05	+0.04
1 × 9 V zinc/carbon battery 1 M copper sulfate electrolyte in small sample vial 1 mm dia. copper rod immersed ±30 mm Distance between electrodes ±10 mm Electrode surface sanded	26	−0.27	+0.25

The small-scale setup for the electrowinning of copper is shown in Figure 22.2.

Table 22.3 gives two example experimental results, from which it can be concluded that even fairly short experiments can give adequate results. The consistent observation that the mass loss of the anode exceeds the mass gain of the cathode can be related to the impure copper used for the anode.

The results show that this is a feasible technique to study the effects of plating potential, current, and time on electrolysis. For longer plating times, the mass gain/loss is well within the range of accuracy provided by the DPS.

This is also a very effective visual experiment because students are able to observe one copper electrode getting "thinner" (oxidation) and the other copper electrode getting "fatter" (reduction) (see Figure 22.3).

Apart from conceptual understanding, students also learn that to obtain good quantitative results requires careful attention to details (such as clean, dry electrodes before and after!).

22.7.3
The Role of Sensors, Probes, and the Digital Multimeter in Quantitative Microscale Chemistry

Contemporary developments in electronics and instrumentation have paved the way for measurement of chemical quantities using low-cost methods that do not need to be performed in a typical laboratory. Regardless of scale, chemical sensors or sensing probes are able to recognize the material to be analyzed, and the primary signal received (electrochemical, heat, light absorbed) can be translated into a secondary signal that is easy to record and interpret. The digital multimeter is one such device that can measure a secondary electrical signal and display it as an

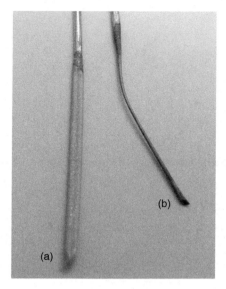

Figure 22.3 Cathode (a) and anode (b) after electrolysis.

electrical quantity, as the following examples illustrate. The microscale applicability of instrumental methods being a given, these examples show how the authors have aimed to make them low cost and therefore widely accessible.

22.7.3.1 Cell Potential Measurements

The use of a six-well microplate (Wellplate 6) has been promoted for setting up half-cells and creating batteries made of three galvanic cells in adjacent wells [20, 21].

The same well plate has also been used to construct zinc and copper half-cells in a single well, using plastic tips to hold the metals and their corresponding 1 M metal sulfate solutions. The potential of the zinc/copper cell was determined using a digital multimeter [20, 21].

Our group has adopted a similar design by using the microchemistry kits together with plastic tips and a digital multimeter to construct a zinc/copper cell and lead/copper cell in large wells of the comboplate (Figure 22.4), as well as to show the effect on cell potential of changing the concentration of the copper sulfate solution in the zinc/copper cell (Figure 22.5).

Learners are also able to explore additivity of cell potentials and the Nernst Equation [22].

The use of the "electrician's multimeter" for carrying out measurements in a wide range of potentiometric experiments has also been advocated by others [23].

pH Measurements pH measurements are specific examples of cell potential measurements. The costly and fragile glass electrode is a barrier to wider access to these measurements and several authors have sought to overcome this. The first

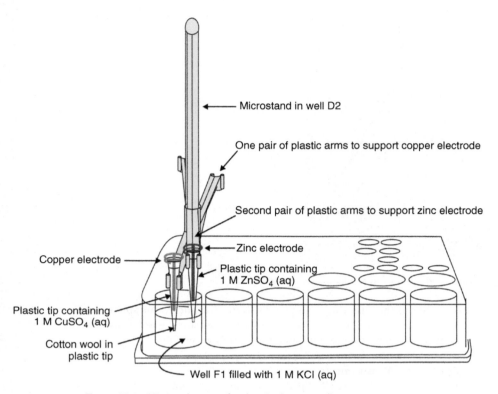

Figure 22.4 Microscale setup for the zinc/copper cell.

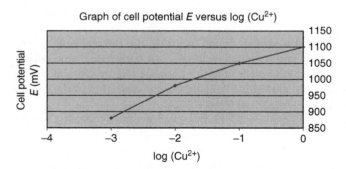

Figure 22.5 Effect of decreasing the concentration of Cu^{2+}(aq) in the Zn/Cu cell using the setup in Figure 24.4. The cell potential is approximately directly proportional to log $[Cu^{2+}$(aq)].

22.7 Case Study III: Can Quantitative Practical Skills Be Learned with Microscale Equipment?

phase of Sane's "Locally Produced Low Cost Equipment" *(LPLCE)* Project in India [24] looked at producing a low-cost pH electrode using carbon rods, but this was not adapted for microscale work at the time. A breakthrough was made in 2005 when the team at University of Santo Tomas, Manila, fabricated a potentiometric hydrogen ion sensor based on the dispersion of quinhydrone in an oil paste [25]. This is lower in cost than glass electrodes because the body of the pH sensor is a 1-ml plastic syringe, and the conducting wire is formed using the stem of the syringe to which a safety pin has been attached. The carbon paste containing quinhydrone becomes the sensing membrane which is packed into the lower end and tip of the syringe. The sensor has provided repeatable results in solutions of varying pH. It is not fragile like the glass electrode, nor does it need sensitive storage conditions. Since its diameter is comparable to that of a 1-ml syringe, it could be used with small volumes of solutions for microscale work.

22.7.3.2 Electrical Conductivity, Light Absorption, and Temperature Measurements

Small-Scale Conductivity/Conductance Meters For many school and first-year university students, measuring solution conductivity means looking at the brightness of a bulb and comparing its intensity to that produced by another solution. There have been other semi-quantitative microscale methods of measuring conductivity, such as flashing LEDs and buzzers (reviewed by Bergantin *et al.* [26]) as well as the portable conductivity meter and probe described by Ganong [27] that can be used for field-testing of smaller volumes of environmental water samples.

In recent years, the quantitative measurement of conductivity has involved the further development of sensing probes that can be adapted for microscale chemistry experiments. Coupling of these probes/sensors with a multimeter allows quantitative measurements of conductivity to be easily made at low cost [23].

Our group's Microconductivity Unit is based on a control box with a conductivity probe connected to a 9 V battery and a digital multimeter. When the probe is lowered into a sample of choice, the output signal is displayed as microamps or milliamps on the multimeter.

The unit can be used with other microchemistry equipment, making use of the large wells of the microwell plate to measure conductivity of electrolyte samples (± 2 ml) (Figure 22.6).

The activities that make use of the unit require the construction of calibration curves [28] (Figure 22.6). With this apparatus, students are able to use conductivity to explore the conductivities of different liquids, study the dissolving of different solids in water, distinguish between strong and weak acids, investigate the factors that affect the conductivity of aqueous solutions, and more.

A special conductance meter has also been developed recently by Sevilla *et al.* at the University of Santo Tomas, Manila [26]. It can be powered using a regulated power supply or two 9 V batteries. The output voltage is also measured with a digital multimeter. The probe consists of a pair of pencil leads set at a fixed distance in a plastic lid that fits into the wells of a microwell plate. A maximum

 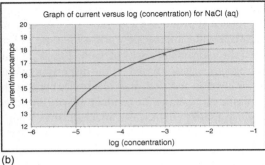

(a) (b)

Figure 22.6 (a) The conductivity probe is lowered into a small volume in a microwell plate, and (b) a calibration curve for NaCl(aq) constructed using current readings obtained when a digital multimeter is coupled with the microconductivity unit.

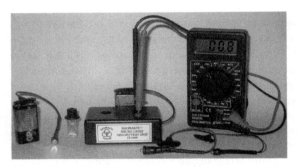

Figure 22.7 Low-cost light absorption unit based on a custom-made control box with digital multimeter.

of 2.0 ml of solution is required in each well for testing. Experiments involving conductance and ionic mobility, effect of varying concentration on conductance, and determination of a dissociation constant (K_a) can be carried out. The relative error obtained in these experiments is acceptable for general chemistry courses and the developers intend evaluating the conductance meter further with students.

Light Absorption and Temperature Measurements The versatility of the multimeter means that it can be successfully used to carry out colorimetric and thermometric measurements when coupled with photodetectors and thermistors, respectively [29]. We have adopted this approach in the development of low-cost units for measuring temperature and light absorption in small-scale experiments. Each unit consists of a control box which can be connected to a digital multimeter and

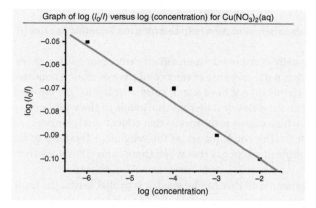

Figure 22.8 Calibration curve – log (I_o/I) versus log (concentration) for $Cu(NO_3)_2$(aq).

which also has a probe that is adapted to measure either light absorption (Light-dependent Resistor (LDR) based) or temperature (thermistor-based).

The Micro Light Absorption Unit (Figure 22.7) is equipped with red, blue and white LEDs. A sample of small volume is placed in a miniature cell which fits into the control box. The output signal is in the form of electrical voltage (mV), and calibration curves need to be compiled (Figure 22.8).

Similarly, the thermistor of the Micro Temperature Unit can be used to find the temperatures of solids, liquids, solutions, and gases/air [30]. Samples can be measured in small vials or in the wells of a microwell plate. The measured voltage depends on the sample temperature. The work is not yet complete and needs further trials to support its use in the classroom.

Sevilla [31] has developed a microscale photometer based on the use of microwells with embedded LEDs and with a digital multimeter to measure signal output. The low-cost microscale electronic thermometer uses a mini probe that contains the thermistor in a glass casing [31]. Various activities accompany the use of this probe, including preparation of cooling curves (e.g., for naphthalene). These experiments need a maximum sample volume of only 3 ml.

All of the examples in this case study indicate that the era where access to quantitative chemical practice was denied because of the expense of traditional instrumentation is being superseded by another where modern, low-cost microscale methods allow learners more experiences reflecting modern practice. While the aim remains the development of quantitative practical skills (and a better understanding of concepts!), the range of available experiences is being greatly extended. This is a substantial advantage of microscale experimentation in chemistry education.

22.8
Case Study IV: Can Microscale Experimentation Help Learning the Scientific Approach?

A regular classroom is usually considered an unsuitable environment for chemistry experimentation. This is partly because of safety and environmental concerns and because regular classrooms do not have water, good ventilation, gas supplies, and perhaps no mains electricity. Besides, there is this image of the white-coated chemist in his laboratory, which clearly is the model that educational planners and teachers want to conform to. The consequence of the religious adherence to the laboratory as a kind of temple of science is that when there is no laboratory there is no experimentation.

When traditional equipment is all that can be used, it is probable that the traditional laboratory is the best way to accommodate this. And it may also be argued that the modeling of the laboratory setting helps orientate learners to laboratories in the real world. Without arguing this possibility, one may simply say the benefits, if any, come at a high price. Also in accepting this idea, one must be mindful of the potential negatives of this exclusivity of science.

Using microscale equipment permits chemistry experiments in the regular classroom, and this facilitates the integration of experimentation with other activities. This surely opens the way to meaningfully "learning the scientific approach" and "gaining a better understanding of theoretical aspects of the subject." It may also lead naturally to adopting alternative laboratory instruction styles (inquiry, discovery, problem-based) [9] (Domin) and to laboratory-driven instruction [32] – except there is no laboratory!

An example of this integration can be taken from such a basic concept as conservation of mass in a chemical change. Two reactant solutions in small vials are placed on the pan of a DPS and their total mass recorded. Then one solution is added to the other and the vials are weighed together again. The final total mass is the same as the initial one. But if it is not, what is the explanation? Perhaps some drops of solution were lost in the transfer? Perhaps the difference is within the precision characteristics of the DPS? Perhaps a gas was one of the reaction products – how can we modify our technique to capture any gas formed? With a variety of reactions being tested around the class, within a short space of time quite a substantial number of observations can be accumulated. This may clarify the fact that great scientific principles are not based upon one observation. Learners may realize that scientists were not always believers in the conservation of mass, and the limitations of their balances made it easy to remain a doubter. Establishing this principle led to the definition of chemical elements by Lavoisier and the atomic theory of Dalton. Molecular modeling (counting the atoms) and writing of balanced chemical equations would be natural related activities in the same classroom. We have developed suites of activities along these lines.

We have used this scientific approach in our teacher development activities. Teachers enjoy it and often express surprise at the experience. Overwhelmingly, this is because they had no previous conception of the scientific approach themselves. Practical activities are generally seen as something apart from theory. Unfortunately, these teachers often have no equipment and have little encouragement to use it if they have.

There will of course be concerns about the potential hazards of working in a normal classroom. These concerns have often been cited for the decline of practical work in schools in developed countries, with the threat of regulations and litigation hanging over teachers. As a counter to this understandable influence, CLEAPSS in the United Kingdom has provided help for teachers in the safe management of practical work. Regulations in the UK require risk assessments and, wherever possible, the use of safer alternative procedures. CLEAPSS has been active in developing such procedures, promoting the use of microscale techniques in this regard [33].

Clearly, teacher competence is the first priority, and all the normal rules of behavior, usually introduced in a laboratory, apply in the classroom. To work safely with chemicals outside of the laboratory surely is an important life skill, which can be seen as an advantage of the classroom activity. Furniture and clothing need to be protected, ventilation must be adequate, aggressive chemicals should be avoided, and so on. Risk assessment should be the norm before starting hands-on activities. All these count as practical skills. These admonitions have greater force outside the laboratory and should more readily translate into appropriate and informed behavior in daily life. Thus, although the abandonment of the laboratory as a necessary temple of science in schools raises questions about maintaining safety, these can be answered positively and the move provides an opportunity for integration with the rest of the curriculum.

22.9
Case Study V: Can Microscale Experimentation Help to Achieve the Aims of Practical Work for All?

22.9.1
The UNESCO-IUPAC/CCE Global Microscience Program and Access to Science Education for All

UNESCO has always had a special interest in providing global access to education, with an emphasis on gender equality [34]. In line with the "Education for All" movement [35], science education for all should mean practical science experiences for all. So it is not surprising that UNESCO recognized the potential of the low-cost, versatile microscale chemistry kits (available and tested already in 1995 [36]), and embarked on the Global Microscience Program (GMP) [7].

Several partners, including IUPAC's Committee on Chemistry Education (CCE), RADMASTE, and the International Organisation for Chemical Sciences in Development (IOCD), forged ties with UNESCO to run introductory microchemistry workshops in over 80 countries around the World. More than half of these took place in Africa. Workshop participants everywhere were positive that the kits could deliver hands-on experimentation for all, and several UNESCO-Associated Centers for Microscience Experiments were established to promote the microscale concept (e.g., *www.microsci.org.za*).

UNESCO continues to promote microscale science experimentation but, regrettably, there has been limited follow-up in many countries that hosted the initial workshops. In some, like Cameroon, large-scale implementation occurred. However, in others the great enthusiasm encountered at the introductory training was not exploited locally [37].

22.9.2
The Global Water Experiment of the 2011 International Year of Chemistry – Learning from the Experience

"Water – A Chemical Solution" was the overall theme of the Global Water Experiment (GWE) – a major initiative undertaken by UNESCO and IUPAC during the 2011 International Year of Chemistry (IYC) to increase awareness among students and the public about the importance of water, access to safe drinking water, and the role that chemistry plays in water quality and purification [38]. Embodied within that theme were narrower aims that bear a close resemblance to the aims of practical work identified by Woolnough and Allsop [3]: "the experiment will give students learning experiences that are engaging and edifying (*motivation*) so that they learn valuable *practical skills* and useful chemistry." "The activities cover important topics and conceptual understanding in science (*better understanding of theoretical aspects*) and ....opportunities to learn important experimental and data gathering skills (*learning the Scientific Approach*)."

The GWE Task Group identified and designed a set of four activities – two exemplifying water quality and two focusing on water purification – that could be carried out by learners at different educational levels. It soon became apparent that the protocols could exclude countries lacking traditional laboratory resources from participation. To give meaning to the global aspirations of IUPAC and UNESCO, therefore, microscale kits were designed specifically for the GWE – one for learners (the Global Water Kit) and the other for the teacher (the School Resource Kit).

These kits were supported by microscale versions of the original procedures, approved by UNESCO and IUPAC, and published on the IYC website [39]. School Packs (SPs) consisting of 10 Global Water Kits and 1 School Resource Kit were constituted to enable classes of 50 learners working in groups to carry out all the GWE activities (Figure 22.9).

Under the auspices of UNESCO, these SPs were distributed to 32 selected member countries around the World. Our group also disseminated the kits extensively (with training) within South Africa through government and private sponsors, and responded to independent requests for SPs in other countries such as the Netherlands, Nigeria, and the Gambia.

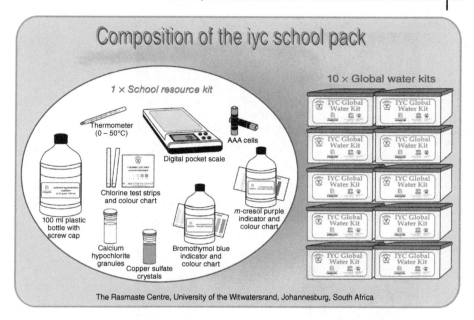

Figure 22.9 The school pack for the GWE.

There is evidence (reports, photographs, videos) that many learners enjoyed the microscale experiences of the GWE (Figure 22.10).

If success is measured as the number of learners participating in the GWE, then it can be claimed that the experiment was indeed successful. The IYC website boasts that 2354 teachers and 128 330 students took part in the GWE. Reports from national coordinators suggest that there may have been more than 2 million participants in fact.

By our own estimate, the microscale kits supplied through UNESCO suggest that around 28 000 learners in less developed countries could have been able to participate [40]. Unfortunately, we have not been able to determine how many did, because of the very scant information available. There is some reporting on the UNESCO website [41], a handful of schools appear on the online GWE schools map [42], and a video has been posted by a girls' school in Jordan showing classroom use of the microscale equipment [43]. It would be good to know what happened elsewhere and whether there has been any legacy activity, as originally intended.

Unfortunately, the lack of follow-up information in countries receiving microscale kits means that we are not learning from the experiences as we should. To achieve the aims of practical work for all, we have to pay more attention to delivery and learning and less to the opening ceremonies.

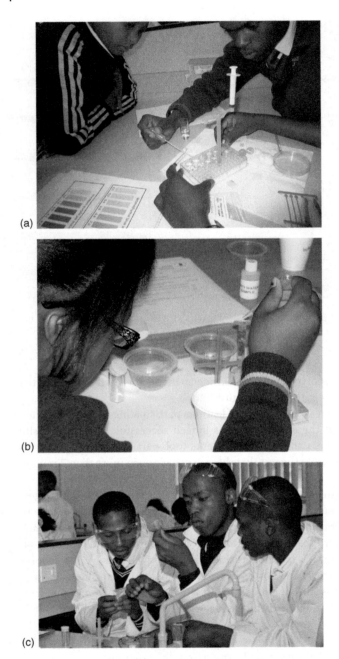

Figure 22.10 (a–c) Learners enjoying the GWE experience through the use of microscale water kits.

22.10
Conclusions

At the start, we recalled the general consensus that experimental work is necessarily part of chemical education, and we drew attention to four aims of this experimental work as identified by Woolnough and Allsop [3]. Our case studies show that all these aims of practical work in science can be achieved in chemistry through microscale experiments. Indeed, they can be achieved at least as well this way as through the use of traditional equipment. Achieving the aims is therefore not a matter of scale. Rather, is it a matter of teacher motivation and competence. These two factors influence all the educational outcomes at all levels.

It remains to weigh the advantages and disadvantages of microscale experimentation. In this regard, there is no contesting that microscale experimentation is less expensive, safer, and less damaging to the environment. These are powerful factors in any educational context. They have particular relevance, however, to developing countries and to the goal of education for all. Providing chemistry education for all clearly entails experimentation for all, and for most of the World this is only feasible through microscale activities.

Change is notoriously slow in the educational field, especially where decision making rests primarily with government rather than with individuals. This is normally the case with primary and secondary education in developing countries and is not uncommon in developed ones. Ministries of Education are plagued by interest groups and advocates of "magic" solutions, and having succumbed to making bad decisions in the past may well be reluctant to make "revolutionary" changes. It seems to us that logic will prevail eventually and this could be accelerated by support from professional chemists (as may be found within universities, academies of science, and chemical societies). The African Academy of Sciences, for example, has taken up this cause, and convened high-level meetings with the Kenyan Ministry of Education, and the International Organisation for Chemical Sciences in Development is supporting this, too. Properly planned pilot projects are perhaps the stepping stone to future wider implementation [44].

References

1. IUPAC Report of the Education Strategy Development Committee (2000) p. 8.
2. Thulstrup, E.W. (1999) in *Science and Environmental Education – Views from Developing Countries* (ed. S. Ware), The World Bank, Washington, DC, pp. 113–127.
3. Woolnough, B. and Allsop, T. (1985) *Practical Work in Science*, Cambridge University Press.
4. Beasley, W. and Chant, D. (1996) *Aust. J. Chem. Educ.*, **41**, 11–16.
5. Pike, R.M. (2006) in *Microscale Chemistry Experimentation for all Ages* (eds M. Hugerat, P. Schwarz, and M. Livneh), The Academic Arab College for Education, Haifa, pp. 13–25.
6. Bradley, J.D. (2001) The UNESCO/IUPAC-CTC global program in microchemistry. *Pure Appl. Chem.*, **73**, 1215–1219.
7. UNESCO *http://www.unesco.org/new/en/ natural-sciences/special-themes/science-education/basic-sciences/microscience/ background* (accessed 4 October 2013).

8. Bradley, J.D., Huddle, P.A., and Sebuyira, M. (2006) in *Microscale Chemistry Experimentation for all Ages* (eds M. Hugerat, P. Schwarz, and M. Livneh), The Academic Arab College for Education, Haifa, pp. 31–37.
9. Domin, D.S. (1999) A review of laboratory instruction styles. *J. Chem. Educ.*, **76**, 543–547.
10. Vermaak, I. (2003) New technologies for effective science education break the cost barrier. Paper presented at the BERA Annual Conference, Edinburgh, Scotland, September 11–13.
11. Michiela, M.R. and Wasike, D.W. (2011) Effectiveness of micro-science kits in teaching primary science in Kenya. *J. Educ. Social Sci.*, **1** (1), 111–116.
12. Abdullah, M., Mohamed, N., and Ismail, Z. (2006) in *Proceedings of the International Science Education Conference ISEC, November* (eds Y.J. Lee, A.L. Tan, and B.T. Ho), Nanyang Technological University, Singapore, pp. 55–65.
13. Richardson, J.N., Stauffer, M.T., and Henry, J.L. (2003) Microscale quantitative analysis of hard water samples using an indirect potassium permanganate redox titration. *J. Chem. Educ.*, **80** (1), 65–67.
14. Abdullah, M., Mohamed, N., and Ismail, Z. (2013) *Chemistry Education and Sustainability in the Global Age*, Springer, pp. 311–320.
15. Schwarz, P. (2006) in *Microscale Chemistry Experimentation for all Ages* (eds M. Hugerat, P. Schwartz, and M. Livneh), The Academic Arab College for Education, Haifa, pp. 334–357.
16. Wloka, K. (2006) in *Microscale Chemistry Experimentation for all Ages* (eds M. Hugerat, P. Schwartz, and M. Livneh), The Academic Arab College for Education, Haifa, pp. 416–424.
17. Worley, R. and Owen, M. (2013) Successful practical work in challenging circumstances – lessons to be learned from Uganda. *Afr. J. Chem. Educ.*, **3** (1), 96–11.3.
18. El-Marsafy, M., Schwarz, P., and Najdoski, M. (2011) *Microscale Chemistry Experiments Using Water and Disposable Materials*, Kuwait Chemical Society, Nuzha, 157 pp, ISBN: 978 608 65076 3 3.
19. Bell, B., Bradley, J., Chikochi, A., Kibasomba, P., Lubango, L., Roberg, C., and Steenberg, E. (2010) Low cost quantitative microscale chemistry. Oral Presentation at 2010 International Conference on Chemical Education.
20. Zhou, N.-H., Habelitz-Tkotz, W., Giesler, D., El-Marsafy, M.K., Schwarz, P., Hugerat, M., and Najdoski, M. (2005) Quantitative microscale chemistry experimentation (volumetry, gravimetry, electrochemistry, thermochemistry). *J. Sci. Educ.*, **6** (2), 84–88.
21. Zhou, N.-H. (2006) in *Microscale Chemistry Experimentation for all Ages* (eds M. Hugerat, P. Schwarz, and M. Livneh), The Academic Arab College for Education, Haifa, pp. 425–434.
22. RADMASTE (2009) Microelectrochemistry Experiments: Manual for Learners and Microelectrochemistry Experiments: Manual for Teachers.
23. Alfonso, R.L., Andres, R.T., and Sevilla, F. III, (1993) Using the electrician's multimeter in the chemistry teaching laboratory. Part 2. Potentiometry and conductimetry. *J. Chem. Educ.*, **70** (7), 580–584.
24. Sane, K.V. (1999) Cost-effective science education in the 21st century – the role of educational technology. *Pure Appl. Chem.*, **71** (6), 999–1006.
25. Tan, J; Lacson, M and Sevilla III, F. (2005) Potentiometric pH sensor based on an oil paste containing quinhydrone. Proceedings of the 2005 Asian Conference on Sensors and the International Conference on New Techniques in Pharmaceutical and Biomedical Research, Kuala Lumpur, Malaysia, September 5–7, 2005, pp 39-42.
26. Bergantin, J., Cleofe, D., and Sevilla, F. (2013) *Chemistry Education and Sustainability in the Global Age*, Springer, pp. 303–310.
27. Ganong, B.R. (2000) Hand-held conductivity meter and probe for small volumes and field work. *J. Chem. Educ.*, **77**, 1606–1608.
28. RADMASTE (2008) RADMASTE Microconductivity: Activities for

Learners Using the RADMASTE Microconductivity Unit.
29. Andres, R.T. and Sevilla, F. III, (1993) Using the electrician's multimeter in the chemistry teaching laboratory. part 1. Colorimetry and thermometry experiments. *J. Chem. Educ.*, **70** (6), 514–517.
30. RADMASTE (2009) Guidelines on the Assembly and Use of the RADMASTE Micro Temperature Unit.
31. Sevilla, F., III, (2010) Low-cost instrumentation for microscale chemistry. Oral Presentation at the Symposium on Green and Microscale Chemistry, Universidad Iberoamericana, Mexico City, *http://www.uia.mx/investigacion/cmqvm/files/simposio2010/Sevilla-1.pdf* (accessed 30 August 2013).
32. Lamba, R.S. (1994) Laboratory-driven instruction in chemistry. *J. Chem. Educ.*, **71**, 1073–1074.
33. CLEAPSS *www.cleapss.org.uk* (accessed 4 October 2013).
34. UNESCO *http://en.unesco.org/themes/education-21st-century* (accessed 4 October 2013).
35. UNESCO *http://www.unesco.org/new/en/education/themes/leading-the-international-agenda/education-for-all/the-efa-movement* (accessed 4 October 2013).
36. Vermaak, I. (1997) Evaluation of cost-effective microscale equipment for a hands-on approach to chemistry practical work in secondary schools. PhD thesis. University of the Witwatersrand.
37. Bell, B. and Bradley, J.D. (2012) Microchemistry in Africa: a reassessment. *Afr. J. Chem. Educ.*, **2** (1), 10–22.
38. Wright, A. and Martinez, J.G. (2010) A global experiment for the international year of chemistry. *Chem. Int.*, **32** (5), 14–17.
39. IYC 2011 (2011) The Global Experiment of the International Year of Chemistry, *http://water.chemistry2011.org/web/iyc/documents* (accessed 28 July 2014).
40. Bell, B., Bradley, J.D., and Steenberg, E. (2012) Learning from the experience – the distribution of low-cost equipment for the global water experiment during the international year of chemistry. Paper Presented by E Steenberg at the 22nd International Conf on Chemical Education, Rome, Italy.
41. UNESCO *http://www.unesco.org/new/en/media-services/single-view/news/global_microscience_and_water_experiment_kits_provide_tools_for_science_education_in_disaster_stricken_areas/#.UlgX_GAaKUk* (accessed 11 October 2013).
42. IYC 2011 The Global Experiment of the International Year of Chemistry, *http://water.chemistry2011.org/web/iyc/the-school-map* (accessed 11 October 2013)
43. International Year of Chemistry 2011 Water a Chemical Solution, UNRWA, Jordan – You Tube, *http://water.chemistry2011.org/web/iyc/videos* (accessed 4 October 2013).
44. RADMASTE (2009) Compendium of innovations and good practices in microscience. Prepared for the UNESCO Office in Abuja under Contract Number ABU/STE/045-09: 3240199599, pp. 102–106.

Part III
The Role of New Technologies

23
Twenty-First Century Skills: Using the Web in Chemistry Education

Jan Apotheker and Ingeborg Veldman

>Where is the life we have lost in living?
>Where is the wisdom we have lost in knowledge?
>Where is the knowledge we have lost in information?
>
>*T.S. Eliot*

23.1
Introduction

Internet is used more and more in education and research. Apart from the Internet smart phones and tablets have emerged which are also being used more and more frequently in education. Peter Atkins recently discussed the changeover from using books toward the use of interactive sites on the Web [1]. He claims that the (text)book has no future. In society the emergence of several type of e-book readers can be seen. Most University courses now use some form of electronic learning environment. Textbooks offer online support for their use in the classroom.

Some years ago a discussion emerged on the function of Internet which was started by Tim O'Reilly. Tim O'Reilly [2] discusses the definition of Web 2.0 in an on line article (Figure 23.1).

It becomes clear that the Web is being used as a platform for the exchange of data and information between people. Many developments are focused on that particular aspect. Especially social platforms like Facebook, twitter, Flickr, Google give people the opportunity to exchange personal data.

In a later article [3] the discussion continues. Instead of being a static medium the Web has become interactive. People have started creating a collective intelligence in which people cooperate to develop new knowledge. One of the best examples of this development is Wikipedia. In Wikipedia self-called experts introduce texts to introduce and exchange information about a subject. Another example is LINUX, where people have collectively developed software that is used widely.

The introduction of the smartphone has connected the Web to our person. Web Blogs have developed, but twitter has taken that a step further. It has led to the

Chemistry Education: Best Practices, Opportunities and Trends, First Edition.
Edited by Javier García-Martínez and Elena Serrano-Torregrosa.
© 2015 Wiley-VCH Verlag GmbH & Co. KGaA. Published 2015 by Wiley-VCH Verlag GmbH & Co. KGaA.

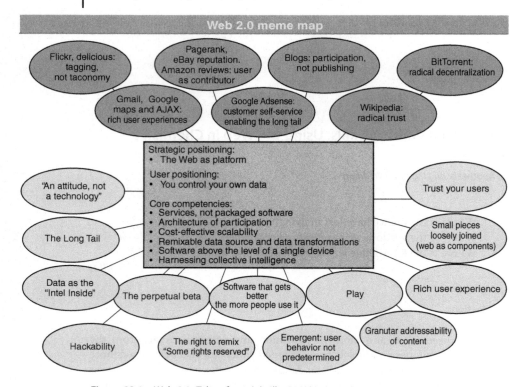

Figure 23.1 Web 2.0. Taken from (o'reilly, 2005).

popularity of twitter, as a type of instant Blog of 140 characters. A lot of data are automatically uploaded to the Web, for example, our location can be traced easily.

The Internet is slowly getting another function. It is storing vast amounts of data, ranging from texts, pictures, photographs, and music, to software. More and more clouds appear in which software applications are now offered online and it is no longer necessary to download complete sets of programs like "Adobe Creative Suite."

Augmented reality is another new development within the Web that couples information to a specific location. Both camera's and smartphones are able to couple time and space together. An application like "layar" uses these data to give information about objects, which is superimposed over a camera view of the object itself (*www.layar.com*). In Figure 23.2 you can see an example of the way the app is used by a real estate agent, giving information about houses for sale.

The new Google glasses are a development along the same lines. In google glass a screen is mounted on glasses. The screen is voice operated and can give different type of information, when connected to the Internet (*http://www.google.com/glass/start/*) (Figure 23.3).

Figure 23.2 Example of the use of layar.

Figure 23.3 Google glass.

In this overwhelming development of uses that have been developed on the Internet since it came into being, education has developed as well. Education has touched on the Internet and within education the use of Internet is becoming more and more a normal procedure. Both in formal and informal education the Internet is used more and more. In this chapter I will discuss some general developments as well as a few specific examples.

23.2
How Can These New Developments Be Used in Education?

Interactivity is a keyword in the description of the Web. The exchange of information is one of the key aspects of the Internet. How can you use the Web in education? In other words what might be the added value of the use of Internet in education. People expect that with the Internet.

- Students will be more motivated, because learning is more "fun"
- Students will perform better
- Students will learn more quickly
- It will be easier to manage the learning process.

There are many developments in the use of the Internet since it came into being that are slowly being introduced within education. In The Netherlands, for example, there are now "Steve Job-schools" using the iPad to provide learning materials.

Still it is not always easy to introduce the use of Internet in a meaningful way in education. Harris *et al.* [4] discuss the problems of introducing Web-based applications into education. They indicate there is a mismatch between the view of developers and the way most practitioners use digital tools:

Researchers emphasize technology uses that support inquiry, collaboration, and reformed practice, whereas many teachers tend to focus on using presentation software, learner-friendly web sites, and management tools to enhance existing practice.

In their discussion they emphasize the techno centric approach toward the design of as opposed to the approaches in which teachers use their pedagogical content knowledge to approach the learning process. This mismatch leads to problems in the use of applications in education.

They have introduced the TPACK (Technological Pedagogical Content Knowledge) framework as an approach for teachers (and designers) to successfully introduce technology into learning activities.

TPACK stands for Technological Pedagogical Content Knowledge. In Figure 23.4 the TPACK framework is described, as it is derived from the usual Pedagogical content knowledge.

Content knowledge is the knowledge about the subject, the paradigm that plays a role within the discipline.

Pedagogical knowledge is the knowledge about the learning process. It enables the teacher to design and use learning activities that fit with the content knowledge they want their students to learn.

Pedagogical content knowledge is the overlap between the two, where teachers are aware of the specific learning problems students have with certain concepts. More specifically PCK gives insight in pre-existing concepts the students generally have and the way these influence the learning process. This is wonderfully illustrated in a picture book for children by Leo Lionni [5]. In this book a tadpole and a goldfish grow up together in a pond. The frog goes into the world and comes back and tells wonderful tales about cows, birds, and people. Lionni illustrates the ideas of the goldfish about the stories of the frog wonderfully. It becomes clear which preconceptual knowledge the goldfish uses and illustrates exactly the learning problems students run into.

Technological knowledge is the knowledge that enables you to understand information technology in such a way that you are able to apply it productively within your teaching. Basically it means you are literate in a technological sense.

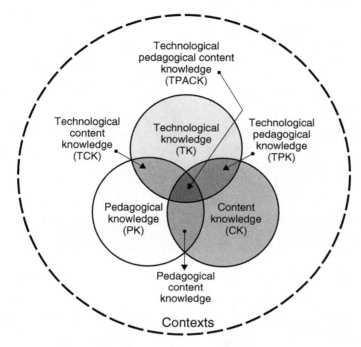

Figure 23.4 TPACK framework (from Ref. [4]).

Technological Pedagogical Knowledge (TPK) means you understand the way technology influences the learning process. Developing TPK implies that you realize the potential benefits and drawbacks of using a particular technology in the learning process. A nice example to illustrate the flexibility needed in using new technology is the change over from the blackboard to the use of smart boards. Smart boards offer an enormous amount of new possibilities in the use of a board in the classroom. It takes technological knowledge as well as pedagogical knowledge to be able to use a smart board in an optimal way (see Figure 23.5).

A smart board combines a projector with an interactive touch screen. It makes the projection of a powerpoint presentation possible. But in addition to the powerpoint you can now also write on the board at the same time. Special software allows for the development of interactive displays. Direct access to Internet is also possible. It is a powerful tool, that gives a lot more possibilities than the normal chalk board. But for optimal use TPACK is a necessity (*https://smarttech.com/Home+Page/Solutions/K-12*) (Figure 23.6).

Technological Content Knowledge deals with the way technology plays a role in the development of content knowledge in the subject being taught. In most science subjects newly developed technology plays a major role in research. A nice example is Scanning Tunneling Microscopy. This has been used to introduce images on a molecular and atomic level. Single molecule research has become

Figure 23.5 Illustration from "fish is fish." Taken from Ref. [6].

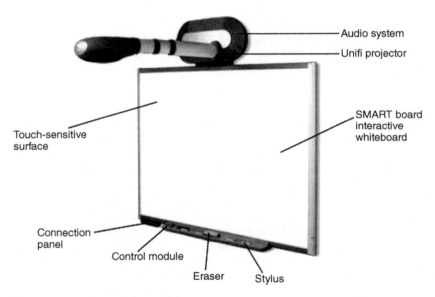

Figure 23.6 A smart board.

possible and feasible by using new technologies. Teachers need to be aware of the new technologies being used.

Technological Pedagogical Content Knowledge is the intersection of all. It is the type of professional knowledge that technological adept teachers use. It combines the pedagogical content knowledge that teachers have with the insight in which technology can play a role in the learning process of students. These Teachers understand the ways in which concepts are best introduced to students and are able to use technology to help students understand the content better.

A number of teachers that have developed this TPACK have developed interactive applets on the Web. A nice example can be found on *http://micro.magnet.fsu.edu/primer/java/jablonski/jabintro/*.

On this URL an applet can be found that illustrates a Jablonski diagram. The interactive possibilities of the applet illustrate the concept of fluorescence and helps students understand the concept better (see Figure 23.7).

The TPACK framework can be used as a tool in the professionalization of teachers (and technological designers). It demonstrates the necessity for the broad development of knowledge and the need for interactivity in the different fields of knowledge. For a truly well developed use of technology interaction between teachers and technological designers is need.

Only that way designs for educational materials using modern media techniques can meet the following criteria:

- Have added value over traditional material
- Be attractive to use
- Be easy to use
- Be easily accessible
- Cater to individual learning wishes.

A number of activities on the Web that have emerged lately, using Technological Pedagogical Content Knowledge illustrate the possibilities the Web offers.

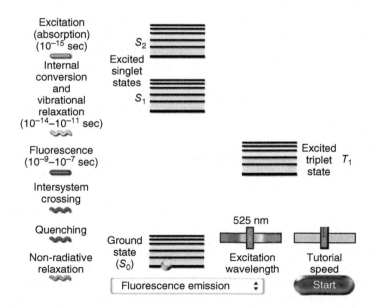

Figure 23.7 Screenshot from an online app about the Jablonski diagram.

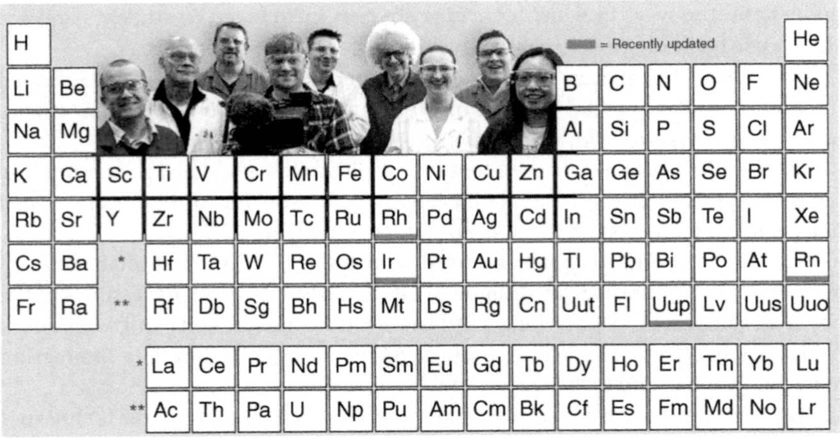

Figure 23.8 Screenshot from periodic videos.

The first I would like to introduce was developed by Martin Poliakov's group at the University of Nottingham, *http://www.periodicvideos.com/*. At first sight (see Figure 23.8) it seems to be just a regular periodic table. By clicking on an element a video about the element is started in which Martin and his team demonstrate specific reactions of the element as well as its properties. It is ideal for students to get a first impression about elements.

In the King's Centre for Visualization in Science Peter Mahaffy and his coworkers of Kings College in Edmonton, Canada have developed a number of interactive applications. The one shown in the screenshot (Figure 23.9) demonstrates the role of carbon dioxide in climate change (*http://www.kcvs.ca/site/index.html*).

The visualization center offers students the opportunity to work interactively with the applets and thus get a better understanding of the chemical issues involved.

The Chemistry portal in Wikipedia (*http://en.wikipedia.org/wiki/Portal:Chemistry*) is a true portal as it opens up a wide range of information. The easy part here is that clicking on a subject immediately yields an enormous amount of information (Figure 23.10).

For students this opens a wide world of information about chemistry. Because of the structure of the web site it is easy to obtain information and explanations about subjects not directly understood.

23.3
MOOCs (Massive Open Online Courses)

Quite some universities have started to put their education on line. Most of that material is freely available for others. A list of available courses can be found on *http://www.mooc-list.com*. Basically A MOOC (massive open online course) is a

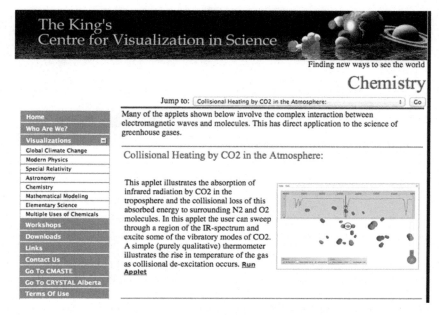

Figure 23.9 Screenshot from the King's Centre for Visualization in Science.

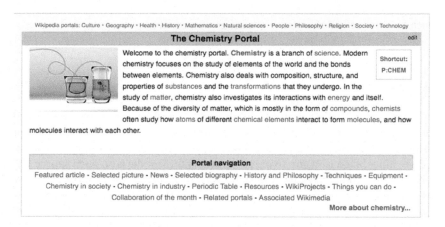

Figure 23.10 A screenshot of the opening screen of the chemistry portal on Wikipedia.

course which is available for all students. Exams can be taken and credits are recognized by universities. Whole communities of students and teachers can be formed around a MOOC.

Learning platforms are often used as a base for eLearning. In these courses the platform is used as the sole means of communication between students and teachers. A nice example is *https://www.edx.org/* in which the University of Harvard, MIT, and other universities work together in offering online courses in science

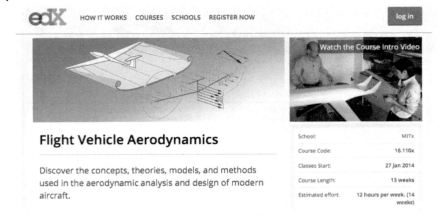

Figure 23.11 Screenshot from edx.

(see Figure 23.11). You can take a course do the exam and get credits at your own university. A number of the courses are focused on Advanced Placement students. It gives these high school students an opportunity to start on courses that have enough level to be challenging. These courses use advanced ways of making interactivity possible between the program and the user. Virtual lab space is offered as well as video instruction. Especially the virtual labs are a nice way of interacting. Both for physics and chemistry options are impressive.

23.4 Learning Platforms

In practice almost all secondary schools and university courses use some type of Learning Platform or Virtual Learning Environment (VLE) like "Blackboard" or "Moodle." This platform is used to communicate information about the course, lecture notes, exercises, and so on, to the students. Moodle is an open source system, which is updated regularly (*http://docs.moodle.org/27/en/Features*) It allows teachers and students to exchange information. It allows teachers to give all information about a course on the one hand. On the other hand students can hand in home work, review each others papers, or work together on Wiki's. Because the system is open source it is updated regularly and offers features that are requested by its users. The site offers an enormous amount of information for prospective users. It is a powerful tool for education.

Some of these universities have offered information in a more systematic way. One example is the Kahn Academy, which offers short instruction videos about different subjects (*https://www.khanacademy.org/*). Also YouTube (*www.youtube.com*) contains a large number of videos from different sources containing lectures given. Most of the material is not directly usable in the classroom. Normally it needs to be adapted.

On of my favorite online courses is offered by iTunes university. It is the physics I: classical Mechanics course taught by Walter Lewin. Not because it is a course that takes advantage of all possibilities the Web has to offer. Basically it is just a video of a lecture series he taught at MIT, with introductory mechanics as a subject. The teacher in this case is the most important factor. Even on tape his enthusiasm for physics, his drive, his excellent structure as well as the well-chosen spectacular demonstration experiments are the main reason for his popularity. These courses have been seen and enjoyed by many. On iTunes a similar course taught by Walter Lewin can be found, discussing electricity and mathematics.

It immediately demonstrates the problems with an on-line course. It lacks the direct face-to-face contact between teacher and student, which is important for the student as well as the teacher. This limits the usefulness of on-line courses. Blended learning in which on Information Technology is used together with face-to-face instruction seems to be very effective [7, 8].

23.5
Online Texts versus Hard Copy Texts

During the development of new material for chemistry in The Netherlands part of that material was shared on the Web. On *http://www.nieuwescheikundeinbedrijf.nl* this material is still available. One of the problems of that material is, that it is not very suited for use on line. An example of that problem, taken from another source, is given in figure 23.12.

Main problem is that in order to read the text you have to scroll down. Students and teachers did not find that convenient. What happened very often that the

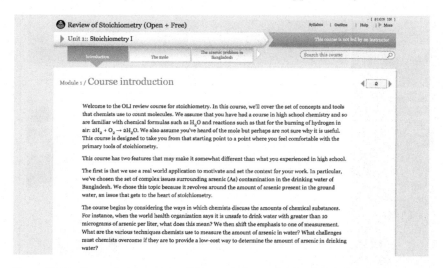

Figure 23.12 Screenshot from *https://oli.cmu.edu/jcourse/lms/students/syllabus.do?section=90d4f80280020ca601d90fc0ad6dd9c7*.

teachers downloaded the text as a pdf document, printed these, and used the print outs in the classroom.

With the introduction of a new curriculum in the Netherlands in September 2013 new textbooks appeared. A number of teachers that had worked in a pilot preparing the new curriculum were unhappy with the way the renewal of the curriculum was presented in the new textbooks. They wanted to publish their own material and make it available to other teachers. They found a publisher that was willing to publish the material on line. Voster content publishes on line material for all subjects in secondary schools on line. This material is freely available for all schools (Figure 23.13).

This material has been edited to make it suitable for direct use on the Web. It included the use of the 5 E model developed by Roger Bybee *et al.* [9]. The steps in this model: Engage, Explore, Explain, Elaborate, and Evaluate are used explicitly throughout the different modules. The material was edited using some of the possibilities that on line material offer such as access to video material. It also uses on line questions with immediate feedback to the students. Chemistry material for the whole of secondary education is now available on line through this source.

Two units have been produced in English, for the island Saba in the Caribbean with which the Netherlands has special ties. These can be found on:

- http://www.studioscheikunde.nl/havovwo_bb/Module_Equilibrium/index.html
- http://www.studioscheikunde.nl/havovwo_bb/Module_Green%20Chemistry/index.html
- http://www.studioscheikunde.nl/havovwo_bb/Module_How can we eat healthily/index.html.

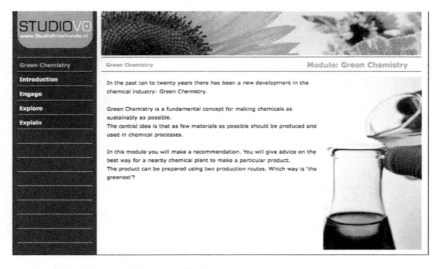

Figure 23.13 Screenshot from stercollecties.

23.6
Learning Platforms/Virtual Learning Environment

Since the beginning in 1985 the Web has been used for communication between groups of people. Once it evolved that function has become more and more important. Communication between teachers and students is one of the important things that Internet can facilitate.

Book publishers have created (and still do) possibilities of communication between teachers and students. One main example is handing in homework. Another example is an online assessment.

Learning Platforms have been created in which a number of services were created for education. Teachers are given a solid platform in which they can organize their courses, Students are given a platform in which their learning is structured, in which they can publicize their work, receive on line feedback, and so on.

The possibilities within virtual environments have increased enormously during the past 10 years. They have become very user friendly, more specifically they make blended learning possible. In blended learning the learning is structured by a teacher, using face to face contact in lectures and lessons on the one hand and the possibilities of the learning environment on the other.

Face to face interaction is one of the most important tools in teaching. It allows for direct feedback and interaction between teacher and learner. A teacher can see the body language of the learner and uses that to interpret the impact of his lecture. The learner has the opportunity to interact directly with the teacher.

The Internet is used to continue the learning process. At present the possibilities of a VLE include the use of short introduction videos, the use of short on line multiple-choice tests. Most important function of the learning platform however is the exchange of information between student and teacher.

The platform is used to schedule the activities for the students. It can offer direct information, like power points, lecture notes, exercises, and so on. It is possible to give students assignments that they can hand in on line. It offers the possibility for students to work together using a Wiki page, a page comparable to a page in Wikipedia. In that case each student is able to adjust the page. That way a complete page or design can be created. Groups of students can be created within the platform. Within the groups on line discussion in a chat session is possible, file exchange is possible as well as a discussion board available.

Teachers and tutors are able to give direct feedback to material handed in by students. It can be graded and grades are immediately available for the students. Final grades can easily be exported as files to be interchanged with other administrations.

Moreover teachers have the opportunity to check papers that are handed in on plagiarism, using "ephorus." Ephorus is an online program embedded in a VLE that checks a text and compares it with other texts on the Web. If similarities are found these are reported. They have the opportunity to give direct feedback.

Teachers can make tests online that are checked directly. This is an important feature to be used in something as the flipped classroom (see Chapter 13).

Figure 23.14 The exhibit: form follows function.

Students can be contacted individually or as groups.

Teachers have the option to structure their course by displaying the timetable for the course, but also by giving specific information about one particular lesson or learning activity (see Figure 23.14).

The number of tools available in Learning Platforms has grown lately. Most additions have been based on requests from teachers. An example is the use of portfolios to present work done during a course, but there are many other examples like working with videos. Usually the video will be uploaded to you tube, but the description and the feedback can be given within the VLE.

In most schools and universities the Learning Platform has become a fixture and integrated part of a course. In schools it is often integrated with the school administration, so that student information can be communicated back and forth, especially grading information.

23.7
The Use of Augmented Reality in (In)Formal Learning

New developments and new possibilities are forever emerging in technology. Communication on the Web is growing and growing. New opportunities for the use of the Web using smart phones are being published. On the one hand teachers do not want to use their students to use smart phones in the classroom on the other hand it brings the Internet in the classroom. Another new aspect is the development of apps for smartphones that can be used in the classroom.

One of the new developments in Internet technology is augmented realty. It is a new technique in which information or 3D images are superimposed over a camera view of the surroundings. A nice example is the app "Layar" in which based on the location of the phone and the direction it is held extra information is given (scan Figure 23.15 with the app "layar" for an example).

The way in which these developments can be used both in formal and informal education is best demonstrated by a detailed description of a successful example. At the University of Groningen a number of coherent activities were developed focused at demonstrating the molecular world underneath the macro world. Both activities focused at informal learning as well as activities focused on the classroom were developed.

Figure 23.15 Starting page of a course in blackboard.

23.8 The Development of Might^y/Macht^{ig}

Science LinX, a small-scale science center at the university of Groningen, plays a pivotal role in the outreach policy, linking formal to informal learning. Together with an online platform (*http://www.sciencelinx.nl*), Science LinX uses blended learning methods to make the invisible visible in the cutting-edge science of the faculty: hands-on exhibit experiences, expos, in-depth workshops, speed dates with scientists, science café's fusing with "virtual" serious games, and simulations. The Science LinX challenge is to translate a seemingly growing but superficial interest in science and technology among the lay audience into deep involvement in research and student education.

A special interest group within Science LinX is teachers at secondary schools. Interaction with these teachers has a dual purpose. The monthly meetings that are organized for the teachers offer the opportunity for in-service training of the teachers in which both pedagogy and scientific content are part of the training. In "Studiestijgers" as this particular activity is called teachers work together with pedagogical experts, researchers from university as well as experts from the science center.

Science LinX is also special because it has the opportunity to work together with both staff and students of the master "education/communication in science." In almost all activities of Science LinX students and staff of the master is involved.

This way Science LinX combines the expertize described in the framework TPACK needed to develop material that can be used in both formal and informal education. In the following the development of an exhibit using augmented reality will be described, from which a tool used in education was derived.

In spring 2009, Science LinX was asked to develop a project to communicate risks and possibilities of nanotechnology to pupils. At the same time, this project produced a quest and discovery of major possibilities of AR to connect not only to teenagers, but also to broader audiences at Arts and Science festivals.

This project gave Science LinX the opportunity to realize an earlier developed concept design for an interactive exhibit on the so-called "Powers of Ten," based on the 1968 American documentary short film by Charles and Ray Eames (*http://www.powersof10.com/film*), exploring the relative scale of the Universe in factors of ten. For the Science LinX exhibit, the medium AR was selected to give a new dimension of immersion to this concept. The original plan evolved into the permanent and mobile exhibit "MIGHT-^{y}" ("MACHT-^{tig}" in Dutch), which is now part of the Science LinX exhibition (see *http://www.rug.nl/sciencelinx/interactieveplattegrond/plattegrond!fullscreen?a-a*).

23.9 The Evolution of MIGHT-^{y}

The goal of the exhibit "MIGHT-^{y} " is to visualize the different length scales on which you can do life science and natural science research, with an accent on the

Nano scale. Pupils encounter on the spectrum of the 40 length scales (10^{24} (100 million light-years) – 10^{-15} (1 fm)) as many as 13 examples of research done in departments of the Faculty of Mathematics and Natural Sciences. In this way a cross section of research areas is presented, giving more insight into the broad range of career options pupils have within our faculty.

The exhibit was designed to be both challenging as well as interesting. Visitors should be attracted to the display. It should also be more playful without losing sight of the content. Julian Oliver's spatial memory game "levelHead" (*http://julianoliver.com/levelhead/*) just seemed perfect to do the job. Julian Oliver, a New Zealand born artist, free software developer, teacher, and writer, has exhibited his electronic artworks in many museums, festivals, and galleries throughout Europe, The Americas, and the South Pacific since beginning his career in 1996. His "levelHead" became a spatial memory game from femtometer to light-years in the context of the Nano LinX project. In the game an avatar is walking through rooms. By tilting a cube the direction the avatar takes can be changed. That way he can walk from one cube to the next cube.

23.10
Game Play

Players of the MIGHT-^{y} exhibit encounter two pairs of AR glasses and two cubes with markers on the desk. After reading the instructions and hearing instruction from the explainer, it's clear that each player has to wear a pair of AR glasses and hold one of the two cubes. It's also clear they should start with the cube interaction (see Figure 23.16).

Player 1 explores cube 1 in which a white virtual man can be navigated by slowly tilting the cube in different directions over stairs, through doors, to other rooms. Each room is provided with graphical clues to the related length scale. Simultaneously, player 2 explores cube 2 in which he'll find thirteen different 3D research objects corresponding to a cross section of research areas on thirteen different length scales.

Both players have to collaborate in order to make a perfect "MATCH": combining, for example, the side of cube 1, which shows the white virtual man in the nanomotor room of 10^{-9}, with the side of cube 2, which shows the 3D nanomotor object, makes a "MATCH." As a response to this "mix and match" principle, the white man from within cube 1 is "beamed up" to one of the beams or towers of the exhibit with an even more lively animated version of his 3D research object, about which he tells his passionate research story. The players learn in this "nanomotor" casus from Dr. Wesley R. Browne (group leader Synthetic Organic Chemistry of the Stratingh Institute for Chemistry) more about the molecular scale, the way lightning fuels the motor, and the possibilities of combining a great number of these nanomotors to change the properties of materials (see Figure 23.17).

After a "MATCH" and meeting with the only 4 in. representation of Dr. Wesley R. Browne, it's easy for player 1 with cube 1 to navigate the virtual white man to

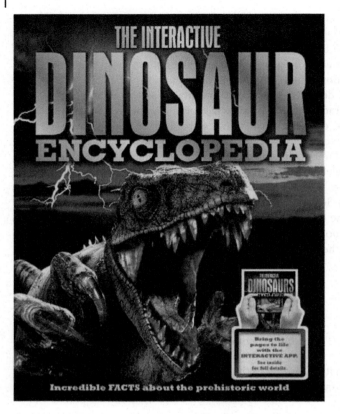

Figure 23.16 Example of "layar." Taken from *http://static.layar.com/website/Layar-Edu-Overview.pdf*

an adjacent room. The room with the representation of the smaller object 10^{-14} (not all of the 40 length scales are covered with content at the moment) will lead to meeting Dr. Maaijke Mevius whose research focuses on cosmic radiance. The room with the representation of the bigger object 10^{-8} will lead to meeting PhD student Jan Willem de Vries, who presents groundbreaking research on drug targeting. A challenging and funny way of informal "zapping" is constructed in this way. The player is in control of which kind of research he wants to learn more about and which researcher he wants to meet virtually (*http://youtu.be/zRGXZdGYfBI*).

23.11
Added Reality and Level of Immersion

AR is a technology which adds an extra layer of computer-generated or synthetic information (visuals or sound) on top of the real world, in real time. It allows the user to see the real world, with virtual objects superimposed upon or composited

Figure 23.17 Two player mode as default.

with the real world. Therefore, AR supplements reality, rather than completely replacing it [10]

While the capability to deliver augmented reality experiences has been around for decades, it is only very recently that those experiences have become easy and portable. AR is becoming an emerging edutainment platform in museums and science centers [11]. Combining AR with nano technological content has been done before in an AR nano manipulator for students to learn nanophysics [12], but this still is an exclusive combination.

In the MIGHT-^{y} exhibit we use the more traditional "marker"-technique. Markers with black and white patterns from a specific library are put on each side of the two cubes and on each of the towers of the exhibit. The markers are used to calculate the positioning of the virtual layers of information and digital objects.

Although the AR solution is not yet a vandal-proof solution nor very consumer friendly because of its sensitive non-ergonomic construction, we believe it's an attractive way to experience the different length scales for teenagers, our main focus. With this device the extra information is mixed with your own perception of the world. The virtual images appear in the cubes and on the beams of the exhibit. A Faculty building with coffee bar, students walking by, and classmates are in the same view as the added digital layer of content. So you could see your classmates and some professors, only 4 in. high, in one view.

An extra level of immersion is reached by making use of "chroma key". All researchers were filmed in front of a blue screen in a studio setting. The challenge for the researchers was, of course, to act normal in this aquarium-like habitat with nothing to hold and having to stand still (otherwise they could fall down from the beam of the exhibit!). Background could be filtered, so the allover effect

Figure 23.18 After a 'MATCH' researchers come alive.

Figure 23.19 MY facilities.

is an even more realistic presence of the researcher on the building blocks of the exhibit (see Figure 23.18).

Depending on the circumstances, it's of course possible to change the AR glasses for webcam solutions (Figure 23.19).

In the exhibit "MIGHT-^{y}" every personal research object on which the researchers comment is brought back to one universal size, although experiencing the relative size of things in relation to your own size is still challenging. Still, it's real data that the player sees. The 3D research objects have a level of "aura": although it's digital and compressed, still it's the real thing. The researchers you meet in the exhibit have studied these datasets for hours and hours, so what's the deal to play this game and listen to their stories with some boundaries of glasses and rules of play to manage!

It was also a no-go to choose actors. Science LinX only shows the real faces behind cutting-edge science. Actors would make filming easier, as they are used to working in a studio and talking on camera, but we wanted the researchers of our faculty to come alive. Having real 3D research objects and researchers also upgrades the level of immersion in this exhibit, as we've talked about earlier in this paper.

In the cross section of researchers, we chose in general a good balance between younger and more established role models. Of course it's nice to see some examples of research by young adults who are just a couple of years ahead of secondary school pupils. More established role models give a more long-term view of what they could achieve as scientists.

To appeal to girls, female role models were also carefully selected; for example, associate Prof. Dr. Roberta Croce, a Rosalind Franklin Fellow. She presents her research on molecular mechanisms of the light reactions of photosynthesis. Much research is done in the Netherlands, for example, by the VHTO (national expert girls/women and science/technology *http://www.vhto.nl*) on how to increase the involvement of girls in science and technology education. Many elements are important, but one of them is to provide female role models. Likewise, a mix of Dutch and English speaking researchers seemed a good design choice to follow, to give a realistic image of our faculty.

Interesting about this AR exhibit is that it has proven to be capable of triggering the interest of not very technologically interested people. After the first sneak previews and usability tests at the High Performance Computing and Visualization Centre, the first submersion of the exhibit into the Nano LinX program with visits of a couple of school classes and the official opening of the exhibit with all of the collaborators, the exhibit crossed borders and was put to the limit by thousands of visitors to the yearly performing arts festival "Noorderzon."

For the first time, a 10 m × 10 m × 10 m science pavilion with the characteristic face of Einstein, and composed of science pics, was present at this major festival (see Figure 23.15). A cube full of arts and science guided about 700 visitors each of the 10 days from micro to macrocosm. The pavilion "Qu3" was the ultimate proof that science is not at all boring, but exciting. In the midst of planetarium, DNA bar, video filter, instant messaging plants, bird flocking simulations, and wipers which also simulated these patterns, robotic koi, and ultrasonic experiments like "bat sounds," MIGHT-^{y} took care of instant virtual researchers in just one hundredth part of the pavilion, the size of the "levelHead" cube.

A mobile version and setup were necessary to do the job, as the permanent version in the entrance of our faculty building is too robust and heavy. There was a glitterbox for the two PC's, speakers and cabling, with on top of it a nicely designed tabletop with a hole in the middle to lead the cables of the AR glasses. It was dressed up with the total setup of beams and cubes with markers, and accompanied by two big LCD screens on both sides of the exhibit to make the content visible for the large numbers of visitors.

Science LinX first explorations of AR with MIGHT-^{y} were successful and has had follow-up applications during the Night of Art and Science of the University of Groningen. AR has proven to be a great medium to present science and art in interactive, inspiring, unconventional ways. And with the possibilities of combining it with architectural objects and cityscapes, even using the GPS based AR techniques and different options of actual showing the content on, for example, smart phone, the boundaries of the medium for science communication seem infinite.

The app Molecular City forms the first follow up. Visualize the chemical building blocks of a PVC drainpipe, iron fences, and caffeine in the showcase of a shop, all with your iPhone or Android app (see below).

Within the process of developing the AR exhibit MIGHT-^{y} it became clearer that although you could really add more reality to one spot in the exhibition; namely, add real researchers of 13 different institutes of the faculty at one spot, still the technique seems to be challenging and sometimes a bit overwhelming. Keep in mind that technology shouldn't overrule the content. On the other hand, don't fear some relapses during the design process, and keep in mind that flexible programming can help get the right balance in interaction, game play, and content in the end.

Optimizing game play in order to make sure the players do meet their favorite scientists in the palm of their hands provides a one-of-a-kind science center experience. Furthermore, every aspect of design has been critically verified against the actual content. "Form follows function" is our motto (see Figure 23.14). But we still need more exploration of what the best possible ways to implement these kind of emerging technologies for the benefits of our own goals are.

23.12
Other Developments

Other exhibits have been developed using the basic ideas of virtual reality. One of these was the "nanophone." This exhibit was used during a temporary display in the night of arts and sciences (*http://www.denachtvankunstenwetenschap.nl/en/*). It consisted of an old phone that rings when you walk past it. When you pick up the phone this triggers a program on a pc connected to the phone. A voice gives instructions and explanations. The person holding the phone sees himself on the screen. By manipulating an object in front of the webcam a 3D image appears in the screen, superimposed over the camera view. Again the background of the object and the explanation are focused on research taking place at the University.

Finally the films used in might^y have been used in a Web application. In order to keep the playful element within this Web application a cube was designed, that contained different markers. The cube can be manipulated so that different markers will be displayed (see Figure 23.20).

Figure 23.20 The layout for the cube.

After loading the url *www.sciencelinx.nl/ar* a page is displayed using the web cam on the computer. Once the camera recognizes a marker a clip about the research corresponding to the relevant power of ten is displayed. See Figure 23.21.

23.13
Molecular City in the Classroom

All matter in our environment consists of small particles, molecules. The molecular world is the world of chemistry. The molecular world is a world of beauty, generally well appreciated by chemists. In order to demonstrate and share this world a application is designed in which the chemical world behind everyday-objects is shown. All sorts of knowledge is explained within the app. By using modern techniques like augmented reality and smart phone technology it is made possible to discover this world in 3D animations.

The application uses the same type of technology as is used in Might^y. The app was first used during a night of arts and sciences, where markers were placed in shop windows (see Figure 23.22). When the app was loaded and the camera of the smart phone recognized the marker a 3D image of a molecule was displayed over the camera view (see Figure 23.23). At the same time an audio message was played. In the app information was given about the molecule (see Figure 23.24). In Table 23.1 the chosen substances are indicated. When the marker was recognized by the app a 3D image of the molecule will be superimposed over the camera image. Also an audio message is played on the phone. For ethanol, for example, the text would be: "you make me surrender." On a different page an image related to the molecule. When clicking on the image information about the molecule is

588 | *23 Twenty-First Century Skills: Using the Web in Chemistry Education*

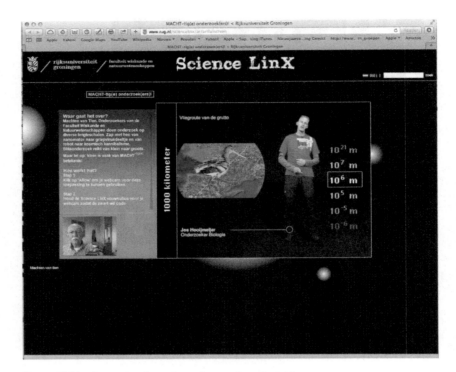

Figure 23.21 Screenshot from the online version of mighty.

Figure 23.22 Qu3 science cube with MIGHT-y exhibit.

Table 23.1 Molecules used in the app molecular city.

Molecule	Location
Caffeine	Café
Aspirin	Apothecary
Ethanol	One of the bars in the vicinity
Bombykol (feromone)	Perfumery
Bucky ball	Sporting goods shop
Testosterone	Erotic goods shop
Iron	Train station
Water	Water bottle
Sugar	Ice cream parlor
Carbon dioxide	Parking garage
Vitamin C	Vegetable store/fruit market
Cellulose	Tree
Urea	Public toilets
Lead chromate	Museum
Diamond	Jewelry shop
Capsaicin	Waga wama (we wok restaurant)
Allyl methyl sulfide (garlic)	Restaurant
Cellulose nitrate (celluloid)	Movie theater
Nicotine	Tobacco shop
Methane	Gas lighter

given. There are English, German, and Dutch versions of the app. A booklet with a description, including all markers is available on *www.molecularcity.nl*.

People were very enthusiastic about the application and tried to find all markers. The app with markers has since been used during several conferences and normally generated a lot of attention (Figure 23.25–23.27).

The molecules in Table 23.1 were used by chemistry teachers in a local school to introduce organic chemistry. The students were asked to design and make models of the molecules in het table. The models of the molecules needed to be large enough so they could be displayed in an exhibition. The students came up with many interesting ideas to make the molecules. The molecule ethanol, for example, was made from cans that contained light alcoholic beverages which are favorite among students. In Figure 23.28 you can see the sign that was designed, as an explanation for the model the students made of caffeine. In Figure 23.29 you can see part of the display in the church, showing bombykol, methane, and caffeine.

This application is perfectly suited to be used as an introduction to organic chemistry at any level. It presents molecules out of the everyday world to students. When these molecules were used to introduce naming organic compounds we experienced no more problems then was the case when a more traditional way of introducing naming organic compounds was used.

Figure 23.23 The nanophone in action.

Figure 23.24 Mobile MIGHT-y on Noorderzon performing arts festival.

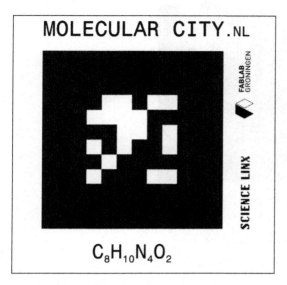

Figure 23.25 Marker for caffeine.

Caffeine

Caffeine is the stimulating substance in coffee, Red Bull, but also tea. One gram of tea contains even more caffeine than one gram of coffee. However, to make a pot of tea only 4 grams is used and for a pot of coffee more than 80 grams, so a cup of coffee contains a lot more caffeine. Caffeine is addictive. Moreover, you get used to it; you need more to reach the same effect. An overdose of caffeine results in all kinds of side effects, such as sweating, trembling, nervousness, fear, etc.

Figure 23.26 Text for badgepage caffeine.

Figure 23.27 Image of molecule caffeine.

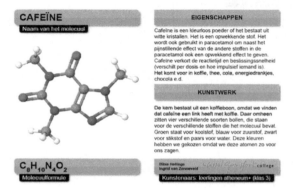

Figure 23.28 Sign for molecules.

Figure 23.29 Caffeine, right hand corner.

23.14
Conclusion

In this chapter I have discussed some general principles regarding the use of ICT-applications in education. One major problem with the design of ICT-applications is that the people having enough knowledge of technology do not know enough about pedagogy, while on the other hand people in education do not have enough knowledge about technology. One of the ways this can be solved is by bringing together a development team in which specialists from different fields are brought together. A number of examples developed at the University of Groningen have been discussed to demonstrate what the result of such a development could be.

One of the most important conclusions is that the use of ICT-applications will only be successful when it has an added value. Teachers and students need to experience this added value otherwise they will not invest the effort needed to implement the application within their education.

The Internet has many opportunities for aiding the learning process of students. It has the opportunity to intersect formal and informal learning. On the one hand it caters to the interest it arouses on the net, on the other hand it offers opportunities for in line learning.

In the end though face to face contact between teacher and student will be needed. A teacher can determine at which stage the student is in a learning process and guide him forward. Face to face interaction, where body language may be read is indispensible in the teacher/learner interaction.

For textbooks there is another story. I am afraid I have to agree with Peter Atkins, that textbooks will ultimately disappear. Developments like Kindle where books are read from a tablet will take over from material printed on paper or some other surface. But books and explaining texts will remain, next to Internet developments.

It will be interesting to see how fruitful the interaction between educators and Internet experts will develop into new materials and possibilities. Practically anything will be possible in the future. One small example that has not yet been discussed, because I just learned about it recently is doing on line experiments. During a day time class UCLA is looking through the telescope at the Weizman Institute in Israel where it is night at the same time. The same happens vice versa.

References

1. Atkins, P.W. (2013) The future of the book. *Chem. Int.*, **35** (2), 3–6.
2. O'Reilly, T. (2005) What is Web 2.0. Design Patterns and Business Models for the Next Generation of Software, 30 September 2005, Message posted to http://oreilly.com/web2/archive/what-is-web-20.html (accessed 30 October 2013).
3. O'Reilly, T. and Batelle, J. (2009) Web Squared: Web 2.0 Five Years On, Message posted to http://www.web2summit.com/web2009/public/schedule/detail/10194 (accessed 30 October 2013).
4. Harris, J., Mishra, P., and Koehler, M. (2009) Teachers' technological pedagogical content knowledge and learning

activity types: curriculum-based technology integration reframed. *J. Res. Technol. Educ.*, **41** (4), 393–416.
5. Lionni, L. (1970) *Fish is Fish*, Dragonfly Books, New York.
6. Classroom 2.0 http://www.classroom20.com/profiles/blogs/fish-is-fish (accessed 30 October 2013).
7. Cherry, L.D. (2010) Blended Learning: An Examination of Online Learning's Impact on Face-to-Face Instruction in High School Classrooms. ProQuest LLC, http://www.proquest.com/en-US/products/dissertations/individuals.shtml (accessed 30 July 2014).
8. Keengwe, J. and Wilsey, B.B. (2012) Online graduate students' perceptions of face-to-face classroom instruction. *Int. J. Inf. Commun. Technol. Educ.*, **8** (3), 45–54.
9. Bybee, R.W., Powell, J.C., and Towbridge, L. (2007) *Teaching Secondary School Science, Strategies for Developing Science Literacy*, 9th edn, Pearrson, Upper Saddle River, NJ.
10. Azuma, R. (1997) A survey of augmented reality. *Presence Teleoper. Virtual Environ.*, **6** (4), 355–385, http://www.cs.unc.edu/~azuma/ARpresence.pdf (accessed 30 October 2013).
11. Bimber, O. and Raskar, R. (2005) *Spatial Augmented Reality. Merging Real and Virtual Worlds*, A.K. Peters Ltd, Wellesley, MA.
12. Marchi, F., Marliere, S., Florens, J.L., Luciani, L., and Chevrier, J. (2008) in *Transactions in Edutainment IV* (eds Z. Pan, A.D. Cheok, W. Müller, X. Zhang, and K. Wong), Springer, pp. 157–175.

24
Design of Dynamic Visualizations to Enhance Conceptual Understanding in Chemistry Courses

Jerry P. Suits

24.1
Introduction

Chemistry students have difficulty learning chemistry for a variety of reasons: For example, many of them tend to fragment knowledge and then memorize those fragments or algorithms [1–3]; also, they tend to focus on the surface features of a visual representation [2, 4]. Conversely, when scientists study a complex research problem, they tend to engage in meaningful and productive learning strategies. Scientists tend to use mental images and to use multiple representations of phenomena when solving complex problems and they can switch from one representation to another [2, 5]. Thus, the goal of dynamic visualizations is to get chemistry students to visualize on the computer screen (i.e., external representations) the same types of representations of chemical phenomena that chemists mentally envision (i.e., internal representations).

24.1.1
Design of Quality Visualizations

An overall principle of dynamic visualizations is that students' focus should be on the essential aspects of the visualization (e.g., the chemical species participating in a chemical reaction) rather than getting distracted by peripheral aspects (e.g., brightly colored red and white balls that represent water solvent molecules). This principle is described in a book that Michael Sanger and I recently edited that introduces an overview of this topic as covered over 17 chapters [6]. Also, in a NSF-funded project [7], Loretta Jones and I assembled a team of cross-disciplinary experts to explore this topic. We described good design principles in molecular visualizations, "violations" of these principles, and looked at how visualizations can produce student misconceptions.

The design and use of dynamic visualizations should focus on creating interactive learning environments that provide a spectrum of goals, which range from explaining concepts to exploring chemical phenomena. To accomplish these goals, the instructional sequence should begin by stating the learning objectives, which are sensitive to students' background knowledge of chemistry topics. It should culminate with assessments that go beyond mere recall, while encouraging students to develop their own mental models [8]. Throughout this sequence, instruction should support appropriate levels of student interactivity while providing cues to direct their attention, feedback to guide their learning, and scaffolding to facilitate their processing of information. Animations are often used to help students understand and explain abstract chemistry concepts, while simulations allow student exploration of phenomena and their representations.

24.1.2
Mental Models and Conceptual Understanding

One of the best means of determining whether students possess a conceptual understanding of chemistry is the extent to which they can correctly express their mental models for solving relatively complex chemistry problems. A *mental model* is a set of mental representations and the mental operations that adapt these representations to a particular problem [9]. Mental models help students organize their experiences into meaningful cognitive structures that allow them to select and transform information and to form hypotheses [10]. Chemists typically use visual images [5] to organize chemical facts into a mental model. This model in turn can solve a real-world problem: What if a Christmas tree ornament is not working properly? In this case it is a "bubble light" [11], which contains a volatile liquid in a sealed container heated by a small incandescent bulb at the bottom with an empty space above the liquid. A problem occurs when the lights are turned "on," but some of the ornaments do not bubble. A quick on-line search for "bubble light" does not help because it yields several misconceptions (e.g., the liquid is water or keep the bulb light upright to start it bubbling). However, chemists know several facts about this matter: (i) the empty space contains vapor, (ii) when an enclosed vapor is heated, the volume cannot expand, (iii) the incandescent bulb is the heat source, (iv) heating an enclosed vapor increases its pressure, (v) boiling point is when the vapor pressure of a liquid equals the external pressure, and (vi) the kinetic molecular theory of gases explains the relationships among facts (i)–(v). Thus, the solution to this problem is to turn the bubble light upside down, which allows the vapor phase to be directly heated by the bulb. Consequently, this increases pressure until it equals the vapor pressure of the liquid and results in its boiling point. Finally, turn the bulb upright and it will bubble continuously. Overall, a chemist can construct a mental model by using mental representations, and mental operations to build a holistic solution to a relatively complex problem.

Conceptual understanding in chemistry requires the acquisition and integration of multiple representations [12] at three different levels [13]:

- *macroscopic level* is the most concrete level of representation (e.g., laboratory observations of chemical phenomena);

- *submicroscopic level* is an intermediate level of representation on the concrete to abstract scale (i.e., to show how entities such as atoms/ions/molecules interact in chemical processes); and
- *symbolic level* is the most abstract level of representation (i.e., to show relationships among abstractions, such as chemical formulas, equations, and mathematical expressions).

Chemical educators believe that students truly understand chemistry only when they can successfully use these three levels of representations and when they can "see" the links among them (see Figure 24.1) [14]. A mental model of these three levels of chemical knowledge and their interconnections is essential in order for a student to develop a conceptual understanding for a given chemical topic.

A titration can be used to illustrate these three levels. If a chemist is titrating an unknown diprotic acid: the *macroscopic* representation features the glassware and chemicals needed to perform the titration experiment (see Figure 24.2a); the *symbolic* level consists of abstractions such as the balanced chemical equation, mathematical calculations, and the resulting plot of volume of titrant versus pH, (Figure 24.2b); and the *submicroscopic* level uses icons (e.g., for H_2O) to show the underlying molecular interactions between a proton and hydroxide ion (see Figure 24.2c). The chemist can interconnect these three levels and essentially fuse then into chemical knowledge that is well organized and coherent (i.e., their mental model of this chemical phenomenon). However, students often misrepresent their chemical knowledge at the submicroscopic level as shown when they do not understand that aqueous protons and hydroxide ions (i.e., two highly reactive species) react to form water (i.e., a very stable species) in a very exothermic manner (see Figure 24.2d).

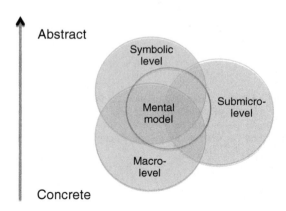

Figure 24.1 A mental model plays a central role in helping students interconnect the three levels of chemical representations for a chemical phenomenon. Figure 1 from [14], © 2010, Taylor&Francis, Ltd. http:www.tandfonline.com)

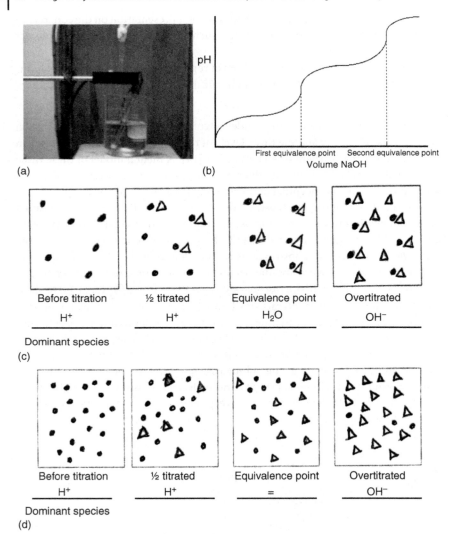

Figure 24.2 (a) Macroscopic titration of an acid using a pH probe. (b) Titration curve for a weak diprotic acid. (c) This is a hand-drawn sketch showing the correct H⁺ and OH⁻ interactions in the four major stages of an acid–base titration. (d) Student's misrepresentation of the H⁺ and OH⁻ interactions in an acid–base titration. Note that both H⁺ and OH⁻ are unreacted at equivalence point.

24.2
Advances in Visualization Technology

In recent years a number of advances in innovative visualization technology have been developed to support students as they strive to understand chemistry concepts and processes. The most popular application is probably a set of 125

PhET interactive simulations that Lancaster *et al.* [15] have designed, developed, and classroom-tested. Their goal was to produce dynamic simulations that enhance student conceptual understanding of chemistry (see Figure 24.3). These simulations were designed to encourage students' interactions with chemical phenomena in such a way that they could develop their inquiry skills. They also feature linked multiple representations, and connections to real world applications. Furthermore, these simulations can help students develop their *mental models* of chemical processes at the molecular level [15, 16]. The PhET simulations are readily available for chemistry instructors to use with their students at ... *https://phet.colorado.edu/en/simulations/category/chemistry*.

Ashe and Yaron [17] have described how analogy-based simulations can use familiar objects in a concrete system to help students learn about the interactions of abstract entities in a chemical system. For example, the use of bouncing balls on the vibrating steps of a stairway can show the interactions of atoms and molecules in a Boltzmann distribution. The designer of the simulation should identify the key features of the chemical system and then set up a one-to-one correspondence with those features of the concrete system. The student could then observe the interactions of objects in the concrete system and then match them to the corresponding visualizations in the abstract, chemical system [17]. For example, to help students understand the progress of reaction diagram for an endothermic chemical reaction, the concrete system can be shown as bouncing balls on platforms

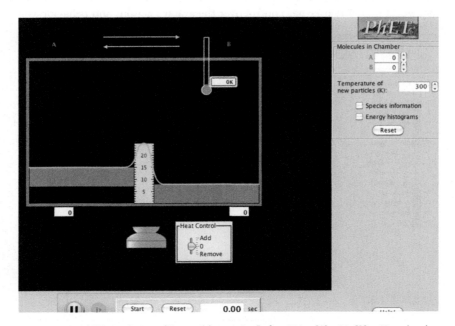

Figure 24.3 PhET simulation of "reversible reaction" after 581 s: [A] = 21, [B] = 40 molecules; initial [A] = [B] = 30 molecules. (Source: *https://phet.colorado.edu/en/simulation/reversible-reactions.*)

of three different heights (i.e., lowest to highest to a middle height). Thus, this system shows interactions of objects (i.e., bouncing balls) on different mechanical potential energies ($PE = m \times g \times h$), which correspond to different chemical potential energies in a chemical system. The best way to show both systems is to have them side-by-side (or one on top of the other). Also, it is important to keep these visualizations as simple as possible so students do not get confused or overwhelmed by the complexity of the two systems [17]. The main advantage of using an analogy-based simulation is that the concrete system helps students visualize a familiar situation, and then they can literally see how the abstract, chemical system exhibits the same processes or properties. Furthermore, as students explore these dynamic simulations, they can internalize them as mental simulations of phenomena where they are using their imagination to organize visual images and to develop useful *mental models* [17, 18]. These mental models, in turn, allow them to use qualitative reasoning skills to solve complex chemistry problems.

Winkelmann [19] has embedded the use of interactive simulations in a virtual laboratory environment (Figure 24.4) [18] that is embedded within a virtual world. A *virtual world* consists of a large network of three-dimensional, visual surroundings where many users (students) can interact with each other and with virtual objects (e.g., chemicals, glassware, buildings, land, cities, and so on). *Second Life* is a popular virtual-world environment, and it becomes more believable when the interactions within it are more like those in real life [20]. Each user creates their own *avatar*, which is a graphical representation of the user's body in the virtual world [21]. The user has control over the avatar's actions and behavior as it

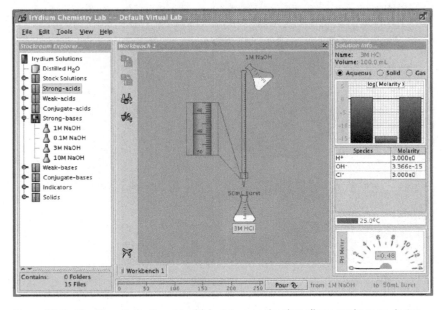

Figure 24.4 The ChemCollective virtual lab. A java applet that allows students to design and carry out their own experiments (*www.chemcollective.org*).

explores this imaginary but shared alternate reality. For example, a student can control the actions of their avatar in a chemistry laboratory by having it prepare and mix chemicals together, and then the student can observe the consequences of this action. They can also use their avatar to communicate with other students' avatars and this discussion could be used to decide how to perform a virtual experiment (i.e., procedural knowledge) and how to make inferences from the observations of the results (i.e., conceptual knowledge). When students assume the identity of their avatar, which they have created as an imaginary character [19], they can change their point-of-view and immerse their thoughts and actions in a virtual world. Thus, when students explore a virtual world, they can modify the learning goal to include their own purposes for learning a chemistry topic. This personalization of learning can help them organize their experiences in a meaningful way that helps them transform and internalize information. This learning process is called the development of *mental models* [9, 10, 22].

Keeney-Kennicutt and Merchant [23] have expanded upon Winkelmann's uses of a virtual world to include a variety of virtual activities, including office hours, videos, simulations, games, quizzes, and interactions with virtual chemical species. In order for a chemist to understand the shape of molecules, they have the ability to visualize the arrangement of atoms in 3D space. These researchers did a study where students in the treatment group visualized and learned about 3D chemical structures (i.e., (VSEPR) Valence-Shell Electron-Pair Repulsion theory) within a virtual environment. The control group learned and performed the same set of tasks exclusively on paper. Results showed that those in the virtual-world treatment group were better able to interpret the corresponding 2D representations on paper. Additional benefits of the 3D virtual group include enhanced visualization skills, immersion (i.e., experience of being completely surrounded in a virtual world), and collaborative learning with other avatars/students [23].

When students are given a 2D printed image of a molecule, many of them have difficulty visualizing its 3D shape [23–26]. The challenge is greater when their task is to explore the various structural features of a molecule and its roles in characterizing a 3D chemical reaction. Designers have developed and used a variety of visualization tools to help students (and researchers) understand the shapes of 3D molecules [24, 25]. Specifically, *Jmol* is a visualization tool that is an open-source Java viewer [27–29], which allows students to observe and manipulate molecules, biomolecules, and crystals (see Figure 24.5). In addition, these tools can help students understand reaction pathways (e.g., S_N2 organic reactions) in terms of both the 3D changes in molecular shape and the corresponding progress of reaction diagrams [24]. With regard to the more complex reactions of biomolecules (i.e., proteins, nucleic acids, carbohydrates, and lipid membranes), visualization tools can provide students with 3D views of binding sites (see Figure 24.6), while allowing them to view them from different perspectives and to literally see an animation of the binding of a substrate and its subsequent change(s) in configuration [28]. Overall, if students are to understand 3D molecules and processes, they must internalize mental images that correspond and complement those of the external

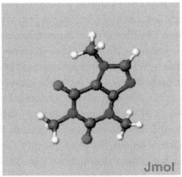

○ Cyclopropane ◉ Caffeine

Figure 24.5 Caffeine molecule in *Jmol*. *http://jmol.sourceforge.net/demo/jssample0/*.

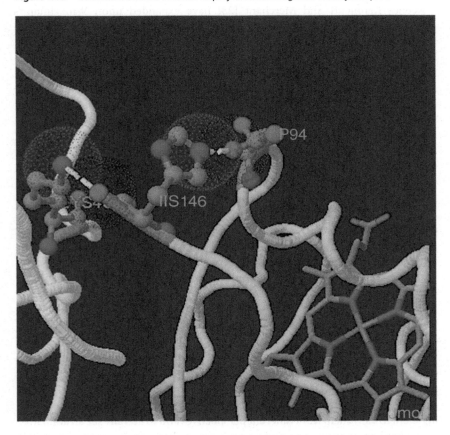

Figure 24.6 A Jmol screenshot of a fragment of transcription factor TFIIIA forming three consecutive zinc finger motifs (cartoons), bound to a stretch of DNA (spacefilled) (*submitted by Angel Herráez*). *http://jmol.sourceforge.net/screenshots/screenshot_hemoglobin.png*.

representations shown by the 3D technologies. When they are engaged in this process, they are constructing a *spatial mental model* of a visual representation [30].

Martin and Mahaffy [31] have developed a set of visualization tools to help students study and actively explore the complex phenomena involved in climate change. Climate change is an interdisciplinary subject that directly requires knowledge of geology, meteorology, chemistry, and physics. It exists as a dynamic and complex system where the various components interact with each other on multiple, interrelated levels. For example, to understand how accumulation of "greenhouse gases" causes tropospheric warming in this system, students must synthesize knowledge of how electromagnetic radiation in the IR spectrum is absorbed by gaseous CO_2 molecules (and other greenhouse gases), and how this absorption produces vibrational modes that allows these molecules to trap, store, and transmit thermal energy to N_2 and O_2 (non-greenhouse gases), which in turn increases the temperature of the troposphere. Thus, scientists possess elaborate mental models of these processes but most students lack the content knowledge and imagination (i.e., a set of visual images in the mind that are interconnected together) needed to engage in this process. Martin, Mahaffy, and their colleagues [31, 32] have developed pedagogic tools that allow students to actively learn from interactive visualizations that are presented using guided inquiry strategies. They have developed, implemented, and assessed a set of seven interactive simulations, which were designed to help students visualize and hence understand the evidence for climate-change systems. These simulations allow students "to construct their own *mental models* that are built upon fundamental science concepts" [31, p. 436], of chemistry and physics.

In summary, all of these innovative visualization technologies, explicitly or implicitly, advocate a learning outcome where students develop their mental models. This common thread is not a coincidence because the development of interactive simulations is based on experts' (e.g., chemists') mental models of complex phenomena. The experts externalize this model in order to provide many opportunities for users (i.e., students) to input and to interact with the model via the visualization technology they are developing [8–10, 22, 30]. In other words, if the assessment used to gauge student understanding was a multiple-choice test, then much would be lost about what the students really understand. One purpose of the remaining sections in this chapter is to show several examples of how chemistry instructors can assess whether or not their students really understand what they experienced after interacting with those visual technologies. Another purpose is to illustrate which design features of high-quality visualization can help students developmental models that progressively move toward becoming viable scientific models of phenomena.

24.3
Dynamic Visualizations and Student's Mental Model

Alicia Courville, my first graduate student, and I designed and evaluated a multimedia module to help students gain qualitative understandings of the gas laws [33]. The course was first semester general chemistry for college science majors. We developed a dynamic animation of an expanding automobile air bag as shown before (Figure 24.7a) and after inflation (Figure 24.7b). This animation gave students both visual and auditory (sound of rapidly expanding gas) feedback. Students calculated their answers to a liter-to-gram gas stoichiometric problem and then got the animated feedback as shown in Figure 24.7c (i.e., expanding air bag in lower right-hand corner) [34]. This animated feedback shows the air bag

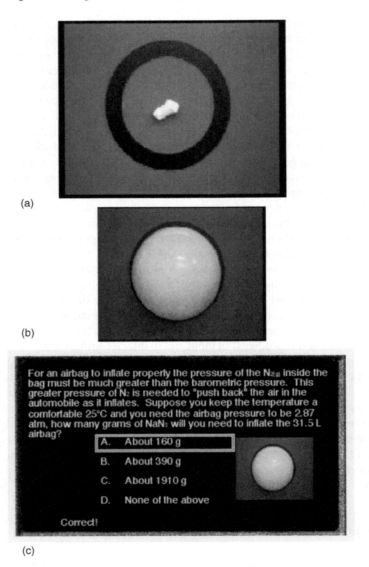

Figure 24.7 Initial and final frames of an air bag movie [33]. (a) Before inflation, (b) after proper inflation, and (c) gas law problem with air bag animated "feedback".

expanding as it is inflated. Normally, after a student enters a calculated answer, they are only given corrective feedback (i.e., you are correct or you missed it) or diagnostic feedback (i.e., you did one step incorrectly, ...). Conversely, our multimedia module used realistic feedback: "You must calculate the proper amount of sodium azide to put into an automobile's air bag so that it will inflate to the proper volume. If you calculate too little, then the air bag with underinflate; too much and you might crush the driver." When a student calculated correctly, answer A in Figure 24.7c, the air bag inflated to the black circle, which represented the steering wheel. Thus, students could visualize the real-world consequences of their answers, and they could see the links between chemistry and its real-world applications.

About 30 students in the class interacted with the multimedia module and their feedback was very positive. For example, several students appreciated the visual features of the multimedia module: "I learned how to visualize the problem" and "It helped give me a visual picture of the calculations." Other students liked the multimedia aspects of the module. We used voices of young people as the source of feedback (e.g., "Crushed by the air bag; poor, poor people!"). One student stated, "The dude talking was funny and it explained before it gave a problem." Still other students liked the real world applications, for example, "The Questions about the air bag w/the information that went along with it." Overall, we found that the module did help students do significantly better on the hour examination that covered gas law calculations.

One student in this class apparently benefitted greatly from her interactions with the visual and auditory aspects of the multimedia module on the gas laws [34]. On her three previous hour examinations, her average score was 69.7%, which was slightly below the average for this chemistry class, which was taught using traditional instruction. However, on the last hour examination, her score was 96% on the gas laws – the highest score in the class. The link between her excellent exam performance and her interactions with the gas laws module is illustrated by her answer to an exam question about an expanding bubble. This question had two parts: it was a multiple-choice question that included a blank space for students to show their work in solving the problem (Figure 24.8a). Although not required, she drew a detailed sketch of a boy, "Cajun Charlie," who was chewing bubble gum and was engaged in the process of blowing a bubble (Figure 24.8a). She also correctly answered the quantitative gas flow problem (i.e., solving for the moles of gas). She used multiple representations [12, 13] in her answer: she performed the mathematical calculations (i.e., symbolic level), sketched the image of a boy blowing a bubble (i.e., macroscopic level), and included the molecular interactions (i.e., submicroscopic level). She used the kinetic molecular theory of gases to draw the molecules moving in random directions at different and rapid velocities, which collide with the inner surface of the bubble to increase the pressure and hence increase the volume of the bubble. In other words, she used a mental simulation similar to an external simulation as shown in Figure 24.8b. Overall, the drawing plus "quantitative explanations" represents her mental model of the phenomenon

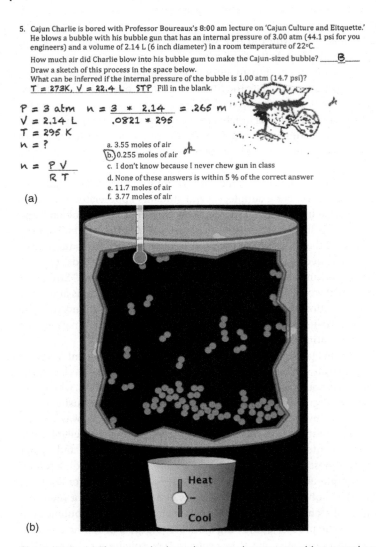

Figure 24.8 (a) This example shows how a student expressed her mental model in response to a multiple-choice gas law problem on a major examination. (b) Animation of expanding N_2 gas http://phet.colorado.edu/en/simulation/states-of-matter.

[14] as a key indicator of her conceptual understanding of the problem. This student's detailed drawing of "Cajun Charlie" (Figure 24.8a) clearly indicates that she had artistic talent, which is consistent with her preferred visual learning style as determined in our study [34].

24.4
Simple or Realistic Molecular Animations?

One of the design decisions that developers of molecular animations must make is "Should the animation be a simple or a realistic representation of a molecular process?" The advantage of realistic animations is that they portray these processes the way chemists envision them at the molecular level. Most chemists believe that the dissolution of sodium chloride should show the randomness of multiple collisions between solvent molecules and the vibrating ionic lattice of NaCl(s) [35]. Likewise, the aqueous environment should be littered with hundreds of water molecules that are hydrogen-bonded together as triads. Unfortunately, when such a realistic animation is shown to students, the information is too complex. Their senses are often oversaturated by too much information being presented at the same time, which produces a cognitive overload [36]. This cognitive overload can destroy their ability to understand the molecular process being portrayed. Also, the chemists' view is not a "replication of reality" but rather it is a scientific model. These models are the best way to understand molecular processes; however, with more and more scientific publications, these processes are getting more complex and the scientific models are getting more abstract and complicated.

Conversely, animations can show molecular processes in simple terms, which convey only one concept per animation. When properly done, this approach allows students to focus on the *key features* of the animation [35]. This guides their mental "visual channels" to the features that a chemist would envision, which allows students to truly understand the process. When students receive verbal information through their "auditory channels," then their attention can be directed to the relevant visual features in a way that is similar to how chemists envision these processes. Cognitive capacity is actually expanded when information is presented in both visual and auditory modes [37]. Sanger and Rosenthal [38] studied whether the sequence of animations should be from simple to realistic, or vice versa. They found that students provided better particulate-level explanations when they had experienced the complex-to-simple sequence. They recommended using this sequence because the more complex animation can get students' attention, then the simpler one can help them give better explanations for chemical reactions.

In our research studies, we allow students to control whether the animation is simple or realistic by using a slider, which changes the degree of complexity between the two extremes: "simplest" $\longleftrightarrow$ "most complex" [39]. As shown in Figure 24.9a,b, the initial position of the slider (i.e., the default) could be set at either extreme. In this case the animation illustrates a complete acid-base titration curve by showing the ratio of hydronium ions (i.e., H_3O^+) to hydroxide ions (i.e., OH^-) at each stage of the titration. Thus, using Sanger and Rosenthal's [38] results – the students could begin by watching the more complex animation (Figure 24.9b) in its entirety. Next, they can be prompted to move the slider to the left to view a simpler animation (Figure 24.9a). In this case, learner control over the degree of complexity of an animation allows them to understand that

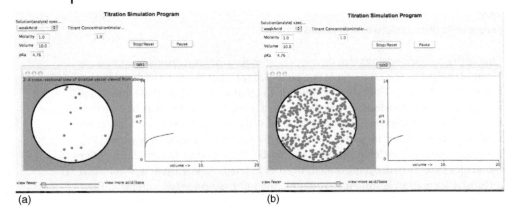

Figure 24.9 Illustration of how a slider can show two different extremes – Part (a) shows a simpler animation and part (b) a more complex animation of a titration process, which shows the ratio of H_3O^+ (red dots) to OH^- ions (blue dots) during the "acid dominant" stage of a titration. The slider is the horizontal line at the bottom of a and b. (a) Simpler animation and (b) more complex animation.

there are intermediate levels of complexity between these two extremes. Also, the simpler animation helps them to get a feel for the interactions of a particular particle (e.g., an H^+ ion) with other particles (e.g., repulsive collisions with other H^+ ions or attractive collisions with OH^- ions).

24.5
Continuous or Segmented Animations?

When chemists contemplate the nature of a chemical reaction, they probably visualize it as being a continuous process, which proceeds from the initial state through the intermediate states (e.g., the transition state or sub-steps within a multistep chemical reaction) and then the final state. Conversely, when teaching a chemical reaction, most chemists emphasize the individual states of the process (e.g., initiation, propagation, and termination of an organic chemical reaction). Many molecular animations of chemical reactions are presented as continuous visual displays, which are often accompanied by verbal descriptions of the chemical structures and their functions.

When students view a continuous animation of a chemical reaction, they must process a large amount of new information that can overload their cognitive capacities. That is, each frame in the animation shows new changes in structures and functions that must be processed and interrelated together with the previous frames. The overall effect is that it is difficult for students to make sense of this hodge-podge of information. They need to focus on each chemical structure individually and to see how it interacts with other structures. When these mental fragments accumulate, students' cognitive capacities can be overloaded, which result in poor understanding of the overall process [40]. The resolution of this problem is to develop a strategy that breaks up the continuous process into a set of distinctive steps, which are meaningful in terms of understanding the overall chemical process. In cognitive psychology, this process is called *segmentation*, and each segment is an organized set of information [37, 41, 42]. Segmented animations allow students to understand each segment before moving on and hence they can understand the entire process without overloading their cognitive capacities. Finally the segmented animation should be paused after each meaningful segment (as determined by chemists) to allow the student to respond to questions about what they just experienced.

24.6
Individual Differences and Visualizations

Chemists' mental models are used to convey visual information to depict chemical phenomena, and verbal information to explain the meaning of the visual information. Thus, students can use pen and paper to draw (in blank boxes) and explain (on blank lines) their mental models for a chemical process – before, during, and after the process occurs. In our research studies, we have found that students who can only draw the correct visual information may or may not understand the meaning of their drawings. This is because they may have just copied what they saw in their lecture notes or the textbook. Likewise, students who appear to explain chemical phenomena may just be reciting or paraphrasing something they heard in lecture [43].

24.6.1
Self-Explanations and Spatial Ability

When students *self-explain* they are elaborating upon the presented information by relating it to their prior knowledge, which can be use to make inferences, and then integrate it with information they already possess [43, p. 297]. Constructivist learning theory is consistent with this self-explanation process [43]. When students translate the instructional text or oration into their own words, this process allows them to understand the meaning that underlies an instructional message. In one study, students who were presented with diagrams generated more self-explanations than those who were presented with textual information on the same topic [43]. The process of self-explanation can help students with good or weak background knowledge of chemistry (i.e., high- and low-prior knowledge, respectively). Chi [43] found that high-prior knowledge (PK) students were able to use their self-explanations to repair their existing mental models and thus to improve their learning outcomes [44]. She found that Low-PK students used their self-explanations to fill in gaps in their missing knowledge. Overall, there is a close link between students' abilities to self-explain the instructional material and their ability to construct and repair their mental models.

Spatial ability (SpA) is the ability to mentally manipulate an object by rotating, twisting, or inverting it to change its orientation in space [45]. A systematic review of the role of SpA in contributing to students' success has found that students who have high-SpA can benefit more from dynamic visualizations than those with low-SpA [45]. Consequently, high-SpA students can use their *visuospatial abilities* to understanding chemical concepts better and to solve complex problems because they can spatially organize and represent the conceptual information presented in those problems [46]. Specifically, when solving problems, high-SpA students tend to draw preliminary figures and sketches even when not required by the questions [47]. Conversely, when solving problems, low-SpA students tend to draw fewer figures or to draw incorrect figures. Chemistry animations helped both low- and high-SpA students gain a better understanding of chemistry concepts [48]. However, the gain was greater for the high-SpA students, which may be because they were able to verbalize their visual images and then translate these verbalizations into spatial mental models [49].

24.6.2
Individual Differences and Visualization Studies

In our research studies, we encouraged students to draw their visual images on a blank template (we use the PowerPoint print option to print a blank three-slide handout) both before and after they experience the animation or simulation [39, 50–54]. Thus, these drawings allowed them to express their *mental models* in an explicit manner that helped them become aware of their conceptions. In addition, they needed to explain their drawings using their own words (i.e., self-explanations). In first semester general chemistry, David Falvo and I studied the effectiveness of a molecular-level animation [35] with respect to four treatment groups (control, arrows, labeled particles, and labels plus arrows), SpA, and gender [55]. One of these treatment groups scored significantly lower than the others – the one that received arrows, which served as signals to show the movement of particles. As expected the high-spatial group scored significantly higher than did the low-spatial group on a posttest of their content knowledge covered in the animation. Surprisingly, female students scored higher than the males on the posttest. We attributed this result to the high-quality instructional guidance that was featured in the animation. Apparently, females may have used this guidance to create better mental models and to use these models as prediction tools [56].

In a follow-up study [57], we used the same molecular-level animation [35], but with a different set of treatments (i.e., pause, pause and pace, and self-explain) to determine their effects on students' content knowledge. We divided this knowledge into three subtests: structures (NaCl as a solid then dissolved), function (dissolved ions attracted to water), and transfer. As expected, the high-spatial group scored significantly higher on the structure posttest questions but not on function nor transfer questions. An interesting aptitude-treatment interaction occurred on the structure posttest questions with respect to SpA: (i) there was no significant difference between high- and low-spatial groups in either the control or "pause and pace" groups, while (ii) in the "pause" group, high-spatial students slightly outscored low-spatial students, and (iii) in the "self-explain" group, high-spatial students scored much higher than did low-spatial students. These results are consistent with the findings of other studies, where low-spatial students need much more instructional support to reduce the cognitive load they are experiencing with a multifaceted molecular level animation. Conversely, high-spatial students are more likely to internalize an animation as a "mental animation," which they can mentally play and re-play as a means of acquiring deeper conceptual understanding while "repairing" their mental models [43].

24.7
Simulations: Interactive, Dynamic Visualizations

A simulation uses a mathematical or logical model to simplify a real-world phenomenon [58]. The goal of running a simulation is to help the user understand the underlying model and scientific concepts [59, 60]. Chemistry simulations are dynamic visualizations of chemical phenomena where users manipulate the variables of a chemical system and then run the simulation in order to observe its outcome on the dependent variable. As experts, chemists can use simulations to deepen their understanding without the need for an external animation to show the outcome. This is because they can use their visual imagery [5] to generate a mental simulation [61] of the chemical system they are exploring. For example, if a chemist wanted to use a "hard core" simulation to visualize the titration curve for an unfamiliar diprotic acid, H_2X (aq), then they could input either the acid's name or chemical formula, and the simulation would output the values of pK_{a1} and pK_{a2}. Next, the chemist would decide what concentration of base to use in their imaginary titration, for example, 0.100 M NaOH, and then they could mentally visualize the shape of the resulting titration curve, see Figure 24.2b. Conversely, a chemistry student enrolled in a college-level general chemistry course would only see two numbers – one for each pK_a value. If a Vulcan mind meld (Star Trek; Mr. Spock's telepathy) were possible, then the chemist's mind would yield a visual titration graph with two Sigmoid-shaped curves (e.g., Figure 24.2b) while the student's mind would reveal only a "blank screen."

24.7.1
Pedagogic Simulations

If students are to engage in meaningful learning from a simulation, then an extra layer must be added to the chemist's "hard core" simulation. The resulting software is a *pedagogic simulation,* and the extra layer is called the "instructional overlay or interface" [62], which supports and guides the student's learning processes. A pedagogic simulation is highly interactive because a student inputs a value for each variable, and then the simulation changes to reflect these particular experimental conditions. In other words, interactivity is a "two-way street" where the actions of the student changes the simulation outcome, and the outcome, in turn, changes the student's perceptions [42, 63]. This interactivity is enhanced when students are encouraged to use a *predict-observe-explain strategy* in their exploration of the simulated phenomenon [44]. After the student selects values for the input variables (i.e., parameters), they should be encouraged to *predict* the outcome. Next, they run the simulation to *observe* the outcome, which they should be able to *explain* in terms of the appropriate chemical concepts. After a student predicts the outcome for their simulation, it can be shown in one of several formats: as a textual message, mathematic or graphical solution, or as an animation (e.g., Figure 24.10a,b). An animation is the preferred format because it is a dynamic visualization, which shows how scientists visualize scientific phenomena [59]. Not surprisingly, students usually prefer seeing an animated outcome [64] rather than reading about it or looking for a pattern in the resulting data table. If the student inputs values for variables that result in a nonproductive outcome, then they should get an appropriate feedback message. For example, if the task is to determine the melting point of an unknown liquid organic compound [65], and the student decides to heat the liquid sample, then a prompting message should state, "You are about to heat a liquid sample. This will not give you the melting point." This statement encouraged students to incorporate new information as mental representations that can restructure their mental models as they adapt to the situations described and depicted in an interactive simulation [66].

Without the appropriate instructional interface, many students would only engage in random, trial-and-error exploration, which can interfere with their ability to manipulate the variables in a simulation [63]. For example, suppose an acid-base titration simulation is programmed to ask the user to input the "number of acidic protons." The intended numerical response might be 1, 2, or 3. However, what if a student enters an unanticipated answer, like 6.02×10^{23}? Thus

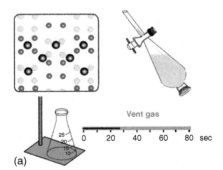
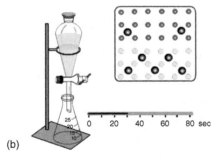

Figure 24.10 Screen-shots from the animation of the organic extraction simulation. In part (a) the separatory funnel is being shaken, and aqueous-organic phases are inter-mixing. In part (b) the layers have separated and the caffeine molecules are mostly in the organic (bottom) layer. (a) Funnel during shaking and (b) funnel after mixing.

the program would "crash," and the student would get frustrated because they would not know what happened or why it happened. To avoid this unfortunate situation, the simulation must anticipate a variety of responses that students input when they are selecting values for each variable. If the student response is anticipated, then an appropriate feedback message can be given to them. An unanticipated answer should trigger a specific statement to give them a direct hint (e.g., Choose 1, 2, or 3 protons/formula).

How should students provide evidence that they have internalized the intended chemistry concepts? As previously described (Section 24.1.2), a student's mental model can be revealed when they express it through multiple representations [67]. This shows they have mental images of the chemical phenomenon that can be verbally explained on a blank template (i.e., three sets of horizontal lines to the right of three boxes). Also, this process can expose their misconceptions when they attempt to represent the phenomenon in sketches and words. In a research study that Sister Nicole Kunze and I conducted, we interviewed a student after she had experienced all three levels of the titration phenomenon [52] as shown in Figure 24.2. We asked her to draw the H^+ and OH^- particles at the submicroscopic (particulate) level in their proper ratio as they exist in the various stages of the titration as shown in her drawings in Figure 24.2d. Although this student established the proper ratio of these chemical species (e.g., H^+ dominates in the flask before the OH^- is added from the buret), she did not understand their interaction when both were present in the flask. However, when given the key (Figure 24.2c), she was able to verbalize the differences between her conceptions/misconceptions (Figure 24.2d) and those shown in the key (Figure 24.2b) [50, 52]. She stated that, "The moles of acid are still there. You make water molecules instead of just counting dots to triangles. I didn't really think about that. I had acid disappearing and base forming and that's not what really happens. *It forms water.* Cause over here (on the key), it's still 6 and 6, 6 acid, but now it's just water. It's still 6 and 6 but you have extra base. It makes a lot more sense [52, p. 1930]." In other words, for her it was an "aha moment" because she had just seen the "missing piece" in the puzzle; it now made sense to her because she could visualize the phenomena as the "whole picture" rather than as fragments that did not make sense. Conversely, other students whose mental models expressed multiple misconceptions were not able to truly understand the phenomenon [50, 52].

24.7.2
An Organic Pre-Lab Simulation

In the laboratory portion of an organic chemistry course, instruction often begins with organic techniques, such as an organic extraction of caffeine from tea-leaves. Saksri Supasorn and I [54] designed a pre-lab simulation to help students understand both the macroscopic procedures and observations that occur during an extraction, and the underlying submicroscopic chemical interactions. In this simulation, students selected values for two parameters of an organic extraction process, and then they observed the efficiency of the extraction process. The goal was to determine the optimal extraction of caffeine from an aqueous tea solution and it was subdivided into two subgoals, where the students explored:

- whether to use only one extraction of caffeine with a large volume of organic solvent (i.e., CH_2Cl_2) or to use smaller volumes and doing multiple extractions, and
- how long they needed to shake the aqueous/organic mixture in a separatory funnel to achieve optimal separation of the caffeine from its aqueous solution.

Thus, the students' task was, "... to design your own extraction to obtain the most efficient result [54, p. 172]."

To assess student learning of the organic extraction process, we asked them to draw and explain their macroscopic and molecular (submicroscopic) representations as expressed in their mental models. One of their challenges was "how to represent the caffeine molecule," which is a complex molecule as show in its Jmol model (see Figure 24.5) [27]. Evaluation of their mental models allowed us to assess the extent to which they understood this process as shown by fewer misconceptions and more of the correct conceptions [39]. The extraction features that most distinguished between good and poor mental models (MMs) were that (i) before mixing good MMs placed most of the caffeine in the aqueous layer and (ii) then they showed that after mixing most of the caffeine was transferred to the organic layer. For example, before interacting with the simulation, Student X's MM incorrectly represented the process by putting the organic layer, which is more dense, on top and then showing "all" of the caffeine molecules, rather than "most," being transferred to the organic layer after being mixed (Figure 24.11a). On the student's post-simulation mental model, he/she corrected both of these misconceptions (see darkened shading and writing). On the other hand, some students, such as Student Y did not correct their misconceptions after interacting with the simulation (Figure 24.11b). Specifically, Student Y initially inverted the two layers without representing the caffeine molecules at all. On their post-simulation MM, they correctly put the organic layer on the bottom while showing that all of the caffeine was transferred to the organic layer. Thus, this student did not alleviate their misconceptions.

The perceptions of most all of the organic students toward the simulation were very positive with most of them either agreeing or strongly agreeing that it helped them to understand the concepts of extraction [54]. The following quotes from students show that they liked the interactivity and visual characteristics of the simulations. For example, Student X stated that "I am also a visual learner, so I liked seeing the experiment, it helps me visualize what we will be doing in the <real> experiment and understand what will be happening [54, p. 177]." And Student Y echoed this perception, "Since the simulation was interactive, I was able to learn the information better. I think this is a great tool for visual learners. This would be a great tool for Org Chem before lab. Instead of just reading out of a lab book, I would also like this simulation tool [54, p. 177]." Quotes from these two students show that the simulation was effective in getting them to visualize the chemical phenomenon of an organic extraction.

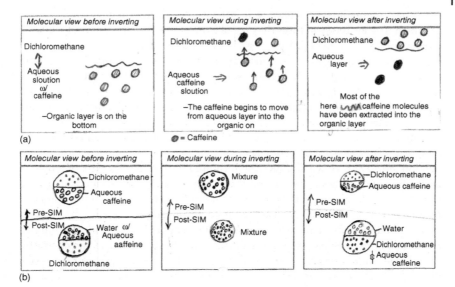

Figure 24.11 (a) Example of Student X's pre-simulation mental model as modified by the darkened circles and writing during the post-simulation (i.e., *before inverting*: vertical double-headed arrow added plus phrase at the bottom; *during inverting*: darkened circle above wavy line; *after inverting*: two darkened circles below wavy line plus "most of the" replaced "all" the caffeine molecules). (b) Example of Student Y's pre-simulation (below this mid-line) and post-simulation (above the line) mental models; note that this student did not include the caffeine molecule in their drawings.

24.8
Conclusions and Implications

In this chapter, my goal has been to show how dynamic visualizations can be designed to help students generate their own viable mental models of chemical phenomena. Chemistry is a complex subject to study because it has three distinctly different levels of chemical representations – symbolic, macroscopic, and submicroscopic levels (See Section 24.1). Unfortunately, many students do not see the relationships among these levels because they are still struggling to try to understand each one as an individual entity. Chemists have developed several innovative visualization technologies to help students overcome these barriers (see Section 24.2). Animations can help by focusing students' attention on the key features of a chemical process, and by showing the relationships among these multiple representations (Section 24.3). Chemists tend to favor realistic animations that are very complex; however, these can overwhelm students with too much information (i.e., cognitive overload). Students prefer simpler animations that focus on the relevant process. A slider allows them to select either a simpler or more-realistic animation (see Section 24.4). A more complex animation can be segmented into smaller parts that are meaningful to chemists. Students should be prompted to explain each segment in their own words before they move on to the next segment (see Section 24.5). Afterwards, the instructor should encourage them to draw the structures they recall from each of the three stages – before, during, and after a chemical process. They should explain each of these drawings in their own words. The combination of student-generated drawings and their self-explanations can help them develop coherent mental models that show their conceptual understanding (see Section 24.6). These mental models can also expose misconceptions so that students experience some discomfort and thus are "open" to accepting more scientifically correct conceptions.

Simulations are highly interactive dynamic visualizations in which the student establishes the experimental conditions by selecting values for each variable (see Section 24.7). This process allows them to predict then observe the outcome of their simulation. If they can explain this outcome, especially at the molecular level, then they are developing a coherent mental model of the chemical phenomenon being studied. All students need an instructional interface to help them construct their mental model of a phenomenon. Flexibility should be designed into interactive simulations to help students with different abilities and background chemistry knowledge. High-spatial students benefit when they have opportunities to sketch drawings and to self-explain their visual experiences. Conversely, low-spatial students need instructional guidance to focus their attention on the relevant aspects of the visualization.

Acknowledgments

The author would like to thank Loretta Jones, Michael Sanger, Saksri Supasorn, Sister Nicole Kunze, Alicia Courville, and Steve Monroney for their various contributions to this chapter. I also thank Bryce Hach, the Hach Scientific Foundation, and Loretta's NSF grant (NSF Award # 0440103).

References

1. Cracolice, M.S., Deming, J.C., and Ehlert, B. (2008) Concept learning versus problem solving: a cognitive difference. *J. Chem. Educ.*, **85** (6), 873.
2. Kozma, R.B. and Russell, J. (1997) Multimedia and understanding: expert and novice responses to different representations of chemical phenomena. *J. Res. Sci. Teach.*, **34** (9), 949–968.
3. Zoller, U. (2002) Algorithmic, LOCS and HOCS (chemistry) exam questions: performance and attitudes of college students. *Int. J. Sci. Educ.*, **24** (2), 185–203.
4. Chi, M.T., Feltovich, P.J., and Glaser, R. (1981) Categorization and representation of physics problems by experts and novices. *Cognit. Sci.*, **5** (2), 121–152.
5. Kleinman, R.W., Griffin, H.C., and Kerner, N.K. (1987) Images in chemistry. *J. Chem. Educ.*, **64**, 766–770.
6. Suits, J.P. and Sanger, M.J. (2013) in *Pedagogic Roles of Animations and Simulations in Chemistry Courses*, ACS Symposium Series, vol. **1142** (eds J.P. Suits and M.J. Sanger), American Chemical Society, Washington, DC, pp. 1–13.
7. Jones, L., Honts, J., Tasker, R., Tversky, B., Suits, J., Falvo, D., and Kelly, R. (2008) Molecules in Motion: Design Principles for Molecular Animations. Website for NSF Grant #04-40103, *artsci.drake.edu/honts/molviz/* (acccessed 20 December 2013).

8. Reiber, L.P. (2005) in *The Cambridge Handbook of Multimedia Learning* (ed. R.E. Mayer), Cambridge University Press, Cambridge, pp. 549–567.
9. Merrill, M.D. (2002) in *The Instructional use of Learning Objects* (ed. D.A. Wiley), AIT & AECT, Washington, DC, pp. 261–280.
10. Kinshuk, T.L. and Patel, A. (2006) in *The International Handbook of Virtual Learning Environments*, Vol. 14 (eds J. Weiss, J. Nolan, J. Hunsinger, and P. Trifonas,) Springer, Dordrecht, The Netherlands, pp. 395–425.
11. Wikipedia *http://en.wikipedia.org/wiki/Bubble_light* (accessed 20 December 2013).
12. Ainsworth, S. (1999) The functions of multiple representations. *Comput. Educ.*, **33**, 131–152.
13. Johnstone, A.H. (1993) The development of chemistry teaching: a changing response to changing demand. *J. Chem. Educ.*, **70**, 701–705.
14. Devetak, I. and Glazar, S.A. (2010) The influence of 16-year-old students' gender, mental abilities, and motivation on their reading and drawing submicrorepresentations achievements. *Int. J. Sci. Educ.*, **32**, 1561–1593.
15. Lancaster, K., Moore, E.B., Parson, R., and Perkins, K. (2013) in *Pedagogic Roles of Animations and Simulations in Chemistry Courses*, ACS Symposium Series, vol. **1142** (eds J.P. Suits and M.J. Sanger), American Chemical Society, Washington, DC, pp. 97–126.
16. Kelly, R.M. and Jones, L.L. (2008) Exploring how different features of animations of sodium chloride dissolution affect students' explanations. *J. Sci. Educ. Technol.*, **16** (5), 413–429.
17. Ashe, C.A. and Yaron, D.J. (2013) in *Pedagogic Roles of Animations and Simulations in Chemistry Courses*, ACS Symposium Series, vol. **1142** (eds J.P. Suits and M.J. Sanger), American Chemical Society, Washington, DC, pp. 367–388.
18. Yaron, D., Karabinos, M., Lange, D., Greeno, J.G., and Leinhardt, G. (2010) The ChemCollective: virtual labs for introductory chemistry courses. *Science*, **328** (5978), 584–585.
19. Winkelmann, K. (2013) in *Pedagogic Roles of Animations and Simulations in Chemistry Courses*, ACS Symposium Series, vol. **1142** (eds J.P. Suits and M.J. Sanger), American Chemical Society, Washington, DC, pp. 161–179.
20. Papagiannakis, G., Kim, H.-S., and Magnenat-Thalmann, N. (2005) Believability and presence in mobile mixed reality environments. IEEE VR2005 Workshop on Virtuality Structures, pp. 1–4, *http://www.academia.edu/389736/Believability_and_Presence_In_Mobile_Mixed_Reality_Environments*
21. Roussos, M., Johnson, A., Moher, T., Leigh, J., Vasilakis, C., and Barnes, C. (1999) Learning and building together in an immersive virtual world. *Presence Teleoper. Virtual Environ.*, **8** (3), 247–263, *http://www.evl.uic.edu/tile/NICE/NICE/PAPERS/PRESENCE/presence.html* (accessed 24 July 2014).
22. Trickett, S.B. and Trafton, J.G. (2007) "What if...": the use of conceptual simulations in scientific reasoning. *Cognit. Sci.*, **31** (5), 843–875.
23. Keeney-Kennicutt, W.L. and Merchant, Z.H. (2013) in *Pedagogic Roles of Animations and Simulations in Chemistry Courses*, ACS Symposium Series, vol. **1142** (eds J.P. Suits and M.J. Sanger), American Chemical Society, Washington, DC, pp. 181–204.
24. Fleming, S. (2013) in *Pedagogic Roles of Animations and Simulations in Chemistry Courses*, ACS Symposium Series, vol. **1142** (eds J.P. Suits and M.J. Sanger), American Chemical Society, Washington, DC, pp. 389–409.
25. Barak, M. (2013) in *Pedagogic Roles of Animations and Simulations in Chemistry Courses*, ACS Symposium Series, vol. **1142** (eds J.P. Suits and M.J. Sanger), American Chemical Society, Washington, DC, pp. 273–291.
26. Charistos, N.D., Tsipis, C.A., and Sigalas, M.P. (2005) 3D molecular symmetry Shockwave: a web application for interactive visualization and three-dimensional perception of molecular symmetry. *J. Chem. Educ.*, **82** (11), 1741.
27. Jmol Jmol: An Open-Source Java Viewer for Chemical Structures in 3D,

http://www.jmol.org/ (accessed 24 July 2014).

28. Herraez, A. (2006) Biomolecules in the computer: Jmol to the rescue. *Biochem. Mol. Biol. Educ.*, **34** (4), 255–261.

29. Hanson, R.M. (2010) Jmol – a paradigm shift in crystallographic visualization. *J. Appl. Crystallogr.*, **43**, 1250–1260.

30. Fiore, S.M. and Schooler, J.W. (2002) How did you get here from there? Verbal overshadowing of spatial mental models. *Appl. Cognit. Psychol.*, **16**, 897–910.

31. Martin, B. and Mahaffy, P. (2013) in *Pedagogic Roles of Animations and Simulations in Chemistry Courses*, ACS Symposium Series, vol. **1142** (eds J.P. Suits and M.J. Sanger), American Chemical Society, Washington, DC, pp. 411–440.

32. McKenzie, L., Versprille, A., Towns, M., Mahaffy, P., Martin, B., and Kirchhoff, M. (2013) Visualizing the chemistry of climate change (VC3Chem): online resources for teaching and learning chemistry through the rich context of climate science. Paper presented at the American Geophysical Union, Fall Meeting 2013, abstract #ED31E-04, http://adsabs.harvard.edu/abs/2013AGUFMED31E..04M (accessed 24 July 2014).

33. Suits, J.P. and Courville, A.A. (1999) in *Mathematics/Science Education and Technology 1999* (ed. D.A. Thomas), AACE, Charlottesville, VA, pp. 531–536.

34. Suits, J.P., Soileau, M., and Pease, R. (2002) Use of interactive response technology to support and assess student visualization of chemistry topics. Proceedings of the CILT (NSF Award# 0085951) Seed Grant Project Entitled Assessment of Problem-Solving Competence via Student-Constructed Visual Representations of Scientific Phenomena, (PI J.P. Suits), http://cilt.concord.org/seedgrants/assessments/Suits_Final_Report_2002.pdf (accessed 31 December 2013).

35. Tasker, R. and Dalton, R. (2006) Research into practice: visualisation of the molecular world using animations. *Chem. Educ. Res. Pract.*, **7** (2), 141–159.

36. Sweller, J., van Merrienboer, J.J.G., and Paas, F. (1998) Cognitive architecture and instructional design. *Educ. Psychol. Rev.*, **10**, 251–296.

37. Mayer, R.E. (2005) in *The Cambridge Handbook of Multimedia Learning* (ed. R.E. Mayer), Cambridge University Press, Cambridge, pp. 169–182.

38. Rosenthal, D.P. and Sanger, M.J. (2013) in *Pedagogic Roles of Animations and Simulations in Chemistry Courses*, ACS Symposium Series, vol. **1142** (eds J.P. Suits and M.J. Sanger), American Chemical Society, Washington, DC, pp. 313–340.

39. Suits, J.P. and Jones, L.L. (2006) Design of dynamic visualizations to promote understanding of chemistry concepts. Paper presented at the 19th International Conference on Chemical Education, Seoul, Korea.

40. Hasler, B.S., Kersten, B., and Sweller, J. (2007) Learner control, cognitive load and instructional animation. *Appl. Cognit. Psychol.*, **21**, 713–729.

41. Spanjers, I., van Gog, T., and van Merriënboer, J. (2010) A theoretical analysis of how segmentation of dynamic visualizations optimizes students' learning. *Educ. Psychol. Rev.*, **23**, 1–13.

42. Betrancourt, M. (2005) in *The Cambridge Handbook of Multimedia Learning* (ed. R.E. Mayer), Cambridge University Press, Cambridge, pp. 287–296.

43. Fonseca, B.A. and Chi, M.T.H. (2011) in *Handbook of Research on Learning and Instruction* (eds R.E. Mayer and P.A. Alexander), Routledge, New York, pp. 296–321.

44. Monaghan, J.M. and Clement, J. (2000) Algorithms, visualization, and mental models: high school students' interactions with a relative motion simulation. *J. Sci. Educ. Technol.*, **9** (4), 311–325.

45. Höffler, T.N. (2010) Spatial ability: its influence on learning with visualizations—A meta-analytic review. *Educ. Psychol. Rev.*, **22**, 245–269.

46. Wu, H.-K. and Shah, P. (2004) Exploring visuospatial thinking in chemistry learning. *Sci. Educ.*, **88**, 465–492.

47. Pribyl, J.R. and Bodner, G.M. (1987) Spatial ability and its role in organic chemistry: a study of four organic

courses. *J. Res. Sci. Teach.*, **24** (3), 229–240.
48. Yang, E., Andre, T., Greenbowe, T.J., and Tibell, L. (2003) Spatial ability and the impact of visualization/animation on learning electrochemistry. *Int. J. Sci. Educ.*, **25** (3), 329–349.
49. Hegarty, M., Kriz, S., and Cate, C. (2003) The roles of mental animations and external animations in understanding mechanical systems. *Cognit. Instr.*, **21** (4), 325–360.
50. Suits, J.P. (2010) Chemical representations and mental models: how do visualizations help students. Invited speaker: Paper presented at the Pacifichem-2010 Conference, Honolulu.
51. Suits, J.P. and Diack, M. (2002) Instructional design of scientific simulations and modeling software to support student construction of perceptual to conceptual bridges. *World Conf. Educ. Multimedia Hypermedia Telecommun.*, **2002** (1), 1904–1909.
52. Suits, J.P., Kunze, N., and Diack, M. (2005) Use of microcomputer-based laboratory experiments to integrate multiple representations of scientific phenomena. *World Conf. Educ. Multimedia Hypermedia Telecommun.*, **2005** (1), 1924–1931.
53. Suits, J.P. and Srisawasdi, N. (2013) in *Pedagogic Roles of Animations and Simulations in Chemistry Courses*, ACS Symposium Series, vol. **1142** (eds J.P. Suits and M.J. Sanger), American Chemical Society, Washington, DC, pp. 241–271.
54. Supasorn, S., Suits, J.P., Jones, L.L., and Vibuljun, S. (2008) Impact of a pre-laboratory computer simulation of organic extraction on comprehension and attitudes of undergraduates. *Chem. Educ. Res. Pract.*, **9**, 169–181.
55. Falvo, D.A. and Suits, J.P. (2009) Gender and spatial ability and the use of specific labels and diagrammatic arrows in a micro-level chemistry animation. *J. Educ. Comput. Res.*, **41** (1), 83–102.
56. Barnea, N. and Dori, Y.J. (1999) High-school chemistry students' performance and gender differences in a computerized molecular modeling learning environment. *J. Sci. Educ. Technol.*, **8** (4), 257–271.
57. Falvo, D.A., Urban, M.J., and Suits, J.P. (2011) Exploring the impact of and perceptions about interactive, self-explaining environments in molecular-level animations. *C.E.P.S. J.*, **1** (4), 45–61.
58. de Jong, T. (2011) in *Handbook of Research on Learning and Instruction* (eds R.E. Mayer and P.A. Alexander), Routledge, New York, pp. 446–466.
59. Wieman, C.E., Adams, W.K., and Perkins, K.K. (2008) PhET: simulations that enhance learning. *Science*, **322** (5902), 682–683.
60. Greca, I.M. and Moreira, M.A. (2002) Mental, physical, and mathematical models in the teaching and learning of physics. *Sci. Educ.*, **86** (1), 106–121.
61. Hegarty, M. and Waller, D. (2005) in *The Cambridge Handbook of Visuospatial Thinking* (eds P. Shah and A. Miyake), Cambridge University Press, Cambridge, pp. 121–169.
62. Reigeluth, C.M. and Schwartz, E. (1989) An instructional theory for the design of computer-based simulations. *J. Comput.-Based Instr.*, **16** (1), 1–10.
63. Plass, J.L., Homer, B.D., and Hayward, E.O. (2009) Design factors for educationally effective animations and simulations. *J. Comput. High. Educ.*, **21** (1), 31–61.
64. Reiber, L.P. (1996) Animation as feedback in a computer-based simulation: representation matters. *Educ. Technol. Res. Dev.*, **44** (1), 5–22.
65. Suits, J.P. and Lagowski, J.J. (1981) Design of computer-simulated experiments to enhance the problem-solving abilities of science students. Proceedings of the National Educational Computing Conference: NECC-81 (eds. D. Harris and L. Nelson-Heern), pp. 184–189.
66. Vosniadou, S. (2007) The cognitive-situative divide and the problem of conceptual change. *Educ. Psychol.*, **42** (1), 55–66.
67. Schnotz, W. and Kurschner, C. (2008) External and internal representations in the acquisition and use of knowledge: visualization effects on mental model construction. *Instr. Sci.*, **36**, 175–190.

25
Chemistry Apps on Smartphones and Tablets
Ling Huang

25.1
Introduction

The past decade witnessed dramatic development and exponential growth of mobile computing. More and more chemistry computer software applications, or "apps," can be found on portable computers such as smartphones and tablets. The landscape of chemical education has been changed and new horizon emerged with the interactive touchscreen platforms. Powerful graphic processors and multicore parallel computing processors extended the boundary of computational power which used be associated with desktop computers in the 1990s and the early years of this century. The competitions among vendors along with fast dropping prices accelerated the adoption of smartphones and tablet computers among high school and college students, teachers, and chemistry professionals. As students and teachers are already spending a lot of time on mobile devices such as smartphones and tablets, using them as an educational tools for Chemistry learning and research became natural and convenient. This chapter will cover many different ways of using Chemistry apps, precautions on adoptions, and various platforms. The readers will hopefully get a better understanding of how Chemistry apps work, their capabilities and their limitations. This review is also a large expansion of our previous review on smartphone-based Chemistry apps [1].

According to a recent Nielsen report, smartphone ownership has increase from 18% in Q3 of 2009 to 62% in Q2 of 2013 [2]. Smartphones can serve as powerful and convenient educational tools on a mobile platform, which potentially encourages learning. A study by StudyBlue found that students who used a mobile application to study spent on average 40 min or more a week on studying [3].

A recent paper by Williams and Pence [4] presented the benefits of using smart phones and similar Internet capable devices in the classroom. In two other articles, Williams *et al.* noted several specific apps for drug discovery [5] and highlighted Chemspider app as a powerful handheld chemical search engine [6]. Several universities are beginning to implement mobile Chemistry apps into their curricula

Chemistry Education: Best Practices, Opportunities and Trends, First Edition.
Edited by Javier García-Martínez and Elena Serrano-Torregrosa.
© 2015 Wiley-VCH Verlag GmbH & Co. KGaA. Published 2015 by Wiley-VCH Verlag GmbH & Co. KGaA.

[7], there are also many universities that provide online guides to mobile applications [7–9]. One web site describes a handful of paid apps that are suitable for professional chemists or chemical engineers [7] while the others put more emphasis on the convenient access to chemistry-related journals online [8, 9]. There is even a Chemical Mahjong game app to make learning chemistry more fun and interesting [10]. This review intends to cover the majority of the free and popular apps available on smartphones (Figure 25.1), iPods (both referred to as *smartphones* from here on) or tablet computers from different perspectives such as disciplines (Table 25.1), functionalities, and target users (Table 25.2).

According to a Pew Research Center report, 34% American adults own tablet computer in May 2013, up from 3% in May 2011 [11]. The increasing tablet ownership also makes it possible for more educational apps to be accessible on lighter and interactive touchscreen portable devices. Compared to smartphones, the increased screen size on tablets (usually 7–11-in. diagonally) provides more area for students to draw chemical structures, to view and interact with 3-D models, and to read a longer article. The added space trades off with portability as the students tend to carry smartphones with them in their pockets or backpack, to the classrooms or to the labs, which make the mobile apps more accessible. The recent appearance of phablets bridges the gaps between portability and larger screen size and creates more opportunities for the adoption of Chemistry apps.

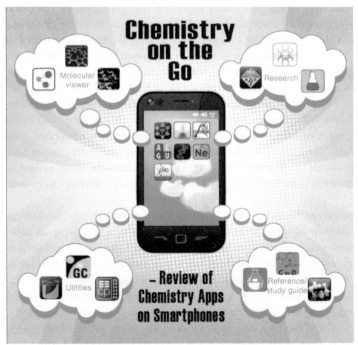

Figure 25.1 Mobile Chemistry apps on Smartphones. Graphics credit: Alan Dans.

Table 25.1 Chemistry apps by disciplines.

Discipline	Android	iOS
General and inorganic	3D Molecular Models, AtomDroid, Chemical Compound Information, ChemistryAid, Chemical Dictionary, Chemistry Acronyms, Chemistry Cheat Sheets, Chemistry Formula Calc., Chemistry Mobile, Convert Pad, Elements, Molecular Viewer 3D, Periodic Droid, Periodic Table, W Chemistry Handbook, Chemistry Quiz, Chemistry Speed Quiz, Formulas Lite, Solution Calculator Lite, Table of Elements	iElements, Chemistry Formula Practice, The Chemical Touch Lite, Chem Pro:Chemistry Tutor, ChemCrafter, CMM Molar Mass Calculator, eDistribution, Elemental, Gas Laws HD Lite, iGasLaw, LabCal, Lewis Dots, MCAT Chemistry Flashcards, Mendeley, Mobile Hyperchem Free, Molarity, NOVA Elements, QuickElem, The Elements: A Visual Exploration, VideoScience
Organic	AtomDroid, Chemistry by Design[a], ChemDoodle Mobile[a], Molecular Viewer 3D, W Chemistry Handbook, ChemSpider Search, Jmol Molecular Visualization, Organic Chemistry-Alkanes, Organic Chemistry Terms, Organic Chemistry Visualized, TLC Timer	Chem3D, Chemistry By Design[a], ChemDoodle Mobile[a], ChemDraw, ChemSpider, Elemental, Green Solvent, Molecules, iSpartan, Named Reactions, Organic Name Reactions, Learn Organic Chemistry Nomenclature, MCAT Chemistry Flashcards, MobReg, OCE, Organic Chemistry Nomenclature Quizillator, Organic Chemistry Test Bank, ReactionFlash, SPRESImobile
Analytical	Chemistry Helper, EMD PTE[a], PubChem Mobile, Scholarley, W Chemistry Handbook, ChemSpider[a], Solution Calculator Lite	EMD PTE[a], ChemMobi (by Accelrys Inc.), ChemSpider[a], GC Calc, Green Solvent, LC Calc, MolPrime, Molecules, Molarity, MS iCalc
Physical	Chemistry Mobile, Convert Pad, W Chemistry Handbook	Atoms in Motion, Insensitive (by Klaus Boldt), PyMOL
Biochemistry	AtomDroid, Biochemistry, ESmol, Molecular Viewer 3D, NDKmol, Promega[a], Chemistry of Life	+Amino-, Amino Acid Tutor, Biochem Euchre Deck, Buffer Calc, Chem3D, DailyCalcs, Promega[a], Gene Link, Genetic Code, iMolview Lite[a], iProtein, Molecules, NutriBiochem[a], Qiagen, RCSB PDB Mobile

a) Denotes apps available on both Android and iOS.

Table 25.2 Chemistry apps by target user groups.

Level	Android	iOS
High school or lower	EMD PTE[a], Periodic Droid, Promega[a], ChemistryAid, Chemistry Quiz, Table of Elements	ChemCrafter, Chemistry Formula Practice, EMD PTE[a], iElements, Molecules, Promega[a], The Chemical Touch Lite, The Elements: A Visual Exploration, NOVA Elements, QuickElem, VideoScience
Undergraduate introductory courses	3D Molecular Models, *ChemDoodle Mobile[a], Atomdroid, Chemical Dictionary, ChemistryAid, Chemistry Cheat Sheets, Chemistry Formula Calc., Chemistry Helper, ConvertPad, EMD PTE[a], Periodic Droid, Periodic Table, W Chemistry Handbook, Chemistry of Life, Chemistry Quiz, Chemistry Speed Quiz, Elements, Formulas Lite, Organic Chemistry-Alkanes, Organic Chemistry Terms, Organic Chemistry Visualized, Solution Calculator Lite	Amino Acid Tutor, ChemDoodle Mobile[a], Chem Pro:Chemistry Tutor, CMM Molar Mass Calculator, eDistribution, Elements Test2, Elemental, EMD PTE[a], Gas Laws HD Lite Genetic Code, iElements, Molecules, iGasLaw, LabCal, Learn Organic Chemistry Nomenclature, Lewis Dots, MCAT Chemistry Flashcards, Mobile Hyperchem Free, Molarity, OCE, Organice Chemistry Nomenclature Quizillator, Organic Chemistry Test Bank
Undergraduate upper level courses	Atomdroid, Biochemistry, ChemDoodle Mobile[a], Chemical Compound information, Chemistry Acronyms, Chemistry Helper, Chemistry by Design[a], Chemistry Mobile, ConvertPad, Molecular Viewer 3D, Promega[a], PubChem, W Chemistry Handbook, ChemSpider[a], ChemSpider Search, Jmol Molecular Visualization, TLC Timer	Buffer Calc, Green Solvent, ChemMobi, ChemSpider[a], MolPrime, Named Reactions, Organic Named Reactions, Promega[a], ChemDoodle Mobile[a], Chemistry by Design[a], Chem3D, ChemDraw, Genetic Code, iMolview Lite[a], MobReg, NutriBiochem[a], iSpartan, PyMOL, RCSB PDB Mobile
Graduate level and professional chemist	ACS Mobile[a], Atomdroid, Biochemistry, ChemDoodle Mobile[a], Chemistry by Design[a], ESMol, NDKMol, Molecular Viewer 3D, Promega[a], PubChem, Scholarley	ACS Mobile[a], Buffer Calc, Chem Doodle Mobile[a], Chemistry By Design[a], ChemMobi, ChemSpider, ChemSpider, GC Calc, Gene Link, Green Solvent, Insensitive, iProtein, LC Calc, MolPrime, Promega[a], Mendeley, MobReg, MS iCalc, Qiagen, PyMOL, RCSB PDB Mobile, ReactionFlash, SPRESImobile

a) Denotes apps available on both Android and iOS.

25.2 Operating Systems and Hardware

The iOS and Android platforms both provide a multitude of applications that can be downloaded directly onto the phone. These mobile applications, or "apps," have a wide range of functionalities and cover many disciplines. The rapid development of cloud computing technology also speeds up the adoption of these mobile apps as chemical education tools or collaborative learning platforms [12] as more programs can be accessed through the "cloud" and large amount of chemical data or structural information can be stored in the "cloud." Collaboration through the interconnection of multiple chemistry apps was recently demonstrated as a new chemoinformatics tool to increase work efficiency [13], which can be utilized to enhance chemistry learning experience to a new level. Because of the current low market share for Windows Mobile Operation System, there are very much fewer Windows-based mobile chemistry apps. Consequently this chapter will only discuss apps found in the two dominant systems, Android and iOS.

Majority of the apps discussed in this chapter are free or of very low cost (<$10), which makes them affordable to students and teachers compared to conventional desktop software which can easily cost a few hundred US dollars. Some apps such as Chemdraw and iSpartan, are the tablet version of the corresponding desktop software. They tablet version might have reduced capability as they are limited by the tablets' computing power, memory constraints and graphics speed. For basic 3D rendering and modeling, however, the tablet apps do a great job, especially at teaching the computational chemistry concepts such molecular geometry, structural orientation, and electron distribution. We found more free apps on the Android platform than on the iOS system. Some of the paid iOS apps possess greater quality or more functionalities, compared to the free trial versions on either systems. The more expensive ones, on the other hand, are not necessarily the better apps. Most of the apps featured in this chapter received three star or above rating from users. The adoption rate is expanding very rapidly as we are writing, most Chemistry apps have at least 1000 installations and some achieved several millions.

Most smartphones and tablets purchased today carry at least a dual-core central processing unit (CPU) with 1–2 GB of memory. More quad-core or octa-core CPUs are entering the market which provides the multitasking capabilities for running apps. The graphic processing power of these portable computers often rivals or exceeds some of the graphical processing desktop computers of just less than a decade ago. The hardware combination provides Chemistry apps enough computing power and storage space so interactive tasks can be performed with simple finger flips and molecular structures can be rendered and viewed in high resolutions. The students or researchers can forgo heavy chemistry dictionaries and CRC handbooks. Instead complex chemical information can reach the readers in a matter of seconds.

Smartphones or iPods have small screens ranging from 4- to 5-in. diagonally, which fit perfectly in a pocket or a handbag. The enhanced portability makes them easy to transport and to take to the field or the lab. The high availability makes them great educational tools in Chemistry lecture rooms or labs. The small screen size with limited resolution and space sometimes make the information hard to read and the rendering of large molecular structures nearly impossible, compared to larger screens found on tablets, laptop, and desktop computer.

Tablet computers can often provide more functionality as the increased screen size (7- to 10-in. diagonally) provides more pixels for texts with larger fonts and pictures with clear boundaries. These are extremely important for reading large amount of Chemical literature and for observing and manipulating a large molecular structure such as protein. The larger size of a tablet, however, adds to the weight and volume, which can reduce the portability. The usually heavier price tag of a tablet also makes it less affordable for students.

25.3
Chemistry Apps in Teaching and Learning

Here we loosely break down all Chemistry apps into nine categories based on their functionalities: Molecular Viewers and Modeling Apps, Molecular Drawing Apps, Periodic Table Apps, Literature Research Apps, Lab Utility Apps, Apps for Teaching and Demonstration, Gaming apps, Chemistry course apps, and Test-prep Apps. Under the Lab Utility Apps category, further branching is used to separate the utilities into five sub-categories: Flashcard apps, Dictionary apps, Search Engine apps, Calculator apps, and Instrumental apps. The classification system is very crude as readers will soon discover that there are a lot of apps carrying overlapping functionalities and hybrid Chemistry apps are seeing increasing adoptions. Some apps are becoming comprehensive Chemistry tools covering several disciplines with the capabilities across several categories or subcategories. The frequent updates by the app developers often expand the functionalities of their software to make the apps more powerful and encompass more territories in chemical study.

25.3.1
Molecular Viewers and Modeling Apps

With the dramatic and rapid improvement of CPU computing power, especially graphic computing power on smartphone and tablet devices, the rotation and rendering of 3-D molecular structures can be easily and smoothly achieved on durable touch-screens with the one-finger swiping or two-finger zooming motions. Less than a decade ago, these functions were reserved for high-end graphics processing workstations with sophisticated software engines that cost thousands of US dollars. Today apps such as "Atomdroid" [14] (Android, by CCB Goettingen) and "Molecules" (iOS, by Sunset Lake Software) can create stunning and interactive 3-D renderings of molecular structures. Both apps show ball-and-stick models of energy-minimized 3-D structures as default. Many apps in this category are developed by software giants or professional developers that are the creators of popular desktop versions of software. So in general, the overall quality of all the molecular viewer apps is very high compared to several other categories.

iOS apps found under this categories include but are not limited to "Chem3D" (by Perkin-Elmer), "iProtein"(by Eidogen-Sertanty), "Molecular HyperChem Free" (by Hypercube, Inc.), "Molecules," "iSpartan" ($19.99 by Wavefunction, Inc), and "PyMOL" (by Schrödinger, Inc). Android apps, mostly free, include "3D Molecular Models" (by MOBSOLUTIONS4ALL), "3D Molecule View" (by Afanche Technologies), Atomdroid, ESMol, NDKMol (both by Biochem_fan), Jmol Molecular Visualization (by Bob Hanson), Molecular Viewer 3D (by Adam Hogan), Organic Chemistry-Alkanes, and Organic Chemistry Visualized (both Budgietainment). iMolView Lite (by Molsoft) can be run on both iOS and Android systems.

Molecular viewer apps can be a useful tool in the classroom. For example, in an organic chemistry class, they serve as supplements to molecular model kit. The advantage of the apps is that a student can easily switch between viewing a space-filling model and ball and stick model. The other advantage from apps is that larger moleculars structures can be manipulated. Similarly, these types of apps can be used in biochemistry and general chemistry, to look at structures. These models can also be used for drug-discovery research, particularly in the investigation of the binding between a drug molecule and a target protein. The viewer tools can be used to evaluate the binding pocket and specifically the type of inter-molecular force involved in binding (i.e., pi-pi stacking and hydrogen-bonding). Even for research involving complex molecules such as enzymes, these molecular viewer apps can help the researchers to visualize molecular structures from different angles. Synthetic organic chemists can use these apps to "touch" and orient the target molecules and explore sites for further modification. The 3-D models provide more realistic pictures of a molecule, which can help introduce accurate parameters for further reactions.

"Molecules" can display both proteins and DNA. It also comes with a search function that utilizes PubChem and Protein Data Bank (PDB). Once the molecule is found, the structure can be downloaded directly onto the device. Several of the Android molecular viewer apps also share the search and download functionalities. In some cases, app users can draw their custom molecules in a desktop or laptop computer and upload their structures to the smartphones for display and rendering. Care must be taken to ensure the compatibility of file formats.

Besides ball-and-stick models, "Atomdroid" (Figure 25.2a) can also display skeletal models along with calculated total energy. Many of the display parameters can be fine-tuned to optimize the presentation and maneuvering speed. In our Android test device, the default speed of two-finger zoom-in could be too sensitive to complete, making the molecule disappear on the screen. Several Android apps such as "Atomdroid," "Molecular Viewer 3D," "ESmol," and "NDKmol" can download protein files from PDB and display the complex protein structures with stunning detail. This could be a valuable tool for teaching protein structures and interactions between small molecules and proteins. The protein

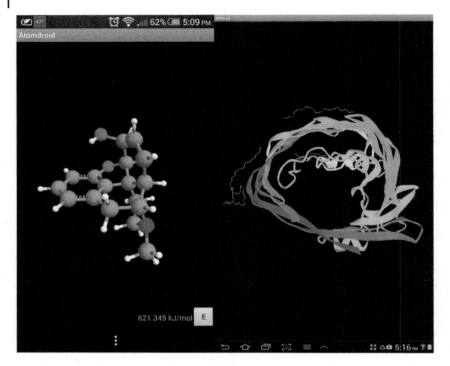

Figure 25.2 "Atomdroid" with a small molecular structure on a 4.7-in. smartphone screen (left) and "NDKmol" with a large protein structure on a 10-in. tablet (right).

structures displayed on a bigger screen usually presents better detail as shown in Figure 25.2b.

"ESmol" can convert polymer structures into beautiful ribbon models. It is also able to show nucleic acids in strands, ladder, or skeletal models. Besides biomolecules such as DNA, RNA, and proteins, "ESmol" can even display polymers and crystals in packing mode. If the file size is bigger than 3 MB, "NDKmol" from the same app developer should be used in the place of "ESmol." Both can handle PDB file formats and load MDL MOL (= SDF) format from SD card or directly from PubChem. "Molecular Viewer 3D" works with PDB, CML, MDL SD, Sybyl Mol2 file formats. The app can display custom-drawn structures or downloaded molecules along with 243 library molecules. Open-source software such as "Avogadro" could be used to convert file formats from Chemdraw or Chemsketch to improve the compatibility with Molecular Viewer apps. "iMolview," available on both iOS and Android, renders large biomolecules such as proteins and DNA with ease, along with additional database search functions that can be a great reference tool for Biochemistry students.

"iSpartan" is the tablet version of "Spartan," a popular chemical modeling software used on desktop computers as chemical education tool. The table version preserves basic molecular rendering capabilities found on the desktop version with some limitations due to the lower computing power associated with tablet hardware and memory storage. The interactive interface on "iSpartan" makes it more appealing than the desktop version where a mouse or keyboard is necessary for the manipulation or drawing of molecules.

Equipped with these molecular viewer apps, students can access simple or complex structures from a device out of their pockets and manipulate the structures to better understand bonding and steric effects. The convenient viewing can be used to assign NMR chemical shifts and to study the reactive sites for Organic Chemistry. In one of our research projects, for example, "Atomdroid" and "Molecules" apps were used to predict the change in NMR chemical shift values when there are subtle changes of derivatization on core structures of synthetic cannabinoids, all of which were used as "designer drugs". With apps such as ESmol and NDKmol, biomolecules become easier to "touch" and "feel" as large molecular structures can be enlarged for detailed observations, which is hard to achieve with physical models. Although some of the free apps in this category contain limited functionality (e.g., can only render alkanes) the paid versions usually have broader capabilities. Some of the free apps such as "Atomdroid" with quick updates and improvement provide professional computational capabilities and stunning graphics. With strong competition, the apps in this category improve their quality dramatically and at high frequencies. They are highly recommended as replacement for expensive desktop versions of computational softwares on conventional PC or Mac computers.

As the fast improvement of mobile computer processing power and video processing capability, complex biomolecules such as proteins and DNA can be visualized, touched, and rotated to provide deeper understanding of molecular biochemistry. Their biophysics can now be studied in a visual manner. After using these apps, the students can appreciate hydrogen-bonding between C-G and A-T base pairs and why the former is a stronger binding thanks to three hydrogen-bonds. The intermolecular forces that are at play in the formation of secondary and tertiary protein structures and enzyme-substrate binding can be revealed with zoomed-in 3-D molecular images. This would also help General Chemistry students understand the importance of intermolecular forces in a biological context.

25.3.2
Molecular Drawing Apps

With the touch screen on smartphones and tablets, 2-D molecular structures can be drawn to precision with one or two fingers. Small molecules can be drawn on "MolPrime" (iOS and android, by Molecular Materials Informatics, Inc.), which was covered in a previous review [5], and "ChemDoodle Mobile" (iOS and Android, by iChemLabs, LLC.). "ChemDoodle Mobile" is an easy tool for sketching molecules to show energy-minimized 2-D structures and to calculate simple NMR spectra. NMR and property predictions work very well for small molecules containing organic elements. The drawn structures can be saved only in the paid version of the apps.

The limited screen sizes (ranging from 3 to 4.7 in. diagonally) on smartphones or iPods, however, made drawing complex molecular structures extremely challenging with these structure drawing apps. The drawing often requires dexterous hand maneuvers and frequent zooming in and out. For simple molecules often encountered in General Chemistry, these apps are suitable for the learning of bonding, molecular geometry, molecular polarity, and Lewis structures.

Tablet computers provide much larger screen sizes (7–10-in. diagonally) which are essential for the display of large 2-D structures. With more real estate for users to touch and tap, an accurate structure can be generated with the precise virtual "push" of icons representing elemental symbols, single or double chemical bonds, and aromatic rings. The larger touchscreen area made these apps great for drawing molecules, sometimes more conveniently than on a desktop or laptop computers with mouse or keyboard control. Conventional drawing software such as ChemDraw ($9.99 by Perkin-Elmer) can be found on iOS-based tablets with identical functions. Many students and teachers are already familiar with ChemDraw (Figure 25.3), which made the app adoption process very simple. On some Android devices, these apps could be a little sluggish due to the lower sensitivity of some touch screens. Other free iOS-based chemical structure sketch tools include "Elemental" (Dotmatics Limited) which provides adequate functions for drawing simple small molecules commonly seen in Chemistry classrooms and labs.

With these drawing apps, Organic Chemistry students can practice drawing 2-D molecular structures following their professors on a blackboard. The students can also ask better questions remotely with clearly drawn structures. After the drawing

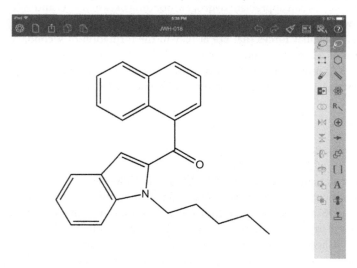

Figure 25.3 ChemDraw tablet interface.

practice, the students will make fewer errors in drawing organic molecules with wrong bonding and charges as many software will auto-correct or alert if a structure incorrectly drawn. Even general chemistry students can improve their understanding of concepts such as molecular geometry, the octet rule, Lewis structure, and bond orders and be well-prepared for Organic Chemistry.

25.3.3
Periodic Table Apps

Periodic table of chemical elements is an important tool used in virtually all chemical education materials. Conventional textbook-based teaching and poster demonstration of the table only provide students with one-way 2-D representations of the chemical periodicity, atomic orbitals, groups, periods, sizes, electronegativity, and other properties. The invention of mobile apps added several dimensions and interactivity for the teaching of periodic table. Several of the utility apps mentioned below, for example, "Chemistry Helper" and "ChemMobile," contain brief introduction and/or link to a periodic table. They provide simple functions directly related to their major functions of the app such as lab reference albeit in a limited scale.

Besides the comprehensive reference apps, there are about a dozen high-quality dedicated periodic table apps that are rich with many useful elemental properties. On iOS, users can find "iElements" (by "SusaSoftX," "Elements" on Android), QuickElem (by Quick Learning LLC), The Chemical Touch Lite (by Christopher J. Fennell), "The Elements: A Visual Exploration" ($13.99, by Touch Press); on Android, "Chemistry Helper," "Periodic Droid"(by DroidLa), "Periodic Table" (by Socratica), "Chemik" (by BK Advance), and "Table of Elements" (by Nomadrobot). "EMD PTE" (by Merck KGaA) can be found on both iOS and Android.

"EMD-PTE" stands out in this category, winning with clean high-definition or HD resolution and rich functions (Figure 25.4a). It carries a calculation tool of atomic weight percentage within a compound. Every element has a history of discovery, which can be found after tapping on the elemental symbol within the app. The accompanied atomic property data are most comprehensive among all periodic table apps. A small drawback of this app is that the small buttons could be hard to touch correctly on a smaller screen. Overall this app has the highest rating.

"Periodic Droid" lists elements by atomic numbers. It comes with a quiz function for studying and reviewing various elemental properties such radii, atomic weights, symbols, physical properties, and history. The elements can also be listed using the order of 16 different properties. "Periodic Table" app is tailored more toward learning General Chemistry and reinforcing various aspects of elemental periodicity.

"iElements" provides good periodic table, with a lot of information on each element such as its name, symbol, atomic number, phase, density, melting point, boiling point, heats of fusion and vaporization, specific heat, oxidation states, ionization energies, electronegativity, covalent, atomic, and van der Waals (VDW) radii. The app also provides a Wikipedia link that opens up in the browser. When an

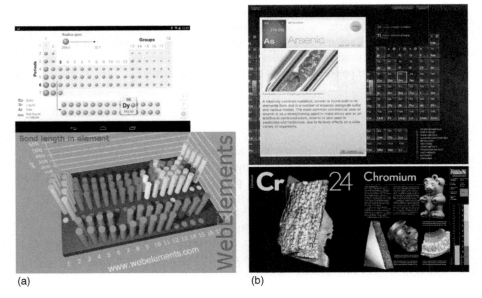

Figure 25.4 (a) "Nova Elements" and (b) "Elements: A Visual Exploration."

element is chosen only the symbol and atomic number appear, for all other information, you have to click "more info" and an element page comes up. Although the element group and location are given, and the table is color-coded by type (halogen, metalloid, etc.) these colors are left for the user to figure out. This could, however, be used as an educational tool, so that students can figure out the grouping for themselves for a review session.

"The Chemical Touch Lite" is another practical periodic table reference. This app gives a good amount of information on the chemical and physical properties of each element, including oxidation states and electronegativity. The table is color-coded depending on the selected chemical or physical property, each property has its own color-coded scale. Another helpful option is the ability to change between different temperature units. For a beginner this table might be a bit confusing as the possible oxidation states are not labeled clearly and the units are not very inherently obvious. This is still a good reference tool for someone with some familiarity with the periodic table.

"The Elements: A Visual Exploration" (Figure 25.4b), the only paid app in this category, provides high-quality multimedia back story about each element. It makes the learning of elements highly "visual" and interactive. The teachers can craft different exercises based on this app to lead student to explore and discover elemental properties and applications of materials. The anecdotes and visual history also help student to get familiar with the discovery process and usage of each element.

These apps also works very well with tablet screens where more elements can be displayed in one screen which makes it less likely for the user to click on the wrong element or property. The larger screens on tablets also depicts the trend across one group or one period more clearly than on smaller screens found on smartphones. These interactive periodic table apps truly transcend the teaching methods that used to be based on conventional textbooks. The enhanced interactivity and easy switching among different elements and properties proved to be several steps ahead on even Web-based periodic tables found on laptop or desktop computer browsers.

Students in General Chemistry who started learning the periodic table can take advantage of these multimedia apps to gain a deeper understanding of periodicity of elements, the properties of different groups, the trend going in a row or in a column in terms of size, electronegativity, valence electrons, reactivity, and so on. All of these information can be accumulated in one or two apps for easy access in the classroom or outside. The rich back story behind each element can be discovered and explored with simple clicks on elemental icons to help student build an intimidate relationship with each element, which enhance their memory and interest in Chemistry. For Inorganic Chemistry student, the transition metals and rare earth elements can be easily displayed in an interactive table that can be pinched and zoomed-in to get a better understanding of positions and related properties.

25.3.4
Literature Research Apps

Because of the great portability from smartphones and tablets, pdf-based chemical literature can be searched, read, organized, and annotated by users on the move, with the easy manual touches. Several apps such as ACS Mobile (iOS and Android, by American Chemical Society), PubChem Mobile (Android, by CRinUS), PubMed Search (Android, by YMED), provide mobile alternatives to website-based search engines. These apps, however, received mixed ratings from users as they only provide marginal functional upgrades compared to conventional Web search engines which can be accessed through the mobile browser. Several of these apps require journal subscription for full-article access. The major advantage comes from the portability and interactive features lacking from conventional desktop versions of the search engine.

Another group of literature organization apps emerged with quick adoption from scholars in different disciplines. "Mendeley" (iOS) is a powerful yet free tool for the organization of research papers from different sources and in different formats (Figure 25.5). The equivalent app on Android is "Scholarley." With these apps, users can search, add, group, and annotate research papers in different libraries. The search can be based on author, year, journal name, type, title, and so on. Now a Chemistry student or teacher can read papers and annotate them when commuting on a train, sharing them among colleagues with the manual touch of several buttons.

Chemistry students can utilize these apps to efficiently search and keep track of research literature. They can write reviews with better searching functions provided in app in addition to conventional Web-based search engines. It's also

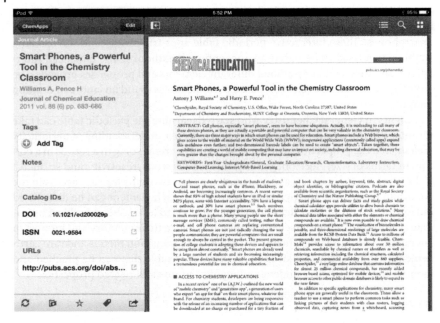

Figure 25.5 Mendeley: a powerful research literature organizer.

much easier to store and organize papers with "Mendeley" and other tools such as "Zotero". Educators can also use these apps to teach chemoinformatics, specifically how to find and process information and how to navigate through massive research databases and online journal libraries. As increasing number of resources provide free or low-cost access to academic community, the emergence of these literature research apps will play an increasingly important role in the access of chemical information for the chemical education community and beyond.

25.3.5
Lab Utility Apps

This category contains the biggest number of apps as developers found it most useful to transform conventional bulky reference books, lab manuals, or instrumental protocols into paperless electronic format accessible on the palm. A lot of these utility apps are must-haves for students working in the lab, reviewing chemical concepts, or operating an advanced chemical instrument. They dramatically reduced the size of reference materials and made the searching process much easier. These apps will also contribute greatly to the revolution of "flipped classroom" discussed in other parts of this book. As pointed out earlier, the portability and rapidly dropping price of smartphones coupled with cheap or free Chemistry apps made these handheld devices the perfect fit for Chemistry students and teachers, serving as lab assistants. They can give the researchers quick access to the information they need whenever and wherever they want. Here all these utility apps are roughly divided into the following five subcategories: Flashcard apps, Dictionary apps, Search Engine apps, Calculator apps, and Instrumental apps. The division is very coarse as there are many overlapping functions. Several apps discussed in one subcategory contain functions under other categories. The portability of smartphones provides a unique strength in chemistry laboratories.

The following apps can be used in various teaching and research labs as reference checkers as mentioned above and also as practical operational tools for the execution of experiments or simply for improving research protocols.

25.3.5.1 **Flashcard Apps**

A lot of students remember flash cards for memorizing amino acid symbols and names when study biochemistry. In Advanced Organic Chemistry, students are also asked to memorize named reactions, which can be recorded on flash cards for self-test and reinforcing the memory. Now by flashcard apps such as "+Amino-" (iOS, by Susan Andryk), "Amino Acid Tutor"(iOS, by Ivan Antonov), "Biochem Euchre Deck" (iOS and Android, by Centre for Mobile Education and Research), "ReactionFlash" (iOS and Android, by Reed Elsevier Properties SA), "Named Reactions"(iOS and Android, by Synthetiq Solutions), "Organic Chemistry Essentials" (iOS, by Christopher Palian), "Organic Chemistry Terms"(Android, by Mobile Study Cards), "Chemistry By Design"[15] (iOS and Android, by Office of Instruction and Assessment, University of Arizona), and "Chemistry of Life" (Android, by JASS). These apps attempt to replace conventional paper flash cards and provide students and researchers with more flexibility and portability. Due to the flipping mechanism, however, these apps still cannot fully replace conventional cards as only one card is displayed on the screen at any time. These apps work best on smartphones or iPod touch as the small screen can mimic the size of paper flash cards. The apps also add extra searching functions and linking capability among different cards through software connections.

"+Amino-," "Amino Acid Tutor," "Biochem Euchre Deck," and "Chemistry of Life" are apps specifically designed for Biochemistry students to review biochemical concepts, names, and acronyms. The other apps deal with Organic Chemical terms or reactions. Students can study these concepts anywhere they want and sometimes can even test their own mastery by flipping through the virtual cards.

"Green Solvent" (iOS, by Molecular Materials Informatics, Inc.) has a variety of solvents grouped by their functional groups. It provides each solvent's safety rating in a flashcard format and links to several mobile structure search apps so users can make a conscious selection of solvent when presented with multiple options (Figure 25.6). For example, when either methanol or acetonitrile can be used as effective mobile phase for HPLC separation, "Green Solvent" app can be used to pick the greener one. Greener solvents can also be selected for sample extraction and glassware rinsing. This app will greatly enhance students' awareness about green chemistry, knowledge on the toxicity, and environmental impact of organic solvents. It will help students to avoid hazardous reagents and select safe alternatives. The teaching of waste disposal can also be simplified with this color-coded app.

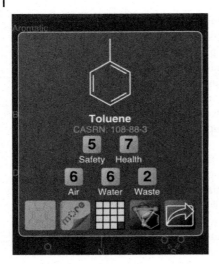

Figure 25.6 "Green Solvents" displays hazard information of chemical solvents.

25.3.5.2 Dictionary/Reference Apps

Students in Chemistry often remember the pain and suffering of carrying hefty CRC handbooks, Merck Index, or other Chemical dictionary or reference books in order to find the a constant, some physical property, or a translation of the term. With the invention of Dictionary apps, the effort became a breeze. Thousands of pages are now condensed in a small, delicate and light touchscreen phone or tablet that can be hold with one hand and searched with the other. The terms, parameters and reactions come out of the app within seconds, replacing the tedious efforts of navigating through a thick dictionary until you hit the correct page and hunting among ant-sized fonts. With two-finger pinching, the text can be enlarged, sometimes accompanied by colorful illustrations to help the understanding of a jargon or a structure.

Apps in this subcategory: "Genetic code" (by Ivan Antonov), "Lab Assistant" (by Visual Ventures), "Organic Named Reactions" (by Indiana University), "Promega" (by Promega), "SPRESImobile" (by Eidogen-Sertanty), "Biochemistry" (by Helen Ginn) are apps found in iOS platform. "Chemical Dictionary" (by Smart Applications), "Chemistry Aid" (by Airwang), "Chemistry Acronyms" (by FIZ Chemie Berlin), "Chemistry Helper" (Adam Hogan, Figure 25.7), "Chemistry Mobile" (by Qan), "Formulas Lite" (by Abhishek Kumar), "W. Chemistry Handbook" (by Dilthiumlabs), "Organic Chemistry Database" (by Freezing Heat Software), "ChemInform Acronyms" (by FIZ Chemie Berlin), "Biochemistry Dictionary" (by Ain Dictionary Developers), "Chemistry Lab Suite" (by Rpor) can be found on Android platform. "NutriBiochem" (by Centre for Mobile Education and Research), a nutritional biochemistry dictionary can be found on both platforms.

Several of these apps contain much more than a searchable dictionary. They have genetic code translators, calculator functions, periodic table, lab protocols, recipes for making buffers or standards, and so on. These apps can be used in all

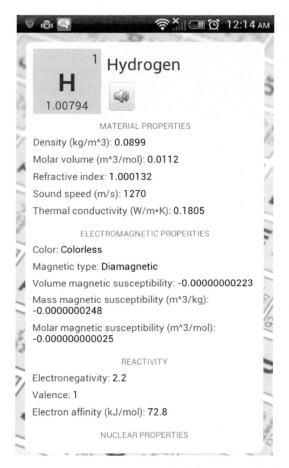

Figure 25.7 "Chemistry Helper" (credit: Adam Hogan).

undergraduate Chemistry labs including General Chemistry, Organic Chemistry, Biochemistry, and Quantitative Chemical Analysis. For example, "W. Chemistry Handbook" contains properties and calculators important for solution chemistry, which becomes an indispensable tool for lab chemists and students in General, Biochemistry, and Organic labs. These apps can be used in conjunction with other apps in the utility apps category to achieve the goal of comprehensive lab assistance. The combined proper usage of these apps can greatly enhance efficiency for students working in Chemistry labs and increase the safety and proper waste disposal.

25.3.5.3 Search Engine Apps

Besides dictionaries, a mobile app can sometimes replace powerful search engines such as Google because more specialized search can be performed in proprietary or chemical specific databases. The tailored search often is quicker and yields instant and specific answers. These apps provide specific and customized tools for searching chemicals using common names, IUPAC names, formula, structure, and various database identifiers. Notable apps in this category includes but are not limited to "ChemMobi" (iOS, Symyc Technologies), "ChemSpider" (iOS and Android, by Royal Society of Chemistry), RCSB PDB (iOS and Android, by RCSB PDB), "mobReg" (iOS and Android, by Eidogen-Sertanty) is a search engine for organic reagents. ChemSpider, a powerful compound search engine [6], can be used as an app to discover structural information, properties, MSDS, related literature, and vendors of chemicals. RCSB PDB (Figure 25.8) is the mighty search engine for protein structures and sequences, along with associated literature. Some of these apps, however, do not fully replace the search engines used in a mobile Web browser. User feedbacks suggest that in order to save app storage space, it might be simpler to run these search engines within browser rather than running them as a single app. The only exception is when a user has to search a specific database frequently for quick access to essential information.

Figure 25.8 RCSB PDB home screen.

25.3.5.4 Calculator Apps

The existence of these handy calculator apps eliminated the need for carrying an extra calculator as most students with smartphones can use their phone for scientific calculations and beyond. These apps can handle most calculations used in General Chemistry, Quantitative Analysis, or Biochemistry Labs. They can calculate molar masses, buffer compositions, convert units, dilution factors, recipes for standard solutions, molarity, melting point, and so on.

In iOS ecosystem, used can find "CMM Molar Mass Calculator" (by Alexandre Perez), "Buffer Calc" (by Wiley), "Converter" (by Ritesh Ranjan), "DailyCalcs" (by Invitrogen), "Dilution" (by Stefano Peruzzi), "LabCal" (by Harald Tammen at iSheepSoft), "Gene Link" (by Ivan Antonov, Figure 25.9); while in Android ecosystem, "Chemistry Formula Calc Lite" (by Vincent Programming), "ConvertPad" (by Sunny Moon), "Solution Calculator Lite" (by *Cooloy.com*), "Chemistry Calculator" (by BLogLR), "Chemical Solutions" (by Amayuki) are frequently mentioned and used. "Molarity" (by Sigma-Aldrich) can be found in both systems.

"Convert Pad" can convert a lot of units and properties in Physical Chemistry, besides its other conversion abilities. This tool can be used in Physical Chemistry labs for quick calculations.

Figure 25.9 "Gene Link" by "Ivan Antonov."

25.3.5.5 Instrumental Apps

The portability of smartphones provides a unique strength in Chemistry laboratories. The following apps can be used in various teaching and research labs as reference checkers as mentioned above and also as practical operational tools for the execution of experiments or simply for improving research protocols. When the experiments involve instrumental control, the parameters can be optimized using in-app calculations.

Agilent has dedicated "LC Calc" and "GC Calc" apps for iOS devices. "LC Calc" can be used to optimize liquid chromatography column conditions by tweaking the correlations among mobile phase flow rate, column dimensions, and back pressure. A well-designed set of separation parameters can be applied to improve resolution and speed. Similar to "LC Calc," "GC Calc" can be used to optimize gas chromatography parameters. Both are beneficial tools for separation scientist as well as students learning chromatographic separations in Instrumental Analysis.

In Organic Labs, our students have been using "TLC Timer" (Android, by Chemovix) to estimate the amount of time needed for full separation and advancing of the solvent front on a thin layer chromatography plate. It can be used by Organic Lab student to manage the timing of developing each TLC plate and by Organic research students to estimate separation time for their synthetic products and impurities.

For more advanced instruments such as NMR, there is a sophisticated app called *"Insensitive"* (iOS, by Klaus Boldt) which can be used to simulate the quantum mechanical models that are used to describe NMR experiment. "ChemDoodle Mobile" can also be used to predict NMR spectra of simple small molecules. "Chemistry Helper" contains IR and NMR tables for peak assignments.

There are also tools for analyzing mass spectra. "MS iCalc" is an iOS app developed by AB Sciex that is dedicated to mass spectrometry calculations. A similar tool in Android exists as "Mass Spectrometry Peaks" app (by Rpor).

These lab utility app tools can be great virtual assistants to Chemistry lab students who work in General Chemistry, Organic Chemistry, Analytical Chemistry, and Biochemistry Labs. Instead of carrying heavy CRC handbooks, a calculator with reference notes, textbooks, or lab manuals, a student only needs to take out a smartphone from pocket to perform all these functions in a lab. The ease of use increases student's confidence at the virtual resources available in chemistry labs and enhance their safety awareness and encourage them to dispose of hazardous waste in a responsible manner. Better designs of experiments and correct preparation of solutions and standards can be expected with the help of these utility apps. Even note-taking in Organic Lab can occur on iPads to replace paper and pens [16]. In summary, these utility apps will dramatically enhance Chemistry Lab students' learning experience and let the student help themselves with the mobile devices. The students can also get extended help outside lab with the constant readiness and ease of access offered by these utility apps.

25.3.6
Apps for Teaching and Demonstration

Chemistry instructors often have to demonstrate dynamic concepts involving particle movements and electron distribution. Apps like "Atoms in Motion" (by Atoms in Motion), "eDistribution" (Arthur Emidio Teixeira), "GasLaws HD" (by T. J. Fletcher), "iGasLaw" (by Cognitive Efficiency), "Lewis Dots" (by Carlo Yuvienco) can help instructors effectively demonstrate these concepts using interactive tools, animations, and movies. All of these apps are iOS-based and more suitable for tablets. After lecture, students can use these tools to review the dynamic concepts and practice aligning electrons around atoms and molecular structures. Students can also tweak parameters (P, V, or T) to observe the results on the gas molecules while getting a better understanding of the ideal gas law.

For high school or entry-level Chemistry coursed videos from "NOVA Elements" (iOS, by PBS, Figure 25.4) and "Video Science" (iOS, by Object Enterprises) can be used to introduce concepts of periodic table and basic chemical knowledge in vivid video formats (Figure 25.10).

Besides the Organic Chemistry utility apps mentioned above, "Learn Organic Chemistry Nomenclature" (iOS, by Aaron M. Hartel) is a helpful tool for student to understand Organic nomenclature. Other Chemistry educational tools include "Chemistry Cheat Sheets" (Android, by *NadsTech.com*), Chemist Free (iOS, by THIX, virtual reaction demo on tablets), "Perfect Chemistry Lite" (Android, by

Figure 25.10 "Video Science" app: a library of demonstration videos.

RanVic Labs). van der Kolk et al. recently demonstrated the application of mobile apps, on both smartphones and tablets, in a Food Chemistry Lab class with some vivid examples [17]. Sometimes, lab specific app is not required for Chemistry lab instruction as demonstrated in a recent report in which camera on smartphones can be used as a simple colorimeter [18].

Many aforementioned apps in other categories can be used for classroom demonstration when appropriate connection device to video projector are used. Usually a simple VGA or HDMI adaptor can be used to project the app screen to a larger screen to be shown to the students. Instead of looking at a static blackboard or whiteboard, or Powerpoint slides, Chemistry lecture students can view vivid and interactive movement of molecules, animations, chemical events in real time while the instructors are manipulating the mobile device screens. The teachers can receive immediate feedback from the students after their viewing and make adjustment on the touch-screen to get a better image. If enough units are available, students can also try to use the mobile devices themselves to get "augmented reality" experience in the learning process of chemistry [19], including the gaming experience described in the next section.

These Chemistry demo apps increase safety in chemistry labs or classrooms, reduce costs for chemicals, glassware, and other supplies. There is also possibility for the teacher to conveniently yet virtually explore the impact of multiple variables such as pH, T, and concentration on the reaction outcome, which is hard or costly to realize in a live demonstration.

Except specific apps, many teaching materials can be accessed free of charge through online media libraries such as iTune and Youtube on a smartphone or tablet. For example, iTune offers "Chemistry Demonstration Series" (*https://itunes.apple.com/us/itunes-u/chemistry-demonstration-series/id428371923?mt=10*) and "Chemistry Lab Procedures" (*https://itunes.apple.com/us/itunes-u/chemistry-lab-procedures-podcast/id386988270?mt=10*) while Youtube has short Chemistry informational videos produced by American Chemical Society and Royal Society of Chemistry and many more demonstrations that can be accessed on a smartphone or tablet. These tools provide detailed and vivid description of Chemistry concepts and lab techniques. The students can view the Chemistry in action if the teachers are not doing a real-time demonstration. The free resources provides helpful teaching materials and reduces the cost and hazard associated with real classroom demonstrations. The best part is that the students can view the video as many times as they want.

25.3.7
Gaming Apps

"Mahjong Chem" (iOS and Android, by Stetson University) is a well-designed game app (Figure 25.11) to help students reinforce general chemistry concepts through the interesting game of mahjong [10]. Another recent gaming app for iPad, "ChemCrafter" (Figure 25.12) creates virtual chemistry experiments in an interactive and fun way for users to "buy" from chemical inventory and create their

25.3 Chemistry Apps in Teaching and Learning | 643

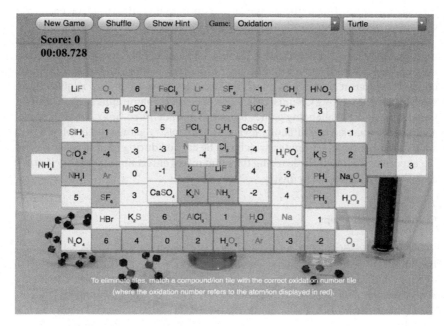

Figure 25.11 Mahjong Chemistry game.

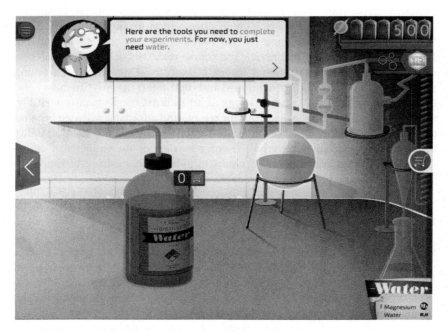

Figure 25.12 "Chemcrafter": A Virtual Experiment Game.

own chemical reactions. The graphical design is well crafted to attract younger users who likes cartoons and interactive games.

Younger students of Chemistry often got bored in traditional lecture settings or at reading Chemistry textbooks. Playing a Chemistry-themed game is a relaxing experience which can often motivate students to explore and utilize chemical concepts or to test themselves General Chemistry knowledge they acquired previously (as in "Mahjong Chemistry"). Other Chemistry game apps such as "Chemistry Games" use challenging games to motivate students to "max out your score" and passing increasingly difficulty levels. The chemistry games can be played on a smartphone or tablet computer, which can be carried during road trips. The gaming experience extends the learning time and mixes entertainment with self-study and review of General Chemistry.

25.3.8
Chemistry Courses Apps

Since the invention of the Internet, the landscape of Chemistry classrooms has changed profoundly. Many chemical educators devoted to the revolution of flipped classrooms. Many mass open online courses (MOOC) of Chemistry are offered online to non-conventional students and disadvantaged students throughout the world. To facilitate the access of online courses, many mobile apps have been developed to provide interactive platforms for online chemical education. Several notable apps include "iTunes U," "Coursera," and "Khan Academy" apps. They provide directly access to online courses taught by famous professors in top-ranking research universities or non-profit organizations (e.g. Khan Academy). Users of these online course apps can pace their studying according to their ability and review the course materials at their convenience, which encourages non-traditional students (such as senior citizens, working students, students from under-developed and rural areas) to learn Chemistry and improve their chemical literacy. As long as an Internet connection is available, the students can use mobile platforms for remote and affordable learning because the cost of the devices are becoming more and more affordable. The portability of these devices drastically lowered the entry barrier for the students of online Chemistry courses. Many publishers start to publish Chemistry e-textbooks to lower the costs and provides convenience..

25.3.9
Test-Prep Apps

Students often get anxious before Chemistry exams as digesting books or notes could be a cumbersome task that lacks efficiency. Fortunately there are a dozen test-prep apps that can greatly enhance the efficiency of "cramming." If used wisely students can dramatically increase their test scores by effectively reviewing the chemical knowledge to prepare for standardized exams.

Under iOS the following test-practice apps can be discovered: "Chemistry Formula Practice" (by Carolina Biological Supply), "Chem Pro: Chemistry Tutor" (by iHelpNYC), "Elements Test2" (by Douglas Roeper), "MCAT Flashcards" (by PALIATech), "Organic Chemistry Nomenclature Quizillator" (by Mobilesce Inc), and "Organic Chemistry Test Bank Lite" (by PALIANTech). Their purposes are self-explanatory. Android system also contains many "quiz apps": "Chemistry Quiz" (by MaksimApps), "Chemistry Quiz" (by Brett Plummer), "Chemistry Speed Quiz" (by Hidden Button Apps), "Periodic Table Quiz" (by Paridae), and "Amino Acid Quiz" (by Next-Gen EduChem).

Many of these apps offer tailored review of specific branches of chemical concepts to assist Chemistry students in the preparation of standard exams. These low-cost or even free apps can act as tutors to help student review chemistry and self-evaluate in order to find out their weaknesses in the understanding of specific chemical concepts. For example, "Organic Chemistry Nomenclature Quizillator" uses quizzes to test the students' mastering of proper nomenclature of organic compounds. Majority of these test-prep apps aid students with General Chemistry, Organic Chemistry, and Biochemistry. Coupled with the other apps described in this chapter, the test-prep apps can serve as enhancement tools in the complete process of Chemistry learning. Students can utilize the test-prep apps to evaluate their learning outcome and make conscience adjustment of their learning strategy and focus.

25.3.10
Apps are Constantly Changing

Most of the Chemistry apps discussed in this chapter have been updated many times and software bugs have been fixed to insure the best compatibility with various devices. The iOS apps only run on Apple devices such as iPhones, iPods, iPads, and iPad-mini, which has very good compatibility as the hardware is tailored to iOS systems and app development. The Android system is more open-source and sometimes a small number of apps might not operate seamlessly due to hardware incompatibility. However, all the developers are diligent at improving their software so software bugs are usually fixed quite efficiently after negative feedback. Thus the app ratings are constantly changing with increasing app adoption and developers' update. Due to various reasons, some apps could drop out of the iTune or Google Play market and new apps are added every day. The icons or names are often changed to reflect updates. This review does not intend to cover the very latest of all the Chemistry apps. Instead, our readers can use this review as a jumping board for the full exploration of the world of mobile Chemistry computing. You are also encouraged to keep track of the latest development of your favorite Chemistry apps so new capabilities can be discovered and drawbacks can be acknowledged to take full advantage of these useful tools for chemical education and research. The readers are cautioned to proceed with the information presented below as it might be outdated by the time this book goes to press.

25.4
Challenges and Opportunities in Chemistry Apps for Chemistry Education

Despite the advantages of educational Chemistry apps described above, the mobile Chemistry computing and applications also faces a lot of challenges in the adoption of this still nascent technology. Readers have to keep in mind the following drawbacks in order to fully realize the potentials of these apps in chemical education and to avoid setbacks. Some of the following will also help chemistry apps beginners better prepare for obstacles in the course of adoption.

Because the Chemistry apps are cohabiting with all the other apps on smartphones and tablet computers, students often get distracted in classroom or labs by social networking apps, unrelated online videos, and non-Chemistry games [20]. The instructor has to provide effective gate-keeping or goal-setting so students will be focused on Chemistry learning or problem solving. Some programs that adopted iPads for Chemistry learning use software to block the access of unrelated contents, however with added cost and mixed results. Fortunately recent research suggested that when students are more engaged with learning activities when smartphones are used in classroom [4, 17, 19] as young students are naturally attracted by new technologies and multimedia presentations which are already an important part of their daily life.

> In wet chemistry labs where hazardous chemicals are handled, contamination can be an issue when it comes to touching touch-screen with gloved hands containing chemicals. Caution must be made to avoid using Chemistry apps with dirty hands [16].

Although the prices for smartphones and tablets are coming down quickly, there are still a lot of students, especially in disadvantaged communities, that cannot afford these expensive electronic gadget. As electronics vendors manufacture low-cost devices, the adoption of smartphone is rapidly accelerating in developing countries, which fuels the usage of Chemistry apps discussed here. Fortunately most apps discussed here are free with a couple high-quality apps charging from $2 to $20, which make them quite affordable. The free apps, however, are always ad-supported and sometimes only offer limited functionality.

During the adoption of Chemistry apps, there is always a learning curve for the instructor to climb as many apps are essentially the mobile version of complicated desktop software that require considerable amount of experience in Computational Chemistry. Though rare, some apps have compatibility issues and a few apps could make the mobile operating system vulnerable. A lot of the apps discussed here require reliable WIFI Internet access, so there must be investment in Web access infrastructure though many support offline operation. Not all apps of the same functions are available on both iOS and Android systems, which makes the quality of various apps unequal. The users must carefully select and test apps before using them for education or learning. The user feedbacks should always be checked before installation to avoid incompatibility or other issues.

Even the perfect chemistry apps cannot fully replace the real-world hands-on experience of handling chemistry glassware, viewing and touching the crystals after synthesis, looking at the color change of pH test paper, catching the endpoint of a titration by the direct observation of indicator color change, smelling the ammonia gas, and getting a first-hand experience on the potential risks of different chemicals. Chemical education is a subject of doing and experimentation, not virtual reality and passive viewing. Safety and environmental aspects can be learned through apps but direct experience is mandatory for the true understanding of chemical knowledge. As pointed out in Wu *et al.*'s review [19], augmented reality (with apps or other technology) should be treated merely as a helpful tool by educators to enhance educational experience. These should not replace the educators' role in leading the learning process. The Chemistry mobile apps should be combined organically with other pedagogical tools such as conventional paper media, live demonstrations, and even the old-school lecturing to reach educational goals.

25.5
Conclusions and Future Perspective

Despite the minor drawbacks mentioned above, mobile Chemistry apps are becoming indispensable tools utilized in chemical education and research. At the time of this review, the total number of installation for all Chemistry apps is well above 100 million and is still increasing as more students and educators are finding the apps as useful tools that can be readily pulled out of their pockets or backpacks. The apps along with the multimedia features introduced in this review will potentially attract students to obtain interactive and effective learning experience in the fields of Science, Technology, Engineering, and Mathematics (STEM). With increasing computing power from quad-core or octa-core CPU, powerful graphic processing units, faster internet access, and increasing memory sizes, the enhanced hardware built in current smartphones or tablet computers will add more interactive functions to mobile Chemistry apps that are quickly replacing conventional mouse-and-keyboard-operated software. The apps that run with accelerated speeds and 3-D graphics will transition more smoothly.

The combined benefits of portability, interactive nature, lowering prices, and increased speed create more possibilities for future app development, which creates new windows for chemical education and research. At the current moment, these Chemistry apps still serve as alternative teaching tools to complement conventional lecture style teaching of chemical concepts. With the gradual adoption and learning of the capabilities from these powerful apps in the chemical education community, innovative ways of teaching and learning Chemistry, not only in the classrooms but also in labs, can be explored to yield further benefits and strengthen the knowledge base developed through conventional teaching methods.

Sounds, pictures, videos, interactive 2-D or 3-D models, and text-based searches can all assist the grasp of chemical or biochemical fundamentals and the applications through multimedia channels. The combination of all these senses can be provided in a tiny device that fits in the pocket or a backpack. The same device can be used to learn, to memorize, to test, to play games, to search, to provide important safety information, all for the purpose of learning chemistry, carrying out chemistry experiments, and conducting original research. The combination of all these media will hopefully enhance the retention of chemical knowledge in students' cognition.

As students carry around apps on smartphones and tablets that contains Chemistry apps, they can wisely manage their study time and go back to the studying materials at any time at their convenience. This flexibility with schedule will not only benefit full-time students who wish to study outside the classrooms, but also enrich and facilitate the learning process for part-time, working students and students from rural, remote areas, or disadvantaged communities.

Many universities started to develop distance learning programs to suit the needs for students who cannot commute to schools on a regular basis or who would like to save transportation costs. Besides traditional platform such as Web browser on desktop or laptop computers, mobile device such as smartphones and tablets are taking a bigger market share in the distance learning market and providing enhanced portability for distance learners. The Chemistry MOOCs saw increasing enrollment all over the world, partially thanks to the rapid spreading or low-cost tablets or smartphones.

The Web 2.0 revolution, coupled with the exponential growth of apps, fueled the integration of mobile Chemistry learning with social media and social networks. Students and educators can exchange ideas, ask and answer questions, discuss, and collaborate online, all with their smartphones or tablets. There are Chemical Education interests group on Linkedin, Facebook, Twitter, ResearchGate, Reddit and other social networking platforms, which all have apps. Many research apps such as Mendeley also provide functions for sharing and collaboration. The boundary between different categories of Chemistry apps is becoming blurred and many multifunctional apps take advantage of the tight integration of several capabilities that benefit students across sub-disciplines in Chemistry, which increases the adoption rates among student users. Many college students are testing more sophisticated and single-function apps (such as apps discussed in Section 25.3.5.5) and find these useful tools even after graduation and when they enter work force. Graduate students and other professional chemists can also benefit from these advanced apps developed by instrument vendors when they are operating complex advanced instrumentations.

As mentioned in Section 25.4, chemical educators face the challenge of adapting to changing landscape of technological platforms built upon the mobile computing systems. The constant emergence of new Chemistry apps that incorporate multiple functions described above requires the continuous learning efforts of all teachers, sometimes from their younger colleagues or even students. The apps alone cannot solve all the problems and obstacles faced by students in the learning of Chemistry. The students still need the instructors' guidance on the proper usage of these mobile software tools. They still need the direction on where to focus when navigating through massive library of chemical information and literature.

Hopefully this review can provide some leads for interested chemical educators and enthusiastic chemistry students.

The future of Chemistry apps on tablet and smartphones will see the continuous growth in user base along with the increasing adoption of mobile computing technologies. More chemical educators will experiment these apps in classroom and in laboratories with constant feedback from students and improvements from developers. The competition in certain categories such as periodic tables and chemical test-prep apps is getting heated and the winners will take a larger market share in chemical education.

While small developer will continue to code niche-application apps that tackle specific problems or challenges in chemical education (e.g., "Insensitive" NMR app). Many large vendors take advantage of their success in conventional platforms and modify their products into mobile apps (e.g., "iSpartan," "ChemDraw") which provides limited functionalities and has received mixed reviews. In the next 10 years, there will be more transition from bulky desktop computers to app platforms with the improving hardware capability of mobile CPUs approaching their laptop or even desktop counterparts.

The new generation of Chemistry students grow up in the new era of iPhones, iPads, Android phones, and tablets. They will be the leading force in the adoption of Chemistry apps. Their spontaneous efforts will drive the adoption among educators. As Windows mobile system is taking market share, more Chemistry apps will be installed on Windows systems and integrated with other windows-based software with which many users are already familiar. The future of Chemistry apps is limitless as the demand for better technology for chemical learning is endless. Peer-to-peer collaborative learning models and augmented reality will keep enriching the learning experience, which makes the process of acquiring Chemistry concepts a painless and enjoyable one.

References

1. Libman, D. and Huang, L. (2013) Chemistry on the go: review of chemistry apps on smartphones. *J. Chem. Educ.*, **90** (3), 320–325.
2. Smith, J. and Tse, D. (2013) Getting Started with Mobile: What Marketers Need to Know, *http://www.nielsen.com/us/en/reports/2013/whats-next--getting-started-with-mobile-what-marketers-need-to-k.html* (accessed 15 December 2013).
3. StudyBlue Infographic *http://www.studyblue.com/projects/infographic-mobile-studying-online-flashcards-on-smartphones/* (accessed 15 December 2013).
4. Williams, A.J. and Pence, H.E. (2011) Smart phones, a powerful tool in the chemistry classroom. *J. Chem. Educ.*, **88**, 683–688.
5. Williams, A.J., Ekins, S., Clark, A.M., Jack, J.J., and Apodaka, R.L. (2011) Mobile apps for chemistry in the world of drug discovery. *Drug Discovery Today*, **16**, 928–939.
6. Pence, H.E. and Williams, A.J. (2010) ChemSpider: an online chemical information resource. *J. Chem. Educ.*, **87**, 1123–1124.
7. Stanford University Library (0000) The Mobile Chemist and Chemical Engineer, *http://lib.stanford.edu/*

swain-library/mobile-apps-chemists-chemical-engineers (accessed 15 December 2013).
8. University of Chicago Library (0000) Chemistry on Mobile Devices, *http://guides.lib.uchicago.edu/content.php?pid=65132&sid=1703522* (accessed 15 December 2013).
9. Indiana University Bloomington (0000) Mobile Chemistry, Indiana University Bloomington Chemistry Library Blog, *https://blogs.libraries.iub.edu/libchem/category/mobile-chemistry/* (accessed 15 December 2013).
10. Crossairt, T.J. and Grubbs, W.T. (2011) Chemical mahjong. *J. Chem. Educ.*, **88**, 841–842.
11. Zickuhr, K. (2013) Tablet Ownership 2013, *http://pewinternet.org/~/media/Files/Reports/2013/PIP_Tablet%20ownership%202013.pdf* (accessed 15 December 2013).
12. Bennett, J. and Pence, H.E. (2011) Managing laboratory data using cloud computing as an organizational tool. *J. Chem. Educ.*, **88**, 761–763.
13. Clark, A.M., Ekins, S., and Williams, A.J. (2012) Redefining cheminformatics with intuitive collaborative mobile apps. *Mol. Inf.*, **31**, 569–584.
14. Feldt, J., Mata, R.A., and Dieterich, J.M. (2012) Atomdroid: a computational chemistry tool for mobile platforms. *J. Chem. Inf. Model.*, **52**, 1072–1078.
15. Draghici, C. and Njardarson, J.T. (2012) Chemistry by design: a web-based educational flashcard for exploring synthetic organic chemistry. *J. Chem. Educ.*, **89**, 1080–1082.
16. Amick, A.W. and Cross, N. (2014) An almost paperless organic chemistry course with the use of iPads. *J. Chem. Educ.*, **91** (5), 753–756.
17. van der Kolk, K., Hartog, R., Beldman, G., and Gruppen, H. (2013) Exploring the potential of smartphones and tablets for performance support in food chemistry laboratory classes. *J. Sci. Educ. Technol.*, **22**, 984–992.
18. Kehoe, E. and Penn, R.L. (2013) Introducing colorimetric analysis with camera phones and digital cameras: an activity for high school or general chemistry. *J. Chem. Educ.*, **90**, 1191–1195.
19. Wu, H.-K., Lee, S.W.-Y., Chang, H.-Y., and Liang, J.-C. (2013) Current status, opportunities and challenges of augmented reality in education. *Comput. Educ.*, **62**, 41–49.
20. Banchero, S. and Philips, E.E. (2013) Schools Learn Tablets' Limits, *http://online.wsj.com/news/articles/SB10001424052702304500404579129812858526576* (accessed 15 December 2013).

26
E-Learning and Blended Learning in Chemistry Education
Michael K. Seery and Christine O'Connor

26.1
Introduction

The past decade has seen a surge in the use of a variety of learning technologies in higher education. The increased accessibility of the Internet, the variety of content-authoring tools, and the rising expectations of students have combined to lead to the generation of an ever-growing repository of learning materials in institutional virtual learning environments (VLEs), behind textbook publisher paywalls, and openly on the Internet. There is no shortage of content.

The ubiquity of content has altered the emphasis of the lecture room from content provider toward models of understanding and applying content to the context of the disciplines. Lecturers and their students can now draw on this external material to support and supplement their in-class teaching and learning. In addition, the classroom discussions can move beyond the lecture halls to online for such as discussion boards and social media. Technology has made new assessment models available, which allows students to work through their understanding of material at their own pace. Therefore, the approach to teaching and the consequent skills required of a lecturer are changing dramatically. With no shortage of opportunities, the difficulty for faculty is how best to integrate learning technologies into their curriculum. In this chapter, we aim to present a series of learning technologies, some better known than others, with an emphasis on how they may be usefully incorporated into the twenty-first century classroom. We advocate these technologies by basing them on what is known about how people learn. In order to be useful to practitioners in higher education, we supplement them with examples from the chemistry practice literature.

When designing the use of learning technologies, the key aspect is to engage learners at the digital chalk face by creating ways to motivate them in using the technology and *creating an environment with added value*. This sense of value is an important concept to consider. If an approach has value to a lecturer, it is likely to be more deeply embedded into curriculum delivery, and in turn more likely to be used by and useful to students. Thus e-learning can be considered as "learning facilitated and supported through the use of information and communication

Chemistry Education: Best Practices, Opportunities and Trends, First Edition.
Edited by Javier García-Martínez and Elena Serrano-Torregrosa.
© 2015 Wiley-VCH Verlag GmbH & Co. KGaA. Published 2015 by Wiley-VCH Verlag GmbH & Co. KGaA.

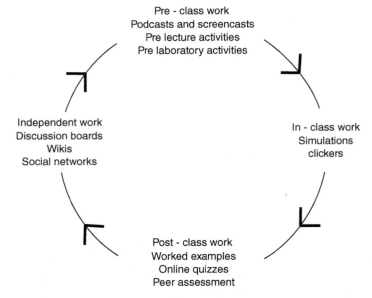

Figure 26.1 The blended learning cycle: incorporating ICT into the curriculum.

technologies" [1]. This definition is useful, as it presents a framework for how *e-learning* may be integrated into our curriculum delivery. In what has become known as *blended learning*, the curriculum delivery aspect of our teaching – the act of teaching, assessment, and feedback – can be complemented and supported by what we do in the online component of our teaching. The classroom extends to the online space, and in turn the online space informs the classroom, thus creating a blended learning cycle (Figure 26.1). A more holistic approach to curriculum delivery in a blended learning environment is not just to consider the efficiency of the acquisition of knowledge and skills but also to enable the learners to take control of their learning so that they become independent learners [2]. The provision of resources, formative assessments, and support frameworks are a means to enable learners to achieve this status. To emphasize this point, the term "technology-enhanced learning" has emerged in more recent years in place of "application of technology to learning" [1]. Several popular models of e-learning design incorporate these principles, including Gagné's nine principles [3], the ADDIE (Analysis, design, development, implementation, and evaluation) model [4], and Salmon's five stages [5].

26.2
Building a Blended Learning Curriculum

Why should a practitioner include any given technology in their teaching? Decisions about whether a given approach is useful will likely depend on a given practitioner's teaching scenario, and perhaps even their own subdiscipline of chemistry. The preeminent factor in deciding whether to include any technology into the curriculum delivery is what value it will have in enabling teaching or facilitating learning. Activities and resources that are deeply embedded in curriculum delivery are more likely to be seen as valuable and more likely to be used by students and promoted by lecturers.

The suggestions below represent a selection of common methods of including technology in the blended learning curriculum. The technology usage is considered in the context of where it can play a role in curriculum delivery, and is listed along with some emerging areas that are developing for each technology in the future. Examples of some curriculum delivery issues are summarized (Table 26.1), and a range of technologies are surveyed in this chapter.

Table 26.1 Some common issues that arise in teaching and learning chemistry and the potential role of technology in helping to address these issues.

Curriculum delivery issue	Example of technology that can be used to address issue	Future directions of this technology
Chemistry is a technical subject and learners can become overwhelmed in lectures or laboratory classes	Online pre-lecture/lab activities can present some or all of the material beforehand, allowing time in lecture/laboratory to be devoted to more in-depth discussion/interaction	Flipped lectures, where the content delivery aspect of lectures is moved in its entirety to before the lecture class, are being trialed in many chemistry lecture classes. This allows for more active learning to take place in the lecture room
Lecturer wishes to gauge whether students understand a topic in any given lecture	Clickers are widely used to examine student understandings at a given moment in a lecture	In-class polling no longer requires hardware, as several apps are now available
Students wish to test their own understanding of topics under study	Quizzes can be used to allow students check their understanding and gain feedback	Response-adaptive quizzes branch the students progression through a quiz depending on answer. Peer authoring of quizzes develop students' understanding of topics through question authoring
Students find basic problem-solving strategies difficult	Worked examples can allow students learn how to approach problems of progressive difficulty	Inclusion of answer-specific tutorial videos to address difficulties based on individual student requirements
Little time in lecture for discussion and learners who may work through material at a different pace	Discussion boards allow for tutorial questions to be discussed in advance, with more complex problems left to the tutorial	Social networking tools such as Twitter are being explored for presenting key topics and facilitating on-the-go discussion

26.3
Cognitive Load Theory in Instructional Design

As well as consideration of the pedagogical basis for technology-enhanced learning, the second consideration is the framework for the design, development, and implementation of multimedia and e-resources. There is a lot of research devoted to the design of online resources for optimum instruction, which can be considered under the umbrella of cognitive load theory (CLT) [6, 7]. CLT provides a basis for how learners receive and process new information, and purports that there is a maximum capacity that learners can process during any given learning event. This capacity or working memory space is governed by three types of load. The *intrinsic load* is the difficulty of the material. With new material, learners need to process new terminology and relationships, and this takes a certain amount of mental effort. The *extraneous load* is the effort required to extract the information from the learning materials. Finally, the *germane load* is the effort required for processing new information and integrating it into the long-term memory. Therefore in the context of CLT, the ultimate goal in effective instruction is to manage the intrinsic load and extraneous load effectively so that there is capacity for germane load, or, in other words, learning (Figure 26.2).

CLT can be used as a framework for the design and implementation of many of the teaching initiatives described below. Providing some information in advance of a class or laboratory (pre-lecture and pre-laboratory activities) can reduce the intrinsic load during class time. Structuring approaches to problem solving using worked examples can reduce the load associated with the combination of theory required to address a problem and the recall of the steps required in the algorithmic element of the calculations.

Furthermore, the design of resources can influence the extraneous load. For novice learners, effective design of instructional material means students can quickly identify the primary objectives of instruction. Materials that make it difficult to identify the learning topic or objective, or present the information in

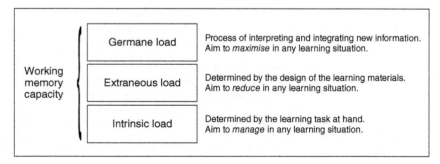

Figure 26.2 The working memory capacity can be considered in terms of cognitive load which consists of three types. The aim of instruction is to maximize germane load so as to facilitate learning. (Based on Ref. [6].)

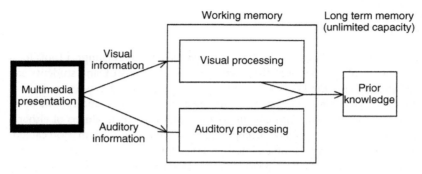

Figure 26.3 The information processing model. Using both visual and auditory learning materials when presenting material online can facilitate more efficient processing as they utilize different channels. (Based on Clarke and Mayer [7].)

an unclear manner, result in an increase in extraneous load. It should be noted, however, that as learners move from novice to expert in a discipline (for example, early to late undergraduate), increasing the extraneous load can have a positive effect on learning.

The design of learning materials for use in online or computer-based instruction has been extensively discussed by Clarke and Mayer [7], who have developed a series of principles of instruction. Briefly, in this context, learners perceive new information through auditory and visual channels – what they hear and what they see. Therefore effective instructional materials will use both these channels to convey new information (Figure 26.3). For example, a lot of information presented as text might be better presented as audio, with the associated visual consisting of a diagram or key equation, displayed coincident with the audio commentary. Finally, the integration of new learning materials with the long-term memory becomes more effective with the level of prior knowledge. Therefore, linking new information with prior knowledge facilitates learners in integrating new knowledge.

26.4
Examples from Practice

Several approaches to blending online work with classroom work with the aim of enhancing the overall curriculum delivery are described below. Common issues that arise in teaching and learning chemistry are used to demonstrate how incorporating particular concepts or technologies may assist in particular teaching and learning scenarios. Exemplars from practice-based literature are highlighted so that readers can get a sense of how they may incorporate any given approach into their own teaching.

26.4.1
Podcasts and Screencasts

Podcasts and screencasts can be very useful audio and visual support tools for a variety of learning situations. They can be hosted on the VLE and students can access this information in class or online as they require. Podcasts are audio files that can be played on a computer or portable device such as a smartphone or an MP3 player. Screencasts combine audio and visuals to create videos of either a computer screen or a video recording. Podcasts and screencasts may be categorized into substitutional, supplemental, and student-specific [8]. A substitutional resource is a podcast or screencast that covers something already given to students. A typical example is a lecture that is delivered to students but is also captured electronically for students to review. A supplemental resource is something that provides additional information to what is delivered to students. A typical example would be some worked-out questions that relate to what was talked about in a lecture. Student-specific resources are often individual to students, and cover resources such as student-generated resources or feedback on assignments.

While all three types of resources have educational value, it is clear that effort spent on substitutional resources is essentially duplicating what is already done in contact time with students. While these resources are popular with students, who like the option of rewatching entire lectures available to them, many reports suggest that students simply skip to a point of difficulty in the lecture. In fact, a recent article suggests that there was a negative correlation between students rewatching lecture capture and exam performance – these students were essentially repeating what was a passive experience (watching a lecture) and not actively working on their understanding of the topic at hand [9].

Podcasts or screencasts can be of any length, and there is little work done on defining optimum lengths. Perhaps more important than the length of the cast is what is expected of students while listening or watching. A 10-min video on a complex topic will feel like a long time if the student is not doing anything. Therefore, podcasts and screencasts can be associated with worksheets or activities that will structure the time on task. In order to enhance the reusability of the recording, it is best to avoid mentioning dates or specific details relating to any class group. Names, rooms, and module codes change frequently! The broadcast usually opens with an introduction of what will be covered, and what is expected of the students. The main body covers the topic of the broadcast, with regular prompts to remind students what they should be doing. With podcasts, verbal signposts are important: summarizing what has been said and what is next at regular intervals [10, 11]. Podcasts can be produced using audio recording software, and Audacity® is a freely available open source tool [12].

Screencasts are audio-visual presentations that cover lecture notes and/or material generated by the lecturer or the students. Although the first reports of videos in chemistry education date to 1957 [13], there has been a meteoric rise in the amount of screencasts published online on sites such as YouTube (e.g., *http://www.youtube.com/user/periodicvideos*) and iTunesU. Common software for production of screencasts include Echo 360 [14], Camtasia Studio® [15], and ProfCast [16]. While lecture capture screencasts just offer students a reviewing of the lecture, there is a broad diversity in application of supplemental screencasts, such as covering a conceptually difficult topic, presenting worked out examples, or providing an overview of a laboratory experiment. A useful approach is to prepare screencasts that respond to areas students find difficult. It is possible to identify areas of difficulty from student responses in class or through online quizzes, and generate tutorial video clips to address these topics. Formal evaluation of this approach has found that these tutorials are a useful tool in improving students' understanding of a topic and in mastering associated problems on the content [17]. A broad range of chemistry screencasts on an array of chemistry topics have been collected [18, 19]. Student-specific podcasts include using screencasts to write feedback on student work. Students can review their feedback on their work listening to the lecturer's comments and watching the annotations. This has been shown to be a more engaging way to illustrate ways to improve work for final submission [20].

Some tools are emerging that enable reuse of screencasts and videos already available on YouTube or other video hosting sites. These include eduCanon [21] which facilitates the embedding of multiple-choice questions into existing videos and EDpuzzle [22] which allows existing videos to be cropped, to add new voiceover, and to embed quizzes. The benefit of these Web applications are that, in many instances, useful existing resources need only minor changes to make them useful learning resources in an individual lecturer's curriculum.

One issue for consideration with regard to podcasts and screencasts is that, unlike information on a piece of paper, it can take more mental effort to glance back at what has just being covered. Therefore, while watching a video or listening to a podcast, there can be a cognitive load demand in remembering some information that was just presented while processing information in a given moment. This is known as the *transient information effect*. In order to alleviate this load, information should be segmented, and where animations are used, incorporation of interactive elements to allow students determine the pace of the animation has been found to be beneficial [23].

26.4.2
Preparing for Lectures and Laboratory Classes

Chemistry lectures can present a significant amount of new information, and often in multiple representations (symbolic, molecular, macroscopic) [24, 25]. Therefore, novice learners may become quickly overwhelmed during the lecture time and either disengage or resort to surface learning. One effective strategy to avoid this situation is to prepare the student by presenting some of the information in advance, in the form of pre-lecture activities.

Pre-lecture activities are some form of work students do in preparation for lectures. This can be as simple as reading an assigned section of the textbook in advance [26]. However, harnessing a technological method offers several advantages: the material can be tailor-made or made specific for the lecture; it is possible to know whether students complete the activity; and an embedded quiz can both help students identify areas of difficulty and alert the lecturer about topics that are proving difficult across a class group [27].

Several examples of pre-lecture activities are available in the literature. These range from presenting one or two key topics in advance of the lecture to presenting the entire lecture online in advance (a phenomenon now used in the emerging trend of "flipped lectures," see Chapter 13). Asking students to read some text in advance and then completing an online quiz can be beneficial to students, although, while students liked the concept, there is a perception that they are being penalized in the quiz for not knowing material that they had not yet been taught in the formal in-class session [28]. Pre-lecture activities were also tried where students completed some preparatory work and a quiz at the start of a lecture. Their implementation led to an improvement in grades of students who had not done chemistry at school [29]. This concept was extended by Seery and Donnelly, who described their use of pre-lecture activities for introductory chemistry. Students watched a pre-lecture activity online prior to the lecture. These were designed to introduce the key concepts and terminology of the lecture. Their implementation removed the gap in performance that had been previously observed between students who had completed chemistry before and those who had not. In keeping with the principles of blending in-class and online, the importance of building on the pre-lecture activities within the lecture was highlighted [30].

While pre-lecture activities are still a relatively new concept in chemistry, pre-laboratory work is a long established practice. Learning in laboratories poses difficulties, as the laboratory protocol adds an extra cognitive demand on learners, who may also be novice learners on the underlying chemistry upon which the laboratory class is based. Learning difficulties in the lab are well documented [31, 32], and students who are overwhelmed may resort to just coping with what they can process, and follow the list of instructions in their manual, resulting in little learning [33, 34]. Reducing some of the new information learners are presented within the laboratory class may help ameliorate the load of learning.

Pre-laboratory exercises online offer students the flexibility of being able to complete the work in their own time at their own pace. Designed well, they can motivate students to read the required laboratory material in advance of the class, and provide immediate feedback on their responses [35]. The design principles of a pre-laboratory simulation and cognitive load considerations have been described fully in a study on organic laboratory simulations [36], and the incorporation of pre-laboratory activities demonstrated a dramatic enhancement in physical chemistry problem-solving ability in the laboratory [37].

A temptation in the design of pre-laboratory activities is to focus on the experimental procedure or "how to" protocol instead of the conceptual basis for the laboratory practical. However, these resources are better used in presenting the theoretical basis to the laboratory class, demonstrating how this theory will be tested, and leaving the practical protocol to the laboratory itself, where it can be explained "just in time" [38]. Focusing the pre-laboratory work on non-numeric theory requires students to understand the conceptual basis of the laboratory, rather than completing a task-oriented exercise of getting the right answer [39]. Consequently, videos focusing on laboratory techniques have a useful role in the laboratory class, and some interesting recent work has highlighted the potential of videos available as mobile podcasts that students can bring into the laboratory class. These prove more efficient for students wishing to gain information about at-hand laboratory protocols than pre-laboratory materials do, which is likely due to the fact that they are on hand when required [40].

26.4.3
Online Quizzes

Online quizzes have become popular in third-level education. They are a reusable resource that can be used to carry out individual formative assessment of students and allow immediate feedback or grading. As technology develops, the use of online quizzes is becoming more widespread and more tailored to what individual students need. Four categories, listed in Table 26.2, are described below.

Responsive quizzes are the standard type of quiz used in chemistry education. As well as a lecturer authoring questions independently, most chemistry text books from large publishing houses supply relevant question databases corresponding to the topics covered in the book chapters. These can be incorporated into the VLE to reinforce learning from the lecture, introduce concepts, test the individuals' progress with the topic, and provide feedback on their answer. An issue with using textbook databanks is that many answers may now be on popular internet answer sites. With most VLEs, it is now feasible to create and upload in-house question banks. With algorithmic problems, large numbers of different questions can be created easily so that each student is provided with a unique set. With questions that require multiple stages, it is possible to give students tailored feedback, based on what stage they answered incorrectly. This process can be partially automated

Table 26.2 Some categories of quizzes that can be used to test and promote student understanding.

Quiz type	Format
Responsive	Feedback informs students whether they are correct or not, and may provide explanations
Response-adaptive	In addition to student feedback, subsequent questions depend on the performance in any particular question. This allows the questions to be tailored to the needs of individual students
Worked examples	Questions are presented in blocks, with each stage requiring more input from students until eventually they can complete a question independently
Student-authored	Students author their own questions, answers, and distractors and can rate and answer other students' questions

by generating large question banks and feedback answers using Microsoft Excel software, so that students receive individual questions and tailored feedback with comparatively little authoring time required [41].

The instantaneous feedback available from online quizzes is especially beneficial to novice learners, and a common approach is to use a database of questions as homework for early undergraduate stages. These can structure the learners' study of a topic and provide scaffolding as they work through the important core topics in an introductory chemistry course. As well as providing feedback, these quizzes can be augmented by providing fully worked solutions and direct students on the basis of their response to targeted tutorial assistance. Such an approach has been shown to assist students in their learning of a topic and their engagement with their introductory courses [42]. Additional resources are available on mobile phone apps, such as Chemistry Quiz (Figure 26.4).

Quizzes can also be used as a pragmatic way to address differences that may arise within large class groups due to differences in prior knowledge, misconceptions held by students from different learning backgrounds, and different gender. Some studies have indicated that there is a difference between female and male achievement in general chemistry courses, and the inclusion of quizzes has helped to address this gap. A recent report advocating this approach demonstrated that, with the inclusion of online quizzes in the assessment schedule, scores across the entire class group increased, and the difference gap between female and male

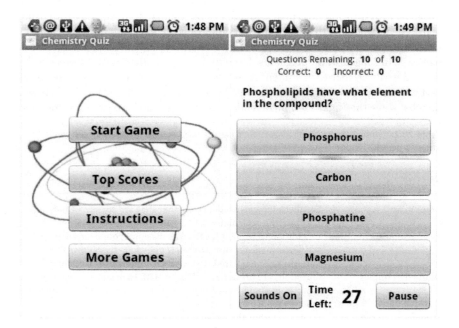

Figure 26.4 Chemistry Quiz App (*http://bit.ly/1qPaFxv*).

students narrowed. This is attributed to the improvement of students' active learning time on the task of completing the quizzes [43]. Online quizzes can also be used in checking students' understanding of core concepts prior to a lecture or safety aspects prior to a laboratory, and thus are often included as part of a larger element of curriculum delivery. Examples include their use to check students' understanding of core terminology prior to a lecture [30] and their use in preparatory laboratory exercises, which ensures students have awareness of the safety considerations of a practical class.

While online quizzes are commonplace, an emerging trend is response-adaptive quizzes. These quizzes incorporate the feedback and/or hints associated with traditional quizzes, but in addition changes the sequence of questions asked on the basis of the response to each particular question. In practice, this results in learners who have difficulty with a topic being redirected through additional questions on that topic to stimulate their understanding, while learners who have demonstrated understanding of a topic progress onto the next stage. Therefore, adaptive systems aim to individualize online quizzes to the particular learning needs of each student. Their use has been shown to additionally increase the performance of students compared to the increase already observed with students who used responsive quizzes only [44].

In recent years, more consideration has been given to student authoring of quiz questions. This has the dual advantage of requiring students to understand a topic well enough so that they are in a position to generate a question, answer, and distractors, and building up a databank of questions for all students to use. This can all be facilitated by PeerWise [45], a freely available online tool where students to generate their own questions and answers and complete questions submitted by their peers. The questions take the format of multiple-choice questions, and a detailed guide of how students develop and evaluate questions is available [46]. PeerWise is gaining popularity in chemistry education with reports published for biochemistry [47] and organic chemistry [48]. A common theme is that the skills involved in developing questions and evaluating the questions of others promote higher order thinking skills. Typically, students are asked to author a set of questions and answer some of their peers' questions. Gamification is introduced by creating league tables based on the number of questions completed, authored, and so on, and these can increase student engagement by adding an element of competition to the process.

26.4.4
Worked Examples

Another feature to consider with the use of quizzes is the incorporation of worked examples. Worked examples are grounded in CLT. As a novice learner approaches a problem, he or she must simultaneously try to understand the chemical theory associated with the problem and draw on this theory to identify the steps required to solve the problem. This can lead to the learner becoming overloaded, with students resorting to the surface learning approach by simply learning off the associated equations or protocol. Worked examples aim to introduce the steps to problem solving in a structured way, so that the method of approach can be introduced gradually, allowing the learner to master the approach to the problem [49]. They can be introduced into the curriculum in various ways. One is to show worked examples in class and follow up with weekly online quizzes aligned with the content covered in class [50]. Another is to deliver the worked examples online as part of the quiz, with iterative worked examples removing increasing amounts of support until the learner can independently solve a problem type. This approach, called *fading*, allows learners to build confidence in their learning and decrease the level of support as they improve their problem-solving skills. As an illustration, a problem requiring three stages to answer would involve three worked examples being presented, with each one removing one step from the end of the process (Figure 26.5). This process is easily facilitated using the quizzing feature on VLEs. This approach demonstrated improvement in the learners' problem-solving skills [51].

26.4.5
Clickers

Personal response systems, or clickers, are a popular tool in teaching, especially in large-enrolment introductory chemistry classes, although their use across different levels and sizes of class groups has been beneficial [52]. While chemistry has been slower to adopt clickers than physics [53, 54], the last 5 years has seen a surge in the number of reports of clickers in chemistry classrooms [55].

Clickers can be used for a variety of purposes and in a variety of ways. A simple poll involves asking a class group a multiple-choice question, and the class

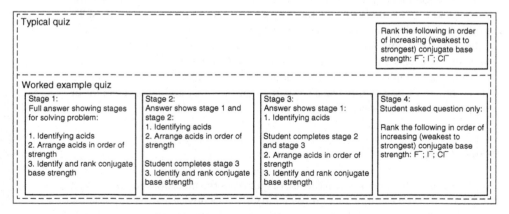

Figure 26.5 A model of worked example compared to direct quizzing. Students are prepared for their approach to the quiz question by a series of stages with less information being provided at each step until they are able to complete the problem independently. (Question used derived from Ref. [51].)

response can be used to identify a topic of difficulty and prompt further explanation [56]. Simple polling can be expanded by using predict–observe–explain type scenarios, where students are asked to predict an outcome using a poll and, following a discussion, re-polling to see if there is a change in the type of response [57, 58]. This approach best mirrors the "peer instruction" model proposed by Mazur for physics. Another aspect of simple polling is to use two-stage questioning, where students are asked to respond to a question and then asked which reason best matches why they chose that answer [59]. This approach is especially useful for identifying misconceptions.

As technology developed, the limits of what clickers can be used for have been pushed. Student understanding of organic chemistry mechanisms can be examined using the number sequence entry on clickers, where students have to, for example, enter the sequence of events that occur in a given reaction mechanism [60, 61]. More recent work describes the use of clickers for enabling peer debate and student-centered discussion in the context of how to address chemistry-specific learning outcomes [48]. Use of clickers has also been found to enhance student retention [62]. The use of clickers is no longer restricted to using the proprietary personal response hardware, with mobile apps now proving popular. Examples include the Socrative app [63], Responseware app [64], and Votapedia [65]. Plickers [66] requires only the lecturer to have a smart phone (Figure 26.6).

26.4.6
Online Communities

An often underused aspect of using online learning is the extension of the classroom to an online space where students and lecturers can continue interacting beyond the lecture hour. As well as lecturer–student communication, these online spaces allow for the development of an online community of peers. With increasing Web technologies available, these online social interactions are becoming more commonplace, and are moving into the education domain.

One of the longest established online spaces for communities of learners is a discussion board, a space that facilitates an online conversation that can occur with different participants contributing at different times. While discussion boards are a common component of VLEs, they are typically underutilized, with their use commonly restricted to supporting fully online modules [67]. In a blended learning context, there are some reports on the use of discussion boards in chemistry education. A common first approach to the use of discussion boards is to distribute class materials, address homework queries, or sharing laboratory data [68]. This offers the advantage of reducing the number of individual queries a lecturer might receive about these topics. Some suggestions for prompting student participation with discussion boards include (i) providing timely responses and feedback to questions posed; (ii) creating an environment where students feel comfortable (including allowing anonymous posting); (iii) setting an example by posting sample queries and responses; and (iv) setting etiquette boundaries early on as

Figure 26.6 (a,b) Screenshots from Plickers App.

required, including personal professionalism [69]. An important point noted by Markwell is that "lurking" – viewing a discussion board without posting – is also an effective way for students to learn. The ability of students to be able to focus on one particular task or topic at a time on the discussion board also resonates with effective application of technology in the context of CLT.

Along with tips from practice, more detailed analysis and coding of discussion board postings in discussion boards that have shown high levels of interaction has also led to some useful implications for practice. These include the lecturer being a visible presence by regularly posting messages, summarizing previous posts and prompting further thought, challenging student responses in a constructive manner, requesting responses, and offering support and encouragement [70].

More recently, social networks such as Facebook and Twitter have been used to create communities to support the teaching of a particular module. In the case of Twitter, a recent study with a large first-year chemistry group demonstrated that the inclusion of Twitter promoted the use of questions and answers (more than oral interaction) in the classroom. The tweets from the class were displayed on a screen in the class. However, the authors of this study felt that it could interrupt the flow of the lecture [71]. A recent survey carried out on 9–17-year-olds by the National School boards association in the United States revealed that this cohort of students spent almost as much time on social networking sites as they did watching TV a week, attributing much of their online activity to creative elements such as creating content, sharing information, and blogging [72]. It was observed that the students have a natural interest in the technology platforms, which motivates them to learn and share content. Professional chemical associations such as the American Chemical Society (ACS) and the Royal Society of Chemistry (RSC) have acknowledged the role of social networking sites and have created a set of Web 2.0 initiatives such as the Mobile App for "molecule of the week" by ACS and the RSC "Chemistry World" and "Education in Chemistry" Blogs. Nature.com is another very useful social networking site with Natures Chemistry Blog, The Sceptical Chymist. The *Journal of the American Chemical Society* (JACS), ACS, and the RSC are all active on Facebook [73]. The social networking platforms are useful for both communicating with students and academics but also as networking sites for graduates, for example, RSC, ACS, LinkedIn.

26.5 Conclusion: Integrating Technology Enhanced Learning into the Curriculum

Having discussed the underlying framework of technology-enhanced learning and considered several examples from practice, the final consideration is how to begin to develop a blended learning (or e-learning if fully online) module. A useful approach is to consider the blended learning cycle for each lecture in the module, and consider what supports, assessment, and feedback are required. For example, a module with a series of lectures may involve developing some screencasts to prepare students for lectures, a homework quiz to allow students to check their understanding after the lecture, and a discussion board for facilitating an online tutorial. Particular lectures in the sequence may require a supplemental screencast to tease out a topic that is conceptually challenging. These technologies should only be used as required in the curriculum, as they are time consuming for lecturers to develop and for students to complete. If they are required in the curriculum, this means that they merit some reward for completion in the form of assessment marks.

The time to develop a fully blended curriculum should not be underestimated, and in a busy work environment it is likely that progressively developing different elements of the technology-enhanced curriculum will be more successful than attempting to do it all at once. These can be implemented in an order of priority determined by the lecturer's own experience of what is most needed to support learners on the module.

The usefulness of a particular resource or approach should be monitored periodically. These include simple monitoring techniques such as tracking usage and surveying students. More nuanced monitoring can include examining quiz marks or exam answers to see if there are topics still causing difficulties that are already supported – questioning if the resource is adequate or if another approach is needed. Are the students more engaged in their class work? Has retention increased in the module? In other words, does the technology incorporated into the module address the original problem identified? There is no single approach to embedding technology, as each learning scenario will differ, with different topics, students, and so on. Therefore, the question worth revisiting regularly is whether technology is enhancing the learning of your students. This should ensure that the level of blended learning in the module is appropriate.

References

1. Pachler, N. and Daly, C. (2011) *Key Issues in e-Learning: Research and Practice*, Continuum International, London.
2. Amiel, T. and Reeves, T.C. (2008) Design-based research and educational technology: rethinking technology and the research agenda. *Educ. Technol. Soc.*, **11** (4), 29–40.
3. Gagné, R.M. (1985) *The Conditions of Learning and Theory of Instruction*, 4th edn, Holt, Rinehart & Winston, New York.
4. Morrison, G.R., Ross, S.M., Kemp, J.E., and Kalman, H. (2010) *Designing Effective Instruction*, 6th edn, John Wiley & Sons, Ltd, Chichester.
5. Salmon, G. (2011) *E-Moderating: The Key to Teaching and Learning Online*, 3rd edn edn, Routledge, New York.
6. Sweller, J. (2008) Human Cognitive Architecture, in *Handbook of Research on Educational Communications and Technology*, 3rd edn (eds J.M. Spector, M.D. Merrill, J. van Merrienboer, and M.P. Driscoll), Routledge, New York.
7. Clarke, R.C. and Mayer, R.E. (2008) *E-Learning and the Science of Instruction*, 2nd edn, Pfeiffer (Wiley), San Francisco, CA.
8. McGarr, O. (2009) A review of podcasting in higher education: its influence on the traditional lecture. *Australas. J. Educ. Technol.*, **25** (3), 309–321.
9. Revell, K.D. (2014) A comparison of the usage of tablet PC, lecture capture, and online homework in an introductory chemistry course. *J. Chem. Educ.* doi: 10.1021/ed400372x
10. Seery, M. (2012) Podcasting: support and enrich chemistry education. *Educ. Chem.*, **49** (2), 19–22.
11. Salmon, G. and Edirisingha, P. (2008) *Podcasting for Learning in Universities*, Open University Press, Maidenhead.
12. Audacity http://audacity.sourceforge.net/ (accessed 20 November 2013).
13. Blonder, R., Jonatan, M., Bar-Dov, Z., Benny, N., Rap, S., and Sakhnini, S. (2013) Can You Tube it? Providing chemistry teachers with technological tools and enhancing their self-efficacy beliefs. *Chem. Educ. Res. Pract.*, **14** (3), 269–285.
14. Echo360 http://www.echo360.com/ (accessed 20 November 2013).
15. Techsmith® http://www.techsmith.com/camtasia.html (accessed 20 November 2013).
16. Profcast http://www.profcast.com (accessed 20 November 2013).
17. He, Y., Swenson, S., and Lents, N. (2012) Online video tutorials increase learning of difficult concepts in an undergraduate analytical chemistry course. *J. Chem. Educ.*, **89** (9), 1128–1132.
18. Royal Society of Chemistry, http://www.rsc.org/learn-chemistry/resource/res00001339/chemistry-vignettes
19. Read, D. and Lancaster, S. (2012) Unlocking video: 24/7 learning for the iPod generation. *Educ. Chem.*, **49** (4), 13–16.
20. Haxton, K.J. and McGarvey, D.J. (2011) Screencasting as a means of providing timely, general feedback on assessment. *New Dir.*, 7, 18–21.
21. eduCanon http://www.educanon.com/ (accessed 20 November 2013).
22. EDPuzzle http://www.edpuzzle.com/ (accessed 20 November 2013).
23. Wong, A., Leahy, W., Marcus, N., and Sweller, J. (2012) Cognitive load theory, the transient information effect

and e-learning. *Learn. Instr.*, **2012** (22), 449–475.
24. Johnstone, A.H. (2000) Teaching of chemistry – logical or psychological? *Chem. Educ. Res. Pract. Eur.*, **1** (1), 9–15.
25. Taber, K.S. (2013) Revisiting the chemistry triplet: drawing upon the nature of chemical knowledge and the psychology of learning to inform chemistry education. *Chem. Educ. Res. Pract.*, **14** (2), 156–168.
26. Kristine, F.J. (1985) Developing study skills in the context of the general chemistry course: the prelecture assignment. *J. Chem. Educ.*, **62** (6), 509–510.
27. Seery, M. (2012) Jump-starting lectures. *Educ. Chem.*, **49** (5), 22–25.
28. Collard, D.M., Girardot, S.P., and Deutsch, H.M. (2002) From the textbook to the lecture: improving prelecture preparation in organic chemistry. *J. Chem. Educ.*, **79** (4), 520–523.
29. Sirhan, G., Gray, C., Johnstone, A.H., and Reid, N. (1999) Preparing the mind of the learner. *Univ. Chem. Educ.*, **3** (2), 43–47.
30. Seery, M.K. and Donnelly, R. (2012) The implementation of pre-lecture resources to reduce in-class cognitive load: a case study for higher education chemistry. *Br. J. Educ. Technol.*, **43** (4), 667–677.
31. Hofstein, A. and Lunetta, V.N. (2004) The laboratory in science education: foundations for the twenty-first century. *Sci. Educ.*, **88** (1), 28–54.
32. Bennett, S.W., Seery, M.K., and Sovegjarto-Wigbers, D. (2009) Practical work in higher level chemistry education, in *Innovative Methods in Teaching and Learning Chemistry in Higher Education* (eds I. Eilks and B. Byers), Royal Society of Chemistry, London.
33. Sweller, J. (1988) Cognitive load during problem solving: effects on learning. *Cogn. Sci.*, **12**, 257–285.
34. Johnstone, A.H. (1997) Chemistry teaching – Science or alchemy? *J. Chem. Educ.*, **74**, 262–268.
35. Chittleborough, G.D., Mocerino, M., and Treagust, D.F. (2007) Achieving greater feedback and flexibility using online pre-laboratory exercises with non-major students. *J. Chem. Educ.*, **84** (5), 884–888.
36. Supasorn, S., Suits, J.P., Jones, L.L., and Vibuljan, S. (2008) Impact of a pre-laboratory organic-extraction simulation on comprehension and attitudes of undergraduate chemistry students. *Chem. Educ. Res. Pract.*, **9** (2), 169–181.
37. Avramiotis, S. and Tsaparlis, G. (2013) Using computer simulations in chemistry problem solving. *Chem. Educ. Res. Pract.*, **14** (3), 297–311.
38. van Merrienboer, J.J.G., Kirschner, P.A., and Kester, L. (2003) Taking the load off a learner's mind: instructional design for complex learning. *Educ. Psychol.*, **38** (1), 5–13.
39. Winberg, T.M. and Berg, C.A.R. (2007) Students' cognitive focus during a chemistry laboratory exercise: effect of a computer-simulated prelab. *J. Res. Sci. Teach.*, **44**, 1108–1133.
40. Powell, C.B. and Mason, D.S. (2013) Effectiveness of podcasts delivered on mobile devices as a support for student learning during general chemistry laboratories. *J. Sci. Educ. Technol.*, **22** (2), 148–170.
41. Ashworth, S.H. (2013) Generating large question banks of graded questions with tailored feedback and its effect on student performance. *New Dir.*, **9** (1), 55–59.
42. Freasier, B., Collins, G., and Newitt, P. (2003) A web-based interactive homework quiz and tutorial package to motivate undergraduate chemistry students and improve learning. *J. Chem. Educ.*, **80** (11), 1344–1347.
43. Richards-Babb, M. and Jackson, J.K. (2011) Gendered responses to online homework use in general chemistry. *Chem. Educ. Res. Pract.*, **12** (4), 409–419.
44. Eichler J. F., Peeples, J. (2013) Online homework put to the test: a report on the impact of two online learning systems on student performance in general chemistry, *J. Chem. Educ.*, **90** (9), 1137–1143.
45. Peerwise http://peerwise.cs.auckland.ac.nz/ (accessed 20 November 2013).

46. Bates, S. and Galloway, R. (2013) Student-generated assessment. *Educ. Chem.*, **50** (1), 18–21.
47. Bottomley, S. and Denny, P. (2011) A participatory learning approach to biochemistry using student authored and evaluated multiple-choice questions. *Biochem. Mol. Biol. Educ.*, **39** (5), 352–361.
48. Ryan, B.J. (2013) Line up, line up: using technology to align and enhance peer learning and assessment in a student centred foundation organic chemistry module. *Chem. Educ. Res. Pract.*, **14** (3), 229–238.
49. Crippen, K.J. and Brooks, D.W. (2009) Applying cognitive theory to chemistry instruction: the case for worked examples. *Chem. Educ. Res. Pract.*, **10** (1), 35–41.
50. Crippen, K.J. and Earl, B.L. (2007) Impact of web-based worked examples and self-explanation on performance, problem-solving, and self-efficacy. *Comput. Educ.*, **49**, 809–821.
51. Behmke, D.A. and Atwood, C.H. (2013) Implementation and assessment of Cognitive Load Theory (CLT) based questions in an electronic homework and testing system. *Chem. Educ. Res. Pract.*, **14** (3), 247–256.
52. Sevian, H. and Robinson, W.E. (2011) Clickers promote learning in all kinds of classes—small and large, graduate and undergraduate, lecture and lab. *J. Coll. Sci. Teach.*, **40**, 14–18.
53. Mazur, E. (1997) *Peer Instruction: A User's Manual*, Prentice Hall, Upper Saddle River, NJ.
54. MacArthur, J.R. and Jones, L.L. (2008) A review of literature reports of clickers applicable to college chemistry classrooms. *Chem. Educ. Res. Pract.*, **9** (3), 187–195.
55. MacArthur, J.R. (2013) How will classroom response systems "cross the chasm"? *J. Chem. Educ.*, **90** (3), 273–275.
56. King, D.B. (2008) Using clickers to identify the muddiest points in large chemistry classes. *J. Chem. Educ.*, **88** (11), 1485–1488.
57. Wagner, B.D. (2009) A variation on the use of interactive anonymous quizzes in the chemistry classroom. *J. Chem. Educ.*, **86** (11), 1300–1303.
58. COFA http://online.cofa.unsw.edu.au/learning-to-teach-online/ltto-episodes?view=video&video=265 (accessed 20 November 2013).
59. Chandrasegaran, A.L., Treagust, D.F., and Mocerino, M. (2007) The development of a two-tier multiple-choice diagnostic instrument for evaluating secondary school students' ability to describe and explain chemical reactions using multiple levels of representation. *Chem. Educ. Res. Pract.*, **8** (3), 293–307.
60. Flynn, A.B. (2011) Developing problem-solving skills through retrosynthetic analysis and clickers in organic chemistry. *J. Chem. Educ.*, **88** (11), 1496–1500.
61. Ruder, A.R. and Straumanis, S.M. (2009) A method for writing open-ended curved arrow notation questions for multiple-choice exams and electronic-response systems. *J. Chem. Educ.*, **86** (12), 1392–1396.
62. Gebru, M.T., Phelps, A.J., and Wulfsberg, G. (2012) Effect of clickers versus online homework on students' long-term retention of general chemistry course material. *Chem. Educ. Res. Pract.*, **13** (3), 325–329.
63. Socrative http://www.socrative.com/ (accessed 20 November 2013).
64. Turning Technologies http://www.rwpoll.com/ (accessed 20 November 2013).
65. CSIRO http://bit.ly/evotapedia (accessed 10 December 2013).
66. Plickers www.plickers.com (accessed 10 December 2013).
67. Dori, Y.J. and Barak, M. (2003) A Web-based chemistry course as a means to foster freshman learning. *J. Chem. Educ.*, **80** (9), 1084–1092.
68. Paulisse, K.W. and Polik, W.F. (1999) Use of WWW discussion boards in chemistry education. *J. Chem. Educ.*, **76** (5), 704–707.
69. Markwell, J. (2005) Using the discussion board in the undergraduate biochemistry classroom. *Biochem. Mol. Biol. Educ.*, **33** (4), 260–264.

70. Slocum, L.E., Towns, M.H., and Zielinski, T.J. (2004) Online chemistry modules: interaction and effective faculty facilitation. *J. Chem. Educ.*, **81** (7), 1058–1065.
71. Cole, M.L., Hibbert, D.B., and Kehoe, E.J. (2013) Students' perceptions of using Twitter to interact with the course instructor during lectures for a large enrolment chemistry course. *J. Chem. Educ.*, **90** (5), 671–672.
72. National School Boards Association (2007) *Creating and Connecting: Research and Guidelines on Online Social—and Educational—Networking*, National School Boards Association.
73. Martinez, J.C. (2010) Chemistry 2.0: xreating online communities. *Chem. Int.*, **32**, 4.

27
Wiki Technologies and Communities: New Approaches to Assessing Individual and Collaborative Learning in the Chemistry Laboratory

Gwendolyn Lawrie and Lisbeth Grøndahl

27.1
Introduction

The shift from traditional expository teaching to learner-centered collaborative, active learning environments has catalyzed a parallel increase in the use of Web 2.0 technologies in instructors' pedagogies. These technologies represent social, participatory information-sharing, communication, and collaboration media (including networking sites such as Twitter, Facebook, and LinkedIn) that align with students' own worlds. As part of the emergence of "pedagogy 2.0" [1, 2], Wikis and Blogs in particular have been successfully translated into platforms that enhance learning environments. Wikis are fully editable Web pages [3] that enable collaborative co-construction and revision of online content as an open-editing platform. There is a wide range of Wiki tools available for instructors to consider, including commercial sites tailored for educational applications that require a subscription for access including Wikispaces and PBWorks (URLs supplied in Ref. [4]) and Wikis hosted in their own institution's learning management system such as Blackboard, Sakai, and Moodle. Wikis have been used extensively for the co-construction of information, with Wikipedia representing the most reputed example, including a branch of this Wiki encyclopedia dedicated to chemistry (The Chemistry Portal). There are multiple active and dynamic chemistry information and resource-based Wikis, including ChemWiki: the dynamic chemistry E-Textbook (hosted by the University of California, Davis, URL in Ref. [4]) and LearnChemistry Wiki (hosted by the Royal Society of Chemistry, URL in Ref. [4]).

The potential of a Wiki technology as a viable platform for assessing student learning gains and capturing these broader outcomes is appealing and is considered in this chapter in a particular constructivist learning context of the chemistry laboratory. Learning outcomes from the chemistry laboratory experience are traditionally measured through experimental reports or worksheets, but these genres exclude evaluation of broader learning outcomes such as collaboration and communication, which represent the core elements of working in a research or commercial laboratory.

Chemistry Education: Best Practices, Opportunities and Trends, First Edition.
Edited by Javier García-Martínez and Elena Serrano-Torregrosa.
© 2015 Wiley-VCH Verlag GmbH & Co. KGaA. Published 2015 by Wiley-VCH Verlag GmbH & Co. KGaA.

It would be the rare or lousy chemists who completed a laboratory experiment without continuously recording their observations and data as they progressed. The majority of good chemists typically complete this process using their laboratory notebook, where data collected for the variables currently under investigation have been displayed crudely as a table or a graph along with provisional decisions and planning for their next experiment. Students acquire these skills across the whole of their undergraduate experience in chemistry majors, but the intrinsic role of good lab record keeping as a professional practice may not be explicitly assessed in individual units of study or courses.

In this chapter, assessment of the laboratory learning in relation to the applicability of Wikis as an example of technology-enhanced learning is considered. However, the use of a Wiki is not limited to the chemistry laboratory environment and the affordances of Wikis discussed in this chapter can easily be translated or extended into other chemistry or science learning contexts and environments.

27.2
Shifting Assessment Practices in Chemistry Laboratory Learning

Traditional laboratory experiments have long been regarded as the cornerstones of chemistry courses across the curriculum [5, 6]. Originally, these experiences were intended to either reinforce the acquisition of concepts encountered in lectures by providing experiences of macroscopic phenomena that students need to relate back to molecular-level explanations, or develop students' practical manipulative skills. However, student learning in these environments is much more complex. Kirschner and Meester [7] reviewed the literature and identified 120 learning objectives for science practical activities, subsequently distilling out 8 common general student-centered laboratory learning objectives (left-hand column in Figure 27.1).

These form core objectives, but it is widely recognized that there are many more transferable learning objectives that are valued and achievable within the laboratory context [8, 9]. Of course, learning objectives should be considered hand in hand with strategies to assess the associated learning outcomes. The traditional assessment of student laboratory learning is through the completion of a worksheet, a laboratory report, a portfolio, or other form of scientific communication such as a poster or oral presentation [5, 10]. Indeed, numerous studies examine how students learn in the laboratory but there are fewer published ones that focus on the optimal format for student assessment in the laboratory, particularly in the tertiary context. To illustrate this point, for the purpose of this chapter, we reviewed the 113 articles published in the "Laboratory Experiment" section of the *Journal of Chemical Education* (volume 89, 2012) in a single year. The level of study was assigned through the keywords cited by their authors and the basis of the experiment (traditional expository, inquiry/discovery, or other) was assigned on the basis of the format of instructions provided to students: whether the students worked individually or collaboratively and the

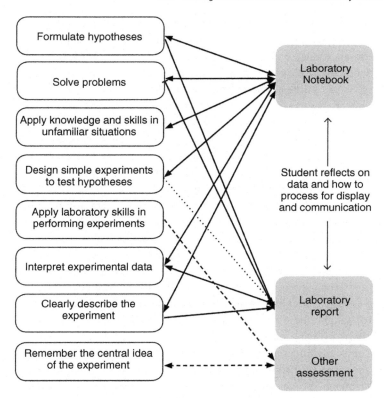

Figure 27.1 Core learning objectives in the science laboratory related to the primary form of assessment that measures these student outcomes (a dotted line indicates that communication of the outcome would need to be explicitly required in the assessment, and a dashed line indicates that the outcome is assessed by another unspecified mode).

format of assessment, if cited. Three clear levels of experiment emerged: (i) high school/introductory/first-year/general chemistry (25%), (ii) second-year/upper division (27%), and (iii) upper division/graduate (48%). Seventy-nine percent of experiments were categorized as traditional or expository laboratory exercises, while 10% were inquiry- or discovery-based (the remaining experiments were dry labs, project-based, or case studies). Seventy-nine percent of articles stated whether students worked primarily in pairs or groups (53%) or individually (20%) or a combination of both (6%). The mode of assessing the outcomes of the laboratory experiment were categorized as either uncited (33%), report (39%), post-lab questions (17%), or other (11%) which included oral presentations. Very little detail was evident in terms of the expected structure or content of laboratory reports provided and this absence of literature perhaps reflects the challenges for an instructor in assessing student learning within the laboratory environment beyond observing their manipulative skills and competencies.

It is difficult to make individual student thinking visible without engaging in a "think-aloud" activity such as a viva or oral exam where they explain their procedures, observations, and interpretations. A completed worksheet only provides evidence that a sequence of processes has been completed particularly during prescribed procedures, characteristic of a traditional "cookbook" exercise. Collectively written group reports or oral presentations represent collaborative products but the assignment of individual roles in the latter is fraught with difficulty in the assignment of marks or grades [11]. A key question is, therefore: "when students submit individual reports or lab notebooks at the end of a session, having completed an experiment collaboratively with their peers, how does an instructor measure their individual learning outcomes?"

In the last two decades, there has been a strong movement toward more active-learning inquiry and undergraduate research experiences in the laboratory because there is evidence that these foster greater learning gains, confidence, engagement, student ownership, metacognition, and extended lab memory [8, 12–15]. Students complete less structured experimental processes in these environments and the outcomes are indeterminate, so the instructors are seeking better strategies to gain insight into their thinking through aligned assessment.

Students are taught the structure of the genre of the scientific laboratory report as early as primary science education, following the general sequence: introduction, aims, method, results, discussion, and conclusion. It has been found that posing questions to scaffold and support students in their communication of their experimental processes and outcomes is effective using the Science Writing Heuristic (SWH) [16] which supported improved learning outcomes from the laboratory [15, 17]. The laboratory report, however, does not provide evidence of the iterative nature of experimentation in inquiry laboratories where students make decisions and plan future experiments based on their observations and data. Indeed, the production of a laboratory report implicitly requires that students record their observations and collect data throughout the laboratory sessions, meaning that the laboratory notebook and scientific report are intrinsically interlinked and, in combination, represent authentic scientific practice (right-hand column in Figure 27.1).

In 1971, Latreille and Chavance [18] re-published the translation of a page from Grignard's lab notebook record of his first discovery of the formation of an organomagnesium compound (dated around 1989). They identified this as an "excellent lab record" and proposed that the page be given to students along with their advice to improve their own record keeping (Box 27.1). The benefits [19] and formal requirements of keeping a laboratory notebook are well documented [20] and still resonate with contemporary guidelines for how professional scientific laboratory notebooks should be structured. Their importance has increased because these documents represent accurate records of experimental procedures that may be used as legal evidence to establish intellectual property.

BOX 27.1 Advice on Keeping an Excellent Lab Notebook Inspired by Francois Grignard's Record of His Discovery of the Formation of a Grignard Reagent [18]

1) The chemist describes immediately everything that he does and observes.
2) He does not say anything that he cannot prove.
3) What he crosses out should remain readable.
4) The description must be such that if the writer should suddenly be called *away*, then any other chemist would be able to continue or duplicate the same experiment.
5) The notebook should be written in the first person and in the present tense.
6) The author's name should be clearly written on the paper in the order that credit may be given where credit is due.
7) Each page should be numbered and there should be no blank spaces so that the date can be established without a doubt.

Recently, educators have been encouraged to "overcome the inertia of the traditional school lab report" [21] and adopt new ways to enable students to communicate their experimental outcomes. As an assessment tool, the laboratory notebook represents a dynamic record of a student's experience in the laboratory, and entries can be categorized as defining, descriptive, designing, verbal or visual reporting, or interpreting data [22]. As shown in Figure 27.1, the laboratory notebook enables assessment of "apply knowledge and skills in unfamiliar situations" that cannot readily be assessed in the format of a laboratory report that is completed post experimental investigation. Also, assessment of "design simple experiments to test a hypothesis" will only be achieved if students are explicitly required to justify their design strategy as part of the report structure and criteria.

So what strategies are available for instructors to support students' acquisition of skills in laboratory record keeping while, in parallel, assessing their individual learning gains from inquiry-based laboratory experiments? One option is to adopt a technology-enhanced tool that provides a dynamic record and history of the processes that students apply in their record-keeping along with their changes in appraisal of their experimental outcomes. This tool is a Wiki!

27.3
Theoretical and Learning Design Perspectives Related to Technology-Enhanced Learning Environments

Instructional design to incorporate technology with the aim of enhancing the learning environment should be applied to enhance student-centered activities while improving the flow of the task and student engagement. Successful implementation is evident when the benefits can be iterated and the outcome does not simply represent a conventional learning activity that has been mirrored through being hosted on a technology platform. To explore this, an instructor should ask him/herself: are the students able to achieve the same outcomes without the presence of the technology as they do with the technology?

Indeed, there are multiple theoretical perspectives, models, and frameworks that have been developed to describe how technology may enhance students' learning processes and outcomes including, but not limited to, social constructivism, activity theory, communities of practice, and actor-network theory. Grainne Conole provides both a summary of theoretical perspectives and extensive strategies for instructors who aim to design for learning using technology and subsequently evaluate the impact [23]. A common thread between many theoretical perspectives is that the instructor desires to generate a collaborative and constructive environment. The conversational framework for collaborative technologies [24] provides a mechanism to explore and represent how the complexity of learning through technology-enhanced collaboration can be made more visible while simultaneously examining the rigor of the activity. In mapping the activities and relationships evident in lab Wikis to this conversational framework [25], several points are highlighted where students may iterate on the provision of formative feedback from the instructor, their team members, and laboratory tutors who were also part of their learning community during an inquiry laboratory experiment (Figure 27.2).

Through the Wiki, the instructor provides the scaffolding for students to progress effectively across a sequence of sessions, and the Wiki makes the peer interactions more visible. An example Wiki page is provided in Figure 27.3,

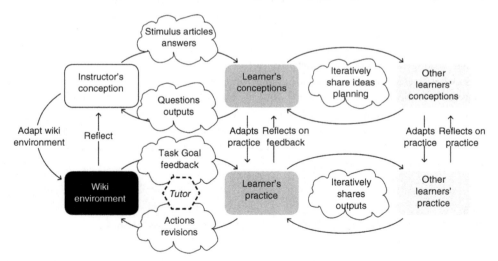

Figure 27.2 The modified conversational framework for collaborative technologies applied to represent the relationships involved in a Wiki lab notebook activity adapted from [24].

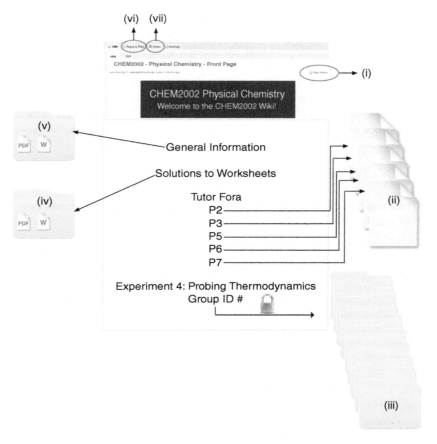

Figure 27.3 An example course Wiki illustrating the underlying structure that can be linked to the front page (shown center). Features include (i) the history function which occurs on every page in the Wiki, (ii) pages dedicated to each experiment in the course where the tutor responds to student questions and shares information, (iii) folders with restricted membership enabling assignment of groups and including the instructor and tutors, (iv) instructor-provided resources, (v) instructor-provided information, (vi) pages and files function enabling navigation of the subsections of the Wiki, and (vii) user function where membership can be controlled.

indicating how the structure supports the interactions shown in Figure 27.2. The instructor provides resources including stimulus material, task instructions, supporting information, and assessment criteria in dedicated folders linked to the Wiki front page (Figure 27.3(iv,v)). Membership of the Wiki or any subfolders can be restricted to the instructor, team members, and the practical tutor (Figure 27.3(iii)). The tutors can have a dedicated forum for each experiment in a course (Figure 27.3ii) where students can post questions or the tutors can provide feedback and their marking schemes.

In tasks where the entire Wiki is assigned to an individual group, each student member is able to add to, edit, or structure the group Wiki as part of collaborative activities (e.g., for a virtual lab notebook) and the instructor can find evidence of these processes through the Wiki history function (this function is highlighted as (i) in the Wiki shown in Figure 27.3) establishing a chronological record of contributions, processes, and outcomes. The instructor and the tutor support constructive collaboration between students during iterative experimental processes throughout the duration of the task by provision of timely feedback.

27.4
Wiki Learning Environments as an Assessment Platform for Students' Communication of Their Inquiry Laboratory Outcomes

Online collaborative environments such as Wikis enable students to collaborate, create, communicate, edit, and link content either synchronously or asynchronously [2, 26–30]. It is important to consider the educational affordances of a Wiki for each specific task, and technology-enhanced learning is most effective when that technology is a low threshold for student participation in terms of their information, communication, or technology skills.

Laboratory learning encompasses a number of key activities (Figure 27.4) and, to be considered a viable tool, the Wiki environment should enhance each of these. Limitations in, or absence of, one or more of these activities will require a supplementary method for facilitating collaboration and assessment of students in the broader learning outcomes in the laboratory. This quickly increases the complexity of the task and is likely to deter faculty from adopting this tool. While

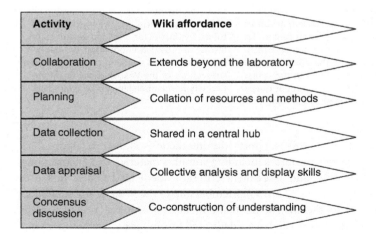

Figure 27.4 Activities that students complete during a laboratory experience, indicating the affordances that a Wiki environment can provide in enhancing that experience.

there are multiple choices in Wiki tools, many do not support communication and display of scientific experimental data because of deficiencies in formatting, fonts, and embedding. In particular, the symbolic and graphical basis of chemistry, in particular, requires ready inclusion of superscripts, subscripts, symbols, mathematical equations, tabulation, and graphs in the pages. Students also need opportunity to become familiar with the Wiki environment prior to any higher stakes assessment and this is achievable by embedding smaller Wiki activities in a course as orientation. A Wiki platform can be used in multiple ways to support learning in the laboratory, and specific examples are provided below.

27.4.1
Co-Construction of Shared Understanding of Experimental Observations

Wikis represent a constructivist learning environment that is particularly suited to fostering the development of shared understanding between participants. Engaging students in Wiki environments can involve the requirement of submitting a contribution to a discussion perhaps as an "icebreaker" for group work. For example, to increase their scientific literacy, three groups of seven to eight students enrolled in a third-level tertiary nanoscience course in a research-intensive Australian university were required to develop a synchronous description of the same scanning electron microscopy image of a diatom (sourced from the Web [31]). This was completed during a 40-min period within a workshop session in the course. They entered their descriptors (supported by a glossary of electron microscopy terms provided in class) into the Wiki comments field and submitted until there was a collective agreement on the final description. The evolution of the sophistication of their description was evident during the course of the task with early descriptors including *"bulges out symmetrically," "two wide points of approximately 10 µm and a pinched region in the middle," "peanut boat,"* and *"a peanut shape, with few layers on the outshell and planety organized holes inside the layer."* While each group began with very different descriptions, all three groups arrived at a remarkable consensus (Table 27.1).

Indeed, the synchronous discussion enabled students to challenge and question their descriptors, for example, one student in group 3 challenged the group with *"we could change the word brittle for porous,"* or exclude observations that could not be verified, for example, *"made of different elements due to the different shades of gray."*

27.4.2
Enhancing the Role of Tutors in the Wiki Laboratory Community

As illustrated by the conversational framework for collaborative technologies, the presence and role of tutors in the Wiki provides an important source of feedback to students beyond face-to-face contact in the laboratory. Once students leave the laboratory session, they are often required to collate their results and

Table 27.1 Final statements developed collaboratively and synchronously by three separate groups of seven to eight students through a Wiki discussion to establish a collectively agreed description of a scanning electron microscope image.

Group	Final description
1	The image above depicts a ~28.5 μm long peanut shaped structure with a maximum width of ~11.1 μm. The structural features of the image include a clathrate layer comprising holes of ~0.3 μm, macroporous, divided in two by a linear component with frill-like sides traveling the length of the structure. Beneath this layer appears to be other clathrate layers of similar structure
2	An elongated ovoid structure of approximately 30 μm, with a maximum width of 12 μm and a pinch in the middle having a diameter of approximately 8 μm. The shape is symmetrical along the length and across the 8 μm center. Internally, it has a regular macroporous honeycomb (clathrate) structure, with the length being divided by a fissure flanked by crimped walls
3	This structure has the properties of a core shell with approximately 30 μm in length and aspect ratio of 0.41176. There is radial symmetry and a C2 axis of rotation. A honeycombed structure which seemed brittle, mounted on a sturdy looking backbone structure. Elongated with a partially platy-clathrate structure

interpret them in traditional laboratories and also make decisions and plan future experiments in inquiry laboratories. The Wiki offers an alternative to each student e-mailing the tutor with his or her individual question; it can represent a hub for a tutor forum (traditional experiments) or source of additional feedback (collaborative inquiry experiments).

A tutor forum for traditional laboratory settings offers an efficiency that tutors themselves appreciate as they are often asked the same question repeatedly and this reduces both e-mails and questions in the laboratory. Interviews with laboratory tutors who participated in a course Wiki tutor forum for the first time resulted in the observations provided in Box 27.2 (G.A. Lawrie and L. Grøndahl, unpublished data):

BOX 27.2 Affordances of a Wiki Lab Tutor Forum

- Students are able to review posted responses, reducing the frequency of questions asked in the laboratory.
- Students have better access to the tutor and this increased the quality of the questions they posted (though they often expected the tutor to be online 24/7).
- Tutors are able to post last-minute modifications to procedures or instructions if required, which reduces repetition in the laboratory increasing time for other activities.

- Tutors are able to post their marking criteria and general advice in relation to data processing that students frequently find difficult, such as error analysis.
- Tutors are able to review other tutors' entries, which makes provision of advice more efficient (no repetition and ability to cross-reference between tutors).

The interactions recorded in the tutor forum also benefit the instructor who gains feedback on which student questions are asked most frequently and which are often invisible in the laboratory setting as a result of the direct interaction between the tutor and student.

27.5 Practical Examples of the Application of Wikis to Enhance Laboratory Learning Outcomes

The benefit in adopting the Wiki environment, both as a platform and an assessment tool, is evident through the advantage it offers over traditional formats of assessment. Because this technology tool has only recently been applied to the context of science laboratory learning, there are only a few examples of good practice that have emerged to date to inspire instructors. These different applications are summarized as case studies below and include chemistry, physics, and engineering laboratory course exemplars, and they have been categorized to illustrate the potential utility and applicability of a Wiki.

27.5.1 Supporting Collaborative Discussion of Experimental Data by Large Groups of Students during a Second-Level Organic Chemistry Inquiry Experiment

In this case study [32], set in the context of a second-level organic chemistry course at a mid-western US institution serving 12 000 tertiary students, Chem-Wiki was developed by the authors and was hosted on the platform Wikispaces [4] to provide a collaborative domain for the creation of content. Students, working in teams of six to eight, could share and discuss their data generated through their inquiry-based experiments prior to preparing a report. This online collaborative domain enabled students to reflect on their data and make decisions between laboratory sessions. Each group also generated a glossary of chemical terminology, procedures, and concepts in their own words, which were compiled to make a course-specific dictionary. Individual students were responsible for preparing separate sections of the laboratory report (samples can be viewed through the link provided as reference 15 in [32]).

27.5.2
Virtual Laboratory Notebook Wiki Enhancing Laboratory Learning Outcomes from a Collaborative Research-Style Experiment in a Third-Level Nanoscience Course

Working in groups of three to five, students in the third-level tertiary nanoscience course referred to in Section 27.4.1 used a Wiki hosted through PBWorks [4] (Figure 27.5a) as their virtual laboratory notebook. They completed all the processes of planning, method development, data processing and display, reflection, and discussion iteratively in the Wiki over several weeks culminating in a discussion of their results [25]. The Wiki provided a very effective environment in which students could collaborate with substantial evidence of gains in scientific literacy as a result of the process. Students identified multiple benefits including shared vision and consensus in laboratory experiments *"everyone could see what everyone else was thinking,"* and improved collaboration in the laboratory and access to data afterwards *"having all the data available to all group members at once"* rather than waiting for a group member to share their manually recorded observations.

The role of the tutor in providing feedback or encouraging students to reflect on their data, again, was particularly important during inquiry laboratory activities as evidenced through the tutor interaction with a group in their Wiki outside the laboratory (Figure 27.5b). This reinforces the relationships described by the conversational framework, which support laboratory learning [25].

27.5.3
Scaffolding Collaborative Laboratory Report Writing through a Wiki

As discussed earlier, a post-experiment laboratory report translates experimental planning, implementation, and critical appraisal of resulting data into a scientific communication of the students' understanding of the processes and outcomes. Wikis can enhance the collaborative writing process through careful scaffolding, and two separate strategies have been reported to engage students in the collective writing process. In the first [33], set in the context of a second-level mechanical engineering technology laboratory in a private US college, each student in a group of four takes responsibility for submitting one out of six sections of the report to the Wiki (Dashboard Wiki) hosted in the institution's learning management system. A sequence has been developed that requires increasingly higher order thinking and interlinking to previous submitted sections: Appendix (including raw data), Method, Results, Analysis, Introduction, and Conclusion. The Appendix and Conclusion sections represented group submissions and, while collaboration and peer editing are encouraged throughout, giving each student responsibility for submission of a section builds in interdependency between team members. There were seven reports required across the course, and students reported gains in their learning through peer reviewing and editing in the Wiki.

In the second study [34], SWH was successfully adapted to an (unidentified) Wiki online environment. Set in the context of a general physics traditional laboratory at a Taiwanese university, groups of two to three freshman students worked

27.5 Practical Examples of the Application of Wikis to Enhance Laboratory Learning Outcomes

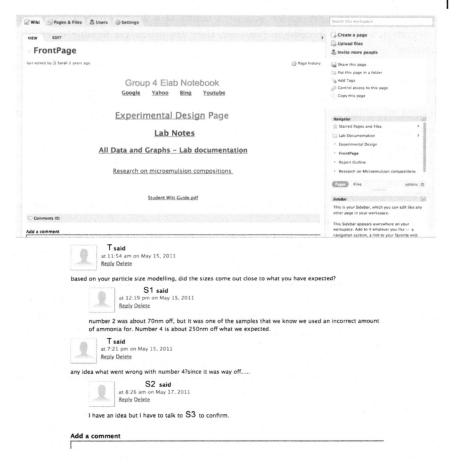

Figure 27.5 (a) Screen capture of the PBWorks Wiki platform showing the front page of a group laboratory notebook. (b) Screen capture example of how the tutor (T) interaction with the students (S1 and S2) can continue beyond the face-to-face interactions within the laboratory session by encouraging reflection on data post-lab.

collaboratively to answer questions that scaffolded the report format. For example, the question "What is the difference between what we did and what was planned in the pre-lab report?" scaffolded writing the procedure section and "How do we analyze the experimental data and what are the results?" scaffolded the results section (including text, graphs, tables).

The choice of Wiki tool will depend on multiple factors. Commercial sites such as Wikispaces and PBWorks offer initial free accounts for instructors to trial (with limited student membership) and provide extensive technical support. Class e-mail lists can be easily uploaded to these sites to establish Wiki membership, and pages can be set to have restricted membership (ideal for groups) and a range of editing rights. Wikis hosted within learning management systems such as Blackboard can offer easier access in terms of a single student login and integration with grade centers; however, they are often less flexible in editing options.

Clearly, the benefit of the Wiki is in the shared space providing students time to work collaboratively and reflectively beyond the laboratory environment. The product of these activities is easily accessed by the instructor and provides greater insight into collaborative processes. There may be instances where the instructor seeks to delve further into the data that can be accessed through the Wiki environment and this represents a transition into the domain of learning analytics.

27.6
Emerging Uses of Wikis in Lab Learning Based on Web 2.0 Analytics and Their Potential to Enhance Lab Learning

Instructors often seek methods for evaluating the engagement of their students and individual learning outcomes from collaborative tasks where they have worked in pairs, trios, or larger groups. The Wiki platform offers the ability to gain remarkable insight into student activity in either individual or collaborative tasks through the following dimensions: student products/artifacts, students' perceptions, and individual or group processes [29]. Empirical data, collected in multiple forms as a function of time from online environments (including learning management systems) are increasingly being used to measure student engagement in tasks [29, 35]. In general, analytics in online platforms enable both facile measurement of the timing and categorization of student contributions to a shared domain [36] and mapping the structural arrangement of the components of the student-generated Wiki [30]. Caution is advised, as there is evidence that the digital artifacts in a Wiki cannot be regarded as an independent measure of collaborative activity because students may collaborate face-to-face in parallel to discuss their Wiki contents with one student taking responsibility to write a collective stance (G.A. Lawrie and L. Grøndahl, unpublished data).

27.6.1
Evaluating Student Participation and Contribution as Insight into Engagement

In constructive collaborative environments, such as Wikis, instructors can quantify individual student contribution both in timing and the number of characters (words) contributed [36]. For example, in Figure 27.6, the frequency and quantity of individual contributions by each member are displayed for two separate practical groups that completed the same task on different days in a second-level physical chemistry laboratory in an Australian university (G.A. Lawrie and L. Grøndahl, unpublished data). As part of the assessment, students were required to engage in an online discussion and interpretation of their qualitative observations of an inquiry experiment to develop a molecular

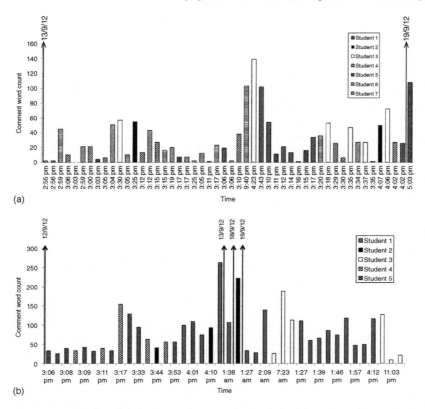

Figure 27.6 Display of the timing and extent of individual students' contributions to the group's discussion. (a) Group A (one final entry 6 days after lab session) and (b) group B (new entries continued beyond 1 week after the laboratory session).

explanation. The task assessment criteria required that each student contribute at least two entries to the discussion to encourage participation (Table 27.2).

When comparing these two groups, the difference in engagement becomes very evident to the instructor through the timing and extent of their contributions and collaboration [29]. A group that complied with the criteria is shown in Figure 27.6a, and there is a strong interactive pattern of participation by all individual group members. For a second group (Figure 27.6b), the discussion exchanges are initially only between two students in the group. A third student within this group (Student 1), who had engaged in discussion late but had met the criteria, continued to monitor the Wiki discussion after the deadline had passed on 12/9/12. Student 1 interacted with student 3, who had only commenced his or her contribution to their group discussion 1 week after the experimental session.

The collection of empirical data through the Wiki provides one measure of student engagement in terms of activity and participation; another is to directly compare the outcomes for students who complete a collaborative task in a Wiki with

Table 27.2 Example of assessment criteria that encourages students to participate in a collaborative discussion of observations made during an inquiry experiment.

Marks	3	2	1
Participation in online discussion on Wiki	A minimum of two comments (at least one sentence each) has been submitted to the Wiki and contributions to the discussion in the Wiki task evident	A minimum of two comments has been submitted to the Wiki and participation in the Wiki online discussion is evident	One comment (at least one sentence) has been submitted to the Wiki for each of the stations
Scientific contribution	Contribution to establishing the link between macroscopic observations and molecular-level processes made in the laboratory for each of the stations Evidence of *significant* contribution to the development of the overall understanding of underpinning thermodynamic principles and chemical concepts	Contribution to establishing the link between macroscopic observations and molecular-level processes made in the laboratory for each of the stations Evidence of a *satisfactory* contribution to the development of the overall understanding of underpinning thermodynamic principles and chemical concepts	Contribution to establishing the link between macroscopic observations and molecular-level processes made in the laboratory for each of the stations *Limited* contribution to the development of the overall understanding of underpinning thermodynamic principles and chemical concepts

those who complete the same task as individuals [37]. This type of research has not yet been considered for the application of Wikis in the context of science learning environments and would represent an interesting study.

27.6.2
Categorizing the Level of Individual Student Understanding

As students generate explanations of experimental data, the level of their understanding can be explored by characterizing their contributions to a collaborative

Comparing versions of Experiment 4 Post-lab
Showing changes between May 17, 2011 at 8:20:57 am (crossed-out) and May 17, 2011 at 9:33:33 am (underlined)

> ~~Purification~~Experiment 4 Pre-lab Purification of Silica Particles
> ~~12 Falcon tubes containing suspended silica particles were prepared for the deposition of poly-electrolyte.~~
> ~~6x tubes of150.0nm particles~~
> ~~6x tubes of570.0nm particles~~
> ~~Of these above mentioned tubes, 3x tubes of 150.0nm particles& 1x tube of570.0nm particleswere found to be contaminated by an unknown substance, which caused irreversible agglomeration of the particles. This led to the omission of the affected tubes from further participation in the experiment.~~
> ~~(MOVED TO IN-LAB)~~
> The purification process ~~didn't occur without hiccup, however at then end of the lab~~occurred with a few problems, but we were ~~able~~stillable to achieve the silica
> The major issue we had was the contamination of 4 of our samples. We believe the contamination occurred by the tubes falling over in the sonication water bath. Some of the lids may not have been fastened tightly enough and we suspect that some unknown chemical has entered the tubes and reacted with the silica particles.
> ~~Even with~~Tubes of material left after contamination:
> 3x 150.0nm particles
> 5x 570.0nm particles
> However, it is likely thatour experiment will proceed as per normal, as no additional changes are neccessary to the ~~loss of 30%~~overall plan of our ~~sample, we believe there is enough silica remaining~~experiment.To ensure that no additional samples will be contaminated, care must be taken to ~~continue~~check that the ~~experiment as we had initially planned it. The only issue now is whether or not time permits.~~lids of the tubes are tight and secure.
> Preparation of the Poly-Electrolytes
> (not finished yet)

Figure 27.7 Screen capture of an example of a Wiki student laboratory notebook page where the history function reveals the nature of edits comparing two time points. In this case, correction of language between students to develop a more scientific description is shown.

Wiki. The history function makes comparison between time points and individual student edits on each page possible: an example of the comparison is provided in Figure 27.7. In this example, the first student has written a reflective entry regarding the in-lab activities, and then a second student has edited these to attempt a more scientific communication of their processes.

Detailed evaluation using the history function can be a time-consuming process, which may be of more interest to the educational researcher or a practitioner who wishes to evidence the efficacy of their teaching initiative, rather than an instructor simply implementing the initiative in their course. Indeed, currently it is not practical to apply these analytics to very large class enrolments, but the pace with which tools for analysis of online text are being developed is so rapid that it is feasible that suitable tools will be realized in the very near future.

An alternative to the history function is to export the student entries from the Wiki and manually assign categories. An example is shown in Figure 27.8, where students in groups of five to seven have contributed to a discussion of their observations made synchronously during an inquiry laboratory activity and with the opportunity to continue asynchronously for 1 week after the session (described in Section 27.5.1 (G.A. Lawrie and L. Grøndahl, unpublished data)). The instructor

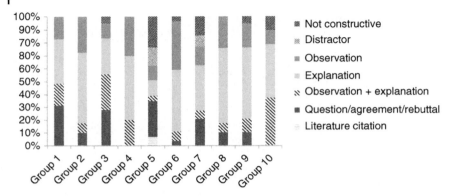

Figure 27.8 Categorization of the type of individual student entries in a group discussion forum.

can compare the patterns and types of individual contributions between groups across the course, so that disruptive behavior can be easily identified (as observed for group 5 in Figure 27.8). Students were required to develop a molecular-level explanation for their macroscopic observations, and the most common strategy was to make a statement of their observations followed by an attempt at an explanations. Higher level explanations of their observations were developed by most groups, but Group 5 clearly failed to engage in the task appropriately, with only 50% of entries contributing constructively to the discussion (these entries were of a high standard). Several individual group members submitted distracting, irrelevant, and nonconstructive comments, representing the remaining 50%, which disrupted the flow of the collaborative process (perhaps attributable as Wiki vandalism [36]).

27.7
Conclusion

It is clear that Wikis have the potential to add a useful dimension to an instructor's "pedagogy 2.0" toolbox as a platform to gain deeper insight into students' experimental and collaborative processes during inquiry laboratory activities in particular. This tool also has the potential to enhance student-centered activities by extending interactions between students, tutors, and instructors from the face-to-face laboratory sessions into a blended environment. Key affordances of Wikis in supporting assessment of laboratory learning that have been considered using examples (G.A. Lawrie and L. Grøndahl, unpublished data) [25, 32–34] in this chapter include the following:

- The role of group discussion to construct a shared understanding of experimental observations related to the underlying chemical concepts. The additional

benefit is individual students' acquisition and application of correct language and terminology;
- Collaborative laboratory record-keeping in the genre of a virtual laboratory notebook where experimental design processes, observations, collation, and processing of data and display of results are co-constructed;
- Collectively constructed written products such as an experimental discussion or laboratory report in which experimental outcomes are evaluated and described through collaborative authorship, editing, and peer review;
- Enhanced relationships with tutors through discussion for enabling guidance, Socratic questioning, and encouragement of groups of students to reflect on their processes and data.

While the Wiki has been considered in the context of laboratory learning in this chapter, the affordances and limitations of this technology tool are widely applicable to other chemistry education contexts and can be translated into the learning environments of other science disciplines. There are numerous studies that report enhanced learning outcomes from use of Wikis as a pedagogical tool, particularly in supporting learning in collaborative group activities [2, 26–30, 37]. While the affordances are evident, there are always limitations and these include nonparticipation by individual students due to social factors, technology literacy reasons [38], disruptive behaviors [36], or simply task avoidance. These can be addressed through the careful design of task instructions and assessment coupled with active monitoring by instructors or tutors and intervention strategies during tasks.

References

1. Jimoyiannis, A., Tsiotakis, P., Roussinos, D., and Siorenta, A. (2013) Preparing teachers to integrate Web 2.0 in school practice: toward a framework for pedagogy 2.0. *Australas. J. Educ. Technol.*, **29**, 248–267.
2. Su, F. and Beaumont, C. (2010) Evaluating the use of a wiki for collaborative learning. *Innovations Educ. Teach. Int.*, **47**, 417–431.
3. Augar, N., Raitman, R., and Zhou, W. (2004) Teaching and learning online with wikis. Beyond the Comfort Zone: Proceedings of the 21st ASCILITE Conference (eds R. Atkinson, C. McBeath, D. Jonas-Dwyer and R. Phillips), 95–104, *http://www.ascilite.org.au/conferences/perth04/procs/augar.html* (accessed 31 July 2014).
4. Web URLs for wiki sites: Wikispaces *https://www.wikispaces.com/*; PBWorks *http://www.pbworks.com/*; ChemWiki *http://chemwiki.ucdavis.edu/*; LearnChemistry Wiki *http://www.rsc.org/learn-chemistry/wiki/Main_Page* (accessed 31 July 2014).
5. National Research Council (2006) *America's Lab Report: Investigations in High School Science. Committee on High School Science Laboratories: Role and Vision* (eds S.R. Singer, M.L. Hilton, and H.A. Schweingruber), Board on Science Education, Center for Education, Division of Behavioral and Social Sciences and Education, The National Academies Press, Washington, DC. pp. 76–77.
6. Hofstein, A. and Lunetta, V.N. (2004) The laboratory in science education. Foundations for the twenty-first century. *Sci. Educ.*, **88**, 28–54.
7. Kirschner, P.A. and Meester, M.A.M. (1988) The laboratory in higher science education: problems, premises and objectives. *High. Educ.*, **17**, 81–98.

8. Johnstone, A.H. and Al-Shuaili, A. (2001) Learning in the laboratory; some thoughts from literature. *Univ. Chem. Educ.*, **5**, 42–51.
9. Reid, N. and Shah, I. (2007) The role of laboratory work in university chemistry. *Chem. Educ. Res. Pract.*, **8**, 172–185.
10. Prades, A. and Rodriguez Espinar, S. (2010) Laboratory assessment in chemistry: an analysis of the adequacy of the assessment process. *Assess. Eval. High. Educ.*, **35**, 449–461.
11. King, P.E. and Behnke, R.R. (2005) Problems associated with evaluating student performance in groups. *Coll. Teach.*, **53**, 57–61.
12. Kipnis, M. and Hofstein, A. (2008) The inquiry laboratory as a source for development of metacognitive skills. *Int. J. Sci. Math. Educ.*, **6**, 601–627.
13. Szteinberg, G.A. and Weaver, G.C. (2013) Participants' reflections two and three years after an introductory chemistry course-embedded research experience. *Chem. Educ. Res. Pract.*, **14**, 23–35.
14. Hunter, A.-B., Laursen, S.L., and Seymour, E. (2006) Becoming a scientist: the role of undergraduate research in students' cognitive, personal and professional development. *Sci. Educ.*, **91**, 36–74.
15. Brickman, P., Gormally, C., Armstrong, N., and Hallar, B. (2009) Effects of inquiry-based learning on students' science literacy skills and confidence. *Int. J. Sch. Teach. Learn.*, **3** (2), 1–22, http://academics.georgiasouthern.edu/ijsotl/v3n2.html (accessed 31 July 2014).
16. Poock, J.R., Burke, K.A., Greenbowe, T.J., and Hand, B.M. (2007) Using the science writing heuristic in the general chemistry laboratory to improve students' academic performance. *J. Chem. Educ.*, **84**, 1371–1379.
17. Akkus, R., Gunel, M., and Hand, B. (2007) Comparing an inquiry-based approach known as the science writing heuristic to traditional science teaching practices: are there differences? *Int. J. Sci. Educ.*, **29**, 1745–1765.
18. Latreille, H. and Chavance, M. (1971) An excellent laboratory notebook. *J. Chem. Educ.*, **48**, 846–847.
19. MacNeil, J. and Falconer, R. (2010) When learning the hard way makes learning easy: building better lab note-taking skills. *J. Chem. Educ.*, **87**, 703–704.
20. Eisenberg, A. (1982) Keeping a laboratory notebook. *J. Chem. Educ.*, **59**, 1045–1046.
21. Moskovitz, C. and Kellog, D. (2011) Inquiry-based writing in the laboratory course. *Science*, **332**, 919–920.
22. Hanauer, D.I. (2013) A genre analysis of student microbiology laboratory notebooks, in *Applied Linguistics and Literacies for STEM: Founding Concepts, Methodologies and Research Projects*, Chapter 5 (eds M.J. Curry and D.I. Hanauer), John Benjamins Publishing Company, Philadelphia, PA.
23. Conole, G. (2013) Designing for learning in an open world, in *Explorations in the Learning Sciences, Instructional Systems and Performance Technologies*, vol. **4**, Springer Science + Business Media LLC.
24. Laurillard, D. (2009) The pedagogical challenges to collaborative technologies. *Int. J. Comput.-Support. Collaborat. Learn.*, **4**, 5–20.
25. Lawrie, G., Grøndahl, L., Boman, S., and Andrews, T. (2014) Wiki lab notebooks: supporting student learning in collaborative inquiry laboratory experiments Submitted to. *J. Sci. Educ. Tech.*
26. Wheeler, S., Yeomans, P., and Wheeler, D. (2008) The good, the bad and the wiki: evaluating student-generated content for collaborative learning. *Br. J. Educ. Technol.*, **39**, 987–995.
27. Larusson, J.A. and Alterman, R. (2009) Wikis to support the "collaborative" part of collaborative learning. *Int. J. Comput.-Support. Collaborat. Learn.*, **4**, 371–402.
28. Witney, D. and Smallbone, T. (2011) Wiki work: can using wikis enhance student collaboration for group assignment tasks? *Innovations Educ. Teach. Int.*, **48**, 101–110.
29. Roussinos, D. and Jimoyiannis, A. (2013) Analysis of students' participation patterns and learning presence in a wiki-based project. *Educ. Media Int.*, **50** (4), 306–24.

30. Trentin, G. (2009) Using a wiki to evaluate individual contribution to a collaborative learning project. *J. Comput. Assist. Learn.*, **25**, 43–55.
31. University of Guam Image Sourced from Professor Christopher Lobban's Research Web Page, *http://university.uog.edu/botany/lobban/lobban.htm* (accessed May 2014).
32. Elliott, E.W. and Fraiman, A. (2010) Using Chem-Wiki to increase student collaboration through online lab reporting. *J. Chem. Educ.*, **87**, 54–56, (Reference 15: *http://orion.neiu.edu/~chemdept/Pfraiman.html* (accessed May 2014)).
33. Ge, C. (2012) Application of Wikis with scaffolding structure in laboratory reporting. *Am. J. Eng. Educ.*, **3**, 89–104.
34. Lo, H.-C. (2013) Design of online report writing based on constructive and cooperative learning for a course on traditional general physics experiments. *Educ. Technol. Soc.*, **16**, 380–391.
35. Ferguson, R. (2012) Learning analytics: drivers, developments and challenges. *Int. J. Technol. Enhanc. Learn.*, **4**, 304–317.
36. Meisher-Tal, H. and Gorsky, P. (2010) Wikis: what students do and do not do when writing collaboratively. *Open Learn.:J. Open, Distance e-Learn.*, **25**, 25–35.
37. Neumann, D.L. and Hood, M. (2009) The effects of using a wiki on student engagement and learning of report writing skills in a university statistics course. *Australas. J. Educ. Technol.*, **25**, 382–398.
38. Naismith, L., Leet, B.-H., and Pilkington, R.M. (2011) Collaborative learning with a wiki: differences in perceived usefulness in two contexts of use. *J. Comput. Assist. Learn.*, **27**, 228–242.

28
New Tools and Challenges for Chemical Education: Mobile Learning, Augmented Reality, and Distributed Cognition in the Dawn of the Social and Semantic Web

Harry E. Pence, Antony J. Williams, and Robert E. Belford

28.1
Introduction

To understand the impact the social and semantic Web will have on chemical education and the practice of science, it is imperative to look at the nature of the evolution of scientific communications. An underlying concept of this chapter is that late twentieth/early twenty-first century advances in ICTs (information and communication technologies) are changing the way our society shares, communicates, and manipulates information, and we are entering a new era of societal communications that will impact the practice of science and education. In fact, recent advances in ICTs have been so fast that, perhaps for the first time in human history, the youth tend to use ICTs with greater expertise than their parents' generation, a phenomenon that has come to be known as the *second-level digital divide* [1], and is of great relevance to educators. In this chapter, we identify some of the emergent technologies related to the practice of the chemical sciences, along with some of the associated cognitive issues of relevance to chemical educators as they grapple with this new information landscape and related educational paradigms.

The pre-Web document-centric world of scientific communications can be classified as the Gutenberg era, after Johannes Gutenberg who invented the "modern printing press," which in itself represented an ICT revolution. Prior to Gutenberg, each page of a document had to be individually created, making the process very expensive and labor intensive. By using movable typeset, Gutenberg could instantly generate multiple pages as the need arose, and this led to what could be called the *first era of mass communication* [2]. Today's peer-reviewed scientific publication paradigm evolved out of the constraints of these technologies, which, for reasons of economics and distribution, required validation prior to setting the type and printing the document.

Because the early Web appeared at a time when document-centric forms of communication were dominant, one can ask a logical question: are documents the only form of communication appropriate for Web-based ICTs? Actually, this is not the first time a technology changed the nature of communications and, as

Chemistry Education: Best Practices, Opportunities and Trends, First Edition.
Edited by Javier García-Martínez and Elena Serrano-Torregrosa.
© 2015 Wiley-VCH Verlag GmbH & Co. KGaA. Published 2015 by Wiley-VCH Verlag GmbH & Co. KGaA.

Paul Smart points out, early writing took the form of a continuous script type of structure that mimicked the oral traditions that the written text grew out of. It took time for the printed publication to evolve into a form most suited to the technologies of the printed word (in contrast to the spoken word) [3]. He goes on to extend this evolution of communications to the early document-centric Web, where "Web pages" mimic the printed publications that dominated scientific and educational communications in the latter part of the twentieth century. He suggests that the document-centric Web may not support extended cognitive processes in the manner a data-centric semantic Web could. These are the core issues this chapter grapples with, and the "dawn of the social and semantic Web" represents this adoption of human communications to the new ICTs that are becoming prevalent in the twenty-first century.

The first section of this chapter, "The Semantic Web and the Social Semantic Web" introduces the concept Web 3.0 (semantic Web) with a primary focus on technologies of relevance to the chemical sciences. The next section, "Mobile Devices in Chemical Education" presents an anytime/anywhere paradigm shift in digital communications from desktop computers to the more ubiquitous mobile computing devices, which is followed by an introduction to "Smartphone Applications for Chemistry." These mobile technologies enable augmented reality applications, which are discussed in "Teaching Chemistry in Virtual and Augmented Space." The topics then shift to material on "The Role of the Social Web: Crowdsourced Open Online Databases, Application Programming Interfaces (APIs), and Open Notebook Science." The next section, "Distributed Cognition, Cognitive Artifacts, and the Second-Level Digital Divide" steps back from the technologies and looks at the material from a social-situated learning cognitive sciences perspective and brings forth some of the issues facing teachers in the classroom. The final section of this chapter, "The Future of Chemical Education" attempts to provide guidance to contemporary educators in this new and evolving information landscape of the early twenty-first century.

28.2
The Semantic Web and the Social Semantic Web

The World Wide Web (WWW) was developed less than three decades ago by Berners-Lee at the CERN laboratories in Switzerland in December 1990 [4]. The addition of *graphical user interfaces* (GUIs) and *universal resource locators* (URLs) with *hypertext markup language* (HTML) created a more useful Web of Documents that could display information using different colors and fonts as well as including pictures and sounds. Soon, the first search engines came into existence, and Google was created in 1997. The next development during the time between 2000 and 2010 was the Web of Social Networks, which is sometimes called *Web 2.0* or the *Read/Write Web*.

Web 2.0 added improvements in information sharing, creativity, and collaboration by the formation of Web-based social communities that shared Blogs, Wikis, video, or images. The size of the Web continued to expand at an incredible pace throughout this time. Maurice de Kunder estimated that, as of September 2013, Google's Web *index* contained at least 1.55 billion *pages*, and since many Web pages are located behind firewalls that make them inaccessible to methods used for this count, the total size of the WWW is much larger.

Having this much information available is both a blessing and a curse. It is quite likely that almost anything that anyone would want to know is somewhere on the Web, but that does not guarantee that what is wanted will be easy to find. Search engines often return hundreds of thousands of hits for a simple search, and the results may be a mixture of sites that are relevant to the search as well as many others of little or no value. As early as 2001, Berners-Lee *et al.* were publicly describing an expansion of the WWW concept that would offer even greater capabilities than what already existed. At that time, most of the Web's content was document-centered, designed to be read by humans, and not in a format readily understood by computers. The Berners-Lee group proposed to add a computer-readable structure to the Web pages in order to make Web searches more focused and to allow software agents to carry out more sophisticated tasks for users. If this is to happen, the Web must consist of structured information that obeys sets of inference rules, which allow computers to cooperate better with humans. This idea was called the *Semantic Web* [5]. Many groups have been working toward accomplishing this goal, and recently there have been a number of steps in this direction.

One way to create the types of data relationships analogous to the Semantic Web is by the use of human evaluators who will create knowledge maps that show how different concepts are related. In 2012, Google announced on its official Blog that it had added a function called *Knowledge Graph* to its regular search results [6]. Knowledge Graph recognizes that searches are not just strings of words, but a particular concept, which is logically linked to other concepts. Now, many results of a Google search include a sidebar, called a *Knowledge Graph*, which suggests several different meanings for the search phrase as well as other searches that may be related. For example, when searching for Robert Woodward, an accompanying Knowledge Graph reveals that there are two well-known persons named Robert Woodward, the organic chemist and the investigative reporter of Deep Throat fame. The sidebar provides a short vita for each one that can be used to narrow the search (see Figure 28.1). Google implemented the Knowledge Web in 2010 based on a site called *Freebase*, a massive public database semantically structured to link together useful information. Freebase depends on individual volunteers from the Web community creating metadata that connects information in semantic ways. Freebase now contains over 500 million entities or objects, which are connected by over 3.5 billion links, and it continues to expand.

Robert Burns Woodward

Chemist

Robert Burns Woodward was an American organic chemist. He is considered by many to be the preeminent organic chemist of the twentieth century, having made many key contributions to the subject, especially ... Wikipedia

Born: April 10, 1917, Boston, MA

Died: July 8, 1979, Cambridge, MA

Education: Massachusetts Institute of Technology

Awards: Nobel Prize in Chemistry, Copley Medal, National Medal of Science for Physical Science

Get updates about Robert Burns Woodward   Keep me updated

People also search for

Roald Hoffmann | William von Eggers Doering | Elias James Corey | Gilbert Stork | Albert Eschenm...

Figure 28.1 Google knowledge graph for Robert Woodward.

The Royal Society of Chemistry's (RSC) Project Prospect integrates chemical names from its journal publications from 2008 to 2010 with a chemical database called *ChemSpider* (see later in this chapter for a discussion of ChemSpider) [7]. The names of compounds that have been linked are highlighted in the HTML version of the journal articles so that readers can click on the name to obtain data on the compound and also to link to articles referencing the same compound in RSC journals covering a decade of publications or to other related information sources. This initial work, focused on the period 2008–2010, provided experience both in the field of text-mining and in the validation of chemical names and terms to map to the appropriate associated data.

At present, the RSC is working on its DERA project, Data Enhancing the RSC Archive [8], which expands on their initial text-mining work and will extend into the extraction of physicochemical parameters and spectral data [9]. The work is investigating the extraction of spectral data in two forms: the conversion of textual forms of spectral data (e.g., NMR peak lists of chemical shifts) into spectral curve representations, and the conversion of spectral figures contained within papers and supplementary info into spectral curves [9]. It is estimated that there are tens of thousands of spectral curves, specifically NMR and IR data, contained within the RSC publication archive, and it is hoped that these can be converted into a more meaningful and searchable form. At the time of writing, over 100 000 articles in the RSC archive have already been marked up for chemical compounds, and work is progressing across the rest of the archive.

The ChemEd DL (digital library) WikiHyperGlossary project automates the mark-up of digital documents and Web pages, linking words within a user-specified glossary to database content associated with the glossary [10]. These can be canonical (non-editable) definitions, such as those of IUPAC glossaries, but also socially generated (Wiki-editable) definitions with embedded multimedia content targeting up to five levels of background knowledge [11]. If the word is a chemical, the WikiHyperGlossary associates an InChI (International Chemical Identifier) (discussed in detail below), which extracts chemical information through the ChemSpider API interfaces (see below). At *hyperglossary.org*, the development site for ChemEd DL WikiHyperGlossary, a molecular editor is also connected to a molecule's InChI that enables a semantic framework, where molecules in a paper can be changed into other molecules, whose physicochemical data can be obtained via the same ChemSpider API.

Although developments such as Knowledge Base, WikiHyperGlossary, and Project Prospect can be very valuable, they fall short of the original vision that the Semantic Web should accumulate computer-readable descriptive information, called *metadata*, which permits the unambiguous identification of objects on the Web. Tim Berners-Lee argues that, " ... a goal of the Web was that, if the interaction between person and hypertext could be so intuitive that the *machine-readable* information space gave an accurate representation of the state of people's thoughts, interaction, and work patterns, the machine analysis could become a very powerful management tool, seeing patterns in user's work and facilitating working together through the typical problems which beset the management of large organizations [12]." The computer-readable descriptive information permits the unambiguous identification and manipulation of "objects" on the Web by computers.

According to The Worldwide Web Consortium (W3C), the goal of the Semantic Web is

> ... to create a universal medium for the exchange of data. It is envisaged to smoothly interconnect personal information management, enterprise application integration, and the global sharing of commercial, scientific and cultural data. Facilities to put machine-understandable data on the Web are quickly becoming a high priority for organizations, individuals and communities. The Web can reach its full potential only if it becomes a place where data can be shared and processed by automated tools as well as by people. For the Web to scale, tomorrow's programs must be able to share and process data even when these programs have been designed totally independently [13].

These "objects" may refer to a piece of equipment, a chemical compound, a particular scientist, an analytical procedure, or any other kind of content. The metadata from these objects is then linked in such a way that it permits new kinds of questions to be asked. Just as URLs designate specific Web pages, the Semantic Web uses universal resource identifiers (URIs) to identify objects on the Web and their properties. In chemistry, this requires a precise way to identify chemical compounds. For example, The IUPAC InChI is a universal identifier based on the molecular structure of compounds [14] and provides a string of characters that encode a chemical structure. The key component of InChI-enabled communication is an Open Source software package that encodes a chemical structure into a string of letters and numbers called an *identifier*. The InChI is less arbitrary and less ambiguous than other commonly used compound identifiers, such as systematic chemical names, the Chemical Abstracts Service registry number (CAS number), or Simplified Molecular-Input Line-Entry System (SMILES) [15]. In the future, more links will be made based on chemical structures as publishers, chemical suppliers, database producers, and chemists embed InChIs into their documents, programs, and Internet resources. The URI allows information on an object to be combined from all over the Web with the confidence that the object described is identical, and this is a basic requirement for the Semantic Web. InChIs associated with URIs will enable a chemical semantic Web.

The Resource Description Framework (RDF) uses the URI to not only identify an entity on the Web but also to describe how that entity relates to other possible Web entities. It organizes this information into triples of subject, predicate, and object. For example, if the subject is a chemical compound, the predicate might be a particular property of that compound, and the object is the value of that property. It is also possible that the subject is a particular scientist, the predicate is a relationship, such as a coworker, and the object is another scientist. It might also correlate different names for the same compound, so that the subject is "isopropyl_alcohol," the predicate is "has systematic name," and the object is "propan-2-ol." A set of these triple structures that shows the interrelationships between a number of different types of Web entities is called an *RDF schema* [15]. This works for not only Web pages but also for videos, sounds, images, and texts. Since objects with the same URI are identical, this makes it possible to correlate information from different databases on the Web. According to Gruber, "A specification of a representational vocabulary for a shared domain of discourse – definitions of classes, relations, functions, and other objects – is called *ontology* [16]."

Currently, the main use for RDF is in combination with what are called *RSS feeds*, where RSS stands for *rich site summary* or more commonly *really simple syndication*. The RSS software is called a *feed reader*, *aggregator*, or *RSS reader* and uses standard Web formats to gather and publish frequently updated information (commonly called a "*Web feed*") from various types of Web sites, Blogs, videos, audios, news headlines, and Web pages. The RSS feed can include not only full or summarized text, but also useful metadata, such as reference information. The human user selects which feeds to subscribe to and, with an appropriate reader, can receive human-readable summaries. Pence and Pence have described the use of RSS feeds for teaching information literacy to undergraduate chemistry students [17].

The Semantic Web promises not just better searches but also an improved way to evaluate information from the Web. The RDF will provide information on the provenance of data, that is, the origin of a given piece of information and how it was obtained. Unfortunately, not all laboratories are equally trustworthy, nor is all equipment equally accurate. As Taylor *et al.* point out, "knowing who did what, where, and when allows the selection of the best available data for the task at hand and the verification of the validity of that result [15]." Normally, this type of verification requires careful examination by a trained expert in the particular field. Providing the information for the computer to make at least an initial assessment is more time-efficient and may well be more enlightening than assessing the validity of results on a case-by-case basis.

Taylor *et al.* have used Semantic Web technologies to improve computer access to chemical data by using URIs and an RDF framework to add metadata, which simplifies search and sharing for molecular information [15]. Figure 28.2 shows how a RDF triple structure can be used to remove ambiguity caused by the varying methods for identifying a particular compound. There are several projects in biology, geography, and chemistry that are attempting to apply the Semantic Web to science problems using extensible markup language (XML) to set up rules to describe data that is platform-independent and computer-readable [15]. The most popular of these efforts in chemistry is based on chemical markup language (CML), which describes molecules at the atomic level. CML is a data-file format that can represent complex information structures, and uses generic XML software to interface with relational or object-oriented databases. CML is based on the RDF triples that can be used for both conventional searches as well as for recognizing data relationships in chemical applications. For example, the RDF/RSS approach has been applied to chemical problems as described by a series of papers

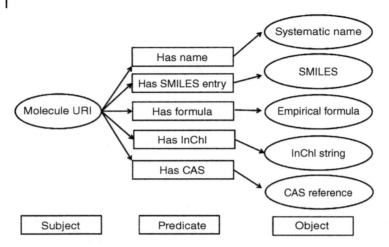

Figure 28.2 Use of triples to identify a specific molecule. (Adapted from Ref. [8].)

that attempt to apply RDF structure using CML [18]. (The reference is to the most recent of these papers, which is a point of access to previous work.)

Schoefeggera *et al.* explain that there are several shortcomings of semantic search systems [19]:

- Formal semantic search structures may not accurately correspond to each user's individual and subjective perception of a knowledge domain.
- Semantic Search engines often cannot keep up with exchanges in external environments and users' definitions.
- The way users view the semantics of a Web page may not correspond to the intention of the original authors.

The development of the WWW has not been limited to pages of information. There has also been an explosion in the number of social networking sites (SNSs) that are social, conversational, and highly participatory. These sites includes not only Blogs and Wikis, but also a broad spectrum of personal interactions occurring on "friending" sites such as Facebook, Orkut, LinkedIn, and Twitter; content-sharing sites such as YouTube, Flickr, Pinterest, and Stumbleupon; and gaming sites such as Everquest and World of Warcraft. These sites share many features in common, including creation of networks of friends, comment boards, recommendations, and media uploading. These SNS sites demonstrate that the Internet is not just about content but also about connecting people. Unfortunately, social networks are good at connecting people but often do not identify what interests the people in a group have in common. Breslin and Decker argue that SNSs need to develop object-oriented sociality, which would describe how people are connected in terms of their professional or avocational interests; that is, social networks need a semantic component [20]. They suggest that object-centered Web environments have come to situate and stabilize human selves and define individual identity just as much as communities or families did in the past.

At the same time, others are suggesting that one way to respond to the shortcomings of the Semantic Web is to depend more upon human interaction, not just computer readability, by having independent users assign keywords (or tags) to Web resources of interest. These tags create a new form of classification, which is not based upon predefined categories, such as an ontology, which have been determined by some central authority. The new form of labeling Web pages based on user-created tags is called a *folksonomy* [19]. It is suggested that folksonomies are more closely aligned with the individual users and more responsive to changes in the user's perception of the information environment.

This idea that social interactions on the Web, such as folksonomies, should be combined with the machine-readable data characteristic of the Semantic Web is called the *Social Semantic Web* [21]. Thus, two largely independent developments on the WWW, namely the Semantic Web and the Social Web, are merging to create what is called the *Social Semantic Web*. The goal is to create an object-centered social Web, which would be a more semantically rich form of knowledge representation based on both human interactions and computer-readable connections. As the online interactions become increasingly similar to what interests people in everyday life, Web interrelations should begin to reflect more accurately what happens in the real world [20]. This requires the expansion of the entity connections envisioned in the Semantic Web to include the Web of human connections, which is a critical component of knowledge.

The Social Web offers a way to connect people and their interests, which is analogous to the manner in which data is described in RDF. The value and relevance of a piece of information can be measured by the reputation and the number of users who tag it. There are already several Web sites that are designed to forge these new connections, including NEPOMUK – The Social Semantic Desktop [22], the Friend-of-a-Friend Project [23], and the SemanticMediaWiki [24]. Object-centered networks like these can significantly enrich the Semantic Web by connecting human or community activity with specific objects in the semantic space. For example, if a group knows that John Smith is working on a particular class of compounds that is also of interest to them, it would be possible to keep abreast of what Smith is doing with a query, "What content has #JohnSmith interacted with recently?"

The Semantically-Interlinked Online Communities Initiative [25] represents one attempt to integrate the personal profiles available from the Friend-of-a-Friend project with the RDF knowledge organizing formats of the Simple Knowledge Organization System [26] to link people and activities identified on Blogs, discussion boards, and other postings. This initiative is particularly interesting because it plans to expand beyond text-based activities to include the audio and video content, which is becoming more and more popular on the Web.

One of the most ambitious projects intended to combine the ideas of the Social and Semantic Webs for chemistry is the Open Pharmacological Concept Triple Store (Open PHACTS) [27]. This project is a European Innovative Medicines Initiative, bringing together pharmaceutical companies, academics, publishers, and commercial and open database providers in the chemistry and life sciences domains. The intention of the project is to utilize semantic Web technologies to gather together, integrate, and deliver pharmacological and physicochemical data from public domain chemistry and biology databases via a semantic Web platform. The project has already produced the first beta release as the Open PHACTS explorer [28], providing access to the associated data of over 500 million triples. This is done via an Open API that has been used to deliver a series of other tools including visualization and querying tools tuned to the needs of the various user communities [29]. (The API concept is discussed later in this chapter.) This approach demonstrates directly the value of providing access to open APIs, as the community has the ability to take advantage of the data assembled during the project to deliver the necessary solutions for the user base.

The primary challenge in modern drug design is the complexity of the aggregated chemical and biological data. Increasingly, public-domain data from protein databases, genetic data banks, and chemical screening assay datasets are used by the pharmaceutical industry as part of their informatics datasets. The majority of current scientific advances therefore result from collective and international collaborative efforts. In addition to the authors and creators of datasets, other contributors and consumers of scientific information can be mobilized as curators, ranging from students to medical practitioners to patients. Such a *crowdsourced* approach is, of course, a valuable approach to the creation of knowledge, as demonstrated by the success of Wikipedia. One of the most important requirements to enable such efforts is that the data and information generated are preserved in a stable, unambiguous, trustworthy, and computer-readable state. Semantic Web technologies provide open standards that simplify this process, and can significantly contribute to data and information interoperability. This type of approach to building open APIs (see below) and datasets of value to the educational community would provide the necessary data that developers need to deliver appropriate applications to the users that are built on the data. Thus, Open PHACTS is intended to combine Semantic and Social Web components to facilitate improvements in drug discovery in academia and industry, while supporting both open innovation and in-house nonpublic drug discovery research.

28.3
Mobile Devices in Chemical Education

Until recently, it has been assumed that when the Semantic Web became a reality it would be accessed primarily by means of a Web browser on a desktop or laptop computer. Now this expectation is changing rapidly because of the popularity of mobile devices, especially smartphones. A *smartphone* is generally described as a mobile phone that has a complete operating system available for application developers and a Web browser that creates a viable alternative for accessing the Web. Because mobile devices, like smartphones, tablets, and mini tablets, are portable and are becoming ubiquitous, users can access the Web at any time or place. In addition, the internal navigation functions of the phone provide copious information about the user that can be incorporated into the Semantic Web to make it more personal. This combination creates a virtual Personal Information Space that can have an impact on society, including chemical education, which may be even greater than the changes brought about by the personal computer.

Mobile devices are frequently banned in classrooms where the faculty often view them as a distraction, even though research is showing they are becoming an integral part of the lives for many students. For example, the Speak Up Research Project organized by Project Tomorrow from 2003 to 2013 surveyed over 3 million K-12 students, teachers, and parents in the United States, Australia, and Canada to find their views on the use of technology within education [30]. What they have consistently found was "that digital learners wish that their school based education experiences more closely replicated how they are using technology outside of schools." Table 28.1 (used by permission from Project Tomorrow) shows the pervasive use of smartphones by high school students in this survey.

Mobile devices are moving personal computing from the traditional desktop paradigm to a new ubiquitous environment where information is always readily available. College students have enthusiastically adopted smartphones and are using them almost constantly [31]. As of May 2013, the Pew Research Center reports that 91% of American adults have a cell phone. They further report that 56% of American adults have a smartphone, and of special interest to educators is

Table 28.1 Students' use of emerging technologies to self-direct and support schoolwork.

	Grade 6 students (%)	Grade 9 students (%)	Grade 12 students (%)
Texting with classmates about assignments	39	65	67
Using Facebook to collaborate with classmates on a school project	19	35	40
Taking photos of school assignments of materials using my mobile device	11	28	35
Watch a video I find online to help with homework	29	30	33
Using Twitter to communicate or to follow others	7	20	25
Communicate with classmates using a webcam Skype or online chat	20	29	26
Using a mobile app to keep school work organized	15	24	25
Texting with my teacher	7	11	20

© Project Tomorrow 2013.

that 80% of Americans 18–29 years of age have a smartphone [32]. This is affecting personal behavior. Many young people strongly prefer texting, instead of email, and a recent summary of statistics about mobile usage indicates that in 2013 more than 90% of all humans who took a picture did so using a smartphone camera rather than traditional film cameras [33]. The camera on a smartphone can be used to save and share information. Many students have enthusiastically adapted to using their cell phone camera for tasks such as capturing diagrams from the whiteboard, making videos of demonstrations, and quickly making copies of a classmates notes to compensate for a class absence.

Banning the use of cell phones in class is by no means the best option for teachers who wish to make sure that students use their cell phones for educational purposes rather than spending their class time texting or Web browsing. Howard Rheingold, an early pioneer in online communities who is credited with originating the term virtual community, argues that students have too little control over where their devices lead their thoughts and argues it is important to help them learn to manage their attention [34]. He suggests that students recognize that multitasking during class with their devices almost always means giving less attention to the teacher and other students, a situation Rheingold calls *continuous partial attention*. He is using classroom exercises to help his students focus their attention, such as designating some portions of class as "technology on" and other times as "technology off." This helps the students to use their technology constructively and to recognize that there are some times when technology is not appropriate. Of course, this means that the instructor must make sure that students are given constructive ways to use their phones when the technology is permitted.

On the positive side, Warschauer [35] has studied the results of using laptop computers in classrooms, and found that this produced the following:

1) More just-in-time learning
2) More autonomous, individualized learning
3) A greater ease of conducting research
4) More empirical investigation
5) More in-depth learning.

Warschauer's article was based on laptop use in middle and high school classes, and it seems likely that these would be equally applicable for a college classroom where smartphones were in use.

Successful mobile learning requires teachers to find ways to integrate smartphones into the educational process while minimizing the potential for distraction. Banning smartphones is like teaching a class in the library and telling students that they are forbidden to use any of the books. Because smartphones are so portable and have become such an integral part of the lives of young people that they represent a constantly available access point for all of the information on the WWW. Ramsey Musallam, AP Chemistry teacher at Sacred Heart Cathedral Preparatory in San Francisco, uses cell phones to communicate with his students. During class, he enhances the day's material by sending relevant questions, quizzes, polls, and other material to their cell phones [36]. Thus, he diminishes their social use of the phone by integrating their device into the academic experience. It should also be possible to give the students short research questions to answer using their phones. In classes where not all of the students have access to smartphones, it would be relatively easy to form small groups that share a phone. This would, of course, require some course redesign, but anyone who thinks that teaching in the twenty-first century will look exactly like the practice 20 years ago is going to have a rude awakening.

Another way to provide more individualized learning is by means of podcast[1] on chemical topics. Abilene Christian University has issued iPhones or iPod Touches to all faculty and students in order to gain experience with mobile learning. Powell did her doctoral research partially on the use of podcasts designed for smartphones for general chemistry laboratory instruction [37]. She is currently working with Professor Sutherlin, both presently at Abilene Christian University, to expand this type of podcast to courses in biochemistry and general science for preservice teachers. A high school teacher, Laura McDonald, is using cell phones as an alternative to personal response systems (often called *clickers*) in her classroom [38]. Clearly, instructors are just beginning to explore the new ways that smartphones can be used to supplement traditional classroom teaching.

Henrik Berggren, CEO and cofounder of a company that markets an e-book reader, reports that people with smartphones spend more time reading per book and read books more often with their smartphones than e-book readers or tablets [39]. One factor that may help explain this unexpected popularity of reading on a smartphone is that most users carry their devices with them everywhere, which may also suggest something about the changing character of the science classroom. Sparrow *et al.* have pointed out that most people now have constant access to information because so much of what they want to know is available online [40]. As a result, digital devices serve as an external memory which helps to keep track of the large amount of information that is used on a day-to-day basis. It seems probable that the ubiquitous smartphones will amplify this trend. This should impact science teaching in two different ways. First, traditional teaching methods that emphasize memorization are based on practices that are becoming more obsolete every day, and, second, students will need training in the best ways to take advantage of the new ways to access information.

When the semantic Web becomes a reality, it will combine with smartphone portability to create exciting capabilities for knowledge creation. Semantic technologies will improve search results based on an individual history of past searches and will suggest conceptual relationships that had not previously been considered. This will be done on a mobile device that is always at hand. This partnership will enable a new type of mental symbiosis, where the device does not simply support the thoughts of the user but recommends new directions that may lead to unexpected insights.

1) A **podcast** is a series of audio, video, or PD files subscribed to and downloaded or streamed online to a computer or mobile device.

28.4
Smartphone Applications for Chemistry[2]

The power of the smartphone does not just rest upon its portability, but also upon the profusion of applications that are available [41]. The online Web browser included on most smartphones is sufficient to search and run programs on the WWW, but even more powerful is the type of software application, called an *app* or a *mobile app*, that is available for smartphones, tablets, or other mobile devices. These apps are designed to do a specific job, such as give directions or find the location of some desired service. There are literally millions of apps available, and there is an expectation that whatever one might wish to do, there is "an app for that." It is helpful to separate a discussion of apps that are generally useful for teaching from those that are specific for chemistry. Regardless, most apps can be downloaded easily at no charge or purchased for much less that the cost of software for a laptop or desktop computer. Unfortunately, there are so many apps available that this discussion can only scratch the surface in either category.

A number of smartphone apps are generally useful in the classroom, such as those that allow a teacher to use a smartphone to link pictures of his or her students with class rosters, log observed data, capture notes from a whiteboard, scan documents, or do concept mapping [42]. Some of the most useful types of apps are those that allow documents to be saved or shared. Cloud-based programs, such as Dropbox and Sugarsync, allow a user to save or share a document on a smartphone or tablet (or even a desktop computer) instead of printing it or sending it by email. This can be used to coordinate multiple participants on a single project, to allow students to turn in their reports to an instructor more conveniently than by email or paper, or simply to save material that is of interest to be read later. Some apps, such as GoodReader or Evernote, permit a user to read files in the pdf format as well as make notes or add highlights that are incorporated into the document. One of the most useful sources of chemical information that is available for both the iPhone and the iPad is the ChemSpider app. This free app provides access to terabytes of data on physical and chemical properties, molecular structures, spectral data, synthetic methods, safety information, and systematic nomenclature for over 30 million unique chemical compounds sourced and linked out to over 500 separate data sources on the Web [43]. The ChemSpider database is discussed in more detail later in this chapter. There are quite a few apps available that are free or cost less than 10 dollars [44].

Many campuses have created apps that allow students and faculty to access the local library resources, either for immediate reading or to be saved and read later. Twiss-Brooks has created an excellent list of chemistry apps, including links to several publishers who have apps to connect to their journals [45]. Apps such as Chemfacts, the Elements StudyBuddy, ChemSpider, or Amino Acid Reference make reference material conveniently available when viewed on a portable device,

---

2) This topic is also covered in Chapter 27 of this book, Incorporating Apps for Teaching and Learning Chemistry.

such as a smartphone. Visualization of molecular structures for millions of substances is available with ChemMobi or the RCSB Protein Data Bank [46]. This means that in-class exercises can be based on journal articles and references that were previously available only if the class were physically happening in the library. The resulting learning environment should be supportive of more in-time and in-depth learning. Classroom instruction should look more like research and less like the rote transfer of information. As an example of how even chemistry gaming may be popularized, efforts have been made to ensure that the Spectral Game is optimized for the mobile interface by ensuring the usage of HTML5 spectral display widgets [47].

The plethora of available apps may make it difficult to find one that is appropriate for the job at hand, let alone finding the one that is the best. Two smartphone operating systems are currently widely popular, Android and Apple iOS, and some apps work on only one or the other, not both. Libman and Huang have done a very valuable review of some chemistry apps and offer "an objective opinion" on the usefulness of each. They classified apps in several different ways, including specific chemical disciplines (college chemistry classes), functionalities (what do they do), and target users (who will find them to be advantageous) [48]. They reviewed about 30 popular and mostly free apps, which can be used to support learning chemistry or may serve as reference or research tools. Table 28.2 from Libman and Huang classifies some popular apps by the typical chemistry classes where they may be beneficial. Of course, this does not completely solve the problem because

Table 28.2 Chemistry apps for the smartphone by chemistry classes [30].

Discipline	Android	iOS
General and inorganic	AtomDroid, Chemistry Cheat Sheets, Chemistry Mobile, Convert Pad, Molecular Viewer 3D, Periodic Droid, Periodic Table, W Chemistry Handbook	iElements, The Chemical Touch Lite
Organic	AtomDroid, Chemistry By Design[a], ChemDoodle Mobile[a], Molecular Viewer 3D, W Chemistry Handbook	Chemistry By Design[a], ChemSpider[a], ChemDoodle Mobile[a], Green Solvent, Molecules, Named Reactions, Organic Name Reactions
Analytical	Chemistry Helper, EMD PTE[a], PubChem Mobile, W Chemistry Handbook, ChemSpider[a]	EMD PTE[a], ChemMobi (by Accelrys Inc.), ChemSpider[a], GC Calc, Green Solvent, LC Calc, MolPrime Molecules
Physical	Chemistry Mobile, Convert Pad, W Chemistry Handbook	Insensitive (by Klaus Boldt)
BioChemistry	AtomDroid, ESmol, Molecular Viewer 3D, NDKmol, Promega[a]	Amino Acid Tutor, Buffer Calc, Promega, Genetic Code

a) Denotes apps that are available on both Android and iOS.

new apps are constantly being released and Libman and Huang's analysis does not include the apps for less popular smartphone operating systems, such as that from Microsoft.

There are also specialized apps that are probably more useful for advanced courses and for research than for general teaching. Ekins *et al.* have reported on a free Green Solvents mobile app for the Apple iOS phones [49]. This app accesses a solvent guide created by the American Chemical Society Green Chemistry Institute Pharmaceutical Roundtable. Such a service is particularly valuable for those working in green chemistry when essential information is locked up in sites that have limited availability because of pay walls. In a similar fashion, mobile devices are increasingly being used instead of laptops or desktops for drug discovery [50]. The use of mobile phone apps for specific tasks in the chemistry laboratory is still at a very early stage, but in the coming years developments in this area may be of considerable interest to many research chemists. Finally, several available apps are good sources for information about chemical safety hazards, and there is a Web page that lists a variety of apps that are related to industrial hygiene [51]. An effort to assemble a categorized list of scientific apps that can be supported by crowdsourced participation by the community has been the assembly of the SciMobileApps Wiki [52] by Williams [53]. They made use of the community's familiarity with the Media Wiki platform to encourage developers and users of mobile scientific apps to post their own articles and commentary. It is hoped that the list will continue to be updated and for further community participation as the listings are easier to navigate than the existing iTunes listings relative to segregation of scientific content.

If the future of smartphones in the classroom is difficult to predict, their use in research is even less clearly defined, but it seems likely that there will be a special place for smartphones in the chemistry laboratory as well as in the chemistry classroom. This is especially true as electronic lab notebooks (ELNs) make their way into the laboratory on tablets and smartphones. Early efforts in this area are already under way. One example is the Notelus iPad app [54], which also makes use of the ChemSpider platform for linking to chemistry data. An increasing number of commercial ELN applications are also developing mobile interface applications for scientists to work in the lab.

There seems to be little diminution in the rate at which new apps are being created, which means that more game-changing variations will be on the way soon, and that it will become increasingly difficult to find the best app for a given procedure in chemistry. There is, however, every indication that the increased power of apps will make them an integral part of the laboratory of the future.

28.5
Teaching Chemistry in a Virtual and Augmented Space

Virtual space refers to objects that exist only as bits and bytes in the memory of a computer rather than in real life. Perhaps the simplest way to use virtual space for teaching in an undergraduate course is to employ one of the commercial services, which allows a class to save files on an offsite computer. For example, Abrams has combined a storage facility called *DropBox* with the course management system called *Blackboard* to allow students to share research data for inquiry-driven environmental laboratory exercises [55]. He notes that a major drawback of this approach is that only one student at a time can enter data. At campuses where many students have Google accounts, it is possible to avoid this bottleneck by using one of the forms of Google Drive to store documents, spreadsheets, or slides [56]. Spaeth and Black have used Google Groups as a basis for cooperative learning in laboratory courses [57], Soulsby has reported the use of Google Groups to distribute NMR data to all members of a class [58], and Bennett and Pence have used Google Docs to collect student research results from undergraduate organic laboratories [59].

Milgram and Kishino describe a spectrum of possible computer interactions that range from traditional reality, through digitally enriched environments, to the physical world enhanced by virtual components, to true virtual worlds [60]. A virtual world is one in which the user, represented by a computer-generated image called an *avatar*, moves through a relatively realistic looking landscape, which also exists only in the computer memory. Examples of this would be the many Massively Multiplayer Online Role-playing Games (MMORPGs), such as World of Warcraft or virtual worlds, such as Second Life. The intermediate stage of physical world enhanced by virtual components would be a teleconference, where several people at different locations can view each other on television sets and exchange information using telecommunications. Augmented reality (AR) adds computer-generated information in the form of text, sound, video, or graphics to a live view of a real-world environment.

Several years ago, education seemed ready to rapidly expand in the virtual world called *Second Life* [61]. Although it is possible to participate in many of the activities of Second Life with a minimal outlay of funds [62], creating a permanent presence to offer courses requires a significant commitment of time and money. Despite this, a number of higher education organizations leased sites, and some began to offer courses almost a decade ago. This initial enthusiasm did not last however, and the combination of a steep learning curve to use the system, rising costs to lease virtual land for projects, and the financial stringencies faced by many colleges during the economic downturn of the late 2000s forced the termination of many of these efforts.

As seen in Figure 28.3, Second Life allows the creation of molecular models that appear to be life-sized in comparison with the *avatar* that represents the user. Simulations of reactions at the molecular level allow for a very intimate connection to the world at the atomic level. On the right in Figure 28.3 is a periodic table that is clickable so that the user, as his or her avatar, can explore elemental properties. It is also possible to develop games and simulations. Lang and Bradley have reviewed some of the ways in which a virtual world, like Second Life, can contribute to chemical education, and conclude, "The open, immersive, and highly visual three-dimensional platform combined with the ease of content creation and scripting

Figure 28.3 An avatar viewing a chemistry display in Second Life.

makes Second Life a valuable tool for conducting chemistry research, education, and collaboration [63]." More recently, Keeney-Kennicutt and coworkers have created a suite of virtual laboratories in Second Life that compare favorably with their real-life counterparts [64]. For some experiments, the students doing the virtual laboratory seemed to do even better than those who did the same procedure in real life. Many of the most common approaches to online learning are variations of the physical world enhanced by virtual components. Currently, a major topic of discussion in the online learning is the educational value of Massive Open Online Courses or MOOCs. The term MOOC was first used in 2008 to describe an online course called *Connectivism and Connective Knowledge* taught by Dave Cormier, Stephen Downes, and George Siemens [65], but the concept has now become both popular and controversial [66]. At this point, it is probably too early to predict what role MOOCs will play in the future of chemical education, but a number of different chemistry topics are being taught using this approach, ranging in level from introductory chemistry to statistical molecular thermodynamics.

Another innovative attempt to use the physical world enhanced by virtual components for teaching chemistry was the Physical Chemistry Online (PCOL) Consortium [67]. This effort was an attempt to support faculty who were teaching physical chemistry at small departments by sharing resources and supporting collaborative learning among the students at these institutions. The cooperating faculty, which eventually included 13 different institutions, developed 14 different modules that allowed students at geographically separated institutions to work together. Over 30 faculty and institutions were ultimately involved in designing and using PCOL materials. Over the life of the project, many different technologies were employed to support student collaboration, including email, online discussion boards, course management platforms, and Wikis [68].

AR provides computer-generated information in the form of text, sound, video, or graphics, which is added to a live view of a real-world environment. Typically, the image of the real world is provided by the camera on a mobile device such as a smartphone, and information overlay is determined from an app on the device that connects to the WWW and downloads the required information. AR can be used to deliver information in environments beyond the traditional classroom, such as field trips or at historic sites. Figure 28.4 shows a piece of paper with a QR code becoming a link to a video about the element calcium. The combination of the widespread availability of the WWW with the prevalent use of mobile devices is creating new opportunities to use AR for education and training.

The two main types of AR that are currently in common use are Marker-based AR and Markerless AR. Marker-based AR places a physical label, such as a quick response (QR) barcode, on an object or location. A smartphone, tablet, or other mobile device can either obtain data directly from the label or else use a URL from the label to accesses a Web site that contains more information. The label is often a two-dimensional (2D) barcode, which is a complicated pattern of squares, dots, hexagons, or other geometric patterns. These labels can represent much more information than the more familiar one-dimensional code of multiple straight lines. Figure 28.5 compares a one-dimensional (1D) barcode with a two-dimensional barcode. Currently the most common 2D coding pattern in the United States is called the QR code. A 2D barcode can encode several thousand characters, which is much more than a 1D barcode and more than enough to represent the URL of a Web site. A QR code can direct a phone to access a Web site, dial a phone number, or send a text message.

There are several advantages to using a 2D barcode to connect to a Web page. It is quicker than typing a long URL, especially on a cell phone, and, if the Web page is updated, the modification is available immediately to the users who click on the barcode. Markered applications are the easiest way to create a simple form of augmented reality. (Note: Some argue that the use of physical labels does not

Figure 28.4 Augmented reality on a smartphone showing a QR code connecting to an online video.

Figure 28.5 One-dimensional barcode (left) and two-dimensional barcode (right).

represent true augmented reality, but it is convenient to include the topic here because it is similar to markered AR in many ways.) An object that includes a 2D barcode is called a *smart object* because it can connect to digital information on the Web. Thus, 2D barcodes make objects in the real world clickable like a Web page. There are several free online sites available that allow an individual to create 2D barcodes for Web site addresses and various apps that work with all the major types of smartphones to open a Web link from a barcode [69]. Tolliver-Nigro gives more suggestions about using QR codes, including phone compatibilities for a number of free sites that will create these codes [70].

Markerless AR does not use a visible marker, like a QR code, but instead data from the mobile device's global positioning system (GPS), compass, accelerometer, and/or gyroscope determines the location and orientation of the device and connects to online information, which is to be superimposed on the image of the location. Several companies have created markerless phone apps that combine camera views with digital data, including Layer and Junaio. These companies use location-based AR to provide information about nearby schools, museums, restaurants, transportation, and health care to be added to the live camera image. Thus far, markerless AR seems to be used mostly by commercial firms for advertising and libraries [71] rather than by chemical educators.

Bujak *et al.* have surveyed the literature on AR learning and on this basis suggest several arguments to justify why it should be useful [72]. Although their article focuses on mathematics education, their conclusions seem to be equally applicable to chemistry. They suggest that an effective learning environment should minimize any extraneous tasks that may increase the cognitive load on the learner. There is some research that indicates that the combination of virtual learning objects with familiar physical spaces improves information recall. When acquiring procedural knowledge, it is helpful for students to see the instructions integrated with the materials being manipulated. And finally, Bujak *et al.* predict that AR should help to clarify the connection between abstract concepts, physical objects, and symbols [72]. As Johnstone pointed out many years ago, one of the fundamental problems in chemical instruction is the conflating of these three different types of knowledge [73].

In a recent review, Merchant *et al.* reported on a meta-analysis of 69 studies that examined the results of using games, simulations, and virtual worlds in K-12 and higher education settings. These authors found that, whereas all three techniques were generally effective in improving educational outcomes, the learning gains for games were about twice as large as simulations or virtual worlds [74]. They found that games and virtual worlds were equally effective for three kinds of learning outcomes: knowledge-based, abilities-based, or skill-based, but they concluded that simulations were less successful at developing skills. They explained this by pointing out that skill acquisition is a gradual process that requires more extensive practice. It would be reasonable to expect that students may spend more time in games and virtual worlds because a game may have multiple levels and the open structure of virtual worlds offers more opportunity for continued involvement. Thus, the lower results for simulations may simply reflect the fact that students are spending more time in that type of learning environment. They also identified several areas where more research is needed.

Benedict and Pence have reported on a number of different ways to use QR codes for teaching chemistry [75]. As long as most of the students in a class have smartphones or tablet computers, small groups can be created, each of which includes at least one student with a mobile device. This allows everyone to access videos during the lecture and repeat portions of the video at will rather than just seeing them on a large screen. The images can create a discussion focus around which the groups can share impressions and analysis. QR codes in the laboratory give students easy access to instructional videos showing how to use a piece of equipment or do an experiment. Adding QR codes to homework and class worksheets allows the students to see the procedures related to the calculations and easily review any activities that may be confusing. These researchers found that students liked using worksheets that included QR codes to access videos.

Historically, many chemistry departments have used hallway displays to provide supplemental chemical information and to kindle more interest in chemistry. AR creates an easy way to develop a situated (or locative) narrative that is tied to some specific physical or geographic location, such as a hallway or display case. A Web page is available that shows QR codes on a periodic table that link to videos and other information about each of the elements [76]. Figure 28.6 shows some of the QR codes from this table, which can be scanned with a mobile device that has the appropriate app. Young people love to use their mobile devices, and so providing QR codes that connect passersby to chemical information is an excellent advertisement for chemistry. See Figure 28.4 for an example of this use of augmented reality. Another possible basis for a hallway display would be the poster created by Bonifació, which uses QR codes to connect to information about the Nobel Prize winners in chemistry [77].

Vernier Technologies have recently introduced the LabQuest 2 "Connected Science System," where students can connect their mobile devices and smartphones directly to data streams resulting from a variety of scientific probes (pH meters, spectrometers, barometers, etc.). Students can manually type in an IP address or scan a QR code on the Labquest 2 and instantly view and download their data, or,

Figure 28.6 (a,b) Selected QR codes that link to videos and information about the elements (a portion of the complete periodic table which is available at http://www.flickr.com/photos/periodicvideos/5915143448/sizes/o/in/photostream/.

with the Labquest app, even analyze the results [78]. These devices can even generate their own 802.11 Wi-Fi signals and function as mobile hot spots, allowing access to the data in remote locations where there is no network.

Barcode-labeled smart objects can be very useful in the chemistry laboratory. One advantage of having a barcode linked to instructions for an instrument is that they could be updated on the Web whenever necessary. This means that, instead of having multiple versions of the instructions or not being able to find the instructions when they are needed, the latest version is always available and up to date. A bottle of chemicals with a barcode could redirect the user to a safety data sheet (SDS), formerly called a *Material Safety Data sheet*, or to other chemical and physical information. Even a simple page of laboratory instructions could now connect to a video showing how a procedure should be done. The U.S. National Library of Medicine offers a number of resources for health and safety information, including an app and Web pages optimized for mobile devices [79].

Although QR codes are the easiest way to introduce AR, any image can be used as a label as long as appropriate software is available to read that particular format. One way to do this is to design software which will respond to a set of labels that will create not just a single image but will combine labels on several different cards to create a 3D model of a molecule. For example, Fjeld and Voegtli have created a system based on a set of interactive tools which allows a student to choose elements from a booklet menu, which can be combined to create 3D molecular models [80]. Bringing together multiple element labels triggers the creation of a custom-designed molecular model with the composition defined by the labels. Art Olson of the Scripps Research Institute has used a 3D printer to create models of molecular structures, both targets, such as the HIV protease enzyme, and the drugs that bond to these targets. Olson added AR labels to each molecular model so that, when viewed on a computer with appropriate software, the resulting image will show how well the drugs fit their targets, and simultaneously test how well the two molecules are likely to bond on a chemical level [81]. Table Figure 28.7 is a screen capture from Olson's video.

Figure 28.7 Screen capture from Olson's video *http://www.youtube.com/watch?v=gZxK6j4JTHQ&feature=youtu.be*.

There are several commercial software systems available for purchase which will provide a set of AR labels that can create complex molecules. Sponholtz Productions has introduced markered AR cards that will simulate 3D molecular models when viewed with a computer webcam [82], and a company called DAQRI sells Elements 4D, a set of laser-etched wooden blocks that have a different chemical element on each side. When the webcam of a computer that has the accompanying augmented reality DAQRI software interacts with these blocks they can be used to create chemical reactions and new molecules in a virtual space [83]. There are several products like this at various stages of commercial development. Some of the better examples show how easy it is to add and subtract atoms to form molecules to create a chemical reaction [84] or to view a 3D sodium chloride lattice [85].

As noted earlier in this chapter, most college students now have access to smartphones, and young people increasingly view their phone as an integral part of their lives. This represents an opportunity to create a unique type of interactive chemistry learning experience. This suggests that augmented reality applications may soon become more common in the chemistry classroom and laboratory. Beyond the developments discussed previously, it would be a simple matter to supplement the images in chemistry textbooks by adding marker-based AR codes that could allow the student to view videos and animations. Much of chemistry is based on the 3D shape of molecules and anything that gives students a more complete understanding of the spatial relationships that determine properties and structure would be highly beneficial.

There are several interesting developments in the combination of AR with wearable devices, such as smart watches, but probably the most intriguing development for educators is AR glasses. In the future, phones may be replaced or supplemented by glasses that overlay an AR image onto any surface. There are already military applications of this technology, called *head-up displays (HUDs)*, being used for modern aircraft flight controls and even for combat soldiers. Although *Wired Magazine* argues that, "walking around with a camera mounted on the side of your face at all times makes you look dorky," it is important to make a distinction between general use and educational use [86]. Imagine the educational value if one could view a complex operation, such as a surgery or performing a chemical reaction, just as it was visible through the eyes of an expert performing the process. Google already has announced a version called *Google Glass*, which is currently available to a limited number of users but is soon expected be in commercial production [87]. It is possible for AR glasses to not only provide step-by-step instructions for doing a process but even highlight the locations and required tools, as shown in a video from BMW AG [88]. Although the current AR glasses clearly do not meet the required safety standards for laboratory goggles, this might well be a future development.

Although it is not included in Milgram's listing of interactions between the real and virtual worlds, one interaction that may become very important for chemists is the Internet of Things. The Internet of Things represents the expansion of the Web into the physical world by means of inexpensive microprocessor sensors that embed identification, sensing, and/or measuring capabilities into physical objects ranging from automobiles to appliances to scientific instruments. Atzori *et al.* suggest that the Internet of Things is, "a world-wide network of interconnected objects uniquely addressable, based on standard communication protocols [89]." They go on to point out that various groups may view this development from either a perspective of "Things," "the Internet," or "Semantic Connections," and so each group works on the basis of somewhat different perspectives.

Those who focus on the "thing" aspect emphasize inventorying and managing things, ranging from everyday objects to people, through some identification tags, such as QR codes, near-field communication (NFC) barcodes, or radio-frequency identification (RFID) chips. A lost object could be found easily by checking its position on a computer. This global visibility of objects could represent a revolutionary development, but it is by no means the whole story of the Internet of Things. Those who focus on the Internet facet of this concept argue that we are moving from anytime, anyplace connectivity for individual persons to similar connectivity for anything. Objects would use the Internet to communicate with humans and with each other in ways that are beneficial to human activity. Finally, the semantic vision of the Internet of Things recognizes that the number of objects involved would become very large, and so semantic solutions would be necessary to locate, interconnect, and organize information from this Web of objects.

The Internet of Things creates a new form of virtual space where physical objects have a digital representation that is discernable only through computer mediation. Gartner says the Internet of Things' installed base will grow much faster than the number of computers and tablets combined and will be 26 billion units by 2020, a 30-fold increase from today [90]. Using augmented reality connectivity, an automobile could remind its owner that it is low on gasoline, or a personal health monitor could remind a person that it was time to take medication. It is still too early to predict how chemical educators might use these new capabilities, but they promise to open up a new connection between the physical and virtual worlds.

28.6
The Role of the Social Web

The methods by which scientists from all over the world can contribute to an open Web of science have been led by the shift toward mass collaboration platforms of many types. Whether acknowledged or not, standard social networking tools such as Facebook and Twitter are mass collaboration platforms. Even with only 140 characters to exchange and distribute information, individuals can quickly share information, data, and knowledge by tweeting, retweeting, and sharing off the originating platform. People on Facebook can post about a scientific article, magazine article, or an observation in their own lab, and this information is broadcast almost instantly and can certainly cascade to an international audience very quickly. Twitter is already host to RealTimeChemistry [91], a community of scientists that post about what they are up to in real time, including sharing photos, videos, and so on.

Most people would probably identify the most successful crowdsourced collaborative platform for sharing knowledge as Wikipedia [92]. By providing a content management system for the masses, the world has been allowed to contribute large amounts of data, information, and knowledge in a manner that allows for instant contribution, validation, and engaging discussions regarding the "facts." As a result, an incredible wealth of contents hosted in Wikipedia can be used, reused, repurposed, and integrated very simply into other platforms [93]. There are well over 10 000 compound pages now on individual chemicals and drugs, and masses of details regarding chemical reactions, minerals, industrial processes, and individual scientists. Since the content is all openly licensed, it can be moved onto other sites for reuse. For example, the ScientistsDB platform [94], which was seeded with scientists from Wikipedia, and the Wikipedia compound pages are integrated with the ChemSpider database (see below) using calls against the open programming interface available for Wikipedia.

While Wikipedia is an ideal example of a mass collaboration platform, there are other examples more dedicated to the domain of chemistry that are worth mentioning. For example, while Wikipedia as a content management system does hold chemical structures, they are held in the form of images. However, scientists require data in formats that they can use in their software tools of choice. For example, chemical structures are best handled in connection table formats [95], and spectral data in data interchange standards such as JCAMP [96]. Popular databases used by chemists for sourcing chemical data include PubChem [97], ChemSpider (see below), DrugBank [98], and Learn Chemistry [99]. Spectral data can be accessed via systems such as NMRShiftDB [100] but only in the form of lists of assignments associated with the chemical structure. Interactive spectral data for NMR, IR, Raman, mass spectrometry, and others can also be accessed and downloaded for further manipulation from ChemSpider.

The ChemSpider platform actually holds a unique position as a chemistry database because users can contribute their own data, up to tens of thousands of individual compounds in a separate file if they wish, and share their own property and spectral data [101]. In this way ChemSpider is a crowdsourcing platform for the entire chemical community which also encourages scientists to participate in the validation of the data with interactive capabilities to add and remove synonyms, flag errors in the data, and provide other diverse feedback on any record. As a result, ChemSpider is a hybrid of a cheminformatics platform for the hosting of chemistry data, integrated to a content management system for diverse data types and delivering Wikipedia-like crowdsourced contributions. The platform offers powerful domain-specific searches but with chemical structure, substructure, and property searches, as well as access to a number of property prediction services. What makes ChemSpider extremely powerful is how it both utilizes and delivers access to its data and tools via a set of APIs.

An API defines the manner by which software components can integrate and interact with each other. Commonly, the API includes specifications to initiate specific routines, access defined data structures, trigger object classes, and so on. The vast majority of APIs of interest to the domain of cheminformatics and accessing chemistry data are based on specifications of remote calls against a particular resource. For Web-based resources, the API is generally a set of HTTP requests along with a definition of the structure of response messages in XML or, increasingly common, in the JavaScript Object Notation (JSON) format. Web-based APIs have moved from Simple Object Access Protocol (SOAP) based Web services to representational state transfer (REST) Web resources. Some of this shift is related to the Semantic Web and the RDF discussed earlier.

While many databases today allow downloading of data, it is the provision of APIs that greatly enable integration and utilization of the system. For example, the PubChem service provides access to an extremely powerful API, the Power User Gateway [102], to access its data and can support many hundreds of thousands of calls in a day. The ChemSpider database offers access to an expansive set of Web service calls, allowing users to source data from the Web site [103]. A ChemSpider record (e.g., example, see Figure 28.8) will contain the chemical structure itself and a set of previously predicted molecular properties, including algorithmically generated systematic names, InChI, SMILES, molecular formula, and molecular mass.

Each page also assembles a series of so-called infoboxes and can provide data such as links to chemical vendors and other external Web sites. While some of these are direct links with specific URLs written into the database, many of these infoboxes use calls against services to retrieve information on the fly. Some of these services are as follows:

- Wikipedia – the header to the article is retrieved on the fly and displayed. A link to the entire record for reading or editing is provided.

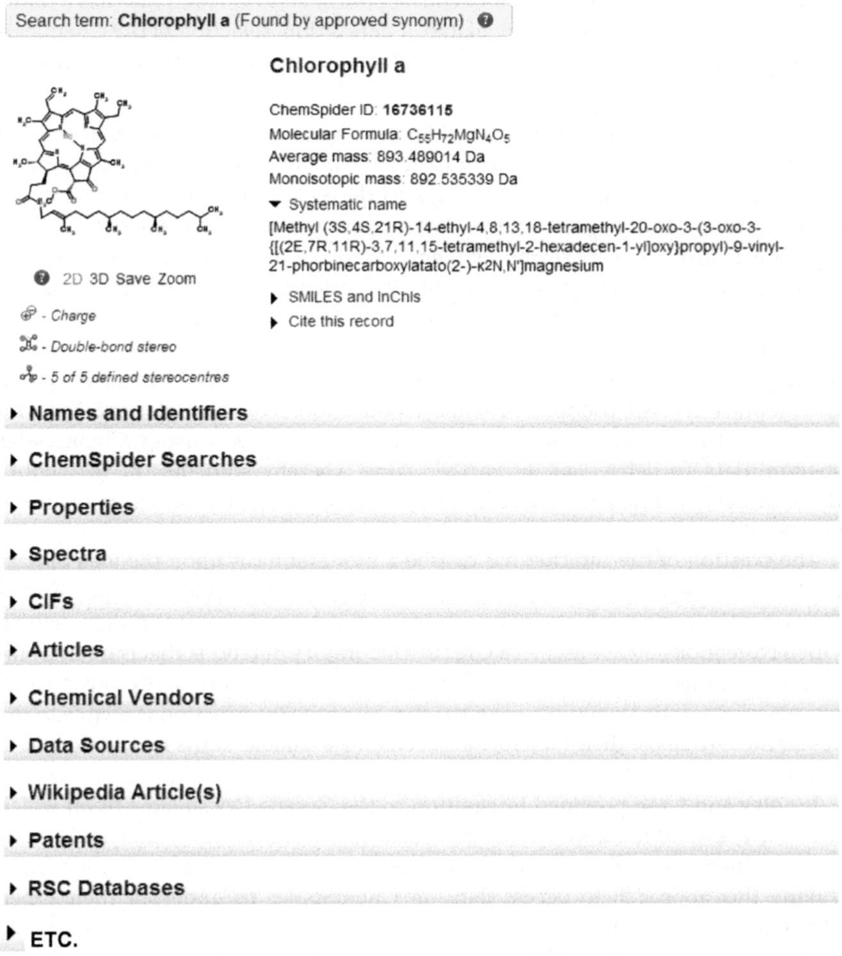

Search term: **Chlorophyll a** (Found by approved synonym)

Chlorophyll a

ChemSpider ID: **16736115**
Molecular Formula: $C_{55}H_{72}MgN_4O_5$
Average mass: 893.489014 Da
Monoisotopic mass: 892.535339 Da

▼ Systematic name
[Methyl (3S,4S,21R)-14-ethyl-4,8,13,18-tetramethyl-20-oxo-3-(3-oxo-3-{[(2E,7R,11R)-3,7,11,15-tetramethyl-2-hexadecen-1-yl]oxy}propyl)-9-vinyl-21-phorbinecarboxylatato(2-)-κ2N,N']magnesium

▶ SMILES and InChIs
▶ Cite this record

2D 3D Save Zoom
- Charge
- Double-bond stereo
- 5 of 5 defined stereocentres

▶ **Names and Identifiers**

▶ **ChemSpider Searches**

▶ **Properties**

▶ **Spectra**

▶ **CIFs**

▶ **Articles**

▶ **Chemical Vendors**

▶ **Data Sources**

▶ **Wikipedia Article(s)**

▶ **Patents**

▶ **RSC Databases**

▶ **ETC.**

Figure 28.8 Partial header for a typical ChemSpider record of cholesterol a.

- Google Patents, Google Books, and Google Scholar – all chemical names associated with a compound and validated to be of high quality are passed to the Google API and can provide the list of books, articles, and patents containing information about the chemical. Since the search is performed by the Google search engine, the results set is rank-ordered according to their algorithm, and therefore provides the most relevant data in the first few hits.
- PubMed – as with the Google service searches, the chemical names are used to search articles held in PubMed.
- RSC databases – chemical names are again used as the input source to perform searches against a series of RSC databases.

- Chemicalize [104] – while each chemical published to ChemSpider has a number of molecular properties calculated at deposition, selection of the Chemicalize tab pushes the molecular connection table out to the prediction services and calculates a series of other properties on the fly.

These examples demonstrate ChemSpider utilizing other services to enrich its platform. However, since ChemSpider itself provides Web services, these have been used by a number of groups to derive value from the ChemSpider data platform. Mass spectrometry instrument vendors have integrated their software platforms to perform searches based on monoisotopic mass and/or molecular formula. While it is possible to perform such searches across the entire ChemSpider database, commonly they focus their searches on specific data sources such as human metabolites. This approach has been used to develop structure identification approaches as described by Little *et al.* [105]. ChemSpider is also useful in chemical education. For example, the Spectral Game discussed earlier as a mobile app was developed using ChemSpider spectral data to deliver an interactive game whereby students try to identify the consistency between an NMR spectrum and one of a series of chemical compounds.

The creation of the Internet has caused a new emphasis upon the idea of having free and open access to scientific information. One aspect of this Open Access Science Movement is the creation of Open Notebook Science. Bradley defined Open Notebook Science as, " … a URL to a laboratory notebook that is freely available and indexed on common search engines. It does not necessarily have to look like a paper notebook but it is essential that all of the information available to the researchers to make their conclusions is equally available to the rest of the world [106]."

Bradley and Lang provided integration to the Google Docs platform on which they have assembled a series of molecular spreadsheets [107] that make calls against ChemSpider data as the basis of a series of molecular calculations [108]. Taking this approach to access data on ChemSpider to underpin their Open Notebook Science projects, the molecular structure handling, data curation, and ongoing updates of data can be handled using a centralized community platform for chemistry. They have also utilized the ChemSpider database and API as the basis of data curation, including experimental solubility validation and the development of a curated melting point dataset. Access to the data and related prediction algorithms have been made available as Web services [109], and the resulting API has been used to deliver mobile accessibility to the services [110].

While Open Notebook Science in academia is only just starting to expand in influence, it lags behind the shift toward ELNs in industry. The adoption of ELNs has been primarily due to the need to support intellectual property protection but with the additional needs of efficiency in research driven by access to searching of historical data not locked up in a standard paper notebook [111]. The largest installed bases of ELNs in the chemical and pharmaceutical industry are primarily commercial in nature, and there have been a number of corporate investments in developing in-house solutions. There are also a number of open source notebooks available to the community. The choice as to whether to purchase, reuse open source solutions, or build an electronic notebook in the academic community is likely driven primarily by cost constraints, with very few commercial ELN solutions installed to date. One specific example is at the University of Cambridge in the United Kingdom [112]; however, Elliott, an expert in the field of ELNs, cites very low adoption of ELNs in academia and comments that there is a "general unwillingness to change and adopt ELN very early on in teaching [111]."

Despite the slow adoption of ELNs in academia, there have been certain high-profile efforts to encourage adoption including the UK-based Dial-a-Molecule project, a project that utilized a commercial ELN during the initial phases [113]. The project is presently investigating the implementation of chemistry-specific widgets into the Open Source LabTrove [114] generic ELN in order to provide support for the chemistry laboratory. This includes working with the RSC to take advantage of their ChemSpider API to access chemical compound information in order to create a stoichiometry table and populate the database with analytical data to access through the ELN [115]. This approach once again shows the power of open APIs to allow software to integrate and utilize the value of the platforms, in this case for providing access to chemical compound data. This project is part of an overall effort presently active in the United Kingdom to provide access to a Chemical Database Service [116] and associated data repository whereby students will be able to store their lab data (chemical compounds, synthetic procedures, analytical data, and associated meta data) as either private, public, or embargoed data. It is to be expected that, with the increasing mandates of the funding agencies to release scientific data to the community, this and other services will ultimately result in the release of an increasing amount of open data to the community. While this is a UK-centric initiative, the expectations that academic scientists will utilize ELNs for the management of data appears to be increasing in momentum, and there is likely to be increasing adoption of such platforms in the foreseeable future.

28.7
Distributed Cognition, Cognitive Artifacts, and the Second Digital Divide

The first sections of this chapter provided an overview of current and evolving social and semantic Web technologies that would be of interest to early twenty-first century chemical educators, and this section will look at a variety of cognitive and pedagogic implications these have for chemical education and the practice of science. The focus will be on how these technologies are changing the fundamental (cognitive) artifacts used to represent, communicate, and manipulate (chemical) information, and how these impact the actual tasks performed in a cognitive process. One of the basic premises introduced in this section is that the problem-solving schema used by chemists in the practice of science and education are dependent on the cognitive artifacts used in this representation, communication, and manipulation of the information germane to the problem. The earlier sections of this chapter introduced numerous types of new digital-enabled cognitive artifacts, which may not only enable the development of new schema in the pursuit of science but appear to be more easily adopted by the youth than by practicing scientists and educators. This latter aspect has huge ramifications for chemical education and is resulting in a new challenge in the university classroom that some are calling the *second-level digital divide*.

The common denominator that transcends all of these technologies is that they are based on software agents mediating human cognitive processes. Implicit in this computer–human interaction is a digital artifact–human distribution of tasks associated with cognitive processes. This warrants a look at this material from the perspective of distributed cognitive science, where, as stated by Hollan *et al.*, "[the] cognitive process is delimited by the functional relationships among the elements that participate in it, rather than the spatial collocation of the elements [117]." They further clarify, "Whereas traditional views look for cognitive events in the manipulation of symbols inside individual actors, distributed cognition looks for a broader class of cognitive events and does not expect all such events to be encompassed by the skin or skull of an individual." It needs to be understood that the principles developed in distributed cognition are not based on human–computer interactions, but because the computer is external to the human, any cognitive task where a human interacts with a computer is a distributed cognitive process.

Holland recognizes three types of distributed cognitive processes:

1) Cognitive processes may be distributed across the members of a social group.
2) Cognitive processes may involve coordination between internal and external (material or environmental) structures.
3) Processes may be distributed through time in such a way that the products of earlier events can transform the nature of later events.

McGarry states that, in distributed cognition, cognitive processes are seen to "manifest (themselves) in propagations of representational states between human and nonhuman agents across a variety of media" and that "Distributed cognition is a recasting of cognitive science to account for the cognitive properties exhibited by associations of people and artifacts ... [118]" In investigating cognition–artifact relations, Heersmink states, "It is then better, for explanatory reasons, to see agent and artifact as one cognitive system with a distributed informational architecture [119]." Nemeth sees distributed cognition as the "shared awareness of goals, plans, and details that no one single individual grasps" and cognitive artifacts as objects that are part of distributed cognitive systems. To better understand the importance of distributed cognition to the earlier material in this chapter on the social and semantic Web, one needs to develop a deeper understanding of cognitive artifacts.

Humans are masters in the use of tools for survival as they adapt to the myriad of ever changing environments. Norman points out that "Many artifacts make us stronger or faster, or protect us from the elements or predators, or feed and clothe us. And many artifacts make us smarter, increasing cognitive capabilities, and making possible the modern intellectual world [120]." Norman uses the example of the notepad-and-pencil "to-do list" to introduce a cognitive artifact that functions as a memory aid, and then points out that the cognitive artifact does not extend memory. Instead, it changes the nature of the cognitive task, in this case from rote memorization to writing down what you want, bringing the notepad with you, and looking at it while shopping. This realization that artifacts change the nature of human cognitive tasks is worth restating. It is absolutely critical for chemical educators to understand that these artifacts are not enhancing established cognitive processes, but they are changing the tasks performed in the cognitive process. This is in essence the origin of the second-level digital divide, which will be discussed in the conclusion of this section.

Heersmink recently created a taxonomy of cognitive artifacts that considers technologies as "artifactual" (in contrast to techniques) and therefore limits the concept to physical objects [121]. This approach is limiting in that it does not allow semantic Web technologies to be artifactual, and so we will take a broader definition to include nonphysical objects and entities but follow Heersmink's taxonomy in not considering techniques to be artifacts. He distinguishes two genera of cognitive artifacts: representational (in both functional or information senses) and non-representational or ecological. Common representational artifacts include printed documents, lab notebooks, Web pages, online databases, ontologies, and functional entities such as calculators and keyboards. Nonrepresentational cognitive artifacts are ecological objects that could be used to enable a cognitive process, like the stars being used for maritime navigation, or consistently keeping car keys in a specified location (memory aid), or possibly Web browsers and API services that enable the digital landscape. Note that neither the stars nor the car keys were created with cognitive purposes in mind, but they acquire cognitive roles through agent-based behavioral patterns.

Although symbolic systems, such as the linguistic and mathematical symbols used in verbal and analytical reasoning, are typically not considered to be cognitive artifacts, they are intricately related to cognitive artifacts, as the actual nature of the symbol and its value as a representation are defined by the artifacts used to generate and communicate it. This has profound consequences for the practice of science, which today uses 2D print-based symbolic representations which are not unlike prehistoric cave drawings. For example, in predicting reaction mechanisms, the organic chemist uses arrow-pushing schema based on Lewis dot structure-based symbolic representations generated through the pen-and-paper types of cognitive artifacts. That is, the actual schema taught by the educator and used by the scientist is ultimately defined by both the functional artifact used to generate and the informational artifact used to communicate the symbolic representations that the schema is based on. (You could not have a Lewis dot structure-based schema without a Lewis dot structure.) This limits the growth of science to those representations supported by the artifacts employed in their cognitive processes.

As tomorrow's scientists employ new cognitive artifacts based on the types of semantic Web and digital technologies introduced in this chapter, one can expect new CML-type multidimensional symbolic representations to evolve, enabling higher levels of cognitive scaffolding that will generate new schemas for solving problems and understanding chemical processes. This connection of symbolic representations to cognitive artifacts gives new meaning to Lavoisier's preface in *Traité Elémentaire de Chimie* (translation by Robert Kerr – Edinburgh, 1790).

> " … we cannot improve the language of any science without at the same time improving the science itself; neither can we, on the other hand, improve a science, without improving the language or nomenclature which belongs to it [122]."

What are the issues associated with scientists adopting new cognitive artifacts in their pursuit of science? Jones and Nemeth used cognitive ethnography to investigate how cognitive artifacts were used in two work environments that are currently being influenced by digital technologies: acute healthcare and scientific research [123]. With respect to the contemporary practice of science, they identified the role of distributed cognition in the scientific discovery process and the current dependence of distributed cognition on a subset of cognitive artifacts, which they identify as information artifacts, specifically the laboratory notebook and the printed article. From these, scientists would extract "information objects" from multiple contexts that they would relate to the context of their work. Through this ethnographic study, it was noted that the work practices of the scientists contained "habituated stable behaviors" that could be traced to the technologies that were prevalent when they were students. In the context of their research, the two most common methods for information acquisition were digital: Google and PubMed. But when scientists cognitively used that information, they would print the documents deemed to contain "information objects" of importance to their research. Once printed, they would bring them to research meetings and discuss them, employing the distributed cognitive intellect of their research group.

In their work on cognitive artifacts in scientific work, Jones and Nemeth noted "a substantial gap between the current ecology of information resources and the potential for supporting cognitive demands of distributed discovery [123]." This behavior can be understood from an information processing theory perspective, where the scientists' "habituated stable behaviors" represented an automation of tasks, thus reducing their cognitive load and freeing their working memory to focus on the research needs at hand. Here lies one of the major challenges that educators need to be aware of: that is, unlike novices, experts tend to have "habituated stable behaviors" that can interfere with their adaption of new technologies and the cognitive artifacts associated with those technologies.

It might help to put this into a historic/evolutionary context to understand why in the age of the social Web this is so important for chemical educators to understand. Sterelny states that "intergenerational social learning profoundly shapes our minds and lives" and argues a sort of niche construction based on cultural inheritance (in contrast to genetic), resulting in "cognitive tools and [the] assembling [of] other informational resources that support and scaffold intelligent action [124]." She goes on to state, "The cognitive competence of generation $N+1$ individually and collectively depends on cognitive provisioning by generation N. The most critical, mind-and-brain-shaping environmental supports for cognition are these cumulatively built, collectively provided tools for thinking, tools that are provided to many or all of a generation by many or all of the previous generation." Here lies one of the primary challenges for contemporary educators, as today, due to the speed with which digital technologies are evolving, this is one of the few times in human history where the cognitive artifacts used by the youth (generation $N+1$) are unfamiliar to the majority of the elders (generation N).

That is, many educators, like the scientists Jones and Nemeth studied, have "habituated stable behaviors" based on the use of different cognitive artifacts than the ones their students use. It needs to be understood that this intergenerational social learning that Sterelny describes need not be from one individual of a parent's generation to child (as in the case of an apprentice learning a trade), although she argues that in primitive societies it was from parent to child. What is of interest is that today, with social Web technologies, the youth can learn to use new technologies from someone who lives anywhere in the world, including peers from their own generation. This can result in rapid intragenerational (in contrast to intergenerational) learning where through peer-to-peer interaction students gain mastery of digital cognitive artifacts their teachers are not only unfamiliar with, but in fact whose prior "expertise" interferes with their adaption of these technologies that their students (who lack the associated "habituated stable behaviors") find so easy to adopt. This leads to a special challenge for chemical educators, as many of the advances in cheminformatics will be more difficult for them to master than their students.

This phenomenon of the spread of cognitive artifacts through intragenerational learning is evidenced by the findings of research projects such as Project Tomorrow [125] that runs the Speak Up Research Project, where from 2003 to 2013 they surveyed over 3 million K-12 students, teachers, and parents in the United States, Australia, and Canada to find their views on the use of technology within education. What they have consistently measured was "that digital learners wish that their school based education experiences more closely replicated how they are using technology outside of schools [126]." Nagel quotes Julie Evans, the CEO of Project Tomorrow: "In fact, students tell us that they have to power down to go to school, and then, at the end of the school day, they power back up again – a real disconnect in the way students are viewing technology from the adults in their educational lives [127]."

Here is an excerpt (printed with permission) from their 2013 national report, "From Chalkboards to Tablets: The Emergence of the K-12 Digital Learner."

> Year after year, students in our focus groups remind us that their dissatisfaction with using technology at their school is not about the quantity or

quality of the equipment or resources; it is about the unsophisticated use of those tools by their teachers, which they believe is holding back their learning potential. The comparison of the students' perspectives on obstacles to technology use at school from 2003 to 2012 reflects this new reality which some are calling the second level digital divide.

It needs to be understood that the first cohorts of students that Project Tomorrow studied in 2003 are just now starting to enter our colleges and universities. These students are natively pioneering the use of cognitive artifacts based on social and semantic Web technologies that change the nature of the cognitive task from that employed by their elders. This disconnect between the students and their teachers is one of the primary origins of the second-level digital divide, and represents one of the greatest challenges facing chemical educators in the dawn of the social and semantic Web.

28.8
The Future of Chemical Education

As described in this chapter, chemists have many new tools for working with information. Crowdsourced online databases, such as ChemSpider, provide easy access to a wealth of chemical properties, especially when they are combined with access by smartphones. Projects such as Open PHACTs are already beginning to employ social semantic Web techniques to make available information that had formerly been isolated in multiple separate information silos. Electronic notebooks facilitate the sharing of data with multiple researchers who need not be in the same geographical area. There is clearly a need to integrate these new tools into the traditional chemical curriculum. In 1996, Holmes and Warden reported on the "first Web-centric course in Chemical Information, CI Studio, which was delivered on the World Wide Web and integrated electronic and print sources with Internet-accessible references [128]." Moy *et al.* have taught their students the value of crowdsourcing by having them edit Wikipedia entries [129]. Pence and Pence recommend that students be taught how to develop a personalized information management system, including the use of Really Simple Syndication (RSS), professional connections on Twitter and LinkedIn, and the use of social tagging [130].

Part of the solution may rest in looking at this problem from the perspective of distributed cognition, and recognizing that today's students bring not only enthusiasm in the use of technology to today's college classroom but also considerable skills in the use of digital cognitive artifacts, many of which are unfamiliar to their instructors. Students have learned how to share their skills with peers through social Web technologies that have enabled rapid intragenerational learning. This has created one of the greatest challenges for contemporary educators, namely the second-level digital divide. That is, students are using different digital technologies in their everyday cognitive activities than those that their teachers employ. Perhaps the most important thing for faculty to recognize is how cognitive artifacts change the actual tasks involved in cognitive processes, and how little instruction their students need to undertake complex educational tasks with these new technologies. As the students' expertise with these technologies often outstrips that of the professor, the professor's job becomes one of defining the problem and pointing the class toward possible solutions. Students are ready and eager to become creators, not just consumers, of knowledge. As expertise becomes more widely distributed throughout the classroom, professors must learn to share the job of knowledge creation with the students.

Many students today have the technical ability to create content and images and may be accustomed to doing this, but few of them understand how to use images effectively for learning. Many teachers may also need to revise their pedagogy. Jones has surveyed the recent research on how visualizations of molecular structure and dynamics are used in modern chemistry classes and concluded, "pedagogy designed for classroom instruction using books and paper needed to be redesigned for instruction using technology, particularly when interactive technologies were used [131]." She explains that until recently there has been insufficient interaction between cognitive psychologists studying how learners interact with visualizations and chemical educators who are using these tools in the classroom. Jones concludes that, "we still need to know how best to design and use instructional technology," and offers an extensive list of possible topics for educational applications of scientific visualizations. Unfortunately, the pace of change in technology seems to be far outstripping the developments in pedagogy. Since technology shows no tendency to slow down, there is certainly a need to move more rapidly in discovering the best ways to use these new tools for teaching.

One essential fact to recognize is that young people, and even those not so young, are thoroughly committed to the use of mobile devices. Google has recently published a study which found that people commonly consume media on more than one screen, television, smartphone, tablet, and personal computer, and often they work with several different screens simultaneously [132]. They report that, "smartphones are the backbone of our media interactions. They have the highest number of user interactions per day and serve as the most common starting point for interactions across multiple screens." Despite continuing reports that multitasking makes people less effective [133], the Google study finds that people feel more effective when using multiple screens. Chemistry instructors who attempt to fight this change in popular behavior will probably not be very successful. It is much better to attempt to redirect this situation toward more useful goals by incorporating smartphones into their instructional plans.

The combination of a smartphone plus appropriate apps is clearly becoming an important component of chemistry teaching, learning, and research. Since the prices for both devices and phone connectivity are likely to continue to decrease, the percentage of students who have smartphones in class will probably increase. This means that those in a chemistry classroom will routinely carry in their pockets the equivalent of a major research library. In fact, if Google Glass or some similar personal augmented reality device becomes popular, this access may be as clear as the glasses on their faces. What will the chemistry classroom look like in a decade? An important first step is to stop thinking of mobile devices as being just phones, but as really powerful and portable devices for linking into the Internet of Everything. Because they are so portable and so personally linked to the user, it is quite probable that they may change education more profoundly than did the desktop and laptop computer revolutions. Now is the time to begin planning how to incorporate these new capabilities into the way students learn and are taught.

AR represents a new way to create more engaging and immersive learning environments. This capability is especially valuable for chemical education, where spatial reasoning and visual interpretation are significant challenges. AR offers many new opportunities for innovation and meshes well with the current public interest in social networks. This approach also is desirable as a way to integrate mobile access into traditional methods of chemical education. Currently, the most popular way for chemical educators to explore AR technology is based on QR codes, because they can be developed with minimal investment of time and money. Commercial availability of markered AR and AR glasses, such as Google Glass, represents another avenue for development that will probably become more popular in the near future. The greatest challenge here is not just understanding the technology but creating new learning environments that will make the best use of these applications.

Although this chapter has focused on technological developments of the social and semantic Web that may affect chemical education and addressed some of their implications from the perspective of cognitive sciences, one cannot ignore how these technologies enable new modes of student data collection and emphasize the importance of learning analytics. The use of data that is collected by design is a key component to making decisions, and there will be many complications involved with understanding the cognitive processes associated with the use of social and semantic Web technologies in education and science. Educational data mining is clearly beyond the scope of this chapter, and the interested reader can check Siemens article on "Learning Analytics" for a recent review of this topic [134].

In 1995, George Long *et al.* summarized their report on one of the early online chemical education conferences sponsored by the American Chemical Society Committee on Computers in Chemical Education by saying, "As the computer and other new technological or pedagogical tools are added to our curricula, as is happening now at an incredible rate, the way students think about chemistry will change. We, as educators, will no longer be able to rely solely on the old methods to educate and evaluate students [135]." That advice is as relevant today as it was then.

We dedicate this chapter to the memory of our friend and colleague Jean-Claude Bradley. JC's contributions to the vision of what can result from increased openness and crowdsourced participation in the chemical sciences made him one of the leading voices in many of the concepts presented in this chapter.

References

1. SpeakUp Anonymous Project Speak-Up 2012 National Report: From Chalkboards to Tablets: The Emergene of the K-12 Digital Learner (2013), http://www.tomorrow.org/speakup/SU12_DigitalLearners_StudentsTEXT.html (accessed December 30 2013).
2. Williams, B.S. (2010) in *Enhancing Learning with Online Resources, Social Networking and Digital Libraries*, ACS Symposium Series Books, vol. 1060 (eds R.E. Belford, J.W. Moore, and H.E. Pence), ACS, Washington, DC, pp. 95–114.
3. Smart, P.R. (2012) The web-extended mind. *Metaphilosophy*, **43** (4), 426–445.
4. Peter, I. History of the World Wide Web, (2004) http://www.nethistory.info/History%20of%20the%20Internet/web.html (accessed 20 November 2013).
5. Berners-Lee, T., James Hendler, J., and Lassila, O. (2001) The Semantic Web, http://www.cs.umd.edu/~golbeck/LBSC690/SemanticWeb.html (accessed 22 November 2013).
6. Google Google Introducing the Knowledge Graph: Things, Not Strings, (2012) http://googleblog.blogspot.com/2012/05/introducing-knowledge-graph-things-not.html (accessed 5 November 2013).
7. RSC RSC Semantic Publishing, http://www.rsc.org/publishing/journals/projectprospect/ (accessed 22 November 2013).
8. Batchelor, C., Karapetyan, K., Psenichov, A., Sharpe, D., Steele, J., Tkchenko, V., and Williams, A.J. Digitally Enhancing the RSC Archive, (2013) http://www.slideshare.net/AntonyWilliams/digitally-enabling-the-rsc-archive (accessed 30 December 2013).
9. Williams, A.J., Batchelor, C., Brouwer, W., and Tkachenko, V. Digitizing Documents to Provide a Public Spectroscopy Database, (2013) http://www.slideshare.net/AntonyWilliams/digitizing-documents-to-provide-a-public-spectroscopy-database (accessed 30 December 2013).
10. Belford, R.A. et al. About the WikiHyperGlossary, (2011) http://whg.chemeddl.org (accessed 14 May 2014).
11. Belford, R.E., Bauer, M.A., Berleant, D., Holmes, J.L., and Moore, J.W. (2012) ChemEd DL WikiHyperGlossary: connecting digital documents to online resources, while coupling social to canonical definitions within a glossary. *G. Didattica Cult. Soc. Chim. Ital. (Rivista CnS)*, **34** (3), 46–50.
12. Berners-Lee, T. The World Wide Web: Past, Present and Future, (1996) http://www.w3.org/People/Berners-Lee/1996/ppf.html (accessed 2 December 2013).
13. W3C World Wide Web Consortium (W3C) Semantic Web Activity Statement, (2013) http://www.w3.org/2001/sw/Activity (accessed 2 December 2013).
14. The IUPAC International Chemical Identifier (InChI), (2011) http://www.iupac.org/home/publications/e-resources/inchi.html (accessed 1 January 2014).
15. Taylor, K.R., Gledhill, R.J., Essex, J.W., and Frey, J.G. (2006) Bringing chemical data onto the semantic web. *J. Chem. Inf. Model.*, **46** (3), 939–952.
16. Gruber, T.R. (1993) A translation approach to portable ontology specifications. *Knowledge Acquis.*, **5** (2), 199–220.
17. Pence, L.E. and Pence, H.E. (2009) Accessing and managing scientific literature: using RSS in the classroom. *J. Chem. Educ.*, **86** (1), 41–44.
18. Adams, N., Winter, J., Murray-Rust, J.P., and Rzepa, H.S. (2008) Chemical markup, XML and the world-wide web. 8. Polymer markup language. 8. *J. Chem. Inf. Comput. Sci.*, **48** (11), 2118–2128.
19. Schoefeggera, K., Tammet, T., and Granitzerc, M. (2013) A survey on socio-semantic information retrieval. *Comput. Sci. Rev.*, **8**, 25–46.
20. Breslin, J. and Decker, S. (2007) The future of social networks on the internet: the need for semantics. *Internet Comput.*, **11** (6), 86–90.

21. Morville, P. (2005) *Ambient Findability: What We Find Changes Who We Become*, O'Reilly Media, Sebastopol, CA.
22. Bernardi, A. NEPOMUK – The Social Semantic Desktop – P6-027705, http://nepomuk.semanticdesktop.org (accessed 16 May 2014).
23. The Friend of a Friend (FOAF) Project, http://www.foaf-project.org (accessed 15 May 2014).
24. Semantic Media Wiki, http://semantic-mediawiki.org/ (accessed 14 May 2014).
25. Semantically-Interlinked Online Communities The SIOC Initiative, (2013) http://sioc-project.org/ (accessed 21 May 2014).
26. Introduction to SKOS, http://www.w3.org/2004/02/skos/intro (accessed 15 May 2013).
27. Williams, A.J. et al. (2012) Open PHACTS: semantic interoperability for drug discovery. *Drug Discovery Today*, **17** (21–22), 1188–1198.
28. Open PHACTS Explorer, (2013) http://www.openphacts.org/explorer (accessed 16 May 2014).
29. Open PHACTS Discovery Platform, https://dev.openphacts.org/ (accessed 16 May 2014).
30. From Chalkboards to Tablets: The Emergence of the K-12 Digital Learner, http://www.tomorrow.org/speakup/SU12_DigitalLearners_StudentsTEXT.html (accessed 11 December 2013).
31. Katz, J.E. (2002) *Perpetual Contact: Mobile Communication, Private Talk, Public Performance*, Cambridge University Press, Cambridge.
32. Brenner, J. Pew Internet: Mobile, (2013) http://pewinternet.org/Commentary/2012/February/Pew-Internet-Mobile.aspx (accessed 4 November 2013).
33. Mobile Statistics, (2013) http://iminmarketer.com/mobile-statistics/ (accessed 1 November 2013).
34. Rheingold, H. Attention Literacy, (2009) http://www.sfgate.com/cgi-bin/blogs/rheingold/detail?entry_id=38828 (accessed 31 October 2013).
35. Warschauer, M. (2009) Information literacy in the laptop classroom. *Teach. Coll. Rec.*, **190** (11), 2511–2540.
36. Barseghian, T. (2012) How Teachers Make Cell Phones Work in the Classroom, http://blogs.kqed.org/mindshift/2012/05/how-teachers-make-cell-phones-work-in-the-classroom/ (accessed 18, October, 2014).
37. Powell, C. (2010) Podcast effectiveness as scaffolding support for students enrolled in first-semester general chemistry laboratories. PhD dissertation. University of North Texas, Denton, TX.
38. McDonald, L.M. (2010) Anecdotal Uses of Facebook, Google Calendar, and Cell Phones in a High School Classroom, http://science.widener.edu/svb/cccenews/fall2010/paper4.html (accessed 28 October 2013).
39. Cheredar, T. (2013) Surprisingly, People Spend More Time Reading Books on Smartphones than Tablets, http://venturebeat.com/2013/08/21/surprisingly-people-spend-more-time-reading-books-on-smartphones-than-tablets/ (accessed 21 August 2013).
40. Sparrow, B. et al. (2011) Google effects on memory: cognitive consequences of having information at our fingertips. *Science*, **333** (3604), 776–778.
41. Williams, A.J. (2010) Mobile Chemistry – Chemistry in Your Hands and in Your Face. Chemistry World, http://www.rsc.org/chemistryworld/Issues/2010/May/MobileChemistryChemistryHandsFace.asp (accessed 25 July 2014).
42. Young, J.R. (2011) 6 top smartphone apps to improve teaching, research, and your life. The Chronicle of Higher Education, http://chronicle.com/article/6-Top-Smartphone-Apps-to/125764/ (accessed 25 July 2014).
43. Pence, H.E. and Williams, A.J. (2010) ChemSpider: an online chemical information resource. *J. Chem. Educ.*, **87** (11), 1123–1124.
44. AppAdvice (2014) Mind Mapping Apps, http://appadvice.com/appguides/show/mind-mapping-apps (accessed 21 May 2014).
45. Twiss-Brooks, A. (2013) Chemistry on Mobile Devices, http://guides.lib.uchicago.edu/content.php?pid=65132&sid=1703522 (accessed 12 November 2013).

46. Williams, A.J. and Pence, H.E. (2011) Smart phones, a powerful tool in the chemistry classroom. *J. Chem. Educ.*, **88** (6), 683–686.
47. Bradley, J.-C., Lancashire, R.J., Lang, A., and Williams, A.J. (2009) The spectral game: leveraging open data and crowdsourcing for education. *J. Cheminf.*, **1**, 9, http://www.jcheminf.com/content/1/1/9 (accessed 6 August 2014).
48. Libman, D. and Huang, L. (2013) Chemistry on the go: review of chemistry apps on smartphones. *J. Chem. Educ.*, **90** (3), 320–325.
49. Elkins, S., Clark, A.M., and Williams, A.J. (2012) Open drug discovery teams: a chemistry mobile app for collaboration. *Mol. Inf.*, **31** (8), 585–597.
50. Williams, A.J., Ekins, S., Clark, A.M., Jack, J.J., and Apodaca, R.L. (2011) Mobile apps for chemistry in the world of drug discovery. *Drug Discovery Today*, **16**, 928–939.
51. EHS Freeware, (2014) http://www.ehsfreeware.com/ihinfo.htm (accessed 16 May 2014).
52. Williams, A. J. (2011) SciMobileApps Wiki, www.scimobileapps.com (accessed 16 May 2014).
53. Williams, A.J. (2011) Mobile Chemistry and the SciMobileApps Wiki, http://www.slideshare.net/Antony Williams/mobile-chemistry-and-the-scimobileapps-wiki-october-2011-version (accessed 2 January 2014).
54. University of Southampton (2013) Notelus, https://itunes.apple.com/us/app/notelus/id593269701?mt=8 (accessed 14 May 2014).
55. Abrams, N.M. (2012) Combining cloud networks and course management systems for enhanced analysis in teaching laboratories. *J. Chem. Educ.*, **89** (4), 482–486.
56. Google Drive http://www.google.com/drive/apps.html?usp=ad_search&gclid=CIqS0ez3qrsCFepaMgod3GcAvw (accessed 1 January 2013).
57. Spaeth, A.D. and Black, R.S. (2012) Google docs as a form of collaborative learning. *J. Chem. Educ.*, **89** (8), 1078–1079.
58. Soulsby, D. (2012) Using cloud storage for NMR data distribution. *J. Chem. Educ.*, **89** (8), 1007–1011.
59. Bennett, J. and Pence, H.E. (2011) Managing laboratory data using cloud computing as an organizational tool. *J. Chem. Educ.*, **88** (6), 761–763.
60. Milgram, P. and Kishino, F.A. (1994) Taxonomy of mixed reality visual displays. *IEICE Trans. Inf. Syst.* (Special Issue on Networked Reality), **E77-D** (12), 1321–1329.
61. Castronova, E. (2007) *Exodus to the Virtual World*, Palgrave Macmillan, New York.
62. Pence, H.E. (2007-2008) The homeless professor in second life. *J. Educ. Technol. Syst.*, **36** (2), 171–177.
63. Lang, A. and Bradley, J.-C. (2009) Chemistry in second life. *Chem. Cent. J.*, **3**, 14.
64. Keeney-Kennicutt, W. and Winkelmann, K. (2013) What Can Students Learn from Virtual Labs? http://www.ccce.divched.org/P9Fall2013 CCCENL (accessed 12 December 2013).
65. Downes, S. (2008) Connectivism & Connective Knowledge. The Daily, (Sept. 15 2008), http://connect.downes.ca/archive/08/09_15_thedaily.htm (accessed 14 December 2013).
66. Pence, H.E. (2013-2014) Are MOOCs a solution or a symptom? *J. Educ. Technol. Syst.*, **42** (2), 131–132.
67. Saunder, D. *et al.* (2000) Phsical chemistry online: maximizing your potential. *Chem. Educ.*, **5**, 77–82.
68. Cole, R.S. (2013) in *Current Status of PCOL* (ed. H.E. Pence).
69. QR Code Generator, http://www.the-qrcode-generator.com/ (accessed 14 May 2014).
70. Tolliver-Nigro, H. (2009) Making the most of quick response codes. *Seybold Rep.: Anal. Publ. Technol.*, **9** (21), 2–8.
71. Pence, H.E. (2013) in *The Handheld Libary* (eds T.A. Peters and L. Bell), Libraries Unlimited, Santa Barbara, CA, pp. 133–142.
72. Bujak, K.R. (2013) A psychological perspective on augmented reality in the mathematics classroom. *Comput. Educ.*, **68**, 536–544.

73. Johnstone, A.H. (1991) Why is science difficult to learn? Things are seldom what they seem. *J. Comput. Assisted Learn.*, **7**, 75–83.
74. Merchant, Z., Goetz, E.T., Cifuentes, L., Keeney-Kennicutt, W., and Davis, T.J. (2014) Effectiveness of virtual reality-based instruction on students' learning outcomes in K-12 and higher education: a meta-analysis. *Comput. Educ.*, **70**, 29–40.
75. Benedict, L. and Pence, H.E. (2012) Teaching chemistry using student-created videos and photo blogs accessed with smartphones and two-dimensional barcodes. *J. Chem. Educ.*, **89** (4), 492–496.
76. The Periodic Table Bar Codes, *http://www.flickr.com/photos/ periodicvideos/5915143448/sizes/o/in/ photostream/* (accessed 14 May 2014).
77. Bonifácio, V.D.B. (2013) Offering QR-code access to information on nobel prizes in chemistry, 1901–2011. *J. Chem. Educ.*, **90** (10), 1401–1402.
78. Vernier Connected Science System, *http://www.vernier.com/products/ wireless-solutions/connected-science-system/* (accessed 21 May 2014).
79. Gallery of Mobile Apps and Sites, (2014) *http://www.nlm.nih.gov/mobile/* (accessed 2 January 2014).
80. Fjeld, M. and Voegtli, B.M. (2012) Augmented Chemistry: An Interactive Educational Workbench, *http://nguyendangbinh.org/Proceedings/ ISMAR/2002/papers/ismar_fjeld_chem.pdf* (accessed 28 October 2013).
81. Olson, A. Augmented Reality for Chemists, (2011) *http://www.youtube.com/ watch?v=gZxK6j4JTHQ&feature= youtu.be* (accessed 25 October 2013).
82. Sponholtz Productions, (2010) *http://www.sponholtzproductions.com/ index.html* (accessed 16 May 2014).
83. Carney, M. (2013) DAQRI Launches Elements 4D on Kickstarter, Hopes to Introduce Augmented Reality to the Masses Through Education, *http://pandodaily.com/2013/07/24/ daqri-elements-4d-kickstarter/* (accessed 9 October).
84. Klinker, G.J. Augmented Chemical Reactions, *http://www.youtube. com/watch?v=aPd8fr46bng* (accessed 14 May 2014).
85. Van-Bossuyt, B. (2010) Augmented Reality in Chemistry: An NaCl Lattice, *https://www.youtube.com/watch?v=dj7f_ PEknK4* (accessed 14 May 2014).
86. Wolsen, M. (2013) Guys Like This Could Kill Google Glass Before It Ever Gets Off the Ground, *http://www.wired.com/business/2013/05/ inherent-dorkiness-of-google-glass/* (accessed 29 November 2013).
87. Von Abo, R. (2014) Welcome to a World through Glass, *http://www.bizcommunity.com/Article/ 196/16/114723.html* (accessed 1 January 2014).
88. BMW Augmented Reality, *http://www.youtube.com/watch?v= P9KPJlA5yds* (accessed 1 January 2014).
89. Atzori, L., Iera, A., and Morabito, G. (2010) The internet of things: a survey. *Comput. Networks*, **54**, 2287–2805.
90. Rivera, J. and van der Meulen, R. (2013) Gartner Says the Internet of Things Installed Base Will Grow to 26 Billion Units By 2020, *http://www.gartner.com/newsroom/id/ 2636073* (accessed 18 December 2013).
91. Twitter Doctor_Galactic Real Time Chem, (2013) *https://twitter.com/RealTime Chem* (accessed 15 May 2014).
92. Hannay, T. Comparing Wikipedia and Britannica, (2005) *http://blogs.nature.com/ nascent/2005/12/comparing_ wikipedia_and_britan_ 1.html* (accessed 14 May 2014).
93. Walker, M. (2010) in *Enhancing Learning with Online Resources, Social Networking, and Dgital Libraries*, vol. 1060 (eds R.E. Belford, J.W. Moore, and H.E. Pence), American Chemical Society, Washington, DC, pp. 79–92.
94. Williams, A.J., The Vision for the ScientistsDB Wiki, (2012) *http://www. scientistsdb.com/index.php?title= Main_Page* (accessed 1 January 2014).
95. Wikipedia Wikipedia Chemical Table File, (2014) *http://en.wikipedia.org/wiki/*

Chemical_table_file (accessed 15 May 2014).
96. IUPAC Homepage of the IUPAC CPEP Subcommittee on Electronic Data Standards, (2007) *http://www.jcamp-dx.org/*). (accessed 15 May 2014).
97. PubChem, *http://pubchem.ncbi. nlm.nih.gov/* (accessed 15 May 2014).
98. DrugBank: Open Data Drug Bank and Drug Target Database, (2014) *http://www.drugbank.ca* (accessed 15 May 2014).
99. RSC LearnChemistry: Enhancing Chemistry Learning and Teaching, (2014) *http://www.rsc.org/learn-chemistry* (accessed 15 May 2014).
100. About nmrshiftdb2, (2014) *http://nmrshiftdb.nmr.uni-koeln.de/* (accessed 1 January 2014).
101. Williams, A.J. (2010) in *Enhancing Learning with Online Resources, Social Networking, and Dgital Libraries*, vol. 1060 (eds R.E. Belford, J.W. Moore, and H.E. Pence), American Chemical Society, Washington, DC, pp. 23–39.
102. NCBI The PubChem Power User Gateway, *http://pubchem.ncbi.nlm. nih.gov/pug/pughelp.html* (accessed 15 May 2014).
103. ChemSpider API *http://www. chemspider.com/AboutServices.aspx* (accessed 2 January 2014).
104. Chemicalize, (2014) *http://www. chemicalize.org/* (accessed 15 May 2014).
105. Little, J.L., Williams, A.J., Pshenichnov, A., and Tkachenko, V. (2012) Identification of "known unknowns" utilizing accurate mass data and ChemSpider. *J. Am. Soc. Mass. Spectrom.*, **23** (1), 179–185.
106. Bradley, J.-C. Open Notebook Science, (2006) *http://drexel-coas-elearning.blogspot.com/2006/09/open-notebook-science.html* (accessed 17 December 2013).
107. Bradley, J.-C. Shining a Light on Chemical Properties with Open Notebook Science and Open Strategies, (2012) *http://www.slideshare.net/jcbradley/ bradley-acs2012* (accessed 2 January 2014).
108. Williams, A.J. et al. (2012) Feeding and Consuming Data to Support Open Notebook Science via the ChemSpider Platform, *http://www.slideshare.net/Antony Williams/feeding-and-consuming-data-to-support-open-notebook-science-via-the-chem-spider-platform* (accessed 17 December 2013).
109. Bradley, J.-C. and Lang, A. (2014) Open Notebook Science Web Services, *http://onswebservices.wikispaces.com/* (accessed 15 May 2014).
110. Bradley, J.-C. (2011) Melting Point Prediction for MMDS by Way of Open Notebook Science, *http://cheminf20.org/2011/06/09/ melting-point-prediction-for-mmds-by-way-of-open-notebook-science/* (accessed 17 December 2013).
111. King, A. (2013) Notebooks Go Digital, *http://www.rsc.org/chemistryworld/2013/ 05/electronic-lab-notebook-review* (accessed 1 January 2014).
112. King, A. (2011) Waving goodbye to the paper lab book. *Chem. World*, **2011**, 46–49.
113. Dial-a-Molecule::An EPSRC Grand Challenge Network, (2014) *http://www.dial-a-molecule.org/wp/* (accessed 2 January 2014).
114. TheFreyGroup Lab Trove *http://www.labtrove.org/*) (accessed 15 May 2013).
115. Day, A. (2013) Create Stoichiometry Tables with New ChemSpider Widget, *http://www.chemspider.com/blog/create-stoichiometry-tables-with-new-chemspider-widget.html* (accessed 1 January 2014).
116. RSC National Chemical Database Service, (2014) *http://cds.rsc.org* (accessed 14 May 2014).
117. Hollan, J.D., Hutchins, E., and Kirsh, D. Distributed cognition: towards a new foundation for human-computer interaction research. (2000) *ACM Trans. Comput.-Hum. Interact.*, 7 (2), 175.
118. McGarry, B. (2005) Things to think with: understanding interactions with artefacts in engineering design. PhD dissertation. University of Queensland, Australia.

119. Heersmink, R. (2012) in *Mind and Artifact: A Multidimensional Matrix for Exploring Cognition-Artifact Relations in the 5th AISB Symposium on Computing and Philosophy* (eds J.M. Bishop and Y.J. Erden), PhilPapers, Birmingham, pp. 54–61, http://philpapers.org/rec/HEEMAA (accessed 20 October, 2014).
120. Norman, D.A. (1991) in *Designing Interaction: Psychology at the Human-Computer Interface* (ed. J. Carroll), Cambridge University Press, Cambridge, p. 17.
121. Heersmink, R. (2013) A taxonomy of cognitive artifacts: function, information, and categories. *Rev. Philos. Psychol.*, **4** (30), 465–481.
122. Lavoisier, A. (2009) *The Project Gutenberg EBook of Elements of Chemistry* (translated by R. Kerr), http://www.gutenberg.org/files/30775/30775-h/30775-h.htm (accessed 7 January 2014).
123. Jones, P.H. and Nemeth, C.P. (2005) in *Ambient Intelligence for Scientific Discovery* (ed. Y. Cai), Springer-Verlag, New York, pp. 152–183.
124. Sterelny, K. (2010) Minds: extended or saffolded? *Phenomenol. Cognit. Sci.*, **9**, 465–481.
125. Project Tomorrow, (2014) http://www.tomorrow.org (accessed 16 May 2014).
126. From Chalkboards to Tablets: The Emergence of the K-12 Digital Learner, (2013) http://www.tomorrow.org/speakup/SU12_DigitalLearners_StudentsTEXT.html (accessed 19 October, 2014).
127. Nagel, D. Students as Free Agent Learners, (2009) *T.H.E. Journal*, http://thejournal.com/articles/2009/04/24/students-as-free-agent-learners.aspx (accessed 2 January 2014).
128. Holmes, C.O. and Warden, J.T. (1996) CIStudio: a worldwide web-based, interactive chemical information course. *J. Chem. Educ.*, **73** (4), 325–331.
129. Moy, C.L. et al. (2010) Improving science education and understanding through editing Wikipedia. *J. Chem. Educ.*, **87** (11), 1159–1162.
130. Pence, L.E. and Pence, H.E. (2010) in *Enhancing Learning with Online Resources, Social Networking, and Digital Libraries*, vol. 1060 (eds R.E. Belford, J.W. Moore, and H.E. Pence), American Chemical Society, Washington, DC, pp. 115–127.
131. Jones, L.L. (2013) How multimedia-based learning and molecular visualization change the landscape of chemical education research. *J. Chem. Educ.*, **90** (12), 1571–1576.
132. The New Multiscreen World by Google, (2012) http://www.slideshare.net/smobile/the-new-multiscreen-world-by-google-14128722 (accessed 21 December 2013).
133. Kleiman, J. (2013) How Multitasking Affects your Brain (and your Effectiveness at Work), http://www.forbes.com/sites/work-in-progress/2013/01/15/how-multitasking-hurts-your-brain-and-your-effectiveness-at-work/ (accessed 21 December 2013).
134. Siemens, G. (2013) Learning analytics: the emergence of a discipline. *Am. Behav. Sci.*, **57** (10), 1380–1400, http://ehis.ebscohost.com.ezproxy.oneonta.edu:2048/eds/detail?vid=3&sid=09121ac6-0df5-4001-841f-d5c0abe09a67@sessionmgr110&hid=105&bdata=JnNpdGU9ZWRzLWxpdmU=#db=bah&AN=90161126 (accessed 30 July 2014).
135. Long, G., Pence, H.E., and Zielinski, T.J. (1993) New tools vs. old methods: a description of the ChemConf '93 discussion. *Comput. Educ.*, **24** (4), 259–269.

Index

3D representations, visualization 601–3
4C/ID (four-component instructional design) approach, competency-based undergraduate curriculum 91
AAAS *see* American Association for the Advancement of Science

a

acceleration program, gifted learners 477
acid, TKRS searches 59–60, 65–7
acid–base chemistry, human activity 16, 18–20
acid–base reactions
- Broensted concept 406, 409–12
- misconceptions 405–12
- proton transfer 405–12
acids/metals experiments, guided-inquiry-based laboratories 311–14
active-learning assignments, classroom 238
active-learning inquiry, laboratory learning 674, 678–81
active learning pedagogies 296
- *see also* problem-based learning (PBL); service-learning
activity system, competency-based teaching 82–3
ADI model *see* argument-driven instructional (ADI) model
air bag/gas laws, dynamic visualization 603–6
air quality 32–3
American Association for the Advancement of Science (AAAS)
- real-world chemistry 280
- recommendations 280, 282
America's Lab Report, high-order learning skills 519

analytical/environmental chemistry projects, service-learning 291–2
ANAPOGIL: Process-Oriented Guided Inquiry Learning in Analytical Chemistry, chemistry education research 167–9
anarchistic model of problem solving, problem solving research 193–9
animations
- *see also* visualization
- chemistry education research 161–3
- continuous/segmented 608–9
- molecular animations 607–8
- simple/realistic 607–8
- student-generated animations 225–6
- visualization 607–9
Anthropocene Epoch, human activity 14–16
apps, chemistry *see* chemistry apps
AR *see* augmented reality
Argument-Driven Inquiry, chemistry education research 160
argument-driven instructional (ADI) model 81
argumentation
- chemistry education research 159–60
- chemistry laboratory 528–31
- epistemic practice 527–8
- high-order learning skills 526–31
- learning skills 529–31
- nature of 526–31
- Toulmin's model 528–9
argumentation and discourse, problems with language 436
argumentation and evidence, *Real Work* 222

ASCI (Attitude toward the Subject of Chemistry Inventory), chemistry education research 164
atomic structure, TKRS searches 60–1, 65–7
atoms first, human activity 6–7
augmented reality (AR) 708–17, 727–8
– game play 581–2
– Internet 566–7, 579–90
– LabQuest 2 app 713–14
– layar 566–7, 579, 582
– Marker-based AR and Markerless AR 711–12
– Might-^{y}/Mach^{tig} 580–90
– modeling apps 714–15
– molecular viewers 714–15
– QR (quick response) codes 711–14
– Second Life 709–10
– wearable devices 715–16
authentic learning
– community-based learning (CBL) 363
– community-based research (CBR) 363
– *Real Work* 206–9
authentic learning experiences, *Real Work* 203–6
authentic materials, *Real Work* 243–4
authentic tasks, *Real Work* 206–9
authentic texts and evidence 228–32
– Course-Based Undergraduate Research Experiences (CURE) 230–1
– generating questions 230
– interdisciplinary research-based projects 231–2
– literature seminars 229–30
– literature summaries 228–9
– public science courses 230
– *Real Work* 228–32

b

balls problem, problem solving research 182–4
barcodes, QR (quick response) codes 711–14
Berners-Lee, Tim 694–5, 697
Big Ideas in chemistry 30
bioinorganic chemistry course
– problem-based learning (PBL) 288–9
– service-learning 293–4, 295
Blackboard, Virtual Learning Environment (VLE) 574
blended learning 651–66
– added value 651–2
– clickers 662–3, 664
– cognitive load theory (CLT) 654–5

– curriculum 652–3, 661, 662, 665–6
– curriculum integration 665–6
– defining 322
– discussion boards 653, 663–4
– examples 655–65
– Facebook 664–5
– flipped classrooms 322, 653, 658
– gamification 661
– information processing model 654–5
– iTunesU 656–7
– laboratory classes 657–9
– lectures 657–9
– online communities 663–5
– online quizzes 659–62
– PeerWise 235, 661
– personal response systems 662–3, 664
– podcasts 656–7
– pre-laboratory activities 658–9
– pre-lecture activities 657–8
– preparation for lectures and laboratory classes 657–9
– role in science education 652–3
– social networking 664–5
– Twitter 664–5
– YouTube 656–7
boiling point experiments, cognitive conflict strategy 449, 454, 465–6
books, popular, lifelong learning 135–6
branches of chemistry 469–70
broadcast media, lifelong learning 140–1
Broensted concept, acid–base reactions 406, 409–12

c

calculator apps 639
calibrated peer review (CPR), *Real Work* 221, 240–1
carbon cycle 32
carbon dioxide
– combustion connection 32–3
– context-based learning (CBL) 272–3
– physical and chemical equilibria 272–3
cards problem, problem solving research 182
carrying out chemistry, human activity 10–14
cartoons, lifelong learning 136–7
CBL *see* community-based learning; context-based learning
CBR *see* community-based research
cell potential measurements, microscale experimentation 549–51
challenging issues, teaching 107–13

ChemCollective virtual laboratory, visualization 600
ChemEd DL (digital library) 697
chemical education
– *vs* chemistry education 3
– future 726–8
chemical equations
– chemistry education research 155, 156
– strengths/shortcomings 31–2
chemical equilibrium
– misconceptions 401–5
– TKRS searches 61–2, 65–7
chemical markup language (CML) 699–700
chemical switches example, context-based learning (CBL) 273–5
chemical symbols, history and development of chemical language 423–5
chemical tetrahedron, context-based learning (CBL) 270–1
'Chemie im Kontext'
– context-based learning (CBL) 261–3
– school-level 261–3
chemistry apps 621–49, 706–8
– calculator apps 639
– challenges 646–7
– chemistry courses apps 644
– Chemistry Quiz App 660
– ChemSpider app 697, 706–7, 708, 717–20
– demonstration and teaching apps 641–2
– dictionary/reference apps 636–7
– flashcard apps 635–6
– future perspectives 647–9
– gaming apps 642–4
– hardware 625–6
– instrumental apps 640
– interactive applets 571–2
– lab utility apps 634–40
– LabQuest 2 app 713–14
– literature research apps 633–4
– modeling apps 626–9, 714–15
– Molecular City app 587–92
– molecular drawing apps 629–31
– molecular viewers 626–9, 714–15
– operating systems 625–6
– opportunities 646–7
– Periodic Table apps 631–3
– search engine apps 637–8
– teaching and demonstration apps 641–2
– test-prep apps 644–5
– updating apps 645
chemistry concepts, planetary boundaries 16–17

chemistry courses apps 644
chemistry education
– *vs* chemical education 3
– tetrahedral 4–5
chemistry education projects, service-learning 292–3
chemistry education research 151–74
– *see also* new teaching methods; research-based teaching; teaching strategies
– ANAPOGIL: Process-Oriented Guided Inquiry Learning in Analytical Chemistry 167–9
– animations 161–3
– Argument-Driven Inquiry 160
– argumentation 159–60
– ASCI (Attitude toward the Subject of Chemistry Inventory) 164
– chemical equations 155, 156
– CHEMX (Chemistry Expectations Survey) 163
– CLASS-Chem (Colorado Learning Attitudes about Science Survey) 163
– CLUE: Chemistry, Life, the Universe, and Everything 169–70
– concept inventories 158–9
– connecting research to practice 154–65, 171–2
– cycle of pedagogical research 172–4
– demonstrations, interactive 166–7
– Group Assessment of Logical Thinking (GALT) 164
– implementation 171–2
– instructional practice 155–6, 171–4
– instruments 163–5
– interactive demonstrations 166–7
– logical thinking tests 164
– misconceptions, students' 154–7
– problem solving 161
– process-oriented guided inquiry learning (POGIL) 160
– professional knowledge/development 171–4
– representations 161–3
– research-based teaching practice 165–70
– Science Writing Heuristic 160
– simulations 161–3
– stoichiometry 155–6
– student argumentation 159–60
– student discourse 159–60
– student learning research 153–4
– student response systems (SRSs) 157–8
– Test of Logical Thinking (TOLT) 164
Chemistry in Context 34

chemistry literacy *see* scientific literacy
Chemistry Quiz App   660
ChemSpider app   697, 706–7, 708, 717–20
CHEMX (Chemistry Expectations Survey), chemistry education research   163
citizen science, lifelong learning   143–4
CLASS-Chem (Colorado Learning Attitudes about Science Survey), chemistry education research   163
classroom response systems, Peer Instruction   333–5, 338
clickers, blended learning   662–3, 664
climate change, human activity   9–10, 16–18, 32–3
CLT *see* cognitive load theory
CLUE: Chemistry, Life, the Universe, and Everything, chemistry education research   169–70
CML *see* chemical markup language
cognition levels, gifted learners   477–8
cognitive accommodation, cognitive conflict strategy   451–2
cognitive artifacts   721–6
cognitive conflict recognition/resolution   459–61
cognitive conflict strategy   447–67
– boiling point experiments   449, 454, 465–6
– changing conceptions   452–5
– cognitive accommodation   451–2
– cognitive conflict recognition/resolution   459–61
– cognitive dissonance   451–5
– constructivism   452, 462
– defining   448
– discrepant events   448, 453–9, 461–5
– disequilibrium   451–2, 453
– examples of cognitive conflict situations   449–51
– logic processes   459–61
– misconceptions   451–2
– origins   451–3
– pitfalls   459
– predict–observe–explain (POE) technique   456, 462–3
– research literature   461–5
cognitive cycle, gifted learners   477–8
cognitive dissonance, cognitive conflict strategy   451–5
cognitive load theory (CLT)
– blended learning   654–5
– transient information effect   657
cognitive sequence, gifted learners   477–8

cognitive skills
– constructivism   304–5
– inquiry-based student-centered instruction   304–5
– learning cycle   304–5
collaboration
– future trends   117
– learning communities   116
– teacher learning   116, 117
– Wikis   681
collaborative identification, team-based learning   223–4
collective writing
– laboratory learning   682–4
– Wikis   682–4
color, understanding, context-based learning (CBL)   273–5
combustion, carbon dioxide connection   32–3
comics, lifelong learning   136–7
command words, problems with language   435–6
community-based learning (CBL)   345–67
– *see also* service-learning
– authentic learning   363
– barriers   360–4
– benefits   353–60
– chemistry education   349–53
– clarity of purpose   360
– community partnerships   361
– current trends   364–6
– curriculum   353
– defining   346
– developments   366
– e-learning   365–6
– economic uncertainty   364–5
– ethical issues   360
– future trends   364–6
– geographic spread   364
– graduate attributes   354
– guidelines   352–3
– high-impact educational practices   354–6
– institutional commitment   363
– online learning   365–6
– PARE (preparation, action, reflection, and evaluation) model   353
– peer-reviewed accounts   349–51
– personal development   354
– reciprocity   359
– reflection   363–4
– regulatory issues   360
– resources   352–3
– scholarship   365
– sustainability   362

– vocabulary   345–9
community-based research (CBR)   345–67
– authentic learning   363
– barriers   360–4
– benefits   353–60
– chemistry education   349–53
– clarity of purpose   360
– community partnerships   361
– current trends   364–6
– curriculum   353
– developments   366
– e-learning   365–6
– economic uncertainty   364–5
– ethical issues   360
– future trends   364–6
– geographic spread   364
– guidelines   352–3
– high-impact educational practices   354–6
– institutional commitment   363
– online learning   365–6
– PARE (preparation, action, reflection, and evaluation) model   353
– peer-reviewed accounts   349–51
– reciprocity   359
– reflection   363–4
– regulatory issues   360
– resources   352–3
– scholarship   365
– sustainability   362
– vocabulary   345–9
competency, defining   82
competency-based teaching   81–3
– activity system   82–3
– characteristics   83
competency-based undergraduate curriculum   83–92, 93
– 4C/ID (four-component instructional design) approach   91
– competency area analysis   86–7, 92
– competency area modeling   89–90, 92
– competency area synthesis   88–9, 92
– Entrepreneurship Education and Training (EET)   91
– structure   84–6
computer-based technologies
– *see also* chemistry apps; information and communication technology (ICT); Internet; simulations; visualization
– learning communities   116
– lifelong learning   141–3
– teacher learning   111–13, 116
concept development, inquiry-based student-centered instruction   308–10

concept inventories, chemistry education research   158–9
ConcepTests, Peer Instruction   330–9
conceptual integration   375–92
– compartmentalization of learning   387–8
– concepts as public knowledge systems   377–8
– concepts, nature of   375–7
– conceptual coherence   381–3, 386–7
– conceptual inductive effect   380–1
– conceptual structure   379–80, 389
– expertise   389–90
– impeding learning   388–9
– implications   390–2
– in learning   385–6
– models/modeling   383–5
– multiple models   383–5
– Personal Construct Theory (PCT)   378–9
– research directions   391–2
conceptual understanding
– macroscopic level   596–7
– mental models   596–8
– submicroscopic level   597
– symbolic level   597
conductivity/conductance meters, microscale experimentation   551–2
connecting research to practice, chemistry education research   154–65, 171–2
connections
– connecting the dots   46–8
– responsibilities   27–48
– stories   27–48
– transforming thinking   27–48
connectivism, lifelong learning   141–3
constructivism
– cognitive conflict strategy   452, 462
– cognitive skills   304–5
– curriculum   74–6
– inquiry-based student-centered instruction   304–6
– learning cycle   304–5
content-based teaching   47–8
content of ideas, lifelong learning   129–30
context-based learning (CBL)   259–76
– applying chemical knowledge   267–8
– approaches   269–75
– carbon dioxide example   272–3
– chemical switches example   273–5
– chemical tetrahedron   270–1
– 'Chemie im Kontext'   261–3
– color, understanding   273–5
– design   263–9
– design of tasks   269, 271–2
– differentiated tasks   271–2

context-based learning (CBL) (contd.)
- effects of learning 263–9
- empirical background 260–1
- empirical study 263–9
- feedback 275
- future studies 276
- goals 269–75
- human activity 20–3
- implications 275
- magnetism, understanding 273–5
- meaningful learning 260–1
- motivation 260–1
- need for further insights 263–9
- school-level 261–3
- strategies, context-based tasks 265–7
- theoretical background 260–1
- university level 269–75

context-based tasks
- strategies to approach 265–7
- transferring knowledge 265–7

context-based teaching 47–8
- future trends 117
- teacher learning 107–9, 117

convergent assignments
- *Real Work* 209–18
- team learning 215–16

Course-Based Undergraduate Research Experiences (CURE), *Real Work* 230–1, 243

CPR *see* calibrated peer review

craft model, teacher learning 100

creativity
- divergent explanations 240
- gifted learners 479–80
- promoting 479–80

crowdsourcing, Semantic Web 702

cultural context, gifted learners 485

CURE *see* Course-Based Undergraduate Research Experiences

curriculum
- *see also* teacher learning
- blended learning 652–3, 661, 662, 665–6
- community-based learning (CBL) 353
- community-based research (CBR) 353
- competency-based undergraduate curriculum 83–92, 93
- constructivism 74–6
- employers' influences 78
- globalization influences 78
- innovation 74–8
- new teaching methods 78–83
- research on student learning 74–6
- school-level formal chemistry education 123–5

- SENCER (Science Education for New Civic Engagements and Responsibilities) 27–8, 37–9, 47–8
- traditional undergraduate 73–4

curriculum development, gifted learners 477–84

curriculum integration, blended learning 665–6

curriculum reform
- experimental experience 498–502
- problem-based group-organized project work 498–9
- second semester project work 499–502

cycle of pedagogical research 172–4

d

Dale pyramid, learning approaches 75–6

Data Enhancing the RSC Archive (DERA) 697

demonstration and teaching apps 641–2

demonstrations, interactive, chemistry education research 166–7

DERA *see* Data Enhancing the RSC Archive

development of chemistry *see* history and development of chemical language; history and development of chemistry

developments in teaching
- experimental experience 494–7
- lessons learned 497
- tertiary level 495–6
- upper secondary level 494–5

diaries, *Reflective Diaries*, guided-inquiry-based laboratories 315–16

dictionary/reference apps 636–7

digital environments, lifelong learning 141–3

digital multimeters, microscale experimentation 548–53

discourse, student, chemistry education research 159–60

discrepant events, cognitive conflict strategy 448, 453–9, 461–5

discussion boards, blended learning 653, 663–4

disequilibrium, cognitive conflict strategy 451–2, 453

disruptive innovation, flipped classrooms 319, 341

distributed cognition 722, 724

Distributed Drug Discovery (D^3) project, *Real Work* 231–2

divergent assignments
- *Real Work* 209–18

- team learning   216–18
divergent tasks, for training organic chemistry peer facilitators   217–18
diversity of students
- language issues   437–41
- second language learners (SLLs)   437–40
domain-specificity of giftedness   472–4
Dora/Flora story   203–6
dynamic visualization   611–15
- *see also* simulations
- mental models   603–6

e
e-learning
- *see also* blended learning
- community-based learning (CBL)   365–6
- community-based research (CBR)   365–6
- defining   651–2
Educational Quality Improvement Program (EQUIP 1), active learning pedagogies   296
EET *see* Entrepreneurship Education and Training
Ehime University Science Innovation Program   482–4
eight-balls problem, problem solving research   182–4
electrical conductivity, microscale experimentation   551–2
electron transfer, misconceptions   414–16
electronic homework systems   225, 235–6, 237
electronic lab notebooks (ELNs)   708, 720–1
electrowinning of copper, microscale experimentation   546–8
ELLs *see* English language learners
ELNs *see* electronic lab notebooks
employers' influences
- curriculum   78
- skills   78
energy, TKRS searches   58–9, 65–7
engagement/involvement
- evaluating   684–8
- flipped classrooms   322
- Wikis   684–8
English language learners (ELLs), problems with language   437–40
enrichment program, gifted learners   477
Entrepreneurship Education and Training (EET), competency-based undergraduate curriculum   91
environmental/analytical chemistry projects, service-learning   291–2

Ephorus plagiarism-detection tool   577
epistemic practice
- argumentation   527–8
- high-order learning skills   527–8
equations, chemical *see* chemical equations
EQUIP 1: *see* Educational Quality Improvement Program
ethylene, TKRS searches   62–3, 65–7
evolving nature of chemistry, influences on teaching   77
experimental experience   489–514
- analysis of project reports   502–3
- analysis, project reports   502–3
- benefits   497
- chemistry point of view   511–12
- curriculum reform   498–502
- design, project reports   509
- developments in teaching   494–7
- elements of experimental work   503–11
- experimental, project reports   509–10
- freshman students   497, 502, 510–11
- inquiry emphasis   493–4
- instruction styles   492–4
- interpretation, project reports   510–11
- objectives, project reports   508–9
- practical work   489–92
- problem-based group-organized project work   498–9
- problem orientation   498
- project-based learning   512–14
- project-organization   499–502
- project reports   502–11
- results, project reports   510
- second semester project work   499–503
- teaching   489–92
experimental observations, Wikis   679
experimental optimization, team-based learning   224
expert model, teacher learning   100
expert style, gifted learners   473–4
explanatory knowledge, *Real Work*   219–20
extension of formal education opportunities   125–9
extraneous load, cognitive load theory (CLT)   654–5

f
face-to-face teams, team-based learning   222–3
Facebook, blended learning   664–5

filtering information
- inquiry-based student-centered instruction 305
- learning cycle 305

flaschcard apps 635–6

flipped classrooms 319–41
- *see also* Peer Instruction
- agile approach 327–8
- big ideas 321–2
- blended learning 322, 653, 658
- criticism 339–40
- defining 320–1
- disruptive innovation 319, 341
- engagement/involvement 322
- examples 325–8
- future of education 341
- history 323
- Just-in-Time Teaching (JiTT) 326
- Khan Academy 324
- Learning Platforms 577–8
- lectures 326–7
- methods 325–9
- myths 326–9
- Peer Instruction 329–39
- pitfalls 336–8
- preparedness 327
- principles 328–9
- prior knowledge 321–2
- protocol 328–9
- self-regulation 322
- student attitudes 329
- student-centered pedagogy 320, 322
- technology dependency 324–5
- *vs* traditional classrooms 323–4
- videos 323–4, 326
- YouTube 321

Flora/Dora story 203–6

formal education, lifelong learning 125–6

four-card problem, problem solving research 182

Freebase 695

freshman students, experimental experience 497, 502, 510–11

future of education
- chemical education 726–8
- flipped classrooms 341
- Peer Instruction 341

future perspectives, chemistry apps 647–9

future studies, context-based learning (CBL) 276

future trends
- collaboration 117
- community-based learning (CBL) 364–6
- community-based research (CBR) 364–6
- context-based teaching 117
- research-based teaching 117
- teacher learning 116–18

g

GALT (Group Assessment of Logical Thinking), chemistry education research 164

game play
- augmented reality (AR) 581–2
- informal education 581–2

gamification, blended learning 661

gaming apps 642–4

gas laws/air bag, dynamic visualization 603–6

general chemistry 6, 7, 9, 40–1, 46–7
- problem solving research 184–6

generating questions, *Real Work* 230

germane load, cognitive load theory (CLT) 654–5

gifted learners 469–86
- acceleration program 477
- cognition levels 477–8
- cognitive cycle 477–8
- cognitive sequence 477–8
- creativity 479–80
- cultural context 485
- curriculum development 477–84
- domain-specificity of giftedness 472–4
- education programs 477
- effects of gifted education on students 480, 482–4
- Ehime University Science Innovation Program 482–4
- enrichment program 477
- expert style 473–4
- gifted behavioral checklist in science 474–5
- gifted styles in science 473–4
- identifying 472–7
- implementation of chemistry education 477–84
- innovation 479–80, 482–4
- IQ Intelligence Quotients tests 472
- meeting needs 482–4
- natural selection model 474–7
- Nobel Prize in chemistry from 1901 to 2012: 470–2
- opportunities 485
- science contests 480–2
- solid style 473–4
- spontaneous style 473–4
- studying beyond the classrooms 480–2

glassware and equipment
- information overload   306–7
- inquiry-based student-centered instruction   306–7
global warming, TKRS searches   57, 65–7
Global Water Experiment (GWE), UNESCO-IUPAC/CCE Global Microscience Program   556–7
globalization, curriculum influences   78
glocalization   51–70
goals, learning *see* learning goals
Google Books   719
Google Glass   566–7, 727–8
Google Patents   719
Google Scholar   719
graduate attributes, community-based learning (CBL)   354
graphic novels, lifelong learning   137–40
gravimetric measurements, microscale experimentation   546–8
green chemistry   39–41
Grignard's lab notebook record, laboratory learning   674–5
Group Assessment of Logical Thinking (GALT), chemistry education research   164
group discussions, Wikis   687–8
guided-inquiry-based laboratories   310–16
- *see also* inquiry-based student-centered instruction
- assessment   314–16
- metals/acids experiments   311–14
- *Reflective Diaries*   315–16
guided peer review and revision   221–2
GWE *see* Global Water Experiment

h

hard copy texts *vs* online texts   575–6
heterogeneity, problems with language   422, 437–9
high-impact educational practices
- community-based learning (CBL)   354–6
- community-based research (CBR)   354–6
high-order learning skills   517–36
- *see also* laboratory projects
- *America's Lab Report*   519
- argumentation   526–31
- asking questions   531–2
- development   519–22
- epistemic practice   527–8
- goals, laboratory studies   522–3
- high school chemistry laboratory   517–36

- inquiry-based teaching   521–2
- inquiry-type chemistry laboratories   523–6, 529–31, 533–6
- metacognition   523–6, 527
- *National Science Education Standards*   521
- *Nuffield Curriculum Projects*   517, 518
- questioning skills   531–2
- using laboratories   522–3
higher order cognitive skills (HOCS)   80
history and development of chemical language   423–8
- chemical symbols   423–5
- systematic nomenclature   425–8
history and development of chemistry   469–70
- branches of chemistry   469–70
- Nobel Prize in chemistry from 1901 to 2012:   470–2
HOCS *see* higher order cognitive skills
home work, *vs Real Work*   203–6
human activity   5–17
- acid–base chemistry   16, 18–20
- Anthropocene Epoch   14–17
- atoms first   6–7
- carrying out chemistry   11–14
- climate change   9–10, 16–18, 32–3
- context-based learning (CBL)   21–3
- learning and teaching chemistry   5–9
- ocean acidification   16, 18–20
- rich contexts   18–19, 20–23
- sustainability   8, 14, 17, 27–8, 37–41
- visualization   9–10

i

ICT *see* information and communication technology
imagination, powers of   11–12, 30, 34–5
InChI compound identifier   698
informal education
- augmented reality (AR)   579–90
- game play   581–2
- lifelong learning   126–7
- Massive Open Online Courses (MOOCs)   126–7
- Might-^{y}/Mach^{tig} 580–90
- Science LinX   580
information and communication technology (ICT)
- *see also* blended learning; computer-based technologies
- influences on teaching   76–7
information overload, glassware and equipment   306–7

information processing model
- blended learning   654–5
- inquiry-based student-centered instruction   308
- long-term memory   305–6, 308
innovation
- curriculum   74–8
- disruptive innovation   341
- Ehime University Science Innovation Program   482–4
- flipped classrooms   341
- gifted learners   479–80, 482–4
- promoting   479–80
inquiry-based learning, laboratory learning   674, 678–81
inquiry-based student-centered instruction   301–17
- *see also* guided-inquiry-based laboratories
- cognitive skills   304–5
- concept development   308–10
- constructivism   304–6
- filtering information   305
- glassware and equipment   306–7
- guided-inquiry-based laboratories   310–16
- inductive approach   303–4
- information overload   306–7
- information processing model   308
- inquiry-based instruction   303–4
- laboratory projects   310–16
- learning cycle   304–7
- lecturing/lecture notes   301–2
- long-term memory   305–6, 308
- 'scientists' *vs* 'technicians'   303–4
inquiry-based teaching   80
- *National Science Education Standards*   521–2
inquiry emphasis
- experimental experience   493–4
- instruction style   493–4
inquiry-type chemistry laboratories
- high-order learning skills   523–6, 529–31, 533–6
- metacognition   523–6, 527
instruction styles
- experimental experience   492–4
- inquiry emphasis   493–4
- practical work   492–3
instructional design, Peer Instruction   334–6
instructional materials, student-generated   225–8, 232, 244–5
instructional practice, chemistry education research   155–6, 171–4

instructional technologies   224–8, 242
- electronic homework systems   225
- learning by design   224–5
- student-generated animations   225–6
- student-generated metaphors   227–8
- student-generated video-blogs   226–7
- student-generated videos   225
- Wiki environment   227
- Wikipedia editing   227
instrumental apps   640
instruments, chemistry education research   163–5
interaction
- learning communities   116
- teacher learning   116
interactive applets   571–2
- Jablonski diagram   571–2
- periodic videos   572, 656
interactive demonstrations, chemistry education research   166–7
interactive lectures   79
interactive model
- Interconnected Model of Teacher Professional Growth   101–2
- teacher learning   101–2
interactive simulations   611–14
interactivity, Internet   567–72
Interconnected Model of Teacher Professional Growth, teacher learning   101–2
interdisciplinary research-based projects, *Real Work*   231–2
Internet   565–93
- *see also* augmented reality (AR); chemistry apps; online learning; Semantic Web
- augmented reality (AR)   566–7, 579–90
- Blackboard   574
- Ephorus plagiarism-detection tool   577
- interactive applets   571–2
- interactivity   567–72
- iTunes university   575
- Khan Academy   324, 574
- layar   566–7, 579, 582
- Learning Platforms   574–5, 577–9
- lifelong learning   141–3
- Molecular City app   587–92
- Moodle   574
- online texts *vs* hard copy texts   575–6
- Science LinX   580–90
- Semantic Web   694–702
- Technological Pedagogical Content Knowledge (TPACK)   568–71
- Virtual Learning Environment (VLE)   574–5, 577–9

- Web 2.0: 565–6, 694–5
- Wikipedia 227, 292–3, 572, 573, 717–18
- YouTube 574

intrinsic load, cognitive load theory (CLT) 654–5
inverted classrooms *see* flipped classrooms
invisible, 'seeing' the 28–34
involvement/engagement, flipped classrooms 322
ionic bonding, misconceptions 397–401
IQ Intelligence Quotients tests, gifted learners 472
iTunes university 575
iTunesU, blended learning 656–7

j

Jablonski diagram, interactive applet 571
JiTT *see* Just-in-Time Teaching
Jmol visualization tool 601–2
Just-in-Time Teaching (JiTT), flipped classrooms 326

k

Khan Academy 574
- flipped classrooms 324

knowledge base
- pedagogical content knowledge (PCK) 102–6
- subject matter knowledge (SMK) 102–6
- teacher learning 102–6

l

lab utility apps 634–40
laboratory classes, blended learning 657–9
laboratory learning
- *see also* practical work
- active-learning inquiry 674, 678–81
- assessment 672–5
- collaboration 681
- collective writing 682–4
- experimental observations 679
- Grignard's lab notebook record 674–5
- inquiry-based learning 674, 678–81
- objectives 672–3
- scaffolding collaborative laboratory report writing 682–4
- Science Writing Heuristic (SWH) 674
- virtual laboratory notebook 682
- Wikis 675–89

laboratory notebooks, Wikis 672–6, 678, 682–3, 687
laboratory practical work 489–91
- *see also* microscale experimentation

laboratory projects
- *see also* high-order learning skills
- guided-inquiry-based laboratories 310–16
- inquiry-based student-centered instruction 310–16
- metals/acids experiments 311–14
- safety teams 239
- team-based learning 223

LabQuest 2 app, augmented reality (AR) 713–14
language 421–42
- *see also* scientific literacy
- argumentation and discourse 436
- chemical symbols 423–5
- command words 435–6
- dimensions 429–30
- diversity of students 437–41
- English language learners (ELLs) 437–40
- heterogeneity 422, 437–9
- history and development of chemical language 423–8
- literacy reduction factors 422
- logical connectives 434–5
- non-native speakers 437–40
- nontechnical words 433–4
- problems 430–7
- readability of text 436–7
- role in science education 428–30
- second language learners (SLLs) 437–40
- special-needs students 440–1
- systematic nomenclature 425–8
- technical words/terms 430–3, 434

layar, augmented reality (AR) 566–7, 579, 582
learning about science
- *vs* learning science 203–9
- *vs* learning to be a scientist 203–9

learning and teaching chemistry, human activity 5–9
learning approaches, Dale pyramid 75–6
learning by design, instructional technologies 224–5
learning communities
- collaboration 116
- computer-based technologies 116
- interaction 116
- professional knowledge/development 114–16
- teacher learning 114–16

learning cycle
- cognitive skills 304–5
- constructivism 304–5
- filtering information 305

learning cycle (*contd.*)
- inquiry-based student-centered instruction   304–7
learning goals   21, 40, 41
- superpowers   33–7
Learning Platforms
- flipped classrooms   577–8
- Internet   574–5, 577–9
learning research *see* chemistry education research
learning science
- *vs* learning about science   203–9
- *vs* learning to be a scientist   203–9
learning to be a scientist
- *vs* learning about science   203–9
- *vs* learning science   203–9
lectures
- blended learning   657–9
- flipped classrooms   326–7
- pre-lecture activities   657–8
lecturing/lecture notes   301–2
lifelong learning   123–46
- books, popular   135–6
- broadcast media   140–1
- cartoons   136–7
- citizen science   143–4
- comics   136–7
- computer-based technologies   141–3
- connectivism   141–3
- content of ideas   129–30
- digital environments   141–3
- emphases   127–9
- extension of formal education opportunities   125–9
- formal education   125–6
- four-stage model   127–9
- graphic novels   137–40
- informal education   126–7
- Internet   141–3
- key aspects   125
- magazines   134–5
- Massive Open Online Courses (MOOCs)   142
- media   133–46
- media selection criteria   133, 145–6
- newspapers   134–5
- novels, graphic   137–40
- opportunities   123–9, 144–6
- pedagogy   131–2
- 'popular' books   135–6
- presentation of ideas   130–1
- printed media   134–40
- radio   140–1
- school-level formal chemistry education   123–5
- science centers   134
- science museums   133–4
- social circumstances   131–2
- structure   127–9
- television   140–1
lifelong research-oriented teachers   113–14
light absorption, microscale experimentation   552–3
literacy reduction factors, scientific literacy   422
literacy, scientific *see* scientific literacy
literature research apps   633–4
literature seminars, *Real Work*   229–30
literature summaries, *Real Work*   228–9
logic processes
- cognitive conflict recognition   459–61
- cognitive conflict resolution   459–61
logical connectives, problems with language   434–5
logical thinking tests, chemistry education research   164
long-term memory
- information processing model   305–6, 308
- inquiry-based student-centered instruction   305–6, 308
looking up   44–6

m
macro level (reality) misconceptions   417
macro *vs* micro, microscale experimentation   542–3
macroscopic level
- conceptual understanding   596–7
- mental models   596–7
magazines, lifelong learning   134–5
magnetism, understanding, context-based learning (CBL)   273–5
Marker-based AR and Markerless AR, augmented reality   711–12
Massive Open Online Courses (MOOCs)   79, 572–4, 710
- informal education   126–7
- lifelong learning   142
meaningful learning   224–5
- context-based learning (CBL)   260–1
- *vs* rote learning   203–4
media, lifelong learning   133–46
- broadcast media   140–1
- citizen science   143–4
- digital environments   141–3
- printed media   134–40

- science centers 134
- science museums 133–4
- selection criteria 133, 145–6
media literacy, scientific literacy 53–4
melamine, TKRS searches 63–4, 65–7
mental models
- conceptual understanding 596–8
- dynamic visualization 603–6
- macroscopic level 596–7
- submicroscopic level 597
- symbolic level 597
- visualization 596–606, 609–15
metacognition
- high-order learning skills 523–6, 527
- inquiry-type chemistry laboratories 523–6, 527
- student-generated instructional materials 244–5
metals/acids experiments, guided-inquiry-based laboratories 311–14
metals, descriptive chemistry, problem-based learning (PBL) 285–6
microscale experimentation 539–59
- advantages 542–3
- aims of practical work 540–3, 555–7, 559
- benefits 542–4
- case study 542–58
- cell potential measurements 549–51
- conductivity/conductance meters 551–2
- digital multimeters 548–53
- electrical conductivity 551–2
- electrowinning of copper 546–8
- Global Water Experiment (GWE) 556–7
- gravimetric measurements 546–8
- light absorption 552–3
- macro vs micro 542–3
- microtitration 544–6
- pH measurements 549–51
- probes 548–53
- quantitative 548–53
- roots 541
- scale 542–3
- scientific approach, learning the 554–5
- sensors 548–53
- temperature measurements 552–3
- UNESCO-IUPAC/CCE Global Microscience Program 541, 555–7
- volumetric analysis 544–6
microtitration, microscale experimentation 544–6
Might-y/Mach^{tig}
- augmented reality (AR) 580–90

- informal education 580–90
misconceptions 395–419
- acid–base reactions 405–12
- best practice to challenge 416–17
- chemical equilibrium 401–5
- cognitive conflict strategy 451–2
- electron transfer 414–16
- integrating into instruction 418–19
- ionic bonding 397–401
- macro level (reality) 417
- oxygen transfer 413–14
- preconcepts 395–6
- proton transfer 405–12
- redox reactions 411–16
- representational level (symbolic level) 417
- school-made misconceptions 395–7
- submicro level (mental and concrete models) 417
misconceptions, students', chemistry education research 154–7
mobile devices 702–5
- see also chemistry apps
models/modeling
- conceptual integration 383–5
- modeling apps 626–9, 714–15
- multiple models 383–5
- teaching about 109–11
molecular animations, visualization 607–8
Molecular City app 587–92
molecular drawing apps 629–31
molecular viewers, chemistry apps 626–9, 714–15
MOOCs see Massive Open Online Courses
Moodle, Virtual Learning Environment (VLE) 574
Motivated Strategies for Learning Questionnaire (MSLQ) 243–4
MSLQ see Motivated Strategies for Learning Questionnaire

n

nano, TKRS searches 64–7
National Science Education Standards
- high-order learning skills 521
- inquiry-based teaching 521–2
natural selection model, gifted learners 474–7
new teaching methods
- see also chemistry education research
- activity system 82–3
- competency-based teaching 81–3
- curriculum influences 78–83
- higher order cognitive skills (HOCS) 80

new teaching methods (*contd.*)
- inquiry-based teaching 80
- interactive lectures 79
- Massive Open Online Courses (MOOCs) 79, 126–7, 142, 572–4
- problem-based teaching 80
- process-oriented guided inquiry learning (POGIL) 80, 160
- research-based teaching 80–1

newspapers, lifelong learning 134–5
Nobel Prize in chemistry from 1901 to 2012: 470–2
non-majors, chemistry for 34, 37, 39
non-native speakers, problems with language 437–40
nontechnical words, problems with language 433–4
notebooks
- electronic lab notebooks (ELNs) 708, 720–1
- Grignard's lab notebook record 674–5
- laboratory notebooks, Wikis 672–6, 678, 682–3, 687

novels, graphic, lifelong learning 137–40
Nuffield Curriculum Projects, high-order learning skills 517, 518

o

ocean acidification, human activity 16, 18–20
odors, industrial 41–4
online communities, blended learning 663–5
online learning
- *see also* Internet
- community-based learning (CBL) 365–6
- community-based research (CBR) 365–6
- online texts *vs* hard copy texts 575–6
- online quizzes, blended learning 659–62

open data 698, 701–2, 717, 720–1
Open Notebook Science 720
Open Pharmacological Concept Triple Store (Open PHACTS) 701–2
opportunities
- chemistry apps 646–7
- extension of formal education opportunities 125–9
- gifted learners 485
- lifelong learning 125–9, 144–6
- teacher learning 113–16

organic chemistry
- divergent tasks for training organic chemistry peer facilitators 217–18
- problem-based learning (PBL) 287–8

- problem solving research 186–92
- simulations 613–15
- virtual problem-based learning (VPBL) 287–8

organizing information, problem solving research 195–9
oxygen transfer, misconceptions 413–14

p

PARE (preparation, action, reflection, and evaluation) model 353
PBL *see* problem-based learning
PBWorks Wiki platform 682–3
PCK *see* pedagogical content knowledge
PCOL *see* Physical Chemistry Online
PCT *see* Personal Construct Theory
pedagogic simulation 611–13
- predict–observe–explain (POE) technique 612

pedagogical content knowledge (PCK)
- models/modeling, teaching about 109–11
- teacher learning 102–6, 109–13

pedagogy, lifelong learning 131–2
Peer Instruction
- *see also* flipped classrooms
- classroom response systems 333–5, 338
- ConcepTests 330–9
- discussion 337
- flipped classrooms 329–39
- future of education 341
- goals 330
- instructional design 334–6
- pitfalls 336–8
- research 336
- workflow 332–4

peer presentation, review and critique 218–22
peer review and critique, conceptual weaknesses 240–1
peer-to-peer instruction, *Real Work* 220
PeerWise, blended learning 235, 661
Periodic Table apps 631–3
periodic videos, interactive applet 572, 656
Personal Construct Theory (PCT), conceptual integration 378–9
personal development, community-based learning (CBL) 354
personal response systems, blended learning 662–3, 664
pH measurements, microscale experimentation 549–51
PhET interactive simulations 598–9

physical chemistry, problem solving research 197–9
Physical Chemistry Online (PCOL)   710
plagiarism, Ephorus plagiarism-detection tool   577
planetary boundaries, chemistry concepts 16–17
podcasts   236–8, 239, 705
– blended learning   656–7
POE technique *see* predict–observe–explain technique
POGIL *see* process-oriented guided inquiry learning
Polya's model, problem solving research   195
'popular' books, lifelong learning   135–6
practical work
– *see also* laboratory learning; microscale experimentation
– aims   540–3, 555–7, 559
– benefits   497
– chemistry education   489–91
– experimental experience   489–92
– goals   492–3
– instruction styles   492–3
– laboratory   489–91
– reasons for   490–1
pre-laboratory activities, blended learning 658–9
predict–observe–explain (POE) technique
– cognitive conflict strategy   456, 462–3
– pedagogic simulation   612
preparedness, flipped classrooms   327
presentation of ideas, lifelong learning 130–1
print textbooks   233–5
printed media, lifelong learning   134–40
prior knowledge, flipped classrooms   321–2
probes, microscale experimentation 548–53
problem-based group-organized project work
– curriculum reform   498–9
– experimental experience   498–9
problem-based learning (PBL)   280–90
– bioinorganic chemistry course   288–9
– content   289–90
– history   281
– metals, descriptive chemistry   285–6
– options   282–3
– organic chemistry   287–8
– problems   285–9
– process   281–3
– research preparation   282

– tasks   281–3
– thermochemistry   286–7
– virtual PBL   283–5, 287–8
problem-based teaching   80
problem orientation, experimental experience   498
problem solving, chemistry education research   161
problem solving research   181–200
– anarchistic model of problem solving 193–9
– eight-balls problem   182–4
– four-card problem   182
– general chemistry   184–6
– organic chemistry   186–92
– organizing information   195–9
– physical chemistry   197–9
– Polya's model   195
– problem-solving mindset   193
– Purdue Visualization of Rotation (ROT) Test   184–6
– reasons for   181–4
– spatial ability tests   184–7
– successful problem solvers' characteristics 199–200
– synthesis problems   187–93
– trial and error strategy   182–3, 184, 195–6, 198
Process-Oriented Guided Inquiry Learning in Analytical Chemistry (ANAPOGIL), chemistry education research   167–9
process-oriented guided inquiry learning (POGIL)   80
– chemistry education research   160
professional knowledge/development   22–3
– chemistry education research   171–4
– Interconnected Model of Teacher Professional Growth   101–2
– learning communities   114–16
– teacher learning   99–108
project-based learning   512–14
– *see also* experimental experience
project-organization, experimental experience   499–502
project reports   502–11
– analysis   502–3
– design   509
– experimental   509–10
– experimental experience   502–11
– interpretation   510–11
– objectives   508–9
– results   510
proton transfer
– acid–base reactions   405–12

proton transfer (*contd.*)
- misconceptions   405–12
public health   37–9
public science courses, *Real Work*   230
PubMed   719
Purdue Visualization of Rotation (ROT) Test, problem solving research   184–6

q

QR (quick response) codes   711–14
quantitative microscale experimentation   548–53
questioning skills, high-order learning skills   531–2
quizzes, online, blended learning   659–62

r

radio, lifelong learning   140–1
RDF *see* Resource Description Framework
readability of text, problems with language   436–7
Real Work
- active-learning assignments, classroom   238
- argumentation and evidence   222
- authentic learning   206–9
- authentic learning experiences   203–6
- authentic materials   243–4
- authentic tasks   206–9
- authentic texts and evidence   228–32
- calibrated peer review (CPR)   221, 240–1
- convergent assignments   209–18
- Course-Based Undergraduate Research Experiences (CURE)   230–1, 243
- creativity, divergent explanations   240
- defining   203–6
- Distributed Drug Discovery (D^3) project   231–2
- divergent assignments   209–18
- electronic homework systems   225, 235–6, 237
- explanatory knowledge   219–20
- generating questions   230
- guided peer review and revision   221–2
- *vs* home work   203–6
- instructional technologies   242
- interdisciplinary research-based projects   231–2
- learning from   239–45
- literature seminars   229–30
- literature summaries   228–9
- peer presentation, review and critique   218–22
- peer review and critique, conceptual weaknesses   240–1
- peer-to-peer instruction   220
- podcasts   236–8, 239
- print textbooks   233–5
- public science courses   230
- safety teams, laboratory   239
- situated cognition   20, 203, 207
- situated learning   206–9
- student-generated instructional materials   225–8, 232, 244–5
- team-based learning   222–4
- team learning, achievement gains   241–2
- tutor learning   220
- web-based textbooks   233–5
- Wiki textbooks   232–3
real-world chemistry   17, 37–9, 41–4, 279–80
- *see also* problem-based learning (PBL)
- American Association for the Advancement of Science (AAAS)   280
redox reactions, misconceptions   412–16
Reflective Diaries, guided-inquiry-based laboratories   315–16
representational level (symbolic level) misconceptions   417
representations, chemistry education research   161–3
research
- *see also* chemistry education research; problem solving research
- conceptual integration   391–2
- student learning   153–4
research-based teaching   80–1
- *see also* chemistry education research
- future trends   117
- lifelong research-oriented teachers   113–14
- research-based teaching practice   165–70
- teacher learning   117
research on student learning, curriculum   74–6
research preparation, problem-based learning (PBL)   282
Resource Description Framework (RDF)   698–9
responsibilities
- connections   27–48
- SENCER (Science Education for New Civic Engagements and Responsibilities)   27–8, 37–9, 47–8
- stories   27–48
- transforming thinking   27–48

retrosynthetic analysis   213
rich contexts, human activity   13–14, 17–23
ROT test, Purdue Visualization of Rotation (ROT) Test   184–6
rote learning, *vs* meaningful learning   203–4
RSC databases   719
RSS feeds   699

s

safety teams, laboratory   239
scaffolding collaborative laboratory report writing, Wikis   682–4
school-level
– 'Chemie im Kontext'   261–3
– context-based learning (CBL)   261–3
– formal chemistry education   123–5
science centers, lifelong learning   134
science contests, gifted learners   480–2
Science LinX, informal education   580–90
science museums, lifelong learning   133–4
Science Writing Heuristic, chemistry education research   160
Science Writing Heuristic (SWH), laboratory learning   674
scientific approach, learning the   554–5
scientific literacy   52–5
– see also language
– defining   52
– elements   53
– literacy reduction factors   422
– media literacy   53–4
– teaching keywords-based recommendation system (TKRS) searches   55–70
scientific terms, teaching keywords-based recommendation system (TKRS) searches   55–70
'scientists' *vs* 'technicians', inquiry-based student-centered instruction   303–4
screencasts, blended learning   656–7
search engine apps   637–8
search systems shortcomings, Semantic Web   700–1
second language learners (SLLs), problems with language   437–40
second-level digital divide   693, 721–3, 725–7
Second Life, augmented reality (AR)   709–10
see blended learning, screencasts   656–7
'seeing' the invisible   28–34
self-explanations, visualization   609–10

self-regulation
– flipped classrooms   322
– student-generated instructional materials   244–5
Semantic Web   694–702
– Berners-Lee, Tim   694–5, 697
– ChemEd DL (digital library)   697
– chemical markup language (CML)   699–700
– crowdsourcing   702
– Data Enhancing the RSC Archive (DERA)   697
– Freebase   695
– goal   698
– InChI compound identifier   698
– Open Pharmacological Concept Triple Store (Open PHACTS)   701–2
– Resource Description Framework (RDF)   698–9
– RSS feeds   699
– search systems shortcomings   700–1
– Semantically-Interlinked Online Communities Initiative   701
– Social Semantic Web   701
– WikiHyperGlossary   697
SENCER (Science Education for New Civic Engagements and Responsibilities), national curriculum reform project   27–8, 37–9, 47–8
sensors, microscale experimentation   548–53
service-learning   290–5
– see also community-based learning (CBL)
– analytical/environmental chemistry projects   291–2
– benefits   294
– bioinorganic chemistry course   293–4, 295
– chemistry education projects   292–3
– defining   346
– scope   290–1
– Wikipedia editing   292–3
simulations
– chemistry education research   161–3
– interactive   611–14
– organic chemistry   613–15
– pedagogic simulation   611–13
– PhET interactive simulations   598–9
– visualization   598–9, 611–15
situated cognition, *Real Work*   20, 203, 207
situated learning, *Real Work*   206–9
skills, required, employers' influences   78
SLLs *see* second language learners

smartphone devices 702–5
– *see also* chemistry apps
SMK *see* subject matter knowledge
social circumstances, lifelong learning 131–2
social networking, blended learning 664–5
Social Semantic Web 701
– role 717–21
solid style, gifted learners 473–4
spatial ability, visualization 609–10
spatial ability tests, problem solving research 184–7
special-needs students, problems with language 440–1
spontaneous style, gifted learners 473–4
SRSs *see* student response systems
stoichiometry, chemistry education research 155–6
stories 27–48
– connections 27–48
– responsibilities 27–48
– transforming thinking 27–48
strategies, context-based tasks 265–7
strategies, teaching *see* teaching strategies
student argumentation, chemistry education research 159–60
student-centered learning *see* inquiry-based student-centered instruction
student-centered pedagogy, flipped classrooms 320, 322
student communication, Wikis 678–81
student discourse, chemistry education research 159–60
student-generated animations 225–6
student-generated instructional materials 225–8, 232, 244–5
student-generated metaphors 227–8
student-generated video-blogs 226–7
student-generated videos 225
student learning research 153–4
– *see also* chemistry education research
student numbers, influences on teaching 77–8
student response systems (SRSs), chemistry education research 157–8
student understanding, Wikis 686–8
students' misconceptions, chemistry education research 154–7
subject matter knowledge (SMK)
– models/modeling, teaching about 109–11
– teacher learning 102–6, 109–11
submicro level (mental and concrete models) misconceptions 417

submicroscopic level
– conceptual understanding 597
– mental models 597
successful problem solvers' characteristics 199–200
super-learning environments 34–7
superpowers, learning goals 33–7
sustainability
– human activity 8, 14, 17, 27–8, 37–41
– TKRS searches 57–8, 65–7
SWH *see* Science Writing Heuristic
symbolic level
– conceptual understanding 597
– mental models 597
synthesis problems, problem solving research 187–93
systematic nomenclature, history and development of chemical language 425–8

t
tablet devices 702–5
– *see also* chemistry apps
teacher learning 99–108
– *see also* curriculum
– challenges 113–16
– challenging issues, teaching 107–13
– collaboration 116, 117
– complex/reciprocal processes 100–2
– computer-based technologies 111–13, 116
– context-based teaching 107–9, 117
– craft model 100
– elements 99–100
– empowering teachers 107–13
– expert model 100
– future trends 116–18
– interaction 116
– interactive model 101–2
– Interconnected Model of Teacher Professional Growth 101–2
– knowledge base 102–6
– learning communities 114–16
– lifelong research-oriented teachers 113–14
– models/modeling, teaching about 109–11
– opportunities 113–16
– pedagogical content knowledge (PCK) 102–6, 109–13
– professional knowledge/development 99–108
– research-based teaching 117

– subject matter knowledge (SMK) 102–6, 109–11
– Technological Pedagogical Content Knowledge (TPACK) 111–13
teaching and demonstration apps 641–2
teaching challenging issues, teacher learning 107–13
teaching developments
– experimental experience 494–7
– lessons learned 497
– tertiary level 495–6
– upper secondary level 494–5
teaching keywords-based recommendation system (TKRS) searches 55–70
– implications for chemistry education 68–70
teaching strategies 151–74
– approaches 151–2
– focuses 151–2
– phases 151–2, 154, 156–7
team-based learning
– collaborative identification 223–4
– experimental optimization 224
– face-to-face teams 222–3
– laboratory projects 223
– *Real Work* 222–4
– virtual teams 223
team learning
– achievement gains 241–2
– convergent assignments 215–16
– divergent assignments 216–18
technical words/terms, problems with language 430–3, 434
Technological Pedagogical Content Knowledge (TPACK) 568–71
– Jablonski diagram 571
– teacher learning 111–13
technology dependency, flipped classrooms 324–5
technology-enhanced learning
– *see also* blended learning; chemistry apps
– theoretical perspectives 675–8
television, lifelong learning 140–1
temperature measurements, microscale experimentation 552–3
Test of Logical Thinking (TOLT), chemistry education research 164
test-prep apps 644–5
tetrahedral chemistry education, visual metaphor 4
textbooks
– print textbooks 233–5
– web-based textbooks 233–5
– Wiki textbooks 232–3

theoretical perspectives, technology-enhanced learning 675–8
thermochemistry, problem-based learning (PBL) 286–7
TKRS searches *see* teaching keywords-based recommendation system searches
TOLT (Test of Logical Thinking), chemistry education research 164
Toulmin's model of argumentation 528–9
TPACK/TPCK *see* Technological Pedagogical Content Knowledge
traditional classrooms, *vs* flipped classrooms 323–4
traditional undergraduate curriculum 73–4
transferring knowledge, context-based tasks 265–7
transforming thinking
– connections 27–48
– responsibilities 27–48
– stories 27–48
transient information effect, cognitive load theory (CLT) 657
trial and error strategy, problem solving research 182–3, 184, 195–6, 198
tutor learning, *Real Work* 220
tutors' role, Wikis 679–81
Twitter, blended learning 664–5

u

undergraduate research 40–1, 81
– Course-Based Undergraduate Research Experiences (CURE) 230–1, 243
UNESCO-IUPAC/CCE Global Microscience Program
– Global Water Experiment (GWE) 556–7
– microscale experimentation 541, 555–7
United States Agency for International Development (USAID), active learning pedagogies 296

v

videos
– *see also* YouTube
– flipped classrooms 324–5, 326
– iTunesU 656–7
– periodic videos, interactive applet 572
– student-generated video-blogs 226–7
– student-generated videos 225
– transient information effect 657
virtual laboratory notebook, Wikis 682
Virtual Learning Environment (VLE) 574–5, 577–9, 708–17
– *see also* blended learning

virtual problem-based learning (VPBL) 283–5
– organic chemistry 287–8
virtual teams, team-based learning 223
virtual worlds
– ChemCollective virtual laboratory 600
– visualization 600–1
visualization 595–616
– *see also* animations
– 3D representations 601–3
– animations 607–9
– ChemCollective virtual laboratory 600
– conceptual understanding 596–8
– design 595–6
– dynamic visualization 603–6, 611–15
– gas laws/air bag 603–6
– human activity 8–9
– individual differences 609, 610–11
– Jmol visualization tool 601–2
– mental models 596–606, 609–15
– molecular animations 607–8
– PhET interactive simulations 598–9
– self-explanations 609–10
– simulations 598–9, 611–15
– spatial ability 609–10
– technology 598–603
– virtual worlds 600–1
Visualization of Rotation (ROT) Test, problem solving research 184–6
VLE *see* Virtual Learning Environment
volumetric analysis, microscale experimentation 544–6
VPBL *see* virtual problem-based learning

w

wearable devices, augmented reality (AR) 715–16
Web, 2.0: 565–6
– *see also* Internet; Semantic Web
web-based textbooks 233–5
Wiki environment 227
Wiki textbooks 232–3
WikiHyperGlossary 697
Wikipedia 572, 573, 717–18
– editing 227, 292–3
– service-learning 292–3
Wikis 671–89
– collaboration 681
– collective writing 682–4
– emerging uses 684–8
– engagement/involvement, evaluating 684–8
– examples 681–4
– experimental observations 679
– group discussions 687–8
– laboratory learning 675–89
– laboratory notebooks 672–6, 678, 682–3, 687
– PBWorks Wiki platform 682–3
– scaffolding collaborative laboratory report writing 682–4
– shared understanding 679
– student communication 678–81
– student participation, evaluating 684–8
– student understanding 686–8
– tools 682–4
– tutors' role 679–81
– virtual laboratory notebook 682
workflow, Peer Instruction 332–4

y

YouTube 113, 574, 642, 714–15
– blended learning 656–7
– flipped classrooms 321

CPSIA information can be obtained
at www.ICGtesting.com
Printed in the USA
LVOW02*1421170217
524629LV00006B/30/P

9 783527 336050